A Companion to Analysis

A Second First and First Second Course in Analysis

A Companion to Analysis
A Second First and First Second Course in Analysis

T. W. Körner

Graduate Studies
in Mathematics
Volume 62

American Mathematical Society
Providence, Rhode Island

2000 *Mathematics Subject Classification*. Primary 26–01.

For additional information and updates on this book, visit
www.ams.org/bookpages/gsm-62

Library of Congress Cataloging-in-Publication Data

Körner, T. W. (Thomas Williamm), 1946–
A companion to analysis : a second first and first second course in analysis / T. W. Körner.
p. cm. — (Graduate studies in mathematics, ISSN 1065-7339 ; v. 62)
Includes bibliographical references and index.
ISBN 0-8218-3447-9 (alk. paper)
1. Mathematical analysis. I. Title. II. Series.

QA300.K589 2003
515—dc22 2003062905

∞ The paper used in this book is acid-free and falls within the guidelines
established to ensure permanence and durability.
Visit the AMS home page at http://www.ams.org/

10 9 8 7 6 5 4 3 2 1 09 08 07 06 05 04

[Archimedes] concentrated his ambition exclusively upon those speculations which are untainted by the claims of necessity. These studies, he believed, are incomparably superior to any others, since here the grandeur and beauty of the subject matter vie for our admiration with the cogency and precision of the methods of proof. Certainly in the whole science of geometry it is impossible to find more difficult and intricate problems handled in simpler and purer terms than in his works. Some writers attribute it to his natural genius. Others maintain that phenomenal industry lay behind the apparently effortless ease with which he obtained his results. The fact is that no amount of mental effort of his own would enable a man to hit upon the proof of one of Archimedes' theorems, and yet as soon as it is explained to him, he feels he might have discovered it himself, so smooth and rapid is the path by which he leads us to the required conclusion.

Plutarch *Life of Marcellus* [Scott-Kilvert's translation]

It may be observed of mathematicians that they only meddle with such things as are certain, passing by those that are doubtful and unknown. They profess not to know all things, neither do they affect to speak of all things. What they know to be true, and can make good by invincible argument, that they publish and insert among their theorems. Of other things they are silent and pass no judgment at all, choosing rather to acknowledge their ignorance, than affirm anything rashly.

Barrow *Mathematical Lectures*

For [A. N.] Kolmogorov mathematics always remained in part a sport. But when ... I compared him with a mountain climber who made first ascents, contrasting him with I. M. Gel′fand whose work I compared with the building of highways, both men were offended. ' ... Why, you don't think I am capable of creating general theories?' said Andreĭ Nikolaevich. 'Why, you think I can't solve difficult problems?' added I. M.

V. I. Arnol′d in *Kolmogorov in Perspective*

Contents

Introduction

In his autobiography [**12**], Winston Churchill remembered his struggles with Latin at school. ' ... even as a schoolboy I questioned the aptness of the Classics for the prime structure of our education. So they told me how Mr Gladstone read Homer for fun, which I thought served him right.' 'Naturally' he says 'I am in favour of boys learning English. I would make them all learn English; and then I would let the clever ones learn Latin as an honour, and Greek as a treat.'

This book is intended for those students who might find rigorous analysis a treat. The content of this book is summarised in Appendix J and corresponds more or less (more rather than less) to a recap at a higher level of the first course in analysis followed by the second course in analysis at Cambridge in 2004 together with some material from various methods courses (and thus corresponds to about 60 to 70 hours of lectures). Like those courses, it aims to provide a foundation for later courses in functional analysis, differential geometry and measure theory. Like those courses also, it assumes complementary courses such as those in mathematical methods and in elementary probability to show the practical uses of calculus and strengthen computational and manipulative skills. In theory, it starts more or less from scratch, but the reader who finds the discussion of Section 1.1 baffling or the ϵ, δ arguments of Section 1.2 novel will probably find this book unrewarding. I assume a fair degree of algebraic fluency and, from Chapter 4 onwards, some exposure to linear algebra.

This book is about mathematics for its own sake. It is a guided tour of a great but empty Opera House. The guide is enthusiastic but interested only in sight-lines, acoustics, lighting and stage machinery. If you wish to see the stage filled with spectacle and the air filled with music you must come at another time and with a different guide.

Although I hope this book may be useful to others, I wrote it for students to read either before or after attending the appropriate lectures. For this reason, I have tried to move as rapidly as possible to the points of difficulty, show why they are points of difficulty and explain clearly how they are overcome. If you understand the hardest part of a course then, almost automatically, you will understand the easiest. The converse is not true.

In order to concentrate on the main matter in hand, some of the simpler arguments have been relegated to exercises. The student reading this book *before* taking the appropriate course may take these results on trust and concentrate on the central arguments which are given in detail. The student reading this book *after* taking the appropriate course should have no difficulty with these minor matters and can also concentrate on the central arguments. I think that doing at least some of the exercises will help students to 'internalise' the material, but I hope that even students who skip most of the exercises can profit from the rest of the book.

I have included further exercises in Appendix K. Some are standard, some form commentaries on the main text and others have been taken or adapted from the Cambridge mathematics exams. None are 'makeweights', they are all intended to have some point of interest. Sketches of some solutions are available from the home pages given on page xiii. I have tried to keep to standard notations, but a couple of notational points are mentioned in the index under the heading 'notation'.

I have not tried to strip the subject down to its bare bones. A skeleton is meaningless unless one has some idea of the being it supports, and that being in turn gains much of its significance from its interaction with other beings, both of its own species and of other species. For this reason, I have included several sections marked by a ♡. These contain material which is not necessary to the main argument but which sheds light on it. Ideally, the student should read them but not study them with anything like the same attention which she devotes to the unmarked sections. There are two sections marked ♡♡ which contain some, very simple, philosophical discussion. It is entirely intentional that removing the appendices and the sections marked with a ♡ more than halves the length of the book.

It is an honour to publish with the AMS; I am grateful to the editorial staff for making it a pleasure. I thank Dr Gunther Leobacher for computer generating the figures for this book. I owe particular thanks to Jorge Aarao, Brian Blank, Johan Grundberg, Jonathan Partington, Ralph Sizer, Thomas Ward and an anonymous referee, but I am deeply grateful to the many other people who pointed out errors in and suggested improvements to earlier versions of this book. My e-mail address is `twk@dpmms.cam.ac.uk` and I shall try to keep a list of corrections accessible from an AMS page at `www.ams.org/bookpages/gsm-62` as well as from my home page page at `http://www.dpmms.cam.ac.uk/~twk/`.

I learned calculus from the excellent *Calculus* [**13**] of D. R. Dickinson and its inspiring author. My first glimpse of analysis was in Hardy's *Pure Mathematics* [**24**] read when I was too young to really understand it. I learned elementary analysis from Ferrar's *A Textbook of Convergence* [**18**] (an excellent book for those making the transition from school to university, now, unfortunately, out of print) and Burkill's *A First Course in Mathematical Analysis* [**10**]. The books of Kolmogorov and Fomin [**31**] and, particularly, Dieudonné [**14**] showed me that analysis is not a collection of theorems but a single coherent theory. Stromberg's book *An Introduction to Classical Real Analysis* [**48**] lies permanently on my desk for browsing. The expert will easily be able to trace the influence of these books on the pages that follow. If, in turn, my book gives any student half the pleasure that the ones just cited gave me, I will feel well repaid.

Cauchy began the journey that led to the modern analysis course in his lectures at the École Polytechnique in the 1820's. The times were not propitious. A reactionary government was determined to keep close control over the school. The faculty was divided along fault lines of politics, religion and age whilst physicists, engineers and mathematicians fought over the contents of the courses. The student body arrived insufficiently prepared and then divided its time between radical politics and worrying about the job market (grim for both staff and students). Cauchy's course was not popular[1].

Everybody can sympathise with Cauchy's students who just wanted to pass their exams and with his colleagues who just wanted the standard material taught in the standard way. Most people neither need nor want to know about rigorous analysis. But there remains a small group for whom the ideas and methods of rigorous analysis represent one of the most splendid triumphs of the human intellect. We echo Cauchy's defiant preface to his printed lecture notes.

> As to the methods [used here], I have sought to endow them with all the rigour that is required in geometry and in such a way that I have not had to have recourse to formal manipulations. Such arguments, although commonly accepted ... cannot be considered, it seems to me, as anything other than [suggestive] to be used sometimes in guessing the truth.

[1] Belhoste's splendid biography [4] gives the fascinating details.

> Such reasons [moreover] ill agree with the mathematical sciences' much vaunted claims of exactitude. It should also be observed that they tend to attribute an indefinite extent to algebraic formulas when, in fact, these formulas hold under certain conditions and for only certain values of the variables involved. In determining these conditions and these values and in settling in a precise manner the sense of the notation and the symbols I use, I eliminate all uncertainty. ... It is true that in order to remain faithful to these principles, I sometimes find myself forced to depend on several propositions that perhaps seem a little hard on first encounter .... But, those who will read them will find, I hope, that such propositions, implying the pleasant necessity of endowing the theorems with a greater degree of precision and restricting statements which have become too broadly extended, will actually benefit analysis and will also provide a number of topics for research, which are surely not without importance.

I dedicate this book to the memory of my parents in gratitude for many years of love and laughter.

Chapter 1

The Real Line

1.1. Why do we bother?

It is surprising how many people think that analysis consists in the difficult proofs of obvious theorems. All we need know, they say, is what a limit is, the definition of continuity and the definition of the derivative. All the rest is 'intuitively clear'[1].

If pressed they will agree that the definition of continuity and the definition of the derivative apply as much to the rationals $\mathbb{Q}$ as to the real numbers $\mathbb{R}$. If you disagree, take your favorite definitions and examine them to see where they require us to use $\mathbb{R}$ rather than $\mathbb{Q}$. Let us examine the workings of our 'clear intuition' in a particular case.

What is the integral of t^2? More precisely, what is the general solution of the equation

$$(*) \qquad g'(t) = t^2?$$

We know that $t^3/3$ is a solution but, if we have been well taught, we know that this is not the general solution since

$$(**) \qquad g(t) = \frac{t^3}{3} + c,$$

with c any constant, is also a solution. Is $(**)$ the most general solution of $(*)$?

If the reader thinks it is the most general solution then she should ask herself why she thinks it is. Who told her and how did they explain it?

[1] A good example of this view is given in the book [**9**]. The author cannot understand the problems involved in proving results like the intermediate value theorem and has written his book to share his lack of understanding with a wider audience.

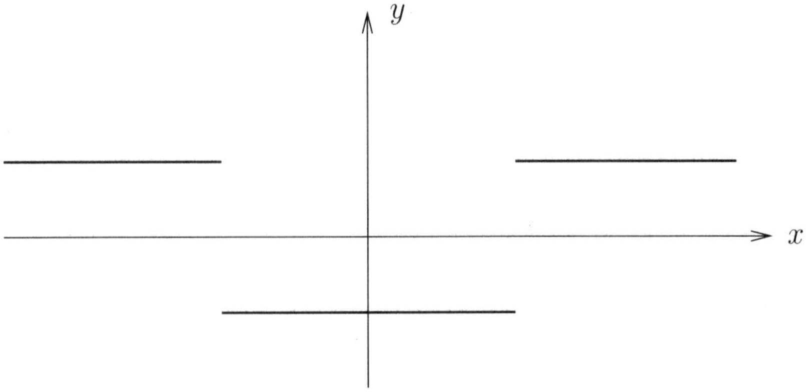

Figure 1.1. A continuous and differentiable function over the rationals

If the reader thinks it is not the most general solution, then can she find another solution?

After a little thought she may observe that if $g(t)$ is a solution of $(*)$ and we set

$$f(t) = g(t) - \frac{t^3}{3}$$

then $f'(t) = 0$ and the statement that $(**)$ is the most general solution of $(*)$ reduces to the following theorem.

Theorem 1.1. (Constant value theorem.) *If $f : \mathbb{R} \to \mathbb{R}$ is differentiable and $f'(t) = 0$ for all $t \in \mathbb{R}$, then f is constant.*

If this theorem is 'intuitively clear' over $\mathbb{R}$ it ought to be intuitively clear over $\mathbb{Q}$. The same remark applies to another 'intuitively clear' theorem.

Theorem 1.2. (The intermediate value theorem.) *If $f : \mathbb{R} \to \mathbb{R}$ is continuous, $b > a$ and $f(a) \geq 0 \geq f(b)$, then there exists a c with $b \geq c \geq a$. such that $f(c) = 0$.*

However, if we work over $\mathbb{Q}$ both putative theorems vanish in a puff of smoke.

Example 1.3. *If $f : \mathbb{Q} \to \mathbb{Q}$ is given by*

$$f(x) = -1 \quad \text{if } x^2 < 2,$$
$$f(x) = 1 \quad \text{otherwise,}$$

then

(i) f is a continuous function with $f(0) = -1$, $f(2) = 1$, yet there does not exist a c with $f(c) = 0$,

(ii) f is a differentiable function with $f'(x) = 0$ for all x, yet f is not constant.
[We sketch f in Figure 1.1.]

Sketch proof. We have not yet formally defined what continuity and differentiability are to mean. However, if the reader believes that f is discontinuous, she must find a point $x \in \mathbb{Q}$ at which f is discontinuous. Similarly, if she believes that f is not everywhere differentiable with derivative zero, she must find a point $x \in \mathbb{Q}$ for which this statement is false. The reader will be invited to give a full proof in Exercise 1.16 after continuity has been formally defined[2]. ▲

The question 'Is (∗∗) the most general solution of (∗)?' now takes on a more urgent note. Of course, we work in $\mathbb{R}$ and not in $\mathbb{Q}$, but we are tempted to echo Acton ([**1**], end of Chapter 7).

> This example is horrifying indeed. For if we have actually seen one tiger, is not the jungle immediately filled with tigers, and who knows where the next one lurks.

Here is another closely related tiger.

Exercise 1.4. *Continuing with Example 1.3, set $g(t) = t + f(t)$ for all t. Show that $g'(t) = 1 > 0$ for all t but that $g(-8/5) > g(-6/5)$.*

Thus, if we work in $\mathbb{Q}$, a function with strictly positive derivative need not be increasing.

Any proof that there are no tigers in $\mathbb{R}$ must start by identifying the difference between $\mathbb{R}$ and $\mathbb{Q}$ which makes calculus work on one even though it fails on the other. Both are 'ordered fields', that is, both support operations of 'addition' and 'multiplication' together with a relation 'greater than' ('order') with the properties that we expect. I have listed the properties in the appendix on page 357, but only to reassure the reader. We are not interested in the properties of general ordered fields but only in that particular property (whatever it may be) which enables us to avoid the problems outlined in Example 1.3 and so permits us to do analysis.

1.2. Limits

Many ways have been tried to make calculus rigorous and several have been successful. In this book I have chosen the first and most widely used path, via the notion of a limit. In theory, my account of this notion is complete in itself. However, my treatment is unsuitable for beginners and I expect

[2]We use ■ to mark the end of a proof and ▲ to mark the end of an argument which we do not claim to be a full proof.

my readers to have substantial experience with the use and manipulation of limits.

Throughout this section $\mathbb{F}$ will be an ordered field. The reader will miss nothing if she simply considers the two cases $\mathbb{F} = \mathbb{R}$ and $\mathbb{F} = \mathbb{Q}$. She will, however, miss something if she fails to check that everything we say applies to both cases equally.

Definition 1.5. *We work in an ordered field $\mathbb{F}$. We say that a sequence a_1, a_2, … tends to a limit a as n tends to infinity, or more briefly*

$$a_n \to a \text{ as } n \to \infty,$$

if, given any $\epsilon > 0$, we can find an integer $n_0(\epsilon)$ [read 'n_0 depending on ϵ'] such that

$$|a_n - a| < \epsilon \text{ for all } n \geq n_0(\epsilon).$$

The following properties of the limit are probably familiar to the reader.

Lemma 1.6. *We work in an ordered field $\mathbb{F}$.*

(i) The limit is unique. That is, if $a_n \to a$ and $a_n \to b$ as $n \to \infty$, then $a = b$.

(ii) If $a_n \to a$ as $n \to \infty$ and $1 \leq n(1) < n(2) < n(3) \ldots$, then $a_{n(j)} \to a$ as $j \to \infty$.

(iii) If $a_n = c$ for all n, then $a_n \to c$ as $n \to \infty$.

(iv) If $a_n \to a$ and $b_n \to b$ as $n \to \infty$, then $a_n + b_n \to a + b$.

(v) If $a_n \to a$ and $b_n \to b$ as $n \to \infty$, then $a_n b_n \to ab$.

(vi) Suppose that $a_n \to a$ as $n \to \infty$. If $a_n \neq 0$ for each n and $a \neq 0$, then $a_n^{-1} \to a^{-1}$.

(vii) If $a_n \leq A$ for each n and $a_n \to a$ as $n \to \infty$, then $a \leq A$. If $b_n \geq B$ for each n and $b_n \to b$ as $n \to \infty$, then $b \geq B$.

Proof. I shall give the proofs in detail, but the reader is warned that similar proofs will be left to her in the remainder of the book.

(i) By definition:–

Given $\epsilon > 0$ we can find an $n_1(\epsilon)$ such that $|a_n - a| < \epsilon$ for all $n \geq n_1(\epsilon)$.
Given $\epsilon > 0$ we can find an $n_2(\epsilon)$ such that $|a_n - b| < \epsilon$ for all $n \geq n_2(\epsilon)$.

Suppose, if possible, that $a \neq b$. Then setting $\epsilon = |a - b|/3$ we have $\epsilon > 0$. If $N = \max(n_1(\epsilon),\ n_2(\epsilon))$ then

$$|a - b| \leq |a_N - a| + |a_N - b| < \epsilon + \epsilon = 2|b - a|/3,$$

which is impossible. The result follows by reductio ad absurdum.

(ii) By definition,

† Given $\epsilon > 0$ we can find an $n_1(\epsilon)$ such that $|a_n - a| < \epsilon$ for all $n \geq n_1(\epsilon)$.

Let $\epsilon > 0$. Since $n(j) \geq j$ (proof by induction, if the reader demands a proof) we have $|a_{n(j)} - a| < \epsilon$ for all $j \geq n_1(\epsilon)$. The result follows.

(iii) Let $\epsilon > 0$. Taking $n_1(\epsilon) = 1$ we have

$$|a_n - c| = 0 < \epsilon$$

for all $n \geq n_1(\epsilon)$. The result follows.

(iv) By definition, † holds, as does

†† Given $\epsilon > 0$ we can find an $n_2(\epsilon)$ such that $|b_n - b| < \epsilon$ for all $n \geq n_2(\epsilon)$.

Observe that

$$|(a_n + b_n) - (a + b)| = |(a_n - a) + (b_n - b)| \leq |a_n - a| + |b_n - b|.$$

Thus if $\epsilon > 0$ and $n_3(\epsilon) = \max(n_1(\epsilon/2),\ n_2(\epsilon/2))$ we have

$$|(a_n + b_n) - (a + b)| \leq |a_n - a| + |b_n - b| < \epsilon/2 + \epsilon/2 = \epsilon$$

for all $n \geq n_3(\epsilon)$. The result follows.

(v) By definition, † and †† hold. Let $\epsilon > 0$. The key observation is that

$$(1) \qquad |a_n b_n - ab| \leq |a_n b_n - a_n b| + |a_n b - ab| = |a_n||b_n - b| + |b||a_n - a|.$$

If $n \geq n_1(1)$, then $|a_n - a| < 1$ so $|a_n| < |a| + 1$ and (1) gives

$$(2) \qquad |a_n b_n - ab| \leq (|a| + 1)|b_n - b| + |b||a_n - a|.$$

Thus setting[3] $n_3(\epsilon) = \max\big(n_1(1),\ n_1(\epsilon/(2(|b|+1))),\ n_2(\epsilon/(2(|a|+1)))\big)$ we see from (2) that

$$|a_n b_n - ab| < \epsilon/2 + \epsilon/2 = \epsilon$$

for all $n \geq n_3(\epsilon)$. The result follows.

(vi) By definition, † holds. Let $\epsilon > 0$. We observe that

$$(3) \qquad \left|\frac{1}{a_n} - \frac{1}{a}\right| = \frac{|a - a_n|}{|a||a_n|}.$$

Since $a \neq 0$ we have $|a|/2 > 0$. If $n \geq n_1(|a|/2)$ then $|a_n - a| < |a|/2$, so $|a_n| > |a|/2$ and (3) gives

$$(4) \qquad \left|\frac{1}{a_n} - \frac{1}{a}\right| \leq \frac{2|a - a_n|}{|a|^2}.$$

[3]The reader may ask why we use $n_1(\epsilon/(2(|b|+1)))$ rather than $n_1(\epsilon/(2|b|))$. Observe first that we have not excluded the possibility that $b = 0$. More importantly, observe that all we are required to do is to find an $n_3(\epsilon)$ that works, and it is futile to seek a 'best' $n_3(\epsilon)$ in these or similar circumstances.

Thus setting $n_3(\epsilon) = \max(n_1(|a|/2),\ n_1(a^2\epsilon/2))$ we see from (4) that

$$\left|\frac{1}{a_n} - \frac{1}{a}\right| < \epsilon$$

for all $n \geq n_3(\epsilon)$. The result follows.

(vii) The proof of the first sentence in the statement is rather similar to that of (i). By definition, † holds. Suppose, if possible, that $a > A$, that is, $a - A > 0$. Setting $N = n_1(a - A)$ we have

$$a_N = (a_N - a) + a \geq a - |a_N - a| > a - (a - A) = A,$$

contradicting our hypothesis. The result follows by reductio ad absurdum.

To prove the second sentence in the statement we can either give a similar argument, or set $a_n = -b_n$, $a = -b$ and $A = -B$ and use the first sentence.

[Your attention is drawn to part (ii) of Exercise 1.8.] ■

Exercise 1.7. *Prove that the first few terms of a sequence do not affect convergence. Formally, show that if there exists an N such that $a_n = b_n$ for $n \geq N$, then $a_n \to a$ as $n \to \infty$ implies $b_n \to a$ as $n \to \infty$.*

Exercise 1.8. *In this exercise we work within $\mathbb{Q}$. (The reason for this will appear in Section 1.5, which deals with the axiom of Archimedes.)*

(i) Observe that if $\epsilon \in \mathbb{Q}$ and $\epsilon > 0$, then $\epsilon = m/N$ for some strictly positive integers m and N. Use this fact to show, directly from Definition 1.5, that (if we work in $\mathbb{Q}$) $1/n \to 0$ as $n \to \infty$.

(ii) Show, by means of an example, that, if $a_n \to a$ and $a_n > b$ for all n, it does not follow that $a > b$. (In other words, taking limits may destroy strict inequality.)

Does it follow that $a \geq b$? Give reasons.

Exercise 1.9. *A more natural way of proving Lemma 1.6 (i) is to split the argument in two.*

(i) Show that if $|a - b| < \epsilon$ for all $\epsilon > 0$, then $a = b$.

(ii) Show that if $a_n \to a$ and $a_n \to b$ as $n \to \infty$, then $|a - b| < \epsilon$ for all $\epsilon > 0$.

(iii) Deduce Lemma 1.6 (i).

(iv) Give a similar 'split proof' for Lemma 1.6 (vii).

Exercise 1.10. *Here is another way of proving Lemma 1.6 (v). I do not claim that it is any simpler, but it introduces a useful idea.*

(i) Show from first principles that, if $a_n \to a$, then $ca_n \to ca$.

(ii) Show from first principles that, if $a_n \to a$ as $n \to \infty$, then $a_n^2 \to a^2$.

(iii) Use the relation $xy = ((x + y)^2 - (x - y)^2)/4$ together with (ii), (i) and Lemma 1.6 (iv) to prove Lemma 1.6 (v).

The next result is sometimes called the sandwich lemma or the squeeze lemma.

Exercise 1.11. *Suppose* $a_m \geq c_m \geq b_m$ *for all* m. *Then, if* $a_n \to c$ *and* $b_n \to c$, *it follows that* $c_n \to c$ *as* $n \to \infty$.

Suppose $|a_m| \geq |c_m| \geq |b_m|$ *for all* m *and that* $a_n \to c$ *and* $b_n \to c$ *as* $n \to \infty$. *Does it follow that* $c_n \to c$? *Give a proof or counterexample as appropriate.*

1.3. Continuity

Our definition of continuity follows the same line of thought.

Definition 1.12. *We work in an ordered field* $\mathbb{F}$. *Suppose that* E *is a subset of* $\mathbb{F}$ *and that* f *is some function from* E *to* $\mathbb{F}$. *We say that* f *is continuous at* $x \in E$ *if given any* $\epsilon > 0$ *we can find* $\delta_0(\epsilon, x)$ *[read '*δ_0 *depending on* ϵ *and* x*'] with* $\delta_0(\epsilon, x) > 0$ *such that*

$$|f(x) - f(y)| < \epsilon \text{ for all } y \in E \text{ such that } |x - y| < \delta_0(\epsilon, x).$$

If f *is continuous at every point of* E *we say that* $f : E \to \mathbb{F}$ *is a continuous function.*

The reader, who, I expect, has seen this definition before, and is, in any case, a mathematician, will be anxious to move on to see some theorems and proofs. Non-mathematicians might object that our definition does not correspond to their idea of what continuous should mean. If we consult the dictionary we find the definition 'connected, unbroken; uninterrupted in time or sequence: not discrete'. A mathematician would object that this merely defines one vague concept in terms of other equally vague concepts. However, if we rephrase our own definition in words we see that it becomes 'f is continuous if $f(y)$ is close to $f(x)$ whenever y is sufficiently close to x', and this clearly belongs to a different circle of ideas from the dictionary definition.

This will not be a problem when we come to define differentiability since there is no 'common sense' notion of differentiability. In the same way the existence of a 'common sense' notion of continuity need not trouble us provided that whenever we use the word 'continuous' we add under our breath 'in the mathematical sense' and we make sure our arguments make no appeal (open or disguised) to 'common sense' ideas of continuity.

Here are some simple properties of continuity.

Lemma 1.13. *We work in an ordered field* $\mathbb{F}$. *Suppose that* E *is a subset of* $\mathbb{F}$, *that* $x \in E$, *and that* f *and* g *are functions from* E *to* $\mathbb{F}$.

(i) If $f(x) = c$ *for all* $x \in E$, *then* f *is continuous on* E.

(ii) Let us define $f+g: E \to \mathbb{F}$ by $(f+g)(t) = f(t)+g(t)$ for all $t \in E$. If f and g are continuous at x, then so is $f+g$.

(iii) Let us define $f \times g : E \to \mathbb{F}$ by $(f \times g)(t) = f(t)g(t)$ for all $t \in E$. Then if f and g are continuous at x, then so is $f \times g$.

(iv) Suppose that $f(t) \neq 0$ for all $t \in E$. If f is continuous at x, then so is $1/f$.

Proof. Follow the proofs of parts (iii) to (vi) of Lemma 1.6. ∎

By repeated use of parts (ii) and (iii) of Lemma 1.13 it is easy to show that polynomials $P(t) = \sum_{r=0}^{n} a_r t^r$ are continuous. The details are spelled out in the next exercise.

Exercise 1.14. *We work in an ordered field $\mathbb{F}$. Prove the following results.*

(i) Suppose that E is a subset of $\mathbb{F}$ and that $f : E \to \mathbb{F}$ is continuous at $x \in E$. If $x \in E' \subset E$ then the restriction[4] $f|_{E'}$ of f to E' is also continuous at x.

(ii) If $J : \mathbb{F} \to \mathbb{F}$ is defined by $J(x) = x$ for all $x \in \mathbb{F}$, then J is continuous on $\mathbb{F}$.

(iii) Every polynomial P is continuous on $\mathbb{F}$.

(iv) Suppose that P and Q are polynomials and that Q is never zero on some subset E of $\mathbb{F}$. Then the rational function P/Q is continuous on E (or, more precisely, the restriction of P/Q to E is continuous).

The following result is little more than an observation, but will be very useful.

Lemma 1.15. *We work in an ordered field $\mathbb{F}$. Suppose that E is a subset of $\mathbb{F}$, that $x \in E$, and that f is continuous at x. If $x_n \in E$ for all n and $x_n \to x$ as $n \to \infty$, then $f(x_n) \to f(x)$ as $n \to \infty$.*

Proof. Left to reader. ∎

We have now done quite a lot of what is called ϵ, δ analysis, but all we have done is sharpened our proof of Example 1.3. The next exercise gives the details.

Exercise 1.16. *We work in $\mathbb{Q}$. The function f is that defined in Example 1.3.*

(i) Show that the equation $x^2 = 2$ has no solution. (See any elementary text on number theory or Exercise K.1.)

(ii) If $|x| \leq 2$ and $|\eta| \leq 1$, show that $|(x+\eta)^2 - x^2| \leq 5|\eta|$.

[4]Recall that, if $U \supseteq V$ and $f : U \to X$ is a function, we define the restriction of f to V to be the function $f|_V : V \to X$ given by $f|_V(v) = f(v)$ for all $v \in V$.

(iii) If $x^2 < 2$ *and* $\delta = (2-x^2)/6$, *show that* $y^2 < 2$ *whenever* $|x-y| < \delta$. *Conclude that* f *is continuous at* x.

(iv) Show that if $x^2 > 2$ *then* f *is continuous at* x.

(v) Conclude that f *is a continuous function.*

Unless we can isolate the property that distinguishes the rationals from the reals, we can make no progress.

1.4. The fundamental axiom

The key property of the reals, the *fundamental axiom* which makes everything work, can be stated as follows:

The fundamental axiom of analysis. *If* $a_n \in \mathbb{R}$ *for each* $n \geq 1$, $A \in \mathbb{R}$ *and* $a_1 \leq a_2 \leq a_3 \leq \ldots$ *and* $a_n < A$ *for each* n, *then there exists an* $a \in \mathbb{R}$ *such that* $a_n \to a$ *as* $n \to \infty$.

Less ponderously, and just as rigorously, the fundamental axiom for the real numbers says *every increasing sequence bounded above tends to a limit.*

Everything which depends on the fundamental axiom is analysis, everything else is mere algebra[5].

I claim that all the theorems of classical analysis can be obtained from the standard 'algebraic' properties of $\mathbb{R}$ together with the fundamental axiom. I shall start with the intermediate value theorem. (Here $[a, b]$ is the closed interval $[a, b] = \{x \in \mathbb{R} : a \leq x \leq b\}$.)

Theorem 1.17. (The intermediate value theorem.) *If* $f : [a, b] \to \mathbb{R}$ *is continuous and* $f(a) \geq 0 \geq f(b)$, *then there exists a* $c \in [a, b]$ *such that* $f(c) = 0$.

(The proof will eventually be given as Theorem 1.35.)

Exercise 1.18. *Assuming Theorem 1.17, prove the apparently more general result:–*

If $f : [a, b] \to \mathbb{R}$ *is continuous and* $f(a) \geq t \geq f(b)$, *then there exists a* $c \in [a, b]$ *such that* $f(c) = t$.

In less precise terms, 'if a function f is such that small changes in x produce small changes in $f(x)$, then f takes all the values between $f(a)$ and $f(b)$'.

How might this programme of obtaining the intermediate value theorem from the fundamental axiom fail?

[5]This is an analysis textbook. Remember that, to a differential geometer, the implicit function theorem of Section 13.2 is mere analysis.

(1) The reader has higher standards of rigour than I do but can fill in the gaps herself. For example, in the statement of Theorem 1.17, I do not explicitly say that $b \geq a$. Again, I talk about the 'algebraic' properties of $\mathbb{R}$ when, strictly speaking, a set cannot have algebraic properties and I should refer instead to the algebraic properties of $(\mathbb{R}, +, \times, >)$. Such problems may annoy the reader but do not cause the programme to fail.

(2) As is almost certainly the case, my proofs contain errors or have gaps which do not occur in other accounts of the material and which can thus be corrected. In such a case, I apologise, but it is I who have failed and not the programme.

(3) My proofs contain a serious error or have a serious gap which occurs in all accounts of this material. If this can be corrected then the programme survives but looks a great deal more shaky. If a serious error has survived for a century, who knows what other errors may lurk.

(4) All accounts contain an error which cannot be corrected or a gap that cannot be filled. The programme has failed.

We start our attempt with a simple consequence of the fundamental axiom.

Lemma 1.19. *In $\mathbb{R}$ every decreasing sequence bounded below tends to a limit.*

Proof. Observe that if $a_1, a_2, a_3, \ldots$ is a decreasing sequence bounded below then $-a_1, -a_2, -a_3, \ldots$ is an increasing sequence bounded above. We leave the details to the reader as an exercise. ∎

Exercise 1.20. *(i) If $m_1, m_2, \ldots$ is an increasing sequence of integers bounded above, show that there exists an N such that $m_j = m_N$ for all $j \geq N$.*

(ii) Show that every non-empty set $A \subseteq \mathbb{Z}$ bounded above has a maximum. More formally, show that if $A \subseteq \mathbb{Z}$, $A \neq \varnothing$ and there exists a K such that $K \geq a$ whenever $a \in A$, then there exists an $a_0 \in A$ with $a_0 \geq a$ whenever $a \in A$.

1.5. The axiom of Archimedes

Our first genuinely 'analysis' result may strike the reader as rather odd.

Theorem 1.21. (Axiom of Archimedes.)

$$\frac{1}{n} \to 0 \text{ as } n \to \infty$$

Proof. Observe that the $1/n$ form a decreasing sequence bounded below. Thus, by the fundamental axiom (in the form of Lemma 1.19), $1/n$ tends to some limit l. To identify this limit we observe that the limit of a product is a product of the limits (Lemma 1.6 (v)), so

$$\frac{1}{2n} = \frac{1}{2} \times \frac{1}{n} \to \frac{l}{2},$$

and the limit of a subsequence is the limit of the sequence (Lemma 1.6 (ii)), so

$$\frac{1}{2n} \to l$$

as $n \to \infty$. Thus, by the uniqueness of limits (Lemma 1.6 (i)), $l = l/2$, so $l = 0$ and $1/n \to 0$, as required. ■

Exercise 1.22. *[Exercise 1.8 (ii) concerned $\mathbb{Q}$. We repeat that exercise but this time we work in $\mathbb{R}$.] Show, by means of an example, that, if $a_n \to a$ and $a_n > b$ for all n, it does not follow that $a > b$. (In other words, taking limits may destroy strict inequality.)*

Does it follow that $a \geq b$? Give reasons.

Theorem 1.21 shows that there is no 'exotic' real number $\gimel$ say (to choose an exotic symbol) with the property that $1/n > \gimel$ for all integers $n \geq 1$ and yet $\gimel > 0$ (that is, $\gimel$ is smaller than all strictly positive rationals and yet strictly positive). There exist number systems with such exotic numbers (the famous 'non-standard analysis' of Abraham Robinson and the 'surreal numbers' of Conway constitute two such systems); but, just as the rationals are, in some sense, too small a system for the standard theorems of analysis to hold, so these non-Archimedean systems are, in some sense, too big. Eudoxus and Archimedes[6] realised the need for an axiom to show that there is no exotic number $\daleth$ bigger than any integer (i.e. $\daleth > n$ for all integers $n \geq 1$; to see the connection with our form of the axiom consider $\gimel = 1/\daleth$). However, in spite of its name, what was an axiom for Eudoxus and Archimedes is a theorem for us.

Exercise 1.23. *(i) Show that there does not exist a $K \in \mathbb{R}$ with $K > n$ for all $n \in \mathbb{Z}$ by using Theorem 1.21.*

(ii) Show the same result directly from the fundamental axiom.

Exercise 1.24. *(i) Show that if a is real and $0 \leq a < 1$, then $a^n \to 0$ as $n \to \infty$. Deduce that if a is real and $|a| < 1$, then $a^n \to 0$.*

(ii) Suppose that a is real and $a \neq -1$. Discuss the behaviour of

$$\frac{1 - a^n}{1 + a^n}$$

as $n \to \infty$ for the various possible values of a.

[Hint: $(1 - a^n)/(1 + a^n) = (a^{-n} - 1)/(a^{-n} + 1)$.]

[6] This is a simplification of a more complex story.

Here is an important consequence of the axiom of Archimedes.

Exercise 1.25. *(i) Use the fact that every non-empty set of integers bounded above has a maximum (see Exercise 1.20) to show that, if* $x \in \mathbb{R}$*, then there exists an integer* m *such that* $m \leq x < m+1$*. Show that* $|x-m| < 1$*.*

(ii) If $x \in \mathbb{R}$ *and* n *is a strictly positive integer, show that there exists an integer* q *such that* $|x - q/n| < 1/n$*.*

(iii) Deduce Lemma 1.26 below, using the axiom of Archimedes explicitly.

Lemma 1.26. *If* $x \in \mathbb{R}$*, then, given any* $\epsilon > 0$*, there exists a* $y \in \mathbb{Q}$ *such that* $|x-y| < \epsilon$*.*

Thus the rationals form a kind of skeleton for the reals. (We say that the rationals are dense in the reals.)

The reader will probably already be acquainted with the following definition.

Definition 1.27. *If* $a_1, a_2, \ldots$ *is a sequence of real numbers, we say that* $a_n \to \infty$ *as* $n \to \infty$ *if, given any real* K*, we can find an* $n_0(K)$ *such that* $a_n \geq K$ *for all* $n \geq n_0(K)$*.*

Exercise 1.28. *Using Exercise 1.23, show that* $n \to \infty$ *as* $n \to \infty$*.*

Exercise 1.28 shows that two uses of the words 'n tends to infinity' are consistent. It is embarrassing to state the result, but it would be still more embarrassing if it were false. Here is a another simple exercise on Definition 1.27.

Exercise 1.29. *Let* $a_1, a_2, \ldots$ *be a sequence of non-zero real numbers. Show that, if* $a_n \to \infty$*, then* $1/a_n \to 0$*. Is the converse true? Give a proof or counterexample.*

Exercise 1.30. *It is worth noting explicitly that ordered fields may satisfy the axiom of Archimedes but not the fundamental axiom. Show in particular that the rationals satisfy the axiom of Archimedes. (This is genuinely easy, so do not worry if your answer is brief.)*

Exercise 1.31. *The reader may be interested to see an ordered field containing* $\mathbb{Z}$ *which does not satisfy the axiom of Archimedes. We start by considering polynomials* $P(X) = \sum_{n=0}^{N} a_n X^n$ *with real coefficients* a_n*, and form the set* $\mathbb{K}$ *of rational functions* $P(X)/Q(X)$ *where* P *and* Q *are polynomials and* Q *is not the zero polynomial (that is,* $Q(X) = \sum_{m=0}^{M} b_m X^m$ *with* $b_M \neq 0$*.) Convince yourself that, if we use the usual standard formal algebraic rules for manipulating rational functions, then* $\mathbb{K}$ *is a field (that is, it satisfies conditions (A1) to (D) as set out in the axioms on page 357).*

To produce an order on $\mathbb{K}$ we define the set $\mathbb{P}$ to consist of all quotients of the form

$$\frac{\sum_{n=0}^{N} a_n X^n}{\sum_{m=0}^{M} b_m X^m}$$

with $a_N, b_M \neq 0$ and $a_N b_M > 0$. Convince yourself that this is a consistent definition (remember that the same quotient will have many different representations; $P(X)/Q(X) = R(X)P(X)/(R(X)Q(X))$ whenever $R(X)$ is a non-zero polynomial) and that $\mathbb{P}$ satisfies conditions (P1) to (P3). If we say that $P_1(X)/Q_1(X) > P_2(X)/Q_2(X)$ whenever we have $P_1(X)/Q_1(X) - P_2(X)/Q_2(X) \in \mathbb{P}$, condition (P4) is automatically satisfied and $\mathbb{K}$ is indeed an ordered field.

We note that the elements of $\mathbb{K}$ of the form $a/1$ with $a \in \mathbb{R}$ can be identified in a natural way with $\mathbb{R}$. If we make this natural identification, $\mathbb{K}$ contains $\mathbb{Z}$.

To see that the axiom of Archimedes fails, observe that $1/n > 1/X > 0$ for all $n \in \mathbb{Z}$, $n \geq 1$.

Of course, since the axiom of Archimedes fails, the fundamental axiom fails. By examining the proof of Theorem 1.21, show that the $1/n$ form a decreasing sequence bounded below but not tending to any limit.

If the reader knows some modern algebra she will see that our presentation can be sharpened in various ways. (It would be better to define $\mathbb{K}$ using equivalence classes. We should take greater care over checking consistency. The words 'identified in a natural way' should be replaced by 'there is an isomorphism of $(\mathbb{R}, +, \times, >)$ with a subfield of $\mathbb{K}$'.) Such readers should undertake the sharpening as an instructive exercise.

Exercise 1.32. *We shall not make any essential use of the decimal expansion of the real numbers, but it is interesting to see how it can be obtained. Let us write*

$$\mathbb{D} = \{n \in \mathbb{Z} : 9 \geq n \geq 0\}.$$

(i) If $x_j \in \mathbb{D}$, show that $\sum_{j=1}^{N} x_j 10^{-j} < 1$.

(ii) If $x_j \in \mathbb{D}$, show that $\sum_{j=1}^{N} x_j 10^{-j}$ converges to a limit x, say, as $N \to \infty$. Show that $0 \leq x \leq 1$, and that $x = 1$ if and only if $x_j = 9$ for all j.

(iii) If $y \in [0,1]$, show that there exist $y_j \in \mathbb{D}$ such that

$$y - 10^{-N} < \sum_{j=1}^{N} y_j 10^{-j} \leq y$$

and that $\sum_{j=1}^{N} y_j 10^{-j} \to y$ as $N \to \infty$.

(iv) Identify explicitly the use of the axiom of Archimedes in the proof of the last sentence of (ii) and in the proof of (iii).

(v) Suppose that $a_j,\ b_j \in \mathbb{D}$, $a_j = b_j$ *for* $j < M$ *and* $a_M > b_M$. *If* $\sum_{j=1}^{N} a_j 10^{-j} \to a$ *and* $\sum_{j=1}^{N} b_j 10^{-j} \to b$ *as* $N \to \infty$, *show that* $a \geq b$. *Give the precise necessary and sufficient condition for equality, and prove it.*

Exercise 1.33. *It seems, at first sight, that decimal expansion gives a natural way of treating real numbers. It is not impossible to do things in this way, but there are problems. Here is one of them. Let* $0 < a,\ b < 1$ *and* $c = ab$. *If* $\sum_{j=1}^{N} a_j 10^{-j} \to a$, $\sum_{j=1}^{N} b_j 10^{-j} \to b$, *and* $\sum_{j=1}^{N} c_j 10^{-j} \to c$, *find* c_j *in terms of the various* a_k *and* b_k. *(The reader is invited to reflect on this problem rather than solve it. Indeed, one question is 'what would constitute a nice solution?')*

Exercise 1.34. *Here is a neat application of decimal expansion.*

(i) Define $f : [0,1] \to [0,1]$ *as follows. Each* $x \in [0,1]$ *has a unique non-terminating decimal expansion*

$$x = \sum_{j=1}^{\infty} x_j 10^{-j}$$

with the x_j *integers such that* $0 \leq x_j \leq 9$. *If there exists an integer* $N \geq 2$ *such that* $x_{2j} = 1$ *for all* $j \geq N$ *but* $x_{2N-2} \neq 1$ *we set*

$$f(x) = \sum_{j=1}^{\infty} x_{2(j+N)+1} 10^{-j}.$$

Otherwise we set $f(x) = 0$. *Show that, given any* $y \in [0,1]$, *any* $\epsilon > 0$ *and any* $t \in [0,1]$, *we can find an* $x \in [0,1]$ *with* $|x - y| < \epsilon$ *such that* $f(x) = t$. *In other words* f *takes every value in* $[0,1]$ *arbitrarily close to every point.*

(ii) Show that if $0 \leq a < b \leq 1$ *then, given any* t *lying between* $f(a)$ *and* $f(b)$ *(that is to say,* t *with* $f(a) \leq t \leq f(b)$ *if* $f(a) \leq f(b)$ *or with* $f(b) \leq t \leq f(a)$ *if* $f(b) \leq f(a)$*), there exists a* $c \in [a,b]$ *such that* $f(c) = t$. *(Thus the fact that a function satisfies the conclusion of the intermediate value theorem (Exercise 1.18) does not show that it is well behaved.)*

(iii) Find a $g : \mathbb{R} \to [0,1]$ *such that, given any* $y \in \mathbb{R}$, *any* $\epsilon > 0$ *and any* $t \in [0,1]$, *we can find an* x *with* $|x - y| < \epsilon$ *with* $f(x) = t$.

(iv) (This may require a little thought.) Find a $g : \mathbb{R} \to \mathbb{R}$ *such that, given any* $y \in \mathbb{R}$, *any* $\epsilon > 0$ *and any* $t \in \mathbb{R}$, *we can find an* x *with* $|x - y| < \epsilon$ *such that* $f(x) = t$.

Although decimal expansion is a very useful way of representing numbers, it is not the only one. In Exercises K.13 and K.14 we discuss representation by continued fractions.

1.6. Lion hunting

Having dealt with the axiom of Archimedes, we can go on at once to prove the intermediate value theorem.

Theorem 1.35. (The intermediate value theorem.) *We work in $\mathbb{R}$. If $f : [a, b] \to \mathbb{R}$ is continuous and $f(a) \geq 0 \geq f(b)$, then there exists a $c \in [a, b]$ such that $f(c) = 0$.*

Proof. Since the method of proof is important to us, I shall label its three main parts.

Part A: Set $a_0 = a$ and $b_0 = b$. We observe that $f(a_0) \geq 0 \geq f(b_0)$. Now set $c_0 = (a_0 + b_0)/2$. There are two possibilities. Either $f(c_0) \geq 0$, in which case we set $a_1 = c_0$ and $b_1 = b_0$, or $f(c_0) < 0$, in which case we set $a_1 = a_0$ and $b_1 = c_0$. In either case, we have

$$\begin{gathered} f(a_1) \geq 0 \geq f(b_1), \\ a_0 \leq a_1 \leq b_1 \leq b_0, \\ \text{and } b_1 - a_1 = (b_0 - a_0)/2. \end{gathered}$$

Continuing inductively, we can find a sequence of pairs of points a_n and b_n such that

$$\begin{gathered} f(a_n) \geq 0 \geq f(b_n), \\ a_{n-1} \leq a_n \leq b_n \leq b_{n-1}, \\ \text{and } b_n - a_n = (b_{n-1} - a_{n-1})/2, \end{gathered}$$

for all $n \geq 1$.

Part B: We have $a_0 \leq a_1 \leq \cdots \leq a_n \leq b_0$, so that the a_n form an increasing sequence bounded above. By the fundamental axiom there is a real number c, say, such that $a_n \to c$ as $n \to \infty$. Since $a = a_0 \leq a_n \leq b_0$ we have $a \leq c \leq b$. We note also that $b_n - a_n = 2^{-n}(b_0 - a_0)$, so, by the axiom of Archimedes, $b_n - a_n \to 0$ and thus

$$b_n = a_n + (b_n - a_n) \to c + 0 = c$$

as $n \to \infty$.

Part C: Since f is continuous at c and $a_n \to c$, it follows that $f(a_n) \to f(c)$ as $n \to \infty$. Since $f(a_n) \geq 0$ for all n, it follows that $f(c) \geq 0$. A similar argument applied to the b_n shows that $f(c) \leq 0$. Since $0 \leq f(c) \leq 0$, it follows that $f(c) = 0$, and we are done. ■

Exercise 1.36. *(i) Give the complete details in the inductive argument in Part A of the proof of Theorem 1.35 above.*

(ii) Give the details of the 'similar argument applied to the b_n' which shows that $f(c) \leq 0$.

(iii) We use various parts of Lemma 1.6 in our Theorem 1.35. Identify the points where we use Lemma 1.6.

Exercise 1.37. *(i) Think how the argument used to prove Theorem 1.35 applies to* $[a, b] = [0, 1]$, $f(x) = 2-4x^2$. *(You are not asked to write anything, though you may well choose to draw a diagram.)*

(ii) Think also how the argument used to prove Theorem 1.35 applies to $[a, b] = [0, 1]$, $f(x) = (1 - 5x)(2 - 5x)(3 - 5x)$.

The method used to prove Theorem 1.35 is called 'Lion hunting'[7]. The method is also called 'successive bisection', 'bisection search' or simply 'bisection'.

Let us summarise the proof. In Part A we have evidence of a lion in the interval $[a_{n-1}, b_{n-1}]$. We split the interval into two halves $[a_{n-1}, c_{n-1}]$ and $[c_{n-1}, b_{n-1}]$ and show that, since there is evidence of a lion in the interval $[a_{n-1}, b_{n-1}]$, either there is evidence of a lion in $[a_{n-1}, c_{n-1}]$, in which case we take $[a_n, b_n] = [a_{n-1}, c_{n-1}]$, or, if there is no evidence of a lion in $[a_{n-1}, c_{n-1}]$ (this does not mean that there are no lions in $[a_{n-1}, c_{n-1}]$, simply that we do not have evidence of one), then it follows that there must be evidence of a lion in $[c_{n-1}, b_{n-1}]$ and we take $[a_n, b_n] = [c_{n-1}, b_{n-1}]$.

In Part B we use the fundamental axiom of analysis to show that these successive bisections 'close in' on a point c which we strongly suspect of being a lion. Finally in Part C we examine the point c to make sure that it really is a lion. (It might be a wolf or a left handed corkscrew.)

Let us see what goes wrong if we omit parts of the hypotheses of Theorem 1.35. If we omit the condition $f(a) \geq 0 \geq f(b)$, then we cannot even start Part A of the argument. The example $[a, b] = [0, 1]$, $f(x) = 1$ shows that the conclusion may indeed be false.

If we have $f(a) \geq 0 \geq f(b)$ but replace $\mathbb{R}$ by another ordered field for which the fundamental axiom does not hold, then Part A goes through perfectly but Part B fails. Example 1.3 with which we started shows that the conclusion may indeed be false. (Take $[a, b] = [-2, 0]$.)

If we have $f(a) \geq 0 \geq f(b)$ and we work over $\mathbb{R}$ but we do not demand f continuous, then Part C fails. Working over $\mathbb{R}$ we may take $[a, b] = [0, 1]$ and define $f(x) = 1$ for $x \leq 1/3$ and $f(x) = -1$ for $x > 1/3$. Parts A and B work perfectly to produce $c = 1/3$, but there is no lion (that is, no zero of f) at c.

Exercises 1.38 to 1.40 are applications of the intermediate value theorem.

[7]The name probably comes from *A Contribution to the Mathematical Theory of Big Game Hunting* by H. Pétard. This squib is reprinted in [**8**].

Exercise 1.38. *Show that any real polynomial of odd degree has at least one root. Is the result true for polynomials of even degree? Give a proof or counterexample.*

Exercise 1.39. *Suppose that* $g : [0,1] \to [0,1]$ *is a continuous function. By considering* $f(x) = g(x) - x$*, or otherwise, show that there exists a* $c \in [0,1]$ *with* $g(c) = c$*. (Thus every continuous map of* $[0,1]$ *into itself has a fixed point.)*

Give an example of a bijective[8] *continuous function* $k : (0,1) \to (0,1)$ *such that* $k(x) \neq x$ *for all* $x \in (0,1)$*. Give an example of a bijective continuous function* $l : \mathbb{R} \to \mathbb{R}$ *such that* $l(x) \neq x$ *for all* $x \in \mathbb{R}$*.*

Give an example of a bijective (but, necessarily, non-continuous) function $h : [0,1] \to [0,1]$ *such that* $h(x) \neq x$ *for all* $x \in [0,1]$*.*
[Hint: *First find a function* $H : [0,1] \setminus \{0,\ 1,\ 1/2\} \to [0,1] \setminus \{0,\ 1,\ 1/2\}$ *such that* $H(x) \neq x$*.*]

Exercise 1.40. *Every mid-summer day at six o'clock in the morning, the youngest monk from the monastery of Damt starts to climb the narrow path up Mount Dipmes. At six in the evening he reaches the small temple at the peak where he spends the night in meditation. At six o'clock in the morning on the following day he starts downwards, arriving back at the monastery at six in the evening. Of course, he does not always walk at the same speed. Show that, none the less, there will be some time of day when he will be at the same place on the path on both his upward and downward journeys.*

Finally we give an example of lion hunting based on splitting the interval in five equal pieces rather than bisecting it.

Exercise 1.41. *Suppose that we have a sequence* $x_1,\ x_2,\ \ldots$ *of real numbers. Let* $[a_0, b_0]$ *be any closed interval. Show that we can find a sequence of pairs of points* a_n *and* b_n *such that*

$$\begin{aligned}
&\textit{either } x_n \geq b_n + (b_{n-1} - a_{n-1})/5 \textit{ or } x_n \leq a_n - (b_{n-1} - a_{n-1})/5,\\
&\qquad a_{n-1} \leq a_n \leq b_n \leq b_{n-1},\\
&\textit{and } b_n - a_n = (b_{n-1} - a_{n-1})/5,
\end{aligned}$$

for all $n \geq 1$*.*

Show that a_n *and* b_n *tend to some limit* $c \in [a_0, b_0]$*. Show further that, for each* $n \geq 1$*, either* $x_n \geq c + (b_{n-1} - a_{n-1})/5$ *or* $x_n \leq c - (b_{n-1} - a_{n-1})/5$ *and so in particular* $x_n \neq c$*.*

[8] If you do not know the terms injective, surjective and bijective, look at Exercise K.89. The traditional term for surjective is 'onto', but the confusion surrounding the terms 'one-one' and 'one-to-one' is now so deeply embedded that it is unwise to use them.

Thus we cannot write the points of $[a_0, b_0]$ as a sequence. (We say that $[a_0, b_0]$ is uncountable. The reader may know a proof of this via decimal expansions.)

For more on countability and an important extension of this exercise, see Appendix B and Exercise B.7 within that appendix.

1.7. The mean value inequality

Having disposed of one of the three tigers with which we started, by proving the intermediate value theorem, we now dispose of the other two by using the 'mean value inequality'.

Theorem 1.42. (The mean value inequality.) *Let U be the open interval (α, β) on the real line $\mathbb{R}$. Suppose that $K \geq 0$ and that $a, b \in U$ with $b > a$. If $f : U \to \mathbb{R}$ is differentiable with $f'(t) \leq K$ for all $t \in U$, then*

$$f(b) - f(a) \leq (b - a)K.$$

Before we can do this we must define differentiability and the derivative. The reader will almost certainly be familiar with a definition along the following lines.

Definition 1.43. *Let U be an open set in $\mathbb{R}$. We say that a function $f : U \to \mathbb{R}$ is differentiable at $t \in U$ with derivative $f'(t)$ if, given $\epsilon > 0$, we can find a $\delta(t, \epsilon) > 0$ such that $(t - \delta(t, \epsilon), t + \delta(t, \epsilon)) \subseteq U$ and*

$$\left| \frac{f(t+h) - f(t)}{h} - f'(t) \right| < \epsilon$$

whenever $0 < |h| < \delta(t, \epsilon)$.

In Chapter 6 we shall define a more general notion of differentiation and derive many of its properties. For the moment all we need is Definition 1.43.

Exercise 1.44. *Let U be an open interval in $\mathbb{R}$ and suppose functions $f, g : U \to \mathbb{R}$ are differentiable at $t \in U$ with derivatives $f'(t)$ and $g'(t)$. If λ, $\mu \in \mathbb{R}$, show that $\lambda f + \mu g$ is differentiable at t with derivative $\lambda f'(t) + \mu g'(t)$.*

To obtain Theorem 1.42 we prove an apparently weaker result.

Lemma 1.45. *We use the notation and assumptions of Theorem 1.42. If $\epsilon > 0$, then $f(b) - f(a) \leq (K + \epsilon)(b - a)$.*

Proof of Theorem 1.42 from Lemma 1.45. Because the inequality

$$f(b) - f(a) \leq (K + \epsilon)(b - a)$$

holds for all $\epsilon > 0$, we have $f(b) - f(a) \leq K(b - a)$. ■

Proof of Lemma 1.45. We suppose that $f(b)-f(a)>(K+\epsilon)(b-a)$ and use lion-hunting to derive a contradiction. Set $a_0=a$, $b_0=b$. We observe that $f(b_0)-f(a_0)>(K+\epsilon)(b_0-a_0)$. Now set $c_0=(a_0+b_0)/2$. Since

$$\begin{aligned}\big(f(c_0)-f(a_0)-(K+\epsilon)(c_0-a_0)\big)+\big(f(b_0)-f(c_0)-(K+\epsilon)(b_0-c_0)\big)\\ =\big(f(b_0)-f(a_0)-(K+\epsilon)(b_0-a_0)\big)>0,\end{aligned}$$

at least one of the expressions $f(c_0)-f(a_0)-(K+\epsilon)(c_0-a_0)$ and $(f(b_0)-f(c_0)-(K+\epsilon)(b_0-c_0)$ must be strictly positive. If $f(b_0)-f(c_0)-(K+\epsilon)(b_0-c_0)>0$, we set $a_1=c_0$ and $b_1=b_0$. Otherwise, we set $a_1=a_0$ and $b_1=c_0$. In either case, we have

$$\begin{aligned}f(b_1)-f(a_1)&>(K+\epsilon)(b_1-a_1),\\ a_0\leq a_1&\leq b_1\leq b_0,\\ \text{and } b_1-a_1&=(b_0-a_0)/2.\end{aligned}$$

Continuing inductively, we can find a sequence of pairs of points a_n and b_n such that

$$\begin{aligned}f(b_n)-f(a_n)&>(K+\epsilon)(b_n-a_n),\\ a_{n-1}\leq a_n&\leq b_n\leq b_{n-1},\\ \text{and } b_n-a_n&=(b_{n-1}-a_{n-1})/2,\end{aligned}$$

for all $n\geq 1$.

We have $a_0\leq a_1\leq\cdots\leq a_n\leq b_0$, so that the a_n form an increasing sequence bounded above. By the fundamental axiom there is a real number c, say, such that $a_n\to c$ as $n\to\infty$. Since $a=a_0\leq a_n\leq b_0$ we have $a\leq c\leq b$ and similarly $a_N\leq c\leq b_N$ for all N. We note also that $b_n-a_n=2^{-n}(b_0-a_0)$; so, by the axiom of Archimedes, $b_n-a_n\to 0$ and thus

$$b_n=a_n+(b_n-a_n)\to c+0=c$$

as $n\to\infty$.

Since f is differentiable at c, we can find a $\delta>0$ such that $(c-\delta,c+\delta)\subseteq U$ and

$$\left|\frac{f(c+h)-f(c)}{h}-f'(c)\right|<\epsilon/2$$

whenever $0<|h|<\delta$. Thus

$$|f(c+h)-f(c)-f'(c)h|\leq\epsilon|h|/2$$

whenever $|h|<\delta$, and so, since $f'(c)\leq K$,

$$\begin{aligned}f(c+h)-f(c)&\leq(K+\epsilon/2)h &&\text{for } 0\leq h<\delta,\\ f(c)-f(c+h)&\leq-(K+\epsilon/2)h &&\text{for } -\delta\leq h\leq 0.\end{aligned}$$

Since $a_n \to c$ and $b_n \to c$, we can find an N such that $|a_N - c| < \delta$ and $|b_N - c| < \delta$. It follows, first taking $h = a_N - c$ and then $h = b_N - c$, that

$$f(c) - f(a_N) \le (K + \epsilon/2)(c - a_N)$$
$$\text{and } f(b_N) - f(c) \le (K + \epsilon/2)(b_N - c).$$

Thus

$$\begin{aligned} f(b_N) - f(a_N) &= \big(f(b_N) - f(c)\big) + \big(f(c) - f(a_N)\big) \\ &\le (K + \epsilon/2)(b_N - c) + (K + \epsilon/2)(c - a_N) \\ &= (K + \epsilon/2)(b_N - a_N), \end{aligned}$$

contradicting our earlier assumption that $f(b_n) - f(a_n) > (K + \epsilon)(b_n - a_n)$ for all n.

Thus our initial assumption must be wrong, and the theorem is proved. ■

Theorem 1.42 immediately proves Theorem 1.1.

Theorem 1.46. (The constant value theorem.) *Let U be the open interval (α, β) or the real line $\mathbb{R}$. If $f : U \to \mathbb{R}$ is differentiable with $f'(t) = 0$ for all $t \in U$, then f is constant.*

Proof. Let b, $a \in U$ with $b \ge a$. By applying Theorem 1.42 to f with $K = 0$ we see that $f(b) - f(a) \le 0$. By applying Theorem 1.42 to $-f$ with $K = 0$, we see that $f(a) - f(b) \le 0$. Thus $f(a) = f(b)$. But a and b were arbitrary, so f is constant. ■

In Section 1.1 we noted the importance of the following simple corollary.

Theorem 1.47. *Let U be the open interval (α, β) or the real line $\mathbb{R}$. If the functions $f, g : U \to \mathbb{R}$ are differentiable with $f'(t) = g'(t)$ for all $t \in U$, then there exists a constant c such that $f(t) = g(t) + c$ for all $t \in U$.*

Proof. Apply Theorem 1.46 to $f - g$. ■

In *A Tour of the Calculus* [**5**], Berlinski greets this theorem with a burst of rhetoric.

> ... functions agreeing in their derivates, the theorem states, differ on an interval only by a constant. It is the derivative of a real-valued function that like some pulsing light illuminates again the behaviour of the function, enforcing among otherwise anarchic and wayward mathematical objects a stern uniformity of behaviour. Such is the proximate burden of the mean value theorem, which is now revealed to play a transcendental role in the scheme of things.

To which the present author, constrained by the conventions of textbook writing from such active expressions of enthusiasm, can only murmur 'Hear, hear'.

It is fairly easy to see that Theorem 1.42 is equivalent to the following result.

Theorem 1.48. *Let U be the open interval (α, β) or the real line $\mathbb{R}$. Suppose that $a, b \in U$ and $b > a$. If $g : U \to \mathbb{R}$ is differentiable with $g'(t) \geq 0$ for all $t \in U$, then*

$$g(b) - g(a) \geq 0.$$

Exercise 1.49. *(i) By taking $f = -g$ and $K = 0$, prove Theorem 1.48 from Theorem 1.42.*

(ii) By taking $g(t) = Kt - f(t)$, prove Theorem 1.42 from Theorem 1.48.

Thus a function with positive derivative is increasing.

The converse result is 'purely algebraic' in the sense that it does not involve the fundamental axiom.

Lemma 1.50. *If $g : (a, b) \to \mathbb{R}$ is differentiable and increasing on (a, b) then $g'(t) \geq 0$ for all $t \in (a, b)$.*

Exercise 1.51. *Use the definition of the derivative to prove Lemma 1.50. [*Hint: *Show first that, given any $\epsilon > 0$, we have $g'(t) > -\epsilon$.]*

Readers who know the mean value theorem (given as Theorem 4.49 later) may wish to extend Theorem 1.42 as follows.

Exercise 1.52. *Suppose that $a, b \in \mathbb{R}$ and $b > a$. If $f : [a, b] \to \mathbb{R}$ is continuous and f is differentiable on (a, b) with $f'(t) \leq K$ for all $t \in (a, b)$, use Theorem 1.42 on intervals (a_n, b_n) with $a < a_n < b_n < b$ and continuity to show that*

$$f(b) - f(a) \leq (b - a)K.$$

Experience shows that students do not fully realise the importance of the mean value inequality. Readers should take note whenever they use Theorem 1.47 or Theorem 1.48 since they are then indirectly using the mean value inequality.

1.8. Full circle

We began this chapter with an example of an ordered field for which the intermediate value theorem failed. A simple extension of that example shows that just as the fundamental axiom implies the intermediate value theorem, so the intermediate value theorem implies the fundamental axiom.

Theorem 1.53. *Let $\mathbb{F}$ be an ordered field for which the intermediate value theorem holds, that is to say:*

Let $a, b \in \mathbb{F}$ with $b > a$ and set $[a,b] = \{x \in \mathbb{F}\,;\, b \geq x \geq a\}$. If $f : [a,b] \to \mathbb{F}$ is continuous and $f(a) \geq 0 \geq f(b)$, then there exists a $c \in [a,b]$ such that $f(c) = 0$.

Then the fundamental axiom holds. That is to say:

If $a_n \in \mathbb{F}$ for each $n \geq 1$, $A \in \mathbb{F}$, $a_1 \leq a_2 \leq a_3 \leq \ldots$ and $a_n < A$ for each n, then there exists a $c \in \mathbb{F}$ such that $a_n \to c$ as $n \to \infty$.

Proof. Suppose $a_1 \leq a_2 \leq a_3 \leq \ldots$ and $a_n < A$ for all n. Choose $a < a_1$ and $b > A$. Define $f : [a,b] \to \mathbb{F}$ by

$$\begin{aligned} f(x) &= 1 && \text{if } x < a_n \text{ for some } n, \\ f(x) &= -1 && \text{otherwise.} \end{aligned}$$

Since f does not take the value 0, the intermediate value theorem tells us that there must be a point $c \in [a,b]$ at which f is discontinuous.

Suppose that $y < a_N$ for some N, so $\epsilon = a_N - y > 0$. Then, whenever $|x-y| < \epsilon/2$, we have $x \leq a_N - \epsilon/2$, so $f(x) = f(y) = 1$ and $|f(x)-f(y)| = 0$. Thus f is continuous at y. We have shown that $c \geq a_n$ for all n.

Suppose that there exists an $\epsilon > 0$ such that $y \geq a_n + \epsilon$ for all n. Then, whenever $|x-y| < \epsilon/2$, we have $x \geq a_n + \epsilon/2$ for all n, so $f(x) = f(y) = -1$ and $|f(x) - f(y)| = 0$. Thus f is continuous at y. We have shown that, given $\epsilon > 0$, there exists an $n_0(\epsilon)$ such that $c < a_{n_0(\epsilon)} + \epsilon$.

Combining the results of the two previous paragraphs with the fact that the a_n form an increasing sequence, we see that, given $\epsilon > 0$, there exists an $n_0(\epsilon) > 0$ such that $a_n \leq c < a_n + \epsilon$ and so $|c - a_n| < \epsilon$ for all $n \geq n_0(\epsilon)$. Thus $a_n \to c$ as $n \to \infty$, and we are done. ∎

Exercise 1.54. *State and prove results similar to Theorem 1.53 for Theorem 1.47 and Theorem 1.42. (Thus the constant value and the mean value theorem are equivalent to the fundamental axiom.)*

1.9. Are the real numbers unique?

Our statement of Theorem 1.53 raises the question as to whether there may be more than one ordered field satisfying the fundamental axiom. The answer, which is as good as we can hope for, is that all ordered fields satisfying the fundamental axiom are isomorphic. The formal statement is given in the following theorem.

Theorem 1.55. *If the ordered field $(\mathbb{F}, +, \times, >)$ satisfies the fundamental axiom of analysis, then there exists a bijective map $\theta : \mathbb{R} \to \mathbb{F}$ such that, if*

x, $y \in \mathbb{R}$, then

$$\begin{aligned}\theta(x+y) &= \theta(x)+\theta(y),\\ \theta(xy) &= \theta(x)\theta(y),\\ \theta(x) > 0 &\text{ whenever } x > 0.\end{aligned}$$

Exercise 1.56. *Show that the conditions on θ in Theorem 1.55 imply that*

$$\theta(x) > \theta(y) \text{ whenever } x > y.$$

We shall need neither Theorem 1.55 nor its method of proof. For completeness we sketch a proof in Exercises A.1 to A.5 starting on page 359, but I suggest that the reader merely glance at them. Once the reader has acquired sufficient experience in both analysis and algebra, she will find that the proof of Theorem 1.55 writes itself. Until then it is not really worth the time and effort involved.

Chapter 2

A First Philosophical Interlude ♡♡

This book contains two philosophical interludes. The reader may omit both on the grounds that mathematicians should do mathematics and not philosophise about it[1]. However, the reader who has heard Keynes' gibe that 'Practical men who believe themselves to be exempt from any intellectual influences, are usually the slaves of some defunct economist' may wonder what kind of ideas underlie the standard presentation of analysis given in this book.

2.1. Is the intermediate value theorem obvious? ♡♡

It is clear from Example 1.3 that the intermediate value theorem is not obvious to a well trained mathematician. Psychologists have established that it is not obvious to very small children, since they express no surprise when objects appear to move from one point to another without passing through intermediate points. But most other human beings consider it obvious. Are they right?

The Greek philosopher Zeno has made himself unpopular with 'plain honest men' for over 2000 years by suggesting that the world may not be as simple as 'plain honest men' believe. I shall borrow and modify some of his arguments.

[1]My father on being asked by Dieudonné to name any mathematicians who had been influenced by any philosopher, instantly replied 'Descartes and Leibniz'.

There are two ways in which the intermediate value theorem might be obvious: through observation and introspection. Is it obvious through observation? Suppose we go to the cinema and watch the film of an arrow's flight. It certainly looks as though the arrow is in motion passing through all the points of its flight. But, if we examine the film, we see that it consists of series of pictures of the arrow at rest in different positions. The arrow takes up a finite, though large, number of positions in its apparent flight and the tip of the arrow certainly does not pass through all the points of the trajectory. Both the motion and the apparent truth of the intermediate value theorem are illusions. If they are illusory in the cinema, might they not be illusory in real life?

There is another problem connected with the empirical study of the intermediate value theorem. As Theorem 1.53 proves, the intermediate value theorem is deeply linked with the structure of the real numbers. If the intermediate value theorem is physically obvious then the structure of the real numbers should also be obvious. To give an example, the intermediate value theorem implies that there is a positive real number x satisfying $x^2 = 2$. We know that this number $\sqrt{2}$ is irrational. But the existence of irrational numbers was so non-obvious that this discovery precipitated a crisis in Greek mathematics[2].

Of course, it is possible that we are cleverer than our Greek forefathers, or at least better educated, and what was not obvious to them may be obvious to us. Let us try and see whether the existence of $\sqrt{2}$ is physically obvious.

One way of doing this would be to mark out a length of $\sqrt{2}$ metres on a steel rod. We can easily mark out a length of 1.4 metres (with an error of less than .05 metres). With a little work we can mark out a length of 1.41 metres (with an error of less than .005 metres) or, indeed, a length of 1.414 metres (with an error of less than .0005 metres). But it is hard to see why looking at a length of $1.414 \pm .0005$ metres should convince us of the existence of a length $\sqrt{2}$. Of course we can imagine the process continued, but few 'plain honest men' would believe a craftsman who told them that because they could work to an accuracy of $\pm.000\,05$ metres they could therefore work to an accuracy of $\pm.000\,000\,005$ metres 'and so on'. Indeed if someone claimed to have marked out a length of 1.414 213 562 373 095 metres to an accuracy of $\pm 5 \times 10^{-16}$ metres we might point out that the claimed error was less than the radius of a proton.

[2]Unfortunately, we have no contemporary record of how the discovery was viewed, and we can be certain that the issues were looked at in a very different way to that which we see them today. Many modern historians even deny that there was a crisis, but there is substantial indirect evidence for the traditional view.

If we try to construct such a length indirectly as the length of the hypotenuse of a right angled triangle with shorter sides both of length 1 metre we simply transfer the problem to that of producing two lengths of 1 metre (we can produce one length of 1 metre by using the standard metre, but how can we copy it exactly?) and an exact right angle. If we try to construct it by graphing $y = x^2 - 2$ and looking at the intersection with the axis $y = 0$, close inspection of the so-called intersection merely reveals a collection of graphite blobs (or pixels or whatever). Once again, we learn that there are many numbers which almost satisfy the equation $x^2 = 2$ but not whether there are any which satisfy the equation exactly.

Since the intermediate value theorem does not seem to be obvious by observation, let us see whether it is obvious by introspection. Instead of observing the flight of an actual physical arrow, let us close our eyes and imagine the flight of an ideal arrow from A to B. Here a difficulty presents itself. We can easily picture the arrow at A and at B but, if we wish to imagine the complete flight, we must picture the arrow at the half way point A_1 from A to B. This is easy enough but, of course, the same argument shows that we must picture the arrow at the half way point A_2 from A to A_1, at the half way point A_3 from A to A_2 and so on. Thus in order to imagine the flight of the arrow we must see it at each of the infinite sequence of points A_1, A_2, A_3, ... Since an electronic computer can only perform a finite number of operations in a given time, this presents a problem to those who believe that the brain is a kind of computer. Even those who believe that, in some way, the brain transcends the computer may feel some doubts about our ability to picture the arrow at each of the *uncountable* set[3] of points which according to the 'obvious' intermediate value theorem (the intermediate value theorem implies the fundamental axiom by Theorem 1.53 and the fundamental axiom yields the result of Exercise 1.41) must be traversed by the tip of the arrow on its way to the target.

There is another difficulty when we try to picture the path of the arrow. At first, it may seem to the reader to be the merest quibble, but in my opinion (and that of many cleverer people) it becomes more troubling as we reflect on it. It is due to Zeno but, as with his other 'paradoxes', we do not have his own words. Consider the flying arrow. At every instant it has a position, that is, it occupies a space equal to itself. But everything that occupies a space equal to itself is at rest. Thus the arrow is at rest.

From the time of Zeno to the end of the 19th century, all those who argued about Zeno's paradoxes, whether they considered them 'funny little riddles' or deep problems, did not doubt that, in fact, the arrow did have a position and velocity and did, indeed, travel along some path from A to B.

[3]Plausible statement B.10 is relevant here.

Today we are not so sure. In the theory of quantum mechanics it is impossible to measure the position and momentum of a particle simultaneously to more than a certain accuracy. But 'plain honest men' are uninterested in what they cannot measure. It is, of course, possible to believe that the particle has an exact position and momentum which we can never know, just as it is possible to believe that the earth is carried by invisible elephants standing on an unobservable turtle, but it is surely more reasonable to say that particles do not have position and momentum (and so do not have position and velocity) in the sense that our too hasty view of the world attributed to them. Again the simplest interpretation of experiments like the famous two slit experiment which reveal the wavelike behaviour of particles is that particles do not travel along one single path but along all possible paths.

A proper modesty should reduce our surprise that the real world and the world of our thoughts should turn out to be more complicated than we first expected.

> Two things fill the mind with ever-fresh admiration and reverence, the more often and the more enduringly the mind is occupied with them: the starry heaven above me and the moral law within me. [Kant, *Critique of Practical Reason*]

We cannot justify results like the intermediate value theorem by an appeal to our fallible intuition or an imperfectly understood real world, but we can try to prove them from axioms. Those who wish may argue as to whether and in what sense those axioms are 'true' or 'a model for reality', but these are not mathematical problems.

A note on Zeno: We know practically nothing about Zeno except that he wrote a book containing various paradoxes. The book itself has been lost, and we only know the paradoxes in the words of other Greek philosophers who tried to refute them. Plato wrote an account of a discussion between Socrates, Zeno and Zeno's teacher Parmenides, but it is probably fictional. The most that we can hope for is that, like one of those plays in which Einstein meets Marilyn Monroe, it remains true to what was publicly known.

According to Plato:

> Parmenides was a man of distinguished appearance. By that time he was well advanced in years with hair almost white; he may have been sixty-five. Zeno was nearing forty, a tall and attractive figure. It was said that he had been Parmenides' lover. They were staying with Pythadorus .... Socrates and a few others came there, anxious to hear a reading of Zeno's treatise, which the two visitors had brought for the first time to Athens.

Parmenides taught that what is must be whole, complete, unchanging and one. The world may appear to be made of many changing things, but change and plurality are illusions. Zeno says that his book is

> ... a defense of Parmenides argument against those who try to make fun of it by showing that his supposition, that [only one thing exists] leads to many absurdities and contradictions. This book, then, is a retort against those who assert a plurality. It pays them back in the same coin with something to spare, and aims at showing that on a thorough examination, the assumption that there is a plurality leads to even more absurd consequences than the hypothesis of the one. It was written in that controversial spirit in my young days ... [**42**]

Many historians of mathematics believe that Zeno's paradoxes and the discussion of the reasoning behind them were a major factor in the development of the Greek method of mathematical proof, which we use to this day.

Chapter 3

Other Versions of the Fundamental Axiom

Since all of analysis depends on the fundamental axiom, it is not surprising that mathematicians have developed a number of different methods of proof to exploit it. We have already seen the method of 'lion hunting'. In this chapter we see two more: the 'supremum method' and the 'Bolzano–Weierstrass method'.

3.1. The supremum

Just as the real numbers are distinguished among all systems enjoying the same algebraic properties (that is, all ordered fields) by the fundamental axiom, so the integers are distinguished among all systems enjoying the same algebraic properties (that is, all ordered integral domains) by the statement that every non-empty set of the integers bounded above has a maximum.

Well ordering of integers. *If $A \subseteq \mathbb{Z}$, $A \neq \varnothing$ and there exists a K such that $K \geq a$ whenever $a \in A$, then there exists an $a_0 \in A$ with $a_0 \geq a$ whenever $a \in A$.*

(We proved this as a consequence of the fundamental axiom in Exercise 1.20.)

The power of this principle is illustrated by the fact that it justifies the method of mathematical induction.

Exercise 3.1. *(i) State formally and prove the statement that every non-empty set of integers bounded below has a minimum.*

(ii) Suppose that $\mathcal{P}$ is some mathematical property which may be possessed by a positive integer n. Let $P(n)$ be the statement that n possesses the property $\mathcal{P}$. Suppose that

(a) $P(0)$ is true.

(b) If $P(n)$ is true, then $P(n+1)$ is true.

By considering the set

$$E = \{n \in \mathbb{Z} : n \geq 0 \text{ and } P(n) \text{ is true}\},$$

and using (i), show that $P(n)$ is true for all positive integers n.

(iii) Examine what happens to your proof of (ii) when $P(n)$ is the statement $n \geq 4$, when $P(n)$ is the statement $n \leq 4$ and when $P(n)$ is the statement $n = -4$.

Unfortunately, as the reader probably knows, a bounded non-empty set of real numbers need not have a maximum.

Example 3.2. *If $E = \{x \in \mathbb{R} : 0 < x < 1\}$, then E is a non-empty bounded set of real numbers with no maximum.*

Proof. If $a \geq 1$ or if $a \leq 0$, then $a \notin E$, so a is not a maximum. If $0 < a < 1$, then $a < (a+1)/2 \in E$, so a is not a maximum. ∎

However, we shall see in Theorem 3.7 that any non-empty bounded set of real numbers does have a least upper bound (supremum).

Definition 3.3. *Consider a non-empty set A of real numbers. We say that α is a* least upper bound *(or* supremum*) for A if the following two conditions hold.*

(i) $\alpha \geq a$ for all $a \in A$.

(ii) If $\beta \geq a$ for all $a \in A$, then $\beta \geq \alpha$.

Lemma 3.4. *If the supremum exists, it is unique.*

Proof. The only problem is setting out the matter in the right way.

Suppose α and α' are least upper bounds for a non-empty set A of real numbers. Then:

(i) $\alpha \geq a$ for all $a \in A$.

(ii) If $\beta \geq a$ for all $a \in A$, then $\beta \geq \alpha$.

(ii′) $\alpha' \geq a$ for all $a \in A$.

(ii′) If $\beta \geq a$ for all $a \in A$, then $\beta \geq \alpha'$.

By (i), $\alpha \geq a$ for all $a \in A$, so by (ii′), $\alpha \geq \alpha'$. Similarly, $\alpha' \geq \alpha$, so $\alpha = \alpha'$ and we are done. ∎

It is convenient to have the following alternative characterisation of the supremum.

Lemma 3.5. *Consider a non-empty set A of real numbers; α is a supremum for A if and only if the following two conditions hold.*

(i) $\alpha \geq a$ for all $a \in A$.

(ii) Given $\epsilon > 0$, there exists an $a \in A$ such that $a + \epsilon \geq \alpha$.

Proof. Left to the reader. ■

We write $\sup A$ or $\sup_{a \in A} a$ for the supremum of A, if it exists.

Exercise 3.6. *Check that the discussion of the supremum given above carries over to all ordered fields. (There is no need to write anything unless you wish to.)*

Here is the promised theorem.

Theorem 3.7. (Supremum principle.) *If A is a non-empty set of real numbers which is bounded above (that is, there exists a K such that $a \leq K$ for all $a \in A$), then A has a supremum.*

Note that the result is false for the rationals.

Exercise 3.8. *Let us work in $\mathbb{Q}$. Using Exercise 1.16 (ii), or otherwise, show that*

$$\{x \in \mathbb{Q} : x^2 < 2\}$$

has no supremum.

We must thus use the fundamental axiom in the proof of Theorem 3.7. One way to do this is to use 'lion hunting'.

Exercise 3.9. *(i) If A satisfies the hypotheses of Theorem 3.7, show that we can find a_0, $b_0 \in \mathbb{R}$ with $a_0 < b_0$ such that $a \leq b_0$ for all $a \in A$ but $[a_0, b_0] \cap A \neq \varnothing$.*

(ii) Continuing with the discussion of (i), show that we can find a sequence of pairs of points a_n and b_n such that

$$\begin{aligned} a &\leq b_n \text{ for all } a \in A, \\ [a_n, b_n] &\cap A \neq \varnothing, \\ a_{n-1} &\leq a_n \leq b_n \leq b_{n-1}, \\ \text{and } b_n - a_n &= (b_{n-1} - a_{n-1})/2, \end{aligned}$$

for all $n \geq 1$.

(iii) Deduce Theorem 3.7.

Here is another way of using the fundamental axiom to prove Theorem 3.7. (However, if the reader is only going to do one of the two Exercises 3.9 and 3.10, she should do Exercise 3.9.)

Exercise 3.10. *(i) If A satisfies the hypotheses of Theorem 3.7, explain carefully why we can find an integer $r(j)$ such that*

$$r(j)2^{-j} > a \text{ for all } a \in A \text{ but}$$
$$\text{there exists an } a(j) \in A \text{ with } a(j) \geq (r(j)-1)2^{-j}.$$

[Hint: Recall that every non-empty set of the integers bounded below has a minimum.]

(ii) By applying the fundamental axiom to the sequence $(r(j)-1)2^{-j}$ and using the axiom of Archimedes, deduce Theorem 3.7.

We leave it to the reader to define the *greatest lower bound* or *infimum*, written $\inf A$ or $\inf_{a\in A} a$, when it exists.

Exercise 3.11. *(i) Define the greatest lower bound by modifying Definition 3.3. State and prove results corresponding to Lemmas 3.4 and 3.5.*

(ii) Show that

$$\inf_{a\in A} a = -\sup_{a\in A}(-a),$$

provided that either side of the equation exists.

(iii) State and prove a result on greatest lower bounds corresponding to Theorem 3.7. (Part (ii) gives a short proof.)

As an example of how a 'supremum argument' can be used we reprove the intermediate value theorem (Theorem 1.35).

Theorem 3.12. *We work in $\mathbb{R}$. If $f : [a,b] \to \mathbb{R}$ is continuous and $f(a) \geq 0 \geq f(b)$, then there exists a $c \in [a,b]$ such that $f(c) = 0$.*

Proof. Our proof will have three labelled main parts which may be compared with those in the 'lion hunting proof' on page 15.

Part A: Consider the set

$$E = \{x \in [a,b] \, : \, f(x) \geq 0\}.$$

We observe that $f(a) \geq 0$, so $a \in E$ and E is non-empty. Since $x \in E$ implies $x \leq b$, the set E is automatically bounded above.

Part B: Since every non-empty set bounded above has a supremum, E has a supremum, call it c.

Part C: Let $\epsilon > 0$. Since f is continuous at c, we can find a $\delta > 0$ such that if $x \in [a,b]$ and $|x-c| < \delta$ then $|f(x) - f(c)| < \epsilon$. We must consider three possible cases according as $a < c < b$, $c = b$ or $c = a$.

If $a < c < b$, we proceed as follows. Since $c = \sup E$ we can find $x_0 \in E$ such that $0 \leq c - x_0 < \delta$, and so $|f(x_0) - f(c)| < \epsilon$. Since $x_0 \in E$, $f(x_0) \geq 0$ and so $f(c) \geq -\epsilon$. On the other hand, choosing $y_0 = c + \min(b - c, \delta)/2$ we know that $0 \leq y_0 - c < \delta$, and so $|f(x_0) - f(c)| < \epsilon$. Since $y_0 > c$ it follows that $y_0 \notin E$, so $f(y_0) < 0$ and $f(c) < \epsilon$. We have shown that $|f(c)| \leq \epsilon$.

If $c = b$, we proceed as follows. Since $c = \sup E$, we can find $x_0 \in E$ such that $0 \leq c - x_0 < \delta$ and so $|f(x_0) - f(c)| < \epsilon$. Since $x_0 \in E$, we have $f(x_0) \geq 0$, and so $f(c) \geq -\epsilon$. By hypothesis $f(b) \leq 0$, so $|f(c)| \leq \epsilon$.

If $c = a$, we proceed as follows. Since $c = a$, we know that $f(c) \geq 0$. On the other hand, choosing $y_0 = c+\min(b-c, \delta)/2$, we know that $0 \leq y_0-c < \delta$, and so $|f(y_0) - f(c)| < \epsilon$. Since $y_0 > c$ it follows that $y_0 \notin E$, so $f(y_0) < 0$ and $f(c) < \epsilon$. We have shown that $|f(c)| < \epsilon$.

In all three cases we have shown that $|f(c)| \leq \epsilon$ for all $\epsilon > 0$, so $f(c) = 0$. ■

Exercise 3.13. *In both the 'lion hunting' and the 'supremum argument' proofs we end up with a point c where $f(c) = 0$, that is, a 'zero of f'. Give an example of a function $f : [a, b] \to \mathbb{R}$ satisfying the hypotheses of the intermediate value theorem for which 'lion hunting' and the 'supremum argument' give different zeros.*

Let us summarise the proof just given. In Part A we construct a set E on which f has a certain kind of behaviour and show that E is bounded and non-empty. In Part B we use the fact that E is bounded and non-empty to give us a point $c = \sup E$ which is in some sense a 'transition point'. Finally in Part C we examine the point c to make sure that it really has the desired property. Note that we had to examine the cases $c = a$ and $c = b$ separately. This often happens when we use the 'supremum argument'.

Let us see what goes wrong if we omit parts of the hypotheses of Theorem 3.12. If we omit the condition $f(a) \geq 0$, then we cannot even start Part A of the argument.

If we have $f(a) \geq 0$ but replace $\mathbb{R}$ by another ordered field for which it is not necessarily true that every non-empty bounded subset has a supremum, Part A goes through perfectly but Part B fails.

If we have $f(a) \geq 0 \geq f(b)$ and we work over $\mathbb{R}$ but we do not demand f continuous, then Part C fails. Part C also fails if $f(a) \geq 0$ and f is continuous but we do not know that $0 \geq f(b)$.

As a second example of how a 'supremum argument' can be used we reprove Lemma 1.45, from which the mean value inequality (Theorem 1.42) follows.

Lemma 3.14. *Let U be the open interval (α, β) on the real line $\mathbb{R}$. Suppose that $a, b \in U$ and $b > a$. If $f : U \to \mathbb{R}$ is differentiable with $f'(t) \leq K$ for* *all* $t \in U$ *and* $\epsilon > 0$, *then*

$$f(b) - f(a) \leq (b-a)(K+\epsilon).$$

Proof. Consider the set

$$E = \{x \in [a,b] \,:\, f(t) - f(a) \leq (K+\epsilon)(t-a) \text{ for all } a \leq t \leq x\}.$$

We observe that $f(a) - f(a) = 0 = (K+\epsilon)(a-a) \geq 0$, so $a \in E$ and E is non-empty. Since $x \in E$ implies $x \leq b$, the set E is automatically bounded above.

Since every non-empty set bounded above has a supremum, E has a supremum, call it c.

Since f is differentiable at c, we can find a $\delta > 0$ such that $(c-\delta, c+\delta) \subseteq U$ and

$$\left|\frac{f(c+h) - f(c)}{h} - f'(c)\right| < \epsilon/2$$

whenever $0 < |h| < \delta$. Thus

$$|f(c+h) - f(c) - f'(c)h| \leq \epsilon|h|/2$$

whenever $|h| < \delta$, and so, since $f'(c) \leq K$,

$$\begin{aligned} f(c+h) - f(c) &\leq (K+\epsilon/2)h && \text{for } 0 \leq h < \delta, \\ f(c) - f(c+h) &\leq -(K+\epsilon/2)h && \text{for } -\delta \leq h \leq 0. \end{aligned}$$

We must consider three possible cases, according as $a < c < b$, $c = b$ or $c = a$.

If $a < c < b$ we proceed as follows. If $a \leq t < c$, then, by the definition of the supremum, we can find an $x \in E$ with $t < x \leq c$, and so, by the definition of E,

$$f(t) - f(a) \leq (K+\epsilon)(t-a).$$

If $c \leq t < c+\delta$, then, choosing a $t_0 \geq a$ with $c > t_0 > c-\delta$, we have

$$f(t_0) - f(a) \leq (K+\epsilon)(t_0 - a)$$

whilst, by the result of the previous paragraph,

$$\begin{aligned} f(c) - f(t_0) &\leq (K+\epsilon/2)(c-t_0), \\ f(t) - f(c) &\leq (K+\epsilon/2)(t-c), \end{aligned}$$

so, adding the last three inequalities, we obtain

$$f(t) - f(a) \leq (K+\epsilon)(t-a).$$

Thus $f(t) - f(a) \leq (K+\epsilon)(t-a)$ for all $a \leq t < c+\delta$, contradicting the definition of c.

If $c = a$, then we know that $f(t) - f(a) \leq (K + \epsilon/2)(t - a)$ for all $a \leq t < a + \delta$, again contradicting the definition of c. Since we have shown that it is impossible that $c = a$ or $a < c < b$, it follows that $c = b$. Since the supremum of a set need not belong to that set, we must still prove that $b \in E$. However, if we choose a $t_0 \geq a$ with $b > t_0 > b - \delta$, the arguments of the previous paragraph give $f(b) - f(t_0) \leq (K + \epsilon/2)(b - t_0)$ and $f(t_0) - f(a) \leq (K + \epsilon)(t_0 - a)$, so $f(c) \leq (K + \epsilon)(c - a)$. The arguments of the previous paragraph also give $f(t) - f(a) \leq (K + \epsilon)(t - a)$ for $a \leq t < c$, so $c \in E$ and the theorem follows. ∎

We now discuss the relation between the fundamental axiom and the supremum principle. In the proof of Theorem 3.7 we saw that the fundamental axiom together with the usual rules of algebra implies the supremum principle. In the proof of Theorem 3.12 we saw that the supremum principle together with the usual rules of algebra implies the intermediate value theorem. However, Theorem 1.53 tells us that the intermediate value theorem together with the usual rules of algebra implies the fundamental axiom.

Although our argument has been a bit informal, the reader should be satisfied that the following is true. (If she is not satisfied, she may write out the details for herself.)

Theorem 3.15. *Let* $(\mathbb{F}, +, \times, >)$ *be an ordered field. Then the following two statements are equivalent.*

(i) Every increasing sequence bounded above has a limit.

(ii) Every non-empty set bounded above has a supremum.

The next exercise sketches a much more direct proof that the supremum principle implies the fundamental axiom.

Exercise 3.16. *Let* $(\mathbb{F}, +, \times, >)$ *be an ordered field such that every non-empty set bounded above has a supremum. Suppose that* $a_n \in \mathbb{F}$ *for each* $n \geq 1$, $A \in \mathbb{F}$, $a_1 \leq a_2 \leq a_3 \leq \dots$ *and* $a_n < A$ *for each* n. *Write* $E = \{a_n : n \geq 1\}$. *Show that* E *has a supremum,* a *say, and that* $a_n \to a$.

3.2. The Bolzano–Weierstrass theorem

This section is devoted to the following important result.

Theorem 3.17. (Bolzano–Weierstrass.) *If* $x_n \in \mathbb{R}$ *and there exists a* K *such that* $|x_n| \leq K$ *for all* n, *then we can find* $n(1) < n(2) < \dots$ *and* $x \in \mathbb{R}$ *such that* $x_{n(j)} \to x$ *as* $j \to \infty$.

Mathematicians say a sequence converges if it tends to a limit. The Bolzano–Weierstrass theorem thus says that every bounded sequence of reals has a convergent subsequence. Notice that we say nothing about uniqueness; if $x_n = (-1)^n$ then $x_{2n} \to 1$ but $x_{2n+1} \to -1$ as $n \to \infty$.

Exercise 3.18. *(i) Find a sequence $x_n \in [0,1]$ such that, given any $x \in [0,1]$, we can find $n(1) < n(2) < \dots$ such that $x_{n(j)} \to x$ as $j \to \infty$.*

(ii) Is it possible to find a sequence $x_n \in [0,1]$ such that, given any $x \in [0,1]$ with $x \neq 1/2$, we can find $n(1) < n(2) < \dots$ and $x \in \mathbb{R}$ such that $x_{n(j)} \to x$ as $j \to \infty$, but we cannot find $m(1) < m(2) < \dots$ such that $x_{m(j)} \to 1/2$ as $j \to \infty$? Give reasons for your answer.

The proof of the Bolzano–Weierstrass theorem given in modern textbooks depends on an ingenious observation of Donald Newman [**40**].

Lemma 3.19. *If $x_n \in \mathbb{R}$, then at least one of the following two statements must be true.*

(A) There exist $m(1) < m(2) < \dots$ such that $x_{m(j)} \geq x_{m(j+1)}$ for each $j \geq 1$.

(B) There exist $n(1) < n(2) < \dots$ such that $x_{n(j)} \leq x_{n(j+1)}$ for each $j \geq 1$.

Exercise 3.20. *Prove Theorem 3.17 from Lemma 3.19 and the fundamental axiom.*

Proof of Lemma 3.19. Call an integer $m \geq 1$ a 'far seeing integer' if $x_m \geq x_n$ for all $n \geq m$. (The term 'far seeing' is invented for use in this particular proof and is not standard terminology.) There are two possibilities:

(A) There exist infinitely many far seeing integers. Thus we can can find $m(1) < m(2) < \dots$ such that each $m(j)$ is far seeing and so $x_{m(j)} \geq x_{m(j+1)}$ for each $j \geq 1$.

(B) There are only finitely many far seeing integers. Thus there exists an N such that, if $n \geq N$, there exists an $n' > n$ with $x_{n'} > x_n$ and so, in particular, $x_{n'} \geq x_n$. Thus, given $n(j) \geq N$, we can find $n(j+1) > n(j)$ with $x_{n(j)} \leq x_{n(j+1)}$. Proceeding inductively, we obtain $n(1) < n(2) < \dots$ with $x_{n(j)} \leq x_{n(j+1)}$ for each $j \geq 1$. ∎

I admit that the proof above is very clever, but I feel that clever proofs should only be used when routine proofs do not work. Here is a proof of the Bolzano–Weierstrass theorem by lion hunting.

Exercise 3.21. *We assume the hypotheses of Theorem 3.17. Set $[a_0, b_0] = [-K, K]$. Show that we can find a sequence of pairs of points a_n and b_n such* *that*

$$\begin{aligned} &x_m \in [a_n, b_n] \textit{ for infinitely many values of } m, \\ &a_{n-1} \leq a_n \leq b_n \leq b_{n-1}, \\ \textit{and } &b_n - a_n = (b_{n-1} - a_{n-1})/2, \end{aligned}$$

for all $n \geq 1$.

Show that $a_n \to c$ *as* $n \to \infty$ *for some* $c \in [a_0, b_0]$. *Show further that we can find* $m(j)$ *with* $m(j+1) > m(j)$ *and* $x_{m(j)} \in [a_j, b_j]$ *for each* $j \geq 1$. *Deduce the Bolzano–Weierstrass theorem.*

Exercise 3.21 links directly with the origins of the theorem since, according to the notes of his students, Weierstrass[1] used bisection (our 'lion hunting') to prove Theorem 3.17.

Here is another natural proof of Theorem 3.17, this time using a supremum argument.

Exercise 3.22. *We assume the hypotheses of Theorem 3.17. Explain why*

$$y_n = \sup\{x_m : m \geq n\}$$

is well defined. Show that the y_n *form a decreasing sequence bounded below, and conclude that* y_n *tends to a limit* y.

Show carefully that we can find $n(j)$ *with* $n(j+1) > n(j)$ *such that* $|y - x_{n(j)}| < j^{-1}$. *Deduce the Bolzano–Weierstrass theorem.*

The y of Exercise 3.22 is called $\limsup_{n\to\infty} x_n$. More formally, we have the following definition.

Definition 3.23. *If* x_n *is a sequence of real numbers which is bounded above, we write*

$$\limsup_{n\to\infty} x_n = \lim_{n\to\infty} \sup\{x_m : m \geq n\}.$$

Exercise 3.24. *Let* x_n *be a bounded sequence of real numbers.*

(i) Define $\liminf_{n\to\infty} x_n$ *by analogy with* $\limsup$, *taking care to show that* $\liminf_{n\to\infty} x_n$ *exists.*

(ii) Show that $\liminf_{n\to\infty} x_n = -\limsup_{n\to\infty}(-x_n)$.

---

[1]Bernard Bolzano (1781–1848) was a professor of philosophy and religion the University of Prague who seems to have held advanced opinions on every topic on which it was possible to hold advanced opinions. Neither the Austrian Empire nor the Catholic Church approved, and he was suspended on a charge of heresy, put under house arrest and forbidden to publish. (The storm blew over to some extent, but he had to retire to private life and was never allowed to teach again.) He was also interested in the foundations of mathematics and wrote with deep insight on many of the topics discussed in the first three chapters of this book. If his work had come to the attention of the mathematical community when it should, he would now be considered as the cofounder with Cauchy of modern analysis.

(iii) Show that $\liminf_{n\to\infty} x_n \leq \limsup_{n\to\infty} x_n$.

(iv) Show that $\liminf_{n\to\infty} x_n = \limsup_{n\to\infty} x_n$ *if and only if* x_n *tends to a limit.*

(v) If $n(1) < n(2) < \dots$ *and* $x_{n(j)} \to x'$ *as* $j \to \infty$*, show that*

$$\liminf_{n\to\infty} x_n \leq x' \leq \limsup_{n\to\infty} x_n.$$

(vi) If $\liminf_{n\to\infty} x_n \leq x' \leq \limsup_{n\to\infty} x_n$*, does it follow that there exist* $n(1) < n(2) < \dots$ *such that* $x_{n(j)} \to x'$ *as* $j \to \infty$*? Give a proof or counterexample.*

(vii) Show that $y = \limsup_{n\to\infty} x_n$ *if and only if both the following conditions hold.*

(A) Given $\epsilon > 0$*, we can find an* $N(\epsilon)$ *such that* $x_n < y + \epsilon$ *for all* $n \geq N(\epsilon)$.

(B) Given $\epsilon > 0$ *and* N*, we can find* $n(N, \epsilon) \geq N$ *such that* $x_{n(N,\epsilon)} > y - \epsilon$.

State and prove a similar result for lim inf.

Although we mention lim sup from time to time, we shall not make great use of the concept.

The reader will not be surprised to learn that the Bolzano–Weierstrass theorem is precisely equivalent to the fundamental axiom.

Exercise 3.25. *Suppose that* $\mathbb{F}$ *is an ordered field for which the Bolzano–Weierstrass theorem holds (that is, every bounded sequence has a convergent subsequence). Suppose that* a_n *is an increasing sequence bounded above. Use the Bolzano–Weierstrass theorem to show that there exist an* $a \in \mathbb{F}$ *and* $n(1) < n(2) < \dots$ *such that* $a_{n(j)} \to a$ *as* $j \to \infty$*. Show that* $a_n \to a$ *as* $n \to \infty$*, and so* $\mathbb{F}$ *obeys the fundamental axiom.*

We illustrate the use of the Bolzano–Weierstrass theorem by applying it to the familiar example of the intermediate value theorem.

Theorem 3.26. *We work in* $\mathbb{R}$*. If* $f : [a, b] \to \mathbb{R}$ *is continuous and* $f(a) \geq 0 \geq f(b)$*, then there exists a* $c \in [a, b]$ *such that* $f(c) = 0$.

Proof. Our proof will have three labelled main parts which may be compared with those in the 'lion hunting proof' on page 15 and the 'supremum proof' on page 34.

Part A: Without loss of generality, we may suppose $a = 0$ and $b = 1$. Consider the real numbers $f(0)$, $f(1/n)$, $f(2/n)$, ... , $f(1)$. Since $f(0) \geq 0$ and $f(1) \leq 0$, there must exist an integer r with $0 \leq r \leq n - 1$ such that $f(r/n) \geq 0 \geq f((r + 1)/n)$. Set $x_n = r/n$.

Part B: Since the $x_n \in [0,1]$, they form a bounded sequence and so, by the Bolzano–Weierstrass theorem, we can find a $c \in [0,1]$ and $n(1) < n(2) < \dots$ such that $x_{n(j)} \to c$ as $j \to \infty$.

Part C: Since f is continuous and $x_{n(j)} \to c$, it follows that $f(x_{n(j)}) \to f(c)$ as $j \to \infty$. Since $f(x_{n(j)}) \geq 0$, it follows that $f(c) \geq 0$.

By the axiom of Archimedes, $x_{n(j)} + n(j)^{-1} \to c$, so $f(x_{n(j)} + n(j)^{-1}) \to f(c)$ as $j \to \infty$. Since $f(x_{n(j)} + n(j)^{-1}) \leq 0$, it follows that $f(c) \leq 0$. Combining this result with that of the paragraph above, we obtain $f(c) = 0$ as required. ∎

Exercise 3.27. *Produce a version of the proof just given in which we do not assume that a and b take particular values.*

Exercise 3.28. *Extend Exercise 3.13 to cover the 'Bolzano–Weierstrass' argument as well.*

Let us summarise the proof just given. In Part A we construct a sequence of points x_n which look 'more and more promising'. In Part B we use the fact that every bounded sequence has a convergent subsequence to give a point which 'ought to be very promising indeed'. Finally in Part C we examine the point c to make sure that it really has the desired property.

Let us see what goes wrong if we omit parts of the hypotheses of Theorem 3.26. If we omit the condition $f(a) \geq 0$, $f(b) \leq 0$, then we cannot even start Part A of the argument.

If we have $f(a) \geq 0 \geq f(b)$ but replace $\mathbb{R}$ by another ordered field for which the Bolzano–Weierstrass theorem does not hold, then Part A goes through perfectly but Part B fails. As usual, this is shown by Example 1.3.

If we have $f(a) \geq 0 \geq f(b)$ and we work over $\mathbb{R}$ but we do not demand that f is continuous, then Part C fails.

Exercise 3.29. *Let $f : [0,1] \to \mathbb{R}$ Show that, if $f(1) - f(0) \geq L$, then there must exist an integer r with $0 \leq r \leq n-1$ such that $f((r+1)/n) - f(r/n) \geq L/n$.*

By imitating the proof of Theorem 3.26, give a Bolzano–Weierstrass proof of Lemma 1.45 and thus of Theorem 1.42 (the mean value inequality).

Exercise 3.30. *In this exercise you may use the axiom of Archimedes and the fact that any non-empty bounded set of integers has a maximum.*

(i) Let E be a non-empty set of real numbers which is bounded above (that is, there exists a K such that $K \geq x$ whenever $x \in E$). If n is a strictly positive integer, show that there exists an integer r such that

$$\text{there exists an } e \in E \text{ with } e \geq r/n$$
$$\text{but } (r+1)/n \geq x \text{ whenever } x \in E.$$

(ii) Arguing in the manner of the proof of Theorem 3.26, show, using the Bolzano–Weierstrass theorem, that E has a supremum.

3.3. Some general remarks

We have now obtained three different but equivalent forms of the fundamental axiom (the fundamental axiom itself, the existence of a supremum for a non-empty bounded set, and the Bolzano–Weierstrass theorem) and used methods based on these three forms to prove the intermediate value theorem and the mean value inequality (themselves equivalent to the fundamental axiom). I make no excuse for the time we have spent on this programme. All of analysis rests like an inverted pyramid on the fundamental axiom, so it makes sense to study it closely.

For reasons which will become clear in the next chapter, we will rely most strongly on 'Bolzano–Weierstrass' techniques. However, there will be several places where we prefer 'supremum methods'. Exercises K.133 to K.135 show that lion hunting is useful in the theory of integration, and, although it lies outside the scope of this book, it should be remarked that the standard proof of Cauchy's theorem, on which complex analysis is based, relies on lion hunting. There is a fourth method of exploiting the fundamental axiom, based on 'Heine–Borel' or 'compactness', which is not discussed here (see Exercises K.35 to K.37, which are intended to be done after reading Chapter 4) but which, when she does meet it, the reader should consider in the context of this chapter.

Whenever we use one of these techniques it is instructive to see how the others could have been used. (Often this is easy, but occasionally it is not.) It is also worth bearing in mind that, whenever we genuinely need to use one of these methods or some theorem based on them, we are using the basic property of the reals. Everything that depends on the fundamental axiom is analysis — the rest is mere algebra.

However, the reader should also remember that most of the difficulties in analysis are resolved not by the precise manipulation of axioms but by the clear understanding of concepts.

Chapter 4

Higher Dimensions

4.1. Bolzano–Weierstrass in Higher Dimensions

In 1908, Hardy wrote a textbook to introduce the new rigorous analysis (or 'continental analysis' as it was known in a Cambridge more insular than today) to 'first year students at the Universities whose abilities approach something like what is usually described as "scholarship standard" '. Apart from the fact that even the most hardened analyst would now hesitate to call an introduction to analysis *A Course of Pure Mathematics* [**24**], it is striking how close the book is in both content and feel to a modern first course in analysis. (And, where there are changes, it is often not clear that the modern course[1] has the advantage.) One major difference is that Hardy only studied the real line, but later advances in mathematics mean that we must now study analysis in $\mathbb{R}^m$ as well.

We start with some algebra which is probably very familiar to most of my readers. If $\mathbf{x}$, $\mathbf{y} \in \mathbb{R}^m$, we define the inner product (sometimes called the dot product or scalar product) $\mathbf{x} \cdot \mathbf{y}$ of the two vectors by

$$\mathbf{x} \cdot \mathbf{y} = \sum_{j=1}^{m} x_j y_j.$$

(We shall sometimes use the alternative notation $\langle \mathbf{x}, \mathbf{y} \rangle = \mathbf{x} \cdot \mathbf{y}$. Many texts use the notation $\mathbf{x}.\mathbf{y} = \mathbf{x} \cdot \mathbf{y}$.)

Lemma 4.1. *If* $\mathbf{x}$, $\mathbf{y}$, $\mathbf{z} \in \mathbb{R}^m$ *and* $\lambda \in \mathbb{R}$, *then*

(i) $\mathbf{x} \cdot \mathbf{x} \geq 0$ *with equality if and only if* $\mathbf{x} = \mathbf{0}$,

[1]Indeed, anyone who works through the exercises in Hardy as a first course and the exercises in Whittaker and Watson's even older *A Course of Modern Analysis* [**50**] as a second will have had a splendid education in analysis.

(ii) $\mathbf{x} \cdot \mathbf{y} = \mathbf{y} \cdot \mathbf{x}$,

(iii) $(\lambda \mathbf{x}) \cdot \mathbf{y} = \lambda(\mathbf{x} \cdot \mathbf{y})$,

(iv) $(\mathbf{x} + \mathbf{y}) \cdot \mathbf{z} = \mathbf{x} \cdot \mathbf{z} + \mathbf{y} \cdot \mathbf{z}$.

Proof. Direct calculation which is left to the reader. ■

Since $\mathbf{x} \cdot \mathbf{x} \geq 0$ we may define the 'Euclidean norm of $\mathbf{x}$' by

$$\|\mathbf{x}\| = (\mathbf{x} \cdot \mathbf{x})^{1/2},$$

where we take the positive square root.

Lemma 4.2. (The Cauchy–Schwarz inequality.) *If* $\mathbf{x}$, $\mathbf{y} \in \mathbb{R}^m$, *then* $|\mathbf{x} \cdot \mathbf{y}| \leq \|\mathbf{x}\|\|\mathbf{y}\|$.

Proof. If $\|\mathbf{x}\| = 0$, then $\mathbf{x} = \mathbf{0}$, so $\mathbf{x} \cdot \mathbf{y} = 0$ and the inequality is trivial. If not, we observe that

$$\begin{aligned} 0 &\leq (\lambda \mathbf{x} + \mathbf{y}) \cdot (\lambda \mathbf{x} + \mathbf{y}) \\ &= \lambda^2 \|\mathbf{x}\|^2 + 2\lambda \mathbf{x} \cdot \mathbf{y} + \|\mathbf{y}\|^2 \\ &= \left(\lambda \|\mathbf{x}\| + \frac{\mathbf{x} \cdot \mathbf{y}}{\|\mathbf{x}\|}\right)^2 + \|\mathbf{y}\|^2 - \frac{(\mathbf{x} \cdot \mathbf{y})^2}{\|\mathbf{x}\|^2}. \end{aligned}$$

If we now set $\lambda = -(\mathbf{x} \cdot \mathbf{y})/\|\mathbf{x}\|^2$, this gives us

$$0 \leq \|\mathbf{y}\|^2 - \frac{(\mathbf{x} \cdot \mathbf{y})^2}{\|\mathbf{x}\|^2},$$

which, after rearrangement and taking square roots, gives the desired result[2]. ■

Exercise 4.3. *Although the proof just given is fairly detailed, it is a worthwhile exercise to extend it so that all the steps are directly justified by reference to the properties of the inner product given in Lemma 4.1.*

We can now obtain the standard properties of the Euclidean norm.

Lemma 4.4. *If* $\| \ \|$ *is the Euclidean norm on* $\mathbb{R}^m$, *then the following results hold.*

(i) $\|\mathbf{x}\| \geq 0$ *for all* $\mathbf{x} \in \mathbb{R}^m$.

(ii) If $\|\mathbf{x}\| = 0$, *then* $\mathbf{x} = \mathbf{0}$.

[2]The reader may ask why we do not 'simply' say that $\mathbf{x} \cdot \mathbf{y} = \|\mathbf{x}\|\|\mathbf{y}\| \cos\theta$, where θ is the angle between the vectors $\mathbf{x}$ and $\mathbf{y}$. The Cauchy–Schwarz inequality is then 'simply' the statement that $|\cos\theta| \leq 1$. However, our program is to deduce all of analysis from a limited set of statements about $\mathbb{R}$. We have not yet discussed what 'cos' is to be and, even more importantly, what the 'angle between two vectors' is to mean. When we finally reach a definition of angle in Exercise 5.56, the reader will see that we have actually *reversed* the suggested argument of the first two sentences of this footnote. This way of proceeding greatly amuses mathematicians and greatly annoys educational theorists.

(iii) If $\lambda \in \mathbb{R}$ and $\mathbf{x} \in \mathbb{R}^m$, then $\|\lambda\mathbf{x}\| = |\lambda|\|\mathbf{x}\|$.

(iv) (The triangle inequality.) If $\mathbf{x}, \mathbf{y} \in \mathbb{R}^m$ then $\|\mathbf{x}+\mathbf{y}\| \leq \|\mathbf{x}\| + \|\mathbf{y}\|$.

Proof. The triangle inequality can be deduced from the Cauchy–Schwarz inequality as follows:

$$\begin{aligned}\|\mathbf{x}+\mathbf{y}\|^2 &= (\mathbf{x}+\mathbf{y})\cdot(\mathbf{x}+\mathbf{y}) = \|\mathbf{x}\|^2 + 2\mathbf{x}\cdot\mathbf{y} + \|\mathbf{y}\|^2 \\ &\leq \|\mathbf{x}\|^2 + 2\|\mathbf{x}\|\|\mathbf{y}\| + \|\mathbf{y}\|^2 = (\|\mathbf{x}\| + \|\mathbf{y}\|)^2.\end{aligned}$$

The remaining verifications are left to the reader. ■

Exercise 4.5. *(i) By carefully examining the proof of Lemma 4.2, or otherwise, show that we have equality in the Cauchy–Schwarz inequality (that is, we have $|\mathbf{x}\cdot\mathbf{y}| = \|\mathbf{x}\|\|\mathbf{y}\|$) if and only if $\mathbf{x}$ and $\mathbf{y}$ are linearly dependent (that is, we can find real λ and μ, not both zero, such that $\lambda\mathbf{x} = \mu\mathbf{y}$).*

(ii) Show that we have equality in the triangle inequality (that is, $\|\mathbf{x}\| + \|\mathbf{y}\| = \|\mathbf{x}+\mathbf{y}\|$) if and only if we can find positive λ and μ, not both zero, such that $\lambda\mathbf{x} = \mu\mathbf{y}$.

We observe that

$$\|\mathbf{x}-\mathbf{y}\| = \left(\sum_{i=1}^{m}(x_j - y_j)^2\right)^{1/2},$$

which (at least if $m = 1$, $m = 2$, or $m = 3$) is recognisably the distance (more properly, the Euclidean distance) between $\mathbf{x}$ and $\mathbf{y}$. If we set $\mathbf{x} = \mathbf{a} - \mathbf{b}$ and $\mathbf{y} = \mathbf{b} - \mathbf{c}$, then the triangle inequality of Lemma 4.4 (iv) becomes

$$\|\mathbf{a}-\mathbf{c}\| \leq \|\mathbf{a}-\mathbf{b}\| + \|\mathbf{b}-\mathbf{c}\|,$$

which is usually read as saying that the length of one side of a triangle is less than or equal to the sum of the lengths of the other two sides.

Exercise 4.6. *If $\mathbf{x} = (x_1, x_2, \ldots, x_m) \in \mathbb{R}^m$, show that*

$$\max_{1\leq i\leq m} |x_i| \leq \|\mathbf{x}\| \leq \sum_{i=1}^{m} |x_i| \leq m \max_{1\leq i\leq m} |x_i|.$$

Looking at each of the 3 inequalities in turn, find necessary and sufficient conditions for equality.

Exercise 4.7. *We proved the triangle inequality by going via the Cauchy–Schwarz inequality. Try and prove the inequality*

$$\left(\sum_{i=1}^{3}(x_j + y_j)^2\right)^{1/2} \leq \left(\sum_{i=1}^{3} x_j^2\right)^{1/2} + \left(\sum_{i=1}^{3} y_j^2\right)^{1/2}$$

directly.

Although we shall not take the matter to extremes, we shall have a strong preference for coordinate free methods and statements. So far as I am aware, no one has found a set of labelled axes (perhaps carved in stone or beautifully cast in bronze) bearing an attestation from some higher power that these are 'nature's coordinate axes'. Coordinate free statements and methods encourage geometric intuition and generalise more readily.

Maxwell, who played a crucial role in the development of vector methods, wrote in the first chapter of his great *Treatise on Electricity and Magnetism*

> For many purposes of physical reasoning, as distinguished from calculation, it is desirable to avoid explicitly introducing the Cartesian coordinates, and to fix the mind at once on a point of space instead of its three coordinates, and on the magnitude and direction of a force instead of its three components. This mode of contemplating geometrical and physical quantities is more primitive and more natural than the other. (Chapter 1, [**38**])

We now turn towards analysis.

Definition 4.8. *We work in* $\mathbb{R}^m$ *with the Euclidean norm. We say that a sequence* $\mathbf{a}_1, \mathbf{a}_2, \ldots$ *tends to a limit* $\mathbf{a}$ *as* n *tends to infinity, or more briefly*

$$\mathbf{a}_n \to \mathbf{a} \text{ as } n \to \infty,$$

if, given $\epsilon > 0$*, we can find an* $n_0(\epsilon)$ *such that*

$$\|\mathbf{a}_n - \mathbf{a}\| < \epsilon \text{ for all } n \geq n_0(\epsilon).$$

Notice that this shows that Definition 1.5 was about the *distance* between two points and not the absolute value of the difference of two numbers.

We can prove the following results on sequences in $\mathbb{R}^m$ in exactly the same way as we proved the corresponding results for $\mathbb{R}$ (and more general ordered fields $\mathbb{F}$) in Lemma 1.6.

Lemma 4.9. *We work in* $\mathbb{R}^m$ *with the Euclidean norm.*

(i) The limit is unique. That is, if $\mathbf{a}_n \to \mathbf{a}$ *and* $\mathbf{a}_n \to \mathbf{b}$ *as* $n \to \infty$*, then* $\mathbf{a} = \mathbf{b}$.

(ii) If $\mathbf{a}_n \to \mathbf{a}$ *as* $n \to \infty$ *and* $n(1) < n(2) < n(3) \ldots$*, then* $\mathbf{a}_{n(j)} \to \mathbf{a}$ *as* $j \to \infty$.

(iii) If $\mathbf{a}_n = \mathbf{c}$ *for all* n*, then* $\mathbf{a}_n \to \mathbf{c}$ *as* $n \to \infty$.

(iv) If $\mathbf{a}_n \to \mathbf{a}$ *and* $\mathbf{b}_n \to \mathbf{b}$ *as* $n \to \infty$*, then* $\mathbf{a}_n + \mathbf{b}_n \to \mathbf{a} + \mathbf{b}$.

(v) Suppose $\mathbf{a}_n \in \mathbb{R}^m$*,* $\mathbf{a} \in \mathbb{R}^m$*,* $\lambda_n \in \mathbb{R}$ *and* $\lambda \in \mathbb{R}$*. If* $\mathbf{a}_n \to \mathbf{a}$ *and* $\lambda_n \to \lambda$*, then* $\lambda_n \mathbf{a}_n \to \lambda \mathbf{a}$.

Proof. Left to the reader. ∎

Exercise 4.10. *The parts of Lemma 1.6 do not all have corresponding parts in Lemma 4.9. Explain briefly why these differences occur.*

Exercise 4.11. *(i) If $\mathbf{a}_n \to \mathbf{a}$ and $\mathbf{b}_n \to \mathbf{b}$ as $n \to \infty$, show that $\mathbf{a}_n \cdot \mathbf{b}_n \to \mathbf{a} \cdot \mathbf{b}$. (You may find the Cauchy–Schwarz inequality useful.)*

(ii) If $\mathbf{x}$, $\mathbf{y} \in \mathbb{R}^m$ show that

$$\mathbf{x} \cdot \mathbf{y} = (\|\mathbf{x}+\mathbf{y}\|^2 + \|\mathbf{x}-\mathbf{y}\|^2)/4.$$

Prove part (i) following the method of Exercise 1.10.

Lemma 4.9 is, of course, merely algebra and applies to $\mathbb{Q}^m$ as much as to $\mathbb{R}^m$. In order to do analysis we need a more powerful tool, and, in keeping with the spirit of our general programme, we extend the Bolzano–Weierstrass theorem to $\mathbb{R}^m$.

Theorem 4.12. (Bolzano–Weierstrass.) *If $\mathbf{x}_n \in \mathbb{R}^m$ and there exists a K such that $\|\mathbf{x}_n\| \le K$ for all n, then we can find $n(1) < n(2) < \dots$ and $\mathbf{x} \in \mathbb{R}^m$ such that $\mathbf{x}_{n(j)} \to \mathbf{x}$ as $j \to \infty$.*

Once again, 'any bounded sequence has a convergent subsequence'.

Proof. We prove the result for $m = 2$, leaving it to the reader to prove the general result. Let us write $\mathbf{x}_n = (x_n, y_n)$. Observe that, since $\|\mathbf{x}_n\| \le K$, it follows that $|x_n| \le K$. By the Bolzano–Weierstrass theorem for the reals (Theorem 3.17), it follows that there exist a real number x and a sequence $m(1) < m(2) < \dots$ such that $x_{m(k)} \to x$ as $k \to \infty$.

Since $\|\mathbf{x}_{m(k)}\| \le K$, it follows that $|y_{m(k)}| \le K$; so, again by the Bolzano–Weierstrass theorem for the reals, if follows that there exist a real number y and a sequence $k(1) < k(2) < \dots$ such that $y_{m(k(j))} \to y$ as $j \to \infty$.

Setting $n(j) = m(k(j))$ and $\mathbf{x} = (x, y)$, we have

$$\|\mathbf{x}_{n(j)} - \mathbf{x}\| \le |x_{n(j)} - x| + |y_{n(j)} - y| = |x_{m(k(j))} - x| + |y_{m(k(j))} - y| \to 0,$$

and so $\mathbf{x}_{n(j)} \to \mathbf{x}$ as $j \to \infty$. ∎

Exercise 4.13. *Prove Theorem 4.12 for general m.*

Exercise 4.14. *We can also prove Theorem 4.12 by 'multidimensional lion hunting'. In this exercise we again consider the case $m = 2$, leaving the general case to the reader. She should compare this exercise with Exercise 3.21.*

We assume the hypotheses of Theorem 4.12. Consider the square

$$S_0 = [a_0, b_0] \times [a_0', b_0'] = [-K, K] \times [-K, K].$$

Explain why $\mathbf{x}_n \in S_0$ for all n.

Set $c_0 = (a_0 + b_0)/2$, $c_0' = (a_0' + b_0')/2$ *and consider the four squares*

$$T_{0,1} = [a_0, c_0] \times [a_0', c_0'],\ \ T_{0,2} = [c_0, b_0] \times [a_0', c_0'],$$
$$T_{0,3} = [a_0, c_0] \times [c_0', b_0'],\ \ T_{0,4} = [c_0, b_0] \times [c_0', b_0'].$$

Explain why at least one of the squares $T_{0,p}$*, say, must be such that* $x_r \in T_{0,p}$ *for infinitely many values of* r*. We set* $[a_1, b_1] \times [a_1', b_1'] = T_{0,p}$*.*

Show that we can find a sequence of pairs of intervals $[a_n, b_n]$ *and* $[a_n', b_n']$ *such that*

$$\mathbf{x}_r \in [a_n, b_n] \times [a_n', b_n'] \text{ for infinitely many values of } r,$$
$$a_{n-1} \le a_n \le b_n \le b_{n-1},\ \ a_{n-1}' \le a_n' \le b_n' \le b_{n-1}',$$
$$\text{and } b_n - a_n = (b_{n-1} - a_{n-1})/2,\ \ b_n' - a_n' = (b_{n-1}' - a_{n-1}')/2,$$

for all $n \ge 1$*.*

Show that $a_n \to c$ *as* $n \to \infty$ *for some* $c \in [a_0, b_0]$ *and* $a_n' \to c'$ *as* $n \to \infty$ *for some* $c' \in [a_0', b_0']$*. Show further that we can find* $m(j)$ *with* $m(j+1) > m(j)$ *and* $\mathbf{x}_{m(j)} \in [a_j, b_j] \times [a_j', b_j']$ *for each* $j \ge 1$*. Deduce the Bolzano–Weierstrass theorem.*

The proof of Theorem 4.12 involves extending a one-dimensional result to several dimensions. This is more or less inevitable, because we stated the fundamental axiom of analysis in a one-dimensional form. However, the Bolzano–Weierstrass theorem itself contains no reference as to whether we are working in $\mathbb{R}$ or $\mathbb{R}^m$. It is thus a excellent tool for multidimensional analysis.

4.2. Open and closed sets

When we work in $\mathbb{R}$ the intervals are, in some sense, the 'natural' sets to consider. One of the problems that we face when we try to do analysis in many dimensions is that the types of sets with which we have to deal are much more diverse. It turns out that the so-called closed and open sets are both sufficiently diverse and sufficiently well behaved to be useful. This section is devoted to deriving some of their simpler properties. Novices frequently find the topic hard, but eventually the reader will appreciate that this section is a rather trivial interlude in a deeper discussion.

The definition of a closed set is a natural one.

Definition 4.15. *A set* $F \subseteq \mathbb{R}^m$ *is closed if whenever* $\mathbf{x}_n \in F$ *for each* n *and* $\mathbf{x}_n \to \mathbf{x}$ *as* $n \to \infty$ *then* $\mathbf{x} \in F$*.*

Thus a set is closed in the sense of analysis if it is 'closed under the operation of taking limits'. An indication of why this is good definition is given by the following version of the Bolzano–Weierstrass theorem.

Theorem 4.16. *(i) If K is a closed bounded set in $\mathbb{R}^m$, then every sequence in K has a subsequence converging to a point of K.*

(ii) Conversely, if K is a subset of $\mathbb{R}^m$ such that every sequence in K has a subsequence converging to a point of K, then K is a closed bounded set.

Proof. Both parts of the proof are easy.

(i) If $\mathbf{x}_n$ is a sequence in K, then it is a bounded sequence and so, by Theorem 4.12, has a convergent subsequence $\mathbf{x}_{n(j)} \to \mathbf{x}$, say. Since K is closed, $\mathbf{x} \in K$ and we are done.

(ii) If K is not closed, we can find $\mathbf{x}_n \in K$ and $\mathbf{x} \notin K$ such that $\mathbf{x}_n \to \mathbf{x}$ as $n \to \infty$. Since any subsequence of a convergent subsequence converges to the same limit, no subsequence of the $\mathbf{x}_n$ can converge to a point of K.

If K is not bounded, we can find $\mathbf{x}_n \in K$ such that $\|\mathbf{x}_n\| > n$. If $\mathbf{x}$ is any point of $\mathbb{R}^m$, then the inequality

$$\|\mathbf{x}_n - \mathbf{x}\| \geq \|\mathbf{x}_n\| - \|\mathbf{x}\| > n - \|\mathbf{x}\|$$

shows that no subsequence of the $\mathbf{x}_n$ can converge. ■

When working in $\mathbb{R}^m$, the words 'closed and bounded' should always elicit the response 'Bolzano–Weierstrass'. We shall see important examples of this slogan in action in the next section (Theorem 4.41 and Theorem 4.63).

The following remark is sometimes useful.

Exercise 4.17. *(i) If A is a non-empty closed subset of $\mathbb{R}$ with supremum α, then we can find $a_n \in A$ with $a_n \to \alpha$ as $n \to \infty$.*

(ii) If A is a non-empty closed subset of $\mathbb{R}$, then, if $\sup_{a\in A} a$ exists, $\sup_{a\in A} a \in A$.

We turn now to the definition of an open set.

Definition 4.18. *A set $U \subseteq \mathbb{R}^m$ is open if, whenever $\mathbf{x} \in U$, there exists an $\epsilon > 0$ such that, whenever $\|\mathbf{x} - \mathbf{y}\| < \epsilon$, we have $\mathbf{y} \in U$.*

Thus every point of an open set lies 'well inside the set'. The ability to 'look in all directions' plays an important role in many proofs. The first example we shall see occurs in the proof of Rolle's theorem (Theorem 4.52), and the idea will play a key part in the study of complex analysis.

Exercise 4.19. *Consider sets in $\mathbb{R}$. Prove the following results.*

(i) The interval $[a,b] = \{x \in \mathbb{R} : a \leq x \leq b\}$ is closed.

(ii) The interval $(a,b) = \{x \in \mathbb{R} : a < x < b\}$ is open.

(iii) The interval $[a,b) = \{x \in \mathbb{R} : a \leq x < b\}$ is neither open nor closed.

(iv) $\mathbb{R}$ is both open and closed.

Lemma 4.20. *Consider sets in $\mathbb{R}^m$. Let $\mathbf{x} \in \mathbb{R}^m$ and $r > 0$.*

(i) The set $B(\mathbf{x}, r) = \{\mathbf{y} \in \mathbb{R}^m : \|\mathbf{x} - \mathbf{y}\| < r\}$ is open.

(ii) The set $\bar{B}(\mathbf{x}, r) = \{\mathbf{y} \in \mathbb{R}^m : \|\mathbf{x} - \mathbf{y}\| \leq r\}$ is closed.

Proof. This is routine[3]. There is no loss of generality in taking $\mathbf{x} = \mathbf{0}$.

(i) If $\mathbf{z} \in B(\mathbf{0}, r)$, then $\|\mathbf{z}\| < r$ so, if we set $\delta = r - \|\mathbf{z}\|$, it follows that $\delta > 0$. If $\|\mathbf{z} - \mathbf{y}\| < \delta$, the triangle inequality gives

$$\|\mathbf{y}\| \leq \|\mathbf{z}\| + \|\mathbf{z} - \mathbf{y}\| < \|\mathbf{z}\| + \delta = r,$$

so that $\mathbf{y} \in B(\mathbf{0}, r)$. Thus $B(\mathbf{0}, r)$ is open.

(ii) If $\mathbf{y}_n \in \bar{B}(\mathbf{0}, r)$ and $\mathbf{y}_n \to \mathbf{y}$ as $n \to \infty$, then, by the triangle inequality,

$$\|\mathbf{y}\| \leq \|\mathbf{y}_n\| + \|\mathbf{y} - \mathbf{y}_n\| \leq r + \|\mathbf{y} - \mathbf{y}_n\| \to r + 0 = r$$

as $n \to \infty$. Thus $\|\mathbf{y}\| \leq r$, and we have shown that $\bar{B}(\mathbf{0}, r)$ is closed. ∎

We call $B(\mathbf{x}, r)$ the open ball of radius r and centre $\mathbf{x}$. We call $\bar{B}(\mathbf{x}, r)$ the closed ball of radius r and centre $\mathbf{x}$. Observe that the closed and open balls of $\mathbb{R}$ are precisely the closed and open intervals.

The following restatement of the definition helps us picture an open set.

Lemma 4.21. *A subset U of $\mathbb{R}^m$ is open if and only if each point of U is the centre of an open ball lying entirely within U.*

Thus every point of an open set is surrounded by a ball consisting only of points of the set.

The topics of this section are often treated using the idea of *neighbourhoods*. We shall not use neighbourhoods very much, but they come in handy from time to time.

Definition 4.22. *The set N is a neighbourhood of the point $\mathbf{x}$ if we can find an $r > 0$ (depending on both $\mathbf{x}$ and N) such that $B(\mathbf{x}, r) \subseteq N$.*

Thus a set is open if and only if it is a neighbourhood of every point that it contains.

Returning to the main theme, we note the following remarkable fact.

Lemma 4.23. *A subset E of $\mathbb{R}^m$ is open if and only if its complement $\mathbb{R}^m \setminus E$ is closed.*

---

[3]Mathspeak for 'It may be hard the first time you see it but when you look at it later you will consider it to be routine.'

Proof. Again, this is only a question of writing things down clearly. We split the proof into two parts.

Necessity: Suppose that E is open. If $\mathbf{x}_n \in \mathbb{R}^m \setminus E$ for all n and $\mathbf{x}_n \to \mathbf{x}$ as $n \to \infty$, then we claim that $\mathbf{x} \in \mathbb{R}^m \setminus E$. For, if not, we must have $\mathbf{x} \in E$, and so, since E is open, we can find an $\epsilon > 0$ such that, whenever $\|\mathbf{a}-\mathbf{x}\| < \epsilon$, it follows that $\mathbf{a} \in E$. Since $\mathbf{x}_n \to \mathbf{x}$, we can find an N such that $\|\mathbf{x}_N - \mathbf{x}\| < \epsilon$, and so $\mathbf{x}_N \in E$, contradicting the statement that $\mathbf{x}_n \in \mathbb{R}^m \setminus E$. Thus $\mathbf{x} \in \mathbb{R}^m \setminus E$, and we have shown that $\mathbb{R}^m \setminus E$ is closed.

Sufficiency: Suppose that $\mathbb{R}^m \setminus E$ is closed. We show that E is open. For, if not, there must exist an $\mathbf{a} \in E$ such that, given any $\epsilon > 0$, there exists a $\mathbf{y} \notin E$ with $\|\mathbf{y} - \mathbf{a}\| < \epsilon$. In particular, we can find $\mathbf{x}_n \in \mathbb{R}^m \setminus E$ such that $\|\mathbf{x}_n - \mathbf{a}\| < 1/n$. By the axiom of Archimedes, this means that $\mathbf{x}_n \to \mathbf{a}$ as $n \to \infty$, and so, since $\mathbb{R}^m \setminus E$ is closed, $\mathbf{a} \in \mathbb{R}^m \setminus E$, contradicting our assumption that $\mathbf{a} \in E$. Thus E is open. ∎

We observe the following basic results on open and closed sets.

Lemma 4.24. *Consider the collection τ of open sets in $\mathbb{R}^m$.*

(i) $\varnothing \in \tau$, $\mathbb{R}^m \in \tau$.

(ii) If $U_\alpha \in \tau$ for all $\alpha \in A$, then $\bigcup_{\alpha \in A} U_\alpha \in \tau$.

(iii) If $U_1, U_2, \ldots, U_n \in \tau$, then $\bigcap_{j=1}^n U_j \in \tau$.

Proof. This is routine.

(i) Since the empty set contains no points, every point in the empty set has any property we desire (in this case, that of being the centre of an open ball lying within the empty set). Thus the empty set is open. If $\mathbf{x} \in \mathbb{R}^m$ then $B(\mathbf{x}, 1) \subseteq \mathbb{R}^m$. Thus $\mathbb{R}^m$ is open.

(ii) If $\mathbf{x} \in \bigcup_{\alpha \in A} U_\alpha$, then we can find a particular $\alpha(0) \in A$ such that $\mathbf{x} \in U_{\alpha(0)}$. Since $U_{\alpha(0)}$ is open, we can find a $\delta > 0$ such that $B(\mathbf{x}, \delta) \subseteq U_{\alpha(0)}$. Automatically, $B(\mathbf{x}, \delta) \subseteq \bigcup_{\alpha \in A} U_\alpha$. We have shown that $\bigcup_{\alpha \in A} U_\alpha$ is open.

(iii) If $\mathbf{x} \in \bigcap_{j=1}^n U_j$, then $\mathbf{x} \in U_j$ for each $1 \leq j \leq n$. Since each U_j is open, we can find a $\delta_j > 0$, such that $B(\mathbf{x}, \delta_j) \subseteq U_j$ for each $1 \leq j \leq n$. Setting $\delta = \min_{1 \leq j \leq n} \delta_j$, we have $\delta > 0$ (note that this part of the argument requires that we are only dealing with a finite number of open sets U_j) and $B(\mathbf{x}, \delta) \subseteq U_j$ for each $1 \leq j \leq n$. Thus $B(\mathbf{x}, \delta) \subseteq \bigcap_{j=1}^n U_j$, and we have shown that $\bigcap_{j=1}^n U_j$ is open. ∎

Lemma 4.25. *Consider the collection $\mathcal{F}$ of closed sets in $\mathbb{R}^m$.*

(i) $\varnothing \in \mathcal{F}$, $\mathbb{R}^m \in \mathcal{F}$.

(ii) If $F_\alpha \in \mathcal{F}$ for all $\alpha \in A$, then $\bigcap_{\alpha \in A} F_\alpha \in \mathcal{F}$.

(iii) If $F_1, F_2, \ldots, F_n \in \mathcal{F}$, then $\bigcup_{j=1}^n F_j \in \mathcal{F}$.

(Note the difference between (ii) and (iii). In (ii) A is completely arbitrary and could, for example, be uncountable. In (iii) we deal only with a *finite* collection of closed sets.)

Proof. This follows from Lemma 4.24 by repeated use of Lemma 4.23.

(i) Observe that $\varnothing = \mathbb{R}^m \setminus \mathbb{R}^m$ and $\mathbb{R}^m = \mathbb{R}^m \setminus \varnothing$. Now use Lemma 4.24 (i).

(ii) Observe that

$$\bigcap_{\alpha \in A} F_\alpha = \mathbb{R}^m \setminus \bigcup_{\alpha \in A} (\mathbb{R}^m \setminus F_\alpha)$$

and use Lemma 4.24 (ii).

(iii) Observe that

$$\bigcup_{j=1}^{n} F_j = \mathbb{R}^m \setminus \bigcap_{j=1}^{n} (\mathbb{R}^m \setminus F_j)$$

and use Lemma 4.24 (iii). ■

Exercise 4.26. *We proved Lemma 4.24 directly and obtained Lemma 4.25 by complementation. Prove Lemma 4.25 and obtain Lemma 4.24 by complementation.*

Exercise 4.27. *(i) We work in $\mathbb{R}$ and use the usual notation for intervals (see Exercise 4.19 if necessary). Show that*

$$\bigcap_{j=1}^{\infty} (-1 - j^{-1}, 1) = [-1, 1)$$

and conclude that the intersection of open sets need not be open. Why does this not contradict Lemma 4.24?

(ii) Let U_1, U_2, ... be open sets in $\mathbb{R}$ such that $U_1 \supseteq U_2 \supseteq U_3 \supseteq \dots$. Show, by means of examples, that $\bigcap_{j=1}^{\infty} U_j$ may be (a) open but not closed, (b) closed but not open, (c) open and closed or (d) neither open nor closed.

(iii) What result do we get from (iii) by complementation?

(iv) Let $F_j = [a_j, b_j]$ and $F_1 \subseteq F_2 \subseteq F_3 \subseteq \dots$. Show, by means of examples, that $\bigcap_{j=1}^{\infty} F_j$ may be (a) open but not closed, (b) closed but not open, (c) open and closed or (d) neither open nor closed.

(v) Let $a < b$ and $c < d$. Show that, if we work in $\mathbb{R}^2$, $[a, b] \times [c, d]$ is closed, $(a, b) \times (c, d)$ is open and $(a, b) \times [c, d]$ is neither open nor closed.

(vi) Do part (ii) with $\mathbb{R}$ replaced by $\mathbb{R}^2$.

(vii) If A is open in $\mathbb{R}^m$ and B is open in $\mathbb{R}^n$, show that $A \times B$ is open in $\mathbb{R}^{m+n}$. State and prove the corresponding result for closed sets.

Of course, analysis deals with (reasonably well behaved) functions as well as sets. The notion of continuity gives us a natural class of reasonably well behaved functions. The definition carries over unchanged from the one-dimensional case.

Definition 4.28. *Let $E \subseteq \mathbb{R}^m$. We say that a function $\mathbf{f} : E \to \mathbb{R}^p$ is continuous at some point $\mathbf{x} \in E$ if, given $\epsilon > 0$, we can find a $\delta(\epsilon, \mathbf{x}) > 0$ such that, whenever $\mathbf{y} \in E$ and $\|\mathbf{x} - \mathbf{y}\| < \delta(\epsilon, \mathbf{x})$, we have*

$$\|\mathbf{f}(\mathbf{x}) - \mathbf{f}(\mathbf{y})\| < \epsilon.$$

If $\mathbf{f}$ is continuous at every point $\mathbf{x} \in E$, we say that $\mathbf{f}$ is a continuous function on E.

This may be the place to make a comment on vector notation. It is conventional in elementary analysis to distinguish elements of $\mathbb{R}^m$ from those in $\mathbb{R}$ by writing points of $\mathbb{R}^m$ in boldface when printing and underlining them when handwriting[4]. Eventually this convention becomes tedious and, in practice, mathematicians only use boldface when they wish to emphasise that vectors are involved.

Exercise 4.29. *Look at Lemma 1.13 and parts (iii) to (v) of Lemma 4.9, and give the corresponding results for continuous functions. (Thus part (v) of Lemma 4.9 corresponds to the statement that, if $\lambda : E \to \mathbb{R}$ and $\mathbf{f} : E \to \mathbb{R}^p$ are continuous at $\mathbf{x} \in E$, then so is $\lambda\mathbf{f}$.) Prove your statements directly from Definition 4.28.*

Suppose that $E \subseteq \mathbb{R}^m$ and $f : E \to \mathbb{R}$ is continuous at $\mathbf{x}$. Show that, if $f(\mathbf{t}) \neq 0$ for all $\mathbf{t} \in E$, then $1/f$ is continuous at $\mathbf{x}$.

Once again we have the following useful observation.

Lemma 4.30. *Let E be a subset of $\mathbb{R}^m$ and $\mathbf{f} : E \to \mathbb{R}^p$ a function. Suppose that $\mathbf{x} \in E$ and that $\mathbf{f}$ is continuous at $\mathbf{x}$. If $\mathbf{x}_n \in E$ for all n and $\mathbf{x}_n \to \mathbf{x}$ as $n \to \infty$, then $\mathbf{f}(\mathbf{x}_n) \to \mathbf{f}(\mathbf{x})$ as $n \to \infty$.*

Proof. Left to the reader. ■

Another way of looking at continuity, which will become progressively more important as we proceed, is given by the following lemma.

Lemma 4.31. *The function $\mathbf{f} : \mathbb{R}^m \to \mathbb{R}^p$ is continuous if and only if $\mathbf{f}^{-1}(U)$ is open whenever U is an open set in $\mathbb{R}^p$.*

The reader may need to be reminded of the definition

$$\mathbf{f}^{-1}(U) = \{\mathbf{x} \in \mathbb{R}^m : \mathbf{f}(\mathbf{x}) \in U\}.$$

[4] Or placing an arrow over them.

Proof. As with most of the proofs in this section, this is just a matter of writing things down correctly. We split the proof into two parts.

Necessity: Suppose $\mathbf{f}$ is continuous and U is an open set in $\mathbb{R}^p$. If $\mathbf{x} \in \mathbf{f}^{-1}(U)$, then $\mathbf{f}(\mathbf{x}) \in U$. But U is open, so there exists an $\epsilon > 0$ such that $B(\mathbf{f}(\mathbf{x}), \epsilon) \subseteq U$. Since $\mathbf{f}$ is continuous at $\mathbf{x}$, we can find a $\delta > 0$ such that

$$\|\mathbf{f}(\mathbf{x}) - \mathbf{f}(\mathbf{y})\| < \epsilon \text{ whenever } \|\mathbf{x} - \mathbf{y}\| < \delta.$$

We thus have $B(\mathbf{x}, \delta) \subseteq \mathbf{f}^{-1}(U)$. It follows that $\mathbf{f}^{-1}(U)$ is open.

Sufficiency: Suppose that $\mathbf{f}^{-1}(U)$ is open whenever U is an open subset of $\mathbb{R}^p$. Let $\mathbf{x} \in \mathbb{R}^m$ and $\epsilon > 0$. Since $B(\mathbf{f}(\mathbf{x}), \epsilon)$ is open, it follows that $\mathbf{f}^{-1}(B(\mathbf{f}(\mathbf{x}), \epsilon))$ is open. But $\mathbf{x} \in \mathbf{f}^{-1}(B(\mathbf{f}(\mathbf{x}), \epsilon))$, so there exists a $\delta > 0$ such that $B(\mathbf{x}, \delta) \subseteq \mathbf{f}^{-1}(B(\mathbf{f}(\mathbf{x}), \epsilon))$. We thus have

$$\|\mathbf{f}(\mathbf{x}) - \mathbf{f}(\mathbf{y})\| < \epsilon \text{ whenever } \|\mathbf{x} - \mathbf{y}\| < \delta.$$

It follows that f is continuous. ∎

Exercise 4.32. *Show that* $\sin((-5\pi, 5\pi)) = [-1, 1]$. *Now give examples of bounded open sets A in $\mathbb{R}$ such that (a)* $\sin A$ *is closed and not open, (b)* $\sin A$ *is open and not closed, (c)* $\sin A$ *is neither open nor closed, (d)* $\sin A$ *is open and closed. (Observe that $\varnothing$ is automatically bounded.)*

The reader may object that we have not yet derived the properties of sin. *In my view this does not matter if we are merely commenting on or illustrating our main argument. (I say a little more on this topic in Appendix C.) However, if the reader is interested, she should be able to construct a polynomial P such that (a), (b), (c) and (d) hold for suitable A when* $\sin A$ *is replaced by $P(A)$.*

The next exercise gives a simple example of how Lemma 4.31 can be used and asks you to contrast the new 'open set' method with the old 'ϵ, δ' method

Exercise 4.33. *Prove the following result, first directly from Definition 4.28 and then by using Lemma 4.31 instead.*

If $\mathbf{f} : \mathbb{R}^m \to \mathbb{R}^p$ and $\mathbf{g} : \mathbb{R}^p \to \mathbb{R}^q$ are continuous, then so is their composition $\mathbf{g} \circ \mathbf{f}$.

(Recall that we write $\mathbf{g} \circ \mathbf{f}(\mathbf{x}) = \mathbf{g}(\mathbf{f}(\mathbf{x}))$.)

The reader who has been following carefully may have observed that we have only defined limits of sequences. Here is another notion of limit which is probably familiar to the reader.

Definition 4.34. *Let $E \subseteq \mathbb{R}^m$, $\mathbf{x} \in E$ and $\mathbf{a} \in \mathbb{R}^p$. Consider a function $\mathbf{f} : E \setminus \{\mathbf{x}\} \to \mathbb{R}^p$ (or[5] a function $\mathbf{f} : E \to \mathbb{R}^p$). We say that $\mathbf{f}(\mathbf{y}) \to \mathbf{a}$ as*

[5] Thus it does not matter whether $\mathbf{f}$ is defined at $\mathbf{x}$ or not (and, if it is defined, it does not matter what the value of $\mathbf{f}(\mathbf{x})$ is).

$\mathbf{y} \to \mathbf{x}$ through values of $\mathbf{y} \in E$ if, given $\epsilon > 0$, we can find a $\delta(\epsilon, \mathbf{x}) > 0$ such that, whenever $\mathbf{y} \in E$ and $0 < \|\mathbf{x} - \mathbf{y}\| < \delta(\epsilon, \mathbf{x})$, we have

$$\|\mathbf{f}(\mathbf{x}) - \mathbf{a}\| < \epsilon.$$

(We give a slightly more general definition in Exercise K.23.)

Exercise 4.35. *Let $E \subseteq \mathbb{R}^m$, $\mathbf{x} \in E$ and $\mathbf{a} \in \mathbb{R}^p$. Consider a function $\mathbf{f} : E \setminus \{\mathbf{x}\} \to \mathbb{R}^p$. Define $\tilde{\mathbf{f}} : E \to \mathbb{R}^p$ by $\tilde{\mathbf{f}}(\mathbf{y}) = \mathbf{f}(\mathbf{y})$ if $\mathbf{y} \in E \setminus \{\mathbf{x}\}$ and $\tilde{\mathbf{f}}(\mathbf{x}) = \mathbf{a}$. Show that $\mathbf{f}(\mathbf{y}) \to \mathbf{a}$ as $\mathbf{y} \to \mathbf{x}$ through values of $\mathbf{y} \in E$ if and only if $\tilde{\mathbf{f}}$ is continuous at $\mathbf{x}$.*

Exercise 4.36. *After looking at your solution of Lemma 4.29, state and prove the corresponding results for limits.*

Exercise 4.37. *[In this exercise you should start from Definition 1.43.] Let U be an open set in $\mathbb{R}$. Show that a function $f : U \to \mathbb{R}$ is differentiable at $t \in U$ with derivative $f'(t)$ if and only if*

$$\frac{f(t+h) - f(t)}{h} \to f'(t)$$

as $h \to 0$ (through values of h with $t + h \in U$).

Exercise 4.38. *In Chapter 6 we approach the properties of differentiation in a more general manner. However, the reader will probably already have met results like the following which can be proved using Exercises 4.36 and 4.37.*

(i) If $f, g : (a, b) \to \mathbb{R}$ are differentiable at $x \in (a, b)$, then so is the sum $f + g$, and we have $(f + g)'(x) = f'(x) + g'(x)$.

(ii) If $f, g : (a, b) \to \mathbb{R}$ are differentiable at $x \in (a, b)$, then so is the product $f \times g$, and we have $(f \times g)'(x) = f'(x)g(x) + f(x)g'(x)$.

[Hint: $f(x+h)g(x+h) - f(x)g(x) = (f(x+h) - f(x))g(x+h) + f(x)(g(x+h) - g(x))$.]

(iii) If $f : (a, b) \to \mathbb{R}$ is nowhere zero and f is differentiable at $x \in (a, b)$, then so is $1/f$, and we have $(1/f)'(x) = -f'(x)/f(x)^2$.

(iv) State accurately and prove a result along the lines of (ii) and (iii) dealing with the derivative of f/g.

(v) If $c \in \mathbb{R}$, $c \neq 0$ and $f : \mathbb{R} \to \mathbb{R}$ is differentiable at x, show that the function f_c defined by $f_c(t) = f(ct)$ $[t \in \mathbb{R}]$ is differentiable at $c^{-1}x$ and we have $f_c'(c^{-1}x) = cf'(x)$. What happens if $c = 0$?

(vi) Use part (ii) and induction on n to show that if $r_n(x) = x^n$, then r_n is everywhere differentiable with $r_n'(x) = nr_{n-1}(x)$ $[n \geq 1]$. Hence show that every polynomial is everywhere differentiable. If P and Q are polynomials and $Q(t) \neq 0$ for all $t \in (a, b)$, show that P/Q is everywhere differentiable on (a, b).

Exercise 4.39. *(i) Use part (ii) of Exercise 4.38 to show that, if $f, g : (a,b) \to \mathbb{R}$ satisfy the equation $f(t)g(t) = 1$ for all $t \in (a,b)$ and are differentiable at $x \in (a,b)$, then $g'(x) = -f'(x)/f(x)^2$.*

(ii) Explain why we cannot deduce part (iii) of Exercise 4.38 directly from part (i) of this exercise. Can we deduce the result of part (i) of this exercise directly from part (iii) of Exercise 4.38?

(iii) Is the following statement true or false? If $f, g : (a,b) \to \mathbb{R}$ are differentiable at $x \in (a,b)$ and $f(x)g(x) = 1$, then $g'(x) = -f'(x)/f(x)^2$. Give a proof or counterexample.

Exercise 4.40. *From time to time the eagle-eyed reader will observe statements like*

$$\text{`}f(x) \to \infty \text{ as } x \to -\infty\text{'}$$

which have not been formally defined. If this really bothers her, she is probably reading the wrong book (or the right book but too early). It can be considered a standing exercise to fill in the required details.

In Appendix D, I sketch a method used in Beardon's elegant treatment [**2**] *which avoids the need for such repeated definitions.*

4.3. A central theorem of analysis

In this section we prove Theorem 4.44, which says that a real-valued continuous function on a closed bounded set in $\mathbb{R}^m$ is bounded and attains its bounds. This result together with the intermediate value theorem (proved as Theorem 1.35) and the mean value inequality (proved as Theorem 1.42 and later in a more general context as Theorem 6.27) are generally considered to be the central theorems of elementary analysis.

Our next result looks a little abstract at first.

Theorem 4.41. *Let K be a closed bounded subset of $\mathbb{R}^m$ and $\mathbf{f} : K \to \mathbb{R}^p$ a continuous function. Then $\mathbf{f}(K)$ is closed and bounded.*

In other words, the continuous image of a closed bounded set is closed and bounded.

Proof. By Theorem 4.16 (ii), we need only prove that any sequence in $\mathbf{f}(K)$ has a subsequence converging to a limit in $\mathbf{f}(K)$.

To this end, suppose that $\mathbf{y}_n$ is a sequence in $\mathbf{f}(K)$. By definition, we can find $\mathbf{x}_n \in K$ such that $\mathbf{f}(\mathbf{x}_n) = \mathbf{y}_n$. Since K is a closed and bounded subset of $\mathbb{R}^m$, Theorem 4.16 (i) tells us that there exist $n(j) \to \infty$ and $\mathbf{x} \in K$ such that $\mathbf{x}_{n(j)} \to \mathbf{x}$ as $j \to \infty$. Since $\mathbf{f}$ is continuous, $\mathbf{y}_{n(j)} = \mathbf{f}(\mathbf{x}_{n(j)}) \to \mathbf{f}(\mathbf{x}) \in \mathbf{f}(K)$, and we are done. ∎

Exercise 4.42. *Let* $N : \mathbb{R}^m \to \mathbb{R}$ *be given by* $N(\mathbf{x}) = \|\mathbf{x}\|$. *Show that* N *is continuous. Deduce in particular that if* $\mathbf{x}_n \to \mathbf{x}$ *as* $n \to \infty$, *then* $\|\mathbf{x}_n\| \to \|\mathbf{x}\|$.

Exercise 4.43. *(i) Let* A *be the open interval* $(0,1)$. *Show that the map* $f : A \to \mathbb{R}$ *given by* $f(x) = 1/x$ *is continuous but that* $f(A)$ *is unbounded. Thus the continuous image of a bounded set need not be bounded.*

(ii) Let $A = [1,\infty) = \{x \in \mathbb{R} : x \geq 1\}$ *and* $f : A \to \mathbb{R}$ *be given by* $f(x) = 1/x$. *Show that* A *is closed and* f *is continuous but* $f(A)$ *is not closed. Thus the continuous image of a closed set need not be closed.*

(iii) Show that the function $\pi : \mathbb{R}^2 \to \mathbb{R}$ *given by* $\pi(x,y) = x$ *is continuous. (The function* π *is called a projection.) Show that the set*

$$A = \{(x, 1/x) : x > 0\}$$

is closed in $\mathbb{R}^2$ *but that* $\pi(A)$ *is not.*

We derive a much more concrete corollary.

Theorem 4.44. *Let* K *be a closed bounded subset of* $\mathbb{R}^m$ *and* $f : K \to \mathbb{R}$ *a continuous function. Then we can find* $\mathbf{k}_1$ *and* $\mathbf{k}_2$ *in* K *such that*

$$f(\mathbf{k}_2) \leq f(\mathbf{k}) \leq f(\mathbf{k}_1)$$

for all $\mathbf{k} \in K$.

Proof. Since $f(K)$ is a non-empty bounded set, it has a supremum, M, say. Since $f(K)$ is closed, $M \in f(K)$, that is to say, $M = f(\mathbf{k}_1)$ for some $\mathbf{k}_1 \in K$. We obtain $\mathbf{k}_2$ similarly. ∎

In other words, a real-valued continuous function on a closed bounded set is bounded and attains its bounds. Less usefully we may say that, in this case, f actually has a maximum and a minimum. Notice that there is no analogous result for vector-valued functions. Much popular economic writing consists of attempts to disguise this fact (there is unlikely to be a state of the economy in which *everybody* is best off).

Exercise 4.45. *When I was an undergraduate, we used another proof of Theorem 4.44 which used lion hunting to establish that* f *was bounded and then a clever trick to establish that it attains its bounds.*

(i) We begin with some lion hunting in the style of Exercise 4.14. As in that exercise, we shall only consider the case $m = 2$, *leaving the general case to the reader. Suppose, if possible, that* $f(K)$ *is not bounded above (that is, given any* $\kappa > 0$, *we can find an* $\mathbf{x} \in K$ *such that* $f(\mathbf{x}) > \kappa$*).*

Since K *is closed and bounded, we can find a rectangle* $S_0 = [a_0, b_0] \times [a_0', b_0'] \supseteq K$. *Show that we can find a sequence of pairs of intervals* $[a_n, b_n]$

and $[a'_n, b'_n]$ *such that*

$$f(K \cap [a_n, b_n] \times [a'_n, b'_n]) \text{ is not bounded,}$$
$$a_{n-1} \le a_n \le b_n \le b_{n-1},\ a'_{n-1} \le a'_n \le b'_n \le b'_{n-1},$$
$$\text{and } b_n - a_n = (b_{n-1} - a_{n-1})/2,\ b'_n - a'_n = (b'_{n-1} - a'_{n-1})/2,$$

for all $n \ge 1$.

Show that $a_n \to c$ *as* $n \to \infty$ *for some* $c \in [a_0, b_0]$ *and* $a'_n \to c'$ *as* $n \to \infty$ *for some* $c' \in [a'_0, b'_0]$. *Show that* $\mathbf{c} = (c, c') \in K$. *Use the fact that* f *is continuous at* $\mathbf{c}$ *to show that there exists an* $\epsilon > 0$ *such that, if* $\mathbf{x} \in K$ *and* $\|\mathbf{x} - \mathbf{c}\| < \epsilon$, *then* $f(\mathbf{x}) < f(\mathbf{c}) + 1$. *Show that there exists an* N *such that*

$$[a_n, b_n] \times [a'_n, b'_n] \subseteq B(\mathbf{c}, \epsilon)$$

for all $n \ge N$, *and derive a contradiction.*

Hence deduce that $f(K)$ *is bounded above. Show also that* $f(K)$ *is bounded below.*

(ii) Since any non-empty bounded subset of $\mathbb{R}$ *has a supremum, we know that* $M = \sup f(K)$ *and* $m = \inf f(K)$ *exist. We now produce our clever trick. Suppose, if possible, that* $f(\mathbf{x}) \ne M$ *for all* $\mathbf{x} \in K$. *Explain why, if we set* $g(\mathbf{x}) = 1/(M - f(\mathbf{x}))$, $g : K \to \mathbb{R}$ *will be a well defined strictly positive continuous function. Deduce that there exists a real number* $M' > 0$ *such that* $g(\mathbf{x}) \le M'$ *for all* $\mathbf{x} \in K$ *and show that* $f(\mathbf{x}) \le M - 1/M'$ *for all* $\mathbf{x} \in K$. *Explain why this contradicts the definition of* M, *and conclude that there must exist some* $\mathbf{k}_1 \in K$ *such that* $f(\mathbf{k}_1) = M$. *We obtain* $\mathbf{k}_2$ *similarly. (The author repeats the remark he made on page 38 that amusing as proofs like these are, clever proofs should only be used when routine proofs do not work.)*

Our next theorem is just a particular but useful case of Theorem 4.44.

Theorem 4.46. *Let* $f : [a, b] \to \mathbb{R}$ *be a continuous function. Then we can find* $k_1, k_2 \in [a, b]$ *such that*

$$f(k_2) \le f(x) \le f(k_1)$$

for all $x \in [a, b]$.

Later we will use this result to prove Rolle's theorem (Theorem 4.52), from which in turn we shall obtain the mean value theorem (Theorem 4.49).

Theorem 4.44 can also be used to prove the fundamental theorem of algebra, which states that every complex polynomial has a root. If the reader cannot wait to see how this is done, she can look ahead to Section 5.8.

Exercise 4.47. *(i) Prove Theorem 4.46 directly from the one-dimensional version of the Bolzano–Weierstrass theorem. (Essentially just repeat the arguments of Theorem 4.41.)*

(ii) Give an example of a continuous $f : (a, b) \to \mathbb{R}$ *which is unbounded.*

(iii) Give an example of a continuous $f : (a, b) \to \mathbb{R}$ *which is bounded but does not attain its bounds.*

(iv) How does your argument in (i) fail in (ii) and (iii)?

(v) Suppose now we work over $\mathbb{Q}$ *and write* $[a, b] = \{x \in \mathbb{Q} : a \leq x \leq b\}$. *Show that* $f(x) = (1+(x^2-2)^2)^{-1}$ *defines a continuous function* $f : [0, 2] \to \mathbb{Q}$ *which is continuous and bounded but does not attain its upper bound. How does your argument in (i) fail?*

Define a continuous function $g : [0, 2] \to \mathbb{Q}$ *which is continuous and bounded but does not attain either its upper bound or its lower bound. Define a continuous function* $h : [0, 2] \to \mathbb{Q}$ *which is continuous but unbounded.*

We conclude this section with an exercise which emphasises once again the power of the hypotheses 'closed and bounded' when combined with the Bolzano–Weierstrass method. The result is important, but we shall not make much use of it.

Exercise 4.48. *(i) By picking* $x_j \in K_j$ *and then applying the Bolzano–Weierstrass theorem, prove the following result.*

Suppose that $K_1, K_2, \ldots$ *are non-empty bounded closed sets in* $\mathbb{R}^m$ *such that* $K_1 \supseteq K_2 \supseteq \ldots$. *Then* $\bigcap_{j=1}^{\infty} K_j \neq \varnothing$. *(That is, the intersection of a nested sequence of bounded, closed, non-empty sets is itself non-empty.)*

(ii) By considering $K_j = [j, \infty)$, *or otherwise, show that boundedness cannot be dropped from the hypothesis.*

(iii) By considering $K_j = (0, j^{-1})$, *or otherwise, show that closedness cannot be dropped from the hypothesis.*

Exercises K.35 to K.37 discuss a substantial generalisation of Exercise 4.48, called the Heine–Borel theorem.

4.4. The mean value theorem

Traditionally one of the first uses of the theorem that every continuous function on a closed interval is bounded and attains its bounds has been to prove a slightly stronger version of the mean value inequality.

In common with Dieudonné ([**14**], page 142) and Boas ([**8**], page 118), I think that the mean value inequality is sufficient for almost all needs and that the work required to understand the subtleties in the statement and proof of Theorem 4.49 far outweigh any gain.

However, Theorem 4.49 is likely to remain part of the standard analysis course for many years, so I include it here[6].

Theorem 4.49. (The mean value theorem.) *If $f : [a, b] \to \mathbb{R}$ is a continuous function with f differentiable on (a, b), then we can find a $c \in (a, b)$ such that*

$$f(b) - f(a) = (b - a)f'(c).$$

Here are some immediate consequences.

Lemma 4.50. *If $f : [a, b] \to \mathbb{R}$ is a continuous function with f differentiable on (a, b), then the following results hold.*

(i) If $f'(t) > 0$ for all $t \in (a, b)$, then f is strictly increasing on $[a, b]$. (That is, $f(y) > f(x)$ whenever $b \geq y > x \geq a$.)

(ii) If $f'(t) \geq 0$ for all $t \in (a, b)$, then f is increasing on $[a, b]$. (That is, $f(y) \geq f(x)$ whenever $b \geq y > x \geq a$.)

(iii) If $f'(t) = 0$ for all $t \in (a, b)$, then f is constant on $[a, b]$. (That is, $f(y) = f(x)$ whenever $b \geq y > x \geq a$.)

Proof. We prove part (i), leaving the remaining parts to the reader. If $b \geq y > x \geq a$, then the mean value theorem (Theorem 4.49) tells us that

$$f(y) - f(x) = (y - x)f'(c)$$

for some c with $y > c > x$. By hypothesis $f'(c) > 0$, so $f(y) - f(x) > 0$. ■

Exercise 4.51. *Prove Theorem 1.42 from Theorem 4.49.*

The key step in proving Theorem 4.49 is the proof of the special case when $f(a) = f(b)$.

Theorem 4.52. (Rolle's theorem.) *If $g : [a, b] \to \mathbb{R}$ is a continuous function with g differentiable on (a, b) and $g(a) = g(b)$, then we can find a $c \in (a, b)$ such that $g'(c) = 0$.*

The next exercise asks you to show that the mean value theorem follows from Rolle's theorem.

Exercise 4.53. *(i) If f is as in Theorem 4.49, show that we can find a real number A such that, setting*

$$g(t) = f(t) - At,$$

the function g satisfies the conditions of Theorem 4.52.

(ii) By applying Rolle's theorem (Theorem 4.52) to the function g in (i), obtain the mean value theorem (Theorem 4.49). (Thus the mean value theorem is just a tilted version of Rolle's theorem.)

[6]Moreover, having gone to the trouble of proving it, we shall use it freely.

Cauchy produced an interesting variant on the argument of Exercise 4.53, which we give as Exercise K.52.

Exercise 4.54. *The following very easy consequence of Definition 1.43 will be used in the proof of Rolle's theorem. Let U be an open set in $\mathbb{R}$ and let $f : U \to \mathbb{R}$ be differentiable at $t \in U$ with derivative $f'(t)$. Show that if $t_n \in U$, $t_n \neq t$ and $t_n \to t$ as $n \to \infty$, then*

$$\frac{f(t_n) - f(t)}{t_n - t} \to f'(t)$$

as $n \to \infty$.

We now turn to the proof of Rolle's theorem.

Proof of Theorem 4.52. Since the function g is continuous on the closed interval $[a, b]$, Theorem 4.46 tells us that it is bounded and attains its bounds. More specifically, we can find $k_1, k_2 \in [a, b]$ such that

$$g(k_2) \leq g(x) \leq g(k_1)$$

for all $x \in [a, b]$. If both k_1 and k_2 are end points of $[a, b]$ (that is, k_1, $k_2 \in \{a, b\}$) then

$$g(a) = g(b) = g(k_1) = g(k_2)$$

and $g(x) = g(a)$ for all $x \in [a, b]$. Taking $c = (a + b)/2$, we have $g'(c) = 0$ (the derivative of a constant function is zero), and we are done.

If at least one of k_1 and k_2 is not an end point, there is no loss in generality in assuming that k_1 is not an end point (otherwise, consider $-g$). Write $c = k_1$. Since c is not an end point, $a < c < b$ and we can find a $\delta > 0$ such that $a < c - \delta < c + \delta < b$. Set $x_n = c - \delta/n$. Since c is a maximum for g, we have $g(c) \geq g(x_n)$ and so

$$\frac{g(x_n) - g(c)}{x_n - c} \geq 0$$

for all n. Since

$$\frac{g(x_n) - g(c)}{x_n - c} \to g'(c),$$

it follows that $g'(c) \geq 0$. However, if we set $y_n = c+\delta/n$, a similar argument shows that

$$\frac{g(y_n) - g(c)}{y_n - c} \leq 0$$

for all n, and so $g'(c) \leq 0$. Since $0 \leq g'(c) \leq 0$, it follows that $g'(c) = 0$, and we are done. ∎

(We look more closely at the structure of the preceding proof in Exercise K.46.)

In his interesting text [**11**], R. P. Burn writes

> Both Rolle's theorem and the mean value theorem are geometrically transparent. Each claims, with slightly more generality in the case of the mean value theorem, that for a graph of a differentiable function, there is always a tangent parallel to the chord.

My view is that the apparent geometrical transparency is due to our strong intuitive feeling that a function with positive derivative ought to increase — which is precisely what we are ultimately trying to prove[7]. It is because of this struggle between intuition and rigour that the argument of the second paragraph of the proof always brings to my mind someone crossing a tightrope above a pool full of crocodiles. Let me repeat to any reader tempted to modify that argument, *we wish to use Theorem 4.49 to prove that a function with positive derivative is increasing, and so we cannot use that result to prove Theorem 4.49.* If you believe that you have a substantially simpler proof of Rolle's theorem than the one given above, first check it against Exercise K.47 and then check it with a professional analyst[8]. Exercise K.44 gives another use of the kind of argument used to prove Rolle's theorem.

If the reader uses Theorem 4.49, it is important to note that we know nothing about c apart from the fact that $c \in (a, b)$.

Exercise 4.55. *Suppose that k_2 is as in the proof of Theorem 4.52. Show explicitly that, if k_2 is not an end point, $g'(k_2) = 0$.*

Exercise 4.56. *Suppose that $g : \mathbb{R} \to \mathbb{R}$ is differentiable, that $a < b$ and that $g(a) = g(b)$. Suppose k_1 and k_2 are as in the proof of Theorem 4.52. Show that, if $k_1 = a$, then $g'(a) \leq 0$, and show by example that we may have $g'(a) < 0$. State similar results for the cases $b = k_1$ and $a = k_2$.*

Exercise 4.57. *(This exercise should be compared with Lemma 4.50.)*

(i) Suppose that $f : (a, b) \to \mathbb{R}$ is differentiable and increasing on (a, b). Show that $f'(t) \geq 0$ for all $t \in (a, b)$.

(ii) If $f : \mathbb{R} \to \mathbb{R}$ is defined by $f(t) = t^3$, show that f is differentiable and everywhere strictly increasing, yet $f'(0) = 0$.

---

[7]This should not be interpreted as a criticism of Burn's excellent book. He is writing a first course in analysis and is trying to persuade the unwilling reader that what looks complicated is actually simple. I am writing a second course in analysis and trying to persuade the unwilling reader that what looks simple is actually complicated.

[8]Other proofs do exist. See, for example, Exercise K.15.

Exercise 4.58. *I said above that the mean value inequality is sufficient for most purposes. For the sake of fairness, here is an example where the extra information provided by Rolle's theorem does seem to make a difference. Here, as elsewhere in the exercises, we assume that reader knows notations like $F^{(r)}$ for the rth derivative of F and can do things like differentiating a polynomial which have not been explicitly treated in the main text.*

Suppose that $f : \mathbb{R} \to \mathbb{R}$ is n times differentiable and that

$$a < x_1 < x_2 < \cdots < x_n < b.$$

Suppose that P is a polynomial of degree $n-1$ with $P(x_j) = f(x_j)$. (We say that P is an interpolating polynomial for f.) We are interested in the error

$$E(t) = f(t) - P(t)$$

at some point $t \in [a, b]$. Since we already know that $E(x_j) = 0$, we may also assume that t, x_1, x_2, ..., x_n are distinct.

We consider the function

$$F(x) = f(x) - P(x) - E(t) \prod_{j=1}^{n} \frac{x - x_j}{t - x_j}.$$

(i) Show that F vanishes at t, x_1, x_2, ..., x_n and so vanishes at $n + 1$ distinct points in (a, b).

(ii) By using Rolle's theorem (Theorem 4.52), show that F' vanishes at n distinct points in (a, b).

(iii) By repeated use of Rolle's theorem, show that $F^{(n)}$ vanishes at some point $c \in (a, b)$.

(iv) By computing $F^{(n)}$ explicitly, deduce that

$$0 = f^{(n)}(c) - n! E(t) \prod_{j=1}^{n} \frac{1}{t - x_j},$$

and so

$$E(t) = \frac{f^{(n)}(c)}{n!} \prod_{j=1}^{n} (t - x_j).$$

Of course, we know nothing about c, but, if we know that $|f^{(n)}(x)| \leq A$ for all $x \in [a, b]$, we can deduce that

$$|f(t) - P(t)| \leq \frac{A}{n!} \prod_{j=1}^{n} (t - x_j)$$

for all $t \in [a, b]$. (We discuss this matter further in Exercise K.49.)

(v) Deduce the weaker inequality

$$|f(t) - P(t)| \leq A\frac{(b-a)^n}{n!}$$

for all $t \in [a, b]$.

An argument similar to the one just given is used in Exercise K.50 to prove a version of Taylor's theorem.

The reader, if she is very sharp-eyed, may observe that we cannot prove Lemma 4.50 (i) from the mean value inequality[9].

4.5. Uniform continuity

The mean value inequality (Theorem 1.42) is an excellent example of the way that many theorems of analysis convert *local* information into *global* information. We know that $f'(x) \leq K$, so that *locally* the rate of increase of f is no greater than K. We deduce that $f(u) - f(v) \leq K(u - v)$, so that *globally* the rate of increase of f is no greater than K. I remind the reader, once again, that this conversion of local to global fails for $\mathbb{Q}$ and depends on the fundamental axiom of analysis.

The main theorem of this section (Theorem 4.63 on uniform continuity) is another excellent example of the conversion of local information to global. We need a couple of definitions and examples. Recall first from Definition 4.28 what it means for a function to be continuous on a set.

Definition 4.59. *Let* $E \subseteq \mathbb{R}^m$. *We say that a function* $\mathbf{f} : E \to \mathbb{R}^p$ *is* continuous *on* E *if, given any point* $\mathbf{x} \in E$ *and any* $\epsilon > 0$, *we can find a* $\delta(\epsilon, \mathbf{x}) > 0$ *such that, whenever* $\mathbf{y} \in E$ *and* $\|\mathbf{x} - \mathbf{y}\| < \delta(\epsilon, \mathbf{x})$, *we have*

$$\|\mathbf{f}(\mathbf{x}) - \mathbf{f}(\mathbf{y})\| < \epsilon.$$

Now compare Definition 4.59 with our definition of *uniform* continuity.

Definition 4.60. *Let* $E \subseteq \mathbb{R}^m$. *We say that a function* $\mathbf{f} : E \to \mathbb{R}^p$ *is* uniformly continuous *on* E *if, given any* $\epsilon > 0$, *we can find a* $\delta(\epsilon) > 0$ *such that, whenever* $\mathbf{x}, \mathbf{y} \in E$ *and* $\|\mathbf{x} - \mathbf{y}\| < \delta(\epsilon)$, *we have*

$$\|\mathbf{f}(\mathbf{x}) - \mathbf{f}(\mathbf{y})\| < \epsilon.$$

Example 4.62 and Theorem 4.63 depend on understanding what it means for a function not to be uniformly continuous.

Exercise 4.61. *Let* $E \subseteq \mathbb{R}^m$. *Write down the definition of what it means for a function* $\mathbf{f} : E \to \mathbb{R}^p$ *not to be uniformly continuous*[10].

[9] However, the weaker result given in Exercise K.28 is usually sufficient.

[10] Mathematical educationalists call this sort of thing 'finding the contrapositive'.

Example 4.62. *The following three functions are continuous but not uniformly continuous.*

(i) $f_1 : \mathbb{R} \to \mathbb{R}$ *given by* $f_1(x) = x^2$.

(ii) $f_2 : (0,1) \to \mathbb{R}$ *given by* $f_2(x) = x^{-1}$.

(iii) $f_3 : (0,1) \to \mathbb{R}$ *given by* $f_3(x) = \sin(x^{-1})$.

Proof. (i) Suppose that $\delta > 0$. If we take $x > \delta^{-1}$ and $y = x + \delta/2$, we have $|x - y| < \delta$, yet

$$|x^2 - y^2| = y^2 - x^2 = (y + x)(y - x) > 2\delta^{-1}\delta/2 = 1.$$

Thus f_1 is not uniformly continuous.

(ii) and (iii) are left as exercises for the reader. ∎

Theorem 4.63. *Let K be a closed bounded subset of $\mathbb{R}^m$. If $\mathbf{f} : K \to \mathbb{R}^p$ is continuous on K, then $\mathbf{f}$ is uniformly continuous on K.*

Proof. Earlier I said that the words 'closed and bounded' should elicit the response 'Bolzano–Weierstrass'. Let us see how this slogan works here.

If $\mathbf{f}$ is not uniformly continuous, then there must exist an $\epsilon > 0$ such that, given any $\delta > 0$, we can find $\mathbf{x}$, $\mathbf{y} \in K$ such that

$$\|\mathbf{x} - \mathbf{y}\| < \delta \text{ but } \|\mathbf{f}(\mathbf{x}) - \mathbf{f}(\mathbf{y})\| > \epsilon.$$

Thus we can find a sequence of pairs of points $\mathbf{x}_n$, $\mathbf{y}_n \in K$ such that

$$\|\mathbf{x}_n - \mathbf{y}_n\| < 1/n \text{ but } \|\mathbf{f}(\mathbf{x}_n) - \mathbf{f}(\mathbf{y}_n)\| > \epsilon.$$

By the Bolzano–Weierstrass theorem, we can find a $\mathbf{k} \in K$ and a sequence $n(1) < n(2) < n(3) < \dots$ such that $\mathbf{x}_{n(j)} \to \mathbf{k}$ as $j \to \infty$. Since

$$\|\mathbf{y}_{n(j)} - \mathbf{k}\| \le \|\mathbf{y}_{n(j)} - \mathbf{x}_{n(j)}\| + \|\mathbf{x}_{n(j)} - \mathbf{k}\| \to 0 + 0 = 0,$$

it follows that $\mathbf{y}_{n(j)} \to \mathbf{k}$ as $j \to \infty$. Since $\mathbf{f}$ is continuous, it follows that

$$\epsilon < \|\mathbf{f}(\mathbf{x}_{n(j)}) - \mathbf{f}(\mathbf{y}_{n(j)})\| \le \|\mathbf{f}(\mathbf{x}_{n(j)}) - \mathbf{f}(\mathbf{k})\| + \|\mathbf{f}(\mathbf{y}_{n(j)}) - \mathbf{f}(\mathbf{k})\| \to 0$$

as $j \to \infty$. This contradiction proves the theorem. ∎

Exercise 4.64. *Use Theorem 4.63 to prove that a real-valued continuous function on a closed bounded set is bounded. Use the trick given in Exercise 4.45 (ii) to deduce that a real-valued continuous function on a closed bounded set is bounded and attains its bounds (Theorem 4.44).*

Exercise 4.65. *If we work in $\mathbb{Q}$ rather than $\mathbb{R}$, use the function of Example 1.3 to give an example of a function $f : \mathbb{Q} \to \mathbb{Q}$ which is continuous but not uniformly continuous on $[0,2]$. At which stage does our proof of Theorem 4.63 break down?*

We shall use the fact that a continuous function on a closed bounded set is uniformly continuous when we show that every continuous function on a closed interval is integrable (Theorem 8.32). It was in this context that the notion of uniform continuity first arose.

4.6. The general principle of convergence

We know that an increasing sequence of real numbers tends to a limit if and only if the sequence is bounded above. In this section we consider a similarly useful result which applies to sequences in $\mathbb{R}^n$.

Definition 4.66. *We say that a sequence of points* $\mathbf{x}_n \in \mathbb{R}^m$ *is a* Cauchy sequence *if, given any* $\epsilon > 0$*, we can find* $n_0(\epsilon)$ *such that* $\|\mathbf{x}_p - \mathbf{x}_q\| < \epsilon$ *for all* $p, q \geq n_0(\epsilon)$.

Our first lemma is merely[11] algebraic.

Lemma 4.67. *Any convergent sequence in* $\mathbb{R}^m$ *forms a Cauchy sequence.*

Proof. Suppose that $\mathbf{x}_n \to \mathbf{x}$. Let $\epsilon > 0$. By definition, we can find an $n_0(\epsilon)$ such that

$$\|\mathbf{x}_n - \mathbf{x}\| < \epsilon/2 \text{ for all } n \geq n_0(\epsilon).$$

We observe that, by the triangle inequality,

$$\|\mathbf{x}_p - \mathbf{x}_q\| \leq \|\mathbf{x}_p - \mathbf{x}\| + \|\mathbf{x}_q - \mathbf{x}\| < \epsilon/2 + \epsilon/2 = \epsilon$$

for all $p, q \geq n_0(\epsilon)$, so we are done. ■

The converse to Lemma 4.67 is a powerful theorem of analysis.

Theorem 4.68. *Any Cauchy sequence in* $\mathbb{R}^m$ *converges.*

Proof. Suppose that $\mathbf{x}_n \in \mathbb{R}^m$ is a Cauchy sequence. Then, by definition, given any $\epsilon > 0$, we can find $N(\epsilon)$ such that $\|\mathbf{x}_p - \mathbf{x}_q\| < \epsilon$ for all $p, q \geq N(\epsilon)$.

In particular, if $n \geq N(1)$, we have

$$\|\mathbf{x}_n\| \leq \|\mathbf{x}_n - \mathbf{x}_{N(1)}\| + \|\mathbf{x}_{N(1)}\| < 1 + \|\mathbf{x}_{N(1)}\|.$$

It follows that

$$\|\mathbf{x}_n\| \leq \max_{1 \leq r \leq N(1)} \|\mathbf{x}_r\| + 1$$

for all n, and so the sequence $\mathbf{x}_n$ is bounded.

By the Bolzano–Weierstrass theorem, it follows that there are an $\mathbf{x} \in \mathbb{R}^m$ and a sequence $n(1) < n(2) < \dots$ such that $\mathbf{x}_{n(j)} \to \mathbf{x}$ as $j \to \infty$. Thus, given any $\epsilon > 0$, we can find $J(\epsilon)$ such that $\|\mathbf{x}_{n(j)} - \mathbf{x}\| < \epsilon$ for all $j \geq J(\epsilon)$.

[11] Remember that, from the standpoint of this book, any argument, however difficult and complicated it may be, that does not involve the fundamental axiom is 'mere algebra'.

We now observe that if $n \geq N(\epsilon/2)$ and we choose a j such that $j \geq J(\epsilon/2)$ and $n(j) > N(\epsilon/2)$, then

$$\|\mathbf{x}_n - \mathbf{x}\| \leq \|\mathbf{x}_n - \mathbf{x}_{n(j)}\| + \|\mathbf{x}_{n(j)} - \mathbf{x}\| < \epsilon/2 + \epsilon/2 = \epsilon.$$

Thus $\mathbf{x}_n \to \mathbf{x}$ as $n \to \infty$, and we are done. ∎

Exercise 4.69. *Show that, if any subsequence of a Cauchy sequence converges, then the Cauchy sequence converges.*

Combining Theorem 4.68 with Lemma 4.67, we get the general principle of convergence.

Theorem 4.70. (General principle of convergence.) *A sequence in $\mathbb{R}^m$ converges if and only if it is a Cauchy sequence.*

The general principle of convergence is used in the study of infinite sums.

Definition 4.71. *If $\mathbf{a}_j \in \mathbb{R}^m$, we say that $\sum_{j=1}^{\infty} \mathbf{a}_j$ converges to $\mathbf{s}$ if*

$$\sum_{j=1}^{N} \mathbf{a}_j \to \mathbf{s}$$

as $N \to \infty$. We write $\sum_{j=1}^{\infty} \mathbf{a}_j = \mathbf{s}$.

If $\sum_{j=1}^{N} \mathbf{a}_j$ does not tend to a limit as $N \to \infty$, we say that the sum $\sum_{j=1}^{\infty} \mathbf{a}_j$ diverges.

[The definition just given corresponds to how mathematicians talk, but is not how logicians talk. Formally speaking, $\sum_{j=1}^{\infty} \mathbf{a}_j$ either exists, or it does not and cannot converge or diverge. In an effort to get round this some books use locutions like 'the series $\mathbf{a}_j$ converges to the sum $\sum_{j=1}^{\infty} \mathbf{a}_j$'.]

We note that any result on sums can be rewritten as a result on sequences, and vice versa.

Exercise 4.72. *We work in $\mathbb{R}^m$.*

(i) If $\mathbf{s}_n = \sum_{j=1}^{n} \mathbf{a}_j$, then $\sum_{j=1}^{\infty} \mathbf{a}_j$ converges to $\mathbf{s}$ if and only if $\mathbf{s}_n \to \mathbf{s}$ as $n \to \infty$.

(ii) If we write $\mathbf{s}_0 = \mathbf{0}$ and $\mathbf{a}_j = \mathbf{s}_j - \mathbf{s}_{j-1}$, then $\mathbf{s}_n \to \mathbf{s}$ as $n \to \infty$ if and only if $\sum_{j=1}^{\infty} \mathbf{a}_j$ converges to $\mathbf{s}$.

Exercise 4.73. *By writing*

$$\sum_{j=1}^{N} (\mathbf{a}_j + \mathbf{b}_j) = \sum_{j=1}^{N} \mathbf{a}_j + \sum_{j=1}^{N} \mathbf{b}_j,$$

show that, if $\sum_{j=1}^{\infty}\mathbf{a}_j$ *and* $\sum_{j=1}^{\infty}\mathbf{b}_j$ *converge, then so does* $\sum_{j=1}^{\infty}(\mathbf{a}_j+\mathbf{b}_j)$ *and that, in this case,*

$$\sum_{j=1}^{\infty}(\mathbf{a}_j+\mathbf{b}_j)=\sum_{j=1}^{\infty}\mathbf{a}_j+\sum_{j=1}^{\infty}\mathbf{b}_j.$$

Exercise 4.74. *If* $\mathbf{a}_j=\mathbf{b}_j$ *for all* $j\geq M$ *and* $\sum_{j=1}^{\infty}\mathbf{b}_j$ *converges, show that* $\sum_{j=1}^{\infty}\mathbf{a}_j$ *converges. Thus 'the first few terms of an infinite sum do not affect its convergence'.*

Exercise 4.75. *Prove the general principle of convergence for sums. That is to say, prove the following result. The sum* $\sum_{j=1}^{\infty}\mathbf{a}_j$ *converges if and only if, given any* $\epsilon>0$, *we can find* $n_0(\epsilon)$ *such that* $\|\sum_{j=p}^{q}\mathbf{a}_j\|<\epsilon$ *for all* $q\geq p\geq n_0(\epsilon)$.

Exercise 4.76. *(i) Show that if* $\sum_{j=1}^{\infty}\mathbf{a}_j$ *converges, then* $\|\mathbf{a}_n\|\to 0$ *as* $n\to\infty$.

(ii) By observing that $\sum_{j=1}^{n}j^{-1/2}\geq n^{1/2}$, *or otherwise, show that the converse of part (i) is false.*

(iii) Show that, if $\eta>0$, *then* $\sum_{j=1}^{\infty}j^{-1+\eta}$ *diverges. (We will obtain stronger results in Exercises 5.9 and 5.10.)*

We use the general principle of convergence to prove a simple but important result.

Theorem 4.77. *Let* $\mathbf{a}_n\in\mathbb{R}^m$ *for each* n. *If* $\sum_{j=1}^{\infty}\|\mathbf{a}_j\|$ *converges, then so does* $\sum_{j=1}^{\infty}\mathbf{a}_j$.

Proof. Since $\sum_{j=1}^{\infty}\|\mathbf{a}_j\|$ converges, the (trivial, algebraic) necessity part of the general principle of convergence tells that, given any $\epsilon>0$, we can find $n_0(\epsilon)$ such that $\sum_{j=p}^{q}\|\mathbf{a}_j\|<\epsilon$ for all $q\geq p\geq n_0(\epsilon)$. The triangle inequality now tells us that

$$\|\sum_{j=p}^{q}\mathbf{a}_j\|\leq\sum_{j=p}^{q}\|\mathbf{a}_j\|<\epsilon$$

for all $q\geq p\geq n_0(\epsilon)$ and the (profound, analytic) sufficiency part of the general principle of convergence tells us that $\sum_{j=1}^{\infty}\mathbf{a}_j$ converges. ■

Theorem 4.77 is often stated as saying that 'absolute convergence implies convergence'. We formalise this by making a definition.

Definition 4.78. *If* $\mathbf{a}_j\in\mathbb{R}^m$, *we say that the sum* $\sum_{j=1}^{\infty}\mathbf{a}_j$ *is absolutely convergent if* $\sum_{j=1}^{\infty}\|\mathbf{a}_j\|$ *converges.*

Theorem 4.77 has the natural interpretation that if we wander around $\mathbb{R}^m$ taking a sequence of steps of finite total length then we must converge on some point. However, although the result appears 'intuitively evident', the proof ultimately depends on the fundamental axiom of analysis.

Exercise 4.79. *The first proof of Theorem 4.77 that I was given went as follows.*

(i) We work first in $\mathbb{R}$. *Let* $a_n \in \mathbb{R}$ *for each* n. *If* $\sum_{n=1}^{\infty} |a_j|$ *converges, then set*

$$\begin{aligned} a_n^+ &= a_n,\ a_n^- = 0 && \text{if } a_n \geq 0 \\ a_n^+ &= 0,\ a_n^- = -a_n && \text{otherwise.} \end{aligned}$$

Observe that $\sum_{n=1}^{N} a_n^+$ *is an increasing sequence in* N, *and so show that* $\lim_{N\to\infty} \sum_{n=1}^{N} a_n^+$ *exists. Show similarly that* $\lim_{N\to\infty} \sum_{n=1}^{N} a_n^-$ *exists. Establish the equality*

$$\sum_{n=1}^{N} a_n = \sum_{n=1}^{N} a_n^+ - \sum_{n=1}^{N} a_n^-$$

and use it to show that $\lim_{N\to\infty} \sum_{n=1}^{N} a_n$ *exists.*

(ii) By taking real and imaginary parts, use (i) to prove that if $a_n \in \mathbb{C}$ *for each* n *and* $\sum_{j=1}^{\infty} |a_j|$ *converges, then so does* $\sum_{j=1}^{\infty} a_j$.

(iii) Use (i) to prove Theorem 4.77.

Although this is, if anything, an easier proof than the one we gave earlier, it is a bit inelegant and does not generalise to to the context of general normed spaces as discussed in Chapter 10.

Theorem 4.77 is often used in tandem with another simple but powerful tool.

Theorem 4.80. (The comparison test.) *We work in* $\mathbb{R}$. *Suppose that* $0 \leq b_j \leq a_j$ *for all* j. *Then, if* $\sum_{j=1}^{\infty} a_j$ *converges, so does* $\sum_{j=1}^{\infty} b_j$.

Proof. Since $A_n = \sum_{j=1}^{n} a_j$ is an increasing sequence tending to a limit A, we know that $A_n \leq A$ for each n. Thus

$$\sum_{j=1}^{n} b_j \leq \sum_{j=1}^{n} a_j \leq A$$

for each n and $B_n = \sum_{j=1}^{n} b_j$ is an increasing sequence bounded above. By the fundamental axiom, B_n tends to a limit and the result follows. ∎

Exercise 4.81. *We work in* $\mathbb{R}$. *Spell out in detail the proof that, if* $A_1 \leq A_2 \leq \dots$ *and* $A_n \to A$ *as* $n \to \infty$, *then* $A_j \leq A$ *for all* j.

Exercise 4.82. *(This should be routine for the reader. It will be needed in the proof of Lemma 4.83.) We work in* $\mathbb{R}$.

(i) Show that $\sum_{j=0}^{n} r^j = (1 - r^{n+1})/(1 - r)$.

(ii) Deduce that $\sum_{j=0}^{\infty} r^j$ *converges if and only if* $|r| < 1$ *and that, if* $|r| < 1$, *then*

$$\sum_{j=0}^{\infty} r^j = \frac{1}{1-r}.$$

We are now in a position to discuss power series $\sum_{n=0}^{\infty} a_n z^n$ in $\mathbb{C}$. The reader may object that we have not discussed convergence in $\mathbb{C}$, but from our point of view $\mathbb{C}$ is just $\mathbb{R}^2$ with additional algebraic structure. (Alternatively, the reader should be able to supply all the appropriate definitions and generalisations for herself.)

Lemma 4.83. *Suppose that* $a_n \in \mathbb{C}$. *If* $\sum_{n=0}^{\infty} a_n z_0^n$ *converges for some* $z_0 \in \mathbb{C}$, *then* $\sum_{n=0}^{\infty} a_n z^n$ *converges for all* $z \in \mathbb{C}$ *with* $|z| < |z_0|$.

Proof. Since $\sum_{n=0}^{\infty} a_n z_0^n$ converges, the general principle of convergence tells us that $|a_n z_0^n| \to 0$ as $n \to \infty$. In particular, we can find an N such that $|a_n z_0^n| \leq 1$ for all $n \geq N$. Thus, setting $M = 1 + \max_{0 \leq n \leq N} |a_n z_0^n|$, we have $|a_n z_0^n| \leq M$ for all $n \geq 0$.

Now suppose that $|z| < |z_0|$. We observe that

$$|a_n z^n| = |a_n z_0^n| \frac{|z|^n}{|z_0|^n} \leq M r^n,$$

where $r = |z|/|z_0|$. Since $0 \leq r < 1$, the geometric sum $\sum_{n=0}^{\infty} M r^n$ converges. Since $0 \leq |a_n z^n| \leq M r^n$, the comparison test (Theorem 4.80) tells us that $\sum_{n=0}^{\infty} |a_n z^n|$ converges. But, by Theorem 4.77, absolute convergence implies convergence, so $\sum_{n=0}^{\infty} a_n z^n$ converges. ■

Theorem 4.84. *Suppose that* $a_n \in \mathbb{C}$. *Then either* $\sum_{n=0}^{\infty} a_n z^n$ *converges for all* $z \in \mathbb{C}$, *or there exists a real number* R *with* $R \geq 0$ *such that*

(i) $\sum_{n=0}^{\infty} a_n z^n$ *converges if* $|z| < R$,

(ii) $\sum_{n=0}^{\infty} a_n z^n$ *diverges if* $|z| > R$.

We call R the *radius of convergence* of the power series $\sum_{n=0}^{\infty} a_n z^n$. If $\sum_{n=0}^{\infty} a_n z^n$ converges for all z, then, by convention, we say that the power series has infinite radius of convergence and write $R = \infty$. It should be noticed that the theorem says nothing about what happens when $|z| = R$. In Example 5.14 we shall see that the power series may converge for all such z, diverge for all such z, or converge for some points z with $|z| = R$, but not for all.

Proof of Theorem 4.84. If $\sum_{n=0}^{\infty} a_n z^n$ converges for all z, then there is nothing to prove. If this does not occur, there must exist a $z_1 \in \mathbb{C}$ such that

$\sum_{n=0}^{\infty} a_n z_1^n$ diverges. By Lemma 4.83, it follows that $\sum_{n=0}^{\infty} a_n z^n$ diverges whenever $|z| > |z_1|$, and so the set

$$S = \{|z| : \sum_{n=0}^{\infty} a_n z^n \text{ converges}\}$$

is a bounded subset of $\mathbb{R}$. Since $\sum_{n=0}^{\infty} a_n 0^n$ converges, S is non-empty and so has a supremum (see Theorem 3.7) which we shall call R.

By the definition of the supremum, $\sum_{n=0}^{\infty} a_n z^n$ diverges if $|z| > R$, so (ii) holds. Now suppose $|z| < R$. By the definition of the supremum we can find a z_0 with $|z| < |z_0| < R$ such that $\sum_{n=0}^{\infty} a_n z_0^n$ converges. Lemma 4.83 now tells us that $\sum_{n=0}^{\infty} a_n z^n$ converges, and we have shown that (i) holds. ■

Exercise 4.85. *(i) If $a \in \mathbb{C}$, find, with proof, the radius of convergence of $\sum_{n=0}^{\infty} a^n z^n$.*

(ii) Find, with proof, the radius of convergence of $\sum_{n=0}^{\infty} n^n z^n$.

(iii) Conclude that every possible value of R with $R \geq 0$ or $R = \infty$ can occur as a radius of convergence.

Exercise 4.86. *If $\sum_{n=0}^{\infty} a_n z^n$ has radius of convergence R and $|z_1| > R$, show that the sequence $|a_n z_1^n|$ is unbounded.*

(For further exercises on the radius of convergence, look at Exercises K.60 and K.58.)

Theorem 4.84 and its proof (including the steps like Lemma 4.83 which lead up to it) are both beautiful and important. We shall return to it later (for example in Lemma 11.70) and extract still more information from it. Time spent thinking about it will not be wasted. (If you would like an exercise on the proof, look at Exercise K.80.)

We conclude this section and the chapter with a rather less important discussion which the reader may choose to skip or skim. The general principle of convergence is obviously a very strong principle. It is natural to ask whether it is as strong as the fundamental axiom of analysis. The answer is that it is almost as strong but not quite. The general principle of convergence together with the axiom of Archimedes are equivalent to the fundamental axiom of analysis.

Exercise 4.87. *Suppose that $\mathbb{F}$ is an ordered field that satisfies the axiom of Archimedes.*

(i) Show that, if x_n is an increasing sequence bounded above, then, given any positive integer q, there exists an $N(q)$ such that $0 \leq x_n - x_m < 1/q$ for all $n \geq m \geq N(q)$.

(ii) Deduce that any increasing sequence bounded above is a Cauchy sequence.

(iii) Deduce that, if $\mathbb{F}$ satisfies the general principle of convergence, it satisfies the fundamental axiom of analysis.

Later, in Appendix G, as a by-product of more important work done in Section 14.4, we shall obtain an example of an ordered field that satisfies the general principle of convergence but not the axiom of Archimedes (and so not the fundamental axiom of analysis).

Chapter 5

Sums and Suchlike ♡

This chapter contains material on sums which could be left out of a streamlined course in analysis. Much of the material can be obtained as corollaries of more advanced work. However, I believe that working through it will help deepen the reader's understanding of the processes of analysis.

5.1. Comparison tests ♡

How can we tell if a sum $\sum_{n=1}^{\infty} \mathbf{a}_n$ converges?

We have already seen two very useful tools in Theorem 4.77 and Theorem 4.80, which we restate here.

Theorem 5.1. (Absolute convergence implies convergence.) *Let us work in $\mathbb{R}^m$. If $\sum_{j=1}^{\infty} \|\mathbf{a}_j\|$ converges, then so does $\sum_{j=1}^{\infty} \mathbf{a}_j$.*

Theorem 5.2. (The comparison test.) *We work in $\mathbb{R}$. Suppose that $0 \leq b_j \leq a_j$ for all j. Then, if $\sum_{j=1}^{\infty} a_j$ converges, so does $\sum_{j=1}^{\infty} b_j$.*

Comparison with geometric series gives the ratio test.

Exercise 5.3. *We work in $\mathbb{R}$. Suppose that a_n is a sequence of non-zero terms with $a_{n+1}/a_n \to l$ as $n \to \infty$.*

(i) If $|l| < 1$, show that we can find an N such that $|a_{n+1}/a_n| < (1+l)/2$ for all $n \geq N$. Deduce that we can find a constant K such that $|a_n| \leq K((1+l)/2)^n$ for all $n \geq 1$. Conclude that $\sum_{n=1}^{\infty} a_n$ converges.

(ii) If $|l| > 1$, show that $|a_n| \to \infty$ as $n \to \infty$, and so, in particular, $\sum_{n=1}^{\infty} a_n$ diverges.

We can extend the result of Exercise 5.3 by using absolute convergence (Theorem 5.1).

Lemma 5.4. (The ratio test.) *Suppose that* $\mathbf{a}_n$ *is a sequence of non-zero terms in* $\mathbb{R}^m$ *with* $\|\mathbf{a}_{n+1}\|/\|\mathbf{a}_n\| \to l$ *as* $n \to \infty$.

(i) If $|l| < 1$, *then* $\sum_{n=1}^{\infty} \mathbf{a}_n$ *converges.*

(ii) If $|l| > 1$, *then* $\|\mathbf{a}_n\| \to \infty$ *as* $n \to \infty$, *and so, in particular,* $\sum_{n=1}^{\infty} \mathbf{a}_n$ *diverges.*

Exercise 5.5. *Prove Lemma 5.4. What can you say if* $\|\mathbf{a}_{n+1}\|/\|\mathbf{a}_n\| \to \infty$*?*

Exercise 5.6. *If* $a_{2n} = a_{2n-1} = 4^{-n}$, *show that* $\sum_{n=1}^{\infty} a_n$ *converges but* a_{n+1}/a_n *does not tend to a limit. Give an example of a sequence* b_n *with* $b_n > 0$ *such that* $\sum_{n=1}^{\infty} b_n$ *diverges but* b_{n+1}/b_n *does not tend to a limit.*

Notice that Lemma 5.4 says nothing about what happens when $l = 1$. In Exercise 5.9 (ii) the reader is invited to show that, if $l = 1$, we may have convergence or divergence.

The ratio test is a rather crude tool, and the comparison test becomes more effective if we can use other convergent and divergent series besides the geometric series. Cauchy's condensation test provides a family of such series. (However, most people use the integral comparison test, which we obtain later in Lemma 9.6, so the reader need not memorise the result.)

Exercise 5.7. *We work in* $\mathbb{R}$. *Suppose that* a_n *is a decreasing sequence of positive numbers.*

(i) Show that

$$2^N a_{2^N} \geq \sum_{n=2^N}^{2^{N+1}-1} a_n \geq 2^N a_{2^{N+1}}.$$

(ii) Deduce that, if $\sum_{N=0}^{\infty} 2^N a_{2^N}$ *converges to* A, *then*

$$\sum_{m=1}^{M} a_m \leq A$$

for all M. *Explain why this implies that* $\sum_{m=1}^{\infty} a_m$ *converges.*

(iii) Show similarly that if $\sum_{m=1}^{\infty} a_m$ *converges, so does* $\sum_{N=0}^{\infty} 2^N a_{2^N}$.

Tidying up a bit (remember that the convergence of an infinite sum is not affected by the first few terms), we obtain the following lemma.

Lemma 5.8. (Cauchy's condensation test.) *We work in* $\mathbb{R}$. *If* a_n *is a decreasing sequence of positive numbers, then* $\sum_{n=1}^{\infty} a_n$ *and* $\sum_{n=1}^{\infty} 2^n a_{2^n}$ *converge or diverge together.*

The next two exercises use logarithms and powers. These will be formally defined later in Sections 5.6 and 5.7, but we shall run no risk of circularity if we use them in exercises.

Exercise 5.9. *(i) Show that $\sum_{n=1}^{\infty} n^{-\alpha}$ converges if $\alpha > 1$ and diverges if $\alpha \leq 1$.*

(ii) Give an example of a sequence of strictly positive numbers a_n with $a_{n+1}/a_n \to 1$ such that $\sum_{n=1}^{\infty} a_n$ converges. Give an example of a sequence of strictly positive numbers a_n with $a_{n+1}/a_n \to 1$ and $a_n \to 0$ as $n \to \infty$ such that $\sum_{n=1}^{\infty} a_n$ diverges.

(iii) We know from (i) that $\sum_{n=1}^{\infty} n^{-1}$ diverges. Use the inequality of Exercise 5.7 (ii) to give a rough *estimate of the size of N required to give $\sum_{n=1}^{N} n^{-1} > 100$.*

Exercise 5.10. *(i) Show that $\sum_{n=2}^{\infty} n^{-1}(\log n)^{-\alpha}$ converges if $\alpha > 1$ and diverges if $\alpha \leq 1$. Give a* rough *estimate of the size of N required to give $\sum_{n=2}^{N} n^{-1}(\log n)^{-1} > 100$.*

(ii) Show that $\sum_{n=3}^{\infty} n^{-1}(\log n)^{-1}(\log\log n)^{-\alpha}$ converges if $\alpha > 1$ and diverges if $\alpha \leq 1$. Give a rough *estimate of the size of N required to give $\sum_{n=3}^{N} n^{-1}(\log n)^{-1}(\log\log n)^{-1} > 100$.*

(iii) Write down and answer the appropriate part (iii) for this question[1].

As the following exercise makes clear, whatever series we use for a comparison test, there must be some series for which that test fails.

Exercise 5.11. *Suppose a_n is a sequence of positive real numbers such that $\sum_{n=1}^{\infty} a_n$ converges. Show that we can find $N(1) < N(2) < \dots$ such that*

$$\sum_{n=N(j)}^{M} a_n < 4^{-j} \text{ for all } M \geq N(j).$$

If we now set $b_n = a_n$ for $1 \leq n \leq N(1)$ and $b_n = 2^j a_n$ for $N(j) + 1 < n \leq N(j+1)$ $[j \geq 1]$, show that $b_n/a_n \to \infty$ as $n \to \infty$ but $\sum_{n=1}^{\infty} b_n$ converges.

Suppose u_n is a sequence of positive real numbers such that $\sum_{n=1}^{\infty} u_n$ diverges. Show that we can find a sequence of positive real numbers v_n such that $v_n/u_n \to 0$ as $n \to \infty$ but $\sum_{n=1}^{\infty} v_n$ diverges.

5.2. Conditional convergence ♡

If the infinite sum $\sum_{j=1}^{\infty} \mathbf{a}_j$ is convergent but not absolutely convergent, we call it conditionally convergent[2].

The reader should know the following simple test for convergence.

---

[1]Functions with this sort of growth play an important role in certain parts of mathematics, as is indicated by the riddle 'What sound does a drowning number theorist make?' with the answer 'Log log log ... '.

[2]The terminology we use for infinite sums is not that which would be chosen if the subject were developed today, but attempts at reforming established terminology usually do more harm than good.

Lemma 5.12. (Alternating series test.) *We work in $\mathbb{R}$. If we have a decreasing sequence of positive numbers a_n with $a_n \to 0$ as $n \to \infty$, then $\sum_{j=1}^{\infty}(-1)^{j+1}a_j$ converges.*

Further,

$$\left|\sum_{j=1}^{N}(-1)^{j+1}a_j - \sum_{j=1}^{\infty}(-1)^{j+1}a_j\right| \leq |a_{N+1}|$$

for all $N \geq 1$.

The last sentence is sometimes expressed by saying 'the error caused by replacing a convergent infinite alternating sum of decreasing terms by the sum of its first few terms is no greater than the absolute value of the first term neglected'. We give the proof of Lemma 5.12 as an exercise.

Exercise 5.13. *We work in $\mathbb{R}$. Suppose that we have a decreasing sequence of positive numbers a_n (that is, $a_n \geq a_{n+1} \geq 0$ for all n). Set $s_m = \sum_{j=1}^{m}(-1)^{j+1}a_j$.*

(i) Show that the s_{2n} form an increasing sequence and the s_{2n-1} form a decreasing sequence.

(ii) By writing

$$s_{2n} = a_1 - (a_2 - a_3) \cdots - (a_{2n-2} - a_{2n-1}) - a_{2n}$$

show that $s_{2n} \leq a_1$. Show also that $s_{2n-1} \geq 0$. Deduce that $s_{2n} \to \alpha$, $s_{2n-1} \to \beta$ as $n \to \infty$ for some α and β.

(iii) Show that $\alpha = \beta$ if and only if $a_n \to 0$ as $n \to \infty$.

(iv) Suppose now that $a_n \to 0$ as $n \to \infty$ and so $\alpha = \beta = l$. Show that $s_n \to l$ as $n \to \infty$.

(v) Under the assumptions of (iv), show that $s_{2n} \leq l \leq s_{2n+1}$ and deduce that $|l - s_{2n}| \leq a_{2n+1}$. Show similarly that $|l - s_{2n-1}| \leq a_{2n}$ for all n.

(vi) Check that we have indeed proved Lemma 5.12.

We use Lemma 5.12 to give part (iii) of the next example and redeem the promise made on page 70 to exhibit a power series which converges at some point of its circle of convergence but not at all.

Example 5.14. *(i) Show that $\sum_{n=1}^{\infty} nz^n$ has radius of convergence 1 and $\sum_{n=1}^{\infty} nz^n$ diverges for all z with $|z| = 1$.*

(ii) Show that $\sum_{n=1}^{\infty} n^{-2}z^n$ has radius of convergence 1 and $\sum_{n=1}^{\infty} n^{-2}z^n$ converges for all z with $|z| = 1$.

(iii) Show that $\sum_{n=1}^{\infty} n^{-1}z^n$ has radius of convergence 1, and $\sum_{n=1}^{\infty} n^{-1}z^n$ converges if $z = -1$ and diverges if $z = 1$.

Proof. We leave the details to the reader. In each case the fact that the series diverges for $|z| > 1$ and converges for $|z| < 1$ is easily established using the ratio test. ■

Lemma 5.12 is a special case of a test due to Abel.

Lemma 5.15. (Abel's test for convergence.) *Suppose that $\mathbf{a}_j \in \mathbb{R}^m$ and there exists a K such that*

$$\left\| \sum_{j=1}^{n} \mathbf{a}_j \right\| \le K$$

for all $n \ge 1$. If λ_j is a decreasing sequence of real positive numbers with $\lambda_j \to 0$ as $j \to \infty$, then $\sum_{j=1}^{\infty} \lambda_j \mathbf{a}_j$ converges.

Exercise 5.16. *Take $m = 1$ and $\mathbf{a}_j = (-1)^{j+1}$ in Lemma 5.15.*

(i) Deduce the alternating series test (without the estimate for the error).

(ii) By taking $\lambda_j = 1$, show that the condition $\lambda_j \to 0$ cannot be omitted in Lemma 5.15.

(iii) By taking $\lambda_j = (1+(-1)^j)/2j$, show that the condition λ_j decreasing cannot be omitted in Lemma 5.15.

We set out the proof of Lemma 5.15 as an exercise.

Exercise 5.17. *(i) Suppose that $\mathbf{a}_j \in \mathbb{R}^m$ and $\lambda_j \in \mathbb{R}$. We write*

$$\mathbf{S}_n = \sum_{j=1}^{n} \mathbf{a}_j.$$

Show that

$$\sum_{j=p}^{q} \lambda_j \mathbf{a}_j = \sum_{j=p}^{q} \lambda_j (\mathbf{S}_j - \mathbf{S}_{j-1}) = \lambda_{q+1}\mathbf{S}_q + \sum_{j=p}^{q} (\lambda_j - \lambda_{j+1})\mathbf{S}_j - \lambda_p \mathbf{S}_{p-1}$$

for all $q \ge p \ge 1$. This useful trick is known as partial summation *by analogy with integration by parts. Note that $\mathbf{S}_0 = \mathbf{0}$ by convention.*

(ii) Now suppose that the hypotheses of Lemma 5.15 hold. Show that

$$\left\| \sum_{j=p}^{q} \lambda_j \mathbf{a}_j \right\| \le \lambda_{q+1} K + \sum_{j=p}^{q} (\lambda_j - \lambda_{j+1}) K + \lambda_p K = 2\lambda_p K,$$

and use the general principle of convergence to show that $\sum_{j=1}^{\infty} \lambda_j \mathbf{a}_j$ converges.

Abel's test is well suited to the study of power series, as can be seen in part (ii) of the next exercise.

Exercise 5.18. *Suppose that α is real.*

(i) Show that $\sum_{n=1}^{\infty} n^\alpha z^n$ has radius of convergence 1 for all values of α.

(ii) Show that, if $\alpha \geq 0$, $\sum_{n=1}^{\infty} n^\alpha z^n$ diverges (that is, fails to converge) for all z with $|z| = 1$.

(iii) Show that, if $\alpha < -1$, $\sum_{n=1}^{\infty} n^\alpha z^n$ converges for all z with $|z| = 1$.

(iv) Show that, if $-1 \leq \alpha < 0$, $\sum_{n=1}^{\infty} n^\alpha z^n$ converges for all z with $|z| = 1$ and $z \neq 1$, but diverges if $z = 1$.

(v) Identify the points where $\sum_{n=1}^{\infty} n^{-1} z^{2n}$ converges.

So far as I know, it remains an open problem[3] to find a characterisation of those sets $E \subseteq \{z \in \mathbb{C} : |z| = 1\}$ such that there exists a power series $\sum_{n=1}^{\infty} a_n z^n$ of radius of convergence 1 with $\sum_{n=1}^{\infty} a_n z^n$ convergent when $z \in E$ and divergent when $|z| = 1$ and $z \notin E$.

It is important to remember that, if we evaluate a sum like $\sum_{n=1}^{\infty} a_n$ which is convergent but not absolutely convergent, then, as Exercise 5.19 below makes clear, we are, in effect, evaluating '$\infty - \infty$' and the answer we get will depend on the way we evaluate it.

Exercise 5.19. *Let $a_j \in \mathbb{R}$. Write*

$$\begin{aligned} a_j^+ &= a_j,\ a_j^- = 0 && \textit{if } a_j \geq 0, \\ a_j^+ &= 0,\ a_j^- = -a_j && \textit{if } a_j < 0. \end{aligned}$$

(i) If both $\sum_{j=1}^{\infty} a_j^+$ and $\sum_{j=1}^{\infty} a_j^-$ converge, show that $\sum_{j=1}^{\infty} |a_j|$ converges.

(ii) If exactly one of $\sum_{j=1}^{\infty} a_j^+$ and $\sum_{j=1}^{\infty} a_j^-$ converges, show that $\sum_{j=1}^{\infty} a_j$ diverges.

(iii) Show by means of examples that, if both $\sum_{j=1}^{\infty} a_j^+$ and $\sum_{j=1}^{\infty} a_j^-$ diverge, then $\sum_{j=1}^{\infty} a_j$ may converge or may diverge.

(iv) Show that if $\sum_{j=1}^{\infty} a_j$ is convergent but not absolutely convergent, then both $\sum_{j=1}^{\infty} a_j^+$ and $\sum_{j=1}^{\infty} a_j^-$ diverge.

(Exercise K.69 contains some advice on testing for convergence. Like most such advice, it ceases to be helpful at the first point that you really need it.)

In his ground-breaking papers on number theory, Dirichlet manipulated conditionally convergent sums in very imaginative ways. However, he was always extremely careful to justify such manipulations. To show why he took such care he gave a very elegant specific counterexample which is reproduced

[3] See [**32**].

in Exercise K.75. The following more general and extremely striking result is due to Riemann.

Theorem 5.20. *Let $a_j \in \mathbb{R}$. If $\sum_{j=1}^{\infty} a_j$ is convergent but not absolutely convergent, then given any $l \in \mathbb{R}$ we can find a permutation σ of $\mathbb{N}^+ = \{n \in \mathbb{Z} : n \geq 1\}$ (that is, a bijection $\sigma : \mathbb{N}^+ \to \mathbb{N}^+$) with $\sum_{j=1}^{\infty} a_{\sigma(j)}$ convergent to l.*

We give a proof in the next exercise.

Exercise 5.21. *Suppose that $\sum_{j=1}^{\infty} a_j$ is convergent, but not absolutely convergent, and that $l \in \mathbb{R}$. Set*

$$S(0) = \{n \in \mathbb{N}^+ : a_n < 0\},\ T(0) = \{n \in \mathbb{N}^+ : a_n \geq 0\}.$$

We use the following inductive construction.

At the nth step we proceed as follows. If $\sum_{j=1}^{n-1} a_{\sigma(j)} < l$, we let $m_n = \min T(n)$ (that is to say, $m_n = \min_{r \in T(n)} r$) and set $\sigma(n) = m_n$, $T(n+1) = T(n) \setminus \{m_n\}$ and $S(n+1) = S(n)$. If $\sum_{j=1}^{n-1} a_{\sigma(j)} \geq l$, let $m_n = \min S(n)$ and set $\sigma(n) = m_n$, $S(n+1) = S(n) \setminus \{m_n\}$ and $T(n+1) = T(n)$.

(i) Describe the construction in your own words. Carry out the first few steps with $a_n = (-1)^n/n$ and suitable l (for example, $l = 2$ and $l = -1$).

(ii) Show that σ is indeed a permutation.

(iii) If $\kappa > |a|$ for all $a \in S(n) \cup T(n)$, $m > n$ and $S(m) \neq S(n)$, $T(m) \neq T(n)$, show that $|\sum_{j=1}^{m} a_{\sigma(j)} - l| < \kappa$. (It may be helpful to consider what happens when $\sum_{j=1}^{m} a_{\sigma(j)} - l$ changes sign.)

(iv) Conclude that $\sum_{j=1}^{n} a_{\sigma(j)} \to l$ as $n \to \infty$.

Let me add that Example 5.20 is not intended to dissuade you from using conditionally convergent sums, but merely to warn you to use them carefully. As Dirichlet showed, it is possible to be both imaginative and rigorous.

In *A Mathematician's Miscellany* [**36**], Littlewood presents an example which helps understand what happens in Riemann's theorem. Slightly dressed up, it runs as follows.

I have an infinite supply of gold sovereigns and you have none. At one minute to noon I give you 10 sovereigns, at 1/2 minutes to noon I take one sovereign back from you but give you 10 more sovereigns in exchange, at 1/3 minutes to noon I take one sovereign back from you but give you 10 more sovereigns in exchange and so on (that is, at $1/n$ minutes to noon I take one sovereign back from you but give you 10 more sovereigns in exchange [$n \geq 2$]). The process stops at noon. How rich will you be?

As it stands the question is not sufficiently precise. Here are two reformulations.

Reformulation A: I have an infinite supply of gold sovereigns labelled s_1, s_2, ... and you have none. At one minute to noon I give you the sovereigns s_j with $1 \leq j \leq 10$, at 1/2 minutes to noon I take the sovereign s_1 back from you but give you the sovereigns s_j with $11 \leq j \leq 20$, in exchange, at 1/3 minutes to noon I take the sovereign s_2 back from you but give you the sovereigns s_j with $21 \leq j \leq 30$, in exchange and so on (that is, at $1/n$ minutes to noon I take the sovereign s_{n-1} back from you but give you the sovereigns s_j with $10n - 9 \leq j \leq 10n$, in exchange). The process stops at noon. How rich will you be? In this case, it is clear that I have taken all my sovereigns back (I took sovereign s_n back at $1/(n+1)$ minutes to noon), so you are no richer than before.

Reformulation B: I have an infinite supply of gold sovereigns labelled s_1, s_2, ... and you have none. At one minute to noon I give you the sovereigns s_j with $1 \leq j \leq 10$, at 1/2 minutes to noon I take the sovereign s_2 back from you but give you the sovereigns s_j with $11 \leq j \leq 20$, in exchange, at 1/3 minutes to noon I take the sovereign s_4 back from you but give you the sovereigns the s_j with $21 \leq j \leq 30$, in exchange and so on (that is, at $1/n$ minutes to noon I take the sovereign $s_{2(n-1)}$ back from you but give you the sovereigns s_j with $10n - 9 \leq j \leq 10n$, in exchange). The process stops at noon. How rich will you be? In this case, it is clear that I have given you all my odd numbered sovereigns and taken none of them back. You are now infinitely rich.

Exercise 5.22. *Give a reformulation in which you end up with precisely N sovereigns.*

Remark: There is a faint smell of sulphur about Littlewood's example. (What happens if all the gold pieces are indistinguishable[4]?) However, most mathematicians would agree that the original problem was not correctly posed and that reformulations A and B are well defined problems with the answers we have given.

Littlewood's paradox depends on my having an infinite supply of gold sovereigns. In the same way Dirichlet showed that the phenomenon described in Example 5.20 cannot occur if the sum is absolutely convergent. We shall prove this as Lemma 5.26 in the next section.

5.3. Interchanging limits ♡

We often find ourselves wishing to interchange the order of two limiting processes. Among the examples we shall discuss in the course of this book

[4]There is a connection with Zeno's paradoxes. See Chapter II, Section 4 of [**20**] or, in rather less detail (but accompanied by many other splendid paradoxes), [**43**].

are

$$\sum_{i=1}^{\infty}\sum_{j=1}^{\infty} a_{ij} \stackrel{?}{=} \sum_{j=1}^{\infty}\sum_{i=1}^{\infty} a_{ij} \qquad \text{(interchanging the order of summation)}$$

$$\int_a^b \int_c^d f(x,y)\,dx\,dy \stackrel{?}{=} \int_c^d \int_a^b f(x,y)\,dy\,dx \qquad \text{(interchanging the order of integration)}$$

$$\frac{\partial^2 f}{\partial x \partial y} \stackrel{?}{=} \frac{\partial^2 f}{\partial y \partial x} \qquad \text{(changing the order of partial differentiation)}$$

$$\lim_{n\to\infty} \int_a^b f_n(x)\,dx \stackrel{?}{=} \int_a^b \lim_{n\to\infty} f_n(x)\,dx \qquad \text{(limit of integral is integral of limit)}$$

$$\frac{d}{dy}\int_a^b f(x,y)\,dx \stackrel{?}{=} \int_a^b \frac{\partial f}{\partial y}(x,y)\,dx \qquad \text{(differentiation under the integral sign)}$$

(The list is not exhaustive. It is good practice to make a mental note each time you meet a theorem dealing with the interchange of limits.)

Unfortunately, it is not always possible to interchange limits. The reader should study the following example carefully[5]. (A good counterexample may be as informative as a good theorem.)

Exercise 5.23. *Take* $a_{nm} = 1$ *if* $m \leq n$, $a_{nm} = 0$ *otherwise. Write out the values of* a_{nm} *in matrix form (say for* $1 \leq n, m \leq 5$*). Find* $\lim_{n\to\infty} a_{nm}$ *when* m *is fixed and* $\lim_{m\to\infty} a_{nm}$ *when* n *is fixed. Show that*

$$\lim_{m\to\infty}(\lim_{n\to\infty} a_{nm}) \neq \lim_{n\to\infty}(\lim_{m\to\infty} a_{nm})$$

In his *Course of Pure Mathematics* [**24**], Hardy writes.

> If L and L' are two limit operations then the numbers $LL'z$ and $L'Lz$ are not *generally* equal in the strict sense of the word 'general'. We can always, by the exercise of a little ingenuity, find z so that $LL'z$ and $L'Lz$ shall differ from one another. But they are equal generally if we use the word in a more practical sense, viz. as meaning 'in the great majority of such cases as are likely to occur naturally'. In practice, a result obtained by assuming that two limit operations are commutative is *probably* true; at any rate it gives a valuable suggestion of the answer to the problem under consideration. But an answer thus obtained, must in default of further study ... be regarded as suggested only and not proved.

To this I would add that, with experience, analysts learn that some limit interchanges are much more likely to lead to trouble than others.

One particularly dangerous interchange is illustrated in the next example.

[5]In this book we follow the practice, familiar from matrix theory, of writing a_{nm} as an abbreviation for $a_{(n,m)}$.

Exercise 5.24. *Take* $a_{nm} = 1/n$ *if* $m \leq n$, $a_{nm} = 0$ *otherwise. Write out the values of* a_{nm} *in matrix form (say for* $1 \leq n, m \leq 5$*). Show that*

$$\lim_{n\to\infty} \sum_{m=1}^{\infty} a_{nm} \neq \sum_{m=1}^{\infty} \lim_{n\to\infty} a_{nm}.$$

We shall see a similar phenomenon in Example 11.53, and, in Exercise 11.55 following that example, we shall look at a branch of mathematics in which, contrary to Hardy's dictum, the failure of limit interchange is the rule rather than the exception.

Exercise 5.24 involves an 'escape to infinity' which is prevented by the conditions of the next lemma.

Lemma 5.25. (Dominated convergence.) *Suppose that* $c_j \geq 0$ *and* $\sum_{j=1}^{\infty} c_j$ *converges. If* $\mathbf{a}_j(n) \in \mathbb{R}^m$ *and* $\|\mathbf{a}_j(n)\| \leq c_j$ *for all* n*, then, if* $\mathbf{a}_j(n) \to \mathbf{a}_j$ *as* $n \to \infty$*, it follows that* $\sum_{j=1}^{\infty} \mathbf{a}_j$ *converges and*

$$\sum_{j=1}^{\infty} \mathbf{a}_j(n) \to \sum_{j=1}^{\infty} \mathbf{a}_j$$

as $n \to \infty$*.*

Proof. Let $\epsilon > 0$ be given. Since $\sum_{j=1}^{\infty} c_j$ converges, it follows from the general principle of convergence that we can find a $P(\epsilon)$ such that

$$\sum_{j=p}^{q} c_j \leq \epsilon/3$$

for all $q \geq p \geq P(\epsilon)$. It follows that

$$(1) \qquad \sum_{j=P(\epsilon)+1}^{q} \|\mathbf{a}_j(n)\| \leq \sum_{j=P(\epsilon)+1}^{q} c_j \leq \epsilon/3$$

for all $q \geq P(\epsilon)$ and all n. Allowing q to tend to infinity, we have

$$(2) \qquad \sum_{j=P(\epsilon)+1}^{\infty} \|\mathbf{a}_j(n)\| \leq \epsilon/3$$

for all n.

Returning to the inequality (1), but this time letting n tend to infinity whilst keeping q fixed, we obtain

$$\sum_{j=P(\epsilon)+1}^{q} \|\mathbf{a}_j\| \leq \sum_{j=P(\epsilon)+1}^{q} c_j \leq \epsilon/3$$

for all $q \geq P(\epsilon)$. Since an increasing sequence bounded above converges, this shows that $\sum_{j=P(\epsilon)+1}^{\infty} \|\mathbf{a}_j\|$. converges. Since $\sum_{j=P(\epsilon)+1}^{\infty} \mathbf{a}_j$ converges absolutely, it converges, and the same must be true of $\sum_{j=P(\epsilon)+1}^{\infty} \mathbf{a}_j(n)$. We now allow q to tend to infinity to obtain

$$\sum_{j=P(\epsilon)+1}^{\infty} \|\mathbf{a}_j\| \leq \epsilon/3. \tag{3}$$

Since $\mathbf{a}_j(n) \to \mathbf{a}_j$, we can find $M_j(\epsilon)$ such that

$$\|\mathbf{a}_j(n) - \mathbf{a}_j\| \leq \epsilon/(6P(\epsilon))$$

for all $n \geq M_j(\epsilon)$. In particular, taking $M(\epsilon) = \max_{1\leq j\leq P(\epsilon)} M_j(\epsilon)$, we have

$$\left\|\sum_{j=1}^{P(\epsilon)} \mathbf{a}_j(n) - \sum_{j=1}^{P(\epsilon)} \mathbf{a}_j\right\| \leq \epsilon/3 \tag{4}$$

for all $n \geq M(\epsilon)$. Combining the inequalities (2), (3) and (4), we obtain

$$\left\|\sum_{j=1}^{\infty} \mathbf{a}_j(n) - \sum_{j=1}^{\infty} \mathbf{a}_j\right\| \leq \epsilon$$

for $n \geq M(\epsilon)$, so we are done. ■

If we look at Littlewood's sovereigns, the hypothesis in the dominated convergence theorem can be interpreted as saying that we have only $\sum_{j=1}^{\infty} c_j$ sovereigns to play with. It is thus, perhaps, not surprising that we can use the dominated convergence theorem to prove that an absolutely convergent sum can be rearranged in any way we wish without affecting its convergence.

Lemma 5.26. *We work in $\mathbb{R}^m$. If $\sum_{j=1}^{\infty} \mathbf{a}_j$ is absolutely convergent and σ is a permutation of $\mathbb{N}^+$, then $\sum_{j=1}^{\infty} \mathbf{a}_{\sigma(j)}$ is absolutely convergent and*

$$\sum_{j=1}^{\infty} \mathbf{a}_{\sigma(j)} = \sum_{j=1}^{\infty} \mathbf{a}_j.$$

Proof. Define $\mathbf{a}_j(n) = \mathbf{a}_j$ if $j \in \{\sigma(1), \sigma(2), \dots, \sigma(n)\}$, and $\mathbf{a}_j(n) = \mathbf{0}$ otherwise. Then $\|\mathbf{a}_j(n)\| \leq \|\mathbf{a}_j\|$ for all j and n, and

$$\sum_{j=1}^{n} \mathbf{a}_{\sigma(j)} = \sum_{j=1}^{\infty} \mathbf{a}_j(n),$$

so, by the dominated convergence theorem,

$$\sum_{j=1}^{n} \mathbf{a}_{\sigma(j)} \to \sum_{j=1}^{\infty} \mathbf{a}_j.$$

To see that $\sum_{j=1}^{\infty} \mathbf{a}_{\sigma(j)}$ is absolutely convergent, apply the result just obtained with $\|\mathbf{a}_j\|$ in place of $\mathbf{a}_j$. ■

The reader may feel that there is only one 'natural' method of defining $\sum_{r\geq 0}\mathbf{a}_r$, to wit the standard definition

$$\sum_{r\geq 0}\mathbf{a}_r = \lim_{N\to\infty}\sum_{r=0}^{N}\mathbf{a}_r.$$

I would disagree with her even for this case[6], but it is clear that there are several 'natural' methods of defining $\sum_{r,s\geq 0}\mathbf{a}_{rs}$. Among the possibilities are

$$\lim_{N\to\infty}\sum_{r=0}^{N}\sum_{s=0}^{N}\mathbf{a}_{rs},\ \lim_{N\to\infty}\sum_{r=0}^{N}\sum_{s=0}^{r}\mathbf{a}_{(r-s)\,s},\ \text{and}\ \lim_{N\to\infty}\sum_{r^2+s^2\leq N}\mathbf{a}_{rs}$$

(that is, considering sums over squares, triangles or quadrants of circles). Fortunately it is clear that each of these summation methods is a rearrangement of the others, and so, *provided we have absolute convergence*, it does not matter which we use.

We give the details in the next lemma, but the reader who is already convinced can safely ignore them.

Lemma 5.27. *Suppose that E is an infinite set and that E_1, E_2, ... and F_1, F_2, ... are finite subsets of E such that*

(i) $E_1 \subseteq E_2 \subseteq \dots$, $\bigcup_{k=1}^{\infty} E_k = E$,

(ii) $F_1 \subseteq F_2 \subseteq \dots$, $\bigcup_{k=1}^{\infty} F_k = E$.

Suppose further that $\mathbf{a}_e \in \mathbb{R}^m$ for each $e \in E$.

Then, if $\sum_{e\in E_N}\|\mathbf{a}_e\|$ tends to a limit as $N\to\infty$, it follows that each of $\sum_{e\in E_N}\mathbf{a}_e$, $\sum_{e\in F_N}\|\mathbf{a}_e\|$ and $\sum_{e\in F_N}\mathbf{a}_e$ tends to a limit as $N\to\infty$, and

$$\lim_{N\to\infty}\sum_{e\in E_N}\mathbf{a}_e = \lim_{N\to\infty}\sum_{e\in F_N}\mathbf{a}_e.$$

Proof. Using condition (i) and the fact that the X_k are finite, we can enumerate the elements of X as e_1, e_2, ... in such a way that

$$E_N = \{e_1,\, e_2,\, \dots,\, e_{M(N)}\}$$

for some $M(N)$. Thus

$$\sum_{j=1}^{M(N)}\|\mathbf{a}_{e_j}\| = \sum_{e\in E_N}\|\mathbf{a}_e\|$$

[6]Communications engineers sometimes use 'hard summation'

$$\lim_{\delta\to 0+}\sum_{\|\mathbf{a}_r\|>\delta}\mathbf{a}_r.$$

tends to a limit, so $\sum_{j=1}^{M(N)} \|\mathbf{a}_{e_j}\|$ is bounded by some K for all N and, since all terms are positive, $\sum_{j=1}^{n} \|\mathbf{a}_{e_j}\|$ is bounded by K for all n. Thus $\sum_{j=1}^{\infty} \mathbf{a}_{e_j}$ converges absolutely.

We now observe that there is a permutation σ such that

$$\sum_{j=1}^{P(N)} \|\mathbf{a}_{e_{\sigma(j)}}\| = \sum_{e\in F_N} \|\mathbf{a}_e\|$$

for some integer $P(N)$. The required results now follow from Theorem 5.26. ■

Exercise 5.28. *Explain how the previous lemma gives the following result for* $\mathbf{a}_{rs} \in \mathbb{R}^m$.

If any of the three limits

$$\lim_{N\to\infty} \sum_{r=0}^{N}\sum_{s=0}^{N} \|\mathbf{a}_{rs}\|, \quad \lim_{N\to\infty} \sum_{r=0}^{N}\sum_{s=0}^{r} \|\mathbf{a}_{(r-s)s}\|, \quad \textit{or} \quad \lim_{N\to\infty} \sum_{r^2+s^2\le N} \|\mathbf{a}_{rs}\|$$

exists, then so do the other two. Further, under this condition, the three limits

$$\lim_{N\to\infty} \sum_{r=1}^{N}\sum_{s=0}^{N} \mathbf{a}_{rs}, \quad \lim_{N\to\infty} \sum_{r=0}^{N}\sum_{s=0}^{r} \mathbf{a}_{(r-s)s}, \quad \textit{and} \quad \lim_{N\to\infty} \sum_{r^2+s^2\le N} \mathbf{a}_{rs}$$

exist and are equal.

There are two other 'natural' methods of defining $\sum_{r,s\ge 0} \mathbf{a}_{rs}$ in common use, to wit

$$\sum_{r=0}^{\infty}\sum_{s=0}^{\infty} \mathbf{a}_{rs} \text{ and } \sum_{s=0}^{\infty}\sum_{r=0}^{\infty} \mathbf{a}_{rs},$$

or more explicitly

$$\lim_{N\to\infty} \sum_{r=0}^{N} \lim_{M\to\infty} \sum_{s=0}^{M} \mathbf{a}_{rs} \text{ and } \lim_{M\to\infty} \sum_{s=0}^{M} \lim_{N\to\infty} \sum_{r=0}^{N} \mathbf{a}_{rs}.$$

We shall show in the next two lemmas that, *provided we have absolute convergence*, these summation methods are also equivalent.

Lemma 5.29. *Let* $\mathbf{a}_{rs} \in \mathbb{R}^m$ *for* $r,\ s \ge 0$. *The following two statements are equivalent.*

(i) $\sum_{r=0}^{N}\sum_{s=0}^{N} \|\mathbf{a}_{rs}\|$ *tends to a limit as* $N \to \infty$.

(ii) $\sum_{s=0}^{\infty} \|\mathbf{a}_{rs}\|$ *converges for all* $r \ge 0$ *and* $\sum_{r=0}^{\infty}\sum_{s=0}^{\infty} \|\mathbf{a}_{rs}\|$ *converges.*

Further, if statements (i) and (ii) hold, then $\sum_{s=0}^{\infty} \mathbf{a}_{rs}$ *converges for all* $r \ge 0$, *and* $\sum_{r=0}^{\infty}\sum_{s=0}^{\infty} \mathbf{a}_{rs}$ *converges.*

Proof. We first show that (i) implies (ii). Observe that, if $M \geq r$, then

$$\sum_{s=0}^{M} \|\mathbf{a}_{rs}\| \leq \sum_{u=0}^{M}\sum_{s=0}^{M} \|\mathbf{a}_{us}\| \leq \lim_{N\to\infty}\sum_{u=0}^{N}\sum_{s=0}^{N} \|\mathbf{a}_{us}\|.$$

Thus, since an increasing sequence bounded above tends to a limit, it follows that $\sum_{s=0}^{\infty} \|\mathbf{a}_{rs}\|$ converges for all $r \geq 0$. Now observe that, if $N \geq M$,

$$\sum_{r=0}^{M}\sum_{s=0}^{N} \|\mathbf{a}_{rs}\| \leq \sum_{r=0}^{N}\sum_{s=0}^{N} \|\mathbf{a}_{rs}\|.$$

Thus, allowing $N \to \infty$ whilst keeping M fixed (remember that we showed in Lemma 4.9 (iv) that the limit of a finite sum is the sum of the limits), we have

$$\sum_{r=0}^{M}\sum_{s=0}^{\infty} \|\mathbf{a}_{rs}\| \leq \lim_{N\to\infty}\sum_{r=0}^{N}\sum_{s=0}^{N} \|\mathbf{a}_{rs}\|.$$

Thus, since an increasing sequence bounded above tends to a limit, it follows that $\sum_{r=0}^{\infty}\sum_{s=0}^{\infty} \|\mathbf{a}_{rs}\|$ converges.

Now we show that (ii) implies (i). This is easier, since we need only note that

$$\sum_{r=0}^{N}\sum_{s=0}^{N} \|\mathbf{a}_{rs}\| \leq \sum_{r=0}^{N}\sum_{s=0}^{\infty} \|\mathbf{a}_{rs}\| \leq \sum_{r=0}^{\infty}\sum_{s=0}^{\infty} \|\mathbf{a}_{rs}\|$$

and use, again, the fact that an increasing sequence bounded above tends to a limit.

Finally we note that, if (ii) is true, the fact that absolute convergence implies convergence immediately shows that $\sum_{s=0}^{\infty} \mathbf{a}_{rs}$ converges for all $r \geq 0$. We also know that

$$\left\|\sum_{s=0}^{\infty} \mathbf{a}_{rs}\right\| \leq \sum_{s=0}^{\infty} \|\mathbf{a}_{rs}\|,$$

so the comparison test tells us that $\sum_{r=0}^{\infty}\sum_{s=0}^{\infty} \mathbf{a}_{rs}$ converges. ∎

Lemma 5.30. *Let $\mathbf{a}_{rs} \in \mathbb{R}^m$ for r, $s \geq 0$. If the equivalent statements (i) and (ii) in Lemma 5.29 hold, then*

$$\sum_{r=0}^{\infty}\sum_{s=0}^{\infty} \mathbf{a}_{rs} = \lim_{N\to\infty}\sum_{r=0}^{N}\sum_{s=0}^{N} \mathbf{a}_{rs}.$$

Proof. By Lemmas 5.29 and 5.27, we know that both $\sum_{r=0}^{\infty}\sum_{s=0}^{\infty} \mathbf{a}_{rs}$ and $\lim_{N\to\infty}\sum_{r=0}^{N}\sum_{s=0}^{N} \mathbf{a}_{rs}$ exist. We need only prove them equal.

We prove this by using the dominated convergence theorem again. Set $\mathbf{a}_{rs}(N) = \mathbf{a}_{rs}$ if $0 \leq r \leq N$ and $0 \leq s \leq N$, and $\mathbf{a}_{rs}(N) = \mathbf{0}$ otherwise. If r is fixed,

$$\sum_{s=0}^{\infty} \mathbf{a}_{rs}(N) = \sum_{s=0}^{N} \mathbf{a}_{rs} \to \sum_{s=0}^{\infty} \mathbf{a}_{rs}.$$

But we know that

$$\|\sum_{s=0}^{\infty} \mathbf{a}_{rs}(N)\| \leq \sum_{s=0}^{N} \|\mathbf{a}_{rs}(N)\| \leq \sum_{s=0}^{N} \|\mathbf{a}_{rs}\| \leq \sum_{s=0}^{\infty} \|\mathbf{a}_{rs}\|$$

and $\sum_{r=0}^{\infty}\sum_{s=0}^{\infty} \|\mathbf{a}_{rs}\|$ converges. Thus the dominated convergence theorem applied to $\sum_{r=0}^{\infty} \mathbf{b}_r(N)$ with $\mathbf{b}_r(N) = \sum_{s=0}^{\infty} \mathbf{a}_{rs}(N)$ gives

$$\sum_{r=0}^{N}\sum_{s=0}^{N} \mathbf{a}_{rs} = \sum_{r=0}^{\infty}\sum_{s=0}^{\infty} \mathbf{a}_{rs}(N) \to \sum_{r=0}^{\infty}\sum_{s=0}^{\infty} \mathbf{a}_{rs},$$

as required ■

We have proved the following useful theorem.

Lemma 5.31. (Fubini's theorem for sums.) *Let $\mathbf{a}_{rs} \in \mathbb{R}^m$. If any one of the following three objects,*

$$\lim_{N\to\infty} \sum_{r=0}^{N}\sum_{s=0}^{N} \|\mathbf{a}_{rs}\|,\ \sum_{r=0}^{\infty}\sum_{s=0}^{\infty} \|\mathbf{a}_{rs}\|,\ \sum_{s=0}^{\infty}\sum_{r=0}^{\infty} \|\mathbf{a}_{rs}\|$$

is well defined, they all are, as are

$$\lim_{N\to\infty} \sum_{r=0}^{N}\sum_{s=0}^{N} \mathbf{a}_{rs},\ \sum_{r=0}^{\infty}\sum_{s=0}^{\infty} \mathbf{a}_{rs},\ \sum_{s=0}^{\infty}\sum_{r=0}^{\infty} \mathbf{a}_{rs}.$$

Further,

$$\lim_{N\to\infty} \sum_{r=0}^{N}\sum_{s=0}^{N} \mathbf{a}_{rs} = \sum_{r=0}^{\infty}\sum_{s=0}^{\infty} \mathbf{a}_{rs} = \sum_{s=0}^{\infty}\sum_{r=0}^{\infty} \mathbf{a}_{rs}.$$

This theorem allows us to interchange the order of summation of infinite sums (*provided we have absolute convergence*). I attach the name Fubini to it because Fubini proved a general and far-reaching theorem of which this is a special case.

Exercise 5.32. *(i) Set $a_{rr} = 1$ for all $r \geq 1$ and $a_{r\,r-1} = -1$ for all $r \geq 2$. Set $a_{rs} = 0$ otherwise. Write out the matrix with entries a_{rs} where $1 \leq r \leq 4$, $1 \leq s \leq 4$. Show, by direct calculation, that*

$$\sum_{r=1}^{\infty}\sum_{s=1}^{\infty} a_{rs} \neq \sum_{s=1}^{\infty}\sum_{r=1}^{\infty} a_{rs}.$$

(ii) Set $b_{11} = 1$, $b_{1s} = 0$ *for* $s \geq 2$ *and*

$$\begin{aligned} b_{rs} &= 2^{-r+2} && \text{for } 2^{r-2} \leq s \leq 2^{r-1} - 1, \\ b_{rs} &= -2^{-r+1} && \text{for } 2^{r-1} \leq s \leq 2^{r} - 1, \\ b_{rs} &= 0 && \text{otherwise,} \end{aligned}$$

when $r \geq 2$. *Write out the matrix with entries* b_{rs} *where* $1 \leq r \leq 4$, $1 \leq s \leq 8$. *Show by direct calculation that*

$$\sum_{r=1}^{\infty}\sum_{s=1}^{\infty} b_{rs} \neq \sum_{s=1}^{\infty}\sum_{r=1}^{\infty} b_{rs}.$$

Summing over triangles rather than squares and using Lemma 5.27, we obtain another version of Fubini's theorem, which we shall use in the next section.

Lemma 5.33. *If* $\sum_{i=0}^{\infty}\sum_{j=0}^{\infty} \|\mathbf{a}_{ij}\|$ *converges, then writing*

$$\mathbf{s}_n = \sum_{i+j=n} \mathbf{a}_{ij},$$

we have $\sum_{n=0}^{\infty} \mathbf{s}_n$ *absolutely convergent and*

$$\sum_{n=0}^{\infty} \mathbf{s}_n = \sum_{i=0}^{\infty}\sum_{j=0}^{\infty} \mathbf{a}_{ij}.$$

We are no longer so concerned with tracing everything back to the fundamental axiom, but that does not mean that it ceases to play a fundamental role.

Exercise 5.34. *We work in* $\mathbb{Q}$. *Show that we can find* $x_0, x_1, x_2, x_3, \ldots$ *an increasing sequence such that* $x_0 = 0$ *and* $x_j - x_{j-1} \leq 2^{-j}$ *for* $j \geq 1$, *but* x_n *does not tend to a limit.*

Set $x_0 = 0$, $a_{1j} = x_j - x_{j-1}$, $a_{2j} = 2^{-j} - a_{1j}$, *and* $a_{ij} = 0$ *for* $i \geq 3$. *Show that* $a_{ij} \geq 0$ *for all* i, j, *that* $\sum_{i=1}^{\infty} a_{ij}$ *exists for all* j *and* $\sum_{j=1}^{\infty}\sum_{i=1}^{\infty} a_{ij}$ *exists. Show, however, that* $\sum_{j=1}^{\infty} a_{ij}$ *does not exist when* $i = 1$ *or* $i = 2$.

5.4. The exponential function ♡

We have proved many deep and interesting theorems on the properties of continuous and differentiable functions. It is somewhat embarrassing to observe that, up to now, the only differentiable (indeed the only continuous) functions which we could lay our hands on were the polynomials and their quotients. (Even these were obtained as an afterthought in Exercise 4.38. We shall use results from that exercise in this section, but all such results will be proved in a wider context in Chapter 6.) In the next four sections we widen our repertoire considerably.

We start with the function exp. Historically, the exponential function was developed in connection with Napier's marvelous invention of the logarithm as a calculating tool (see Exercise 5.64). However, if by some historic freak, mathematics had reached the state it was in 1850 whilst bypassing the exponential and logarithmic functions, the exponential function might have been discovered as follows.

One way to obtain new functions is to look for solutions to differential equations. Consider one of the simplest such equations, $y'(x) = y(x)$. Without worrying about rigour ('In a storm I would burn six candles to Saint George and half a dozen to his dragon') we might try to find a solution in the form of a power series $y(x) = \sum_{j=0}^{\infty} a_j x^j$.

Plausible Statement 5.35. *The general solution of the equation*

$$\bigstar \qquad y'(x) = y(x),$$

where $y : \mathbb{R} \to \mathbb{R}$ *is a well behaved function, is*

$$y(x) = a \sum_{j=0}^{\infty} \frac{x^j}{j!}$$

with $a \in \mathbb{R}$.

Plausible argument. We seek for a solution of $\bigstar$ of the form $y(x) = \sum_{j=0}^{\infty} a_j x^j$. Assuming that we can treat a power series in the same way as we can treat a polynomial, we differentiate term by term to obtain $y'(x) = \sum_{j=1}^{\infty} j a_j x^{j-1}$. Equation $\bigstar$ thus becomes

$$\sum_{j=0}^{\infty} a_j x^j = \sum_{j=0}^{\infty} (j+1) a_{j+1} x^j,$$

which may be rewritten as

$$\sum_{j=0}^{\infty} (a_j - (j+1) a_{j+1}) x^j = 0.$$

Now, if a polynomial $P(x)$ vanishes for all values of x, all its coefficients are zero. Assuming that the result remains true for power series, we obtain

$$a_j - (j+1) a_{j+1} = 0$$

for all $j \geq 0$, and a simple induction gives $a_j = a_0/j!$. Setting $a = a_0$, we have the result. ▲

Later in the book we shall justify all of the arguments used above (see Theorem 11.73 and Exercise 11.75), but, for the moment, we just use them as a heuristic tool.

We can make an immediate observation.

Lemma 5.36. *The power series*

$$\sum_{j=0}^{\infty} \frac{z^j}{j!}$$

has infinite radius of convergence.

Proof. If $z \neq 0$, then, setting $u_j = z^j/j!$, we have $|u_{j+1}|/|u_j| = |z|/(j+1) \to 0$ as $j \to \infty$. Thus, by the ratio test, $\sum_{j=0}^{\infty} z^j/j!$ is absolutely convergent, and so convergent, for all z. ∎

We may thus define $e(z) = \sum_{j=0}^{\infty} z^j/j!$ for all $z \in \mathbb{C}$. A little playing around with formulae would lead to the key observation.

Lemma 5.37. *If z, $w \in \mathbb{C}$, then*

$$e(z)e(w) = e(z+w).$$

Proof. Observe that $\sum_{r=0}^{\infty} |z^r|/r!$ converges to $e(|z|)$ and $\sum_{s=0}^{\infty} |w^s|/s!$ converges to $e(|w|)$. Thus, writing $a_{rs} = (|z^r|/r!)(|w^s|/s!)$,

$$\begin{aligned}
\sum_{r=0}^{\infty}\sum_{s=0}^{\infty} |a_{rs}| &= \sum_{r=0}^{\infty}\left(\sum_{s=0}^{\infty} \frac{|z^r|}{r!}\frac{|w|^s}{s!}\right) \\
&= \sum_{r=0}^{\infty} \frac{|z^r|}{r!}\left(\sum_{s=0}^{\infty} \frac{|w|^s}{s!}\right) \\
&= \sum_{r=0}^{\infty} \frac{|z^r|}{r!} e(|w|) = e(|z|)e(|w|).
\end{aligned}$$

Thus $\sum_{r=1}^{\infty}\sum_{s=1}^{\infty} a_{rs}$ converges absolutely and, by Lemma 5.33, it follows that, if we write

$$c_n = \sum_{r+s=n} a_{rs},$$

we have $\sum_{n=0}^{\infty} c_n$ absolutely convergent and

$$\sum_{n=1}^{\infty} c_n = \sum_{i=1}^{\infty}\sum_{j=1}^{\infty} a_{ij}.$$

But

$$\begin{aligned}
c_n &= \sum_{r=0}^{n} a_{r(n-r)} = \sum_{r=0}^{n} \frac{z^r}{r!}\frac{w^{n-r}}{(n-r)!} \\
&= \frac{1}{n!}\sum_{r=0}^{n} \binom{n}{r} z^r w^{n-r} = \frac{(z+w)^n}{n!},
\end{aligned}$$

and so

$$e(z)e(w) = \sum_{r=0}^{\infty} \frac{z^r}{r!} \left(\sum_{s=0}^{\infty} \frac{w^s}{s!} \right) = \sum_{r=1}^{\infty} \sum_{s=1}^{\infty} a_{rs} = \sum_{n=1}^{\infty} c_n = e(z+w).$$

■

It should be noticed that the essential idea of the proof is contained in its last sentence. The rest of the proof is devoted to showing that what ought to work does, in fact, work.

Exercise 5.38. (Multiplication of power series.) *The idea of the previous lemma can be generalised.*

(i) Let $\lambda_n, \mu_n \in \mathbb{C}$ *and* $\gamma_n = \sum_{j=0}^{n} \lambda_j \mu_{n-j}$. *Show, by using Lemma 5.33, or otherwise, that, if* $\sum_{n=0}^{\infty} \lambda_n$ *and* $\sum_{n=0}^{\infty} \mu_n$ *are absolutely convergent, so is* $\sum_{n=0}^{\infty} \gamma_n$, *and*

$$\sum_{n=0}^{\infty} \lambda_n \sum_{n=0}^{\infty} \mu_n = \sum_{n=0}^{\infty} \gamma_n.$$

(ii) Suppose that $\sum_{n=0}^{\infty} a_n z^n$ *has radius of convergence* R *and* $\sum_{n=0}^{\infty} b_n z^n$ *has radius of convergence* S. *Explain why both* $\sum_{n=0}^{\infty} a_n w^n$ *and* $\sum_{n=0}^{\infty} b_n w^n$ *converge absolutely whenever* $|w| < \min(R, S)$. *Deduce that, if we write* $c_n = \sum_{r+s=n} a_r b_s$, *then* $\sum_{n=0}^{\infty} c_n w^n$ *converges absolutely and*

$$\sum_{n=0}^{\infty} a_n w^n \sum_{n=0}^{\infty} b_n w^n = \sum_{n=0}^{\infty} c_n w^n.$$

Thus $\sum_{n=0}^{\infty} c_n z^n$ *has radius of convergence at least* $\min(R, S)$ *and, if* $|w| < \min(R, S)$, *the formula just displayed applies. This result is usually stated as 'two power series can be multiplied within their smaller radius of convergence'.*

Exercise 5.39. *Use Exercise 5.38 directly to prove Lemma 5.37.*

Exercise 5.40. *(i) Prove directly that* $\sum_{n=0}^{\infty} z^n$ *has radius of convergence 1 and*

$$\sum_{n=0}^{\infty} z^n = \frac{1}{1-z}$$

for $|z| < 1$.

(ii) Use Exercise 5.38 to show that $\sum_{n=0}^{\infty} (n+1) z^n$ *has radius of convergence at least 1 and that*

$$\sum_{n=0}^{\infty} (n+1) z^n = \frac{1}{(1-z)^2}$$

for $|z| < 1$. *Show that* $\sum_{n=0}^{\infty}(n+1)z^n$ *has radius of convergence exactly* 1.

(iii) It is trivial that $1+z$ *has radius of convergence* ∞. *Use this observation and others made earlier in the exercise to show that, for appropriate choices of* $\sum_{n=0}^{\infty} a_n z^n$ *and* $\sum_{n=0}^{\infty} b_n z^n$, *we can have, in the notation of Exercise 5.38,*

(a) $R = 1$, $S = \infty$, $\sum_{n=0}^{\infty} c_n z^n$ *has radius of convergence* ∞.

(b) $R = 1$, $S = \infty$, $\sum_{n=0}^{\infty} c_n z^n$ *has radius of convergence* 1.

[See also Exercise K.263.]

The next exercise gives an algebraic interpretation of Lemma 5.37.

Exercise 5.41. *Check that* $\mathbb{C}$ *is an Abelian group under addition. Show that* $\mathbb{C}\setminus\{0\}$ *is an Abelian group under multiplication. Show that* $e : (\mathbb{C}, +) \to (\mathbb{C}\setminus\{0\}, \times)$ *is a homomorphism.*

The following estimate is frequently useful.

Lemma 5.42. *If* $|z| < n/2$, *then*

$$\left| e(z) - \sum_{j=0}^{n-1} \frac{z^j}{j!} \right| \leq \frac{2|z|^n}{n!}.$$

Proof. Most mathematicians would simply write

$$\begin{aligned}
\left| e(z) - \sum_{j=0}^{n-1} \frac{z^j}{j!} \right| &= \left| \sum_{j=n}^{\infty} \frac{z^j}{j!} \right| \leq \sum_{j=n}^{\infty} \frac{|z|^j}{j!} \\
&= \sum_{k=0}^{\infty} \frac{|z|^n}{n!} \frac{|z|^k}{(n+1)(n+2)\dots(n+k)} \\
&\leq \frac{|z|^n}{n!} \sum_{k=0}^{\infty} \left(\frac{|z|}{n+1} \right)^k \\
&\leq \frac{|z|^n}{n!} \frac{1}{1-\frac{|z|}{n+1}} = \frac{(n+1)|z|^n}{(n+1-|z|)n!} \\
&\leq \frac{2|z|^n}{n!}.
\end{aligned}$$

■

Exercise 5.43. *A very cautious mathematician*[7] *might prove Lemma 5.42 as follows. Set* $e_m(z) = \sum_{j=0}^{m} \frac{z^j}{j!}$. *Show that, if* $m \geq n$, *then*

$$|e_m(z) - e_{n-1}(z)| \leq \frac{(n+1)|z|^n}{(n+1-|z|)n!}.$$

[7] It is a good idea to be very cautious from time to time.

Deduce that

$$\begin{aligned}|e(z) - e_{n-1}(z)| &\leq |e(z) - e_m(z)| + |e_m(z) - e_{n-1}(z)| \\ &\leq |e(z) - e_m(z)| + \frac{(n+1)|z|^n}{(n+1-|z|)n!}.\end{aligned}$$

By allowing $m \to \infty$, obtain the required result.

We now switch our attention to the restriction of e to $\mathbb{R}$. The results we expect now come tumbling out.

Exercise 5.44. *Consider $e : \mathbb{R} \to \mathbb{R}$ given by $e(x) = \sum_{j=0}^{\infty} x^j/j!$.*

(i) Using Lemma 5.42, show that $|e(h) - 1 - h| \leq h^2$ for $|h| < 1/2$. Deduce that e is differentiable at 0 with derivative 1.

(ii) Explain why $e(x + h) - e(x) = e(x)(e(h) - 1)$. Deduce that e is everywhere differentiable with $e'(x) = e(x)$.

(iii) Show that $e(x) \geq 1$ for $x \geq 0$ and, by using the relation $e(-x)e(x) = 1$, or otherwise, show that $e(x) > 0$ for all $x \in \mathbb{R}$.

(iv) Explain why e is a strictly increasing function.

(v) Show that $e(x) \geq x$ for $x \geq 0$ and deduce that $e(x) \to \infty$ as $x \to \infty$. Show also that $e(x) \to 0$ as $x \to -\infty$.

(vi) Use (v) and the intermediate value theorem to show that $e(x) = y$ has a solution for all $y > 0$.

(vii) Use (iv) to show that $e(x) = y$ has at most one solution for all $y > 0$. Conclude that e is a bijective map of $\mathbb{R}$ to $\mathbb{R}^{++} = \{x \in \mathbb{R} : x > 0\}$.

(viii) By modifying the proof of (v), or otherwise, show that, if P is a polynomial, then $P(x)e(-x) \to 0$ as $x \to \infty$. [We say 'exponential beats polynomial'.]

(ix) By using (viii), or otherwise, show that e is not equal to any function of the form P/Q with P and Q polynomials. [Thus e is a genuinely new function.]

When trying to prove familiar properties of a familiar function, it is probably wise to use a slightly unfamiliar notation. However, as the reader will have realised from the start, the function e is our old friend exp. We shall revert to the mild disguise in the next section, but we use standard notation for the rest of this one.

Exercise 5.45. *(i) Check that $\mathbb{R}$ is an Abelian group under addition. Show that $\mathbb{R}^{++} = \{x \in \mathbb{R} : x > 0\}$ is an Abelian group under multiplication. Show that* $\exp : (\mathbb{R}, +) \to (\mathbb{R}^{++}, \times)$ *is a isomorphism.*

(ii) [Needs a little more familiarity with groups.] Show that $\mathbb{R} \setminus \{0\}$ is an Abelian group under multiplication. By considering the order of the element

$-1 \in \mathbb{R} \setminus \{0\}$, or otherwise, show that the groups $(\mathbb{R}, +)$ and $(\mathbb{R} \setminus \{0\}, \times)$ are not isomorphic.

We can now turn Plausible Statement 5.35 into a theorem

Theorem 5.46. *The general solution of the equation*

$$\bigstar \qquad y'(x) = y(x),$$

where $y : \mathbb{R} \to \mathbb{R}$ is a differentiable function, is

$$y(x) = a\exp(x)$$

with $a \in \mathbb{R}$.

Proof. It is clear that $y(x) = a\exp(x)$ is a solution of $\bigstar$. We must prove there are no other solutions. To this end, observe that, if y satisfies $\bigstar$, then

$$\frac{d}{dx}(\exp(-x)y(x)) = y'(x)\exp(-x) - y(x)\exp(-x) = 0,$$

so, by the mean value theorem, $\exp(-x)y(x)$ is a constant function. Thus $\exp(-x)y(x) = a$ and $y(x) = a\exp(x)$ for some $a \in \mathbb{R}$. ■

Exercise 5.47. *State and prove the appropriate generalisation of Theorem 5.46 to cover the equation*

$$y'(x) = by(x)$$

with b a real constant.

Here is another consequence of Lemma 5.42.

Exercise 5.48. **(e is irrational.)** *Suppose, if possible, that $e = \exp 1$ is rational. Then $\exp 1 = m/n$ for some positive integers m and n. Explain, why, if $N \geq n$,*

$$N!\left(\exp 1 - \sum_{j=0}^{N} \frac{1}{j!}\right)$$

must be a non-zero integer and so

$$N!\left|\exp 1 - \sum_{j=0}^{N} \frac{1}{j!}\right| \geq 1.$$

Use Lemma 5.42 to obtain a contradiction.

Show, similarly, that $\sum_{r=1}^{\infty} \frac{(-1)^{r+1}}{(2r-1)!}$ is irrational.

Most mathematicians draw diagrams frequently, both on paper and in their heads[8]. However, these diagrams are merely sketches. To see this, quickly sketch a graph of $\exp x$.

Exercise 5.49. *Choosing appropriate scales, draw an accurate graph of* exp *on the interval* $[0, 100]$. *Does it look like your quick sketch?*

We conclude this section with a result which is a little off our main track but whose proof provides an excellent example of the use of dominated convergence (Theorem 5.25).

Exercise 5.50. *We work in* $\mathbb{C}$. *Show that if we write*

$$\left(1 + \frac{z}{n}\right)^n = \sum_{j=0}^{\infty} a_j(n) z^j,$$

then $a_j(n)z^j \to z^j/j!$ *as* $n \to \infty$, *and* $|a_j(n)z^j| \leq |z^j|/j!$ *for all* n *and all* j. *Use dominated convergence to conclude that*

$$\left(1 + \frac{z}{n}\right)^n \to e(z)$$

as $n \to \infty$, *for all* $z \in \mathbb{C}$.

5.5. The trigonometric functions ♡

In the previous section we considered the simple differential equation $y'(x) = y(x)$. What happens if we consider the differential equation $y''(x)+y(x) = 0$?

Exercise 5.51. *Proceeding along the lines of Plausible Statement 5.35, show that it is reasonable to conjecture that the general solution of the equation*

★★ $$y''(x) + y(x) = 0,$$

where $y : \mathbb{R} \to \mathbb{R}$ *is a well behaved function, is*

$$y(x) = a \sum_{j=0}^{\infty} \frac{(-1)^j x^{2j}}{(2j)!} + b \sum_{j=0}^{\infty} \frac{(-1)^j x^{2j+1}}{(2j+1)!}$$

with $a, b \in \mathbb{R}$.

A little experimentation reveals what is going on.

Exercise 5.52. *We work in* $\mathbb{C}$. *If we write*

$$c(z) = \frac{e(iz) + e(-iz)}{2}, \quad s(z) = \frac{e(iz) - e(-iz)}{2i},$$

[8] Little of this activity appears in books and papers, partly because, even today, adding diagrams to printed work is non-trivial. It is also possible that it is the process of drawing (or watching the process of drawing) which aids comprehension, rather than the finished product.

show carefully that

$$c(z) = \sum_{j=0}^{\infty} \frac{(-1)^j z^{2j}}{(2j)!}, \quad s(z) = \sum_{j=0}^{\infty} \frac{(-1)^j z^{2j+1}}{(2j+1)!}.$$

We can use the fact that $e(z+w) = e(z)e(w)$ to obtain a collection of useful formulae for s and c.

Exercise 5.53. *Show that if $z, w \in \mathbb{C}$, then*

(i) $s(z+w) = s(z)c(w) + c(z)s(w)$,

(ii) $c(z+w) = c(z)c(w) - s(z)s(w)$,

(iii) $s(z)^2 + c(z)^2 = 1$,

(iv) $s(-z) = -s(z)$, $c(-z) = c(z)$.

We now switch our attention to the restriction of s and c to $\mathbb{R}$.

Exercise 5.54. *Let $c, s : \mathbb{R} \to \mathbb{R}$ be given by $c(x) = \sum_{j=0}^{\infty}(-1)^j x^{2j}/(2j)!$ and $s(x) = \sum_{j=0}^{\infty}(-1)^j x^{2j+1}/(2j+1)!$.*

(i) Using the remainder estimate in the alternating series test (second paragraph of Lemma 5.12), or otherwise, show that $|c(h) - 1| \leq h^2/2$ and $|s(h) - h| \leq |h|^3/6$ for $|h| < 1$. Deduce that c and s are differentiable at 0 with $c'(0) = 0$, $s'(0) = 1$.

(ii) Using the addition formulae of Exercise 5.53 (ii) and (iii) to evaluate $c(x+h)$ and $s(x+h)$, show that c and s are everywhere differentiable with $c'(x) = -s(x)$, $s'(x) = c(x)$.

Suppose that a group of mathematicians who did not know the trigonometric functions were to investigate our functions c and s defined by power series. Careful calculation and graphing would reveal that, incredible as it seemed, c and s appeared to be periodic!

Exercise 5.55. *(i) By using the estimate for error in the alternating series test, show that*

$$c(x) > 0 \text{ for all } 0 \leq x \leq 1.$$

By using a minor modification of these ideas, or otherwise, show that $c(2) < 0$. Explain carefully why this means that there must exist an a with $1 < a < 2$ such that $c(a) = 0$.

(iii) In this part and what follows we make use of the formulae obtained in Exercise 5.53, which tell us that

$$s(x+y) = s(x)c(y) + c(x)s(y), \quad c(x+y) = c(x)c(y) - s(x)s(y),$$
$$c(x)^2 + s(x)^2 = 1, \quad s(-x) = -s(x), \quad c(-x) = c(x)$$

for all x, $y \in \mathbb{R}$. Show that, if $c(a') = 0$ and $c(a'') = 0$, then $s(a' - a'') = 0$. Use the fact that $s(0) = 0$ and $s'(x) = c(x) > 0$ for $0 \leq x \leq 1$ to show that $s(x) > 0$ for $0 < x \leq 1$. Conclude that, if a' and a'' are distinct zeros of c, then $|a' - a''| > 1$. Deduce that $c(x) = 0$ has exactly one solution with $0 \leq x \leq 2$. We call this solution a.

(iv) By considering derivatives, show that s is strictly increasing on $[0, a]$. Conclude that $s(a) > 0$ and deduce that $s(a) = 1$. Show that

$$s(x + a) = c(x), \ \ c(x + a) = -s(x)$$

for all x, and that c and s are periodic with period $4a$ (that is, $s(x + 4a) = s(x)$ and $c(x + 4a) = c(x)$ for all x).

(v) Show that s is strictly increasing on $[-a, a]$, and strictly decreasing on $[a, 3a]$.

(vi) If u and v are real numbers with $u^2 + v^2 = 1$, show that there there is exactly one solution to the pair of equations

$$c(x) = u, \ \ s(x) = v$$

with $0 \leq x < 4a$.

At this point we tear off the thin disguise of our characters and write $\exp(z) = e(z)$, $\sin z = s(z)$, $\cos(z) = c(z)$ and $a = \pi/2$.

Exercise 5.56. *We work in $\mathbb{R}$. Show that, if $|u| \leq 1$, there is exactly one θ with $0 \leq \theta \leq \pi$ such that $\cos\theta = u$.*

Using the Cauchy–Schwarz inequality (Lemma 4.2), show that, if $\mathbf{x}$ and $\mathbf{y}$ are non-zero vectors in $\mathbb{R}^m$, then there is exactly one θ with $0 \leq \theta \leq \pi$ such that

$$\cos\theta = \frac{\mathbf{x} \cdot \mathbf{y}}{\|\mathbf{x}\|\|\mathbf{y}\|}.$$

We call θ the angle between $\mathbf{x}$ and $\mathbf{y}$.

Exercise 5.57. *We work in $\mathbb{C}$ and use the usual disguises except that we write $a = \pi/2$.*

(i) Show that e has period $2\pi i$ in the sense that

$$e(z + 2\pi i) = e(z)$$

for all z, and

$$e(z + w) = e(z)$$

for all z if and only $w = 2n\pi i$ for some $n \in \mathbb{Z}$. State corresponding results for $s,\ c : \mathbb{C} \to \mathbb{C}$.

(ii) If x and y are real, show that

$$e(x + iy) = e(x)(c(y) + is(y)).$$

(iii) If $w \neq 0$, show that there are unique real numbers r and y with $r > 0$ and $0 \leq y < 2\pi$ such that

$$w = re(iy).$$

(iv) If $w \in \mathbb{C}$, find all solutions of $w = re(iy)$ with r and y real and $r \geq 0$.

The traditional statement of Exercise 5.57 (iii) says that $z = re^{i\theta}$, where $r = |z|$ and θ is real. However, we have not defined powers yet, so, for the moment, this must merely be considered as a useful mnemonic. (We will discuss the matter further in Exercise 5.80.)

It may be objected that our definitions of sine and cosine ignore their geometric origins. Exercises K.187 and K.188) give more 'geometric' treatments, but the following points are worth noting.

The trigonometric functions did not arise as part of classical axiomatic Euclidean geometry, but as part of practical geometry (mainly astronomy). An astronomer is happy to consider the sine of 20 degrees, but a classical geometer would simply note that it is impossible to construct an angle of 20 degrees using ruler and compass. Starting with our axioms for $\mathbb{R}$ we can obtain a model of classical geometry, but the reverse is not true.

The natural 'practical geometric' treatment of angle does not use radians. Our use of radians has nothing to do with geometric origins and everything to do with the equation (written in radians)

$$\frac{d}{dx}\sin cx = c\cos cx.$$

Mathematicians measure angles in radians because, for them, sine is a function of analysis. Everyone else measures angles in degrees because, for them, sine is a function used in practical geometry.

In the natural 'practical geometric' treatment of angle it is usual to confine oneself to positive angles less than two right angles (or indeed one right angle). When was the last time you have heard a navigator shouting 'turn -20 degrees left' or 'up 370 degrees'? The extension of sine from a function on $[0, \pi/2]$ to a function on $\mathbb{R}$ and the corresponding extension of the notion of angle is a product of 'analytic' and not 'geometric' thinking.

Since much of this book is devoted to stressing the importance of a 'geometric approach' to the calculus of several variables, I do not wish to downplay the geometric meaning of sine. However, we should treat sine both as a geometric object and as a function of analysis. In this context it matters little whether we start with a power series definition of sine and end up with the parametric description of the unit circle as the path described by the point $(\sin\theta, \cos\theta)$ as θ runs from 0 to 2π, or (as we shall do in

Exercises K.187 and K.188) we start with the Cartesian description of the circle as $x^2 + y^2 = 1$ and end up with a power series for sine.

Exercise 5.58. *Write down the main properties of* cosh *and* sinh *that you know. Starting with a tentative solution of the differential equation $y'' = y$, write down appropriate definitions and prove the stated properties in the style of this section. Distinguish between those properties which hold for* cosh *and* sinh *as functions from $\mathbb{C}$ to $\mathbb{C}$ and those which hold for* cosh *and* sinh *as functions from $\mathbb{R}$ to $\mathbb{R}$.*

5.6. The logarithm ♡

In this section we shall make use of the one-dimensional chain rule.

Lemma 5.59. *Suppose that $f : \mathbb{R} \to \mathbb{R}$ is differentiable at x with derivative $f'(x)$, that $g : \mathbb{R} \to \mathbb{R}$ is differentiable at y with derivative $g'(y)$, and that $f(x) = y$. Then $g \circ f$ is differentiable at x with derivative $f'(x)g'(y)$.*

In traditional notation, $\dfrac{d}{dx}g(f(x)) = f'(x)g'(f(x))$. We divide the proof into two parts.

Lemma 5.60. *Suppose that the hypotheses of Lemma 5.59 hold and, in addition, $f'(x) \neq 0$. Then the conclusion of Lemma 5.59 holds.*

Proof. Since

$$\frac{f(x+h) - f(x)}{h} \to f'(x) \neq 0,$$

we can find a $\delta > 0$ such that

$$\frac{f(x+h) - f(x)}{h} \neq 0$$

for $0 < |h| < \delta$, and so, in particular, $f(x+h) - f(x) \neq 0$ for $0 < |h| < \delta$.

Thus, if $0 < |h| < \delta$, we may write

$$\bigstar \qquad \frac{g(f(x+h)) - g(f(x))}{h} = \frac{g(f(x+h)) - g(f(x))}{f(x+h) - f(x)} \frac{f(x+h) - f(x)}{h}.$$

Now f is differentiable and so continuous at x, so $f(x+h) - f(x) \to 0$ as $h \to 0$. It follows, by using standard theorems on limits (which the reader should identify explicitly), that

$$\frac{g(f(x+h)) - g(f(x))}{h} \to g'(f(x))f'(x)$$

as $h \to 0$, and we are done. ∎

Unfortunately the proof of Lemma 5.60 does not work in general for Lemma 5.59, since we then have no guarantee that $f(x+h)-f(x)\neq 0$, even for small h, and so we cannot use equation ★[9]. We need a separate proof for this case.

Lemma 5.61. *Suppose that the hypotheses of Lemma 5.59 hold and, in addition, $f'(x)=0$. Then the conclusion of Lemma 5.59 holds.*

I outline a proof in the next exercise, leaving the details to the reader.

Exercise 5.62. *We prove Lemma 5.61 by reductio ad absurdum. To this end, suppose that the hypotheses of the lemma hold but the conclusion is false.*

(i) Explain why we can find an $\epsilon > 0$ and a sequence $h_n \to 0$ such that $h_n \neq 0$ and

$$\left|\frac{g(f(x+h_n))-g(f(x))}{h_n}\right| > \epsilon$$

for each $n \geq 0$.

(ii) Explain why $f(x+h_n)\neq f(x)$ for each $n\geq 0$.

(iii) Use the method of proof of Lemma 5.60 to derive a contradiction.

The rather ugly use of reductio ad absurdum in Exercise 5.62 can be avoided by making explicit use of the ideas of Exercise K.23.

Note that, in this section, we only use the special case of the chain rule given in Lemma 5.60. I believe that the correct way to look at the chain rule is by adopting the ideas of Chapter 6 and attacking it directly, as we shall do in Lemma 6.19. We now move on to the main subject of this section.

Since $e:\mathbb{R}\to\mathbb{R}^{++}$ is a bijection (indeed, by Exercise 5.45 a group isomorphism), it is natural to look at its inverse. Let us write $l(x)=e^{-1}(x)$ for $x\in(0,\infty)=\mathbb{R}^{++}$. Some of the properties of l are easy to obtain. (Here and later we use the properties of the function e obtained in Exercise 5.44.)

Exercise 5.63. *(i) Explain why $l:(0,\infty)\to\mathbb{R}$ is a bijection.*

(ii) Show that $l(xy)=l(x)+l(y)$ for all x, $y>0$.

(iii) Show that l is a strictly increasing function.

Exercise 5.64. *No one who went to school after 1960 can really appreciate the immense difference between the work involved in hand multiplication without logarithms and hand multiplication if we are allowed to use logarithms. The invention of logarithms was an important contribution to the*

[9] Hardy's *Pure Mathematics* says 'The proof of [the chain rule] requires a little care' and carries the rueful footnote 'The proofs in many textbooks (and in the first three editions of this book) are inaccurate'. This is the point that the textbooks overlooked.

scientific revolution. When Henry Briggs (who made a key simplification) visited Baron Napier (who invented the idea) 'almost one quarter of an hour was spent, each beholding [the] other ... with admiration before one word was spoke, at last Mr Briggs began.

'My lord, I have undertaken this long Journey purposely to see your Person, and to know by what Engine of Wit or Ingenuity you came first to think of this most excellent Help unto Astronomy, viz., *the* Logarithms; *but, my Lord, being by you found out, I wonder nobody else found it out before, when now known it is so easy.'(Quotation from 9.E.3 of* [**17**].)

(i) As Briggs realised, calculations become a little easier if we use $\log_{10}$ *defined by*

$$\log_{10} x = l(x)/l(10)$$

for $x > 0$. *Show that* $\log_{10} xy = \log_{10} x + \log_{10} y$ *for all* $x, y > 0$ *and that* $\log_{10} 10^r x = r + \log_{10} x$.

(ii) Multiply 1.3245 *by* 8.7893, *correct to five significant figures, without using a calculator.*

(iii) To multiply 1.3245 *by* 8.7893 *using logarithms, one looked up the values of* $\log_{10} 1.3245$ *and* $\log_{10} 8.7893$ *in a table of logarithms. This was quick and easy, giving*

$$\log_{10} 1.3245 \approx 0.1220520, \ \log_{10} 8.7893 \approx 0.9439543.$$

A hand addition, which the reader should do, gave

$$\begin{aligned}\log_{10}(1.3245 \times 8.7893) &= \log_{10} 1.3245 + \log_{10} 8.7893\\ &\approx 0.1220520 + 0.9439543 = 1.0660063.\end{aligned}$$

A quick and easy search in a table of logarithms (or, still easier, a table of inverse logarithms, the so-called antilogarithms) showed that

$$\log_{10} 1.164144 \approx .0660052, \ \log_{10} 1.164145 \approx .0660089$$

so that

$$\log_{10} 11.64144 \approx 1.0660052, \ \log_{10} 11.64145 \approx 1.0660089$$

and, correct to five significant figures, $1.3245 \times 8.7893 = 11.6414$.

(iv) Repeat the exercise with numbers of your own choosing. You may use the '$\log_{10}$*' (often just called '*log*') function on your calculator and the 'inverse* $\log_{10}$*' (often called '*10^x*') but you must do the multiplication and addition by hand. Notice that you need one (or, if you are being careful, two) more extra figures in your calculations than there are significant figures in your answers.*

[There are some additional remarks in Exercises 5.78 and K.93.]

Other properties require a little more work.

Lemma 5.65. *(i) The function $l : (0, \infty) \to \mathbb{R}$ is continuous.*

(ii) The function l is everywhere differentiable with

$$l'(x) = \frac{1}{x}.$$

Proof. (i) We wish to show that l is continuous at some point $x \in (0, \infty)$. To this end, let $\delta > 0$ be given. Since l is increasing, we know that, if

$$e(l(x) + \delta) > y > e(l(x) - \delta),$$

we have

$$l\big(e(l(x) + \delta)\big) > l(y) > l\big(e(l(x) - \delta)\big)$$

and so

$$l(x) + \delta > l(y) > l(x) - \delta.$$

Now e is strictly increasing, so we can find $\eta(\delta) > 0$ such that

$$e(l(x) + \delta) > x + \eta(\delta) > x = l(e(x)) > x - \eta(\delta) > e(l(x) - \delta).$$

Combining the results of the two previous sentences, we see that, if $|x - y| < \eta(\delta)$, then $|l(x) - l(y)| < \delta$. Since δ was arbitrary, l is continuous at x.

(ii) We shall use the result that, if g is never zero and $g(x + h) \to a$ as $h \to 0$, then, if $a \neq 0$, $1/g(x + h) \to 1/a$ as $h \to 0$. Observe that, since l is continuous, we have

$$l(x + h) - l(x) \to 0,$$

and so

$$\frac{l(x + h) - l(x)}{h} = \frac{l(x + h) - l(x)}{e(l(x + h)) - e(l(x))} \to \frac{1}{e'(l(x))} = \frac{1}{e(l(x))} = \frac{1}{x}$$

as $h \to 0$. ■

By using the ideas of parts (iv), (v) and (vi) of Exercise 5.44 together with parts (i) and (iii) of Exercise 5.63 and both parts of Lemma 5.65, we get the following general result.

Exercise 5.66. (One-dimensional inverse function theorem.) *Suppose that $f : [a, b] \to [c, d]$ is continuous and f is differentiable on (a, b) with $f'(x) > 0$ for all $x \in (a, b)$ and $f(a) = c$, $f(b) = d$. Show that f is a bijection, that $f^{-1} : [c, d] \to [a, b]$ is continuous, and that f^{-1} is differentiable on (c, d) with*

$$(f^{-1})'(x) = \frac{1}{f'(f^{-1}(x))}.$$

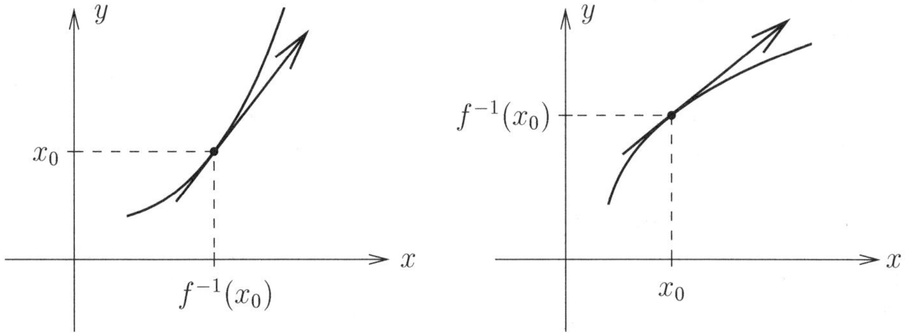

Figure 5.1. The one-dimensional inverse function theorem

We shall give a different proof of this result in a more general (and, I would claim, more instructive) context in Theorem 13.13. Traditionally, the one-dimensional inverse function theorem is illustrated, as in Figure 5.1, by taking the graph $y = f(x)$ with tangent shown at $(f^{-1}(x_0), x_0)$ and reflecting in the angle bisector of the x and y axes to obtain the graph $y = f^{-1}(x)$ with tangent shown at $(x_0, f(x_0))$.

Although the picture is suggestive, this is one of those cases where (at the level of proof we wish to use) a simple picture is inadequate.

Exercise 5.67. *Go through Exercise 5.66 and note where you used the mean value theorem and the intermediate value theorem.*

Exercise 5.68. *(i) Write $A = \{x \in \mathbb{Q} : 2 \geq x \geq 1\}$ and $B = \{x \in \mathbb{Q} : 4 \geq x \geq 1\}$. Define $f : A \to B$ by $f(x) = x^2$. Show that f is strictly increasing on A, that $f(1) = 1$ and $f(2) = 4$, that f is differentiable on A with $f'(x) \geq 2$ for all $x \in A$, and that $f : A \to B$ is injective, yet f is not surjective.*

(ii) Define $f : \mathbb{Q} \to \mathbb{Q}$ by

$$\begin{aligned} f(x) &= x + 1 && \text{for } x < 0,\ x^2 > 2,\\ f(x) &= x && \text{for } x^2 < 2,\\ f(x) &= x - 1 && \text{for } x > 0,\ x^2 > 2. \end{aligned}$$

Show that $f(x) \to -\infty$ as $x \to -\infty$, that $f(x) \to \infty$ as $x \to \infty$, that f is everywhere differentiable with $f'(x) = 1$ for all x, and that $f : \mathbb{Q} \to \mathbb{Q}$ is surjective, yet f is not injective[10].

Initially we defined the exponential and trigonometric functions as maps $\mathbb{C} \to \mathbb{C}$ although we did not make much use of this (they are very important

[10] These examples do not exhaust the ways in which Figure 5.1 is an inadequate guide to what can happen without the fundamental axiom of analysis [**33**].

in more advanced work) and switched rapidly to maps $\mathbb{R} \to \mathbb{R}$. We did nothing of this sort for the logarithm.

The most obvious attempt to define a complex logarithm fails at the first hurdle. We showed that, working over $\mathbb{R}$, the map $\exp : \mathbb{R} \to (0, \infty)$ is bijective, so that we could define log as the inverse function. However, we know (see Exercise 5.57) that, working over $\mathbb{C}$, the map $\exp : \mathbb{C} \to \mathbb{C} \setminus \{0\}$ is surjective but not injective, so no inverse function exists.

Exercise 5.69. *By using the fact that* $\exp 2\pi i = 1 = \exp 0$, *show that there cannot exist a function* $L : \mathbb{C} \setminus \{0\} \to \mathbb{C}$ *with* $L(\exp z) = z$ *for all* $z \in \mathbb{C}$.

However, a one-sided inverse can exist.

Exercise 5.70. *(i) If we set* $L_0(r \exp i\theta) = \log r + i\theta$ *for* $r > 0$ *and* $2\pi > \theta \geq 0$, *show that* $L_0 : \mathbb{C} \setminus \{0\} \to \mathbb{C}$ *is a well defined function with* $\exp(L_0(z)) = z$ *for all* $z \in \mathbb{C} \setminus \{0\}$.

(ii) Let n *be an integer. If we set* $L_n(r \exp i\theta) = L_0(r \exp i\theta) + 2\pi i n$, *show that* $L_n : \mathbb{C} \setminus \{0\} \to \mathbb{C}$ *is a well defined function with* $\exp(L_n(z)) = z$ *for all* $z \in \mathbb{C} \setminus \{0\}$.

(iii) If we set $M(r \exp i\theta) = \log r + i\theta$ *for* $r > 0$ *and* $3\pi > \theta \geq \pi$, *show that* $M : \mathbb{C} \setminus \{0\} \to \mathbb{C}$ *is a well defined function with* $\exp(M(z)) = z$ *for all* $z \in \mathbb{C} \setminus \{0\}$.

The functions L_n and M in the last exercise are not continuous everywhere, and it is natural to ask if there is a continuous function $L : \mathbb{C} \setminus \{0\} \to \mathbb{C}$ with $\exp(L(z)) = z$ for all $z \in \mathbb{C} \setminus \{0\}$. The reader should convince herself, by trying to define $L(\exp i\theta)$ and considering what happens as θ runs from 0 to 2π, that this is not possible. The next exercise crystallises the ideas.

Exercise 5.71. *Suppose, if possible, that there exists a continuous function* $L : \mathbb{C} \setminus \{0\} \to \mathbb{C}$ *with* $\exp(L(z)) = z$ *for all* $z \in \mathbb{C} \setminus \{0\}$.

(i) If θ *is real, show that* $L(\exp(i\theta)) = i(\theta + 2\pi n(\theta))$ *for some* $n(\theta) \in \mathbb{Z}$.

(ii) Define $f : \mathbb{R} \to \mathbb{R}$ *by*

$$f(\theta) = \frac{1}{2\pi}\left(\frac{L(\exp i\theta) - L(1)}{i} - \theta\right).$$

Show that f *is a well defined continuous function, that* $f(\theta) \in \mathbb{Z}$ *for all* $\theta \in \mathbb{R}$, *that* $f(0) = 0$, *and that* $f(2\pi) = -1$.

(iii) Show that the statements made in the last sentence of (ii) are incompatible with the intermediate value theorem, and deduce that no function can exist with the supposed properties of L.

(iv) Discuss informally what connection, if any, the discussion above has with the existence of the international date line.

Exercise 5.71 is not an end but a beginning of much important mathematics. In due course it will be necessary for the reader to understand both the formal proof that, and the informal reasons why, no continuous L can exist.

5.7. Powers ♡

How should we define a^b for $a > 0$ and b any real number? Most people would say that we should first define a^b for b rational and then extend 'by continuity' to non-rational b. This can be done, even with the few tools at our disposal, but it requires hard work to define a^b this way and still more hard work to obtain its properties. When we have more powerful tools at our disposal (uniform convergence and the associated theorems) we shall see how to make this programme work in Exercises K.254 to K.256; but, even then, it requires careful thought.

There are, I think, various reasons why the direct approach is hard.

(1) The first point is mainly psychological. We need to consider a^b as a function of two variables a and b. When we define a^n, we think of the integers n as fixed and a as varying, and the same is true when we define a^b with b rational. However, when we want to define a^b 'by continuity', we think of a as fixed and b as varying.

(2) The second point is mathematical. The fact that a function is continuous on the rationals does not mean that it has a continuous extension to the reals[11]. Consider our standard example, the function $f : \mathbb{Q} \to \mathbb{Q}$ of Example 1.3. We know that f is continuous, but there is no continuous function $F : \mathbb{R} \to \mathbb{R}$ with $F(x) = f(x)$ for $x \in \mathbb{Q}$.

Exercise 5.72. *(i) Prove this statement by observing that, if F is continuous, $F(x_n) \to F(2^{-1/2})$ whenever $x_n \to 2^{-1/2}$, or otherwise.*

(ii) Find a function $g : \mathbb{Q} \to \mathbb{Q}$, which is everywhere differentiable with continuous derivative, such that there is a function $G : \mathbb{R} \to \mathbb{R}$, with $G(x) = g(x)$ for $x \in \mathbb{Q}$, which is everywhere continuous but not everywhere differentiable.

However, the fact that I think something is hard does not prove that it is hard. I suggest that the reader try it for herself. (She may well succeed, all that is required is perseverance and a cool head. I simply claim that the exercise is hard, not that it is impossible.)

Assuming that the reader agrees with me, can we find another approach? We obtained the exponential and trigonometric functions as the solution of

[11] I can still remember being scolded by my research supervisor for making this particular mistake. (The result is true if we replace 'continuity' by 'uniform continuity'. See Exercise K.57.)

differential equations. How does this approach work here? The natural choice of differential equation, if we wish to obtain $y(x) = x^\alpha$, is

$$\bigstar \qquad xy'(x) = \alpha y(x)$$

(Here α is real and $y : (0, \infty) \to (0, \infty)$.)

Tentative solution. We can rewrite ★ as

$$\frac{y'(x)}{y(x)} - \frac{\alpha}{x} = 0.$$

Using the properties of logarithm and the chain rule, this gives

$$\frac{d}{dx}(\log y(x) - \alpha \log x) = 0,$$

so, by the mean value theorem,

$$\log y(x) - \alpha \log x = C,$$

where C is constant. Applying the exponential function and taking $A = \exp C$, we obtain

$$y(x) = A \exp(\alpha \log x),$$

where A is a constant. ▲

Exercise 5.73. *Check, by using the chain rule, that $y(x) = A\exp(\alpha \log x)$ is indeed a solution of ★.*

This strongly suggests that we should define $x^\alpha = \exp(\alpha \log x)$. In order to avoid confusion, we adopt our usual policy of light disguise and investigate the properties of functions $r_\alpha : (0, \infty) \to (0, \infty)$ defined by $r_\alpha(x) = \exp(\alpha \log x)$ [α real].

Exercise 5.74. (Index laws.) *If α, $\beta \in \mathbb{R}$, show that:*

(i) $r_{\alpha+\beta}(x) = r_\alpha(x) r_\beta(x)$ for all $x > 0$.

(ii) $r_{\alpha\beta}(x) = r_\alpha(r_\beta(x))$ for all $x > 0$.

Exercise 5.75. (Consistency.) *Suppose that n, p and q are integers with $n \geq 0$ and $q > 0$. Show that:*

(i) $r_1(x) = x$ for all $x > 0$.

(ii) $r_{n+1}(x) = x r_n(x)$ for all $x > 0$.

(iii) $r_n(x) = \overbrace{x \times x \times \cdots \times x}^{n}$ for all $x > 0$.

(iv) $r_{-n}(x) = \dfrac{1}{r_n(x)}$ for all $x > 0$.

(v) $r_q(r_{p/q}(x)) = r_p(x)$ for all $x > 0$.

Explain briefly why this means that writing $r_{p/q}(x) = x^{p/q}$ is consistent with your previous school terminology.

Exercise 5.76. *Suppose that α is real. Show that:*

(i) $r_\alpha(xy) = r_\alpha(x)r_\alpha(y)$ for all x, $y > 0$.

(ii) $r_0(x) = 1$ for all $x > 0$.

(iii) r_α is everywhere differentiable, and $xr'_\alpha(x) = \alpha r_\alpha(x)$ and $r'_\alpha(x) = \alpha r_{\alpha-1}(x)$ for all $x > 0$.

Exercise 5.77. *(i) If $x > 0$ is fixed, show that $r_\alpha(x)$ is a differentiable function of α with*

$$\frac{d}{d\alpha} r_\alpha(x) = r_\alpha(x) \log x.$$

(ii) If $\alpha > 0$ and α is kept fixed, show that $r_\alpha(x)$ is an increasing function of x. What happens if $\alpha < 0$?

(iii) If $x > 1$ and x is kept fixed, show that $r_\alpha(x)$ is an increasing function of α. What happens if $0 < x < 1$?

(iv) If we write $e = \exp 1$, show that $\exp x = r_e(x)$ (or, in more familiar terms, $\exp x = e^x$).

Exercise 5.78. *Take two rulers A and B marked in centimeters (or some other convenient unit) and lay them marked edge to marked edge. If we slide the point marked 0 on ruler B until it is opposite the point marked x on ruler A, then the point marked y on ruler B will be opposite the point marked $x + y$ on ruler A. We have invented an* adding machine.

Now produce a new ruler A' by renaming the point marked x as 10^x (thus the point marked 0 on A becomes the point marked 1 on A and the point marked 3 on A' becomes the point marked 1000 on A'). Obtain B' from B in the same way. If we slide the point marked 1 on ruler B' until it is opposite the point marked 10^x on ruler A', then the point marked 10^y on ruler B' will be opposite the point marked 10^{x+y} on ruler A'. Explain why, if a, $b > 0$ and we slide the point marked 1 on ruler B' until it is opposite the point marked a on ruler A', then the point marked b on ruler B' will be opposite the point marked ab on ruler A'. We have invented a multiplying machine.

(i) How would you divide a by b using this machine?

(ii) Does the number 10 play an essential role in the device?

(iii) Draw a line segment CD of some convenient length to represent the ruler A'. If C corresponds to 1 and D to 10, draw, as accurately as you can, the points corresponding to $2, 3, \dots, 9$.

The device we have described was invented by Oughtred some years after Napier's discovery of the logarithm and forms the basis for the 'slide rule'. From 1860 to 1960 the slide rule was the emblem of the mathematically competent engineer. It allowed fast and reasonably accurate 'back of an envelope' calculations.

Exercise 5.79. *By imitating the argument of Exercise 5.71, show that there is no continuous function* $S : \mathbb{C} \to \mathbb{C}$ *with* $S(z)^2 = z$ *for all* $z \in \mathbb{C}$. *(In other words, we cannot define a well behaved square root function on the complex plane.)*

Exercise 5.80. *Exercise 5.79 shows, I think, that we cannot hope to extend our definition of* $r_\alpha(x)$ *with* x *real and strictly positive and* α *real to some well behaved* $r_\alpha(z)$ *with* α *and* z *both complex. We can, however, extend our definition to the case when* x *is still real and strictly positive but we allow* α *to be complex. Our definition remains the same:*

$$r_\alpha(x) = \exp(\alpha \log x),$$

but only some of our previous statements carry over.

(i) If $\alpha,\ \beta \in \mathbb{C}$, *show that* $r_{\alpha+\beta}(x) = r_\alpha(x) r_\beta(x)$ *for all* $x > 0$. *Thus part (i) of Exercise 5.74 carries over.*

(ii) Explain carefully why the statement in part (ii) of Exercise 5.74,

$$r_{\alpha\beta}(x) \stackrel{?}{=} r_\alpha(r_\beta(x)),$$

makes no sense (within the context of this question) if we allow α *and* β *to range freely over* $\mathbb{C}$. *Does it make sense and is it true if* $\beta \in \mathbb{R}$ *and* $\alpha \in \mathbb{C}$*? Does it make sense and is it true if* $\alpha \in \mathbb{R}$ *and* $\beta \in \mathbb{C}$*?*

(iii) Find which parts of Exercises 5.76 and 5.77 continue to make sense in the more general context of this question, and prove them.

(iv) Show that, if u *and* v *are real and* $e = \exp(1)$, *then* $\exp(u + iv) = r_{u+iv}(e)$. *We have thus converted the mnemonic*

$$\exp(z) = e^z$$

into a genuine equality.

Exercise 5.81. *According to a well known story*[12], *the Harvard mathematician Benjamin Pierce chalked the formula*

$$e^{i\pi} + 1 = 0$$

on the board and addressed his students as follows.

> *Gentleman, that is surely true, it is absolutely paradoxical; we cannot understand it, and we do not know what it means, but we have proved it, and therefore we know it must be the truth.*

[12]Repeated in Martin Gardner's *Mathematical Diversions.* See also Exercise K.100.

(i) In the context of this chapter, what information is conveyed by the formula

$$\exp(i\pi) + 1 = 0?$$

(What does exp *mean, what does* π *mean and what does* $\exp(i\pi)$ *mean?)*

(ii) In the context of this chapter, what information is conveyed by the formula

$$e^{i\pi} + 1 = 0?$$

There is a superb discussion of the problem of defining x^α in Klein's *Elementary Mathematics from an Advanced Standpoint* [**29**].

5.8. The fundamental theorem of algebra ♡

It is in the nature of a book like this that much of our time is occupied in proving results which the 'physicist in the street' would consider obvious. In this section we prove a result which is less obvious.

Theorem 5.82. (The fundamental theorem of algebra.) *Suppose that* $n \geq 1$, $a_0, a_1, \ldots, a_n \in \mathbb{C}$ *and* $a_n \neq 0$. *Then the equation*

$$a_n z^n + a_{n-1} z^{n-1} + \cdots + a_0 = 0$$

has at least one root in $\mathbb{C}$.

In other words, every polynomial has a root in $\mathbb{C}$.

If the reader believes that this is obvious, then she should stop reading at this point and write down the 'obvious argument'. In fact, Leibniz and other mathematicians doubted the truth of the result. Although d'Alembert, Euler and Lagrange offered proofs of the result, they were unsatisfactory, and the first satisfactory discussion is due to Gauss[13].

The first point to realise is that the 'fundamental theorem of algebra' is in fact a theorem of analysis!

Exercise 5.83. *Suppose* $z = u + iv$ *with* $u, v \in \mathbb{R}$. *If* $z^2 - 2 = 0$, *show that*

$$u^2 - v^2 = 2,$$
$$uv = 0,$$

and deduce that $v = 0$, $u^2 = 2$.

If we write

$$\mathbb{Q} + i\mathbb{Q} = \{x + iy \,:\, x, y \in \mathbb{Q}\},$$

show that the equation $z^2 - 2 = 0$ *has no solution with* $z \in \mathbb{Q} + i\mathbb{Q}$.

[13] See [**30**], Chapter 19, Section 4 and Chapter 25, Sections 1 and 2.

Since $\mathbb{Q} + i\mathbb{Q}$ and $\mathbb{C} = \mathbb{R} + i\mathbb{R}$ share the same algebraic structure, Exercise 5.83 shows that the truth of Theorem 5.82 must depend in some way on the fundamental axiom of analysis. We shall use Theorem 4.44, which states that any continuous function on a closed bounded set in $\mathbb{R}^n$ has a minimum, to establish the following key step of our proof.

Lemma 5.84. *If P is a polynomial, then there exists a $z_0 \in \mathbb{C}$ such that*

$$|P(z)| \geq |P(z_0)|$$

for all $z \in \mathbb{C}$.

We then complete the proof by establishing the following lemma.

Lemma 5.85. *If P is a non-constant polynomial and $|P|$ attains a minimum at z_0, then $P(z_0) = 0$.*

Clearly, Lemmas 5.84 and 5.85 together imply Theorem 5.82. Our proofs of the two lemmas make use of simple results given in the next exercise.

Exercise 5.86. *(i) Let $P(z) = \sum_{j=0}^{n} a_j z^j$ with $n \geq 1$ and $a_n \neq 0$. Show that, if we set $R_0 = 2n|a_n|^{-1}(1 + \max_{0 \leq j \leq n} |a_j|)$, then, whenever $|z| \geq R_0$,*

$$\frac{|a_j|}{|z|^{n-j}} \leq \frac{|a_j|}{R_0} \leq \frac{|a_n|}{2n}$$

for all $0 \leq j \leq n-1$. Hence, or otherwise, show that

$$\left| a_n + \sum_{j=0}^{n-1} \frac{a_j}{z^{n-j}} \right| \geq \frac{|a_n|}{2},$$

and so

$$\left| \sum_{j=0}^{n} a_j z^j \right| \geq \frac{|a_n||z|^n}{2}$$

for all $|z| \geq R_0$.

(ii) By using the result of (i), show that, given any real number $K \geq 0$, we can find an $R(K) > 0$ such that $|P(z)| \geq K$ whenever $|z| \geq R(K)$.

(iii) Let $Q(z) = \sum_{j=k}^{n} b_j z^j$ with $n \geq k \geq 1$ and $b_k \neq 0$. Show that there exists a $\eta_0 > 0$ such that

$$\left| \sum_{j=k+1}^{n} b_j z^j \right| \leq \frac{|b_k||z|^k}{2}$$

for all $|z| \leq \eta_0$.

Proof of Lemma 5.84. We wish to show that, if P is any polynomial, then $|P|$ has a minimum. If P is a constant polynomial there is nothing to prove, so we may suppose that $P(z) = \sum_{j=0}^{n} a_j z^j$ with $n \geq 1$ and $a_n \neq 0$. By Exercise 5.86 (ii), we can find an $R > 0$ such that $|P(z)| \geq |P(0)| + 1$ whenever $|z| \geq R$.

Identifying $\mathbb{C}$ with $\mathbb{R}^2$ in the usual way, we observe that

$$\bar{D}_R = \{z \in \mathbb{C} : |z| \leq R\}$$

is a closed bounded set and that the function $|P| : \mathbb{C} \to \mathbb{R}$ is continuous. Thus we may use Theorem 4.44, which states that a continuous function on a closed bounded set attains its minimum, to show the existence of a $z_0 \in \bar{D}_R$ with $|P(z_0)| \leq |P(z)|$ for all $z \in \bar{D}_R$.

We note, in particular, that $|P(z_0)| \leq |P(0)|$. Thus, if $|z| \geq R$, then

$$|P(z)| \geq |P(0)| + 1 > |P(0)| \geq |P(z_0)|.$$

It follows that $|P(z_0)| \leq |P(z)|$ for all $z \in \mathbb{C}$, as required. ∎

Exercise 5.87. *Define $f : \mathbb{C} \to \mathbb{R}$ by $f(z) = -|z|^2$. Show that f attains a minimum on every set*

$$\bar{D}_R = \{z \in \mathbb{C} : |z| \leq R\}$$

but has no minimum on $\mathbb{C}$. Explain briefly why the proof above works for $|P|$ but not for f.

We must now show that, if z_0 gives the minimum value of the modulus $|P|$ of a non-constant polynomial P, then $P(z_0) = 0$. We start with a collection of remarks intended to simplify the algebra.

Exercise 5.88. *(i) Let P be a non-constant polynomial whose modulus $|P|$ has a minimum at z_0. Show that if $Q(z) = P(z + z_0)$, then Q is a non-constant polynomial whose modulus $|Q|$ has a minimum at 0. Show further that, if $Q(0) = 0$, then $P(z_0) = 0$.*

(ii) Let Q be a non-constant polynomial whose modulus $|Q|$ has a minimum at 0. Show that, for an appropriate $\phi \in \mathbb{R}$, to be defined, the function $R(z) = e^{i\phi}Q(z)$ has $R(0)$ real and positive[14]*. Show that R is a non-constant polynomial whose modulus $|R|$ has a minimum at 0 and that, if $R(0) = 0$, then $Q(0) = 0$.*

(iii) Let R be a non-constant polynomial whose modulus $|R|$ has a minimum at 0 and such that $R(0)$ is real and positive. Explain why we have

$$R(z) = a_0 + \sum_{j=k}^{n} a_j z^j,$$

[14] That is to say, non-negative. (We use 'strictly positive' to mean positive and non-zero.)

where a_0 is real and positive, $k \geq 1$ and $a_k \neq 0$. Set $S(z) = R(e^{i\psi}z)$. Show that, for an appropriate $\psi \in \mathbb{R}$, to be defined,

$$S(z) = b_0 + \sum_{j=k}^{n} b_j z^j,$$

where b_0 is real and positive, $k \geq 1$, and b_k is real and strictly negative (that is, $b_k < 0$).

Most mathematicians would consider the results of Exercise 5.88 to be trivial and use a phrase like 'Without loss of generality we may suppose that $z_0 = 0$ and $P(z) = a_0 + \sum_{j=k}^{n} a_j z^j$, where a_0 is real and positive, $k \geq 1$ and a_k is real and strictly negative' or (better) 'By considering $e^{i\phi}P(e^{i\psi}(z - z_0))$ we may suppose that $z_0 = 0$ and $P(z) = a_0 + \sum_{j=k}^{n} a_j z^j$, where a_0 is real and positive, $k \geq 1$, and a_k is real and strictly negative'.

Proof of Lemma 5.85. We want to show that if P is a non-constant polynomial and z_0 gives a minimum of $|P|$, then $P(z_0) = 0$. Without loss of generality we may suppose that $z_0 = 0$ and $P(z) = a_0 + \sum_{j=k}^{n} a_j z^j$ where a_0 is real and positive, $k \geq 1$ and a_k is real and strictly negative. If $a_0 = 0$ then $P(0) = 0$ and we are done. We suppose that a_0 is strictly positive and seek a contradiction.

By Exercise 5.86, we can find an $\eta_0 > 0$ such that

$$\left| \sum_{j=k+1}^{n} a_j z^j \right| \leq \frac{|a_k z^k|}{2}$$

for all $|z| \leq \eta_0$. Now choose η_1, a real number with $0 < \eta_1 \leq \eta_0$ and $a_0 > |a_k|\eta_1^k/2$ ($\eta_1 = \min(\eta_0, 1, -a_0/(2a_k))$ will do). Remembering that a_0 is real and strictly positive and a_k is real and strictly negative, we see that, whenever η is real and $0 < \eta < \eta_1$, we have

$$\begin{aligned}|P(\eta)| &= \left| a_0 + \sum_{j=k}^{n} a_j \eta^j \right| \leq |a_0 + a_k \eta^k| + \left| \sum_{j=k+1}^{n} a_j \eta^j \right| \\ &\leq |a_0 + a_k\eta^k| + |a_k\eta^k|/2 = a_0 + a_k\eta^k - a_k\eta^k/2 = a_0 + a_k\eta^k/2 < P(0),\end{aligned}$$

contradicting the statement that 0 is a minimum for P. The result follows by reductio ad absurdum. ■

The proof of Theorem 5.82 may look a little complicated, but really it only amounts to a fleshing out of the following sketch argument.

Outline proof of Theorem 5.82. Let P be a non-constant polynomial. Since $|P(z)| \to \infty$ as $|z| \to \infty$, P must attain a minimum. By translation, we may suppose that the minimum occurs at 0. If $P(0) \neq 0$, then

$$P(z) = a_0 + \sum_{j=k}^{n} a_j z^j$$

with $k \geq 1$ and $a_0,\ a_k \neq 0$. Close to zero,

$$P(z) \approx a_0 + a_k z^k.$$

Choosing an appropriate ϕ, we have $|a_0 + a_k(e^{i\phi}\eta)^k| < |a_0|$ whenever η is small and strictly positive, contradicting the statement that $|P|$ attains a minimum, at 0. The result follows by reductio ad absurdum. ▲

Exercise 5.89. *Give an explicit value for ϕ in the outline proof above.*

Exercise 5.90. *We say that z_0 is a* local minimum *of a function $G : \mathbb{C} \to \mathbb{R}$ if we can find a $\delta > 0$ such that $G(z) \geq G(z_0)$ for all z with $|z - z_0| < \delta$. Show that if P is a non-constant polynomial and z_0 is a local minimum of $|P|$, then $P(z_0) = 0$.*

We have already used the strategy of looking for a minimum (or maximum) and then considering the behaviour of the function near that 'extreme' point in our proof of Rolle's theorem (Theorem 4.52). Another example occurs in Exercise K.30 if the reader wishes to try it, and other examples will crop up in this book. The method is very powerful, but we must be careful to establish that an extreme point actually exists (see, as a warning example, the discussion beginning on page 189 of a counterexample due, essentially, to Weierstrass). Notice that our proof required the ability to 'look in all directions'. The minimum had to be in the open set

$$D_R = \{z \in \mathbb{C} \,:\, |z| < R\}$$

and not merely in the set

$$\bar{D}_R = \{z \in \mathbb{C} \,:\, |z| \leq R\}.$$

Exercise 5.91. *This exercise recalls material that is probably familiar from algebra. We work in $\mathbb{C}$.*

(i) Show, by induction on the degree of P, or otherwise, that if P is a non-constant polynomial and $\lambda \in \mathbb{C}$, then there exist a polynomial Q and an $r \in \mathbb{C}$ such that

$$P(z) = (z - \lambda)Q(z) + r.$$

(ii) If P is a non-constant polynomial and $\lambda \in \mathbb{C}$ is such that $P(\lambda) = 0$, then there is a polynomial Q such that

$$P(z) = (z - \lambda)Q(z).$$

(iii) Use the fundamental theorem of algebra and induction on the degree of n to show that any polynomial P of degree n can be written in the form

$$P(z) = a\prod_{j=1}^{n}(z-\lambda_j).$$

(iv) Show that a polynomial of degree n can have at most n distinct roots. What is the minimum number of distinct roots it can have?

(v) If P has real coefficients, show[15] *that $P(z)^* = P(z^*)$ and deduce that, if λ is a root of P, so is λ^*.*

(vi) Use part (v) and induction to show that, if P is a polynomial with real coefficients, then P can be written in the form

$$P(z) = a\prod_{j=1}^{m} Q_j(z),$$

where $a \in \mathbb{R}$ and, for each j, either $Q_j(z) = z + a_j$ with $a_j \in \mathbb{R}$, or $Q_j(z) = z^2 + a_j z + b_j$ with a_j, $b_j \in \mathbb{R}$.

[In the days before mathematicians acquired our present confidence with complex numbers, the fundamental theorem of algebra was given the less general statement that any polynomial with real coefficients could be written as the product of linear and quadratic terms with real coefficients.

It is natural to ask if this restricted result which does not mention complex numbers can be proved without using complex numbers. Gauss's first proof of the restricted result used complex numbers but he later gave a second proof without using complex numbers which depends only on the fact that a real polynomial of odd degree must have a root (Exercise 1.38) and so uses the fundamental axiom in the form of the intermediate value theorem. As might be expected, his proof and its modern successors are rather subtle. The reader is advised to wait until she has studied the rudiments of Galois theory before pursuing these ideas further.]

Exercise 5.92. *Let $P(z) = \sum_{j=0}^{n} a_j z^j$ be a non-constant polynomial with a root at z_0.*

(i) Explain why we can find an $\eta_0 > 0$ such that $P(z) \neq 0$ for all z with $0 < |z - z_0| < \eta_0$.

(ii) If $0 < \eta < \eta_0$, use the fact that a continuous function on a closed bounded set is bounded and attains its bounds to show that there is a $\delta(\eta) > 0$ such that $|P(z)| \geq \delta(\eta) > 0$ for all z with $|z - z_0| = \eta$.

---

[15]We write z^* for the complex conjugate of z. Thus, if x and y are real, $(x+iy)^* = x - iy$. Some authors use $\bar{z}$.

(iii) Continuing with the notations and assumptions of (ii), show that if $Q(z)$ *is a polynomial with* $|P(z) - Q(z)| < \delta(\eta)/2$ *for all* z *with* $|z - z_0| \le \eta$, *then* $|Q|$ *has a local minimum (and so* Q *has a root)* z_1 *with* $|z_1 - z_0| < \eta$.

(iv) Show that given any $\delta > 0$, *we can find an* $\epsilon > 0$ *(depending on* δ, n, a_0, a_1, $\dots$, a_n*) such that, if* $|a_j - b_j| < \epsilon$ *for* $0 \le j \le n$, *then* $\sum_{j=0}^{n} b_j z^j$ *has at least one root* z_1 *with* $|z_0 - z_1| < \delta$.

[Note that this result is not true if we work over $\mathbb{R}$. *The equation* $x^2 = 0$ *has a real root at* 0 *but* $x^2 + \epsilon = 0$ *has no real roots if* $\epsilon > 0$, *however small* ϵ *may be.]*

Exercise 5.93. *(This exercise requires countability and a certain willingness to think like an algebraist.)*

It is sometimes said that we have to introduce $\mathbb{R}$ *in order to provide equations like* $x^2 - 2 = 0$ *with a root. A little thought shows that this is too simple a view of the matter. Recall that a system* $(\mathbb{F}, +, \times)$ *satisfying all the axioms set out in Appendix A except axioms P1 to P4 (the axioms of order) is called a field. If* $(\mathbb{F}, +, \times)$ *is a field and* $\mathbb{G} \subseteq \mathbb{F}$ *is such that*

(a) $0, 1, -1 \in \mathbb{G}$,

(b) if $x, y \in \mathbb{G}$, *then* $x + y, xy \in \mathbb{G}$,

(c) if $x \in \mathbb{G}$ *and* $x \neq 0$, *then* $x^{-1} \in \mathbb{G}$,

then we say that $\mathbb{G}$ *is a subfield of* $\mathbb{F}$. *It easy to see that a subfield is itself a field. In this exercise we show that there is a countable subfield* $\mathbb{L}$ *of* $\mathbb{C}$ *containing* $\mathbb{Q}$ *and such that, if* $a_0, a_1, \dots , a_n \in \mathbb{L}$, *with* $a_n \neq 0$, *then we can find* $a, \lambda_1, \dots , \lambda_n \in \mathbb{L}$ *such that*

$$\sum_{j=0}^{n} a_j z^j = a \prod_{k=1}^{n} (z - \lambda_j)$$

for all $z \in \mathbb{L}$. *In other words, every polynomial with coefficients in* $\mathbb{L}$ *has all its roots in* $\mathbb{L}$. *Here are the steps in the proof.*

(i) If $\mathbb{K}$ *is a countable subfield of* $\mathbb{C}$, *show that the set of polynomials with degree* n *with coefficients in* $\mathbb{K}$ *is countable. Deduce that the set of polynomials* $\mathcal{P}(\mathbb{K})$ *with coefficients in* $\mathbb{K}$ *is countable. Show also that the set* $\mathcal{Z}(\mathbb{K})$ *of roots in* $\mathbb{C}$ *of polynomials in* $\mathcal{P}(\mathbb{K})$ *is countable.*

(ii) If $\mathbb{K}$ *is a subfield of* $\mathbb{C}$ *and* $\omega \in \mathbb{C}$, *we write* $\mathbb{K}(\omega)$ *for the set of numbers* $P(\omega)/Q(\omega)$ *with* $P, Q \in \mathcal{P}(K)$ *and* $Q(\omega) \neq 0$. *Show that* $\mathbb{K}(\omega)$ *is a subfield of* $\mathbb{C}$ *containing* $\mathbb{K}$ *and* ω. *If* $\mathbb{K}$ *is countable, show that* $\mathbb{K}(\omega)$ *is.*

(iii) Let $\mathbb{K}$ *be a subfield of* $\mathbb{C}$ *and* $\boldsymbol{\omega} = (\omega_1, \omega_2, \dots)$, *where* $\omega_j \in \mathbb{C}$. *Set* $\mathbb{K}_0 = \mathbb{K}$ *and define* $\mathbb{K}_n = \mathbb{K}_{n-1}(\omega_n)$ *for all* $n \ge 1$. *If we set* $\mathbb{K}(\boldsymbol{\omega}) = \bigcup_{n=0}^{\infty} \mathbb{K}_n$, *show that* $\mathbb{K}(\boldsymbol{\omega})$ *is a subfield of* $\mathbb{C}$ *containing* $\mathbb{K}$ *and* ω_j *for each* $j \ge 1$. *If* $\mathbb{K}$ *is countable, show that* $\mathbb{K}(\boldsymbol{\omega})$ *is.*

(iv) Let $\mathbb{K}$ *be a countable subfield of* $\mathbb{C}$ *(we could take* $\mathbb{K} = \mathbb{Q}$*). Set* $\mathbb{K}_0 = \mathbb{K}$*. Show by induction, using part (iii), that we may define inductively a sequence* $\mathbb{K}_n$ *of countable subfields of* $\mathbb{C}$ *such that* $\mathbb{K}_n$ *contains* $\mathcal{Z}(\mathbb{K}_{n-1})$ *for each* $n \geq 1$*. If we set* $\mathbb{L} = \bigcup_{n=0}^{\infty} \mathbb{K}_n$*, show that* $\mathbb{L}$ *is a countable subfield of* $\mathbb{C}$ *such that every polynomial with coefficients in* $\mathbb{L}$ *has all its roots in* $\mathbb{L}$*.*

(v) Let $\mathcal{L}$ *be as in (iv). Show that, if* $z \in \mathbb{L}$*, then the conjugate* $z^* \in \mathbb{L}$*. Show that*

$$\mathbb{L} \cap \mathbb{R} = \{\Re z \,:\, z \in \mathbb{L}\}$$

and that $\mathbb{M} = \mathbb{L} \cap \mathbb{R}$ *is an ordered field if we use the usual order on* $\mathbb{R}$*. Is* $\mathbb{M} = \mathbb{R}$*? Why? Is it true that every polynomial with coefficients in* $\mathbb{M}$ *has all its roots in* $\mathbb{M}$*? Why?*

Suppose we define $f : \mathbb{M} \to \mathbb{M}$ *by*

$$\begin{aligned} f(x) &= -1 && \text{if } x^2 < 2, \\ f(x) &= 1 && \text{otherwise.} \end{aligned}$$

Is f *continuous? Why? Does the intermediate theorem (Theorem 1.35) hold with* $\mathbb{R}$ *replaced by* $\mathbb{M}$*? Why?*

[We say that fields like $\mathbb{L}$ *for which every polynomial has a root are 'algebraically closed'. The work we have had to do to obtain an 'algebraically closed'* $\mathbb{L}$ *from* $\mathbb{K}$ *shows the fundamental theorem of algebra in a remarkable light. Although* $\mathbb{R}$ *is not algebraically closed, adjoining a single root* i *of a single equation* $z^2 + 1 = 0$ *to form* $\mathbb{R}(i) = \mathbb{C}$ *produces an algebraically closed field!]*

Chapter 6

Differentiation

6.1. Preliminaries

This section is as much propaganda as technical mathematics and, as with much propaganda, most points are made more than once.

We have already looked briefly at differentiation of functions $f : \mathbb{R} \to \mathbb{R}$. Unfortunately, nature is not one-dimensional and we must consider the more general case of a function $\mathbf{f} : \mathbb{R}^m \to \mathbb{R}^p$. The definition of the derivative in terms of the limit of some ratio is not available, since we cannot divide by vectors.

The first solution that mathematicians found to this problem is via 'directional derivatives' or, essentially equivalently, via 'partial derivatives'. We shall give formal definitions later, but the idea is to reduce a many-dimensional problem to a collection of one-dimensional problems by only examining changes in one direction at a time. Suppose, for example, that $f : \mathbb{R}^m \to \mathbb{R}$ is well behaved. If we wish to examine how f behaves near $\mathbf{x}$, we choose a unit vector $\mathbf{u}$ and look at $f_{\mathbf{u}}(t) = f(\mathbf{x} + t\mathbf{u})$ with $t \in \mathbb{R}$. The function $f_{\mathbf{u}} : \mathbb{R} \to \mathbb{R}$ is 'one-dimensional', and we may look at its derivative

$$f'_{\mathbf{u}}(\mathbf{x}) = \lim_{h \to 0} \frac{f(\mathbf{x} + h\mathbf{u}) - f(\mathbf{x})}{h}.$$

By choosing m unit vectors $\mathbf{u}_j$ at right angles and looking at the associated 'directional derivatives' $f'_{\mathbf{u}_j}(\mathbf{x})$ we can obtain a picture of the way in which f changes.

But, to echo Maxwell,

> ... the doctrine of Vectors ... is a method of thinking and not, at least for the present generation, a method of saving

> thought. It does not, like some more popular mathematical methods, encourage the hope that mathematicians may give their minds a holiday, by transferring all their work to their pens. It calls on us at every step to form a mental image of the geometrical features represented by the symbols, so that in studying geometry by this method we have our minds engaged with geometrical ideas, and are not permitted to call ourselves geometers when we are only arithmeticians. (Page 951, [**39**])

Is there a less 'coordinate bound' and more 'geometric' way of looking at differentiation in many dimensions? If we are prepared to spend a little time and effort acquiring new habits of thought, the answer is yes.

The original discoverers of the calculus thought of differentiation as the process of finding a tangent. If $f : \mathbb{R} \to \mathbb{R}$ is well behaved, then the tangent at x is the line $y = b + a(t - x)$ which touches the curve $y = f(t)$ at $(x, f(x))$; that is, the 'line which most resembles f close to x'. In other words,

$$f(t) = b + a(t - x) + \text{small error}$$

close to x. If we think a little harder about the nature of the 'smallest error' possible, we see that it 'ought to decrease faster than linear', that is,

$$f(t) = b + a(t - x) + E(t)|t - x|$$

with $E(t) \to 0$ as $t \to x$.

Exercise 6.1. *Suppose that $f : \mathbb{R} \to \mathbb{R}$. Show that the following two statements are equivalent.*

(i) $\dfrac{f(t) - f(x)}{t - x} \to a$ *as* $t \to x$.

(ii) $f(t) = f(x) + a(t - x) + E(t)|t - x|$ *with* $E(t) \to 0$ *as* $t \to x$.

Rewriting our equations slightly, we see that f is differentiable at x if

$$f(t) - f(x) = a(t - x) + E(t)|t - x|$$

with $E(t) \to 0$ as $t \to 0$. A final rewrite now tells us that f is differentiable at x if

$$\bigstar \qquad f(x + h) - f(x) = ah + \epsilon(h)|h|,$$

where $\epsilon(h) \to 0$ as $h \to x$. The derivative $f'(x) = a$ is the slope of the tangent at x.

The obvious extension to well behaved functions $f : \mathbb{R}^m \to \mathbb{R}$ is to consider the tangent plane at $(\mathbf{x}, f(\mathbf{x}))$. Just as the equation of a non-vertical line through the origin in $\mathbb{R} \times \mathbb{R}$ is $y = bt$, so the equation of an appropriate plane (or 'hyperplane' if the reader prefers) in $\mathbb{R}^m \times \mathbb{R}$ is $y = \alpha(\mathbf{x})$, where

$\alpha : \mathbb{R}^m \to \mathbb{R}$ is linear. This suggests that we say that f is differentiable at $\mathbf{x}$ if

$$f(\mathbf{x}+\mathbf{h}) - f(\mathbf{x}) = \alpha(\mathbf{h}) + \boldsymbol{\epsilon}(\mathbf{h})\|\mathbf{h}\|,$$

where $\boldsymbol{\epsilon}(\mathbf{h}) \to 0$ as $\mathbf{h} \to \mathbf{0}$. It is natural to call α the derivative of f at $\mathbf{x}$.

Finally, if we consider $\mathbf{f} : \mathbb{R}^m \to \mathbb{R}^p$, the natural flow of our argument suggests that we say that $\mathbf{f}$ is differentiable at $\mathbf{x}$ if we can find a linear map $\alpha : \mathbb{R}^m \to \mathbb{R}^p$ such that

$$\mathbf{f}(\mathbf{x}+\mathbf{h}) = \mathbf{f}(\mathbf{x}) + \alpha(\mathbf{h}) + \boldsymbol{\epsilon}(\mathbf{h})\|\mathbf{h}\|,$$

where $\|\boldsymbol{\epsilon}(\mathbf{h})\| \to 0$ as $\mathbf{h} \to 0$. It is natural to call α the derivative of $\mathbf{f}$ at $\mathbf{x}$.

Important note: It is indeed natural to call α the derivative of $\mathbf{f}$ at $\mathbf{x}$. Unfortunately, it is not consistent with our old definition in the case $m = p = 1$. If $f : \mathbb{R} \to \mathbb{R}$, then, with our new definition, the derivative is the *map* $t \mapsto f'(x)t$ but, with our old, the derivative is the *number* $f'(x)$.

From the point of view we have adopted, the key observation of the one-dimensional differential calculus is that well behaved curves, however complicated they may be globally, behave locally like straight lines, that is, like the simplest curves we know. The key observation of multidimensional calculus is that well behaved functions, however complicated they may be globally, behave locally like linear maps, that is, like the simplest functions we know. It is this observation, above all, which justifies the immense amount of time spent studying linear algebra, that is to say, studying the behaviour of linear maps.

I shall assume that the reader has done a course on linear algebra and is familiar with with the definition and lemma that follow. (Indeed, I have already assumed familiarity with the notion of a linear map.)

Definition 6.2. *We say that a function (or map) $\alpha : \mathbb{R}^m \to \mathbb{R}^p$ is linear if*

$$\alpha(\lambda\mathbf{x} + \mu\mathbf{y}) = \lambda\alpha(\mathbf{x}) + \mu\alpha(\mathbf{y})$$

for all $\mathbf{x}$, $\mathbf{y} \in \mathbb{R}^m$ and λ, $\mu \in \mathbb{R}$.

We shall often write $\alpha\mathbf{x}$ instead of $\alpha(\mathbf{x})$.

Lemma 6.3. *Each linear map $\alpha : \mathbb{R}^m \to \mathbb{R}^p$ is associated with a unique $p \times m$ real matrix $A = (a_{ij})$ such that if $\alpha\mathbf{x} = \mathbf{y}$ then*

$$(\dagger) \qquad y_i = \sum_{j=1}^{m} a_{ij}x_j.$$

Conversely, each $p \times m$ real matrix $A = (a_{ij})$ is associated with a unique linear map $\alpha : \mathbb{R}^m \to \mathbb{R}^p$ by the equation $(\dagger)$.

We shall call A the matrix of α with respect to the standard bases. The point to notice is that, if we take different coordinate axes, we get different matrices associated with the same linear map.

From time to time, particularly in some of the exercises, we shall use other facts about linear maps[1]. The reader should not worry too much if some of these facts are unfamiliar, but she should worry if all of them are.

We now repeat the discussion of differentiation with marginally more generality and precision.

A function is continuous if it is locally approximately constant. A function is differentiable if it is locally approximately linear. More precisely, a function is continuous at a point $\mathbf{x}$ if it is locally approximately constant, with an error which decreases to zero, as we approach $\mathbf{x}$. A function is differentiable at a point $\mathbf{x}$ if it is locally approximately linear, with an error which decreases to zero *faster than linearly*, as we approach $\mathbf{x}$.

Definition 6.4. *Suppose that E is a subset of $\mathbb{R}^m$ and $\mathbf{x}$ a point such that there exists a $\delta > 0$ with the ball $B(\mathbf{x}, \delta) \subseteq E$. We say that $\mathbf{f} : E \to \mathbb{R}^p$ is differentiable at $\mathbf{x}$ if we can find a linear map $\alpha : \mathbb{R}^m \to \mathbb{R}^p$ such that, when $\|\mathbf{h}\| < \delta$,*

$$\bigstar \qquad \mathbf{f}(\mathbf{x} + \mathbf{h}) = \mathbf{f}(\mathbf{x}) + \alpha\mathbf{h} + \boldsymbol{\epsilon}(\mathbf{x}, \mathbf{h})\|\mathbf{h}\|,$$

where $\|\boldsymbol{\epsilon}(\mathbf{x}, \mathbf{h})\| \to 0$ as $\|\mathbf{h}\| \to 0$. We write $\alpha = D\mathbf{f}(\mathbf{x})$ or $\alpha = \mathbf{f}'(\mathbf{x})$.

If E is open and $\mathbf{f}$ is differentiable at each point of E, we say that $\mathbf{f}$ is differentiable on E.

Needless to say, the centre of the definition is the formula $\bigstar$, and the reader should concentrate on understanding the rôle of each term in that formula. The rest of the definition is just supporting waffle. Formula $\bigstar$ is sometimes written in the form

$$\frac{\mathbf{f}(\mathbf{x} + \mathbf{h}) - \mathbf{f}(\mathbf{x}) - \alpha\mathbf{h}}{\|\mathbf{h}\|} \to 0$$

as $\|\mathbf{h}\| \to 0$.

Of course, we need to complete Definition 6.4 by showing that α is unique.

Lemma 6.5. *(i) Let $\gamma : \mathbb{R}^m \to \mathbb{R}^p$ be a linear map and $\boldsymbol{\epsilon} : \mathbb{R}^m \to \mathbb{R}^p$ a function with $\|\boldsymbol{\epsilon}(\mathbf{h})\| \to 0$ as $\|\mathbf{h}\| \to 0$. If*

$$\gamma\mathbf{h} = \boldsymbol{\epsilon}(\mathbf{h})\|\mathbf{h}\|,$$

then $\gamma = 0$, the zero map.

(ii) There is at most one α satisfying the conditions of Definition 6.4.

---

[1]For example, some of the exercises deal with linear maps from $\mathbb{C}^m$ to $\mathbb{C}^p$.

Proof. (i) There are many different ways of setting out this simple proof. Here is one. Let $\mathbf{x} \in \mathbb{R}^m$. If $\eta > 0$, we have

$$\gamma\mathbf{x} = \eta^{-1}\gamma(\eta\mathbf{x}) = \eta^{-1}\boldsymbol{\epsilon}(\eta\mathbf{x})\|\eta\mathbf{x}\| = \boldsymbol{\epsilon}(\eta\mathbf{x})\|\mathbf{x}\|,$$

and so

$$\|\gamma\mathbf{x}\| = \|\boldsymbol{\epsilon}(\eta\mathbf{x})\|\|\mathbf{x}\| \to 0$$

as $\eta \to 0$ through values $\eta > 0$. Thus $\|\gamma\mathbf{x}\| = 0$, and so $\gamma\mathbf{x} = \mathbf{0}$ for all $\mathbf{x} \in \mathbb{R}^m$. In other words, $\gamma = 0$.

(ii) Suppose that we can find linear maps $\alpha_j : \mathbb{R}^m \to \mathbb{R}^p$ such that, when $\|\mathbf{h}\| < \delta$,

$$\bigstar \qquad \mathbf{f}(\mathbf{x}+\mathbf{h}) = \mathbf{f}(\mathbf{x}) + \alpha_j\mathbf{h} + \boldsymbol{\epsilon}_j(\mathbf{x},\mathbf{h})\|\mathbf{h}\|,$$

where $\|\boldsymbol{\epsilon}_j(\mathbf{x},\mathbf{h})\| \to 0$ as $\|\mathbf{h}\| \to 0$ $[j = 1, 2]$.

Subtracting, we see that

$$(\alpha_1 - \alpha_2)\mathbf{h} = \boldsymbol{\epsilon}(\mathbf{x},\mathbf{h}),$$

where

$$\boldsymbol{\epsilon}(\mathbf{x},\mathbf{h}) = \boldsymbol{\epsilon}_2(\mathbf{x},\mathbf{h}) - \boldsymbol{\epsilon}_1(\mathbf{x},\mathbf{h})$$

for $\|\mathbf{h}\| < \delta$. Since

$$\|\boldsymbol{\epsilon}(\mathbf{x},\mathbf{h})\| \leq \|\boldsymbol{\epsilon}_1(\mathbf{x},\mathbf{h})\| + \|\boldsymbol{\epsilon}_2(\mathbf{x},\mathbf{h})\| \to 0$$

as $\|\mathbf{h}\| \to 0$, we can apply part (i) to obtain $\alpha_1 = \alpha_2$. ■

The coordinate free approach can be taken only so far, and to calculate we need to know the matrix A of $\alpha = D\mathbf{f}(\mathbf{x})$ with respect to the standard bases. To find A we have recourse to directional derivatives.

Definition 6.6. *Suppose that E is a subset of $\mathbb{R}^m$ and that we have a function $g : E \to \mathbb{R}$. Suppose further that $\mathbf{x} \in E$ and $\mathbf{u}$ is a unit vector such that there exists a $\delta > 0$ with $\mathbf{x} + h\mathbf{u} \in E$ for all $|h| < \delta$. We can now define a function G from the open interval $(-\delta, \delta)$ to $\mathbb{R}$ by setting $G(t) = g(\mathbf{x} + t\mathbf{u})$. If G is differentiable at 0, we say that g has a directional derivative at $\mathbf{x}$ in the direction $\mathbf{u}$ of value $G'(0)$.*

Exercise 6.7. *Suppose that E is a subset of $\mathbb{R}^m$ and that we have a function $g : E \to \mathbb{R}$. Suppose further that $\mathbf{x} \in E$ and $\mathbf{u}$ is a unit vector such that there exists a $\delta > 0$ with $\mathbf{x} + h\mathbf{u} \in E$ for all $|h| < \delta$. Show that g has a directional derivative at $\mathbf{x}$ in the direction $\mathbf{u}$ of value a if and only if*

$$\frac{g(\mathbf{x} + t\mathbf{u}) - g(\mathbf{x})}{t} \to a$$

as $t \to 0$.

We are interested in the directional derivatives along the unit vectors $\mathbf{e}_j$ in the directions of the coordinate axes. The reader is almost certainly familiar with these under the name of 'partial derivatives'.

Definition 6.8. *Suppose that E is a subset of $\mathbb{R}^m$ and that we have a function $g : E \to \mathbb{R}$. If we give $\mathbb{R}^m$ the standard basis $\mathbf{e}_1$, $\mathbf{e}_2$, ... , $\mathbf{e}_m$ (where $\mathbf{e}_j$ is the vector with jth entry 1 and all other entries 0), then the directional derivative of g at $\mathbf{x}$ in the direction $\mathbf{e}_j$ is called a partial derivative and written $g_{,j}(\mathbf{x})$.*

The recipe for $g_{,j}(\mathbf{x})$ is thus 'differentiate $g(x_1, x_2, \dots, x_j, \dots, x_n)$ with respect to x_j, treating all the x_i with $i \neq j$ as constants'.

The reader would probably prefer me to say that $g_{,j}(\mathbf{x})$ is the partial derivative of g with respect to x_j and write

$$g_{,j}(\mathbf{x}) = \frac{\partial g}{\partial x_j}(\mathbf{x}).$$

I shall use this notation from time to time, but, as I point out in Appendix E, there are cultural differences between the way that applied mathematicians and pure mathematicians think of partial derivatives, so I prefer to use a different notation.

The reader should also know a third notation for partial derivatives:.

$$D_j g = g_{,j}.$$

This 'D' notation is more common than the 'comma' notation and is to be preferred if you only use partial derivatives occasionally or if you only deal with functions $f : \mathbb{R}^n \to \mathbb{R}$. The 'comma' notation is used in tensor analysis and is convenient in the kind of formulae which appear in Section 7.2.

If E is a subset of $\mathbb{R}^m$ and we have a function $\mathbf{g} : E \to \mathbb{R}^p$ then we can write

$$\mathbf{g}(\mathbf{t}) = \begin{pmatrix} g_1(\mathbf{t}) \\ g_2(\mathbf{t}) \\ \vdots \\ g_p(\mathbf{t}) \end{pmatrix}$$

and obtain functions $g_i : E \to \mathbb{R}$ with partial derivatives (if they exist) $g_{i,j}(\mathbf{x})$ (or, in more standard notation, $\dfrac{\partial g_i}{\partial x_j}(\mathbf{x})$). The proof of the next lemma just consists of dismantling the notation so laboriously constructed in the last few paragraphs.

Lemma 6.9. *Let $\mathbf{f}$ be as in Definition 6.4. If we use standard coordinates, then, if $\mathbf{f}$ is differentiable at $\mathbf{x}$, its partial derivatives $f_{i,j}(\mathbf{x})$ exist and the* *matrix of the derivative* $Df(\mathbf{x})$ *is the Jacobian matrix* $(f_{i,j}(\mathbf{x}))$ *of partial derivatives.*

Proof. Left as a strongly recommended but simple exercise for the reader. ■

Notice that, if $f : \mathbb{R} \to \mathbb{R}$, the matrix of $Df(x)$ is the 1×1 Jacobian matrix $(f'(x))$. Notice also that Exercise 6.9 provides an alternative proof of the uniqueness of the derivative (Lemma 6.5 (ii)).

It is customary to point out that the existence of the partial derivatives does not imply the differentiability of the function (see Example 7.32 below), but the main objections to over-reliance on partial derivatives are that it makes formulae cumbersome and stifles geometric intuition. Let your motto be **'coordinates and matrices for calculation, vectors and linear maps for understanding'**.

6.2. The operator norm and the chain rule

We shall need some method of measuring the 'size' of a linear map. The reader is unlikely to have come across this in a standard 'abstract algebra' course, since algebraists dislike using 'metric notions' which do not generalise from $\mathbb{R}$ to more general fields.

Our first idea might be to use some sort of measure like

$$\|\alpha\|_* = \max |a_{ij}|$$

where (a_{ij}) is the matrix of α with respect to the standard bases. However, $\|\alpha\|_*$ has no geometric meaning.

Exercise 6.10. *Show by example that* $\|\alpha\|_*$ *may depend on the coordinate axes chosen.*

Even if we insist that our method of measuring the size of a linear map shall have a geometric meaning, this does not give a unique method. The following chain of ideas gives one method which is natural and standard.

Lemma 6.11. *If* $\alpha : \mathbb{R}^m \to \mathbb{R}^p$ *is linear, there exists a constant* $K(\alpha)$ *such that*

$$\|\alpha\mathbf{x}\| \leq K(\alpha)\|\mathbf{x}\|$$

for all $\mathbf{x} \in \mathbb{R}^m$.

Proof. Since our object is merely to show that some $K(\alpha)$ exists and not to find a 'good' value, we can use the crudest inequalities.

If we write $\mathbf{y} = \alpha\mathbf{x}$, we have

$$\begin{aligned}\|\alpha\mathbf{x}\| = \|\mathbf{y}\| &\leq \sum_{i=1}^{p} |y_i| \\ &\leq \sum_{i=1}^{p}\sum_{j=1}^{m} |a_{ij}||x_j| \leq \sum_{i=1}^{p}\sum_{j=1}^{m} |a_{ij}|\|\mathbf{x}\|.\end{aligned}$$

The required result follows on putting $K(\alpha) = \sum_{i=1}^{p}\sum_{j=1}^{m} |a_{ij}|$. ■

Exercise 6.12. *Use Lemma 6.11 to estimate $\|\alpha\mathbf{x} - \alpha\mathbf{y}\|$ and hence deduce that every linear map $\alpha : \mathbb{R}^m \to \mathbb{R}^p$ is continuous. (This exercise takes longer to pose than to do.)*

Lemma 6.11 tells us that $\{\|\alpha\mathbf{x}\| : \|\mathbf{x}\| \leq 1\}$ is a non-empty subset of $\mathbb{R}$ bounded above by $K(\alpha)$, and so has a supremum.

Definition 6.13. *If $\alpha : \mathbb{R}^m \to \mathbb{R}^p$ is a linear map, then*

$$\|\alpha\| = \sup_{\|\mathbf{x}\|\leq 1} \|\alpha\mathbf{x}\|.$$

Exercise 6.14. *If α is as in Definition 6.13, show that the three quantities*

$$\sup_{\|\mathbf{x}\|\leq 1} \|\alpha\mathbf{x}\|, \quad \sup_{\|\mathbf{x}\|=1} \|\alpha\mathbf{x}\|, \quad \text{and} \quad \sup_{\mathbf{x}\neq\mathbf{0}} \frac{\|\alpha\mathbf{x}\|}{\|\mathbf{x}\|}$$

are well defined and equal.

The 'operator norm' just defined in Definition 6.13 has many pleasant properties.

Lemma 6.15. *Let $\alpha, \beta : \mathbb{R}^m \to \mathbb{R}^p$ be linear maps.*

(i) If $\mathbf{x} \in \mathbb{R}^m$ then $\|\alpha\mathbf{x}\| \leq \|\alpha\|\,\|\mathbf{x}\|$.

(ii) $\|\alpha\| \geq 0$,

(iii) If $\|\alpha\| = 0$ then $\alpha = 0$,

(iv) If $\lambda \in \mathbb{R}$ then $\|\lambda\alpha\| = |\lambda|\|\alpha\|$.

(v) (The triangle inequality) $\|\alpha + \beta\| \leq \|\alpha\| + \|\beta\|$.

(vi) If $\gamma : \mathbb{R}^p \to \mathbb{R}^q$ is linear, then $\|\gamma\alpha\| \leq \|\gamma\|\,\|\alpha\|$.

Proof. I will prove parts (i) and (vi), leaving the equally easy remaining parts as an essential exercise for the reader.

(i) If $\mathbf{x} = \mathbf{0}$, we observe that $\alpha\mathbf{0} = \mathbf{0}$, and so

$$\|\alpha\mathbf{0}\| = \|\mathbf{0}\| = 0 \leq 0 = \|\alpha\|0 = \|\alpha\|\,\|\mathbf{0}\|,$$

as required.

If $\mathbf{x} \neq \mathbf{0}$, we set $\mathbf{u} = \|\mathbf{x}\|^{-1}\mathbf{x}$. Since

$$\|\mathbf{u}\| = \|\mathbf{x}\|^{-1}\|\mathbf{x}\| = 1,$$

we have $\|\alpha\mathbf{u}\| \leq \|\alpha\|$, and so

$$\|\alpha\mathbf{x}\| = \|\alpha(\|\mathbf{x}\|\mathbf{u})\| = \|(\|\mathbf{x}\|\alpha\mathbf{u})\| = \|\mathbf{x}\|\|\alpha\mathbf{u}\| \leq \|\alpha\|\,\|\mathbf{x}\|,$$

as required.

(vi) If $\|\mathbf{x}\| \leq 1$ then, using part (i) twice,

$$\|\gamma\alpha(\mathbf{x})\| = \|\gamma(\alpha(\mathbf{x}))\| \leq \|\gamma\|\|\alpha(\mathbf{x})\| \leq \|\gamma\|\,\|\alpha\|\,\|\mathbf{x}\| \leq \|\gamma\|\,\|\alpha\|.$$

It follows that

$$\|\gamma\alpha\| = \sup_{\|\mathbf{x}\|\leq 1} \|\gamma\alpha(\mathbf{x})\| \leq \|\gamma\|\,\|\alpha\|.$$

■

Exercise 6.16. *(i) Write down a linear map $\alpha : \mathbb{R}^2 \to \mathbb{R}^2$ such that $\alpha \neq 0$ but $\alpha^2 = 0$.*

(ii) Show that we cannot replace the inequality (vi) in Lemma 6.15 by an equality.

(iii) Show that we cannot replace the inequality (v) in Lemma 6.15 by an equality.

Exercise 6.17. *(i) Suppose that $\alpha : \mathbb{R} \to \mathbb{R}$ is a linear map and that its matrix with respect to the standard bases is (a). Show that*

$$\|\alpha\| = |a|.$$

(ii) Suppose that $\alpha : \mathbb{R}^m \to \mathbb{R}$ is a linear map and that its matrix with respect to the standard bases is $(a_1\ a_2\ \dots\ a_m)$. By using the Cauchy–Schwarz inequality (Lemma 4.2) and the associated conditions for equality (Exercise 4.5 (i)), show that

$$\|\alpha\| = \left(\sum_{j=1}^{m} a_j^2\right)^{1/2}.$$

Although the operator norm is, in principle, calculable (we discuss this in Exercises K.108 to K.111), the reader is warned that, except in special cases, there is no simple formula for the operator norm and it is mainly used as a theoretical tool. Should we need to have some idea of its size, extremely rough estimates will often suffice.

Exercise 6.18. *Suppose that $\alpha : \mathbb{R}^m \to \mathbb{R}^p$ is a linear map and that its matrix with respect to the standard bases is $A = (a_{ij})$. Show that*

$$\max_{i,j} |a_{ij}| \leq \|\alpha\| \leq pm \max_{i,j} |a_{ij}|.$$

By using the Cauchy–Schwarz inequality, show that

$$\|\alpha\| \leq \left(\sum_{i=1}^{p}\sum_{j=1}^{m} a_{ij}^2\right)^{1/2}.$$

Show that this inequality implies the corresponding inequality in the previous paragraph.

We now return to differentiation. Suppose that $\mathbf{f} : \mathbb{R}^m \to \mathbb{R}^p$ and $\mathbf{g} : \mathbb{R}^p \to \mathbb{R}^q$ are differentiable. What can we say about their composition $\mathbf{g} \circ \mathbf{f}$? To simplify the algebra, let us suppose that $\mathbf{f}(\mathbf{0}) = \mathbf{0}$, $\mathbf{g}(\mathbf{0}) = \mathbf{0}$ (so $\mathbf{g} \circ \mathbf{f}(\mathbf{0}) = \mathbf{0}$) and ask about the differentiability of $\mathbf{g} \circ \mathbf{f}$ at $\mathbf{0}$. Suppose that the derivative of $\mathbf{f}$ at $\mathbf{0}$ is α and the derivative of $\mathbf{g}$ at $\mathbf{0}$ is β. Then

$$\mathbf{f}(\mathbf{h}) \approx \alpha\mathbf{h}$$

when $\mathbf{h}$ is small ($\mathbf{h} \in \mathbb{R}^m$) and

$$\mathbf{g}(\mathbf{k}) \approx \beta\mathbf{k}$$

when $\mathbf{k}$ is small ($\mathbf{k} \in \mathbb{R}^p$). It ought, therefore, to to be true that

$$\mathbf{g}(\mathbf{f}(\mathbf{h})) \approx \beta(\alpha\mathbf{h}),$$

i.e., that

$$\mathbf{g} \circ \mathbf{f}(\mathbf{h}) \approx (\beta\alpha)\mathbf{h}$$

when $\mathbf{h}$ is small ($\mathbf{h} \in \mathbb{R}^m$). In other words $\mathbf{g} \circ \mathbf{f}$ is differentiable at $\mathbf{0}$ with derivative $\beta\alpha$.

We have been led to formulate the chain rule.

Lemma 6.19. (The chain rule.) *Let U be a neighbourhood of $\mathbf{x}$ in $\mathbb{R}^m$, and V a neighbourhood of $\mathbf{y}$ in $\mathbb{R}^p$. Suppose that $\mathbf{f} : U \to V$ is differentiable at $\mathbf{x}$ with derivative α, that $\mathbf{g} : V \to \mathbb{R}^q$ is differentiable at $\mathbf{y}$ with derivative β, and that $\mathbf{f}(\mathbf{x}) = \mathbf{y}$. Then $\mathbf{g} \circ \mathbf{f}$ is differentiable at $\mathbf{x}$ with derivative $\beta\alpha$.*

In more condensed notation,

$$D(\mathbf{g} \circ \mathbf{f})(\mathbf{x}) = D\mathbf{g}(\mathbf{f}(\mathbf{x}))D\mathbf{f}(\mathbf{x}),$$

or, equivalently,

★★ $$D(\mathbf{g} \circ \mathbf{f})(\mathbf{x}) = (D\mathbf{g}) \circ \mathbf{f}(\mathbf{x})D\mathbf{f}(\mathbf{x}).$$

Proof. We know that

$$\mathbf{f}(\mathbf{x} + \mathbf{h}) = \mathbf{f}(\mathbf{x}) + \alpha\mathbf{h} + \epsilon_1(\mathbf{h})\|\mathbf{h}\|$$

and

$$\mathbf{g}(\mathbf{f}(\mathbf{x}) + \mathbf{k}) = \mathbf{g}(\mathbf{f}(\mathbf{x})) + \beta\mathbf{k} + \epsilon_2(\mathbf{k})\|\mathbf{k}\|,$$

where $\|\boldsymbol{\epsilon}_1(\mathbf{h})\| \to 0$ as $\|\mathbf{h}\| \to 0$ and $\|\boldsymbol{\epsilon}_2(\mathbf{k})\| \to 0$ as $\|\mathbf{k}\| \to 0$. It follows that

$$\mathbf{g} \circ \mathbf{f}(\mathbf{x}+\mathbf{h}) = \mathbf{g}(\mathbf{f}(\mathbf{x}+\mathbf{h})) = \mathbf{g}(\mathbf{f}(\mathbf{x}) + \alpha\mathbf{h} + \boldsymbol{\epsilon}_1(\mathbf{h})\|\mathbf{h}\|),$$

so, taking $\mathbf{k} = \alpha\mathbf{h} + \boldsymbol{\epsilon}_1(\mathbf{h})\|\mathbf{h}\|$, we have

$$\begin{aligned} &\mathbf{g} \circ \mathbf{f}(\mathbf{x}+\mathbf{h}) \\ &\quad = \mathbf{g}(\mathbf{f}(\mathbf{x})) + \beta(\alpha\mathbf{h} + \boldsymbol{\epsilon}_1(\mathbf{h})\|\mathbf{h}\|) + \boldsymbol{\epsilon}_2(\alpha\mathbf{h} + \boldsymbol{\epsilon}_1(\mathbf{h})\|\mathbf{h}\|)\|\alpha\mathbf{h} + \boldsymbol{\epsilon}_1(\mathbf{h})\|\mathbf{h}\|\| \\ &\quad = \mathbf{g} \circ \mathbf{f}(\mathbf{x}) + \beta\alpha\mathbf{h} + \boldsymbol{\eta}(\mathbf{h})\|\mathbf{h}\| \end{aligned}$$

with

$$\boldsymbol{\eta}(\mathbf{h}) = \boldsymbol{\eta}_1(\mathbf{h}) + \boldsymbol{\eta}_2(\mathbf{h}),$$

where

$$\boldsymbol{\eta}_1(\mathbf{h})\|\mathbf{h}\| = \beta\boldsymbol{\epsilon}_1(\mathbf{h})\|\mathbf{h}\|$$

and

$$\boldsymbol{\eta}_2(\mathbf{h})\|\mathbf{h}\| = \boldsymbol{\epsilon}_2(\alpha\mathbf{h} + \boldsymbol{\epsilon}_1(\mathbf{h})\|\mathbf{h}\|)\|\alpha\mathbf{h} + \boldsymbol{\epsilon}_1(\mathbf{h})\|\mathbf{h}\|\|.$$

All we have to do is to show that $\|\boldsymbol{\eta}_1(\mathbf{h})\|$ and $\|\boldsymbol{\eta}_2(\mathbf{h})\|$, and so $\|\boldsymbol{\eta}(\mathbf{h})\| = \|\boldsymbol{\eta}_1(\mathbf{h}) + \boldsymbol{\eta}_2(\mathbf{h})\|$, tend to zero as $\|\mathbf{h}\| \to 0$. We observe first that

$$\|\boldsymbol{\eta}_1(\mathbf{h})\|\mathbf{h}\|\| \le \|\beta\|\|\boldsymbol{\epsilon}_1(\mathbf{h})\|\mathbf{h}\|\| = \|\beta\|\|\boldsymbol{\epsilon}_1(\mathbf{h})\|\|\mathbf{h}\|,$$

so $\|\boldsymbol{\eta}_1(\mathbf{h})\| \le \|\beta\|\|\boldsymbol{\epsilon}_1(\mathbf{h})\| \to 0$ as $\|\mathbf{h}\| \to 0$. Next we observe that

$$\begin{aligned} \|\boldsymbol{\eta}_2(\mathbf{h})\|\mathbf{h}\|\| &= \|\boldsymbol{\epsilon}_2(\alpha\mathbf{h} + \boldsymbol{\epsilon}_1(\mathbf{h})\|\mathbf{h}\|)\|\|\alpha\mathbf{h} + \boldsymbol{\epsilon}_1(\mathbf{h})\|\mathbf{h}\|\| \\ &\le \|\boldsymbol{\epsilon}_2(\alpha\mathbf{h} + \boldsymbol{\epsilon}_1(\mathbf{h})\|\mathbf{h}\|)\|(\|\alpha\mathbf{h}\| + \|\boldsymbol{\epsilon}_1(\mathbf{h})\|\mathbf{h}\|\|) \\ &\le \|\boldsymbol{\epsilon}_2(\alpha\mathbf{h} + \boldsymbol{\epsilon}_1(\mathbf{h})\|\mathbf{h}\|)\|(\|\alpha\| + \|\boldsymbol{\epsilon}_1(\mathbf{h})\|)\|\mathbf{h}\|, \end{aligned}$$

so that

$$\|\boldsymbol{\eta}_2(\mathbf{h})\| \le \|\boldsymbol{\epsilon}_2(\alpha\mathbf{h} + \boldsymbol{\epsilon}_1(\mathbf{h})\|\mathbf{h}\|)\|(\|\alpha\| + \|\boldsymbol{\epsilon}_1(\mathbf{h})\|) \to 0$$

as $\|\mathbf{h}\| \to 0$, and we are done. ■

Students sometimes say that the proof of the chain rule is difficult, but they really mean that it is tedious. It is simply a matter of showing that the error terms $\boldsymbol{\eta}_1(\mathbf{h})\|\mathbf{h}\|$ and $\boldsymbol{\eta}_2(\mathbf{h})\|\mathbf{h}\|$, which ought to be small, actually are. Students also forget the artificiality of the standard proofs of the one-dimensional chain rule (see the discussion of Lemma 5.60 — any argument which Hardy got wrong cannot be natural). The multidimensional argument forces us to address the real nature of the chain rule.

The next result is very simple, but I would like to give two different proofs.

Lemma 6.20. *Let U be a neighbourhood of $\mathbf{x}$ in $\mathbb{R}^n$. Suppose that $\mathbf{f}$, $\mathbf{g} : U \to \mathbb{R}^m$ are differentiable at $\mathbf{x}$. Then $\mathbf{f}+\mathbf{g}$ is differentiable at $\mathbf{x}$ with $D(\mathbf{f}+\mathbf{g})(\mathbf{x}) = D\mathbf{f}(\mathbf{x}) + D\mathbf{g}(\mathbf{x})$.*

Direct proof. By definition

$$\mathbf{f}(\mathbf{x}+\mathbf{h}) = \mathbf{f}(\mathbf{x}) + D\mathbf{f}(\mathbf{x})\mathbf{h} + \epsilon_1(\mathbf{h})\|\mathbf{h}\|$$

and

$$\mathbf{g}(\mathbf{x}+\mathbf{h}) = \mathbf{g}(\mathbf{x}) + D\mathbf{g}(\mathbf{x})\mathbf{h} + \epsilon_2(\mathbf{h})\|\mathbf{h}\|,$$

where $\|\epsilon_1(\mathbf{h})\| \to 0$ and $\|\epsilon_2(\mathbf{h})\| \to 0$ as $\|\mathbf{h}\| \to 0$. Thus

$$\begin{aligned}(\mathbf{f}+\mathbf{g})(\mathbf{x}+\mathbf{h}) &= \mathbf{f}(\mathbf{x}+\mathbf{h}) + \mathbf{g}(\mathbf{x}+\mathbf{h})\\ &= \mathbf{f}(\mathbf{x}) + D\mathbf{f}(\mathbf{x})\mathbf{h} + \epsilon_1(\mathbf{h})\|\mathbf{h}\| + \mathbf{g}(\mathbf{x}) + D\mathbf{g}(\mathbf{x})\mathbf{h} + \epsilon_2(\mathbf{h})\|\mathbf{h}\|\\ &= (\mathbf{f}+\mathbf{g})(\mathbf{x}) + (D\mathbf{f}(\mathbf{x}) + D\mathbf{g}(\mathbf{x}))\mathbf{h} + \epsilon_3(\mathbf{h})\|\mathbf{h}\|\end{aligned}$$

with

$$\epsilon_3(\mathbf{h}) = \epsilon_1(\mathbf{h}) + \epsilon_2(\mathbf{h}).$$

Since

$$\|\epsilon_3(\mathbf{h})\| \le \|\epsilon_1(\mathbf{h})\| + \|\epsilon_2(\mathbf{h})\| \to 0 + 0 = 0$$

as $\|\mathbf{h}\| \to 0$, we are done. ∎

Our second proof depends on a series of observations.

Lemma 6.21. *A linear map $\alpha : \mathbb{R}^n \to \mathbb{R}^m$ is everywhere differentiable with derivative α.*

Proof. Observe that

$$\alpha(\mathbf{x}+\mathbf{h}) = \alpha\mathbf{x} + \alpha\mathbf{h} + \epsilon(\mathbf{h})\|\mathbf{h}\|,$$

where $\epsilon(\mathbf{h}) = \mathbf{0}$, and apply the definition. ∎

As the reader can see, the result and proof are trivial, but they take some getting used to. In one dimension the result says that the map given by $x \mapsto ax$ has derivative $x \mapsto ax$ (or that the tangent to the line $y = ax$ is the line $y = ax$ itself, or that the derivative of the linear map with 1×1 matrix (a) is the linear map with matrix (a)).

Exercise 6.22. *Show that the constant map $\mathbf{f_c} : \mathbb{R}^n \to \mathbb{R}^m$, given by $\mathbf{f_c}(\mathbf{x}) = \mathbf{c}$ for all $\mathbf{x}$, is everywhere differentiable with derivative the zero linear map.*

Lemma 6.23. *Let U be a neighbourhood of $\mathbf{x}$ in $\mathbb{R}^n$ and V a neighbourhood of $\mathbf{y}$ in $\mathbb{R}^m$. Suppose that $\mathbf{f} : U \to \mathbb{R}^p$ is differentiable at $\mathbf{x}$ and $\mathbf{g} : V \to \mathbb{R}^q$ is differentiable at $\mathbf{y}$. Then $U \times V$ is a neighbourhood of $(\mathbf{x},\mathbf{y})$ in $\mathbb{R}^{n+m}$, and the function $(\mathbf{f},\mathbf{g}) : U \times V \to \mathbb{R}^{p+q}$ given by*

$$(\mathbf{f},\mathbf{g})(\mathbf{u},\mathbf{v}) = (\mathbf{f}(\mathbf{u}),\mathbf{g}(\mathbf{v}))$$

is differentiable at $(\mathbf{x},\mathbf{y})$ with derivative $(D\mathbf{f}(\mathbf{x}), D\mathbf{g}(\mathbf{y}))$, where we write

$$(D\mathbf{f}(\mathbf{x}), D\mathbf{g}(\mathbf{y}))(\mathbf{h},\mathbf{k}) = (D\mathbf{f}(\mathbf{x})\mathbf{h}, D\mathbf{g}(\mathbf{y})\mathbf{k}).$$

Proof. We leave some details (such as verifying that $U \times V$ is a neighbourhood of $(\mathbf{x}, \mathbf{y})$) to the reader. The key to the proof is the remark that $\|(\mathbf{h}, \mathbf{k})\| \geq \|\mathbf{h}\|,\ \|\mathbf{k}\|$. Observe that, if we write

$$\mathbf{f}(\mathbf{x}+\mathbf{h}) = \mathbf{f}(\mathbf{x}) + D\mathbf{f}(\mathbf{x})\mathbf{h} + \epsilon_1(\mathbf{h})\|\mathbf{h}\|$$

and

$$\mathbf{g}(\mathbf{y}+\mathbf{k}) = \mathbf{g}(\mathbf{y}) + D\mathbf{g}(\mathbf{y})\mathbf{k} + \epsilon_2(\mathbf{k})\|\mathbf{k}\|,$$

we have

$$\begin{aligned}(\mathbf{f}, \mathbf{g})&((\mathbf{x}, \mathbf{y}) + (\mathbf{h}, \mathbf{k})) \\ &= (\mathbf{f}, \mathbf{g})(\mathbf{x}, \mathbf{y}) + (D\mathbf{f}(\mathbf{x}), D\mathbf{g}(\mathbf{x}))(\mathbf{h}, \mathbf{k}) + \epsilon(\mathbf{h}, \mathbf{k})\|(\mathbf{h}, \mathbf{k})\|,\end{aligned}$$

where

$$\epsilon(\mathbf{h}, \mathbf{k})\|(\mathbf{h}, \mathbf{k})\| = \epsilon_1(\mathbf{h})\|\mathbf{h}\| + \epsilon_2(\mathbf{k})\|\mathbf{k}\|.$$

Using the last equation, we obtain

$$\begin{aligned}\|\epsilon(\mathbf{h}, \mathbf{k})\|\|(\mathbf{h}, \mathbf{k})\| &= \|(\epsilon(\mathbf{h}, \mathbf{k})\|(\mathbf{h}, \mathbf{k})\|)\| = \|(\epsilon_1(\mathbf{h})\|\mathbf{h}\| + \epsilon_2(\mathbf{k})\|\mathbf{k}\|)\| \\ &\leq \|(\epsilon_1(\mathbf{h})\|\mathbf{h}\|)\| + \|(\epsilon_2(\mathbf{k})\|\mathbf{k}\|)\| \leq \|\epsilon_1(\mathbf{h})\|\|(\mathbf{h}, \mathbf{k})\| + \|\epsilon_2(\mathbf{k})\|\|(\mathbf{h}, \mathbf{k})\|.\end{aligned}$$

Thus

$$\|\epsilon(\mathbf{h}, \mathbf{k})\| \leq \|\epsilon_1(\mathbf{h})\| + \|\epsilon_2(\mathbf{k})\| \to 0 + 0 = 0$$

as $\|(\mathbf{h}, \mathbf{k})\| \to 0$. ■

Exercise 6.24. *If* $\mathbf{h} \in \mathbb{R}^n$ *and* $\mathbf{k} \in \mathbb{R}^m$, *show that*

$$\|(\mathbf{h}, \mathbf{k})\|^2 = \|\mathbf{h}\|^2 + \|\mathbf{k}\|^2$$

and

$$\|\mathbf{h}\| + \|\mathbf{k}\| \geq \|(\mathbf{h}, \mathbf{k})\| \geq \|\mathbf{h}\|,\ \|\mathbf{k}\|.$$

We can now give a second proof of Lemma 6.20 using the chain rule.

Second proof of Lemma 6.20. Let $\alpha : \mathbb{R}^n \to \mathbb{R}^{2n}$ be the map given by

$$\alpha(\mathbf{x}) = (\mathbf{x}, \mathbf{x})$$

and $\beta : \mathbb{R}^{2m} \to \mathbb{R}^m$ be the map given by

$$\beta(\mathbf{x}, \mathbf{y}) = \mathbf{x} + \mathbf{y}.$$

Then, using the notation of Lemma 6.23,

$$\mathbf{f} + \mathbf{g} = \beta \circ (\mathbf{f}, \mathbf{g}) \circ \alpha.$$

But α and β are linear, so using the chain rule (Lemma 6.19), we see that $\mathbf{f} + \mathbf{g}$ is differentiable at $\mathbf{x}$ and

$$D(\mathbf{f} + \mathbf{g})(\mathbf{x}) = \beta \circ D(\mathbf{f}, \mathbf{g})(\mathbf{x}, \mathbf{x}) \circ \alpha = D\mathbf{f}(\mathbf{x}) + D\mathbf{g}(\mathbf{x}).$$

■

Exercise 6.25. *Consider the situation described in Lemma 6.23. Write down the Jacobian matrix of partial derivatives for* $(\mathbf{f}, \mathbf{g})$ *in terms of the Jacobian matrices for* $\mathbf{f}$ *and* $\mathbf{g}$. *Write down the matrices associated with* α *and* β *in the proof just given. Compute the Jacobian matrix for* $\mathbf{f} + \mathbf{g}$ *by using matrix multiplication and the chain rule.*

If we only used this idea to prove Lemma 6.20 it would hardly be worth it, but it is frequently easiest to show that a complicated function is differentiable by expressing it as the composition of simpler differentiable functions. (How else would one prove that $x \mapsto \sin(\exp(1 + x^2))$ is differentiable?)

Exercise 6.26. *(i) Show that the function* $J : \mathbb{R}^n \times \mathbb{R}^n \to \mathbb{R}$ *given by the scalar product*

$$J(\mathbf{u}, \mathbf{v}) = \mathbf{u} \cdot \mathbf{v}$$

is everywhere differentiable with

$$DJ(\mathbf{x}, \mathbf{y})(\mathbf{h}, \mathbf{k}) = \mathbf{x} \cdot \mathbf{k} + \mathbf{y} \cdot \mathbf{h}.$$

(ii) Let U *be a neighbourhood of* $\mathbf{x}$ *in* $\mathbb{R}^n$. *Suppose that* $\mathbf{f}$, $\mathbf{g} : U \to \mathbb{R}^m$ *are differentiable at* $\mathbf{x}$. *Show, using the chain rule, that* $\mathbf{f} \cdot \mathbf{g}$ *is differentiable at* $\mathbf{x}$ *with*

$$D(\mathbf{f} \cdot \mathbf{g})(\mathbf{x})\mathbf{h} = \mathbf{f}(\mathbf{x}) \cdot (D(\mathbf{g})(\mathbf{x})\mathbf{h}) + (D(\mathbf{f})(\mathbf{x})\mathbf{h}) \cdot \mathbf{g}(\mathbf{x}).$$

(iii) Let U *be a neighbourhood of* $\mathbf{x}$ *in* $\mathbb{R}^n$. *Suppose that* $\mathbf{f} : U \to \mathbb{R}^m$ *and* $\lambda : U \to \mathbb{R}$ *are differentiable at* $\mathbf{x}$. *State and prove an appropriate result about the function* $\lambda\mathbf{f}$ *given by*

$$(\lambda\mathbf{f})(\mathbf{u}) = \lambda(\mathbf{u})\mathbf{f}(\mathbf{u}).$$

(iv) If you have met the vector product[2] $\mathbf{u} \wedge \mathbf{v}$ *of two vectors* $\mathbf{u}$, $\mathbf{v} \in \mathbb{R}^3$, *state and prove an appropriate theorem about the vector product of differentiable functions.*

(v) Let U *be a neighbourhood of* $\mathbf{x}$ *in* $\mathbb{R}^n$. *Suppose that* $f : U \to \mathbb{R}$ *is non-zero on* U *and differentiable at* $\mathbf{x}$. *Show that* $1/f$ *is differentiable at* $\mathbf{x}$, *and find* $D(1/f)\mathbf{x}$.

6.3. The mean value inequality in higher dimensions

So far our study of differentiation in higher dimensions has remained on the level of mere algebra. (The definition of the operator norm used the supremum and so lay deeper, but we could have avoided this at the cost of using a less natural norm.) The next result is a true theorem of analysis.

[2] Also known as the 'outer product' or 'cross product' and written $\mathbf{u} \times \mathbf{v}$. *Question:* What do you get if you cross a mountaineer with a mosquito? *Answer:* You can't. One is a scaler and the other is a vector.

Theorem 6.27. (The mean value inequality.) *Suppose that U is an open set in $\mathbb{R}^m$ and that $\mathbf{f} : U \to \mathbb{R}^p$ is differentiable. Consider the straight line segment*

$$L = \{(1-t)\mathbf{a} + t\mathbf{b} \,:\, 0 \le t \le 1\}$$

joining $\mathbf{a}$ and $\mathbf{b}$. If $L \subseteq U$ (i.e., L lies entirely within U) and $\|D\mathbf{f}(\mathbf{x})\| \le K$ for all $\mathbf{x} \in L$, then

$$\|\mathbf{f}(\mathbf{a}) - \mathbf{f}(\mathbf{b})\| \le K\|\mathbf{a} - \mathbf{b}\|.$$

Proof. Before starting the proof, it is helpful to note that, since U is open, we can find an $\eta > 0$ such that the extended straight line segment

$$\{(1-t)\mathbf{a} + t\mathbf{b} \,:\, -\eta \le t \le 1+\eta\} \subseteq U.$$

We shall prove our many-dimensional mean value inequality from the one-dimensional version (Theorem 1.42, or if the reader prefers, the slightly sharper Theorem 4.49). To this end, observe that, if $\mathbf{f}(\mathbf{b}) - \mathbf{f}(\mathbf{a}) = \mathbf{0}$, there is nothing to prove. We may thus assume that $\mathbf{f}(\mathbf{b}) - \mathbf{f}(\mathbf{a}) \ne \mathbf{0}$ and consider

$$\mathbf{u} = \frac{\mathbf{f}(\mathbf{b}) - \mathbf{f}(\mathbf{a})}{\|\mathbf{f}(\mathbf{b}) - \mathbf{f}(\mathbf{a})\|},$$

the unit vector in the direction $\mathbf{f}(\mathbf{b}) - \mathbf{f}(\mathbf{a})$. If we now define $g : (-\eta, 1+\eta) \to \mathbb{R}$ by

$$g(t) = \mathbf{u} \cdot \big(\mathbf{f}((1-t)\mathbf{a} + t\mathbf{b}) - \mathbf{f}(\mathbf{a})\big),$$

we see, by using the chain rule or direct calculation, that g is continuous and differentiable on $(-\eta, 1+\eta)$ with

$$g'(t) = \mathbf{u} \cdot (D\mathbf{f}((1-t)\mathbf{a} + t\mathbf{b})(\mathbf{b} - \mathbf{a})).$$

Using the Cauchy–Schwarz inequality (Lemma 4.2) and the definition of the operator norm (Definition 6.13), we have

$$\begin{aligned}
|g'(t)| &\le \|\mathbf{u}\|\|D\mathbf{f}((1-t)\mathbf{a} + t\mathbf{b})(\mathbf{b} - \mathbf{a})\| \\
&= \|D\mathbf{f}((1-t)\mathbf{a} + t\mathbf{b})(\mathbf{b} - \mathbf{a})\| \\
&\le \|D\mathbf{f}((1-t)\mathbf{a} + t\mathbf{b})\|\|\mathbf{b} - \mathbf{a}\| \\
&\le K\|\mathbf{a} - \mathbf{b}\|.
\end{aligned}$$

for all $t \in (0,1)$. Thus, by the one-dimensional mean value inequality,

$$\|\mathbf{f}(\mathbf{a}) - \mathbf{f}(\mathbf{b})\| = |g(1) - g(0)| \le K\|\mathbf{a} - \mathbf{b}\|,$$

as required. ■

Exercise 6.28. *(i) Prove the statement of the first sentence in the proof just given.*

(ii) If g is the function defined in the proof just given, show, giving all the details, that g is continuous and differentiable on $(-\eta, 1+\eta)$ with

$$g'(t) = \mathbf{u} \cdot \big(D\mathbf{f}((1-t)\mathbf{a} + t\mathbf{b})(\mathbf{b} - \mathbf{a})\big).$$

You should give two *versions of the proof, the first using the chain rule (Lemma 6.19) and the second using direct calculation.*

If we have already gone to the trouble of proving the one-dimensional mean value inequality, it seems sensible to make use of it in proving the multidimensional version. However, we could have proved the multidimensional theorem directly without making a one-dimensional detour.

Exercise 6.29. *(i) Reread the proof of Theorem 1.42.*

(ii) We now start the direct proof of Theorem 6.27. As before observe that we can find an $\eta > 0$ such that

$$\{(1-t)\mathbf{a} + t\mathbf{b} \, : \, -\eta \leq t \leq 1+\eta\} \subseteq U,$$

but now consider $\mathbf{F} : (-\eta, 1+\eta) \to \mathbb{R}^p$ by

$$\mathbf{F}(t) = \mathbf{f}((1-t)\mathbf{a} + t\mathbf{b}) - \mathbf{f}(\mathbf{a}).$$

Explain why the theorem will follow if we can show that, given any $\epsilon > 0$, we have

$$\|\mathbf{F}(1) - \mathbf{F}(0)\| \leq K\|\mathbf{a} - \mathbf{b}\| + \epsilon.$$

(ii) Suppose, if possible, that there exists an $\epsilon > 0$ such that

$$\|\mathbf{F}(1) - \mathbf{F}(0)\| \geq K\|\mathbf{a} - \mathbf{b}\| + \epsilon.$$

Show by a lion hunting argument that there exist a $c \in [0,1]$ and $u_n, v_n \in [0,1]$ with $u_n < v_n$ such that $u_n, v_n \to c$ and

$$\|\mathbf{F}(v_n) - \mathbf{F}(u_n)\| \geq (K\|\mathbf{a} - \mathbf{b}\| + \epsilon)(v_n - u_n).$$

(iii) Show from the definition of differentiability that there exists a $\delta > 0$ such that

$$\|\mathbf{F}(t) - \mathbf{F}(c)\| < (K\|\mathbf{a} - \mathbf{b}\| + \epsilon/2)|t - c|$$

whenever $|t - c| < \delta$ and $t \in [0,1]$.

(iv) Prove Theorem 6.27 by reductio ad absurdum.

One of the principal uses we made of the one-dimensional mean value theorem was to show that a function on an open interval with zero derivative was necessarily constant. The reader should do both parts of the following easy exercise and reflect on them.

Exercise 6.30. *(i) Let U be an open set in $\mathbb{R}^m$ such that given any $\mathbf{a}$, $\mathbf{b} \in U$ we can find a finite sequence of points $\mathbf{a} = \mathbf{a}_0, \mathbf{a}_1, \dots, \mathbf{a}_{k-1}, \mathbf{a}_k = \mathbf{b}$ such that each line segment*

$$\{(1-t)\mathbf{a}_{j-1} + t\mathbf{a}_j \,:\, 0 \leq t \leq 1\} \subseteq U$$

$[1 \leq j \leq k]$. Show that, if $\mathbf{f} : U \to \mathbb{R}^p$ is everywhere differentiable on U with $D\mathbf{f}(\mathbf{x}) = \mathbf{0}$, it follows that $\mathbf{f}$ is constant.

(ii) We work in $\mathbb{R}^2$. Let U_1 be the open disc of radius 1 and centre $(-2, 0)$, and let U_2 be the open disc of radius 1 and centre $(2, 0)$. Set $U = U_1 \cup U_2$. Define $f : U \to \mathbb{R}$ by $f(\mathbf{x}) = -1$ for $\mathbf{x} \in U_1$, $f(\mathbf{x}) = 1$ for $\mathbf{x} \in U_2$. Show that f is everywhere differentiable on U with $D(f)(\mathbf{x}) = 0$, but f is not constant.

The reader may ask if we can obtain an improvement to our mean value *inequality* by some sort of *equality* along the lines of Theorem 4.49. The answer is a clear no.

Exercise 6.31. *Let $\mathbf{f} : \mathbb{R} \to \mathbb{R}^2$ be given by $\mathbf{f}(t) = (\cos t, \sin t)^T$. Compute the Jacobian matrix of partial derivatives for $\mathbf{f}$ and show that $\mathbf{f}(0) = \mathbf{f}(2\pi)$ but $D\mathbf{f}(t) \neq 0$ for all t.*

(Although Exercise K.112 is not a counterexample, it points out another problem which occurs when we work in many dimensions.)

It is fairly obvious that we cannot replace the line segment L in Theorem 6.27 by other curves without changing the conclusion.

Exercise 6.32. *Let*

$$U = \{\mathbf{x} \in \mathbb{R}^2 : \|\mathbf{x}\| > 1\} \setminus \{(x, 0)^T : x \leq 0\}.$$

If we take $\theta(\mathbf{x})$ to be the unique solution of

$$\cos(\theta(\mathbf{x})) = \frac{x}{(x^2+y^2)^{1/2}}, \quad \sin(\theta(\mathbf{x})) = \frac{y}{(x^2+y^2)^{1/2}}, \quad -\pi < \theta(\mathbf{x}) < \pi,$$

for $\mathbf{x} = (x, y)^T \in U$, show that $\theta : U \to \mathbb{R}$ is everywhere differentiable with $\|D\theta(\mathbf{x})\| < 1$. (The amount of work involved in proving this depends quite strongly on how clever you are in exploiting radial symmetry.) Show, however, that if $\mathbf{a} = (-1, 10^{-1})^T$, $\mathbf{b} = (-1, -10^{-1})^T$, then

$$|\theta(\mathbf{a}) - \theta(\mathbf{b})| > \|\mathbf{a} - \mathbf{b}\|.$$

It is clear (though we shall not prove it, and, indeed, cannot yet state it without using concepts which we have not formally defined) that the correct generalisation when L is not a straight line will run as follows. 'If L is a well behaved path lying entirely within U and $\|Df(\mathbf{x})\| \leq K$ for all $\mathbf{x} \in L$, then $\|f(\mathbf{a}) - f(\mathbf{b})\| \leq K \times \text{length}\, L$'.

Chapter 7

Local Taylor Theorems

7.1. Some one-dimensional Taylor theorems

By *definition*, a function $f : \mathbb{R} \to \mathbb{R}$ which is continuous at 0 looks like a constant function near 0, in the sense that

$$f(t) = f(0) + \epsilon(t),$$

where $\epsilon(t) \to 0$ as $t \to 0$. By *definition*, again, a function $f : \mathbb{R} \to \mathbb{R}$ which is differentiable at 0 looks like a linear function near 0, in the sense that

$$f(t) = f(0) + f'(0)t + \epsilon(t)|t|,$$

where $\epsilon(t) \to 0$ as $t \to 0$. The next exercise establishes the non-trivial *theorem* that a function $f : \mathbb{R} \to \mathbb{R}$, if it is $n-1$ times differentiable in a neighbourhood of 0 and n times differentiable at 0, looks like a polynomial of degree n near 0, in the sense that

$$f(t) = f(0) + f'(0)t + \frac{f''(0)}{2!}t^2 + \cdots + \frac{f^{(n)}(0)}{n!}t^n + \epsilon(t)|t|^n,$$

where $\epsilon(t) \to 0$ as $t \to 0$.

This exercise introduces several ideas which we use repeatedly in this chapter, so the reader should do it carefully.

Exercise 7.1. *In this exercise we consider functions $f, g : (-a, a) \to \mathbb{R}$, where $a > 0$.*

(i) If f and g are differentiable with $f'(t) \le g'(t)$ for all $0 \le t < a$ and $f(0) = g(0)$, explain why $f(t) \le g(t)$ for all $0 \le t < a$.

(ii) If $|f'(t)| \le |t|^r$ for all $t \in (-a, a)$ and $f(0) = 0$, show that $|f(t)| \le |t|^{r+1}/(r+1)$ for all $|t| < a$.

(iii) If g is n times differentiable with $|g^{(n)}(t)| \leq M$ for all $t \in (-a, a)$ and $g(0) = g'(0) = \cdots = g^{(n-1)}(0) = 0$, show that

$$|g(t)| \leq \frac{M|t|^n}{n!}$$

for all $|t| < a$.

(iv) If g is n times differentiable in $(-a, a)$, and $n+1$ times differentiable at 0 with $g(0) = g'(0) = \cdots = g^{(n)}(0) = g^{(n+1)}(0) = 0$, show, using (iii), that

$$|g(t)| \leq \eta(t)|t|\frac{|t|^n}{n!} = \frac{\eta(t)|t|^{n+1}}{n!},$$

where $\eta(t) \to 0$ as $t \to 0$.

(v) If f is n times differentiable with $|f^{(n)}(t)| \leq M$ for all $t \in (-a, a)$, show that

$$\left| f(t) - \sum_{j=0}^{n-1} \frac{f^{(j)}(0)}{j!} t^j \right| \leq \frac{M|t|^n}{n!}$$

for all $|t| < a$.

(vi) If f is n times differentiable in $(-a, a)$ and $n+1$ times differentiable at 0, show that

$$\left| f(t) - \sum_{j=0}^{n} \frac{f^{(j)}(0)}{j!} t^j \right| \leq \frac{\eta(t)|t|^{n+1}}{n!},$$

where $\eta(t) \to 0$ as $t \to 0$.

Restating parts (v) and (vi) of Exercise 7.1, we get two similar looking but distinct theorems.

Theorem 7.2. (A global Taylor's theorem.) *If $f : (-a, a) \to \mathbb{R}$ is n times differentiable with $|f^{(n)}(t)| \leq M$ for all $t \in (-a, a)$, then*

$$\left| f(t) - \sum_{j=0}^{n-1} \frac{f^{(j)}(0)}{j!} t^j \right| \leq \frac{M|t|^n}{n!}.$$

Theorem 7.3. (The local Taylor's theorem.) *If $f : (-a, a) \to \mathbb{R}$ is $n-1$ times differentiable and $f^{(n)}(0)$ exists, then*

$$f(t) = \sum_{j=0}^{n} \frac{f^{(j)}(0)}{j!} t^j + \epsilon(t)|t|^n,$$

where $\epsilon(t) \to 0$ as $t \to 0$.

We shall obtain other and more precise global Taylor theorems in the course of the book (see Exercise K.50 and Theorem 8.51), but Theorem 7.2 is strong enough for the following typical applications.

Exercise 7.4. *(i) Consider a differentiable function* $e : \mathbb{R} \to \mathbb{R}$ *which obeys the differential equation* $e'(t) = e(t)$ *with the initial condition* $e(0) = 1$. *Quote a general theorem which tells you that, if* $a > 0$, *there exists an* M *with* $|e(t)| \leq M$ *for* $|t| \leq a$. *Show that*

$$\left| e(t) - \sum_{j=0}^{n-1} \frac{t^j}{j!} \right| \leq \frac{M|t|^n}{n!}$$

for all $|t| < a$. *Deduce that*

$$\sum_{j=0}^{n-1} \frac{t^j}{j!} \to e(t)$$

as $n \to \infty$, *and so*

$$e(t) = \sum_{j=0}^{\infty} \frac{t^j}{j!}$$

for all t.

(ii) Consider differentiable functions $s, c : \mathbb{R} \to \mathbb{R}$ *which obey the differential equations* $s'(t) = c(t)$, $c'(t) = -s(t)$ *with the initial conditions* $s(0) = 0$, $c(0) = 1$. *Show that*

$$s(t) = \sum_{j=0}^{\infty} \frac{(-1)^j t^{2j+1}}{(2j+1)!}$$

for all t, *and obtain a similar result for* c.

However, in this chapter we are interested in the *local* behaviour of functions and therefore in the local Taylor theorem. The distinction between local and global Taylor expansion is made in the following very important example of Cauchy.

Example 7.5. *Consider the function* $F : \mathbb{R} \to \mathbb{R}$ *defined by*

$$F(0) = 0,$$
$$F(x) = \exp(-1/x^2) \qquad \textit{otherwise.}$$

(i) Prove by induction, using the standard rules of differentiation, that F *is infinitely differentiable at all points* $x \neq 0$ *and that, at these points,*

$$F^{(n)}(x) = P_n(1/x)\exp(-1/x^2),$$

where P_n *is a polynomial which need not be found explicitly.*

(ii) Explain why $x^{-1}P_n(1/x)\exp(-1/x^2) \to 0$ *as* $x \to 0$.

(iii) Show by induction, using the definition of differentiation, that F *is infinitely differentiable at* 0 *with* $F^{(n)}(0) = 0$ *for all* n. *[Be careful to get this part of the argument right.]*

(iv) Show that

$$F(x) = \sum_{j=0}^{\infty} \frac{F^{(j)}(0)}{j!} x^j$$

if and only if $x = 0$. *(The reader may prefer to say that 'The Taylor expansion of* F *is only valid at* 0*'.)*

(v) Why does part (iv) not contradict the local Taylor theorem (Theorem 7.3)?

[We give a different counterexample making use of uniform convergence in Exercise K.259.]

Example 7.6. *Show that, if we define* $E : \mathbb{R} \to \mathbb{R}$ *by*

$$\begin{aligned} E(x) &= 0 && \text{if } x \leq 0, \\ E(x) &= \exp(-1/x^2) && \text{otherwise,} \end{aligned}$$

then E *is an infinitely differentiable function with* $E(x) = 0$ *for* $x \leq 0$ *and* $E(x) > 0$ *for* $x > 0$

Cauchy gave his example to show that we cannot develop the calculus algebraically but must use ϵ, δ techniques. In later courses the reader will see that his example encapsulates a key difference between real and complex analysis. If the reader perseveres further with mathematics she will also find the function E playing a useful rôle in distribution theory and differential geometry.

A simple example of the use of the local Taylor theorem is given by the proof of (a version of) L'Hôpital's rule in the next exercise.

Exercise 7.7. *If* f, $g : (-a, a) \to \mathbb{R}$ *are* n *times differentiable and*

$$f(0) = f'(0) = \cdots = f^{n-1}(0) = g(0) = g'(0) = \cdots = g^{(n-1)}(0) = 0$$

but $g^{(n)}(0) \neq 0$, *then,*

$$\frac{f(t)}{g(t)} \to \frac{f^{(n)}(0)}{g^{(n)}(0)}$$

as $t \to 0$.

It should be pointed out that the local Taylor theorems of this chapter (and the global ones proved elsewhere) are deep results which depend on the fundamental axiom. The fact that we use mean value theorems to prove them is thus not surprising — we must use the fundamental axiom or results derived from it in the proof.

(Most of my readers will be prepared to accept my word for the statements made in the previous paragraph. Those who are not will need to work through the next exercise. The others may skip it.)

Exercise 7.8. *Explain why we can find a sequence of irrational numbers a_n such that $4^{-n-1} < a_n < 4^{-n}$. We write $I_0 = \{x \in \mathbb{Q} : x > a_0\}$ and*

$$I_n = \{x \in \mathbb{Q} : a_n < x < a_{n-1}\}$$

[$n = 1, 2, 3, \dots$]. Check that, if $x \in I_n$, then $4^{-n-1} < x < 4^{-n+1}$ [$n \geq 1$].

We define $f : \mathbb{Q} \to \mathbb{Q}$ by $f(0) = 0$ and $f(x) = 8^{-n}$ if $|x| \in I_n$ [$n \geq 0$]. In what follows we work in $\mathbb{Q}$.

(i) Show that

$$\frac{f(h) - f(0)}{h} \to 0$$

as $h \to 0$. Conclude that f is differentiable at 0 with $f'(0) = 0$.

(ii) Explain why f is everywhere differentiable with $f'(x) = 0$ for all x. Conclude that f is infinitely differentiable with $f^{(r)} = 0$ for all $r \geq 0$.

(iii) Show that

$$\frac{f(h) - f(0)}{h^2} \to \infty$$

as $h \to 0$. Conclude that, if we write

$$f(h) = f(0) + f'(0)h + \frac{f''(0)}{2!}h^2 + \epsilon(h)h^2,$$

then $\epsilon(h) \nrightarrow 0$ as $h \to 0$. Thus the local Taylor theorem (Theorem 7.3) is false for $\mathbb{Q}$.

7.2. Some many-dimensional local Taylor theorems

In the previous section we used mean value inequalities to investigate the local behaviour of well behaved functions $f : \mathbb{R} \to \mathbb{R}$. We now use the same ideas to investigate the local behaviour of well behaved functions $f : \mathbb{R}^n \to \mathbb{R}$. It turns out that, once we understand what happens when $n = 2$, it is easy to extend the results to general n, and this will be left to the reader.

Here is our first example.

Lemma 7.9. *We work in $\mathbb{R}^2$ and write $\mathbf{0} = (0, 0)$.*

(i) Suppose $\delta > 0$, and that $f : B(\mathbf{0}, \delta) \to \mathbb{R}$ has partial derivatives $f_{,1}$ and $f_{,2}$ with $|f_{,1}(x, y)|, |f_{,2}(x, y)| \leq M$ for all $(x, y) \in B(\mathbf{0}, \delta)$. If $f(0, 0) = 0$,

then

$$|f(x,y)| \leq 2M(x^2+y^2)^{1/2}$$

for all $(x,y) \in B(\mathbf{0},\delta)$.

(ii) *Suppose* $\delta > 0$, *and that* $g : B(\mathbf{0},\delta) \to \mathbb{R}$ *has partial derivatives* $g_{,1}$ *and* $g_{,2}$ *in* $B(\mathbf{0},\delta)$. *Suppose that* $g_{,1}$ *and* $g_{,2}$ *are continuous at* $(0,0)$ *and* $g(0,0) = g_{,1}(0,0) = g_{,2}(0,0) = 0$. *Then, writing*

$$g((h,k)) = \epsilon(h,k)(h^2+k^2)^{1/2},$$

we have $\epsilon(h,k) \to 0$ *as* $(h^2+k^2)^{1/2} \to 0$.

Proof. (i) Observe that the one-dimensional mean value inequality applied to the function $t \mapsto f(x,t)$ gives

$$|f(x,y) - f(x,0)| \leq M|y|$$

whenever $(x,y) \in B(\mathbf{0},\delta)$, and the same inequality applied to the function $s \mapsto f(s,0)$ gives

$$|f(x,0) - f(0,0)| \leq M|x|$$

whenever $(x,0) \in B(\mathbf{0},\delta)$. We now apply a taxicab argument (the idea behind the name is that a New York taxicab which wishes to get from $(0,0)$ to (x,y) will be forced by the grid pattern of streets to go from $(0,0)$ to $(x,0)$ and thence to (x,y)) to obtain

$$\begin{aligned}|f(x,y)| &= |f(x,y) - f(0,0)| = |(f(x,y) - f(x,0)) + (f(x,0) - f(0,0))| \\ &\leq |f(x,y) - f(x,0)| + |f(x,0) - f(0,0)| \leq M|y| + M|x| \\ &\leq 2M(x^2+y^2)^{1/2}\end{aligned}$$

for all $(x,y) \in B(\mathbf{0},\delta)$.

(ii) Let $\epsilon > 0$ be given. By the definition of continuity, we can find a $\delta_1(\epsilon)$ such that $\delta > \delta_1(\epsilon) > 0$ and

$$|g_{,1}(x,y)|,\ |g_{,2}(x,y)| \leq \epsilon/2$$

for all $(x,y) \in B(\mathbf{0},\delta_1(\epsilon))$. By part (i), this means that

$$|g(x,y)| \leq \epsilon(x^2+y^2)^{1/2}$$

for all $(x,y) \in B(\mathbf{0},\delta_1(\epsilon))$, and this gives the desired result. ∎

Theorem 7.10. (Continuity of partial derivatives implies differentiability.) *Suppose* $\delta > 0$, $\mathbf{x} = (x,y) \in \mathbb{R}^2$, $B(\mathbf{x},\delta) \subseteq E \subseteq \mathbb{R}^2$ *and that* $f : E \to \mathbb{R}$. *If the partial derivatives* $f_{,1}$ *and* $f_{,2}$ *exist in* $B(\mathbf{x},\delta)$ *and are continuous at* $\mathbf{x}$, *then, writing*

$$f(x+h,y+k) = f(x,y) + f_{,1}(x,y)h + f_{,2}(x,y)k + \epsilon(h,k)(h^2+k^2)^{1/2},$$

we have $\epsilon(h,k) \to 0$ *as* $(h^2+k^2)^{1/2} \to 0$. *(In other words,* f *is differentiable at* $\mathbf{x}$.*)*

Proof. By translation, we may suppose that $\mathbf{x} = \mathbf{0}$. Now set

$$g(x,y) = f(x,y) - f(0,0) - f_{,1}(0,0)x - f_{,2}(0,0)y.$$

We see that g satisfies the hypotheses of part (ii) of Lemma 7.9. Thus g satisfies the conclusions of part (ii) of Lemma 7.9, and our theorem follows. ■

Although this is not one of the great theorems of all time, it occasionally provides a useful short cut for proving functions differentiable[1]. The following easy extensions are left to the reader.

Theorem 7.11. *(i) Suppose* $\delta > 0$, $\mathbf{x} \in \mathbb{R}^m$, $B(\mathbf{x},\delta) \subseteq E \subseteq \mathbb{R}^m$ *and that* $f : E \to \mathbb{R}$. *If the partial derivatives* $f_{,1}$, $f_{,2}$, $\dots$ $f_{,m}$ *exist in* $B(\mathbf{x},\delta)$ *and are continuous at* $\mathbf{x}$, *then* f *is differentiable at* $\mathbf{x}$.

(ii) Suppose $\delta > 0$, $\mathbf{x} \in \mathbb{R}^m$, $B(\mathbf{x},\delta) \subseteq E \subseteq \mathbb{R}^m$ *and that* $f : E \to \mathbb{R}^p$. *If the partial derivatives* $f_{i,j}$ *exist in* $B(\mathbf{x},\delta)$ *and are continuous at* $\mathbf{x}$ $[1 \le i \le p,\ 1 \le j \le m]$, *then* f *is differentiable at* $\mathbf{x}$.

Ideas similar to those used in the proof of Theorem 7.10 give our next result, which we shall therefore prove more expeditiously. We write

$$f_{,ij}(\mathbf{x}) = (f_{,j})_{,i}(\mathbf{x}),$$

or, in more familiar notation,

$$f_{,ij} = \frac{\partial^2 f}{\partial x_i \partial x_j}.$$

Theorem 7.12. (Second order Taylor series.) *Suppose* $\delta > 0$, $\mathbf{x} = (x,y) \in \mathbb{R}^2$, $B(\mathbf{x},\delta) \subseteq E \subseteq \mathbb{R}^2$ *and that* $f : E \to \mathbb{R}$. *If the partial derivatives* $f_{,1}$, $f_{,2}$, $f_{,11}$, $f_{,12}$, $f_{,22}$ *exist in* $B(\mathbf{x},\delta)$ *and* $f_{,11}$, $f_{,12}$, $f_{,22}$ *are continuous at* $\mathbf{x}$, *then, writing*

$$\begin{aligned} f((x+h,y+k)) &= f(x,y) + f_{,1}(x,y)h + f_{,2}(x,y)k \\ &\quad + (f_{,11}(x,y)h^2 + 2f_{,12}(x,y)hk + f_{,22}(x,y)k^2)/2 + \epsilon(h,k)(h^2+k^2), \end{aligned}$$

we have $\epsilon(h,k) \to 0$ *as* $(h^2+k^2)^{1/2} \to 0$.

Proof. By translation, we may suppose that $\mathbf{x} = \mathbf{0}$. By considering

$$\begin{aligned} &f(h,k) - f(0,0) - f_{,1}(0,0)h - f_{,2}(0,0)k \\ &\qquad - (f_{,11}(0,0)h^2 + 2f_{,12}(0,0)hk + f_{,22}(0,0)k^2)/2, \end{aligned}$$

[1] I emphasise the word *occasionally*. Usually, results like the fact that a differentiable function of a differentiable function is differentiable give a faster and more satisfactory proof.

we may suppose that

$$f(0,0) = f_{,1}(0,0) = f_{,2}(0,0) = f_{,11}(0,0) = f_{,12}(0,0) = f_{,22}(0,0).$$

If we do this, our task reduces to showing that

$$\frac{f(h,k)}{h^2+k^2} \to 0$$

as $(h^2+k^2)^{1/2} \to 0$.

To this end, observe that, if $\epsilon > 0$, the continuity of the given partial derivatives at $(0,0)$ tells us that we can find a $\delta_1(\epsilon)$ such that $\delta > \delta_1(\epsilon) > 0$ and

$$|f_{,11}(h,k)|,\ |f_{,12}(h,k)|,\ |f_{,22}(h,k)| \leq \epsilon$$

for all $(h,k) \in B(\mathbf{0}, \delta_1(\epsilon))$. Using the mean value inequality in the manner of Lemma 7.9, we have

$$|f_{,1}(h,k) - f_{,1}(h,0)| \leq \epsilon|k|$$

and

$$|f_{,1}(h,0) - f_{,1}(0,0)| \leq \epsilon|h|,$$

and a taxicab argument gives

$$\begin{aligned} |f_{,1}(h,k)| &= |f_{,1}(h,k) - f_{,1}(0,0)| \\ &= |(f_{,1}(h,k) - f_{,1}(h,0)) + (f_{,1}(h,0) - f_{,1}(0,0))| \\ &\leq |f_{,1}(h,k) - f_{,1}(h,0)| + |f_{,1}(h,0) - f_{,1}(0,0)| \leq \epsilon(|k|+|h|) \end{aligned}$$

for all $(h,k) \in B(\mathbf{0}, \delta_1(\epsilon))$. (Or we could have just applied Lemma 7.9 with f replaced by $f_{,1}$.) The mean value inequality also gives

$$|f_{,2}(0,k)| = |f_{,2}(0,k) - f_{,2}(0,0)| \leq \epsilon|k|.$$

Now, applying the taxicab argument again, using the mean value inequality and the estimates of the first paragraph, we get

$$\begin{aligned} |f(h,k)| &= |f(h,k) - f(0,0)| = |(f(h,k) - f(0,k)) + (f(0,k) - f(0,0))| \\ &\leq |f(h,k) - f(0,k)| + |f(0,k) - f(0,0)| \\ &\leq \sup_{0\leq s\leq 1} |f_{,1}(sh,k)||h| + \sup_{0\leq t\leq 1} |f_{,2}(0,tk)||k| \\ &\leq \epsilon(|k|+|h|)|h| + \epsilon|k|^2 \\ &\leq 3\epsilon(h^2+k^2). \end{aligned}$$

Since ϵ was arbitrary, the result follows. ∎

Exercise 7.13. *Set out the proof of Theorem 7.12 in the style of the proof of Theorem 7.10.*

We have the following important corollary.

Theorem 7.14. (Symmetry of the second partial derivatives.) *Suppose $\delta > 0$, $\mathbf{x} = (x, y) \in \mathbb{R}^2$, $B(\mathbf{x}, \delta) \subseteq E \subseteq \mathbb{R}^2$ and that $f : E \to \mathbb{R}$. If the partial derivatives $f_{,1}$, $f_{,2}$, $f_{,11}$, $f_{,12}$, $f_{,21}$ $f_{,22}$ exist in $B(\mathbf{x}, \delta)$ and are continuous at $\mathbf{x}$, then $f_{,12}(\mathbf{x}) = f_{,21}(\mathbf{x})$.*

Proof. By Theorem 7.12, we have

$$\begin{aligned} f(x+h, y+k) &= f(x,y) + f_{,1}(x,y)h + f_{,2}(x,y)k \\ &\quad + (f_{,11}(x,y)h^2 + 2f_{,12}(x,y)hk + f_{,22}(x,y)k^2)/2 + \epsilon_1(h,k)(h^2+k^2) \end{aligned}$$

with $\epsilon_1(h,k) \to 0$ as $(h^2+k^2)^{1/2} \to 0$. But, interchanging the rôle of first and second variable, Theorem 7.12 also tells us that

$$\begin{aligned} f(x+h, y+k) &= f(x,y) + f_{,1}(x,y)h + f_{,2}(x,y)k \\ &\quad + (f_{,11}(x,y)h^2 + 2f_{,21}(x,y)hk + f_{,22}(x,y)k^2)/2 + \epsilon_2(h,k)(h^2+k^2) \end{aligned}$$

with $\epsilon_2(h,k) \to 0$ as $(h^2+k^2)^{1/2} \to 0$.

Comparing the two Taylor expansions for $f(x+h, y+k)$, we see that

$$\begin{aligned} f_{,12}(x,y)hk - f_{,21}(x,y)hk &= (\epsilon_1(h,k) - \epsilon_2(h,k))(h^2+k^2) \\ &= \epsilon_3(h,k)(h^2+k^2) \end{aligned}$$

with $\epsilon_3(h,k) \to 0$ as $(h^2+k^2)^{1/2} \to 0$. Taking $h = k$ and dividing by h^2, we have

$$f_{,12}(x,y) - f_{,21}(x,y) = 2\epsilon_3(h,h) \to 0$$

as $h \to 0$, so $f_{,12}(x,y) - f_{,21}(x,y) = 0$ as required. ■

It is possible to produce plausible arguments for the symmetry of second partial derivatives. Here are a couple.

(1) If f is a multinomial, i.e. $f(x,y) = \sum_{p=0}^{P} \sum_{q=0}^{Q} a_{p,q} x^p y^q$, then $f_{,12} = f_{,21}$. But smooth functions are very close to being polynomial, so we would expect the result to be true in general.

(2) Although we cannot interchange limits in general, it is plausible that, if f is well behaved, then

$$\begin{aligned} f_{,12}(x,y) &= \lim_{h\to 0}\lim_{k\to 0} h^{-1}k^{-1}(f(x+h,y+k) - f(x+h,y) - f(x,y+k) + f(x,y)) \\ &= \lim_{k\to 0}\lim_{h\to 0} h^{-1}k^{-1}(f(x+h,y+k) - f(x+h,y) - f(x,y+k) + f(x,y)) \\ &= f_{,21}(x,y). \end{aligned}$$

However, these are merely plausible arguments. They do not make clear the rôle of the continuity of the second derivative (in Example 7.36 we shall see that the result may fail for discontinuous second partial derivatives). More fundamentally, they are *algebraic* arguments and, as the use of the mean value theorem indicates, the result is one of *analysis*. The same kind of argument which shows that the local Taylor theorem fails over $\mathbb{Q}$ (see Example 7.8) shows that it fails over $\mathbb{Q}^2$ and that the symmetry of partial derivatives fails with it (see [**34**]).

If we use the D notation, Theorem 7.14 states that (under appropriate conditions)

$$D_1 D_2 f = D_2 D_1 f.$$

If we write $D_{ij} = D_i D_j$, as is often done, we get

$$D_{12} f = D_{21} f.$$

What happens if a function has higher partial derivatives? It is not hard to guess and prove the appropriate theorem.

Exercise 7.15. *Suppose $\delta > 0$, $\mathbf{x} \in \mathbb{R}^m$, $B(\mathbf{x}, \delta) \subseteq E \subseteq \mathbb{R}^m$ and that $f : E \to \mathbb{R}$. Show that, if all the partial derivatives $f_{,j}$, $f_{,jk}$, $f_{,ijk}$, ... up to the nth order exist in $B(\mathbf{x}, \delta)$ and are continuous at $\mathbf{x}$, then, writing*

$$f(\mathbf{x}+\mathbf{h}) = f(\mathbf{x}) + \sum_{j=1}^{m} f_{,j}(\mathbf{x}) h_j + \frac{1}{2!} \sum_{j=1}^{m} \sum_{k=1}^{m} f_{,jk}(\mathbf{x}) h_j h_k$$
$$+ \frac{1}{3!} \sum_{j=1}^{m} \sum_{k=1}^{m} \sum_{l=1}^{m} f_{,jkl}(\mathbf{x}) h_j h_k h_l + \cdots + \textit{sum up to nth powers} + \epsilon(\mathbf{h}) \|\mathbf{h}\|^n,$$

we have $\epsilon(\mathbf{h}) \to 0$ as $\|\mathbf{h}\| \to 0$.

Notice that you do not have to prove results like

$$f_{,jkl}(\mathbf{x}) = f_{,ljk}(\mathbf{x}) = f_{,klj}(\mathbf{x}) = f_{,lkj}(\mathbf{x}) = f_{,jlk}(\mathbf{x}) = f_{,kjl}(\mathbf{x})$$

since they follow directly from Theorem 7.14.

Applying Exercise 7.15 to the components f_i of a function $\mathbf{f}$, we obtain our full many-dimensional Taylor theorem.

Theorem 7.16. (The local Taylor's theorem.) *Suppose $\delta > 0$, $\mathbf{x} \in \mathbb{R}^m$, $B(\mathbf{x}, \delta) \subseteq E \subseteq \mathbb{R}^m$ and that $\mathbf{f} : E \to \mathbb{R}^p$. If all the partial derivatives $f_{i,j}$, $f_{i,jk}$, $f_{i,jkl}$, ... exist in $B(\mathbf{x}, \delta)$ and are continuous at $\mathbf{x}$, then, writing*

$$f_i(\mathbf{x}+\mathbf{h}) = f_i(\mathbf{x}) + \sum_{j=1}^{m} f_{i,j}(\mathbf{x}) h_j + \frac{1}{2!} \sum_{j=1}^{m} \sum_{k=1}^{m} f_{i,jk}(\mathbf{x}) h_j h_k$$
$$+ \frac{1}{3!} \sum_{j=1}^{m} \sum_{k=1}^{m} \sum_{l=1}^{m} f_{i,jkl}(\mathbf{x}) h_j h_k h_l$$
$$+ \cdots + \textit{sum up to nth powers} + \epsilon_i(\mathbf{h}) \|\mathbf{h}\|^n,$$

we have $\|\boldsymbol{\epsilon}(\mathbf{h})\| \to 0$ as $\|\mathbf{h}\| \to 0$.

The reader will remark that Theorem 7.16 bristles with subscripts, contrary to our announced intention of seeking a geometric, coordinate free view. However, it is very easy to restate the main formula of Theorem 7.16 in a coordinate free way as

$$\mathbf{f}(\mathbf{x}+\mathbf{h}) = \mathbf{f}(\mathbf{x}) + \alpha_1(\mathbf{h}) + \alpha_2(\mathbf{h},\mathbf{h}) + \cdots + \alpha_n(\mathbf{h},\mathbf{h},\ldots,\mathbf{h}) + \epsilon(\mathbf{h})\|\mathbf{h}\|^n,$$

where $\alpha_k : \mathbb{R}^m \times \mathbb{R}^m \cdots \times \mathbb{R}^m \to \mathbb{R}^p$ is linear in each variable (i.e., a k-linear function) and symmetric (i.e., interchanging any two variables leaves the value of α_k unchanged).

Anyone who feels that the higher derivatives are best studied using coordinates should reflect that, if $\mathbf{f} : \mathbb{R}^3 \to \mathbb{R}^3$ is well behaved, then the 'third derivative behaviour' of $\mathbf{f}$ at a single point is apparently given by the $3 \times 3 \times 3 \times 3 = 81$ numbers $f_{i,jkl}(\mathbf{x})$. By symmetry (see Theorem 7.14) only 30 of the numbers are distinct, but these 30 numbers are independent (consider polynomials in three variables for which the total degree of each term is 3). How can we understand the information carried by an array of 30 real numbers?

Exercise 7.17. *(i) Verify the statements in the last paragraph. How large an array is required to give the 'third derivative behaviour' of a well behaved function* $\mathbf{f} : \mathbb{R}^4 \to \mathbb{R}^4$ *at a point? How large an array is required to give the 'fourth derivative behaviour' of a well behaved function* $\mathbf{f} : \mathbb{R}^3 \to \mathbb{R}^3$ *at a point?*

(ii) (Ignore this if the notation is not familiar.) Consider a well behaved function $\mathbf{f} : \mathbb{R}^3 \to \mathbb{R}^3$. *How large an array is required to give* $\operatorname{curl} \mathbf{f} = \nabla \times \mathbf{f} = \nabla \wedge \mathbf{f}$ *and* $\operatorname{div} \mathbf{f} = \nabla \cdot \mathbf{f}$? *How large an array is required to give* $D\mathbf{f}$?

In many circumstances, $\operatorname{curl} \mathbf{f}$ *and* $\operatorname{div} \mathbf{f}$ *give the physically interesting part of* $D\mathbf{f}$, *but physicists also use*

$$(\mathbf{a} \cdot \nabla)\mathbf{f} = \left(\sum_{j=1}^{3} a_j f_{1,j}, \sum_{j=1}^{3} a_j f_{2,j}, \sum_{j=1}^{3} a_j f_{3,j} \right).$$

How large an array is required to give $(\mathbf{a} \cdot \nabla)\mathbf{f}$ *for all* $\mathbf{a} \in \mathbb{R}^3$?

In subjects like elasticity, the description of nature requires the full Jacobian matrix $(f_{i,j})$, *and the treatment of differentiation used is closer to that of the pure mathematician.*

Most readers will be happy to finish this section here[2]. However, some of them[3] will observe that in our coordinate free statement of the local

[2] The rest of this section is marked with a ♡.

[3] Boas notes that 'There is a test for identifying some of the future professional mathematicians at an early age. These are students who instantly comprehend a sentence beginning "Let X be an ordered quintuple $(a, T, \pi, \sigma, \mathcal{B})$ where ... ". They are even more promising if they add, "I never really understood it before." ' ([**8**], page 231.)

Taylor's theorem the 'second derivative behaviour' is given by a *bilinear* map $\alpha_2 : \mathbb{R}^m \times \mathbb{R}^m \to \mathbb{R}^p$, and we defined derivatives in terms of *linear* maps.

Let us be more precise. We suppose $\mathbf{f}$ is a well behaved function on an open set $U \subseteq \mathbb{R}^p$ taking values in $\mathbb{R}^m$. If we write $\mathcal{L}(E, F)$ for the space of linear maps from a finite-dimensional vector space E to a vector space F, then, for each fixed $\mathbf{x} \in U$, we have $D\mathbf{f}(\mathbf{x}) \in \mathcal{L}(\mathbb{R}^m, \mathbb{R}^p)$. Thus, allowing $\mathbf{x}$ to vary freely, we see that we have a function

$$D\mathbf{f} : U \to \mathcal{L}(\mathbb{R}^m, \mathbb{R}^p).$$

We now observe that $\mathcal{L}(\mathbb{R}^m, \mathbb{R}^p)$ is a finite-dimensional vector space over $\mathbb{R}$ of dimension mp; in other words, $\mathcal{L}(\mathbb{R}^m, \mathbb{R}^p)$ can be identified with $\mathbb{R}^{mp}$. We know how to define the derivative of a well behaved function $\mathbf{g} : U \to \mathbb{R}^{mp}$ at $\mathbf{x}$ as a function

$$D\mathbf{g}(\mathbf{x}) \in \mathcal{L}(\mathbb{R}^m, \mathbb{R}^{mp}),$$

so we know how to define the derivative of $D\mathbf{f}$ at $\mathbf{x}$ as a function

$$D(D\mathbf{f})(\mathbf{x}) \in \mathcal{L}(\mathbb{R}^m, \mathcal{L}(\mathbb{R}^m, \mathbb{R}^p)).$$

We have thus shown how to define the second derivative $D^2\mathbf{f}(\mathbf{x}) = D(D\mathbf{f})(\mathbf{x})$. But $D^2\mathbf{f}(\mathbf{x})$ lies in $\mathcal{L}(\mathbb{R}^m, \mathcal{L}(\mathbb{R}^m, \mathbb{R}^p))$ and α_2 lies in the space $\mathcal{E}(\mathbb{R}^m, \mathbb{R}^m; \mathbb{R}^p)$ of bilinear maps from $\mathbb{R}^m \times \mathbb{R}^m$ to $\mathbb{R}^p$. How, the reader may ask, can we identify $\mathcal{L}(\mathbb{R}^m, \mathcal{L}(\mathbb{R}^m, \mathbb{R}^p))$ with $\mathcal{E}(\mathbb{R}^m, \mathbb{R}^m; \mathbb{R}^p)$? Fortunately this question answers itself with hardly any outside intervention.

Exercise 7.18. *Let E, F and G be finite-dimensional vector spaces over $\mathbb{R}$. We write $\mathcal{E}(E, F; G)$ for the space of bilinear maps $\alpha : E \times F \to G$. Define*

$$(\Theta(\alpha)(\mathbf{u}))(\mathbf{v}) = \alpha(\mathbf{u}, \mathbf{v})$$

for all $\alpha \in \mathcal{E}(E, F; G)$, $\mathbf{u} \in E$ and $\mathbf{v} \in F$.

(i) Show that $\Theta(\alpha)(\mathbf{u}) \in \mathcal{L}(F, G)$.

(ii) Show that, if $\mathbf{v}$ is fixed,

$$\big(\Theta(\alpha)(\lambda_1\mathbf{u}_1 + \lambda_2\mathbf{u}_2)\big)(\mathbf{v}) = \big(\lambda_1\Theta(\alpha)(\mathbf{u}_1) + \lambda_2\Theta(\alpha)(\mathbf{u}_2)\big)(\mathbf{v}),$$

and deduce that

$$\Theta(\alpha)(\lambda_1\mathbf{u}_1 + \lambda_2\mathbf{u}_2) = \lambda_1\Theta(\alpha)(\mathbf{u}_1) + \lambda_2\Theta(\alpha)(\mathbf{u}_2)$$

for all λ_1, $\lambda_2 \in \mathbb{R}$ and $\mathbf{u}_1$, $\mathbf{u}_2 \in E$. Conclude that $\Theta(\alpha) \in \mathcal{L}(E, \mathcal{L}(F, G))$.

(iii) By arguments similar in spirit to those of (ii), show that $\Theta : \mathcal{E}(E, F; G) \to \mathcal{L}(E, \mathcal{L}(F, G))$ is linear.

(iv) Show that if $(\Theta(\alpha)(\mathbf{u}))(\mathbf{v}) = \mathbf{0}$ for all $\mathbf{u} \in E$, $\mathbf{v} \in F$, then $\alpha = 0$. Deduce that Θ is injective.

(v) By computing the dimensions of $\mathcal{E}(E, F; G)$ and $\mathcal{L}(E, \mathcal{L}(F, G))$, show that Θ is an isomorphism.

Since our definition of Θ does not depend on a choice of basis, we say that Θ gives a natural isomorphism of $\mathcal{E}(E, F; G)$ and $\mathcal{L}(E, \mathcal{L}(F, G))$. If we use this isomorphism to identify $\mathcal{E}(E, F; G)$ and $\mathcal{L}(E, \mathcal{L}(F, G))$, then $D^2\mathbf{f}(\mathbf{x}) \in \mathcal{E}(\mathbb{R}^m, \mathbb{R}^m; \mathbb{R}^p)$. If we treat the higher derivatives in the same manner, the central formula of the local Taylor theorem takes the satisfying form

$$\mathbf{f}(\mathbf{x}+\mathbf{h}) = \mathbf{f}(\mathbf{x}) + D\mathbf{f}(\mathbf{x})(\mathbf{h}) + \frac{1}{2!}D^2\mathbf{f}(\mathbf{x})(\mathbf{h},\mathbf{h}) + \cdots + \frac{1}{n!}D^n\mathbf{f}(\mathbf{x})(\mathbf{h},\mathbf{h},\dots,\mathbf{h}) + \epsilon(\mathbf{h})\|\mathbf{h}\|^n.$$

For more details, consult sections 11 and 13 of chapter VIII of Dieudonné's *Foundations of Modern Analysis* [**14**], where the higher derivatives are dealt with in a coordinate free way. Like Hardy's book [**24**], Dieudonné's is a masterpiece, but in a very different tradition[4].

7.3. Critical points

In this section we mix informal and formal argument, deliberately using words like 'well behaved' without defining them. Our object is to use the local Taylor formula to produce results about maxima, minima and related objects.

Let U be an open subset of $\mathbb{R}^m$ containing $\mathbf{0}$. We are interested in the behaviour of a well behaved function $f : U \to \mathbb{R}$ near $\mathbf{0}$. Since f is well behaved, the first order local Taylor theorem (which reduces to the definition of differentiation) gives $f(\mathbf{h}) = f(\mathbf{0}) + \alpha\mathbf{h} + \epsilon(\mathbf{h})\|\mathbf{h}\|$, where $\epsilon(\mathbf{h}) \to 0$ as $\|\mathbf{h}\| \to 0$ and $\alpha = Df(\mathbf{0})$ is a linear map from $\mathbb{R}^m$ to $\mathbb{R}$. By a very simple result of linear algebra, we can choose a set of orthogonal coordinates so that $\alpha(x_1, x_2, \dots, x_m) = ax_1$ with $a \geq 0$.

Exercise 7.19. *If $\alpha : \mathbb{R}^m \to \mathbb{R}$ is linear, show that, with respect to any particular chosen orthogonal coordinates,*

$$\alpha(x_1, x_2, \dots, x_m) = a_1x_1 + a_2x_2 + \cdots + a_mx_m$$

for some $a_j \in \mathbb{R}$. Deduce that there is a vector $\mathbf{a}$ such that $\alpha\mathbf{x} = \mathbf{a}\cdot\mathbf{x}$ for all $\mathbf{x} \in \mathbb{R}^m$. Conclude that we can choose a set of orthogonal coordinates so that $\alpha(x_1, x_2, \dots, x_m) = ax_1$ with $a \geq 0$.

In applied mathematics we write $\mathbf{a} = \nabla f$. A longer, but very instructive, proof of the result of this exercise is given in Exercise K.31.

In the coordinate system just chosen

$$f(h_1, h_2, \dots, h_m) = f(\mathbf{0}) + ah_1 + \epsilon(\mathbf{h})\|\mathbf{h}\|,$$

[4]See the quotation from Boas in the previous footnote.

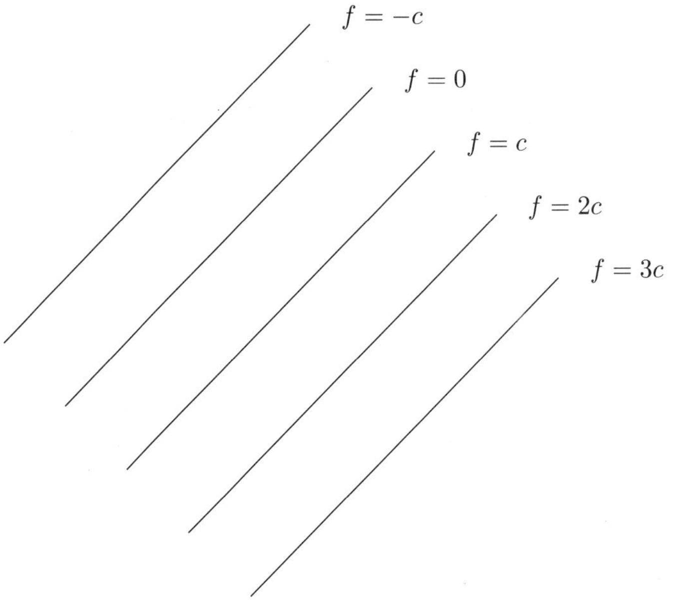

Figure 7.1. Contour lines when the derivative is not zero.

where $\epsilon(\mathbf{h}) \to 0$ as $\|\mathbf{h}\| \to 0$. Thus, speaking informally, if $a \neq 0$ the 'contour lines' $f(\mathbf{h}) = c$ close to $\mathbf{0}$ will look like parallel 'hyperplanes' perpendicular to the x_1 axis. Figure 7.1 illustrates the case $m = 2$. In particular, our contour lines look like those describing a side of a hill but not its peak.

Using our informal insight, we can prove a formal lemma.

Lemma 7.20. *Let U be an open subset of $\mathbb{R}^m$ containing $\mathbf{x}$. Suppose that $f : U \to \mathbb{R}$ is differentiable at $\mathbf{x}$. If $f(\mathbf{x}) \geq f(\mathbf{y})$ for all $\mathbf{y} \in U$, then $Df(\mathbf{x}) = 0$ (more precisely, $Df(\mathbf{x})\mathbf{h} = 0$ for all $\mathbf{h} \in \mathbb{R}^m$).*

Proof. There is no loss in generality in supposing $\mathbf{x} = \mathbf{0}$. Suppose that $Df(\mathbf{0}) \neq \mathbf{0}$. Then we can find an orthogonal coordinate system and a strictly positive real number a such that $Df(\mathbf{0})(h_1, h_2, \ldots, h_n) = ah_1$. Thus, from the definition of the derivative,

$$f(h_1, h_2, \ldots, h_n) = f(\mathbf{0}) + ah_1 + \epsilon(\mathbf{h})\|\mathbf{h}\|,$$

where $\|\epsilon(\mathbf{h})\| \to 0$ as $\|\mathbf{h}\| \to 0$.

Choose $\eta > 0$ such that, whenever $\|\mathbf{h}\| < \eta$, we have $\mathbf{h} \in U$ and $\|\epsilon(\mathbf{h})\| < a/2$. Now choose any real h with $0 < h < \eta$. If we set $\mathbf{h} = (h, 0, 0, \ldots, 0)$, we have

$$f(\mathbf{h}) = f(\mathbf{0}) + ah + \epsilon(\mathbf{h})h > f(\mathbf{0}) + ah - ah/2 = f(\mathbf{0}) + ah/2 > f(\mathbf{0}).$$

∎

The distinctions made in the following definition are probably familiar to the reader.

Definition 7.21. *Let E be a subset of $\mathbb{R}^m$ containing $\mathbf{x}$ and let f be a function from E to $\mathbb{R}$.*

(i) We say that f has a global *maximum at $\mathbf{x}$ if $f(\mathbf{x}) \geq f(\mathbf{y})$ for all $\mathbf{y} \in E$.*

(ii) We say that f has a strict *global maximum at $\mathbf{x}$ if $f(\mathbf{x}) > f(\mathbf{y})$ for all $\mathbf{y} \in E$ with $\mathbf{x} \neq \mathbf{y}$.*

(iii) We say that f has a local *maximum (respectively a strict* local *maximum) at $\mathbf{x}$ if there exists an $\eta > 0$ such that the restriction of f to $E \cap B(\mathbf{x}, \eta)$ has a* global *maximum (respectively a* strict *global maximum) at $\mathbf{x}$.*

(iv) If we can find an $\eta > 0$ such that $E \supseteq B(\mathbf{x}, \eta)$ and f is differentiable at $\mathbf{x}$ with $Df(\mathbf{x}) = 0$, we say that $\mathbf{x}$ is a critical *or* stationary *point*[5] *of f.*

It is usual to refer to the point $\mathbf{x}$ where f takes a (global or local) maximum as a (global or local) maximum, and this convention rarely causes confusion. When mathematicians omit the words local or global in referring to maximum they usually mean the local version (but this convention, which I shall follow, is not universal).

Here are some easy exercises involving these ideas.

Exercise 7.22. *(i) Let U be an open subset of $\mathbb{R}^m$ containing $\mathbf{x}$. Suppose that $f : U \to \mathbb{R}$ is differentiable on U and that Df is continuous at $\mathbf{x}$. Show that, if f has a local maximum at $\mathbf{x}$, then $Df(\mathbf{x}) = 0$.*

*(ii) Suppose that $f : \mathbb{R}^m \to \mathbb{R}$ is differentiable everywhere and E is a closed subset of $\mathbb{R}^m$ containing $\mathbf{x}$. Show that, even if $\mathbf{x}$ is a global maximum of the restriction of f to E, it need not be true that $Df(\mathbf{x}) = 0$. [*Hint: *We have already met this fact when we thought about Rolle's theorem.] Explain informally why the proof of Lemma 7.20 fails in this case.*

(iii) State the definitions corresponding to Definition 7.21 that we need to deal with minima.

(iv) Let E be any subset of $\mathbb{R}^m$ containing $\mathbf{y}$ and let f be a function from E to $\mathbb{R}$. If $\mathbf{y}$ is both a global maximum and a global minimum for f, show that f is constant. What can you say if we replace the word 'global' by 'local'?

[5]In other words, a stationary point is one where the ground is flat. Since flat ground drains badly, the stationary points we meet in hill walking tend to be boggy. Thus we encounter boggy ground at the top of hills and when crossing passes as well as at lowest points (at least in the UK, other countries may be drier or have better draining soils).

We saw above how f behaved locally near $\mathbf{0}$ if $Df(\mathbf{0}) \neq 0$. What can we say if $Df(\mathbf{0}) = 0$? In this case, the second order Taylor expansion gives

$$f(\mathbf{h}) = f(\mathbf{0}) + \beta(\mathbf{h}, \mathbf{h}) + \epsilon(\mathbf{h})\|\mathbf{h}\|^2,$$

where

$$\beta(\mathbf{h}, \mathbf{h}) = \frac{1}{2}\sum_{i=1}^{m}\sum_{j=1}^{m} f_{,ij}(\mathbf{0})h_i h_j$$

and $\epsilon(\mathbf{h}) \to 0$ as $\|\mathbf{h}\| \to 0$. We write $\beta = \frac{1}{2}D^2 f$ and call the matrix $K = (f_{,ij}(\mathbf{0}))$ the Hessian matrix. As we noted in the previous section, the symmetry of the second partial derivatives (Theorem 7.14) tells us that the Hessian matrix is a symmetric matrix and the associated bilinear map $D^2 f$ is symmetric. It follows from a well known result in linear algebra (see, for example, Exercise K.30) that $\mathbb{R}^n$ has an orthonormal basis of eigenvectors of K. Choosing coordinate axes along those vectors, we obtain

$$D^2 f(\mathbf{h}, \mathbf{h}) = \sum_{i=1}^{m} \lambda_i h_i^2,$$

where the λ_i are the eigenvalues associated with the eigenvectors.

In the coordinate system just chosen

$$f(h_1, h_2, \dots, h_m) = f(\mathbf{0}) + \frac{1}{2}\sum_{i=1}^{m} \lambda_i h_i^2 + \epsilon(\mathbf{h})\|\mathbf{h}\|^2,$$

where $\epsilon(\mathbf{h}) \to 0$ as $\|\mathbf{h}\| \to 0$. Thus, speaking informally, if all the λ_i are non-zero, the 'contour lines' $f(\mathbf{h}) = c$ close to $\mathbf{0}$ will look like 'quadratic hypersurfaces' (that is, m-dimensional versions of conics). Figure 7.2 illustrates the two possible contour patterns when $m = 2$. The first type of pattern is that of a summit (if the contour lines are for increasing heights as we approach $\mathbf{0}$) or a bottom (lowest point)[6] (if the contour lines are for decreasing heights as we approach $\mathbf{0}$). The second is that of a pass (often called a saddle). Notice that, for merchants, wishing to get from one valley to another, the pass is the highest point in their journey, but, for mountaineers, wishing to get from one mountain to another, the pass is the lowest point.

When looking at Figure 7.2 it is important to realise that the difference in heights of successive contour lines is not constant. We have taken $f(\mathbf{0}) = 0$ and we have drawn contour lines at heights 0, h, 2^2h, 3^2h, ..., n^2h.

[6]The English language is rich in synonyms for highest points (summits, peaks, crowns, ...) but has few for lowest points. This may be because the English climate ensures that most lowest points are under water.

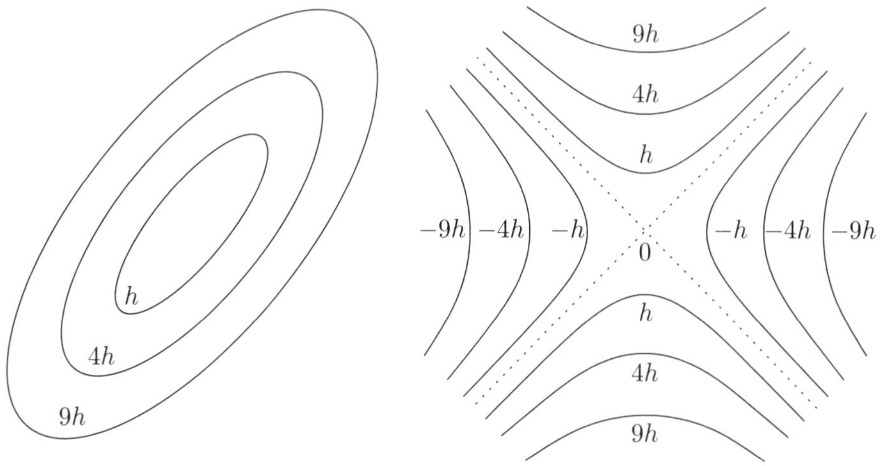

Figure 7.2. Contour lines when the derivative is zero but the second derivative is non-singular

Exercise 7.23. *(i) Redraw Figure 7.2 with contour lines at heights* 0, h, $2h$, $3h$, $\dots$, nh.

(ii) What (roughly speaking) can you say about the difference in heights of successive contour lines in Figure 7.1?

Using our informal insight, we can prove a formal lemma.

Lemma 7.24. *Let* U *be an open subset of* $\mathbb{R}^m$ *containing* $\mathbf{x}$. *Suppose that* $f : U \to \mathbb{R}$ *has second order partial derivatives on* U *and these partial derivatives are continuous at* $\mathbf{x}$. *If* $Df(\mathbf{x}) = 0$ *and* $D^2f(\mathbf{x})$ *is non-singular, then*

(i) f *has a minimum at* $\mathbf{x}$ *if and only if* $D^2f(\mathbf{x})$ *is positive definite.*

(ii) f *has a maximum at* $\mathbf{x}$ *if and only if* $D^2f(\mathbf{x})$ *is negative definite.*

The conditions of the second sentence of the hypothesis ensure that we have a local second order Taylor expansion. In most applications f will be much better behaved than this. We say that $D^2f(\mathbf{x})$ is positive definite if all the associated eigenvalues (that is, all the eigenvalues of the Hessian matrix) are strictly positive, and that $D^2f(\mathbf{x})$ is negative definite if all the associated eigenvalues are strictly negative.

Exercise 7.25. *Prove Lemma 7.24, using the proof of Lemma 7.20 as a model.*

It is a non-trivial task to tell whether a given Hessian is positive or negative definite.

Exercise 7.26. *Let $f(x, y) = x^2 + 6xy + y^2$. Show that $Df(0, 0) = 0$, that all the entries in the Hessian matrix K at $(0, 0)$ are positive and that K is non-singular, but that $D^2 f(0, 0)$ is neither positive definite nor negative definite. (So $(0, 0)$ is a saddle point.)*

Exercise K.117 gives one method of resolving the problem.

Because it is non-trivial to use the Hessian to determine whether a *singular point* (that is, a point $\mathbf{x}$ where $Df(\mathbf{x}) = \mathbf{0}$) is a maximum, a minimum or neither, mathematicians frequently seek short cuts.

Exercise 7.27. *Suppose that $f : \mathbb{R}^m \to \mathbb{R}$ is continuous, that $f(\mathbf{x}) \to 0$ as $\|\mathbf{x}\| \to \infty$ and that $f(\mathbf{x}) > 0$ for all $\mathbf{x} \in \mathbb{R}^m$.*

(i) Explain why there exists an $R > 0$ such that $f(\mathbf{x}) < f(\mathbf{0})$ for all $\|\mathbf{x}\| \geq R$.

(ii) Explain why there exists an $\mathbf{x}_0$ with $\|\mathbf{x}_0\| \leq R$ and $f(\mathbf{x}_0) \geq f(\mathbf{x})$ for all $\|\mathbf{x}\| \leq R$.

(iii) Explain why $f(\mathbf{x}_0) \geq f(\mathbf{x})$ for all $\mathbf{x} \in \mathbb{R}^m$.

(iv) If f is everywhere differentiable and has exactly one singular point $\mathbf{y}_0$, show that f attains a global maximum at $\mathbf{y}_0$.

(v) In statistics we frequently wish to maximise functions of the form

$$f(a, b) = \exp\left(-\sum_{i=1}^{k} (y_i - at_i - b)^2\right)$$

with $\sum_{i=1}^{k} t_i = 0$. Use the results above to find the values of a and b which maximise f.

(vi) We used (v) as an illustrative example, but in this case the result can also be obtained withot calculus. Find a and b by completing the square.

Mathematicians with a good understanding of the topic they are investigating can use insight as a substitute for rigorous verification, but intuition may lead us astray.

Exercise 7.28. *Four towns lie on the vertices of a square of side a. What is the shortest total length of a system of roads joining all four towns? (The answer is given in Exercise K.119, but try to find the answer first before looking it up.)*

The following are standard traps for the novice and occasional traps for the experienced.

(1) Critical points need not be maxima or minima.

(2) Local maxima and minima need not be global maxima or minima.

(3) Maxima and minima may occur on the boundary[7] and may then not be critical points. [We may restate this more exactly as follows. Suppose $f : E \to \mathbb{R}$. Unless E is open, f may take a maximum value at a point $\mathbf{e} \in E$ such that we cannot find any $\delta > 0$ with $B(\mathbf{e}, \delta) \subseteq E$. However well f is behaved, the argument of Lemma 7.20 will fail. For a specific instance see Exercise 7.22.]

(4) A function need not have a maximum or minimum. [Consider $f : U \to \mathbb{R}$ given by $f(x, y) = x$, where $U = B(\mathbf{0}, 1)$ or $U = \mathbb{R}^2$.]

Exercise 7.29. *Find the maxima and minima of the function $f : \mathbb{R}^2 \to \mathbb{R}$ given by*

$$f(x, y) = y^2 - x^3 - ax$$

in the region $\{(x, y) : x^2 + y^2 \leq 1\}$.

Your answer will depend on the constant a.

Matters are further complicated by the fact that different kinds of problems call for different kinds of solutions. The engineer seeks a *global* minimum to the cost of a process. On the other hand, if we drop a handful of ball bearings on the ground they will end up at *local* minima (lowest points), and most people suspect that evolutionary, economic and social changes all involve *local* maxima and minima. Finally, although we like to think of many physical processes as minimising some function, it is often the case that they are really stationarising (finding critical points for) that function. We like to say that light takes a shortest path, but, if you consider a bulb A at the centre of an ellipse, light is reflected back to A from B and B', the two closest points on the ellipse, and from C and C', the two *furthest* points (see Figure 7.3).

We have said that, if $f : \mathbb{R}^2 \to \mathbb{R}$ has a Taylor expansion in the neighbourhood of a point, then (ignoring the possibility that the Hessian is singular) the contour map will look like that in Figure 7.1 or 7.2. But it is very easy to imagine other contour maps, and the reader may ask what happens if the local contour map does not look like that in Figure 7.1 or 7.2. The answer is that the appropriate Taylor expansion has failed, and therefore the hypotheses which ensure the appropriate Taylor expansion must themselves have failed.

Exercise 7.30. *Suppose that $f : \mathbb{R}^2 \to \mathbb{R}$ is given by $f(0, 0) = 0$ and*

$$f(r\cos\theta, r\sin\theta) = rg(\theta)$$

when $r > 0$, where $g : \mathbb{R} \to \mathbb{R}$ is periodic with period 2π. [Informally, we define f using polar coordinates.] Show that, if $g(-\theta) = -g(\theta)$ for all θ,

[7] The highest point of the Netherlands occurs on the boundary of that country.

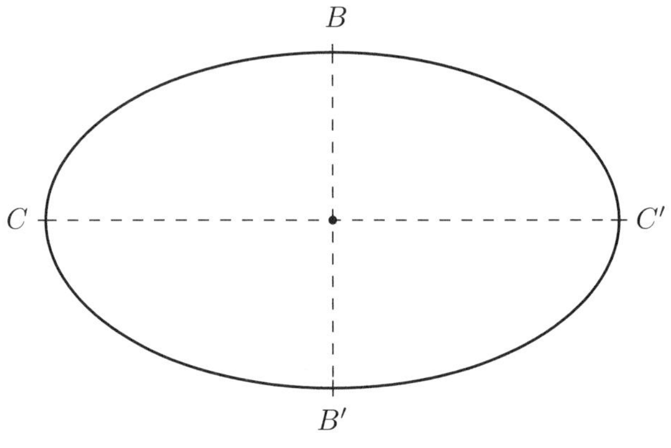

Figure 7.3. Light paths in an ellipse

then f has directional derivatives (see Definition 6.6) in all directions at $(0,0)$.

If we choose $g(\theta) = \sin\theta$, we obtain a contour map like Figure 7.1, but, if $g(\theta) = \sin 3\theta$, we obtain something very different.

Exercise 7.31. *We continue with the notation of Exercise 7.30.*

(i) If $g(\theta) = \sin\theta$, find $f(x,y)$ and sketch the contour lines $f(x,y) = h,\ 2h,\ 3h,\ \dots$ with h small.

(ii) If $g(\theta) = \sin 3\theta$, show that

$$f(x,y) = \frac{y(3x^2 - y^2)}{x^2 + y^2}$$

for $(x,y) \neq 0$. Sketch the contour lines $f(x,y) = h,\ 2h,\ 3h,\ \dots$ with h small.

Example 7.32. *If*

$$f(x,y) = \frac{y(3x^2 - y^2)}{x^2 + y^2} \qquad \text{for } (x,y) \neq (0,0),$$
$$f(0,0) = 0,$$

then f is differentiable except at $(0,0)$, is continuous everywhere, has directional derivatives in all directions at $(0,0)$, but is not differentiable at $(0,0)$.

Proof. By standard results on differentiation (the chain rule, product rule and so on), f is differentiable (and so continuous) except, perhaps, at $(0,0)$. If $u^2 + v^2 = 1$ we have

$$\frac{f(uh,vh) - f(0,0)}{h} \to v(3u^2 - v^2)$$

as $h \to 0$, so f has directional derivatives in all directions at $(0,0)$. Since

$$|f(x,y) - f(0,0)| \leq \frac{4(\max(|x|,|y|))^3}{\max(|x|,|y|))^2} = 4\max(|x|,|y|) \to 0$$

as $(x^2+y^2)^{1/2} \to 0$, f is continuous at $(0,0)$.

Suppose f were differentiable at $(0,0)$. Then

$$f(h,k) = f(0,0) + Ah + Bk + \epsilon(h,k)(h^2+k^2)^{1/2}$$

with $\epsilon(h,k) \to 0$ as $(h^2+k^2)^{1/2} \to 0$, and $A = f_{,1}(0,0)$, $B = f_{,2}(0,0)$. The calculations of the previous paragraph with $v = 0$ show that $f_{,1}(0,0) = 0$, and the same calculations with $u = 0$ show that $f_{,2}(0,0) = -1$. Thus

$$f(h,k) + k = \epsilon(h,k)(h^2+k^2)^{1/2}$$

and

$$\frac{f(h,k)+k}{(h^2+k^2)^{1/2}} \to 0$$

as $(h^2+k^2)^{1/2} \to 0$. Setting $k = h$, we get

$$2^{1/2} = \left|\frac{h+h}{(h^2+h^2)^{1/2}}\right| = \left|\frac{f(h,h)+h}{(h^2+h^2)^{1/2}}\right| \to 0$$

as $h \to 0$, which is absurd. Thus f is not differentiable at $(0,0)$. ■

(We give a stronger result in Exercise C.8 and a weaker but slightly easier result in Exercise 7.34.)

Exercise 7.33. *Write down the details behind the first sentence of our proof of Example 7.32. You will probably wish to quote Lemma 6.20 and Exercise 6.26.*

Exercise 7.34. *If*

$$f(x,y) = \frac{xy}{(x^2+y^2)^{1/2}} \qquad \text{for } (x,y) \neq (0,0),$$
$$f(0,0) = 0,$$

show that f is differentiable except at $(0,0)$, is continuous at $(0,0)$ and has partial derivatives $f_{,1}(0,0)$ and $f_{,2}(0,0)$ at $(0,0)$ but has directional derivatives in no other directions at $(0,0)$. Discuss your results briefly using the ideas of Exercise 7.30.

A further exercise on the ideas just used is given as Exercise K.122.

Emboldened by our success, we could well guess immediately a suitable function to look for in the context of Theorem 7.14.

Exercise 7.35. *Suppose that* $f : \mathbb{R}^2 \to \mathbb{R}$ *is given by* $f(0,0) = 0$ *and*

$$f(r\cos\theta, r\sin\theta) = r^2 \sin 4\theta$$

for $r > 0$. *Show that*

$$f(x,y) = \frac{4xy(x^2 - y^2)}{x^2 + y^2}$$

for $(x,y) \neq 0$. *Sketch the contour lines* $f(x,y) = h,\ 2^2h,\ 3^2h,\ \ldots$ *and compare the result with Figure 7.2.*

Exercise 7.36. *Suppose that*

$$\begin{aligned} f(x,y) &= \frac{xy(x^2 - y^2)}{(x^2 + y^2)} \qquad \text{for } (x,y) \neq (0,0), \\ f(0,0) &= 0. \end{aligned}$$

(i) Compute $f_{,1}(0,y)$, *for* $y \neq 0$, *by using standard results of the calculus.*

(ii) Compute $f_{,1}(0,0)$ *directly from the definition of the derivative.*

(iii) Find $f_{,2}(x,0)$ *for all* x.

(iv) Compute $f_{,12}(0,0)$ *and* $f_{,21}(0,0)$.

(v) Show that f *has first and second partial derivatives everywhere, but* $f_{,12}(0,0) \neq f_{,21}(0,0)$.

It is profoundly unfortunate that Example 7.32 and Exercise 7.36 seem to act on some examiners like catnip on a cat. Multi-dimensional calculus leads towards differential geometry and infinite-dimensional calculus (functional analysis). Both subjects depend on *understanding* objects which we know to be well behaved but which our limited geometric intuition makes it hard for us to comprehend. Counterexamples, such as the ones just produced, which depend on functions having some precise degree of differentiability are simply irrelevant.

At the beginning of this section we used a first order local Taylor expansion and results on linear maps to establish the behaviour of a well behaved function f near a point $\mathbf{x}$ where $Df(\mathbf{x}) \neq 0$. We then used a second order local Taylor expansion and results on bilinear maps to establish the behaviour of a well behaved function f near a point $\mathbf{x}$ where $Df(\mathbf{x}) = 0$ on condition that $D^2 f(\mathbf{x})$ was non-singular. Why should we stop here?

It is not the case that we can restrict ourselves to functions f for which $D^2 f(\mathbf{x})$ is non-singular at all points.

Exercise 7.37. *(i) Let* $A(t)$ *be a* 3×3 *real symmetric matrix with* $A(t) = (a_{ij}(t))$. *Suppose that the entries* $a_{ij} : \mathbb{R} \to \mathbb{R}$ *are continuous. Explain why* $\det A : \mathbb{R} \to \mathbb{R}$ *is continuous. By using an expression for* $\det A$ *in terms* *of the eigenvalues of* A*, show that, if* $A(0)$ *is positive definite and* $A(1)$ *is negative definite, then there must exist a* $c \in (0,1)$ *with* $A(c)$ *singular.*

(ii) Let m *be an odd positive integer,* U *an open subset of* $\mathbb{R}^m$ *and* $\gamma : [0,1] \to U$ *a continuous map. Suppose that* $f : U \to \mathbb{R}$ *has continuous second order partial derivatives on* U*, and that* f *attains a local minimum at* $\gamma(0)$ *and a local maximum at* $\gamma(1)$*. Show that there exists a* $c \in [0,1]$ *such that* $D^2 f(\gamma(t))$ *is singular.*

There is nothing special about the choice of m odd in Exercise 7.37. We do the case $m = 2$ in Exercise K.118, and ambitious readers may wish to attack the general case themselves (however, it is probably only instructive if you make the argument watertight). Exercise K.44 gives a slightly stronger result when $m = 1$.

However, it is only when $Df(\mathbf{x})$ vanishes and $D^2 f(\mathbf{x})$ is singular *at the same point* $\mathbf{x}$ that we have problems, and we can readily convince ourselves (note this is not the same as proving) that this is rather unusual.

Exercise 7.38. *Let* $f : \mathbb{R} \to \mathbb{R}$ *be given by* $f(x) = ax^3 + bx^2 + cx + d$ *with* a*,* b*,* c*,* d *real. Show that there is a* y *with* $f'(y) = f''(y) = 0$ *if and only if one of the following two conditions holds:* $a \neq 0$ *and* $b^2 = 3ac$*, or* $a = b = c = 0$*.*

Faced with this kind of situation, mathematicians tend to use the word *generic* and say 'in the generic case, the Hessian is non-singular at the critical points'. This is a useful way of thinking, but we must remember that:–

(1) If we leave the word generic undefined, any sentence containing the word generic is, strictly speaking, meaningless.

(2) In any case, if we look at any particular function, it ceases to be generic. (A generic function is one without any particular properties. Any particular function that we look at has the particular property that we have looked at it.)

(3) The generic case may be a lot worse than we expect. Most mathematicians would agree that the generic function $f : \mathbb{R} \to \mathbb{R}$ is unbounded on every interval (a, b) with $a < b$, that the generic bounded function $f : \mathbb{R} \to \mathbb{R}$ is discontinuous at every point and that the generic continuous function $f : \mathbb{R} \to \mathbb{R}$ is nowhere differentiable. We should have said something more precise like 'the generic 3 times differentiable function $\mathbf{f} : \mathbb{R}^n \to \mathbb{R}$ has a non-singular Hessian at its critical points'.

So far in this section we have looked at stationary points of f by studying the local behaviour of the function. In this we have remained true to our 17th and 18th century predecessors. In a paper entitled *On Hills and Dales*,

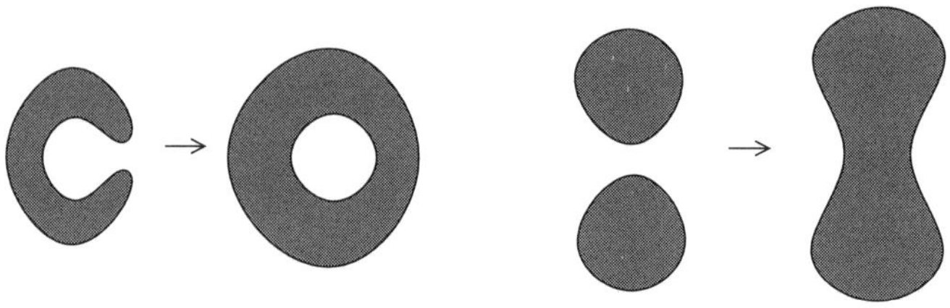

Figure 7.4. Passes vanish under water

Maxwell[8] raises our eyes from the local and shows us the prospect of a global theory.

Plausible Statement 7.39. (Hill and dale theorem.) *Suppose the surface of the moon has a finite number S of summits, B of bottoms and P of passes (all heights being measured from the moon's centre). Then*

$$S + B - P = 2.$$

Plausible Proof. By digging out pits and piling up soil we may ensure that all the bottoms are at the same height, that all the passes are at different heights, but all higher than the bottoms, and that all the summits are at the same height which is greater than the height of any pass. Now suppose that it begins to rain and that the water level rises steadily (and that the level is the same for each body of water). We write $L(h)$ for the number of lakes (a lake is the largest body of water that a swimmer can cover without going on to dry land), $I(h)$ for the number of islands (an island is the largest body of dry land that a walker can cover without going into the water) and $P(h)$ for the number of passes visible when the height of the water is h.

When the rain has just begun and the height h_0, say, of the water is higher than the bottoms, but lower than the lowest pass, we have

$$L(h_0) = B, \ I(h_0) = 1, \ P(h_0) = P. \tag{1}$$

(Observe that there is a single body of dry land that a walker can get to without going into the water, so $I(h_0) = 1$ even if the man in the street would object to calling the surface of the moon with a few puddles an island.) Every time the water rises just high enough to drown a pass, then either

(a) two arms of a lake join so an island appears, a pass vanishes and the number of lakes remains the same, or

(b) two lakes come together so the number of lakes diminishes by one, a pass vanishes and the number of islands remains the same.

We illustrate this in Figure 7.4. In either case, we see that

[8]Maxwell notes that he was anticipated by Cayley.

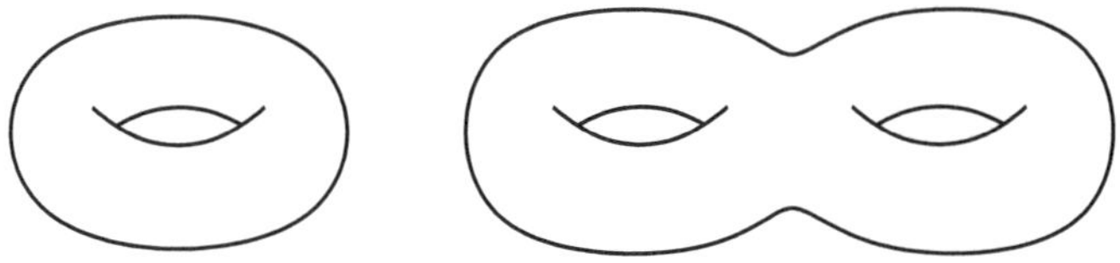

Figure 7.5. One- and two-holed doughnuts

$$I(h) - L(h) + P(h) \text{ remains constant}$$

and so, by equation (1),

$$I(h) - L(h) + P(h) = I(h_0) - L(h_0) + P(h_0) = 1 - B + P. \quad (2)$$

When the water is at a height h_1, higher than the highest pass but lower than the summits, we have

$$L(h_1) = 1, \ I(h_1) = S, \ P(h_1) = 0. \quad (3)$$

(Though the man in the street would now object to us calling something a lake when it is obviously an ocean with S isolated islands.) Using equations (2) and (3), we now have

$$1 - B + P = I(h_1) - L(h_1) + P(h_1) = S - 1$$

and so $B + S - P = 2$. ▲

Exercise 7.40. *State and provide plausible arguments for plausible results corresponding to Plausible Statement 7.39 when the moon is in the shape of a one-holed doughnut, a two-holed doughnut and an n-holed doughnut (see Figure 7.5).*

Notice that local information about the nature of a function at special points provides global 'topological' information about the number of holes in a doughnut.

If you know Euler's theorem (memory jogger: '$V - E + F = 2$'), can you connect it with this discussion?

Exercise 7.41. *The function $f : \mathbb{R}^2 \to \mathbb{R}$ is well behaved (say 3 times differentiable). We have $f(x, y) = 0$ for $x^2 + y^2 = 1$ and $f(x, y) > 0$ for $x^2 + y^2 < 1$. State and provide a plausible argument for a plausible result concerning the number of maxima, minima and saddle points (x, y) for f with $x^2 + y^2 < 1$.*

I find the plausible argument just used very convincing, but it is not clear how we would go about converting it into an argument from first principles (in effect, from the fundamental axiom of analysis). Here are some of the problems we must face.

(1) Do contour lines actually exist (that is, must the points (x, y) with $f(x, y) = h$ lie on nice curves)[9]? We answer this question locally by the implicit function theorem (Theorem 13.21). Our discussion of the solution of differential equations in Section 12.3 sheds some light on the global problem.

(2) 'The largest body of water that a swimmer can cover without going on to dry land' is a vivid but not a mathematical expression. In later work this problem is resolved by giving a formal definition of a connected set.

(3) Implicit in our argument is the idea that a loop divides a sphere into two parts. A result called the Jordan curve theorem gives the formal statement of this idea, but the proof turns out to be unexpectedly hard.

Another, less important, problem is to show that the hypothesis that there are only a 'finite number S of summits, B of bottoms and P of passes' applies to an interesting variety of cases. It is certainly not the case that a function $f : \mathbb{R} \to \mathbb{R}$ will always have only a finite number of maxima in a closed bounded interval. In the same way, it is not true that a moon need have only a finite number of summits.

Exercise 7.42. *Reread Example 7.5. Define $f : \mathbb{R} \to \mathbb{R}$ by*

$$f(x) = (\cos(1/x) - 1)\exp(-1/x^2) \qquad \text{if } x \neq 0,$$
$$f(0) = 0.$$

Show that f is differentiable everywhere and that f has an infinite number of distinct strict local maxima in the interval $[-1, 1]$. (cf. Exercise K.43.)

The answer, once again, is to develop a suitable notion of genericity, but we shall not do so here.

Some will say that there is no need to answer these questions since the plausible argument which establishes Plausible Statement 7.39 is in some sense 'obviously correct'. I would reply that the reason for attacking these questions is their intrinsic interest. Plausible Statement 7.39 and the accompanying discussion are the occasion for us to ask these questions, not the reason for trying to answer them. I would add that we cannot claim to understand Maxwell's result fully unless we can see either how it generalises to higher dimensions or why it does not.

Students often feel that multidimensional calculus is just a question of generalising results from one dimension to many. Maxwell's result shows that the change from one to many dimensions introduces genuinely new phenomena, whose existence cannot be guessed from a one-dimensional perspective.

---

[9]The reader will note that though we have used contour lines as a heuristic tool we have not used them in proofs. Note that, in specific cases, we do not need a general theorem to tell us that contour lines exist. For example, the contour lines of $f(x, y) = a^{-2}x^2 + b^{-2}y^2$ are given parametrically by $(x, y) = (ah^{1/2}\cos\theta, bh^{1/2}\sin\theta)$ for $h \geq 0$.

Chapter 8

The Riemann Integral

8.1. Where is the problem ?

Everybody knows what area is, but then everybody knows what honey tastes like. But does honey taste the same to you as it does to me? Perhaps the question is unanswerable but, for many practical purposes, it is sufficient that we agree on what we call honey. In the same way, it is important that, when two mathematicians talk about area, they should agree on the answers to the following questions:-

(1) Which sets E actually have area?

(2) When a set E has area, what is that area?

One of the discoveries of 20th century mathematics is that decisions on (1) and (2) are linked in rather subtle ways to the question:-

(3) What properties should area have?

As an indication of the ideas involved, consider the following desirable properties for area.

(a) Every bounded set E in $\mathbb{R}^2$ has an area $|E|$ with $|E| \geq 0$.

(b) Suppose that E is a bounded set in $\mathbb{R}^2$. If E is congruent to F (that is, E can be obtained from F by translation and rotation), then $|E| = |F|$.

(c) Any square E of side a has area $|E| = a^2$.

(d) If E_1, E_2, ... are disjoint bounded sets in $\mathbb{R}^2$ whose union $F = \bigcup_{i=1}^{\infty} E_j$ is also bounded, then $|F| = \sum_{i=1}^{\infty} |E_j|$ (so 'the whole is equal to the sum of its parts').

Exercise 8.1. *Suppose that conditions (a) to (d) all hold.*

(i) Let A be a bounded set in $\mathbb{R}^2$ and $B \subseteq A$. By writing $A = B \cup (A \setminus B)$ and using condition (d) together with other conditions, show that $|A| \geq |B|$.

(ii) By using (i) and condition (c), show that, if A is a non-empty bounded open set in $\mathbb{R}^2$, then $|A| > 0$.

We now show that assuming all of conditions (a) to (d) leads to a contradiction. We start with an easy remark.

Exercise 8.2. *If $0 \leq x,\ y < 1$, write $x \sim y$ whenever $x - y \in \mathbb{Q}$. Show that if $x,\ y,\ z \in [0,1)$ we have*

(i) $x \sim x$,

(ii) $x \sim y$ implies $y \sim x$,

(iii) $x \sim y$ and $y \sim z$ together imply $x \sim z$.

(In other words, $\sim$ is an equivalence relation.)

Write

$$[x] = \{y \in [0,1) \ :\ y \sim x\}.$$

(In other words, write $[x]$ for the equivalence class of x.) By quoting the appropriate theorem or direct proof, show that

(iv) $\displaystyle\bigcup_{x \in [0,1)} [x] = [0,1)$,

(v) if $x,\ y \in [0,1)$, then either $[x] = [y]$ or $[x] \cap [y] = \varnothing$.

We now consider a set A which contains exactly one element from each equivalence class.

Exercise 8.3. *(This is easy.) Show that if $t \in [0,1)$, then the equation*

$$t \equiv a + q \mod 1$$

has exactly one solution with $a \in A$, q rational and $q \in [0,1)$.

[Here $t \equiv a + q \mod 1$ means $t - a - q \in \mathbb{Z}$.]

We are now in a position to produce our example. It will be easiest to work in $\mathbb{C}$ identified with $\mathbb{R}^2$ in the usual way and to define

$$E = \{r \exp 2\pi i a \ :\ 1 > r > 0,\ a \in A\}.$$

Since $\mathbb{Q}$ is countable, it follows that its subset $\mathbb{Q} \cap [0,1)$ is countable and we can write

$$\mathbb{Q} \cap [0,1) = \{q_j \ :\ j \geq 1\}$$

with $q_1,\ q_2,\ \ldots$ all distinct. Set

$$E_j = \{r \exp 2\pi i(a + q_j) \ :\ 1 > r > 0,\ a \in A\}.$$

Exercise 8.4. *Suppose that conditions (a) to (d) all hold.*

(i) Describe the geometric relation of E and E_j. Deduce that $|E| = |E_j|$.

(ii) Use Exercise 8.3 to show that $E_j \cap E_k = \varnothing$ if $j \neq k$.

(iii) Use Exercise 8.3 to show that

$$\bigcup_{j=1}^{\infty} E_j = U,$$

where $U = \{z : 0 < |z| < 1\}$.

(iv) Deduce that

$$\sum_{j=1}^{\infty} |E_j| = |U|.$$

Show from Exercise 8.1 (ii) that $0 < |U|$.

(v) Show that (i) and (iv) lead to a contradiction if $|E| = 0$ and if $|E| > 0$. Thus (i) and (iv) lead to a contradiction whatever we assume. It follows that conditions (a) to (d) cannot all hold simultaneously.

Exercise 8.5. *Define E and E_q as subsets of $\mathbb{R}^2$ without using complex numbers.*

The example just given is due to Vitali. It might be hoped that the problem raised by Vitali's example are due to the fact that condition (d) involves infinite sums. This hope is dashed by the following theorem of Banach and Tarski. (It deals with volume, but, of course, length, area and volume are three fingers of the same hand.)

Theorem 8.6. *The unit ball in $\mathbb{R}^3$ can be decomposed into a finite number of pieces which may be reassembled, using only translation and rotation, to form 2 disjoint copies of the unit ball.*

Exercise 8.7. *Use Theorem 8.6 to show that the following four conditions are not mutually consistent.*

(a) Every bounded set E in $\mathbb{R}^3$ has an volume $|E|$ with $|E| \geq 0$.

(b) Suppose that E is a bounded set in $\mathbb{R}^3$. If E is congruent to F (that is to say, E can be obtained from F by translation and rotation), then $|E| = |F|$.

(c) Any cube E of side a has volume $|E| = a^3$.

(d) If E_1 and E_2 are disjoint bounded sets in $\mathbb{R}^3$, then $|E_1 \cup E_2| = |E_1| + |E_2|$.

The proof of Theorem 8.6, which is a lineal descendant of Vitali's example, is too long to be given here. It is beautifully and simply explained in a book [49] devoted entirely to ideas generated by the result of Banach and Tarski[1].

The examples of Vitali and Banach and Tarski show that if we want a well behaved notion of area we will have to say that only certain sets have area. Since the notion of an integral is closely linked to that of area ('the integral is the area under the curve'), this means that we must accept that only certain functions have integrals. It also means that that we must make sure that our definition does not allow paradoxes of the type discussed here.

8.2. Riemann integration

In this section we introduce a notion of the integral due to Riemann. For the moment we only attempt to define our integral for bounded functions on bounded intervals.

Let $f : [a, b] \to \mathbb{R}$ be a function such that there exists a K with $|f(x)| \leq K$ for all $x \in [a, b]$. [To see the connection with 'the area under the curve' it is helpful to suppose initially that $0 \leq f(x) \leq K$. However, all the definitions and proofs work more generally for $-K \leq f(x) \leq K$. The point is discussed in Exercise K.129.] A dissection (also called a partition) $\mathcal{D}$ of $[a, b]$ is a finite subset of $[a, b]$ containing the end points a and b. By convention, we write

$$\mathcal{D} = \{x_0, x_1, \ldots, x_n\} \text{ with } a = x_0 \leq x_1 \leq x_2 \leq \cdots \leq x_n = b.$$

We define the *upper sum* and *lower sum* associated with $\mathcal{D}$ by

$$S(f, \mathcal{D}) = \sum_{j=1}^{n} (x_j - x_{j-1}) \sup_{x \in [x_{j-1}, x_j]} f(x),$$

$$s(f, \mathcal{D}) = \sum_{j=1}^{n} (x_j - x_{j-1}) \inf_{x \in [x_{j-1}, x_j]} f(x).$$

[Observe that, *if the integral* $\int_a^b f(t)\,dt$ *exists*, then the upper sum ought to provide an upper bound and the lower sum a lower bound for that integral.]

Exercise 8.8. *(i) Suppose that* $a \leq c \leq b$. *If* $\mathcal{D} = \{a, b\}$ *and* $\mathcal{D}' = \{a, c, b\}$, *show that*

$$S(f, \mathcal{D}) \geq S(f, \mathcal{D}') \geq s(f, \mathcal{D}') \geq s(f, \mathcal{D}).$$

(ii) Let $c \neq a, b$. *Show by examples that, in (i), we can have either* $S(f, \mathcal{D}) = S(f, \mathcal{D}')$ *or* $S(f, \mathcal{D}) > S(f, \mathcal{D}')$.

---

[1]In more advanced work it is observed that our discussion depends on a principle called the 'axiom of choice'. It is legitimate to doubt this principle. However, anyone who doubts the axiom of choice but believes that every set has volume resembles someone who disbelieves in Father Christmas but believes in flying reindeer.

(iii) Suppose that $a \le c \le b$ and $\mathcal{D}$ is a dissection. Show that, if $\mathcal{D}' = \mathcal{D} \cup \{c\}$, then

$$S(f, \mathcal{D}) \ge S(f, \mathcal{D}') \ge s(f, \mathcal{D}') \ge s(f, \mathcal{D}).$$

(iv) Suppose that $\mathcal{D}$ and $\mathcal{D}'$ are dissections with $\mathcal{D}' \supseteq \mathcal{D}$. Show, using (iii), or otherwise, that

$$S(f, \mathcal{D}) \ge S(f, \mathcal{D}') \ge s(f, \mathcal{D}') \ge s(f, \mathcal{D}).$$

The result of Exercise 8.8 (iv) is so easy that it hardly requires proof. None the less, it is so important that we restate it as a lemma.

Lemma 8.9. *If $\mathcal{D}$ and $\mathcal{D}'$ are dissections with $\mathcal{D}' \supseteq \mathcal{D}$, then*

$$S(f, \mathcal{D}) \ge S(f, \mathcal{D}') \ge s(f, \mathcal{D}') \ge s(f, \mathcal{D}).$$

The next lemma is again hardly more than an observation, but it is the key to the proper treatment of the integral.

Lemma 8.10 (Key integration property). *If $f : [a, b] \to \mathbb{R}$ is bounded and $\mathcal{D}_1$ and $\mathcal{D}_2$ are two dissections, then*

$$\bigstar \qquad S(f, \mathcal{D}_1) \ge S(f, \mathcal{D}_1 \cup \mathcal{D}_2) \ge s(f, \mathcal{D}_1 \cup \mathcal{D}_2) \ge s(f, \mathcal{D}_2).$$

The inequalities $\bigstar$ tell us that, whatever dissection you pick and whatever dissection I pick, your lower sum cannot exceed my upper sum. There is no way we can put a quart into a pint pot[2], and the Banach-Tarski phenomenon is avoided.

Since $S(f, \mathcal{D}) \ge -(b-a)K$ for all dissections $\mathcal{D}$, we can define the *upper integral* as $I^*(f) = \inf_{\mathcal{D}} S(f, \mathcal{D})$. We define the *lower integral* similarly as $I_*(f) = \sup_{\mathcal{D}} s(f, \mathcal{D})$. The inequalities $\bigstar$ tell us that these concepts behave well.

Lemma 8.11. *If $f : [a, b] \to \mathbb{R}$ is bounded, then $I^*(f) \ge I_*(f)$.*

[Observe that, *if the integral $\int_a^b f(t)\,dt$ exists,* then the upper integral ought to provide an upper bound and the lower integral a lower bound for that integral.]

If $I^*(f) = I_*(f)$, we say that f is Riemann integrable and we write[3]

---

[2]Or a litre into a half litre bottle. Any reader tempted to interpret such pictures literally is directed to part (iv) of Exercise K.189.

[3]Some authors believe that the dx in $\int_a^b f(x)\,dx$ is as useless as the vestigial tailbone in human beings and write $\int_a^b f$ or $I_a^b f$ instead. Others see the dx as announcing that x is the dummy variable and point to its utility in expressions like $\int_a^b g(x, y)\,dx$ or value the heuristic expression $\int_a^b f(x)\,dx \overset{?}{\sim} \sum f(x)\delta x$.

$$\int_a^b f(x)\,dx = I^*(f).$$

We write $\mathcal{R}[a,b]$ or sometimes just $\mathcal{R}$ for the set of Riemann integrable functions on $[a,b]$.

Exercise 8.12. *If $k \in \mathbb{R}$, show that the constant function given by $f(t) = k$ for all t is Riemann integrable and*

$$\int_a^b k\,dx = k(b-a).$$

The following lemma provides a convenient criterion for Riemann integrability.

Lemma 8.13. *(i) A bounded function $f : [a,b] \to \mathbb{R}$ is Riemann integrable if and only if, given any $\epsilon > 0$, we can find a dissection $\mathcal{D}$ with*

$$S(f,\mathcal{D}) - s(f,\mathcal{D}) < \epsilon.$$

(ii) A bounded function $f : [a,b] \to \mathbb{R}$ is Riemann integrable with integral I if and only if, given any $\epsilon > 0$, we can find a dissection $\mathcal{D}$ with

$$S(f,\mathcal{D}) - s(f,\mathcal{D}) < \epsilon \text{ and } |S(f,\mathcal{D}) - I| \leq \epsilon.$$

Proof. (i) We need to prove necessity and sufficiency. To prove necessity, suppose that f is Riemann integrable with Riemann integral I (so that $I = I^*(f) = I_*(f)$). If $\epsilon > 0$ then, by the definition of $I^*(f)$, we can find a dissection $\mathcal{D}_1$ such that

$$I + \epsilon/2 > S(f,\mathcal{D}_1) \geq I.$$

Similarly, by the definition of $I_*(f)$, we can find a dissection $\mathcal{D}_2$ such that

$$I \geq s(f,\mathcal{D}_2) > I - \epsilon/2.$$

Setting $\mathcal{D} = \mathcal{D}_1 \cup \mathcal{D}_2$ and using Lemmas 8.9 and 8.10, we have

$$I + \epsilon/2 > S(f,\mathcal{D}_1) \geq S(f,\mathcal{D}) \geq s(f,\mathcal{D}) \geq s(f,\mathcal{D}_2) > I - \epsilon/2,$$

so $S(f,\mathcal{D}) - s(f,\mathcal{D}) < \epsilon$, as required.

To prove sufficiency, suppose that, given any $\epsilon > 0$, we can find a dissection $\mathcal{D}$ with

$$S(f,\mathcal{D}) - s(f,\mathcal{D}) < \epsilon.$$

Using the definition of the upper and lower integrals $I^*(f)$ and $I_*(f)$ together with the fact that $I^*(f) \geq I_*(f)$ (a consequence of our key Lemma 8.10), we already know that

$$S(f,\mathcal{D}) \geq I^*(f) \geq I_*(f) \geq s(f,\mathcal{D}),$$

so we may conclude that $\epsilon \geq I^*(f) - I_*(f) \geq 0$. Since ϵ is arbitrary, we have $I^*(f) - I_*(f) = 0$, so $I^*(f) = I_*(f)$, as required.

(ii) Left to the reader. ■

Exercise 8.14. *Prove part (ii) of Lemma 8.13.*

Many students are tempted to use Lemma 8.13 (ii) as the *definition* of the Riemann integral. The reader should reflect that, without the inequality ★, it is not even clear that such a definition gives a unique value for I. (This is only the first of a series of nasty problems that arise if we attempt to develop the theory without first proving ★, so I strongly advise the reader not to take this path.) We give another equivalent definition of the Riemann integral in Exercise K.128.

It is reasonably easy to show that the Riemann integral has the properties which are normally assumed in elementary calculus.

Lemma 8.15. *If $f, g : [a, b] \to \mathbb{R}$ are Riemann integrable, then so is $f + g$, and*

$$\int_a^b f(x) + g(x)\, dx = \int_a^b f(x)\, dx + \int_a^b g(x)\, dx.$$

Proof. Let us write $I(f) = \int_a^b f(x)\, dx$ and $I(g) = \int_a^b g(x)\, dx$. Suppose $\epsilon > 0$ is given. By the definition of the Riemann integral, we can find dissections $\mathcal{D}_1$ and $\mathcal{D}_2$ of $[a, b]$ such that

$$\begin{aligned} I(f) + \epsilon/4 >& S(f, \mathcal{D}_1) \geq I(f) > s(f, \mathcal{D}_1) - \epsilon/4, \\ I(g) + \epsilon/4 >& S(g, \mathcal{D}_2) \geq I(g) > s(g, \mathcal{D}_2) - \epsilon/4. \end{aligned}$$

If we set $\mathcal{D} = \mathcal{D}_1 \cup \mathcal{D}_2$, then our key inequality ★ and the definition of $I^*(f)$ tell us that

$$I(f) + \epsilon/4 > S(f, \mathcal{D}_1) \geq S(f, \mathcal{D}) \geq I(f).$$

Using this and corresponding results, we obtain

$$\begin{aligned} I(f) + \epsilon/4 >& S(f, \mathcal{D}) \geq I(f) > s(f, \mathcal{D}) - \epsilon/4, \\ I(g) + \epsilon/4 >& S(g, \mathcal{D}) \geq I(g) > s(g, \mathcal{D}) - \epsilon/4. \end{aligned}$$

Now

$$\begin{aligned} S(f + g, \mathcal{D}) &= \sum_{j=1}^{n} (x_j - x_{j-1}) \sup_{x \in [x_{j-1}, x_j]} (f(x) + g(x)) \\ &\leq \sum_{j=1}^{n} (x_j - x_{j-1}) \Big(\sup_{x \in [x_{j-1}, x_j]} f(x) + \sup_{x \in [x_{j-1}, x_j]} g(x) \Big) \\ &= S(f, \mathcal{D}) + S(g, \mathcal{D}), \end{aligned}$$

and similarly $s(f+g,\mathcal{D}) \geq s(f,\mathcal{D}) + s(g,\mathcal{D})$. Thus, using the final inequalities of the last paragraph,

$$\begin{aligned} I(f) + I(g) + \epsilon/2 &> S(f,\mathcal{D}) + S(g,\mathcal{D}) \geq S(f+g,\mathcal{D}) \\ &\geq s(f+g,\mathcal{D}) \geq s(f,\mathcal{D}) + s(g,\mathcal{D}) > I(f) + I(g) - \epsilon/2. \end{aligned}$$

Thus $S(f+g,\mathcal{D}) - s(f+g,\mathcal{D}) < \epsilon$ and $|S(f+g,\mathcal{D}) - (I(f)+I(g))| < \epsilon$. ■

Exercise 8.16. *How would you explain (N.B. explain, not prove) to someone who had not done calculus but had a good grasp of geometry why the result*

$$\int_a^b f(x) + g(x)\,dx = \int_a^b f(x)\,dx + \int_a^b g(x)\,dx$$

is true for well behaved functions. (I hope that you will agree with me that, obvious as this result now seems to us, the first mathematicians to grasp this fact had genuine insight.)

Exercise 8.17. *(i) If $f : [a,b] \to \mathbb{R}$ is bounded and $\mathcal{D}$ is a dissection of $[a,b]$, show that $S(-f,\mathcal{D}) = -s(f,\mathcal{D})$.*

(ii) If $f : [a,b] \to \mathbb{R}$ is Riemann integrable, show that $-f$ is Riemann integrable and

$$\int_a^b (-f(x))\,dx = -\int_a^b f(x)\,dx.$$

(iii) If $\lambda \in \mathbb{R}$, $\lambda \geq 0$, $f : [a,b] \to \mathbb{R}$ is bounded and $\mathcal{D}$ is a dissection of $[a,b]$, show that $S(\lambda f,\mathcal{D}) = \lambda S(f,\mathcal{D})$.

(iv) If $\lambda \in \mathbb{R}$, $\lambda \geq 0$ and $f : [a,b] \to \mathbb{R}$ is Riemann integrable, show that λf is Riemann integrable and

$$\int_a^b \lambda f(x)\,dx = \lambda \int_a^b f(x)\,dx.$$

(v) If $\lambda \in \mathbb{R}$ and $f : [a,b] \to \mathbb{R}$ is Riemann integrable, show that λf is Riemann integrable and

$$\int_a^b \lambda f(x)\,dx = \lambda \int_a^b f(x)\,dx.$$

Combining Lemma 8.15 with Exercise 8.17, we get the following result.

Lemma 8.18. *If $\lambda, \mu \in \mathbb{R}$ and $f, g : [a,b] \to \mathbb{R}$ are Riemann integrable, then $\lambda f + \mu g$ is Riemann integrable and*

$$\int_a^b \lambda f(x) + \mu g(x)\,dx = \lambda \int_a^b f(x)\,dx + \mu \int_a^b g(x)\,dx.$$

In the language of linear algebra, $\mathcal{R}[a,b]$ (the set of Riemann integrable functions on $[a,b]$) is a vector space and the integral is a linear functional (i.e. a linear map from $\mathcal{R}[a,b]$ to $\mathbb{R}$).

Exercise 8.19. *(i) If E is a subset of $[a,b]$, we define the indicator function $\mathbb{I}_E : [a,b] \to \mathbb{R}$ by $\mathbb{I}_E(x) = 1$ if $x \in E$, $\mathbb{I}_E(x) = 0$ otherwise. Show directly from the definition that, if $a \leq c \leq d \leq b$, then $\mathbb{I}_{[c,d]}$ is Riemann integrable and*

$$\int_a^b \mathbb{I}_{[c,d]}(x)\,dx = d - c.$$

(ii) If $a \leq c \leq d \leq b$, we say that the intervals (c,d), $(c,d]$, $[c,d)$, $[c,d]$ all have length $d-c$. If $I(j)$ is a subinterval of $[a,b]$ of length $|I(j)|$ and $\lambda_j \in \mathbb{R}$, show that the step function $\sum_{j=1}^n \lambda_j \mathbb{I}_{I(j)}$ is Riemann integrable and

$$\int_a^b \sum_{j=1}^n \lambda_j \mathbb{I}_{I(j)}\,dx = \sum_{j=1}^n \lambda_j |I(j)|.$$

Exercise 8.20. *(i) If $f, g : [a,b] \to \mathbb{R}$ are bounded functions with $f(t) \geq g(t)$ for all $t \in [a,b]$ and $\mathcal{D}$ is a dissection of $[a,b]$, show that $S(f,\mathcal{D}) \geq S(g,\mathcal{D})$.*

(ii) If $f, g : [a,b] \to \mathbb{R}$ are Riemann integrable functions with $f(t) \geq g(t)$ for all $t \in [a,b]$, show that

$$\int_a^b f(x)\,dx \geq \int_a^b g(x)\,dx.$$

(iii) Suppose that $f : [a,b] \to \mathbb{R}$ is a Riemann integrable function, $K \in \mathbb{R}$ and $f(t) \geq K$ for all $t \in [a,b]$. Show that

$$\int_a^b f(x)\,dx \geq K(b-a).$$

State and prove a similar result involving upper bounds.

(iv) Suppose that $f : [a,b] \to \mathbb{R}$ is a Riemann integrable function, $M \in \mathbb{R}$, $M \geq 0$ and $|f(t)| \leq M$ for all $t \in [a,b]$. Show that

$$\left|\int_a^b f(x)\,dx\right| \leq M(b-a).$$

Although part (iv) is weaker than part (iii), it generalises more easily, and we shall use it frequently in the form

$$|\text{integral}| \leq \text{length} \times \text{sup}.$$

Exercise 8.21. *(i) Let M be a positive real number and $f : [a,b] \to \mathbb{R}$ a function with $|f(t)| \leq M$ for all $t \in [a,b]$. Show that*

$$|f(s)^2 - f(t)^2| \leq 2M|f(s) - f(t)|,$$

and deduce that

$$\sup_{x\in[a,b]} f(x)^2 - \inf_{x\in[a,b]} f(x)^2 \leq 2M(\sup_{x\in[a,b]} f(x) - \inf_{x\in[a,b]} f(x)).$$

(ii) Let $f : [a,b] \to \mathbb{R}$ be a bounded function. Show that, if $\mathcal{D}$ is a dissection of $[a,b]$,

$$S(f^2,\mathcal{D}) - s(f^2\mathcal{D}) \leq 2M(S(f,\mathcal{D}) - s(f,\mathcal{D})).$$

Deduce that, if f is Riemann integrable, so is f^2.

(iii) By using the formula $fg = \frac{1}{4}((f+g)^2 - (f-g)^2)$, or otherwise, deduce that that if $f, g : [a,b] \to \mathbb{R}$ are Riemann integrable, so is fg (the product of f and g). (Compare Exercise 1.10.)

Exercise 8.22. *(i) Consider a function $f : [a,b] \to \mathbb{R}$. We define $f_+, f_- : [a,b] \to \mathbb{R}$ by*

$$\begin{aligned} f_+(t) &= f(t), \quad f_-(t) = 0 && \textit{if } f(t) \geq 0, \\ f_+(t) &= 0, \quad\ \ f_-(t) = -f(t) && \textit{if } f(t) \leq 0. \end{aligned}$$

Check that $f(t) = f_+(t) - f_-(t)$ and $|f(t)| = f_+(t) + f_-(t)$.

(ii) If $f : [a,b] \to \mathbb{R}$ is bounded and $\mathcal{D}$ is a dissection of $[a,b]$, show that

$$S(f,\mathcal{D}) - s(f,\mathcal{D}) \geq S(f_+,\mathcal{D}) - s(f_+,\mathcal{D}) \geq 0.$$

(iii) If $f : [a,b] \to \mathbb{R}$ is Riemann integrable, show that f_+ and f_- are Riemann integrable.

(iv) If $f : [a,b] \to \mathbb{R}$ is Riemann integrable, show that $|f|$ is Riemann integrable and

$$\int_a^b |f(x)|\,dx \geq \left| \int_a^b f(x)\,dx \right|.$$

Exercise 8.23. *In each of Exercises 8.17, 8.21 and 8.22 we used a roundabout route to our result. For example, in Exercise 8.17 we first proved that f^2 is Riemann integrable whenever f is and then used this result to prove that fg is Riemann integrable whenever f and g are. It is natural to ask whether we can give a direct proof in each case. The reader should try to do so. (In my opinion, the direct proofs are not much harder, though they do require more care in writing out.)*

Exercise 8.24. *(i) Suppose that $a \leq c \leq b$ and that $f : [a,b] \to \mathbb{R}$ is a bounded function. Consider a dissection $\mathcal{D}_1$ of $[a,c]$ given by*

$$\mathcal{D}_1 = \{x_0, x_1, \ldots, x_m\} \textit{ with } a = x_0 \leq x_1 \leq x_2 \leq \cdots \leq x_m = c,$$

and a dissection $\mathcal{D}_2$ of $[c,b]$ given by $\mathcal{D}_2 = \{x_{m+1}, x_{m+2}, \ldots, x_n\}$, with $c = x_{m+1} \leq x_{m+2} \leq x_{m+3} \leq \cdots \leq x_n = b$. If $\mathcal{D}$ is the dissection of $[a,b]$ given *by* $\mathcal{D} = \{x_0, x_1, \ldots, x_n\}$*, show that* $S(f, \mathcal{D}) = S(f|_{[a,c]}, \mathcal{D}_1) + S(f|_{[c,b]}, \mathcal{D}_2)$. *(Here* $f|_{[a,c]}$ *means the restriction of* f *to* $[a, c]$*.)*

(ii) Show that $f \in \mathcal{R}[a, b]$ *if and only if* $f|_{[a,c]} \in \mathcal{R}[a, c]$ *and* $f|_{[c,b]} \in \mathcal{R}[c, b]$*. Show also that, if* $f \in \mathcal{R}[a, b]$*, then*

$$\int_a^b f(x)\,dx = \int_a^c f|_{[a,c]}(x)\,dx + \int_c^b f|_{[c,b]}(x)\,dx.$$

In a very slightly less precise and very much more usual notation we write

$$\int_a^b f(x)\,dx = \int_a^c f(x)\,dx + \int_c^b f(x)\,dx.$$

There is a standard convention that we shall follow which says that, if $b \geq a$ and f is Riemann integrable on $[a, b]$, we define

$$\int_b^a f(x)\,dx = -\int_a^b f(x)\,dx.$$

Exercise 8.25. *Suppose* $\beta \geq \alpha$ *and* f *is Riemann integrable on* $[\alpha, \beta]$*. Show that if* a*,* b*,* $c \in [\alpha, \beta]$*, then*

$$\int_a^b f(x)\,dx = \int_a^c f(x)\,dx + \int_c^b f(x)\,dx.$$

[Note that a*,* b *and* c *may occur in any of six orders.]*

However, this convention must be used with caution.

Exercise 8.26. *Suppose that* $b \geq a$*,* $\lambda, \mu \in \mathbb{R}$*, and* f *and* g *are Riemann integrable. Which of the following statements are always true and which are not? Give a proof or counterexample. If the statement is not always true, find an appropriate correction and prove it.*

(i) $\displaystyle\int_b^a \lambda f(x) + \mu g(x)\,dx = \lambda \int_b^a f(x)\,dx + \mu \int_b^a g(x)\,dx.$

(ii) If $f(x) \geq g(x)$ *for all* $x \in [a, b]$*, then* $\displaystyle\int_b^a f(x)\,dx \geq \int_b^a g(x)\,dx.$

Riemann was unable to show that all continuous functions were integrable (we have a key concept that Riemann did not, and we shall be able to fill this gap in the next section). He did, however, have the result of the next exercise. (Note that an increasing function need not be continuous. Consider the Heaviside function $H : \mathbb{R} \to \mathbb{R}$ given by $H(x) = 0$ for $x < 0$, $H(x) = 1$ for $x \geq 0$.)

Exercise 8.27. *Suppose* $f : [a, b] \to \mathbb{R}$ *is increasing. Let* N *be a strictly positive integer and consider the dissection*

$$\mathcal{D} = \{x_0, x_1, \ldots, x_N\} \text{ with } x_j = a + j(b - a)/N.$$

Show that

$$S(f, \mathcal{D}) = \sum_{j=1}^{N} f(x_j)(b-a)/N,$$

find $s(f, \mathcal{D})$ *and deduce that*

$$S(f, \mathcal{D}) - s(f, \mathcal{D}) = (f(b) - f(a))(b-a)/N.$$

Conclude that f *is Riemann integrable.*

Using Lemma 8.18 this gives the following result.

Lemma 8.28. *If* $f : [a, b] \to \mathbb{R}$ *can be written as* $f = f_1 - f_2$ *with* $f_1, f_2 : [a, b] \to \mathbb{R}$ *increasing, then* f *is Riemann integrable.*

At first sight, Lemma 8.28 looks rather uninteresting but, in fact, it covers most of the functions we normally meet.

Exercise 8.29. *(i) If*

$$\begin{aligned} f_1(t) = 0,\ \ f_2(t) &= -t^2 \qquad \text{if } t < 0 \\ f_1(t) = t^2,\ f_2(t) &= 0 \qquad\ \ \ \text{if } t \geq 0, \end{aligned}$$

show that f_1 *and* f_2 *are increasing functions with* $t^2 = f_1(t) - f_2(t)$.

(ii) Show that, if $f : [a, b] \to \mathbb{R}$ *has only a finite number of local maxima and minima, then it can be written in the form* $f = f_1 - f_2$ *with* $f_1, f_2 : [a, b] \to \mathbb{R}$ *increasing.*

Functions which are the difference of two increasing functions are discussed in Exercise K.176, Exercises K.180 to K.184 and more generally in the next chapter as 'functions of bounded variation'. We conclude this section with an important example of Dirichlet.

Exercise 8.30. *If* $f : [0, 1] \to \mathbb{R}$ *is given by*

$$\begin{aligned} f(x) &= 1 \qquad \text{when } x \text{ is rational,} \\ f(x) &= 0 \qquad \text{when } x \text{ is irrational,} \end{aligned}$$

show that, whenever $\mathcal{D}$ *is a dissection of* $[0, 1]$, *we have* $S(f, \mathcal{D}) = 1$ *and* $s(f, \mathcal{D}) = 0$. *Conclude that* f *is not Riemann integrable.*

Exercise 8.31. *(i) If* f *is as in Exercise 8.30, show that*

$$\frac{1}{N}\sum_{r=1}^{N} f(r/N) = 1, \text{ and so } \frac{1}{N}\sum_{r=1}^{N} f(r/N) \to 1 \text{ as } N \to \infty.$$

(ii) Let $g : [0, 1] \to \mathbb{R}$ *be given by*

$$\begin{aligned} g(r/2^n) &= 1 && \text{when } 1 \le r \le 2^n - 1,\ n \ge 1, \text{ and } r \text{ and } n \text{ are integers,} \\ g(s/3^n) &= -1 && \text{when } 1 \le s \le 3^n - 1,\ n \ge 1, \text{ and } s \text{ and } n \text{ are integers,} \\ g(x) &= 0 && \text{otherwise.} \end{aligned}$$

Discuss the behaviour of

$$\frac{1}{N}\sum_{r=1}^{N} g(r/N)$$

as $N \to \infty$ *in as much detail as you consider desirable.*

8.3. Integrals of continuous functions

The key to showing that continuous functions are integrable, which we have and Riemann did not, is the notion of uniform continuity and the theorem (Theorem 4.63) which tells us that a continuous function on a closed bounded subset of $\mathbb{R}^n$, and so, in particular, on a closed interval, is uniformly continuous[4].

Theorem 8.32. *Any continuous function* $f : [a, b] \to \mathbb{R}$ *is Riemann integrable.*

Proof. If $b = a$ the result is obvious, so suppose $b > a$. We shall show that f is Riemann integrable by using the standard criterion given in Lemma 8.13. To this end, suppose that $\epsilon > 0$ is given. Since a continuous function on a closed bounded interval is uniformly continuous, we can find a $\delta > 0$ such that

$$|f(x) - f(y)| \le \frac{\epsilon}{b-a} \text{ whenever } x, y \in [a, b] \text{ and } |x - y| < \delta.$$

Choose an integer $N > (b - a)/\delta$ and consider the dissection

$$\mathcal{D} = \{x_0, x_1, \ldots, x_N\} \text{ with } x_j = a + j(b - a)/N.$$

If x, $y \in [x_j, x_{j+1}]$, then $|x - y| < \delta$, and so

$$|f(x) - f(y)| \le \frac{\epsilon}{b-a}.$$

It follows that

$$\sup_{x\in[x_j,x_{j+1}]} f(x) - \inf_{x\in[x_j,x_{j+1}]} f(x) \le \frac{\epsilon}{b-a}$$

[4]This is a natural way to proceed, but Exercise K.133 shows that it is not the only one.

for all $0 \leq j \leq N-1$, and so

$$\begin{aligned} S(f,\mathcal{D}) - s(f,\mathcal{D}) &= \sum_{j=0}^{N-1}(x_{j+1}-x_j)\Big(\sup_{x\in[x_j,x_{j+1}]} f(x) - \inf_{x\in[x_j,x_{j+1}]} f(x)\Big) \\ &\leq \sum_{j=0}^{N-1} \frac{b-a}{N}\frac{\epsilon}{b-a} = \epsilon, \end{aligned}$$

as required. ■

Slight extensions of this result are given in Exercise I.11. In Exercise K.137 we consider a rather different way of looking at integrals of continuous functions.

Although there are many functions which are integrable besides the continuous functions, there are various theorems on integration which demand that the functions involved be continuous or even better behaved. Most of the results of this section have this character.

Lemma 8.33. *If $f : [a,b] \to \mathbb{R}$ is continuous, $f(t) \geq 0$ for all $t \in [a,b]$ and*

$$\int_a^b f(t)\,dt = 0,$$

it follows that $f(t) = 0$ for all $t \in [a,b]$.

Proof. If f is a positive continuous function which is not identically zero, then we can find an $x \in [a,b]$ with $f(x) > 0$. Setting $\epsilon = f(x)/2$, the continuity of f tells us that there exists a $\delta > 0$ such that $|f(x) - f(y)| < \epsilon$ whenever $|x-y| \leq \delta$ and $y \in [a,b]$. We observe that

$$f(y) \geq f(x) - |f(x) - f(y)| > f(x) - \epsilon = f(x)/2$$

whenever $|x-y| \leq \delta$ and $y \in [a,b]$. If we define $h : [a,b] \to \mathbb{R}$ by $h(y) = f(x)/2$, whenever $|x-y| \leq \delta$ and $y \in [a,b]$, and $h(y) = 0$, otherwise, then $f(t) \geq h(t)$ for all $t \in [a,b]$, and so

$$\int_a^b f(t)\,dt \geq \int_a^b h(t)\,dt > 0.$$

■

Exercise 8.34. *(i) Let $a \leq c \leq b$. Give an example of a Riemann integrable function $f : [a,b] \to \mathbb{R}$ such that $f(t) \geq 0$ for all $t \in [a,b]$ and*

$$\int_a^b f(t)\,dt = 0,$$

but $f(c) \neq 0$.

(ii) If $f : [a, b] \to \mathbb{R}$ is Riemann integrable, $f(t) \geq 0$ for all $t \in [a, b]$ and

$$\int_a^b f(t)\, dt = 0,$$

show that $f(t) = 0$ at every point $t \in [a, b]$ where f is continuous.

(iii) We say that $f : [a, b] \to \mathbb{R}$ is right continuous at $t \in [a, b]$ if $f(s) \to f(t)$ as $s \to t$ through values of s with $b \geq s > t$. Suppose f is Riemann integrable and is right continuous at every point $t \in [a, b]$. Show that if $f(t) \geq 0$ for all $t \in [a, b]$ and

$$\int_a^b f(t)\, dt = 0,$$

it follows that $f(t) = 0$ for all $t \in [a, b]$ with at most one exception. Give an example to show that this exception may occur.

The reader should have little difficulty in proving the following useful related results.

Exercise 8.35. *(i) If $f : [a, b] \to \mathbb{R}$ is continuous and*

$$\int_a^b f(t)g(t)\, dt = 0$$

whenever $g : [a, b] \to \mathbb{R}$ is continuous, show that $f(t) = 0$ for all $t \in [a, b]$.

(ii) If $f : [a, b] \to \mathbb{R}$ is continuous and

$$\int_a^b f(t)g(t)\, dt = 0$$

whenever $g : [a, b] \to \mathbb{R}$ is continuous and $g(a) = g(b) = 0$, show that $f(t) = 0$ for all $t \in [a, b]$. (We prove a slightly stronger result in Lemma 8.61.)

We now prove the fundamental theorem of the calculus, which links the processes of integration and differentiation. Since the result is an important one, it is worth listing the properties of the integral that we use in the proof.

Lemma 8.36. *Suppose λ, $\mu \in \mathbb{R}$, $f, g : [\alpha, \beta] \to \mathbb{R}$ are Riemann integrable and a, b, $c \in [\alpha, \beta]$. The following results hold.*

(i) $\displaystyle\int_a^b 1\, dt = b - a.$

(ii) $\displaystyle\int_a^b \lambda f(t) + \mu g(t)\, dt = \lambda \int_a^b f(t)\, dt + \mu \int_a^b g(t)\, dt.$

(iii) $\displaystyle\int_a^b f(t)\, dt + \int_b^c f(t)\, dt = \int_a^c f(t)\, dt.$

(iv) $\displaystyle\left|\int_a^b f(t)\, dt\right| \leq |b - a| \sup_{0 \leq \theta \leq 1} |f(a + \theta(b - a))|.$

The reader should run through these results in her mind and make sure that she can prove them (note that a, b and c can be in any order).

Theorem 8.37. (The fundamental theorem of the calculus.) *Suppose that $f : (a,b) \to \mathbb{R}$ is a continuous function and that $u \in (a,b)$. If we set*

$$F(t) = \int_u^t f(x)\,dx,$$

then F is differentiable on (a,b) and $F'(t) = f(t)$ for all $t \in (a,b)$.

Proof. Observe that, if $t+h \in (a,b)$ and $h \neq 0$, then

$$\begin{aligned}\left|\frac{F(t+h)-F(t)}{h} - f(t)\right| &= \left|\frac{1}{h}\left(\int_u^{t+h} f(x)\,dx - \int_u^t f(x)\,dx - hf(t)\right)\right| \\ &= \left|\frac{1}{h}\left(\int_t^{t+h} f(x)\,dx - \int_t^{t+h} f(t)\,dx\right)\right| \\ &= \frac{1}{|h|}\left|\int_t^{t+h} (f(x)-f(t))\,dx\right| \\ &\leq \sup_{0\leq\theta\leq 1} |f(t+\theta h) - f(t)| \to 0\end{aligned}$$

as $h \to 0$, since f is continuous at t. (Notice that $f(t)$ remains constant as x varies.) ∎

Exercise 8.38. *(i) Using the idea of the integral as the area under a curve, draw diagrams illustrating the proof of Theorem 8.37.*

(ii) Point out, explicitly, each use of Lemma 8.36 in our proof of Theorem 8.37.

(iii) Let H be the Heaviside function $H : \mathbb{R} \to \mathbb{R}$ given by $H(x) = 0$ for $x < 0$, $H(x) = 1$ for $x \geq 0$. Calculate $F(t) = \int_0^t H(x)\,dx$ and show that F is not differentiable at 0. Where does our proof of Theorem 8.37 break down?

(iv) Let $f(0) = 1$, $f(t) = 0$ otherwise. Calculate $F(t) = \int_0^t f(x)\,dx$ and show that F is differentiable at 0 but $F'(0) \neq f(0)$. Where does our proof of Theorem 8.37 break down?

Exercise 8.39. *Suppose that $f : (a,b) \to \mathbb{R}$ is a function such that f is Riemann integrable on every interval $[c,d] \subseteq (a,b)$. Let $u \in (a,b)$ If we set*

$$F(t) = \int_u^t f(x)\,dx,$$

show that F is continuous on (a,b) and that, if f is continuous at some point $t \in (a,b)$, then F is differentiable at t and $F'(t) = f(t)$.

Sometimes we think of the fundamental theorem in a slightly different way.

Theorem 8.40. *Suppose that $f : (a,b) \to \mathbb{R}$ is continuous, that $u \in (a,b)$ and $c \in \mathbb{R}$. Then there is a unique solution to the differential equation $g'(t) = f(t)$ $[t \in (a,b)]$ such that $g(u) = c$.*

Exercise 8.41. *Prove Theorem 8.40. Make clear how you use Theorem 8.37 and the mean value theorem. Reread Section 1.1.*

We call the solutions of $g'(t) = f(t)$ *indefinite integrals* (or, simply, *integrals*) of f.

Yet another version of the fundamental theorem is given by the next theorem.

Theorem 8.42. *Suppose that $g : (\alpha,\beta) \to \mathbb{R}$ has continuous derivative and $[a,b] \subseteq (\alpha,\beta)$. Then*

$$\int_a^b g'(t)\,dt = g(b) - g(a).$$

Proof. Define $U : (\alpha,\beta) \to \mathbb{R}$ by

$$U(t) = \int_a^t g'(x)\,dx - g(t) + g(a).$$

By the fundamental theorem of the calculus and earlier results on differentiation, U is everywhere differentiable with

$$U'(t) = g'(t) - g'(t) = 0,$$

so, by the mean value theorem, U is constant. But $U(a) = 0$, so $U(t) = 0$ for all t and, in particular, $U(b) = 0$, as required. ■

[**Remark:** In one dimension, Theorems 8.37, 8.40 and 8.42 are so closely linked that mathematicians tend to refer to them all as 'The fundamental theorem of the calculus'. However they generalise in different ways.

(1) Theorem 8.37 shows that, under suitable circumstances, we can recover a function from its 'local average' (see Exercise K.145).

(2) Theorem 8.40 says that we can solve a certain kind of differential equation. We shall obtain substantial generalisations of this result in Section 12.2.

(3) Theorem 8.42 links the value of the derivative f' on the whole of $[a,b]$ with the value of f on the boundary (that is to say, the set $\{a,b\}$). If you have done a mathematical methods course you will already have seen a similar idea expressed by the divergence theorem

$$\int_V \nabla\cdot\mathbf{u}\,dV = \int_{\partial V} \mathbf{u}\cdot d\mathbf{S}.$$

This result and similar ones like Stokes' theorem turn out to be special cases of a master theorem[5] which links the behaviour of the derivative of a certain mathematical object over the whole of some body with the behaviour of the object on the boundary of that body.]

Theorems 8.37 and 8.42 show that (under appropriate circumstances) integration and differentiation are inverse operations and the the theories of differentiation and integration are subsumed in the greater theory of the calculus. Under appropriate circumstances, if the graph of F has tangent with slope $f(x)$ at x, then

$$\begin{aligned}&\text{area under the graph of slope of tangent of } F\\&\quad = \text{area under the graph of } f\\&\quad = \int_a^b f(x)\,dx = \int_a^b F'(x)\,dx = F(b) - F(a).\end{aligned}$$

Exercise 8.43. *Most books give a slightly stronger version of Theorem 8.42 in the following form.*

If $f : [a,b] \to \mathbb{R}$ has continuous derivative, then

$$\int_a^b f'(t)\,dt = f(b) - f(a).$$

Explain what this means (you will need to talk about 'left' and 'right' derivatives), and prove it.

Recalling the chain rule (Lemma 6.19) which tells us that $(\Phi \circ g)'(t) = g'(t)\Phi'(g(t))$, the same form of proof gives us a very important theorem.

Theorem 8.44. (Change of variables for integrals.) *Suppose that $f : (\alpha, \beta) \to \mathbb{R}$ is continuous and $g : (\gamma, \delta) \to \mathbb{R}$ is differentiable with continuous derivative. Suppose further that $g\big((\gamma,\delta)\big) \subseteq (\alpha,\beta)$. Then, if $c,\ d \in (\gamma,\delta)$, we have*

$$\int_{g(c)}^{g(d)} f(s)\,ds = \int_c^d f(g(x))g'(x)\,dx.$$

Exercise 8.45. *(i) Prove Theorem 8.44 by considering*

$$U(t) = \int_{g(c)}^{g(t)} f(s)\,ds - \int_c^t f(g(x))g'(x)\,dx.$$

(ii) Derive Theorem 8.42 from Theorem 8.44 by choosing f appropriately.

(iii) Strengthen Theorem 8.44 along the lines of Exercise 8.43.

[5] Arnol'd calls it the Newton-Leibniz-Gauss-Green-Ostrogradskii-Stokes-Poincaré theorem, but most mathematicians call it the generalised Stokes' theorem or just Stokes' theorem.

(iv) (An alternative proof.) If f is as in Theorem 8.44, explain why we can find an $F : (\alpha, \beta) \to \mathbb{R}$ with $F' = f$. Obtain Theorem 8.44 by applying the chain rule to $F \circ g$.

Because the proof of Theorem 8.44 is so simple and because the main use of the result in elementary calculus is to evaluate integrals, there is a tendency to underestimate the importance of this result. However, it is important for later developments that the reader has an intuitive grasp of this result.

Exercise 8.46. *(i) Suppose that $f : \mathbb{R} \to \mathbb{R}$ is the constant function $f(t) = K$ and that $g : \mathbb{R} \to \mathbb{R}$ is the linear function $g(t) = \lambda t + \mu$. Show by direct calculation that*

$$\int_{g(c)}^{g(d)} f(s)\,ds = \int_c^d f(g(x))g'(x)\,dx,$$

and describe the geometric content of this result in words.

(ii) Suppose now that $f : \mathbb{R} \to \mathbb{R}$ and $g : \mathbb{R} \to \mathbb{R}$ are well behaved functions. By splitting $[c, d]$ into small intervals on which f is 'almost constant' and g is 'almost linear', give a heuristic argument for the truth of Theorem 8.44. To see how this heuristic argument can be converted into a rigorous one, consult Exercise K.133.

Exercise 8.47. *There is one peculiarity in our statement of Theorem 8.44 which is worth noting. We do not demand that g be bijective. Suppose that $f : \mathbb{R} \to \mathbb{R}$ is continuous and $g(t) = \sin t$. Show that, by choosing different intervals (c, d), we obtain*

$$\begin{aligned}\int_0^{\sin\alpha} f(s)\,ds &= \int_0^{\alpha} f(\sin x)\cos x\,dx \\ &= \int_0^{\alpha+2\pi} f(\sin x)\cos x\,dx = \int_0^{\pi-\alpha} f(\sin x)\cos x\,dx.\end{aligned}$$

Explain what is going on.

The extra flexibility given by allowing g not to be bijective is one we are usually happy to sacrifice in the interests of generalising Theorem 8.44.

Exercise 8.48. *The following exercise is traditional.*

(i) Show that integration by substitution, using $x = 1/t$, gives

$$\int_a^b \frac{dx}{1+x^2} = \int_{1/b}^{1/a} \frac{dt}{1+t^2},$$

when $b > a > 0$.

(ii) If we set $a = -1$, $b = 1$ in the formula of (i), we obtain

$$\int_{-1}^{1} \frac{dx}{1+x^2} \stackrel{?}{=} -\int_{-1}^{1} \frac{dt}{1+t^2}.$$

Explain this apparent failure of the method of integration by substitution.

(iii) Write the result of (i) in terms of $\tan^{-1}$ *and prove it using standard trigonometric identities.*

In Sections 5.4 and 5.6 we gave a treatment of the exponential and logarithmic functions based on differentiation. The reader may wish to look at Exercise K.141, in which we use integration instead.

Another result which can be proved in much the same manner as Theorem 8.42 and Theorem 8.44 is the lemma which justifies integration by parts. (Recall the notation $[h(x)]_a^b = h(b) - h(a)$.)

Lemma 8.49. *Suppose that $f : (\alpha, \beta) \to \mathbb{R}$ has continuous derivative and $g : (\alpha, \beta) \to \mathbb{R}$ is continuous. Let $G : (\alpha, \beta) \to \mathbb{R}$ be an indefinite integral of g. Then, if $[a, b] \subseteq (\alpha, \beta)$, we have*

$$\int_a^b f(x)g(x)\, dx = [f(x)G(x)]_a^b - \int_a^b f'(x)G(x)\, dx.$$

Exercise 8.50. *(i) Obtain Lemma 8.49 by differentiating an appropriate U in the style of the proofs of Theorem 8.42 and Theorem 8.44. Quote carefully the results that you use.*

(ii) Obtain Lemma 8.49 by integrating both sides of the equality $(uv)' = u'v + uv'$ and choosing appropriate u and v. Quote carefully the results that you use.

(iii) Strengthen Lemma 8.49 along the lines of Exercise 8.43.

Integration by parts gives a global Taylor theorem with a form that is easily remembered and proved for examination.

Theorem 8.51. (A global Taylor's theorem with integral remainder.) *If $f : (u, v) \to \mathbb{R}$ is n times continuously differentiable and $0 \in (u, v)$, then*

$$f(t) = \sum_{j=0}^{n-1} \frac{f^{(j)}(0)}{j!} t^j + R_n(f, t),$$

where

$$R_n(f, t) = \frac{1}{(n-1)!} \int_0^t (t - x)^{n-1} f^{(n)}(x)\, dx.$$

Exercise 8.52. *By integration by parts, show that*

$$R_n(f,t) = \frac{f^{(n-1)}(0)}{(n-1)!}t^{n-1} + R_{n-1}(f,t).$$

Use repeated integration by parts to obtain Theorem 8.51.

Exercise 8.53. *Reread Example 7.5. If F is as in that example, identify $R_{n-1}(F,t)$.*

Exercise 8.54. *If $f : (-a,a) \to \mathbb{R}$ is n times continuously differentiable with $|f^{(n)}(t)| \leq M$ for all $t \in (-a,a)$, show that*

$$\left| f(t) - \sum_{j=0}^{n-1} \frac{f^{(j)}(0)}{j!}t^j \right| \leq \frac{M|t|^n}{n!}.$$

Explain why this result is slightly weaker than that of Exercise 7.1 (v).

There are several variants of Theorem 8.51 with different expressions for $R_n(f,t)$ (see, for example, Exercise K.50 (vi)). However, although the theory of the Taylor expansion is very important (see, for example, Exercise K.140 and Exercise K.295), these global theorems are not much used in relation to *specific* functions outside the examination hall. We discuss two of the reasons why at the end of Section 11.5. In Exercises 11.82 and 11.84 I suggest that it is usually easier to obtain Taylor series by power series solutions rather than by using theorems like Theorem 8.51. In Exercise 11.85 I suggest that power series are often not very suitable for numerical computation.

8.4. First steps in the calculus of variations ♡

The most famous early problem in the calculus of variations is that of the brachistochrone. It asks for the equation $y = f(x)$ of the wire down which a frictionless particle with initial velocity v will slide from one point (a, α) to another (b, β) (so $f(a) = \alpha$, $f(b) = \beta$, $a \neq b$ and $\alpha > \beta$) in the shortest time. It turns out that that time taken by the particle is

$$J(f) = \frac{1}{(2g)^{1/2}} \int_a^b \left(\frac{1 + f'(x)^2}{\kappa - f(x)} \right)^{1/2} dx,$$

where $\kappa = v^2/(2g) + \alpha$ and g is the acceleration due to gravity.

Exercise 8.55. *If you know sufficient mechanics, verify this. (Your argument will presumably involve arc length, which has not yet been mentioned in this book.)*

This is a problem of minimising which is very different from those dealt with in elementary calculus. Those problems ask us to choose a point x_0 from a one-dimensional space which minimises some function $g(x)$. In Section 7.3

we considered problems in which we sought to choose a point $\mathbf{x}_0$ from an n-dimensional space which minimises some function $g(\mathbf{x})$. Here we seek to choose a function f_0 from an infinite-dimensional space to minimise a function $J(f)$ of functions f.

Exercise 8.56. *We used the words 'infinite-dimensional' somewhat loosely in the previous sentence. However, we can make precise statements along the same lines.*

(i) Show that the collection $\mathcal{P}$ of polynomials P with $P(0) = P(1) = 0$ forms a vector space over $\mathbb{R}$ with the obvious operations. Show that $\mathcal{P}$ is infinite-dimensional (in other words, has no finite spanning set).

(ii) Show that the collection $\mathcal{E}$ of infinitely differentiable functions $f : [0,1] \to \mathbb{R}$ with $f(0) = f(1)$ forms a vector space over $\mathbb{R}$ with the obvious operations. Show that $\mathcal{E}$ is infinite-dimensional.

John Bernoulli published the brachistochrone problem as a challenge in 1696. Newton, Leibniz, L'Hôpital, John Bernoulli and James Bernoulli all found solutions within a year[6]. However, it is one thing to solve a particular problem and quite another to find a method of attack for the general class of problems to which it belongs. Such a method was developed by Euler and Lagrange. We shall see that it does not resolve all difficulties, but it represents a marvelous leap of imagination.

We begin by proving that, under certain circumstances, we can interchange the order of integration and differentiation. (We will extend the result in Theorem 11.62.)

Theorem 8.57. (Differentiation under the integral.) *Let $(a',b') \times (c',d') \supseteq [a,b] \times [c,d]$. Suppose that $g : (a',b') \times (c',d') \to \mathbb{R}$ is continuous and that the partial derivative $g_{,2}$ exists and is continuous. Then, writing $G(y) = \int_a^b g(x,y)\,dx$, we have G differentiable on (c,d) with*

$$G'(y) = \int_a^b g_{,2}(x,y)\,dx.$$

This result is more frequently written as

$$\frac{d}{dy}\int_a^b g(x,y)\,dx = \int_a^b \frac{\partial g}{\partial y}(x,y)\,dx,$$

and interpreted as 'the d clambers through the integral and curls up'. If we use the D notation we get

$$G'(y) = \int_a^b D_2 g(x,y)\,dx.$$

[6]They were giants in those days. Newton had retired from mathematics and submitted his solution anonymously. 'But', John Bernoulli said, 'one recognises the lion by his paw print.'

It may, in the end, be more helpful to note that $\int_a^b g(x,y)\,dx$ is a function of the single variable y, but $g(x,y)$ is a function of the two variables x and y.

Proof. We use a proof technique which is often useful in this kind of situation (we have already used a simple version in Theorem 8.37, when we proved the fundamental theorem of the calculus).

We first put everything under one integral sign. Suppose $y, y+h \in (c,d)$ and $h \neq 0$. Then

$$\begin{aligned}\left|\frac{G(y+h)-G(y)}{h} - \int_a^b g_{,2}(x,y)\,dx\right| &= \frac{1}{|h|}\left|G(y+h)-G(y) - \int_a^b hg_{,2}(x,y)\,dx\right| \\ &= \frac{1}{|h|}\left|\int_a^b \big(g(x,y+h)-g(x,y)-hg_{,2}(x,y)\big)\,dx\right|.\end{aligned}$$

In order to estimate the last integral we use the simple result (Exercise 8.20 (iv))

$$|\text{integral}| \leq \text{length} \times \sup,$$

which gives us

$$\begin{aligned}\frac{1}{|h|}&\left|\int_a^b \big(g(x,y+h) - g(x,y) - hg_{,2}(x,y)\big)\,dx\right| \\ &\leq \frac{b-a}{|h|}\sup_{x\in[a,b]}|g(x,y+h)-g(x,y)-hg_{,2}(x,y)|.\end{aligned}$$

We expect $|g(x,y+h)-g(x,y)-hg_{,2}(x,y)|$ to be small when h is small, because the definition of the partial derivative tells us that $g(x,y+h)-g(x,y) \approx hg_{,2}(x,y)$. In such circumstances, the mean value theorem is frequently useful. In this case, setting $f(t) = g(x,y+t)-g(x,y)$, the mean value theorem tells us that

$$|f(h)| = |f(h)-f(0)| \leq |h| \sup_{0\leq\theta\leq 1}|f'(\theta h)|,$$

and so

$$|g(x,y+h)-g(x,y)-hg_{,2}(x,y)| \leq |h|\sup_{0\leq\theta\leq 1}|g_{,2}(x,y+\theta h)-g_{,2}(x,y)|.$$

There is one further point to notice. Since we are taking a supremum over all $x \in [a,b]$, we shall need to know, not merely that we can make $|g_{,2}(x,y+\theta h)-g_{,2}(x,y)|$ small at a particular x by taking h sufficiently small, but that we can make $|g_{,2}(x,y+\theta h)-g_{,2}(x,y)|$ uniformly small for all x. However, we know that $g_{,2}$ is continuous on $[a,b]\times[c,d]$ and that a function which is continuous on a closed bounded set is uniformly continuous, and this will enable us to complete the proof.

Let $\epsilon > 0$. By Theorem 4.63, $g_{,2}$ is uniformly continuous on $[a, b] \times [c, d]$, and so we can find a $\delta(\epsilon) > 0$ such that

$$|g_{,2}(x, y) - g_{,2}(u, v)| \leq \epsilon/(b - a)$$

whenever $(x - u)^2 + (y - v)^2 < \delta(\epsilon)$ and (x, y), $(u, v) \in [a, b] \times [c, d]$. It follows that, if $y, y + h \in (c, d)$ and $|h| < \delta(\epsilon)$, then

$$\sup_{0 \leq \theta \leq 1} |g_{,2}(x, y + \theta h) - g_{,2}(x, y)| \leq \epsilon/(b - a)$$

for all $x \in [a, b]$. Putting all our results together, we have shown that

$$\left| \frac{G(y + h) - G(y)}{h} - \int_a^b g_{,2}(x, y)\, dx \right| < \epsilon$$

whenever $y, y + h \in (c, d)$ and $0 < |h| < \delta(\epsilon)$, and the result follows. ∎

Exercise 8.58. *Because I have tried to show where the proof comes from, the proof above is not written in a very economical way. Rewrite it more economically.*

A favourite examiner's variation on the theme of Theorem 8.57 is given in Exercise K.147.

Exercise 8.59. *In what follows we will use a slightly different version of Theorem 8.57.*

Suppose $g : [a, b] \times [c, d]$ is continuous and that the partial derivative $g_{,2}$ exists and is continuous. Then, writing $G(y) = \int_a^b g(x, y)\, dx$, we have G differentiable on $[c, d]$ with

$$G'(y) = \int_a^b g_{,2}(x, y)\, dx.$$

Explain what this means in terms of left and right derivatives, and prove it.

The method of Euler and Lagrange applies to the following class of problems. Suppose that $F : \mathbb{R}^3 \to \mathbb{R}$ has continuous second partial derivatives. We consider the set $\mathcal{A}$ of functions $f : [a, b] \to \mathbb{R}$ which are differentiable with continuous derivative and are such that $f(a) = \alpha$ and $f(b) = \beta$. We write

$$J(f) = \int_a^b F(t, f(t), f'(t))\, dt$$

and seek to minimise J, that is, to find an $f_0 \in \mathcal{A}$ such that

$$J(f_0) \leq J(f)$$

whenever $f \in \mathcal{A}$.

In Section 7.3, when we asked if a particular point $\mathbf{x}_0$ from an n-dimensional space minimised $g : \mathbb{R}^n \to \mathbb{R}$, we examined the behaviour of g close to $\mathbf{x}_0$. In other words, we looked at $g(\mathbf{x}_0 + \eta\mathbf{u})$ when $\mathbf{u}$ was an arbitrary vector and η was small. The idea of Euler and Lagrange is to look at

$$G_h(\eta) = J(f_0 + \eta h),$$

where $h : [a,b] \to \mathbb{R}$ is differentiable with continuous derivative and is such that $h(a) = 0$ and $h(b) = 0$ (we shall call the set of such functions $\mathcal{E}$). We observe that G_h is a function from $\mathbb{R}$ and that G_h has a minimum at 0 if J is minimised by f_0. This observation, combined with some very clever, but elementary, calculus gives the celebrated Euler–Lagrange equation.

Theorem 8.60. *Suppose that $F : \mathbb{R}^3 \to \mathbb{R}$ has continuous second partial derivatives. Consider the set $\mathcal{A}$ of functions $f : [a,b] \to \mathbb{R}$ which are differentiable with continuous derivative and are such that $f(a) = \alpha$ and $f(b) = \beta$. We write*

$$J(f) = \int_a^b F(t, f(t), f'(t))\,dt.$$

If $f \in \mathcal{A}$ is such that

$$J(f) \leq J(g)$$

whenever $g \in \mathcal{A}$ then

$$F_{,2}(t, f(t), f'(t)) = \frac{d}{dt}F_{,3}(t, f(t), f'(t)).$$

Proof. We use the notation of the paragraph preceding the statement of the theorem. If $h \in \mathcal{E}$ (that is to say, $h : [a,b] \to \mathbb{R}$ is differentiable with continuous derivative and is such that $h(a) = 0$ and $h(b) = 0$), then the chain rule tells us that the function $g_h : \mathbb{R}^2 \to \mathbb{R}$ given by

$$g_h(\eta, t) = F(t, f(t) + \eta h(t), f'(t) + \eta h'(t))$$

has continuous partial derivative

$$\begin{aligned} g_{h,1}(\eta, t) = h(t)F_{,2}(t, f(t) + \eta h(t), f'(t) + \eta h'(t)) \\ + h'(t)F_{,3}(t, f(t) + \eta h(t), f'(t) + \eta h'(t)). \end{aligned}$$

Thus by Theorem 8.57, we may differentiate under the integral to show that G_h is differentiable everywhere with

$$\begin{aligned} G_h'(\eta) = \int_a^b h(t)F_{,2}(t, f(t) + \eta h(t), f'(t) + \eta h'(t)) \\ + h'(t)F_{,3}(t, f(t) + \eta h(t), f'(t) + \eta h'(t))\,dt. \end{aligned}$$

If f minimises J, then 0 minimises G_h, and so $G_h'(0) = 0$. We deduce that

$$\begin{aligned}0 &= \int_a^b h(t)F_{,2}(t,f(t),f'(t)) + h'(t)F_{,3}(t,f(t),f'(t))\,dt\\ &= \int_a^b h(t)F_{,2}(t,f(t),f'(t))\,dt + \int_a^b h'(t)F_{,3}(t,f(t),f'(t))\,dt.\end{aligned}$$

Using integration by parts and the fact that $h(a) = h(b) = 0$, we obtain

$$\begin{aligned}&\int_a^b h'(t)F_{,3}(t,f(t),f'(t))\,dt\\ &\qquad= \big[h(t)F_{,3}(t,f(t),f'(t))\big]_a^b - \int_a^b h(t)\frac{d}{dt}F_{,3}(t,f(t),f'(t))\,dt\\ &\qquad= -\int_a^b h(t)\frac{d}{dt}F_{,3}(t,f(t),f'(t))\,dt.\end{aligned}$$

Combining the results of the last two sentences, we see that

$$0 = \int_a^b h(t)\left(F_{,2}(t,f(t),f'(t)) - \frac{d}{dt}F_{,3}(t,f(t),f'(t))\right)dt.$$

Since this result must hold for all $h \in \mathcal{A}$, we see that

$$F_{,2}(t,f(t),f'(t)) - \frac{d}{dt}F_{,3}(t,f(t),f'(t)) = 0$$

for all $t \in [a,b]$ (for details see Lemma 8.61 below), and this is the result we set out to prove. ∎

In order to tie up loose ends, we need the following lemma.

Lemma 8.61. *Suppose $f : [a,b] \to \mathbb{R}$ is continuous and*

$$\int_a^b f(t)h(t)\,dt = 0$$

whenever $h : [a,b] \to \mathbb{R}$ is an infinitely differentiable function with $h(a) = h(b) = 0$. Then $f(t) = 0$ for all $t \in [a,b]$.

Proof. By continuity, we need only prove that $f(t) = 0$ for all $t \in (a,b)$. Suppose that, in fact, $f(x) \neq 0$ for some $x \in (a,b)$. Without loss of generality we may suppose that $f(x) > 0$ (otherwise, consider $-f$). Since (a,b) is open and f is continuous, we can find a $\delta > 0$ such that $[x-\delta, x+\delta] \subseteq (a,b)$ and $|f(t) - f(x)| < f(x)/2$ for $t \in [x-\delta, x+\delta]$. This last condition tells us that $f(t) > f(x)/2$ for $t \in [x-\delta, x+\delta]$.

Recall that in Example 7.6 we constructed an infinitely differentiable function $E : \mathbb{R} \to \mathbb{R}$ with $E(t) = 0$ for $t \leq 0$ and $E(t) > 0$ for $t > 0$. Setting $h(t) = E(t-x+\delta)E(-t+x+\delta)$ when $t \in [a,b]$, we see that h is an infinitely differentiable function with $h(t) > 0$ for $t \in (x-\delta, x+\delta)$ and $h(t) = 0$

otherwise (so that, in particular $h(a) = h(b) = 0$). By standard results on the integral,

$$\int_a^b f(t)h(t)\,dt = \int_{x-\delta}^{x+\delta} f(t)h(t)\,dt \geq \int_{x-\delta}^{x+\delta} (f(x)/2)h(t)\,dt$$
$$= \frac{f(x)}{2}\int_{x-\delta}^{x+\delta} h(t)\,dt > 0,$$

so we are done. ■

Exercise 8.62. *State explicitly the 'standard results on the integral' used in the last sentence of the previous proof, and show how they are applied.*

Theorem 8.60 is often stated in the following form. If the function $y : [a,b] \to \mathbb{R}$ minimises J, then

$$\frac{\partial F}{\partial y} = \frac{d}{dx}\frac{\partial F}{\partial y'}.$$

This is concise, but can be confusing to the novice[7].

The Euler–Lagrange equation can only be solved explicitly in a small number of special cases. The next exercise (which should be treated as an exercise in calculus rather than analysis) shows how, with the exercise of some ingenuity, we can solve the brachistochrone problem with which we started. Recall that this asked us to minimise

$$J(f) = \frac{1}{(2g)^{1/2}}\int_a^b \left(\frac{1+f'(x)^2}{\kappa - f(x)}\right)^{1/2} dx.$$

Exercise 8.63. *We use the notation and assumptions of Theorem 8.60.*

(i) Suppose that $F(u,v,w) = G(v,w)$ (often stated as 't does not appear explicitly in $F{=}F(t,y,y')$'). Show that the Euler–Lagrange equation becomes

$$G_{,1}(f(t),f'(t)) = \frac{d}{dt}G_{,2}(f(t),f'(t))$$

and may be rewritten

$$\frac{d}{dt}\big(G(f(t),f'(t)) - f'(t)G_{,2}(f(t),f'(t))\big) = 0.$$

Deduce that

$$G(f(t),f'(t)) - f'(t)G_{,2}(f(t),f'(t)) = c,$$

where c is a constant. (This last result is often stated as $F - y'\dfrac{\partial F}{\partial y'} = c$.)

(ii) (This is not used in the rest of the question.) If $F(u,v,w) = G(u,w)$, show that $G_{,2}(t,f'(t)) = c$ where c is a constant.

[7]It certainly confused me when I met it for the first time.

(iii) By applying (i), show that solutions of the Euler–Lagrange equation associated with the brachistochrone are solutions of

$$\frac{1}{((\kappa - f(x))(1 + f'(x)^2))^{1/2}} = c,$$

where c is a constant. Show that this equation can be rewritten as

$$f'(x) = \left(\frac{B + f(x)}{A - f(x)}\right)^{1/2}.$$

(iv) We are now faced with finding the curve

$$\frac{dy}{dx} = \left(\frac{B + y}{A - y}\right)^{1/2}.$$

If we are sufficiently ingenious (or we know the answer), we may be led to try and express this curve in parametric form by setting

$$y = \frac{A - B}{2} - \frac{A + B}{2}\cos\theta.$$

Show that

$$\frac{dx}{d\theta} = \frac{A + B}{2}(1 + \cos\theta),$$

and conclude that our curve is (in parametric form)

$$x = a + k(\theta - \sin\theta), \ y = b - k\cos\theta$$

for appropriate constants a, b and k. Thus any curve which minimises the time of descent must be a cycloid.

It is important to observe that we have shown that any minimising function satisfies the Euler–Lagrange equation, and not that any function satisfying the Euler–Lagrange equation is a minimising function. Exactly the same argument (or replacing J by $-J$) shows that any maximising function satisfies the Euler–Lagrange equation. Further, if we reflect on the simpler problem discussed in Section 7.3, we see that the Euler–Lagrange equation will be satisfied by functions f such that

$$G_h(\eta) = J(f + \eta h)$$

has a minimum at $\eta = 0$ for some $h \in \mathcal{E}$ and a maximum at $\eta = 0$ for others.

Exercise 8.64. *With the notation of this section, show that, if f satisfies the Euler–Lagrange equations, then $G_h'(0) = 0$ for all $h \in \mathcal{E}$.*

To get round this problem, examiners ask you to 'find the values of f which make J stationary', where the phrase is equivalent to 'find the values of f which satisfy the Euler–Lagrange equations'. In real life, we use physical intuition or extra knowledge about the nature of the problem to

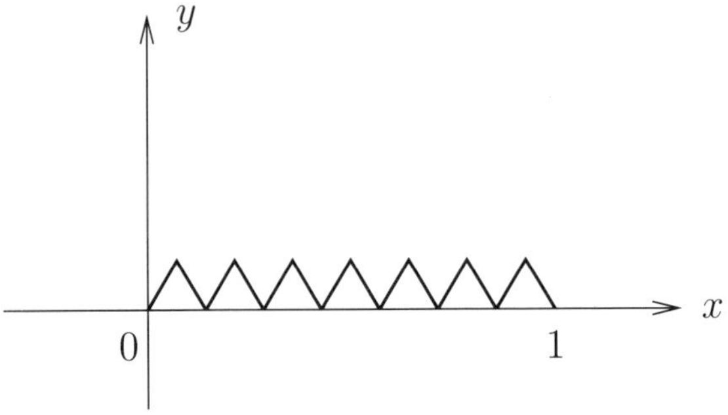

Figure 8.1. A problem for the calculus of variations

find which solutions of the Euler–Lagrange equations represent maxima and which minima.

Mathematicians spent over a century seeking to find an extension to the Euler–Lagrange method which would enable them to distinguish true maxima and minima. However, they were guided by analogy with the one-dimensional (if $f'(0) = 0$ and $f''(0) > 0$ then 0 is minimum) and finite-dimensional cases, and it turns out that the analogy is seriously defective. In the end, Weierstrass produced examples which made it plain what was going on. We discuss a version of one of them.

Consider the problem of minimising

$$I(f) = \int_0^1 (1 - (f'(x))^4)^2 + f(x)^2 \, dx$$

where $f : [0, 1] \to \mathbb{R}$ is once continuously differentiable and $f(0) = f(1) = 0$.

Exercise 8.65. *We look at*

$$G_h(\eta) = I(\eta h).$$

Show that $G_h(\eta) = 1 + A_h\eta^2 + B_h\eta^4 + C_h\eta^8$, *where* A_h, B_h, C_h *depend on* h. *Show that, if* h *is not identically zero, then* $A_h > 0$. *Deduce that* G_h *has a strict minimum at* 0 *for all non-zero* $h \in \mathcal{E}$.

We are tempted to claim that '$I(f)$ has a local minimum at $f = 0$'.

Now look at the function g_n (with$n$ a strictly positive integer) illustrated in Figure 8.1 and defined by

$$g_n(x) = x - \frac{2r}{2n} \qquad \text{for } \left| x - \frac{2r}{2n} \right| \leq \frac{1}{4n},$$

$$g_n(x) = \frac{2r+1}{2n} - x \qquad \text{for } \left| x - \frac{2r+1}{2n} \right| \leq \frac{1}{4n},$$

whenever r is an integer and $x \in [0, 1]$. Ignoring the finite number of points

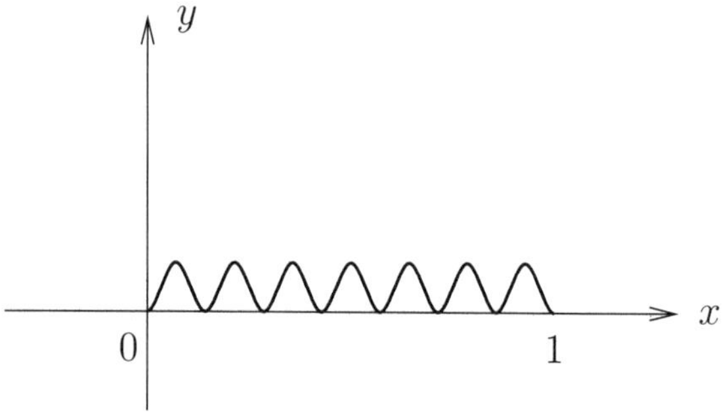

Figure 8.2. The same problem, smoothed

where g_n is not differentiable, we see that $g_n'(x) = \pm 1$ at all other points, and so

$$I(g_n) = \int_0^1 g_n(x)^2\, dx \to 0 \text{ as } n \to \infty.$$

It is clear that we can find a similar sequence of functions f_n which is continuously differentiable by 'rounding the sharp bits' as in Figure 8.2.

The reader who wishes to dot the i's and cross the t's can do the next exercise.

Exercise 8.66. *(i) Let $1/2 > \epsilon > 0$ and let $k : [0,1] \to \mathbb{R}$ be the function such that*

$$\begin{aligned} k(x) &= \epsilon^{-1}x && \text{for } 0 \le x \le \epsilon, \\ k(x) &= 1 && \text{for } \epsilon \le x \le 1-\epsilon, \\ k(x) &= \epsilon^{-1}(1-x) && \text{for } 1-\epsilon \le x \le 1. \end{aligned}$$

Sketch k.

(ii) Let $k_n(x) = (-1)^{[2nx]}k(2nx - 2[nx])$ for $x \in [0,1]$. (Here $[2nx]$ means the integer part of $2nx$.) Sketch the function k_n.

(iii) Let

$$K_n(x) = \int_0^x k_n(t)\, dt$$

for $x \in [0,1]$. Sketch K_n. Show that $0 \le K_n(x) \le 1/(2n)$ for all x. Show that K_n is once differentiable with continuous derivative. Show that $|K_n'(x)| \le 1$ for all x, and identify the set of points where $|K_n'(x)| = 1$.

(iv) Show that there exists a sequence of continuously differentiable functions $f_n : [0,1] \to \mathbb{R}$, with $f_n(0) = f_n(1) = 0$, such that $I(f_n) \to 0$ as $n \to \infty$. (This result is slightly improved in Example K.149.)

This example poses two problems. The first is that in some sense the f_n are close to $f_0 = 0$ with $I(f_n) < I(f_0)$, yet the Euler–Lagrange approach of Exercise 8.65 seemed to show that $I(f_0)$ was smaller than those $I(f)$ with f close to f_0. One answer to this seeming paradox is that, in Exercise 8.65, we only looked at $G_h(\eta) = I(\eta h)$ as η became small, *so we only looked at certain paths approaching* f_0 *and not at all possible modes of approach.* As η becomes small, not only does ηh become small but so does $\eta h'$. However, as n becomes large, f_n becomes small but f'_n does not. In general, when the Euler–Lagrange method looks at a function f it compares it only with functions which are close to f and have derivative close to f'. This does not affect the truth of Theorem 8.60 (which says that the Euler–Lagrange equation is a necessary condition for a minimum), but shows that that the search for a useful sufficient condition is likely to involve new ideas.

Once we have the notion of a metric space we can make matters much clearer. (See Exercises K.223 to K.225.)

Exercise 8.67. *This exercise looks back to Section 7.3. Let* U *be an open subset of* $\mathbb{R}^2$ *containing* $(0,0)$. *Suppose that* $f : U \to \mathbb{R}$ *has second order partial derivatives on* U *and these partial derivatives are continuous at* $(0,0)$. *Suppose further that* $f_{,1}(0,0) = f_{,2}(0,0) = 0$. *If* $\mathbf{u} \in \mathbb{R}^2$ *we write* $G_{\mathbf{u}}(\eta) = f(\eta\mathbf{u})$.

(i) Show that $G'_{\mathbf{u}}(0) = 0$ *for all* $\mathbf{u} \in \mathbb{R}^2$.

(ii) Let $\mathbf{e}_1 = (1,0)$ *and* $\mathbf{e}_2 = (0,1)$ *Suppose that* $G''_{\mathbf{e}_1}(0) > 0$ *and* $G''_{\mathbf{e}_2}(0) > 0$. *Show, by means of an example, that* $(0,0)$ *need not be a local minimum for* f. *Does there exist an* f *with the properties given which attains a local minimum at* $(0,0)$? *Does there exist an* f *with the properties given which attains a local maximum at* $(0,0)$?

(iii) Suppose that $G''_{\mathbf{u}}(0) > 0$ *whenever* $\mathbf{u}$ *is a unit vector. Show that* f *attains a local minimum at* $(0,0)$.

The second problem raised by results like Exercise 8.66 is also very interesting.

Exercise 8.68. *Use Exercise 8.35 to show that* $I(f) > 0$ *whenever* $f : [0,1] \to \mathbb{R}$ *is a continuously differentiable function.*

Conclude, using the discussion above, that the set

$$\{I(f) : f \text{ continuously differentiable}\}$$

has an infimum (to be identified) but no minimum.

Exercise 8.69. *Here is a simpler (but less interesting) example of a variational problem with no solution, also due to Weierstrass. Consider the set* E *of functions* $f : [-1,1] \to \mathbb{R}$ *with continuous derivative and such that*

$f(-1) = -1$, $f(1) = 1$. *Show that*

$$\inf_{f \in E} \int_{-1}^{1} x^2 f'(x)^2 \, dx = 0,$$

but there does not exist any $f_0 \in E$ *with* $\int_{-1}^{1} x^2 f_0'(x)^2 \, dx = 0$.

The discovery that that they had been talking about solutions to problems which might have no solutions came as a severe shock to the pure mathematical community. Of course, examples like the one we have been discussing are 'artificial' in the sense that they have been constructed for the purpose, but unless we can come up with some criterion for distinguishing 'artificial' problems from 'real' problems this takes us nowhere. 'If we have actually seen one tiger, is not the jungle immediately filled with tigers, and who knows where the next one lurks.' The care with which we proved Theorem 4.44 (a continuous function on a closed bounded set is bounded and *attains* its bounds) and Theorem 4.52 (Rolle's theorem, considered as the statement that, if a differentiable function f on an open interval (a, b) attains a maximum at x, then $f'(x) = 0$) are distant echos of that shock. On the other hand, the new understanding which resulted revivified the study of problems of maximisation and led to much new mathematics.

It is always possible to claim that Nature (with a capital N) will never set 'artificial' problems, and so the applied mathematician need not worry about these things. 'Nature is not troubled by mathematical difficulties.' However, a physical theory is not a description of nature (with a small n) but a model of nature which may well be troubled by mathematical difficulties. There are at least two problems in physics where the model has the characteristic features of our 'artificial' problem. In the first, which asks for a description of the electric field near a very sharp charged needle, the actual experiment produces sparking. In the second, which deals with crystallisation as a system for minimising an energy function not too far removed from I, microscopy reveals patterns not too far removed from Figure 8.1!

8.5. Vector-valued integrals

So far we have dealt only with the integration of functions $f : [a, b] \to \mathbb{R}$. The general programme that we wish to follow would direct us to consider the integration of functions $\mathbf{f} : E \to \mathbb{R}^m$ where E is a well behaved subset of $\mathbb{R}^n$. In this section we shall take the first step by considering the special case of a well behaved function $\mathbf{f} : [a, b] \to \mathbb{R}^m$. Since $\mathbb{C}$ can be identified with $\mathbb{R}^2$, our special case contains, as a still more special (but very important) case the integration of well behaved complex-valued functions $f : [a, b] \to \mathbb{C}$.

The definition is simple.

Definition 8.70. *If* $\mathbf{f} : [a,b] \to \mathbb{R}^m$ *is such that* $f_j : [a,b] \to \mathbb{R}$ *is Riemann integrable for each* j*, then we say that* $\mathbf{f}$ *is Riemann integrable and* $\int_a^b \mathbf{f}(x)\,dx = \mathbf{y}$*, where* $\mathbf{y} \in \mathbb{R}^m$ *and*

$$y_j = \int_a^b f_j(x)\,dx$$

for each j.

In other words,

$$\left(\int_a^b \mathbf{f}(x)\,dx\right)_j = \int_a^b f_j(x)\,dx.$$

It is easy to obtain the properties of this integral directly from its definition and the properties of the one-dimensional integral. Here is an example.

Lemma 8.71. *If* $\alpha : \mathbb{R}^m \to \mathbb{R}^p$ *is linear and* $\mathbf{f} : [a,b] \to \mathbb{R}^m$ *is Riemann integrable, then so is* $\alpha\mathbf{f}$ *and*

$$\int_a^b (\alpha\mathbf{f})(x)\,dx = \alpha\left(\int_a^b \mathbf{f}(x)\,dx\right).$$

Proof. Let α have matrix representation (a_{ij}). By Lemma 8.18,

$$(\alpha\mathbf{f})_i = \sum_{j=1}^m a_{ij} f_j$$

is Riemann integrable and

$$\int_a^b \sum_{j=1}^m a_{ij} f_j(x)\,dx = \sum_{j=1}^m a_{ij} \int_a^b f_j(x)\,dx.$$

Comparing this with Definition 8.70, we see that we have the required result. ∎

Taking α to be any orthogonal transformation of $\mathbb{R}^m$ to itself, we see that our definition of the integral is, in fact, coordinate independent. (Remember, it is part of our programme that nothing should depend on the particular choice of coordinates we use. The reader may also wish to look at Exercise K.152.)

Choosing a particular orthogonal transformation, we obtain the following nice result.

Theorem 8.72. *If* $\mathbf{f} : [a,b] \to \mathbb{R}^m$ *is Riemann integrable, then*

$$\left\|\int_a^b \mathbf{f}(x)\,dx\right\| \le (b-a) \sup_{x\in[a,b]} \|\mathbf{f}(x)\|.$$

This result falls into the standard pattern

$$\text{size of integral} \leq \text{length} \times \sup.$$

Proof. If $\mathbf{y}$ is a vector in $\mathbb{R}^m$, we can always find a rotation α of $\mathbb{R}^m$ such that $\alpha\mathbf{y}$ lies along the x_1 axis, that is to say, $(\alpha\mathbf{y})_1 \geq 0$ and $(\alpha\mathbf{y})_j = 0$ for $2 \leq j \leq m$. Let $\mathbf{y} = \int_a^b \mathbf{f}(x)\,dx$. Then

$$\begin{aligned}
\left\|\int_a^b \mathbf{f}(x)\,dx\right\| &= \left\|\alpha\int_a^b \mathbf{f}(x)\,dx\right\| \\
&= \left|\left(\alpha\int_a^b \mathbf{f}(x)\,dx\right)_1\right| \\
&= \left|\int_a^b (\alpha\mathbf{f}(x))_1\,dx\right| \\
&\leq (b-a)\sup_{x\in[a,b]}|(\alpha\mathbf{f}(x))_1| \\
&\leq (b-a)\sup_{x\in[a,b]}\|\alpha\mathbf{f}(x)\| \\
&= (b-a)\sup_{x\in[a,b]}\|\mathbf{f}(x)\|.
\end{aligned}$$

■

Exercise 8.73. *Justify each step in the chain of equalities and inequalities which concluded the preceding proof.*

Exercise 8.74. *Show that the collection $\mathcal{R}$ of Riemann integrable functions $\mathbf{f} : [a,b] \to \mathbb{R}^m$ forms a real vector space with the natural operations. If we write*

$$T\mathbf{f} = \int_a^b \mathbf{f}(x)\,dx$$

and $\|\mathbf{f}\|_\infty = \sup_{t\in[a,b]}\|\mathbf{f}(t)\|$, show that $T : \mathcal{R} \to \mathbb{R}$ is a linear map and $\|T\mathbf{f}\| \leq (b-a)\|\mathbf{f}\|_\infty$.

Chapter 9

Developments and Limitations of the Riemann Integral ♡

9.1. Why go further?

Let us imagine a conversation in the 1880's between a mathematician opposed to the 'new rigour' and a mathematician who supported it. The opponent might claim that the definition of the Riemann integral given in Section 8.2 was dull and gave rise to no new theorems. The supporter might say, as this book does, that definitions are necessary in order that we know when we have proved something and to understand what we have proved when we have proved it. He would, however, have to admit both the dullness and the lack of theorems. Both sides would regretfully agree that there was probably little more to say about the matter.

Twenty years later, Lebesgue, building on work of Borel and others, produced a radically new theory of integration. From the point of view of Lebesgue's theory, Riemann integration has a profound weakness. We saw in Lemma 8.18 and Exercise 8.21 that we cannot leave the class of Riemann integrable functions if we only perform algebraic operations (for example, the product of two Riemann integrable functions is again Riemann integrable). However, we can leave the class of Riemann integrable functions by performing limiting operations.

Exercise 9.1. *Let $f_n : [0,1] \to \mathbb{R}$ be defined by $f_n(r2^{-n}) = 1$ if r is an integer with $0 \leq r \leq 2^n$, $f_n(x) = 0$, otherwise.*

(i) Show that f_n is Riemann integrable.

(ii) Show that there exists an $f : [0,1] \to \mathbb{R}$, which you should define explicitly, such that $f_n(x) \to f(x)$ as $n \to \infty$, for each $x \in [0,1]$.

(iii) Show, however, that f is not Riemann integrable.

[See also Exercise K.153.]

The class of Lebesgue integrable functions includes every Riemann integrable function but behaves much better when we perform limiting operations. As an example, which does not give the whole picture but shows the kind of result that can be obtained, contrast Exercise 9.1 with the following lemma.

Lemma 9.2. *Let $f_n : [a,b] \to \mathbb{R}$ be a sequence of Lebesgue integrable functions with $|f_n(x)| \leq M$ for all $x \in [0,1]$ and all n. If $f_n(x) \to f(x)$ as $n \to \infty$ for each $x \in [0,1]$, then f is Lebesgue integrable and*

$$\int_a^b f_n(x)\,dx \to \int_a^b f(x)\,dx.$$

It is important to realise that mathematicians prize the Lebesgue integral, not because it integrates more functions (most functions that we meet explicitly are Riemann integrable), but because it gives rise to beautiful theorems and, at a deeper level, to beautiful theories way beyond the reach of the Riemann integral.

Dieudonné dismisses the Riemann integral with scorn in [**14**], Chapter VIII:

> It may well be suspected that, had it not been for its prestigious name, this [topic] would have been dropped long ago [from elementary analysis courses], for (with due reverence to Riemann's genius) it is certainly clear to any working mathematician that nowadays such a 'theory' has at best the importance of a mildly interesting exercise in the general theory of measure and integration. Only the stubborn conservatism of academic tradition could freeze it into a regular part of the curriculum, long after it had outlived its historical importance.

Stubborn academic conservatives like the present writer would reply that, as a matter of observation, many working mathematicians[1] do not use and have never studied Lebesgue integration and its generalisation to

[1]Of course, it depends on who you consider to be a mathematician. A particular French academic tradition begins by excluding all applied mathematicians, continues by excluding all supporters of the foreign policy of the United States and ends by restricting the title to pupils of the École Normale Supérieure.

measure theory. Although measure theory is now essential for the study of all branches of analysis and probability, it is not needed for most of number theory, algebra, geometry and applied mathematics.

It is frequently claimed that Lebesgue integration is as easy to teach as Riemann integration. This is probably true, but I have yet to be convinced that it is as easy to learn. Under these circumstances, it is reasonable to introduce Riemann integration as an ad hoc tool to be replaced later by a more powerful theory, if required. If we only have to walk 50 metres, it makes no sense to buy a car.

On the other hand, as the distance to be traveled becomes longer, walking becomes less attractive. We could walk from London to Cambridge, but few people wish to do so. This chapter contains a series of short sections showing how the notion of the integral can be extended in various directions. I hope that the reader will find them interesting and instructive; but, for the reasons just given, she should not invest too much time and effort in their contents, which, in many cases, can be given a more elegant, inclusive and efficient exposition using measure theory.

I believe that, *provided it is not taken too seriously*, this chapter will be useful to those who do not go on to do measure theory, by showing that the theory of integration is richer than most elementary treatments would suggest; and to those who will go on to do measure theory, by opening their minds to some of the issues involved.

9.2. Improper integrals ♡

We have defined Riemann integration for bounded functions on bounded intervals. However, the reader will already have evaluated, as a matter of routine, so-called 'improper integrals'[2] in the following manner:

$$\int_0^1 x^{-1/2}\,dx = \lim_{\epsilon\to 0+}\int_\epsilon^1 x^{-1/2}\,dx = \lim_{\epsilon\to 0+}[2x^{1/2}]_\epsilon^1 = 2,$$

and

$$\int_1^\infty x^{-2}\,dx = \lim_{R\to\infty}\int_1^R x^{-2}\,dx = \lim_{R\to\infty}[-x^{-1}]_1^R = 1.$$

A full theoretical treatment of such integrals with the tools at our disposal is apt to lead into into a howling wilderness of 'improper integrals of the first kind', 'Cauchy principal values' and so on. Instead, I shall give a few typical theorems, definitions and counterexamples from which the reader

[2]There is nothing particularly improper about improper integrals (at least, if they are absolutely convergent, see page 201), but this is what they are traditionally called. Their other traditional name 'infinite integrals' removes the imputation of moral obliquity but is liable to cause confusion in other directions.

should be able to construct any theory that she needs to justify results in elementary calculus.

Definition 9.3. *If $f : [a,\infty) \to \mathbb{R}$ is such that $f|_{[a,X]} \in \mathcal{R}[a,X]$ for each $X > a$ and $\int_a^X f(x)\,dx \to L$ as $X \to \infty$, then we say that $\int_a^\infty f(x)\,dx$ exists with value L.*

Lemma 9.4. *Suppose $f : [a,\infty) \to \mathbb{R}$ is such that $f|_{[a,X]} \in \mathcal{R}[a,X]$ for each $X > a$. If $f(x) \geq 0$ for all x, then $\int_a^\infty f(x)\,dx$ exists if and only if there exists a K such that $\int_a^X f(x)\,dx \leq K$ for all X.*

Proof. As usual we split the proof into two parts, dealing with 'if' and 'only if' separately.

Suppose first that $\int_a^\infty f(x)\,dx$ exists; that is to say, $\int_a^X f(x)\,dx$ tends to a limit as $X \to \infty$. Let $u_n = \int_a^n f(x)\,dx$ when n is an integer with $n \geq a$. Since f is positive, u_n is an increasing sequence. Since u_n tends to a limit, it must be bounded; that is to say, there exists a K such that $u_n \leq K$ for all $n \geq a$. If $X \geq a$ we choose an integer $N \geq X$ and observe that

$$\int_a^X f(x)\,dx \leq \int_a^N f(x)\,dx = u_N \leq K,$$

as required.

Suppose, conversely, that there exists a K such that $\int_a^X f(x)\,dx \leq K$ for all $X \geq a$. Defining $u_n = \int_a^n f(x)\,dx$ as before, we observe that the u_n form an increasing sequence bounded above by K. By the fundamental axiom it follows that u_n tends to a limit L, say. In particular, given $\epsilon > 0$, we can find an $n_0(\epsilon)$ such that $L - \epsilon < u_n \leq L$ for all $n \geq n_0(\epsilon)$.

If X is any real number with $X > n_0(\epsilon) + 1$, we can find an integer n with $n + 1 \geq X > n$. Since $n \geq n_0(\epsilon)$, we have

$$L - \epsilon < u_n \leq \int_a^X f(x)\,dx \leq u_{n+1} \leq L$$

and $|L - \int_a^X f(x)\,dx| < \epsilon$. Thus $\int_a^X f(x)\,dx \to L$ as $X \to \infty$, as required. ∎

Exercise 9.5. *Show that $\int_0^n \sin(2\pi x)\,dx$ tends to a limit as $n \to \infty$ through integer values, but $\int_0^X \sin(2\pi x)\,dx$ does not tend to a limit as $X \to \infty$.*

We use Lemma 9.4 to prove the integral comparison test.

Lemma 9.6. *Suppose $f : [1,\infty) \to \mathbb{R}$ is a decreasing continuous positive function. Then $\sum_{n=1}^\infty f(n)$ exists if and only if $\int_1^\infty f(x)\,dx$ does.*

Just as with sums, we sometimes say that '$\int_1^\infty f(x)\,dx$ converges' rather than '$\int_1^\infty f(x)\,dx$ exists'. The lemma then says '$\sum_{n=1}^\infty f(n)$ converges if and only if $\int_1^\infty f(x)\,dx$ does'.

The proof of Lemma 9.6 is set out in the next exercise.

Exercise 9.7. *Suppose $f : [1, \infty) \to \mathbb{R}$ is a decreasing continuous positive function.*

(i) Show that

$$f(n) \geq \int_n^{n+1} f(x)\, dx \geq f(n+1).$$

(ii) Deduce that

$$\sum_1^N f(n) \geq \int_1^{N+1} f(x)\, dx \geq \sum_2^{N+1} f(n).$$

(iii) By using Lemma 9.4 and the corresponding result for sums, deduce Lemma 9.6.

Exercise 9.8. *(i) Use Lemma 9.6 to show that $\sum_{n=1}^{\infty} n^{-\alpha}$ converges if $\alpha > 1$ and diverges if $\alpha \leq 1$.*

(ii) Use the inequality established in Exercise 9.7 to give a rough *estimate of the size of N required to give $\sum_{n=1}^{N} n^{-1} > 100$.*

(iii) Use the methods just discussed to do Exercise 5.10.

Exercise 9.9. (Simple version of Stirling's formula.) *The ideas of Exercise 9.7 have many applications.*

(i) Suppose $g : [1, \infty) \to \mathbb{R}$ is an increasing continuous positive function. Obtain inequalities for g corresponding to those for f in parts (i) and (ii) of Exercise 9.7.

(ii) By taking $g(x) = \log x$ in part (i), show that

$$\log(N-1)! \leq \int_1^N \log x\, dx \leq \log N!$$

and use integration by parts to conclude that

$$\log(N-1)! \leq N \log N - N + 1 \leq \log N!\ .$$

(iii) Show that $\log N! = N \log N - N + \theta(N)N$, where $\theta(N) \to 0$ as $N \to \infty$.

[A stronger result is proved in Exercise K.156.]

We have a result corresponding to Theorem 4.77.

Lemma 9.10. *Suppose $f : [a, \infty) \to \mathbb{R}$ is such that $f|_{[a,X]} \in \mathcal{R}[a, X]$ for each $X > a$. If $\int_a^\infty |f(x)|\, dx$ exists, then $\int_a^\infty f(x)\, dx$ exists.*

It is natural to state Lemma 9.10 in the form 'absolute convergence of the integral implies convergence'.

Exercise 9.11. *Use the argument of Exercise 4.79 (i) to prove Lemma 9.10.*

Exercise 9.12. *Prove the following general principle of convergence for integrals.*

Suppose $f : [a, \infty) \to \mathbb{R}$ *is such that* $f|_{[a,X]} \in \mathcal{R}[a, X]$ *for each* $X > a$. *Show that* $\int_a^\infty f(x)\,dx$ *exists if and only if, given any* $\epsilon > 0$, *we can find an* $X_0(\epsilon) > a$ *such that*

$$\left| \int_X^Y f(x)\,dx \right| < \epsilon$$

whenever $Y \geq X \geq X_0(\epsilon)$.

Exercise 9.13. *(i) Following the ideas of this section and Section 8.5, provide the appropriate definition of* $\int_a^\infty \mathbf{f}(x)\,dx$ *for a function* $\mathbf{f} : [a, \infty) \to \mathbb{R}^m$.

(ii) By taking components and using Exercise 9.12, or otherwise, prove a general principle of convergence for such integrals.

(iii) Use part (ii) and the method of proof of Theorem 4.77 to prove the following generalisation of Lemma 9.10.

Suppose $\mathbf{f} : [a, \infty) \to \mathbb{R}^m$ *is such that* $\mathbf{f}|_{[a,X]} \in \mathcal{R}[a, X]$ *for each* $X > a$. *If* $\int_a^\infty \|\mathbf{f}(x)\|\,dx$ *exists, then* $\int_a^\infty \mathbf{f}(x)\,dx$ *exists.*

Exercise 9.14. *Suppose* $f : [a, b) \to \mathbb{R}$ *is such that* $f|_{[a,c]} \in \mathcal{R}[a, c]$ *for each* $a < c < b$. *Produce a definition along the lines of Definition 9.3 of what it should mean for* $\int_a^b f(x)\,dx$ *to exist with value* L.

State and prove results analogous to Lemma 9.4 and Lemma 9.10.

Additional problems arise when there are two limits involved.

Example 9.15. *If* $\lambda, \mu > 0$, *then*

$$\int_{-\mu R}^{\lambda R} \frac{x}{1 + x^2}\,dx \to \log\left(\frac{\lambda}{\mu}\right)$$

as $R \to \infty$.

Proof. Direct calculation, which is left to the reader. ∎

A pure mathematician gets round this problem by making a definition along these lines.

Definition 9.16. *If* $f : \mathbb{R} \to \mathbb{R}$ *is such that* $f|_{[-X,Y]} \in \mathcal{R}[-X, Y]$ *for each* $X, Y > 0$, *then* $\int_{-\infty}^\infty f(x)\,dx$ *exists with value* L *if and only if the following condition holds. Given* $\epsilon > 0$, *we can find an* $X_0(\epsilon) > 0$ *such that*

$$\left| \int_{-X}^Y f(x)\,dx - L \right| < \epsilon$$

for all $X, Y > X_0(\epsilon)$.

Exercise 9.17. *Let $f : \mathbb{R} \to \mathbb{R}$ be such that $f|_{[-X,Y]} \in \mathcal{R}[X,Y]$ for each $X, Y > 0$. Show that $\int_{-\infty}^{\infty} f(x)\,dx$ exists if and only if both the integrals $\int_0^{\infty} f(x)\,dx = \lim_{R\to\infty} \int_0^R f(x)\,dx$ and $\int_{-\infty}^0 f(x)\,dx = \lim_{S\to\infty} \int_{-S}^0 f(x)\,dx$ exist. If the integrals exist, show that*

$$\int_{-\infty}^{\infty} f(x)\,dx = \int_{-\infty}^{0} f(x)\,dx + \int_0^{\infty} f(x)\,dx.$$

The physicist gets round the problem by ignoring it. If she is a *real physicist* with correct physical intuition this works splendidly[3], but if not, not.

Speaking broadly, improper integrals $\int_E f(x)\,dx$ work well when they are absolutely convergent, that is to say, $\int_E |f(x)|\,dx < \infty$, but are full of traps for the unwary otherwise. This is not a weakness of the Riemann integral but inherent in any mathematical situation where an object only exists 'by virtue of the cancellation of two infinite objects'. (Recall Littlewood's example on page 79.)

Example 9.18. *Suppose we define the PV (principle value) integral by*

$$\mathcal{PV}\int_{-\infty}^{\infty} f(x)\,dx = \lim_{R\to\infty} \int_{-R}^{R} f(x)\,dx$$

whenever the right hand side exists. Show, by considering Example 9.15, or otherwise, that the standard rule for change of variables fails for PV integrals.

9.3. Integrals over areas ♡

At first sight, the extension of the idea of Riemann integration from functions defined on $\mathbb{R}$ to functions defined on $\mathbb{R}^n$ looks like child's play. We shall do the case $n = 2$, since the general case is a trivial extension.

Let $R = [a,b] \times [c,d]$ and consider $f : R \to \mathbb{R}$ such that there exists a K with $|f(\mathbf{x})| \le K$ for all $\mathbf{x} \in R$. We define a dissection $\mathcal{D}$ of R to be a finite collection of rectangles $I_j = [a_j, b_j] \times [c_j, d_j]$ $[1 \le j \le N]$ such that

(i) $\bigcup_{j=1}^{N} I_j = R$, and

(ii) $I_i \cap I_j$ either is empty or consists of a segment of a straight line $[1 \le j < i \le N]$.

[3]In [**8**], Boas reports the story of a friend visiting the Princeton common room ' ... where Einstein was talking to another man, who would shake his head and stop him; Einstein then thought for a while, then started talking again; was stopped again; and so on. After a while, ... my friend was introduced to Einstein. He asked Einstein who the other man was. "Oh," said Einstein, "that's my mathematician." '

If $\mathcal{D} = \{I_j : 1 \le j \le N\}$ and $\mathcal{D}' = \{I'_k : 1 \le k \le N'\}$ are dissections, we write $\mathcal{D} \wedge \mathcal{D}'$ for the set of non-empty rectangles of the form $I_j \cap I'_k$. If every $I'_k \in \mathcal{D}'$ is contained in some $I_j \in \mathcal{D}$, we write $\mathcal{D}' \succ \mathcal{D}$.

We define the *upper sum* and *lower sum* associated with $\mathcal{D}$ by

$$S(f, \mathcal{D}) = \sum_{j=1}^{N} |I_j| \sup_{\mathbf{x} \in I_j} f(\mathbf{x}),$$

$$s(f, \mathcal{D}) = \sum_{j=1}^{N} |I_j| \inf_{\mathbf{x} \in I_j} f(\mathbf{x}),$$

where $|I_j| = (b_j - a_j)(d_j - c_j)$, the area of I_j.

Exercise 9.19. *(i) Suppose that $\mathcal{D}$ and $\mathcal{D}'$ are dissections with $\mathcal{D}' \succ \mathcal{D}$. Show, using the method of Exercise 8.8, or otherwise, that*

$$S(f, \mathcal{D}) \ge S(f, \mathcal{D}') \ge s(f, \mathcal{D}') \ge s(f, \mathcal{D}).$$

(ii) State and prove a result corresponding to Lemma 8.10.

(iii) Explain how this enables us to define upper and lower integrals and hence complete the definition of Riemann integration. We write the integral as

$$\int_R f(\mathbf{x})\, dA$$

when it exists.

(iv) Develop the theory of Riemann integration on R as far as you can. (You should be able to obtain results like those in Section 8.2 as far as the end of Exercise 8.22.) You should prove that if f is continuous on R then it is Riemann integrable.

We can do rather more than just prove the existence of

$$\int_R f(\mathbf{x})\, dA$$

when f is continuous on the rectangle R.

Theorem 9.20. (Fubini's theorem for continuous functions.) *Let $R = [a, b] \times [c, d]$. If $f : R \to \mathbb{R}$ is continuous, then the functions $F_1 : [a, b] \to \mathbb{R}$ and $F_2 : [c, d] \to \mathbb{R}$ defined by*

$$F_1(x) = \int_c^d f(x, s)\, ds \text{ and } F_2(y) = \int_a^b f(t, y)\, dt$$

are continuous and

$$\int_a^b F_1(x)\, dx = \int_c^d F_2(y)\, dy = \int_R f(\mathbf{x})\, dA.$$

This result is more usually written as

$$\int_a^b \left(\int_c^d f(x,y)\,dy\right)\,dx = \int_c^d \left(\int_a^b f(x,y)\,dx\right)\,dy = \int_R f(\mathbf{x})\,dA,$$

or, simply,

$$\int_a^b \int_c^d f(x,y)\,dy\,dx = \int_c^d \int_a^b f(x,y)\,dx\,dy = \int_{[a,b]\times[c,d]} f(\mathbf{x})\,dA.$$

(See also Exercises K.170, K.172 and K.173.)

We prove Theorem 9.20 in two exercises.

Exercise 9.21. *(We use the notation of Theorem 9.20 in this exercise.) If $|f(x,s)-f(w,s)| \le \epsilon$ for all $s \in [c,d]$, show that $|F_1(x)-F_1(w)| \le \epsilon(d-c)$. Use the uniform continuity of f to conclude that F_1 is continuous.*

For the next exercise we recall the notion of an indicator function $\mathbb{I}_E$ for a set E. If $E \subseteq R$, then $\mathbb{I}_E : R \to \mathbb{R}$ is defined by $\mathbb{I}_E(a) = 1$ if $a \in E$, $\mathbb{I}_E(a) = 0$ otherwise.

Exercise 9.22. *We use the notation of Theorem 9.20. In this exercise interval will mean open, half open or closed interval (that is, intervals of the form, (α,β), $[\alpha,\beta)$, $(\alpha,\beta]$ or $[\alpha,\beta]$) and rectangle will mean the product of two intervals. We say that g satisfies the Fubini condition if*

$$\int_a^b \left(\int_c^d g(x,y)\,dy\right)\,dx = \int_c^d \left(\int_a^b g(x,y)\,dy\right)\,dx = \int_R g(\mathbf{x})\,dA.$$

(i) Show that, given $\epsilon > 0$, we can find rectangles $R_j \subseteq R$ and $\lambda_j \in \mathbb{R}$ such that, writing

$$H = \sum_{j=1}^{N} \lambda_j \mathbb{I}_{R_j},$$

we have $H(\mathbf{x}) - \epsilon \le f(\mathbf{x}) \le H(\mathbf{x}) + \epsilon$ for all $\mathbf{x} \in R$.

(ii) Show by direct calculation that $\mathbb{I}_B$ satisfies the Fubini condition whenever B is a rectangle. Deduce that H satisfies the Fubini condition, and use (i) (carefully) to show that f does.

All this looks very satisfactory, but our treatment hides a problem. If we look at how mathematicians actually use integrals, we find that they want to integrate over sets which are more complicated than rectangles with sides parallel to coordinate axes. (Indeed, one of the guiding principles of this book is that coordinate axes should not have a special role.) If you have studied mathematical methods you will have come across the formula for change of variables[4]

$$\iint_{E'} f(u,v)\,du\,dv = \iint_{E} f(u(x,y),v(x,y)) \left|\frac{\partial(u,v)}{\partial(x,y)}\right| dx\,dy,$$

where

$$E' = \{(u(x,y),v(x,y)) \,:\, (x,y)\in E\}.$$

Even if you do not recognise the formula, you should see easily that any change of variable formula will involve changing not only the integrand but the set over which we integrate.

It is not hard to come up with an appropriate definition for integrals over a set E.

Definition 9.23. *Let E be a bounded set and $f : E \to \mathbb{R}$ a bounded function. Choose $a < b$ and $c < d$ such that $R = [a,b]\times[c,d]$ contains E and define $\tilde{f} : R \to \mathbb{R}$ by $\tilde{f}(\mathbf{x}) = f(\mathbf{x})$ if $\mathbf{x}\in E$, $\tilde{f}(\mathbf{x}) = 0$ otherwise. If $\int_R \tilde{f}(\mathbf{x})\,dA$ exists, we say that $\int_E f(\mathbf{x})\,dA$ exists and*

$$\int_E f(\mathbf{x})\,dA = \int_R \tilde{f}(\mathbf{x})\,dA.$$

Exercise 9.24. *Explain briefly why the definition is independent of the choice of R.*

The most important consequence of this definition is laid bare in the next exercise.

Exercise 9.25. *Let $R = [a,b]\times[c,d]$ and $E \subseteq R$. Let $\mathcal{R}$ be the set of functions $f : R \to \mathbb{R}$ which are Riemann integrable. Then $\int_E f(\mathbf{x})\,dA$ exists for all $f\in\mathcal{R}$ if and only if $\mathbb{I}_E \in \mathcal{R}$.*

If we think about the meaning of $\int_R \mathbb{I}_E(\mathbf{x})\,dA$, we are led to the following definition[5].

Definition 9.26. *A bounded set E in $\mathbb{R}^2$ has Riemann area $\int_E 1\,dA$ if that integral exists.*

Recall that, if $R = [a,b]\times[c,d]$, we write $|R| = (b-a)(d-c)$.

Exercise 9.27. *Show that a bounded set E has Riemann area $|E|$ if and only if, given any ϵ, we can find disjoint rectangles $R_i = [a_i,b_i]\times[c_i,d_i]$*

[4]This formula is included as a memory jogger only. It would require substantial supporting discussion to explain the underlying conventions and assumptions.

[5]Like most of the rest of this chapter, this is not meant to be taken too seriously. What we call 'Riemann area' is traditionally called 'content'. The theory of content is pretty, but was rendered obsolete by the theory of measure.

$[1 \leq i \leq N]$ *and (not necessarily disjoint) rectangles* $R'_j = [a'_j, b'_j] \times [c'_j, d'_j]$ $[1 \leq j \leq M]$ *such that*

$$\bigcup_{i=1}^{N} R_i \subseteq E \subseteq \bigcup_{j=1}^{M} R'_j, \quad \sum_{i=1}^{N} |R_i| \geq |E| - \epsilon \text{ and } \sum_{j=1}^{M} |R'_j| \leq |E| + \epsilon.$$

Exercise 9.28. *Show that, if* E *has Riemann area and* f *is defined and Riemann integrable on some rectangle* $R = [a, b] \times [c, d]$ *containing* E*, then* $\int_E f(\mathbf{x})\, dA$ *exists and*

$$\left| \int_E f(\mathbf{x})\, dA \right| \leq \sup_{x \in E} |f(\mathbf{x})| |E|.$$

In other words,

$$\text{size of integral} \leq \text{area} \times \sup.$$

Our discussion tells us that in order to talk about

$$\int_E f(\mathbf{x})\, dA$$

we need to know not only that f is well behaved (Riemann integrable) but that E is well behaved (has Riemann area). Just as the functions f which occur in 'first mathematical methods' courses are Riemann integrable, so the sets E which appear in such courses have Riemann area, though the process of showing this may be tedious.

Exercise 9.29. *(The reader may just wish to think about how to do this exercise without actually writing down all the details.)*

(i) Show that a rectangle whose sides are not necessarily parallel to the axis has Riemann area and that this area is what we expect.

(ii) Show that a triangle has Riemann area and that this area is what we expect.

(iii) Show that a polygon has Riemann area and that this area is what we expect. (Of course, the answer is to cut it up into a finite number of triangles, but can this always be done?)

However, if we want to go further, it becomes rather hard to decide which sets are nice and which are not. The problem is already present in the one-dimensional case, but hidden by our insistence on only integrating over intervals.

Definition 9.30. *A bounded set* E *in* $\mathbb{R}$ *has Riemann length if, taking any* $[a, b] \supseteq E$*, we have* $\mathbb{I}_E \in \mathcal{R}([a, b])$*. We say then that* E *has Riemann length*

$$|E| = \int_a^b \mathbb{I}_E(t)\, dt.$$

Exercise 9.31. *(i) Explain why the definition just given is independent of the choice of* $[a, b]$.

(ii) Show that

$$\mathbb{I}_{A \cup B} = \mathbb{I}_A + \mathbb{I}_B - \mathbb{I}_A \mathbb{I}_B.$$

Hence show that, if A *and* B *have Riemann length, so does* $A \cup B$. *Prove similar results for* $A \cap B$ *and* $A \setminus B$.

(iii) By reinterpreting Exercise 9.1, show that we can find $A_n \subseteq [0, 1]$ *such that* A_n *has Riemann length for each* n *but* $\bigcup_{n=1}^{\infty} A_n$ *does not.*

(iv) Obtain results like (ii) and (iii) for Riemann area.

It also turns out that the kind of sets we have begun to think of as nice, that is, open and closed sets, need not have Riemann area.

Lemma 9.32. *There exist bounded closed and open sets in* $\mathbb{R}$ *which do not have Riemann length. There exist bounded closed and open sets in* $\mathbb{R}^2$ *which do not have Riemann area.*

The proof of this result is a little complicated, so we have relegated it to Exercise K.174. Fortunately all these difficulties vanish like early morning mist in the sunlight of Lebesgue's theory.

Any belief we may have that we have a 'natural feeling' for how area behaves under complicated maps is finally removed by an example of Peano.

Theorem 9.33. *There exists a continuous map* $f : [0, 1] \to [0, 1] \times [0, 1]$ *which is surjective.*

Thus there exists a curve which passes through every point of a square! A proof depending on the notion of uniform convergence is given in Exercise K.252.

9.4. The Riemann–Stieltjes integral ♡

In this section we discuss a remarkable extension of the notion of integral due to Stieltjes. The reader should find the discussion gives an excellent revision of many of the ideas of Chapter 8.

Before doing so, we must dispose of a technical point. When authors talk about the Heaviside step function $H : \mathbb{R} \to \mathbb{R}$ they all agree that $H(t) = 0$ for $t < 0$ and $H(t) = 1$ for $t > 0$. However, some take $H(0) = 0$, some take $H(0) = 1$ and some take $H(0) = 1/2$. Usually this does not matter, but it is helpful to have consistency.

Definition 9.34. *Let* $E \subseteq \mathbb{R}$ *We say that a function* $f : E \to \mathbb{R}$ *is a right continuous function if, for all* $x \in E$, $f(t) \to f(x)$ *whenever* $t \to x$ *through values of* $t \in E$ *with* $t > x$.

Exercise 9.35. *Which definition of the Heaviside step function makes H right continuous?*

In the discussion that follows, $G : \mathbb{R} \to \mathbb{R}$ will be a right continuous increasing function. (Exercise K.176 sheds some light on the nature of such functions, but is not needed for our discussion.) We assume further that there exist A and B with $G(t) \to A$ as $t \to -\infty$ and $G(t) \to B$ as $t \to \infty$.

Exercise 9.36. *If $F : \mathbb{R} \to \mathbb{R}$ is an increasing function show that the following two statements are equivalent:*

(i) F is bounded.

(ii) $F(t)$ tends to (finite) limits as $t \to -\infty$ and as $t \to \infty$.

We shall say that any finite set $\mathcal{D}$ containing at least two points is a dissection of $\mathbb{R}$. By convention we write

$$\mathcal{D} = \{x_0, x_1, \ldots, x_n\} \text{ with } x_0 < x_1 < x_2 < \cdots < x_n.$$

(Note that we now demand that the x_j are distinct.)

Now suppose $f : \mathbb{R} \to \mathbb{R}$ is a bounded function. We define the *upper Stieltjes G sum* of f associated with $\mathcal{D}$ by

$$\begin{aligned} S_G(f, \mathcal{D}) = (G(x_0) - A) \sup_{t \le x_0} f(t) + \sum_{j=1}^{n} (G(x_j) - G(x_{j-1})) \sup_{t \in (x_{j-1}, x_j]} f(t) \\ + (B - G(x_n)) \sup_{t > x_n} f(t). \end{aligned}$$

(Note that we use half open intervals, since we have to be more careful about overlap than when we dealt with Riemann integration.)

Exercise 9.37. *(i) Define the lower Stieltjes G sum $s_G(f, \mathcal{D})$ in the appropriate way.*

(ii) Show that, if $\mathcal{D}$ and $\mathcal{D}'$ are dissections of $\mathbb{R}$, then $S_G(f, \mathcal{D}) \ge s_G(f, \mathcal{D}')$.

(iii) Define the upper Stieltjes G integral *by $I^*(G, f) = \inf_{\mathcal{D}} S(f, \mathcal{D})$. Give a similar definition for the* lower Stieltjes G integral *$I_*(G, f)$ and show that $I^*(G, f) \ge I_*(G, f)$.*

If $I^*(G, f) = I_*(G, f)$, we say that f is Riemann–Stieltjes integrable with respect to G, and we write

$$\int_{\mathbb{R}} f(x)\, dG(x) = I^*(G, f).$$

Exercise 9.38. *(i) State and prove a criterion for Riemann–Stieltjes integrability along the lines of Lemma 8.13.*

(ii) Show that the set $\mathcal{R}_G$ of functions which are Riemann–Stieltjes integrable with respect to G forms a vector space and the integral is a linear functional (i.e. a linear map from $\mathcal{R}_G$ to $\mathbb{R}$).

(iii) Suppose that $f : \mathbb{R} \to \mathbb{R}$ is Riemann–Stieltjes integrable with respect to G, that $K \in \mathbb{R}$ and that $|f(t)| \leq K$ for all $t \in \mathbb{R}$. Show that

$$\left| \int_{\mathbb{R}} f(x)\, dG(x) \right| \leq K(B - A).$$

(iv) Show that, if $f, g : \mathbb{R} \to \mathbb{R}$ are Riemann–Stieltjes integrable with respect to G, so is fg (the product of f and g).

(v) If $f : \mathbb{R} \to \mathbb{R}$ is Riemann–Stieltjes integrable with respect to G, show that $|f|$ is also and that

$$\int_{\mathbb{R}} |f(x)|\, dG(x) \geq \left| \int_{\mathbb{R}} f(x)\, dG(x) \right|.$$

*(vi) Prove that, if $f : \mathbb{R} \to \mathbb{R}$ is a bounded continuous function, then f is Riemann–Stieltjes integrable with respect to G. [*Hint: *Use the fact that f is uniformly continuous on any $[-R, R]$. Choose R sufficiently large.]*

The next result is more novel, although its proof is routine (it resembles that of Exercise 9.38 (ii)).

Exercise 9.39. *Suppose that F, $G : \mathbb{R} \to \mathbb{R}$ are right continuous increasing bounded functions and λ, $\mu \geq 0$. Show that, if $f : \mathbb{R} \to \mathbb{R}$ is Riemann–Stieltjes integrable with respect to both F and G, then f is Riemann–Stieltjes integrable with respect to $\lambda F + \mu G$ and*

$$\int_{\mathbb{R}} f(x)\, d(\lambda F + \mu G)(x) = \lambda \int_{\mathbb{R}} f(x)\, dF(x) + \mu \int_{\mathbb{R}} f(x)\, dG(x).$$

Exercise 9.40. *(i) If $a \in \mathbb{R}$, show, by choosing appropriate dissections, that $\mathbb{I}_{(-\infty,a]}$ is Riemann–Stieltjes integrable with respect to G and*

$$\int_{\mathbb{R}} \mathbb{I}_{(-\infty,a]}(x)\, dG(x) = G(a) - A.$$

(ii) If $a \in \mathbb{R}$, show that $\mathbb{I}_{(-\infty,a)}$ is Riemann–Stieltjes integrable with respect to G if and only if G is continuous at a. If G is continuous at a, show that

$$\int_{\mathbb{R}} \mathbb{I}_{(-\infty,a)}(x)\, dG(x) = G(a) - A.$$

(iii) If $a < b$, show that $\mathbb{I}_{(a,b]}$ is Riemann–Stieltjes integrable with respect to G and

$$\int_{\mathbb{R}} \mathbb{I}_{(a,b]}(x)\, dG(x) = G(b) - G(a).$$

(iv) If $a < b$, show that $\mathbb{I}_{(a,b)}$ is Riemann–Stieltjes integrable with respect to G if and only if G is continuous at b.

Combining the results of Exercise 9.40 with Exercise 9.38, we see that, if f is Riemann–Stieltjes integrable with respect to G, we may define

$$\int_{(a,b]} f(x)\,dG(x) = \int_{\mathbb{R}} \mathbb{I}_{(a,b]}(x)f(x)\,dG(x)$$

and make similar definitions for integrals like $\int_{(-\infty,a]} f(x)\,dG(x)$

Exercise 9.41. *Show that, if G is continuous and f is Riemann–Stieltjes integrable with respect to G, then we can define $\int_{[a,b]} f(x)\,dG(x)$, and that*

$$\int_{(a,b]} f(x)\,dG(x) = \int_{[a,b]} f(x)\,dG(x).$$

Remark: When we discussed Riemann integration, I said that, in mathematical practice, it was unusual to come across a function that was Lebesgue integrable but not Riemann integrable. In Exercise 9.40 (iv) we saw that the function $\mathbb{I}_{(a,b)}$, which we come across very frequently in mathematical practice, is not Riemann–Stieltjes integrable with respect to any right continuous increasing function G which has a discontinuity at b. In the Lebesgue–Stieltjes theory, $\mathbb{I}_{(a,b)}$ is always Lebesgue–Stieltjes integrable with respect to G. (Exercise K.178 extends Exercise 9.40 a little.)

The next result has an analogous proof to the fundamental theorem of the calculus (Theorem 8.37).

Exercise 9.42. *Suppose that $G : \mathbb{R} \to \mathbb{R}$ is an increasing function with continuous derivative. Suppose further that $f : \mathbb{R} \to \mathbb{R}$ is a bounded continuous function. If we set*

$$I(t) = \int_{(-\infty,t]} f(x)\,dG(x),$$

show that then I is differentiable and $I'(t) = f(t)G'(t)$ for all $t \in \mathbb{R}$.

Using the mean value theorem, in the form which states that the only function with derivative 0 is a constant, we get the following result.

Exercise 9.43. *Suppose that $G : \mathbb{R} \to \mathbb{R}$ is an increasing function with continuous derivative. If $f : \mathbb{R} \to \mathbb{R}$ is a bounded continuous function, show that*

$$\int_{(a,b]} f(x)\,dG(x) = \int_a^b f(x)G'(x)\,dx.$$

Show also that

$$\int_{\mathbb{R}} f(x)\,dG(x) = \int_{-\infty}^{\infty} f(x)G'(x)\,dx,$$

explaining carefully the meaning of the right hand side of the equation.

However, there is no reason why we should restrict ourselves even to continuous functions when considering Riemann–Stieltjes integration.

Exercise 9.44. *(i) If $c \in \mathbb{R}$, define $H_c : \mathbb{R} \to \mathbb{R}$ by $H_c(t) = 0$ if $t < c$, $H_c(t) = 1$ if $t \geq c$. Show, by finding appropriate dissections, that, if $f : \mathbb{R} \to \mathbb{R}$ is a bounded continuous function, we have*

$$\int_{(a,b]} f(x)\, dH_c(x) = f(c),$$

when $c \in (a, b]$. What happens if $c \notin (a, b]$?

(ii) If $a < c_1 < c_2 < \cdots < c_m < b$ and $\lambda_1, \lambda_2, \ldots, \lambda_m \geq 0$, find a right continuous function $G : [a, b] \to \mathbb{R}$ such that, if $f : (a, b] \to \mathbb{R}$ is a bounded continuous function, we have

$$\int_{(a,b]} f(x)\, dG(x) = \sum_{j=1}^{m} \lambda_j f(c_j).$$

Exercise 9.44 shows that Riemann–Stieltjes integration provides a framework in which point masses may be considered along with continuous densities[6].

The reader may agree with this but still doubt whether the extra generality of Riemann–Stieltjes point of view is worth the work involved. The following discussion may help change her mind.

What is a real-valued random variable? It is rather hard to give a proper mathematical definition with the mathematical apparatus available in 1880[7]. However, any real-valued random variable X is associated with a function

$$P(x) = \Pr\{X \leq x\}.$$

Exercise 9.45. *Convince yourself that $P : \mathbb{R} \to \mathbb{R}$ is a right continuous increasing function with $P(t) \to 0$ as $t \to -\infty$ and $P(t) \to 1$ as $t \to \infty$. (Note that, as we have no proper definitions, we can give no proper proofs.)*

Even if we have no definition of a random variable, we do have a definition of a Riemann–Stieltjes integral. So, in a typical mathematician's trick, we turn everything upside down.

Suppose that $P : \mathbb{R} \to \mathbb{R}$ is a right continuous increasing function with $P(t) \to 0$ as $t \to -\infty$ and $P(t) \to 1$ as $t \to \infty$. We say that P is associated

[6]Note that, although we have justified the concept of a 'delta function', we have not justified the concept of 'the derivative of the delta function'. This requires a further generalisation of our point of view to that of distributions.

[7]The Holy Roman Empire was neither holy nor Roman nor an empire. A random variable is neither random nor a variable.

with a real-valued random variable X if

$$\Pr\{X \in E\} = \int_{\mathbb{R}} \mathbb{I}_E(x)\, dP(x)$$

when $\mathbb{I}_E$ is Riemann–Stieltjes integrable with respect to P. (Thus, for example, E could be $(-\infty, a]$ or $(a, b]$.) If the reader chooses to read $\Pr\{X \in E\}$ as 'the probability that $X \in E$', that is up to her. So far as we are concerned, $\Pr\{X \in E\}$ is an abbreviation for $\int_{\mathbb{R}} \mathbb{I}_E(x)\, dP(x)$.

In the same way we *define* the expectation $\mathbb{E}f(X)$ by

$$\mathbb{E}f(X) = \int_{\mathbb{R}} f(x)\, dP(x)$$

when f is Riemann–Stieltjes integrable with respect to P. The utility of this definition is greatly increased if we allow improper Riemann–Stieltjes integrals somewhat along the lines of Definition 9.16.

Definition 9.46. *Let G be as throughout this section. If $f : \mathbb{R} \to \mathbb{R}$ and $R, S > 0$, we define $f_{RS} : \mathbb{R} \to \mathbb{R}$ by*

$$\begin{aligned} f_{RS}(t) &= f(t) && \text{if } R \geq f(t) \geq -S, \\ f_{RS}(t) &= -S && \text{if } -S > f(t), \\ f_{RS}(t) &= R && \text{if } f(t) > R. \end{aligned}$$

If f_{RS} is Riemann–Stieltjes integrable with respect to G for all $R, S > 0$, and we can find an L such that, given $\epsilon > 0$, we can find an $R_0(\epsilon) > 0$ such that

$$\left| \int_{\mathbb{R}} f_{RS}(x)\, dG(x) - L \right| < \epsilon$$

for all R, $S > R_0(\epsilon)$, then we say that f is Riemann–Stieltjes integrable with respect to G with Riemann–Stieltjes integral

$$\int_{\mathbb{R}} f(x)\, dG(x) = L.$$

[As usual, we add a warning that care must be exercised if $\int_{\mathbb{R}} |f(x)|\, dG(x)$ fails to converge.]

Lemma 9.47. (Tchebychev's inequality.) *If P is associated with a real-valued random variable X and $\mathbb{E}X^2$ exists, then*

$$\Pr\{X > a \text{ or } -a \geq X\} \leq \frac{\mathbb{E}X^2}{a^2}.$$

Proof. Observe that

$$x^2 \geq a^2 \mathbb{I}_{\mathbb{R}\setminus(-a,a]}(x)$$

for all x, and so

$$\int_{\mathbb{R}} x^2\, dG(x) \geq \int_{\mathbb{R}} a^2 \mathbb{I}_{\mathbb{R}\setminus(-a,a]}(x)\, dG(x).$$

Thus

$$\int_{\mathbb{R}} x^2\, dG(x) \geq a^2 \int_{\mathbb{R}} \mathbb{I}_{\mathbb{R}\setminus(-a,a]}(x)\, dG(x).$$

In other words,

$$\mathbb{E}X^2 \geq a^2 \Pr\{X \notin (-a,a]\},$$

which is what we want to prove. ■

Exercise 9.48. *(i) In the proof of Tchebychev's theorem we used various simple results on improper Riemann–Stieltjes integrals without proof. Identify these results and prove them.*

(ii) If $P(t) = (\frac{\pi}{2} - \tan^{-1} x)/\pi$, *show that* $\mathbb{E}X^2$ *does not exist. Show that this is also the case if we choose* P *given by*

$$\begin{aligned} P(t) &= 0 && \text{if } t < 1, \\ P(t) &= 1 - 2^{-n} && \text{if } 2^n \leq t < 2^{n+1},\ n \geq 0 \text{ an integer.} \end{aligned}$$

Exercise 9.49. *(Probabilists call this result 'Markov's inequality'. Analysts simply call it a 'Tchebychev type inequality'.) Suppose* $\phi : [0,\infty) \to \mathbb{R}$ *is an increasing continuous positive function. If* P *is associated with a real-valued random variable* X *and* $\mathbb{E}\phi(X)$ *exists, show that*

$$\Pr\{X \notin (-a,a]\} \leq \frac{\mathbb{E}\phi(X)}{\phi(a)}.$$

In elementary courses we deal separately with discrete random variables (typically, in our notation, P is constant on each interval $[n, n+1)$) and continuous random variables[8] (in our notation, P has continuous derivative; this derivative is the 'density function'). It is easy to construct mixed examples.

Exercise 9.50. *The height of water in a river is a random variable* Y *with* $\Pr\{Y \leq y\} = 1 - e^{-y}$ *for* $y \geq 0$. *The height is measured by a gauge which registers* $X = \min(Y, 1)$. *Find* $\Pr\{X \leq x\}$ *for all* x.

Are there real-valued random variables which are not just a simple mix of discrete and continuous? In Exercise K.253 (which depends on uniform convergence) we shall show that there are.

[8]See the previous footnote on the Holy Roman Empire.

The Riemann–Stieltjes formalism can easily be extended to deal with two random variables X and Y by using a two-dimensional Riemann–Stieltjes integral with respect to a function

$$P(x,y) = \Pr\{X \le x,\ Y \le y\}.$$

In the same way we can deal with n random variables $X_1, X_2, \ldots, X_n$. However, we cannot deal with infinite sequences $X_1, X_2, \ldots$ of random variables in the same way. Modern probability theory depends on measure theory.

In the series of exercises starting with Exercise K.180 and ending with Exercise K.186 we see that the Riemann–Stieltjes integral can be generalised further.

9.5. How long is a piece of string? ♡

The topic of line integrals is dealt with quickly and efficiently in many texts. The object of this section is to show why the texts deal with the matter in the way they do. The reader should not worry too much about the details and reserve such matters as 'learning definitions' for when she studies a more efficient text.

The first problem that meets us when we ask for the length of a curve is that it is not clear what a curve is. One natural way of defining a curve is that it is a continuous map $\gamma : [a,b] \to \mathbb{R}^m$. If we do this it is helpful to consider the following examples:

$$\begin{aligned}
&\gamma_1 : [0,1] \to \mathbb{R}^2 \text{ with } \gamma_1(t) = (\cos 2\pi t, \sin 2\pi t),\\
&\gamma_2 : [1,2] \to \mathbb{R}^2 \text{ with } \gamma_2(t) = (\cos 2\pi t, \sin 2\pi t),\\
&\gamma_3 : [0,2] \to \mathbb{R}^2 \text{ with } \gamma_3(t) = (\cos \pi t, \sin \pi t),\\
&\gamma_4 : [0,1] \to \mathbb{R}^2 \text{ with } \gamma_4(t) = (\cos 2\pi t^2, \sin 2\pi t^2),\\
&\gamma_5 : [0,1] \to \mathbb{R}^2 \text{ with } \gamma_5(t) = (\cos 2\pi t, -\sin 2\pi t),\\
&\gamma_6 : [0,1] \to \mathbb{R}^2 \text{ with } \gamma_6(t) = (\cos 4\pi t, \sin 4\pi t).
\end{aligned}$$

Exercise 9.51. *Trace out the curves* γ_1 *to* γ_6. *State in words how the curves* γ_1, γ_4, γ_5 *and* γ_6 *differ.*

Exercise 9.52. *(i) Which of the curves* γ_1 *to* γ_6 *are equivalent and which are not, under the following definitions.*

(a) Two curves $\boldsymbol{\tau}_1 : [a,b] \to \mathbb{R}^2$ *and* $\boldsymbol{\tau}_2 : [c,d] \to \mathbb{R}^2$ *are equivalent if there exist real numbers* A *and* B *with* $A > 0$ *such that* $Ac + B = a$, $Ad + B = b$ *and* $\boldsymbol{\tau}_1(At + B) = \boldsymbol{\tau}_2(t)$ *for all* $t \in [c,d]$.

(b) Two curves $\boldsymbol{\tau}_1 : [a,b] \to \mathbb{R}^2$ *and* $\boldsymbol{\tau}_2 : [c,d] \to \mathbb{R}^2$ *are equivalent if there exists a strictly increasing continuous surjective function* $\theta : [c,d] \to [a,b]$ *such that* $\boldsymbol{\tau}_1(\theta(t)) = \boldsymbol{\tau}_2(t)$ *for all* $t \in [c,d]$.

(c) Two curves $\boldsymbol{\tau}_1 : [a,b] \to \mathbb{R}^2$ *and* $\boldsymbol{\tau}_2 : [c,d] \to \mathbb{R}^2$ *are equivalent if there exists a continuous bijective function* $\theta : [c,d] \to [a,b]$ *such that* $\boldsymbol{\tau}_1(\theta(t)) = \boldsymbol{\tau}_2(t)$ *for all* $t \in [c,d]$.

(d) Two curves $\boldsymbol{\tau}_1 : [a,b] \to \mathbb{R}^2$ *and* $\boldsymbol{\tau}_2 : [c,d] \to \mathbb{R}^2$ *are equivalent if* $\boldsymbol{\tau}_1([a,b]) = \boldsymbol{\tau}_2([c,d])$.

(ii) If you know the definition of an equivalence relation, verify that conditions (a) to (d) do indeed give equivalence relations.

Naturally we demand that 'equivalent curves' (that is, curves which we consider 'identical') should have the same length. I think, for example, that a definition which gave different lengths to the curves described by $\boldsymbol{\gamma}_1$ and $\boldsymbol{\gamma}_2$ would be obviously unsatisfactory. However, opinions may differ as to when two curves are 'equivalent'. At a secondary school level, most people would say that the appropriate notion of equivalence is that given as (d) in Exercise 9.52, and thus the curves $\boldsymbol{\gamma}_1$ and $\boldsymbol{\gamma}_6$ should have the same length. Most of the time, most mathematicians[9] would say that the curves $\boldsymbol{\gamma}_1$ and $\boldsymbol{\gamma}_6$ are not equivalent and that, 'since $\boldsymbol{\gamma}_6$ is really $\boldsymbol{\gamma}_1$ done twice', $\boldsymbol{\gamma}_6$ should have twice the length of $\boldsymbol{\gamma}_1$. If the reader is dubious, she should replace the phrase 'length of curve' by 'distance traveled along the curve'.

The following chain of ideas leads to a natural definition of length. Suppose $\boldsymbol{\gamma} : [a,b] \to \mathbb{R}^m$ is a curve (in other words, $\boldsymbol{\gamma}$ is continuous). As usual, we consider dissections

$$\mathcal{D} = \{t_0, t_1, t_2, \ldots, t_n\}$$

with $a = t_0 \le t_1 \le t_2 \le \cdots \le t_n = b$. We write

$$L(\boldsymbol{\gamma}, \mathcal{D}) = \sum_{j=1}^{n} \|\boldsymbol{\gamma}(t_{j-1}) - \boldsymbol{\gamma}(t_j)\|,$$

where $\|\mathbf{a} - \mathbf{b}\|$ is the usual Euclidean distance between $\mathbf{a}$ and $\mathbf{b}$.

Exercise 9.53. *(i) Explain why* $L(\boldsymbol{\gamma}, \mathcal{D})$ *may be considered as the 'length of the approximating curve obtained by taking straight line segments joining each* $\boldsymbol{\gamma}(t_{j-1})$ *to* $\boldsymbol{\gamma}(t_j)$*'.*

(ii) Show that, if $\mathcal{D}_1$ *and* $\mathcal{D}_2$ *are dissections with* $\mathcal{D}_1 \subseteq \mathcal{D}_2$,

$$L(\boldsymbol{\gamma}, \mathcal{D}_2) \ge L(\boldsymbol{\gamma}, \mathcal{D}_1).$$

[9]But not all mathematicians and not all the time. One very important definition of length associated with the name Hausdorff agrees with the school level view.

Deduce that, if $\mathcal{D}_3$ and $\mathcal{D}_4$ are dissections, then

$$L(\gamma, \mathcal{D}_3 \cup \mathcal{D}_4) \geq \max(L(\gamma, \mathcal{D}_3), L(\gamma, \mathcal{D}_4)).$$

The two parts of Exercise 9.53 suggest the following definition.

Definition 9.54. *We say that a curve $\gamma : [a, b] \to \mathbb{R}^m$ is* rectifiable *if there exists a K such that $L(\gamma, \mathcal{D}) \leq K$ for all dissections $\mathcal{D}$. If a curve is rectifiable, we write*

$$\operatorname{length}(\gamma) = \sup_{\mathcal{D}} L(\gamma, \mathcal{D}),$$

the supremum being taken over all dissections of $[a, b]$.

Not all curves are rectifiable.

Exercise 9.55. *(i) Let $f : [0, 1] \to \mathbb{R}$ be the function given by the conditions $f(0) = 0$, f is linear on $[2^{-n-2}3, 2^{-n}]$ with $f(2^{-n-2}3) = 0$ and $f(2^{-n}) = (n+1)^{-1}$, f is linear on $[2^{-n-1}, 2^{-n-2}3]$ with $f(2^{-n-2}3) = 0$ and $f(2^{-n-1}) = (n+2)^{-1}$ $[n \geq 0]$.*

Sketch the graph of f and check that f is continuous. Show that the curve $\gamma : [0, 1] \to \mathbb{R}^2$ given by $\gamma(t) = (t, f(t))$ is not rectifiable.

(ii) Let $g : [-1, 1] \to \mathbb{R}$ be the function given by the conditions $g(0) = 0$, $g(t) = t^2 \sin |t|^\alpha$ for $t \neq 0$, where α is real. Show that g is differentiable everywhere, but that, for an appropriate choice of α, the curve $\boldsymbol{\tau} : [-1, 1] \to \mathbb{R}^2$ given by $\boldsymbol{\tau}(t) = (t, g(t))$ is not rectifiable.

Exercise 9.56. *(i) By using the intermediate value theorem, show that a continuous bijective function $\theta : [c, d] \to [a, b]$ is either strictly increasing or strictly decreasing.*

(ii) Suppose that $\gamma : [a, b] \to \mathbb{R}^m$ is a rectifiable curve and $\theta : [c, d] \to [a, b]$ is a continuous bijection. Show that $\gamma \circ \theta$ (where $\circ$ denotes composition) is a rectifiable curve and

$$\operatorname{length}(\gamma \circ \theta) = \operatorname{length}(\gamma).$$

(iii) Suppose that $\boldsymbol{\tau} : [-1, 1] \to \mathbb{R}^2$ is given by $\boldsymbol{\tau}(t) = (\sin \pi t, 0)$. Show that $\operatorname{length}(\boldsymbol{\tau}) = 4$. *Comment briefly.*

The next exercise is a fairly obvious but very useful observation.

Exercise 9.57. *Suppose that $\gamma : [a, b] \to \mathbb{R}^m$ is rectifiable. Show that, if $a \leq t \leq b$, then the restriction $\gamma|_{[a,t]} : [a, t] \to \mathbb{R}^m$ is rectifiable. If we write*

$$l_\gamma(t) = \operatorname{length}(\gamma|_{[a,t]}),$$

show that $l_\gamma : [a, b] \to \mathbb{R}$ is an increasing function with $l_\gamma(a) = 0$.

With a little extra effort we can say rather more about l_γ.

Exercise 9.58. *We use the hypotheses and notation of Exercise 9.57.*

(i) Suppose that $\boldsymbol{\gamma}$ has length L and that

$$\mathcal{D} = \{t_0, t_1, t_2, \ldots, t_n\},$$

with $a = t_0 \le t_1 \le t_2 \le \cdots \le t_n = b$, is a dissection such that

$$L(\boldsymbol{\gamma}, \mathcal{D}) \ge L - \epsilon$$

for some $\epsilon > 0$. Show that, if $1 \le j \le n$ and $t_{j-1} \le t \le t' \le t_j$, then

$$0 \le l_{\boldsymbol{\gamma}}(t') - l_{\boldsymbol{\gamma}}(t) \le \|\boldsymbol{\gamma}(t') - \boldsymbol{\gamma}(t)\| + \epsilon.$$

(ii) Use part (i) to show that $l_{\boldsymbol{\gamma}} : [a, b] \to \mathbb{R}$ is continuous.

Exercise 9.59. *[This a just a simple observation obscured by notation.] We use the hypotheses and notation of Exercise 9.57. Let L be the length of $\boldsymbol{\gamma}$. Show that, if $\boldsymbol{\gamma} : [a, b] \to \mathbb{R}^m$ is injective, then setting $\theta = l_{\boldsymbol{\gamma}}$, the map $\theta : [a, b] \to [0, L]$ is bijective and continuous. Explain why this means that θ^{-1} is continuous (see the proof of Lemma 5.65 if necessary). If we set $\boldsymbol{\tau} = \boldsymbol{\gamma} \circ \theta^{-1}$, show that*

$$l_{\boldsymbol{\tau}}(s) = s$$

for $0 \le s \le L$. We say that the curve $\boldsymbol{\tau}$ is the curve $\boldsymbol{\gamma}$ 'reparameterised by arc length'.

If we define $L_{\boldsymbol{\gamma}} : \mathbb{R} \to \mathbb{R}$ by

$$\begin{aligned} L_{\boldsymbol{\gamma}}(t) &= l_{\boldsymbol{\gamma}}(a) && \text{for } t \le a, \\ L_{\boldsymbol{\gamma}}(t) &= l_{\boldsymbol{\gamma}}(t) && \text{for } a < t < b, \\ L_{\boldsymbol{\gamma}}(t) &= l_{\boldsymbol{\gamma}}(b) && \text{for } b \le t, \end{aligned}$$

then, since $L_{\boldsymbol{\gamma}}$ is an increasing function, we can define the Riemann–Stieltjes integral

$$\int_a^b g(t)\, dL_{\boldsymbol{\gamma}}(t)$$

for any continuous function $g : [a, b] \to \mathbb{R}$. It follows that, if $f : \mathbb{R}^m \to \mathbb{R}$ is continuous, we can define the 'integral along the curve' by

$$\int_{\boldsymbol{\gamma}} f(\mathbf{x})\, ds = \int_a^b f(\boldsymbol{\gamma}(t))\, dL_{\boldsymbol{\gamma}}(t).$$

Note that, if the curve $\boldsymbol{\tau}$ is the curve $\boldsymbol{\gamma}$ 'reparameterised by arc length' in the sense of Example 9.59, then

$$\int_{\boldsymbol{\tau}} f(\mathbf{x})\, ds = \int_0^L f(\boldsymbol{\tau}(t))\, dt.$$

We have defined an integral along a curve, but we have not shown how to calculate it. If $\boldsymbol{\gamma}$ is sufficiently smooth, we can proceed as follows.

Exercise 9.60. *Suppose x, $y : [a,b] \to \mathbb{R}$ are continuous functions, $\epsilon > 0$, and A, B are real numbers such that*

$$|(x(t) - x(s)) - A(t-s)|,\ |(y(t) - y(s)) - B(t-s)| \le \epsilon$$

for all t, $s \in [a,b]$. Show that, if $\gamma : [a,b] \to \mathbb{R}^2$ is the curve given by $\gamma(t) = (x(t), y(t))$, then

$$\begin{aligned}\big((\min(|A| - \epsilon, 0))^2 + (\min(|B| - \epsilon, 0))^2\big)(t-s)^2 &\le \|\gamma(t) - \gamma(s)\|^2 \\ &\le \big((|A| + \epsilon)^2 + (|B| + \epsilon)^2\big)(t-s)^2\end{aligned}$$

for all t, $s \in [a,b]$. Deduce that γ is rectifiable and

$$\begin{aligned}\big((\min(|A| - \epsilon, 0))^2 + (\min(|B| - \epsilon, 0))^2\big)^{1/2}(b-a) &\le \operatorname{length}(\gamma) \\ &\le \big((|A| + \epsilon)^2 + (|B| + \epsilon)^2\big)^{1/2}(b-a).\end{aligned}$$

Exercise 9.61. *This exercise uses the ideas of the previous exercise together with the mean value inequality. Suppose that x, $y : [a,b] \to \mathbb{R}$ have continuous derivatives (with the usual conventions about left and right derivatives at end points). Show that the curve $\gamma : [a,b] \to \mathbb{R}^2$ given by $\gamma(t) = (x(t), y(t))$ is rectifiable and, by considering the behaviour of*

$$\frac{l_\gamma(s) - l_\gamma(t)}{s - t}$$

as $s \to t$, show that l_γ is everywhere differentiable on $[a,b]$ with

$$l'_\gamma(t) = (x'(t)^2 + y'(t)^2)^{1/2}.$$

Explain why your proof does not apply to the counterexample in part (ii) of Exercise 9.55.

Exercise 9.62. *Use Exercise 9.43 to compute the lengths of the curves γ_1 to γ_6 defined on page 213.*

Using Exercise 9.43, we see that, if x, $y : [a,b] \to \mathbb{R}^2$ have continuous derivatives (with the usual conventions about left and right derivatives at end points), and we consider the curve $\gamma : [a,b] \to \mathbb{R}^2$ given by $\gamma(t) = (x(t), y(t))$, then, if $f : \mathbb{R}^2 \to \mathbb{R}$ is continuous,

$$\int_\gamma f(\mathbf{x})\, ds = \int_a^b f(x(t), y(t))(x'(t)^2 + y'(t))^{1/2}\, dt,$$

a result coinciding with that we would expect from mathematical methods courses.

Exercise 9.63. *Extend this result to smooth curves $\gamma : [a,b] \to \mathbb{R}^m$, giving as much (or as little) detail as seems to you desirable.*

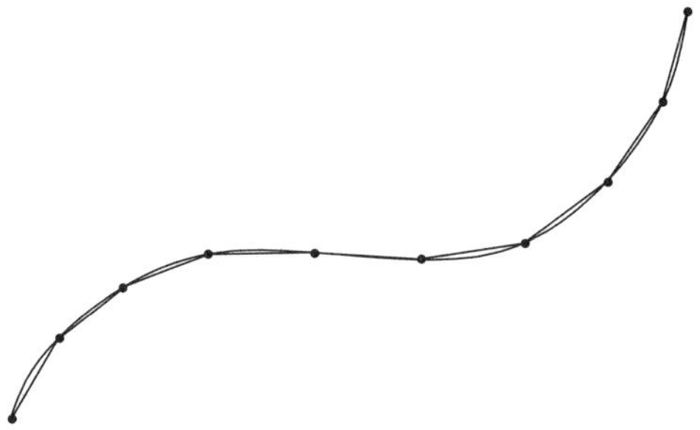

Figure 9.1. Arc length via polygonal paths

This seems highly satisfactory until we read more advanced texts than this and discover that, instead of using the sophisticated general ideas of this section, these advanced texts use a rigorised version of the mathematical methods approach. Why do they do this?

There are various reasons, but one of the most important is that the approaches developed here do not apply in higher dimensions. More specifically, we obtained arc length by 'approximating by polygonal paths' as in Figure 9.1.

In 1890, H. A. Schwarz[10] published an example showing that any naive attempt to find the area of a surface by 'approximating by polyhedral surfaces' must fail. The example provides a good exercise in simple three-dimensional geometry.

Exercise 9.64. (Schwarz's counterexample.) *Split a* 1 *by* 2π *rectangle into* mn *rectangles each* m^{-1} *by* $2\pi n^{-1}$*, and split each of these into four triangles by means of the two diagonals. Bend the large rectangle into a cylinder of height* 1 *and circumference* 2π*, and use the vertices of the* $4mn$ *triangles as the vertices of an inscribed polyhedron with* $4mn$ *flat triangular faces*[11]*. We call the area of the resulting polyhedron* $A(m,n)$*.*

In Figure 9.2 we show one of the nm *rectangles* $ABCD$ *with diagonals meeting at* X *before and after bending. In our discussion, we shall refer only to the system after bending. Let* W *be the mid-point of the arc* AB*,* Y *the mid-point of the chord* AB *and* Z *the mid-point of the line* DA *as shown in Figure 9.3.*

By observing that $WA = XZ$*, or otherwise, show that the area of the triangle* AXD *is* $(2m)^{-1}\sin(\pi/n)$*. Show that* YW *has length* $2(\sin(\pi/2n))^2$ *and deduce, or prove otherwise, that the triangle* XAB *has area given by*

[10] The Schwarz of the Cauchy–Schwarz inequality.

[11] The last two sentences are copied directly from Billingsley's splendid *Probability and Measure* **[6]**, since I cannot see how to give a clearer description than his.

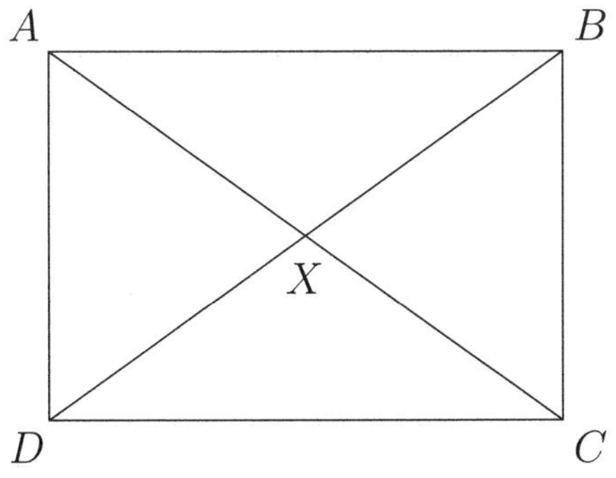

Figure 9.2. Part of Schwarz's polyhedral approximation before bending

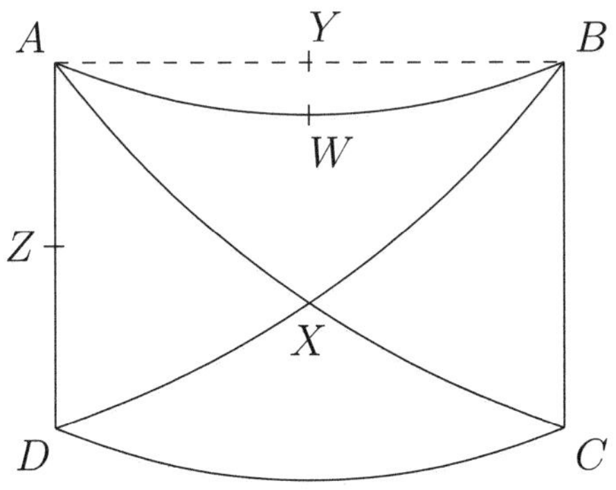

Figure 9.3. Part of Schwarz's polyhedral approximation after bending

$\sin(\pi/n)\big((2m)^{-2} + 4(\sin(\pi/2n))^4\big)$. *Conclude that*

$$A(m,n) = n\sin\frac{\pi}{n}\left(1 + \left(1 + 16m^2\left(\sin\frac{\pi}{2n}\right)^4\right)^{1/2}\right).$$

If we choose $n_j \to \infty$ and $m_j \to \infty$ in such a way that $m_j/n_j^2 \to \lambda$ as $j \to \infty$, what happens to $A(m_j, n_j)$? Can you choose $n_j \to \infty$ and $m_j \to \infty$ so that $A(m_j, n_j) \to \infty$ as $j \to \infty$? Can you choose $n_j \to \infty$ and $m_j \to \infty$ so that $A(m_j, n_j)$ is bounded but does not converge? Can you choose $n_j \to \infty$ and $m_j \to \infty$ so that $A(m_j, n_j) \to \pi$?

Exercise 9.65. *Explain in words and without using calculations why, if we fix n, $A(m,n) \to \infty$ as $m \to \infty$. (Unless you can give a simple geometric account of what is going on, you have not understood Schwarz's example.) Deduce directly that we can choose $n_j \to \infty$ and $m_j \to \infty$ such that $A(m_j, n_j) \to \infty$ as $j \to \infty$. [Thus, if we simply want to show that the* *naive approach fails, we do not need the calculations of the previous exercise. However, if we want more information (for example, necessary and sufficient conditions for $A(m_j, n_j) \to 2\pi$), then we must do the calculations.]*

Once the reader has grasped the point of Exercise 9.65, she may feel that it is obvious what is wrong. However, the problem is not that we cannot see what the area of a cylinder ought to be, but that we cannot think of a simple definition of area which will apply to general surfaces in the same way that our definition of length applied to general curves. Up to now, all attempts to produce a definition of area to parallel our definition of length have failed[12].

If a surface is sufficiently well behaved, it is more or less clear how to frame a notion of approximation by polyhedra which excludes Schwarz's example. But, if a surface is sufficiently well behaved, we can produce a rigorous definition of its area by tidying up the standard mathematical methods definition. If we are going to do this for the area of a surface and its higher-dimensional analogues, it seems a waste of time to have a special theory for length. Life, as the saying goes, is too short to stuff an olive. Our final exercise gives the standard development.

Exercise 9.66. (Standard treatment of line integrals.) *We deal only with curves $\boldsymbol{\gamma} : [a, b] \to \mathbb{R}^m$ which are continuously differentiable (so that, writing*

$$\boldsymbol{\gamma}(t) = (\gamma_1(t), \gamma_2(t), \dots, \gamma_m(t)),$$

we know that $\gamma_j' : [a, b] \to \mathbb{R}$ exists and is continuous). If $f : \mathbb{R}^m \to \mathbb{R}$ is continuous, we **define**

$$\int_{\boldsymbol{\gamma}} f(\mathbf{x})\, ds = \int_a^b f(\gamma_1(t), \gamma_2(t), \dots, \gamma_m(t))(\gamma_1'(t)^2 + \gamma_2'(t)^2 + \dots + \gamma_m'(t)^2)^{1/2}\, dt.$$

Suppose that $\theta : [c, d] \to [a, b]$ is a surjective function with continuous derivative. Explain why $\boldsymbol{\tau} = \boldsymbol{\gamma} \circ \theta$ is a continuously differentiable function from $[c, d]$ to $\mathbb{R}^m$. Show that, if θ is injective, then

$$\int_{\boldsymbol{\tau}} f(\mathbf{x})\, ds = \int_{\boldsymbol{\gamma}} f(\mathbf{x})\, ds,$$

but give an example to show that, if θ is not injective, this may not hold.

We **define** *the length of $\boldsymbol{\gamma}$ to be $\int_{\boldsymbol{\gamma}} 1\, ds$.*

[12]There are other ways of tackling this problem. Once again I refer the reader to the circle of ideas associated with the name of Hausdorff.

Chapter 10

Metric Spaces

10.1. Sphere packing ♡

(In this section and the next we are much more interested in ideas than rigour. We shall use methods and results which go well beyond the scope of this book.)

Human beings prefer order to disorder. Asked to pack a crate of oranges, we do not throw them in at random, but try to pack them in a regular pattern. But choosing patterns requires insight, and sometimes we have no insight. Consider the problem of packing n-dimensional balls in a very large box. (We shall take our balls and boxes to be open, but it should be clear that such details do not matter.)

As an example of a regular packing, let the balls have radius $1/2$ and let us adopt a cubical packing so that the centre of each ball is one of the integer points $\mathbb{Z}^n$. Is this reasonably efficient or not? To answer this question it is helpful to know the volume $V_n(r)$ of an n-dimensional ball of radius r.

Lemma 10.1. *(i) If we write $V_n = V_n(1)$, then $V_n(r) = V_n r^n$.*

(ii) If $a > 0$ and $\mathbb{I}_{[0,a]} : [0, \infty) \to \mathbb{R}$ is defined by $\mathbb{I}_{[0,a]}(t) = 1$ if $t \in [0, a]$, $\mathbb{I}_{[0,a]}(t) = 0$, otherwise, then

$$\int_{\mathbb{R}^n} \mathbb{I}_{[0,a]}(\|\mathbf{r}\|)\, dV = nV_n \int_0^\infty \mathbb{I}_{[0,a]}(r) r^{n-1}\, dr.$$

(iii) If $f : [0, \infty) \to \mathbb{R}$ is given by $f = \sum_{j=1}^k \lambda_j \mathbb{I}_{[0,a_j]}$, then

$$\int_{\mathbb{R}^n} f(\|\mathbf{r}\|)\, dV = nV_n \int_0^\infty f(r) r^{n-1}\, dr.$$

(iv) If $f : [0, \infty) \to \mathbb{R}$ *is such that* $f(r)r^{n+1} \to 0$ *as* $r \to \infty$, *then*

$$\int_{\mathbb{R}^n} f(\|\mathbf{r}\|)\, dV = nV_n \int_0^\infty f(r)r^{n-1}\, dr.$$

(v) Taking $f(r) = \exp(-r^2/2)$ *in (ii), we obtain*

$$(2\pi)^{n/2} = nV_n \int_0^\infty r^{n-1} \exp(-r^2/2)\, dr,$$

and so

$$V_{2n} = \frac{\pi^n}{n!}, \quad V_{2n-1} = \frac{n! 2^{2n} \pi^{n-1}}{(2n)!}.$$

Sketch proof. (i) Use similarity.

(ii) This is a restatement of (i).

(iii) Use linearity of the integral.

(iv) Use an approximation argument.

(v) Using repeated integration,

$$\int_{\mathbb{R}^n} f(\|\mathbf{r}\|)\, dV = \int_{-\infty}^{\infty}\int_{-\infty}^{\infty}\cdots\int_{-\infty}^{\infty} \exp(-(x_1^2 + x_2^2 + \cdots + x_n^2)/2)\, dx_1\, dx_2 \ldots dx_n$$
$$= \int_{-\infty}^{\infty} \exp(-x_1^2/2)\, dx_1 \int_{-\infty}^{\infty} \exp(-x_2^2/2)\, dx_2 \cdots \int_{-\infty}^{\infty} \exp(-x_n^2/2)\, dx_n = (2\pi)^{n/2}.$$

On the other hand, integration by parts gives

$$\int_0^\infty r^{n-1} \exp(-r^2/2)\, dr = (n-2) \int_0^\infty r^{n-3} \exp(-r^2/2)\, dr,$$

so

$$(2\pi)^{n/2} = nV_n \int_0^\infty r^{n-1} \exp(-r^2/2)\, dr$$
$$= n(n-2)V_n \int_0^\infty r^{n-3} \exp(-r^2/2)\, dr = (2\pi)^{(n-2)/2} nV_n/V_{n-2}$$

and $V_n = (2\pi)V_{n-2}/n$. The stated results follow by induction. ▲

Remark: Since we are not seeking rigour in this section, I have only sketched a proof. However, few mathematicians would demand more proof for this result. As I emphasise in Appendix C, we need rigour when we make general statements which must apply to objects we have not yet even imagined. Here, we need a result about a specific property of a specific object (the volume of an n-dimension sphere), so we can have much more confidence in our sketched proof.

If we use a cubical packing, the proportion of space occupied by balls is $2^{-n}V_n$ (think about the ball with center $\mathbf{q}$ and radius $1/2$ inside the cube

$\prod_{j=1}^{n}(q_j - 1/2, q_j + 1/2))$, and Lemma 10.1 shows that this proportion drops very rapidly indeed as n becomes large.

What happens if we 'just place balls wherever we can'. Suppose we are trying to fill the cube

$$C = \{\mathbf{x} : -N - 1/2 < x_j < N + 1/2 \ \ [1 \le j \le n]\}$$

with non-intersecting balls of radius $1/2$, and that we have managed to place m such balls

$$S_k = \{\mathbf{x} : \|\mathbf{x} - \mathbf{y}_k\| < 1/2\} \ \ [1 \le k \le m].$$

Now consider the balls with the same centres and doubled radius

$$\sigma_k = \{\mathbf{x} : \|\mathbf{x} - \mathbf{y}_k\| < 1\} \ \ [1 \le k \le m].$$

A little thought shows that, if $\mathbf{y}$ lies inside the cube

$$C' = \{\mathbf{x} : -N < x_j < N \ [1 \le j \le n]\}$$

and does not lie in any σ_k, then the ball

$$S = \{\mathbf{x} : \|\mathbf{x} - \mathbf{y}\| < 1/2\}$$

lies in C and does not intersect any S_k, so we may add another ball to our collection. Such a point $\mathbf{y}$ will always exist if

$$\operatorname{Vol} C' > \sum_{k=1}^{m} \operatorname{Vol} \sigma_k$$

(since then $\operatorname{Vol} C' > \operatorname{Vol} \bigcup_{k=1}^{m} \sigma_k$, and so $C' \setminus \bigcup_{k=1}^{m} \sigma_k \neq \varnothing$). Thus if

$$\operatorname{Vol} C' > mV_n,$$

we can add extra balls and so, by just filling up available spaces, without following any pattern, we can find at least $\operatorname{Vol} C'/V_n$ disjoint balls of radius $1/2$ in C. Since the volume of a ball of radius $1/2$ is $2^{-n}V_n$ the proportion of space occupied by balls is $2^{-n} \operatorname{Vol} C'/ \operatorname{Vol} C = 2^{-n}(1 + \frac{1}{2N})^{-n}$; and so, for N large, the proportion of space occupied by balls is essentially 2^{-n}, an astounding gain in efficiency over cubical packing.

What morals should we draw? The most immediate is that we do not know what an n-dimensional ball looks like! (Here, 'we' refers to the writer and most of his audience. Specialists in topics like the geometry of Banach spaces do have substantial insight.) The second is that, when we have little insight, trying to impose order on things may well be counter-productive.

10.2. Shannon's theorem ♡

It costs money to send messages. In the old days, telegrams might be charged for by the word. Nowadays, we pay for communication channels which carry so many bits per second (so we are charged by the bit). Suppose that I can afford to send three bits of information to my partner. Then we can make up eight words 000, 001, ... and use them to send eight different messages. For example, 000 could mean 'send money', 001 'come at once', 010 'sell all shares', 011 'flee the country' and so on. Unfortunately, bits are sometimes received wrongly, so 011 could be sent and 001 received with unfortunate consequences. We might, therefore, decide to have only two messages 000 ('all well') and 111 ('flee at once') so that, if at most one digit was received wrongly, my partner could still take the correct action.

Exercise 10.2. *Explain why the last statement is true.*

If I can afford to send 100 bits, then we can still decide to only have two messages 0000...0 and 1111...1; but, if mistakes are rare, this is extremely wasteful. How many different messages can we send and still be reasonably confident that we can tell which message was transmitted even when errors occur? This section gives a simple but very useful model for the problem and solves it.

Consider $X = \{0,1\}^n$ with the rule that, when we add two elements $\mathbf{x}$ and $\mathbf{y}$ of X, the jth component of $\mathbf{x}+\mathbf{y}$ has the form $x_j + y_j$ where $0+0 = 1+1 = 0$ and $1+0 = 0+1 = 1$[1]. Each element $\mathbf{x} \in \{0,1\}^n$ may be considered as a 'word', so X contains 2^n possible words. In transmitting a word over a noisy channel it may become corrupted to

$$\mathbf{x}+\mathbf{e},$$

where each coordinate e_j of the random error $\mathbf{e}$ takes the value 1 with probability p and the value 0 with probability $1-p$, independent of the other coordinates $[0 < p < 1/2]$.

Exercise 10.3. *Why do we not have to consider $p > 1/2$? Why is it hopeless to consider $p = 1/2$?*

If we choose q such that $1/2 > q > p > 0$, then the law of large numbers tells us that, provided n is large enough, it is very unlikely that more than qn of the ϵ_j are non-zero[2]. We may, therefore, simplify our problem to one in which we know that at most qn of the ϵ_j are non-zero.

It is natural to introduce the following notion of the distance $d(\mathbf{x},\mathbf{y})$ between two words (the 'Hamming distance').

---

[1]Formally, we consider the field $\mathbb{Z}_2 = \mathbb{F}_2$ with two elements 0 and 1. $X = \mathbb{F}_2^n$ is then a vector space of dimension n over $\mathbb{F}_2$.

[2]We prove this result directly in Exercise K.191.

Definition 10.4. *If* $\mathbf{x}$, $\mathbf{y} \in X$, *we write*

$$d(\mathbf{x}, \mathbf{y}) = \sum_{j=1}^{n} |x_i - y_i|.$$

Suppose that we only transmit the words $\mathbf{y}_1$, $\mathbf{y}_2$, ... $\mathbf{y}_m$ and that the balls

$$S_k = \{\mathbf{x} : d(\mathbf{x}, \mathbf{y}_k) < qn\} \quad [1 \le k \le m]$$

do not intersect. Since there are no more than qn errors, the received message will lie in one of the balls, S_j, say, and the transmitted message will have been $\mathbf{y}_j$. Thus our system allows us to communicate m distinct messages correctly in spite of the random noise $\mathbf{e}$.

This sounds impressive, but is not very useful unless m is reasonably large. How large can m be? This is clearly a variant of our orange packing problem. The natural way to proceed is to define the volume of a subset E of X to be the number of points in E. Reusing the ideas of our previous argument, we consider the balls with the same centres and doubled radius

$$\sigma_k = \{\mathbf{x} : d(\mathbf{x}, \mathbf{y}_k) < 2qn\} \quad [1 \le k \le m].$$

If $\mathbf{y}$ does not lie in any σ_k, then the ball

$$S = \{\mathbf{x} : d(\mathbf{x}, \mathbf{y}) < qn\}$$

does not intersect any S_k, so we may add another ball to our collection. Such a point $\mathbf{y}$ will always exist if

$$\operatorname{Vol} X > \sum_{k=1}^{m} \operatorname{Vol} \sigma_k,$$

that is, if

$$2^n > \sum_{k=1}^{m} \operatorname{Vol} \sigma_k.$$

The only problem that faces us is to estimate the volume of a ball in this context.

The key turns out to be a simple but, to many people, initially surprising result.

Lemma 10.5. *Suppose* $0 < \lambda < 1/2$.

(i) If $1 \le r \le \lambda n$, *then*

$$\binom{n}{r-1} \le \frac{\lambda}{1-\lambda} \binom{n}{r}.$$

(ii) There exists a constant $A(\lambda)$ such that a ball of radius λn in X has volume at most $A(\lambda)\binom{n}{[\lambda n]}$ and at least $\binom{n}{[\lambda n]}$.

Proof. (i) Observe that

$$\binom{n}{r-1}\Big/\binom{n}{r} = \frac{r}{n+1-r} = \frac{\frac{r}{n}}{1-\frac{r-1}{n}} \leq \frac{\lambda}{1-\lambda}.$$

(ii) Thus, if $r_\lambda = [\lambda n]$, the greatest integer less than λn, we have

$$\begin{aligned}\text{Volume of a ball of radius } \lambda n &= \sum_{r\leq\lambda n}\binom{n}{r}\\ &\leq \sum_{r\leq r_\lambda}\left(\frac{\lambda}{1-\lambda}\right)^{r_\lambda - r}\binom{n}{r_\lambda}\\ &\leq \sum_{m=0}^{\infty}\left(\frac{\lambda}{1-\lambda}\right)^{m}\binom{n}{r_\lambda}\\ &= \frac{1-\lambda}{1-2\lambda}\binom{n}{[\lambda n]}.\end{aligned}$$

Taking $A(\lambda) = (1-\lambda)/(1-2\lambda)$, we have the required result. ■

Exercise 10.6. *(i) If $1 > \lambda > 1/2$, find a good estimate for the volume of a ball of radius λn in X when n is large.*

(ii) Lemma 10.5 says, in effect, that the volume of an appropriate ball in X is concentrated near its 'surface' (that is, those points whose distance from the centre is close to the radius). To what extent is this true for ordinary balls in $\mathbb{R}^m$ when m is large?

In part (i) of Lemma 10.7 we recall a simple form of Stirling's formula (obtained in Exercise 9.9 (iii)). In the rest of the lemma we use it to obtain estimates of the volume $V(\lambda, n)$ of a ball of radius λn in $X = \{0,1\}^n$ when n is large. We use the notation

$$\log_a b = \frac{\log b}{\log a}$$

where a, $b > 0$. (The reader probably knows that $\log_a b$ is called 'the logarithm of b to base a'. She should check the relation $a^{\log_a b} = b$.)

Lemma 10.7. *(i) $\log_e N! = N\log_e N + N + \theta(N)N$, where $\theta(N) \to 0$ as $N \to \infty$.*

(ii) If $0 < \lambda < 1/2$, then

$$n^{-1}\log_e V(\lambda, n) \to -\lambda\log_e\lambda - (1-\lambda)\log_e(1-\lambda)$$

as $n \to \infty$.

(iii) If $0 < \lambda < 1/2$, *then*

$$n^{-1}\log_2 V(\lambda, n) \to -\lambda\log_2\lambda - (1-\lambda)\log_2(1-\lambda)$$

as $n \to \infty$.

Proof. (i) This is Exercise 9.9 (iii).

(ii) In what follows we shall be replacing a real number y by an integer m with $|m-y| \leq 1$. Observe that, if f is well behaved, the mean value inequality gives $|f(y) - f(m)| \leq \sup_{|t-y|\leq 1}|f'(t)|$.

By Lemma 10.5, part (i) of the present lemma, and the remark just made, there exist $\theta_1(n)$, $\theta_2(n) \to 0$ as $n \to \infty$ such that

$$\begin{aligned}
&n^{-1}\log_e V(\lambda, n) = n^{-1}\log_e\binom{n}{[\lambda n]}\\
&= n^{-1}\big(\log_e n! - \log_e(n-[\lambda n])! - \log_e[\lambda n]!\big)\\
&= n^{-1}\big(n\log_e n + n - (n-[\lambda n])\log_e(n-[\lambda n]) - [\lambda n] - [\lambda n]\log_e[\lambda n]\big) + \theta_1(n)\\
&= n^{-1}\big(n\log_e n + n - (n-\lambda n)\log_e(n-\lambda n) - \lambda n - \lambda n\log_e\lambda n\big) + \theta_2(n)\\
&= n^{-1}\big(n\log_e n + n - (1-\lambda)n(\log_e(1-\lambda) + \log_e n) - \lambda n(\log_e\lambda + \log_e n)\big) + \theta_2(n)\\
&= -\lambda\log_e\lambda - (1-\lambda)\log_e(1-\lambda) + \theta_2(n),
\end{aligned}$$

and this is the required result.

(iii) Just apply the definition of $\log_2$. ■

Let us write $H(0) = H(1) = 0$ and

$$H(\lambda) = -\lambda\log_2\lambda - (1-\lambda)\log_2(1-\lambda)$$

for $0 < \lambda < 1$.

Exercise 10.8. *Prove the following results and then sketch the graph of* H.

(i) $H(s) = H(1-s)$ *for all* $s \in [0,1]$.

(ii) $H : [0,1] \to \mathbb{R}$ *is continuous.*

(iii) H *is twice differentiable on* $(0,1)$ *with* $H''(t) < 0$ *for all* $t \in (0,1)$.

(iv) $0 \leq H(t) \leq 1$ *for all* $t \in [0,1]$. *Identify the maximum and minimum points.*

(v) $H(t)/t \to 1$ *as* $t \to 0$ *through positive values.*

Our discussion has shown that we can find $2^{n(1-H(2q))}$ code words all a Hamming distance at least $q'n$ apart. We can immediately interpret this result in terms of our original problem.

Lemma 10.9. *Consider our noisy channel. If* p, *the probability of error in a single bit, is less than* $1/4$, *then, choosing* $p < p' < 1/4$, *we can transmit, with error rate as small as we please, so that information is passed at a rate of* $1 - H(2p')$ *times that of our original channel.*

Shannon pushed the argument a little bit further. Let $\eta > 0$ be small and let N be such that

$$N \times \operatorname{Vol} \text{ball of radius } qn \approx \eta \operatorname{Vol} X.$$

Choose N points $\mathbf{y}_j$ at random in X and let them be code words. There is no reason to expect that the balls

$$S_k = \{\mathbf{x} : d(\mathbf{x}, \mathbf{y}_k) < qn\} \quad [1 \leq k \leq N]$$

do not intersect (and excellent reasons for supposing that they will).

Exercise 10.10. *(This requires some experience with probability.) Write* $Y_{ij} = 1$ *if* S_i *and* S_j *intersect* $[i \neq j]$, $Y_{ij} = 0$, *otherwise. What is* $\mathbb{E}Y_{ij}$*? What is* $\mathbb{E}\sum_{1 \leq j < i \leq N} Y_{ij}$*? If* q *is small, show that* $\mathbb{E}\sum_{1 \leq j < i \leq N} Y_{ij}$ *is large. What does this tell you about the number of intersecting balls?*

If the balls intersect, then, even if less than qn errors are made in transmission, the received message may belong to two or more balls. In such a case we simply agree that our system has failed. How probable is such a failure? Suppose we transmit $\mathbf{y}_1$ and receive $\mathbf{z} = \mathbf{y}_1 + \mathbf{e}$. Since the remaining $\mathbf{y}_j$ with $2 \leq j \leq N$ have been chosen at random, independently of $\mathbf{y}_j$, the probability that $\mathbf{z}$ lies in $\bigcup_{j=2}^N S_j$ has nothing to do with how we defined $\mathbf{z}$ and is just

$$\frac{\operatorname{Vol}(\bigcup_{j=2}^N S_j)}{\operatorname{Vol} X} \leq \frac{\sum_{j=2}^N \operatorname{Vol} S_j}{\operatorname{Vol} X} \approx \eta.$$

Thus the probability of failure of the type discussed in this paragraph can be kept at any specified level η whilst still allowing roughly $\eta \operatorname{Vol} S_1 / \operatorname{Vol} X$ code words.

Exercise 10.11. *How can you reconcile this last result with Exercise 10.10?*

We have thus obtained the full Shannon's theorem.

Theorem 10.12. (Shannon.) *Consider our noisy channel. If* p*, the probability of error in a single bit, is less than* $1/2$*, then, choosing* $p < p' < 1/2$*, we can transmit, with error rate as small as we please, so that information is passed at a rate of* $1 - H(p')$ *times that of our original channel.*

Exercise 10.13. *(For the knowledgeable only.) The statement 'We have thus obtained the full Shannon's theorem' is a bit optimistic, since there are quite a lot of loose ends to tie up. Tie them up.*

In the 50 or so years since Shannon proved his theorem, no one has produced non-random codes to match his random codes for long code words. However, non-random methods are beginning to catch up. For our first problem of sphere packing in $\mathbb{R}^n$, the best known packings in high dimensions are lattice (so very highly ordered) packings. The experts believe that this reflects our ignorance rather than the truth.

10.3. Metric spaces

We discovered our proof of Shannon's theorem (at least in the form of Lemma 10.9) by drawing heavily on analogies with distance and volume in $\mathbb{R}^n$. Many proofs in analysis and elsewhere have been discovered by exploiting analogies between the usual distance in $\mathbb{R}^n$ and 'things that look more or less like distance'.

Recalling that 'once is a trick, twice is a method, thrice a theorem and four times a theory', we seek to codify this insight.

Our first try is to model the properties of Euclidean distance set out in Lemma 4.4.

Definition 10.14. *Let V be a real vector space and $N : V \to \mathbb{R}$ a function. We write $\|\mathbf{x}\| = N(\mathbf{x})$ and say that $\| \ \|$ is a norm if the following conditions hold:*

(i) $\|\mathbf{x}\| \geq 0$ *for all* $\mathbf{x} \in V$.

(ii) If $\|\mathbf{x}\| = 0$, *then* $\mathbf{x} = \mathbf{0}$.

(iii) If $\lambda \in \mathbb{R}$ *and* $\mathbf{x} \in V$, *then* $\|\lambda\mathbf{x}\| = |\lambda|\|\mathbf{x}\|$.

(iv) (The triangle inequality) If $\mathbf{x}, \mathbf{y} \in V$, *then* $\|\mathbf{x} + \mathbf{y}\| \leq \|\mathbf{x}\| + \|\mathbf{y}\|$.

Note that some mathematicians prefer to write $\| \cdot \|$ instead of $\| \ \|$. The '$\cdot$' acts as a placeholder.

Exercise 10.15. *Check that the usual Euclidean norm on $\mathbb{R}^m$ is indeed a norm.*

The appropriate definition when V is a vector space over $\mathbb{C}$, rather than $\mathbb{R}$, runs similarly.

Definition 10.16. *Let V be a complex vector space and $N : V \to \mathbb{R}$ a function: We write $\|\mathbf{x}\| = N(\mathbf{x})$ and say that $\| \ \|$ is a norm if the following conditions hold.*

(i) $\|\mathbf{x}\| \geq 0$ *for all* $\mathbf{x} \in V$.

(ii) If $\|\mathbf{x}\| = 0$, *then* $\mathbf{x} = \mathbf{0}$.

(iii) If $\lambda \in \mathbb{C}$ *and* $\mathbf{x} \in V$, *then* $\|\lambda\mathbf{x}\| = |\lambda|\|\mathbf{x}\|$.

(iv) (The triangle inequality) If $\mathbf{x}$, $\mathbf{y} \in V$, *then* $\|\mathbf{x} + \mathbf{y}\| \leq \|\mathbf{x}\| + \|\mathbf{y}\|$.

We say that $(V, \| \ \|)$ *is a normed space.*

However, we wish to have a notion of distance which applies to spaces which do not have a vector space structure. We seek a definition modelled on those properties of Euclidean distance which do not refer to vector space structures. (Thus you should both *compare* and *contrast* Definition 10.14.)

Definition 10.17. *We say that* (X, d) *is a metric space if* X *is a set and* $d : X^2 \to \mathbb{R}$ *is a function with the following properties:*

(i) $d(x, y) = 0$ *if and only if* $x = y$.

(ii) $d(x, y) = d(y, x)$ *for all* x, $y \in X$.

(iii) (The triangle inequality) $d(x, z) \leq d(x, y) + d(y, z)$ *for all* x, y, $z \in X$.

Exercise 10.18. *Show that, if* (X, d) *is a metric space, then* $d(x, y) \geq 0$ *for all* $x, y \in X$.

It is a remarkable fact that the general notion of a metric space was first introduced by Fréchet in 1906 and the general notion of a normed space by Banach in 1932! (We shall see one reason for the late emergence of the notion of a norm in Theorem 10.45.)

Exercise 10.19. *If* $\| \ \|$ *is a norm over the (real or complex) vector space* V *and we set* $d(\mathbf{x}, \mathbf{y}) = \|\mathbf{x} - \mathbf{y}\|$, *show that* (V, d) *is a metric space.*

The conditions of Definition 10.17 are sometimes called the axioms for a metric space. However, they are not axioms in the same sense as the Fundamental Axiom of Analysis which asserts a fundamental principle of argument which is to be accepted without further proof[3]. Instead they try to isolate the common properties of an interesting class of mathematical objects. (Compare the definition of birds as 'oviparous, warm-blooded, amniotic vertebrates which have their anterior extremities transformed into wings. Metacarpus and fingers carry feathers or quills. There is an intertarsal joint and not more than 4 toes, of which the first is a hallux.' Thus penguins are and dodos were birds but dragonflies and bats are not.)

At first sight, axiom systems like that for metric systems seem merely methods for economising on the number of theorems we have to prove. A theorem which applies to all metric spaces will not need to be proved for each individually. However, a successful axiom system should be a powerful research tool by suggesting appropriate questions. (You will learn more about a penguin by comparing it to a hawk than by comparing it to a frog.) Thus if we have a 'space on which we do analysis' we can ask 'can we make it into a metric space?'. If it is a metric space we can then compare and contrast it with other metric spaces that we know, and this may well suggest new theorems.

[3]For the time being and within the context of a particular system. The axioms of one system may be theorems in another. In Section 14.3 the Fundamental Axiom of Analysis becomes itself an object of discussion.

Although concepts like metric space are intended to apply to important natural systems, we often study 'toy examples' in order to gain insight into what can happen in such a system. Here is such a 'toy example'.

Exercise 10.20. *We work in* $\mathbb{R}^m$ *with the usual Euclidean norm. Show that if* $d(\mathbf{x},\mathbf{y}) = \|\mathbf{x}\| + \|\mathbf{y}\|$ *when* $\mathbf{x} \neq \mathbf{y}$ *and* $d(\mathbf{x},\mathbf{x}) = 0$, *then* $(\mathbb{R}^m, d)$ *is a metric space.*

This metric is called the British Railway non-stop metric. To get from A to B in the fastest time, we travel via London (the origin). Here is another version of the same idea, which we call the British Railway stopping metric.

Exercise 10.21. *We work in* $\mathbb{R}^m$ *with the usual Euclidean norm. If there exists a real* λ *such that* $\lambda\mathbf{x} = \mathbf{y}$, *we write* $d(\mathbf{x},\mathbf{y}) = \|\mathbf{x} - \mathbf{y}\|$. *Otherwise, we set* $d(\mathbf{x},\mathbf{y}) = \|\mathbf{x}\| + \|\mathbf{y}\|$. *Show that* $(\mathbb{R}^m, d)$ *is a metric space.*

Much of the 'mere algebra' which we did for the Euclidean metric carries over with hardly any change to the general metric case. Compare the following definition with Definition 4.8.

Definition 10.22. *Let* (X,d) *be a metric space. If* $a_n \in X$ *for each* $n \geq 1$ *and* $a \in X$, *then we say that* $a_n \to a$ *if, given* $\epsilon > 0$, *we can find an* $n_0(\epsilon)$ *such that*

$$d(a_n, a) < \epsilon \text{ for all } n \geq n_0(\epsilon).$$

We call a the limit of a_n.

Exercise 10.23. *(i) Let* d_1 *be the British Railway non-stop metric on* $\mathbb{R}^m$ *defined in Exercise 10.20. Show that* $\mathbf{x}_n \to \mathbf{x}$ *in* $(\mathbb{R}^m, d_1)$ *if and only if*

(A) there exists an N *such that* $\mathbf{x}_n = \mathbf{x}$ *for all* $n \geq N$, *or*

(B) $\mathbf{x} = \mathbf{0}$ *and* $\|\mathbf{x}_n\| \to 0$ *as* $n \to \infty$.

(ii) Find and prove the corresponding result for the British Railway stopping metric of Exercise 10.21.

If $(V, \|\ \|)$ is a normed space, we say that $\mathbf{x}_n \to \mathbf{x}$ in $(V, \|\ \|)$ if $\mathbf{x}_n \to \mathbf{x}$ in the derived metric. The following result on limits in such spaces is an easy generalisation of Lemma 4.9.

Lemma 10.24. *Let* V *be a real vector space and* $\|\ \|$ *a norm on* V.

(i) The limit is unique. That is, if $\mathbf{a}_n \to \mathbf{a}$ *and* $\mathbf{a}_n \to \mathbf{b}$ *as* $n \to \infty$, *then* $\mathbf{a} = \mathbf{b}$.

(ii) If $\mathbf{a}_n \to \mathbf{a}$ *as* $n \to \infty$ *and* $n(1) < n(2) < n(3)\dots$, *then* $\mathbf{a}_{n(j)} \to \mathbf{a}$ *as* $j \to \infty$.

(iii) If $\mathbf{a}_n = \mathbf{c}$ *for all* n, *then* $\mathbf{a}_n \to \mathbf{c}$ *as* $n \to \infty$.

(iv) If $\mathbf{a}_n \to \mathbf{a}$ *and* $\mathbf{b}_n \to \mathbf{b}$ *as* $n \to \infty$, *then* $\mathbf{a}_n + \mathbf{b}_n \to \mathbf{a} + \mathbf{b}$.

(v) Suppose that $\mathbf{a}_n \in V$, $\mathbf{a} \in V$, $\lambda_n \in \mathbb{R}$, *and* $\lambda \in \mathbb{R}$. *If* $\mathbf{a}_n \to \mathbf{a}$ *and* $\lambda_n \to \lambda$, *then* $\lambda_n \mathbf{a}_n \to \lambda \mathbf{a}$.

Proof. Left to the reader. The reader who looks for the proof of Lemma 4.9 will find herself referred backwards to the proof of Lemma 1.6, but will find the proofs for a general norm on a general vector space just as easy as the proofs for the Euclidean norm in one dimension. ■

Exercise 10.25. *What changes (if any) are necessary in the statements and proofs of Lemma 10.24 if we make* V *be a complex vector space?*

If we consider a general metric, then we have no algebraic structure and the result corresponding to Lemma 10.24 has far fewer parts.

Lemma 10.26. *Let* (X, d) *be a metric space.*

(i) The limit is unique. That is, if $a_n \to a$ *and* $a_n \to b$ *as* $n \to \infty$, *then* $a = b$.

(ii) If $a_n \to a$ *as* $n \to \infty$ *and* $n(1) < n(2) < n(3) \ldots$, *then* $a_{n(j)} \to a$ *as* $j \to \infty$.

(iii) If $a_n = c$ *for all* n, *then* $a_n \to c$ *as* $n \to \infty$.

The proof is again left to the reader.

The material on open and closed sets from Section 4.2 goes through essentially unchanged (and proofs are therefore left to the reader).

Definition 10.27. *Let* (X, d) *be a metric space. A set* $F \subseteq X$ *is closed if, whenever* $x_n \in F$ *for each* n *and* $x_n \to x$ *as* $n \to \infty$, *then* $x \in F$.

A set $U \subseteq X$ *is open if, whenever* $x \in U$, *there exists an* $\epsilon > 0$ *such that, whenever* $d(x, y) < \epsilon$, *we have* $y \in U$.

Example 10.28. *Let* (X, d) *be a metric space. Let* $x \in X$ *and* $r > 0$.

(i) The set $B(x, r) = \{y \in X : d(x, y) < r\}$ *is open.*

(ii) The set $\bar{B}(x, r) = \{y \in X : d(x, y) \leq r\}$ *is closed.*

We call $B(x, r)$ the open ball of radius r and centre x. We call $\bar{B}(x, r)$ the closed ball of radius r and centre x.

Lemma 10.29. *Let* (X, d) *be a metric space. A subset* U *of* X *is open if and only if each point of* U *is the centre of an open ball lying entirely within* U.

Exercise 10.30. *Find the open balls for the two British rail metrics. (For all but one point there is a difference between balls of small and large radius.) Show that the open sets for the British Railway stopping metric of Exercise 10.21 consist of sets of the form* $A \cup B$ *where* A *is empty or*

$A = \{\mathbf{x} : \|\mathbf{x}\| < \delta\}$ *(here* $\| \ \|$ *is the Euclidean norm and* $\delta > 0$*) and* B *is the union of sets of the form* $\{\lambda \mathbf{u} : a < \lambda < b\}$ *where* $0 < a < b$ *and* $\mathbf{u} \neq \mathbf{0}$.

Find the open sets for the British Railway non-stop metric of Exercise 10.20.

Definition 10.31. *The set* N *is a neighbourhood of the point* x *if we can find an* $r > 0$ *such that* $B(x, r) \subseteq N$.

Lemma 10.32. *Let* (X, d) *be a metric space. A subset* U *of* X *is open if and only if its complement* $X \setminus U$ *is closed.*

Lemma 10.33. *Let* (X, d) *be a metric space. Consider the collection* τ *of open sets in* X.

(i) $\varnothing \in \tau$, $X \in \tau$.

(ii) *If* $U_\alpha \in \tau$ *for all* $\alpha \in A$*, then* $\bigcup_{\alpha \in A} U_\alpha \in \tau$.

(iii) *If* $U_1, U_2, \ldots, U_n \in \tau$*, then* $\bigcap_{j=1}^{n} U_j \in \tau$.

Lemma 10.34. *Let* (X, d) *be a metric space. Consider the collection* $\mathcal{F}$ *of closed sets in* X.

(i) $\varnothing \in \mathcal{F}$, $X \in \mathcal{F}$.

(ii) *If* $F_\alpha \in \mathcal{F}$ *for all* $\alpha \in A$*, then* $\bigcap_{\alpha \in A} F_\alpha \in \mathcal{F}$.

(iii) *If* $F_1, F_2, \ldots, F_n \in \mathcal{F}$*, then* $\bigcup_{j=1}^{n} F_j \in \mathcal{F}$.

The new definition of continuity only breaks the chain of translations slightly because it involves *two* metric spaces.

Definition 10.35. *Let* (X, d) *and* (Z, ρ) *be metric spaces. We say that a function* $f : X \to Z$ *is continuous at some point* $x \in X$ *if, given* $\epsilon > 0$*, we can find a* $\delta(\epsilon, x) > 0$ *such that, if* $y \in X$ *and* $d(x, y) < \delta(\epsilon, x)$*, we have*

$$\rho(f(x), f(y)) < \epsilon.$$

If f *is continuous at every point* $x \in X$*, we say that* f *is a continuous function on* X.

The reader may feel that Definition 4.28 is more general than Definition 10.35 because it involves a set E. The following remark shows that this is not so.

Lemma 10.36. *Let* (X, d) *be a metric space and let* $E \subseteq X$*. Let* $d_E : E^2 \to \mathbb{R}$ *be given by* $d_E(u, v) = d(u, v)$ *whenever* $u, v \in E$*. Then* (E, d_E) *is a metric space.*

We conclude this section with more results taken directly from Section 4.2.

Lemma 10.37. *Let (X,d) and (Z,ρ) be metric spaces and suppose that the function $f : X \to Z$ is continuous at $x \in X$. Then, if $x_n \in E$ and $x_n \to x$, it follows that $f(x_n) \to f(x)$.*

Lemma 10.38. *Let (X,d) and (Z,ρ) be metric spaces. The function $f : X \to Z$ is continuous if and only if $f^{-1}(O)$ is open whenever O is open.*

Lemma 10.39. *Let (X,d), (Y,θ) and (Z,ρ) be metric spaces. If $f : X \to Y$ and $g : Y \to Z$ are continuous, then so is their composition $g \circ f$.*

10.4. Norms and the interaction of algebra and analysis

Starting from one metric, we can produce a wide variety of 'essentially equivalent metrics'.

Exercise 10.40. *Let (X,d) be a metric space.*

(i) If we define $d_1(x,y) = \min(1, d(x,y))$, show that (X,d_1) is a metric space. Show further that, if $x_n \in X$ $[n \geq 0]$, then $d_1(x_n, x_0) \to 0$ as $n \to \infty$ if and only if $d(x_n, x_0) \to 0$.

(ii) If we define $d_2(x,y) = d(x,y)^2$, show that (X,d_2) is a metric space. Show further that, if $x_n \in X$ $[n \geq 0]$, then $d_2(x_n, x_0) \to 0$ as $n \to \infty$ if and only if $d(x_n, x_0) \to 0$.

(iii) Let $\alpha > 0$ and define $d_\alpha(x,y) = d(x,y)^\alpha$. For which values of α is (X, d_α) always a metric space? For those values of α, is it always true that, if $x_n \in X$ $[n \geq 0]$, then $d_\alpha(x_n, x) \to 0$ as $n \to \infty$ if and only if $d(x_n, x) \to 0$? Prove your statements.

(iv) (This is trivial but explains why we had to be careful in the statement of (iii).) Give an example of a metric space (X,d) with X an infinite set such that (X, d_α) is a metric space for all $\alpha > 0$.

Because norms reflect an underlying vector space structure, they are much more constrained.

Lemma 10.41. *Let $\| \ \|_1$ and $\| \ \|_2$ be norms on a vector space V. Then the following two statements are equivalent.*

(a) There exist $K, L > 0$ such that $K\|\mathbf{x}\|_1 \geq \|\mathbf{x}\|_2 \geq L\|\mathbf{x}\|_1$ for all $\mathbf{x} \in V$.

(b) If $\mathbf{x}_n \in V$ $[n \geq 0]$, then $\|\mathbf{x}_n - \mathbf{x}_0\|_1 \to 0$ as $n \to \infty$ if and only if $\|\mathbf{x}_n - \mathbf{x}_0\|_2 \to 0$.

Proof. To see that (a) implies (b), observe that, if $\|\mathbf{x}_n - \mathbf{x}_0\|_2 \to 0$, then

$$\|\mathbf{x}_n - \mathbf{x}_0\|_1 \leq L^{-1}\|\mathbf{x}_n - \mathbf{x}_0\|_2 \to 0,$$

and so $\|\mathbf{x}_n - \mathbf{x}_0\|_1 \to 0$ as $n \to \infty$.

To see that (b) implies (a), suppose that (a) is false. Without loss of generality, suppose that there is no $K > 0$ such that $K\|\mathbf{x}\|_1 \geq \|\mathbf{x}\|_2$ for all $\mathbf{x} \in V$. Then we can find $\mathbf{y}_n \in V$ such that

$$\|\mathbf{y}_n\|_2 > n^2\|\mathbf{y}_n\|_1.$$

Setting $\mathbf{x}_0 = \mathbf{0}$ and $\mathbf{x}_n = n^{-1}\|\mathbf{y}_n\|_1^{-1}\mathbf{y}_n$ for $n \geq 1$, we obtain

$$\|\mathbf{x}_n - \mathbf{x}_0\|_1 = \|\mathbf{x}_n\|_1 = 1/n \to 0$$

and

$$\|\mathbf{x}_n - \mathbf{x}_0\|_2 = \|\mathbf{x}_n\|_2 > n,$$

so $\|\mathbf{x}_n - \mathbf{x}_0\|_2 \nrightarrow 0$ as $n \to 0$. Thus (b) is false if (a) is false ■

Exercise 10.42. *Let d_1 and d_2 be metrics on a space X. Consider the following two statements.*

(a) There exist K, $L > 0$ such that $Kd_1(x,y) \geq d_2(x,y) \geq Ld_1(x,y)$ for all x, $y \in X$.

(b) If $x_n \in X$ $[n \geq 0]$, then $d_1(x_n, x_0) \to 0$ as $n \to \infty$ if and only if $d_2(x_n, x_0) \to 0$.

Show that (a) implies (b) but that (b) does not imply (a).

Although we shall not make much use of this, the concepts just introduced have specific names.

Definition 10.43. *(i) Let $K \geq 0$. If (X,d) and (Y,ρ) are metric spaces, we say that a function $f : X \to Y$ is Lipschitz with constant K if we have $\rho(f(x), f(y)) \leq Kd(x,y)$ for all x, $y \in X$.*

(ii) We say that two metrics d_1 and d_2 on a space X are Lipschitz equivalent if there exist $K, L > 0$ such that $Kd_1(x,y) \geq d_2(x,y) \geq Ld_1(x,y)$ for all x, $y \in X$.

Exercise 10.44. *(i) If (X,d) and (Y,ρ) are metric spaces, show that any Lipschitz function $f : X \to Y$ is continuous. Give an example to show that the converse is false.*

(ii) Show that Lipschitz equivalence is indeed an equivalence relation between metric spaces.

The following result is a less trivial example of the interaction between algebraic and metric structure.

Theorem 10.45. *All norms on $\mathbb{R}^n$ are Lipschitz equivalent.*

Proof. We consider the standard Euclidean norm given by

$$\|\mathbf{x}\| = \left(\sum_{j=1}^{n} x_j^2\right)^{1/2}$$

and show that it is equivalent to an arbitrary norm $\| \ \|_*$.

One of the required inequalities is easy to obtain algebraicly. If we write $\mathbf{e}_i$ for the unit vector along the x_i axis, we have

$$\begin{aligned}\|\mathbf{x}\|_* &= \left\|\sum_{j=1}^n x_j\mathbf{e}_j\right\|_* \leq \sum_{j=1}^n |x_j|\|\mathbf{e}_j\|_* \\ &\leq n \max_{1\leq r\leq n}|x_r| \max_{1\leq r\leq n}\|\mathbf{e}_r\|_* \leq (n \max_{1\leq r\leq n}\|\mathbf{e}_r\|_*)\|\mathbf{x}\|.\end{aligned}$$

We have thus shown the existence of a K such that $K\|\mathbf{x}\| \geq \|\mathbf{x}\|_*$.

To obtain the other inequality we use an argument from analysis[4]. Observe that, if we give $\mathbb{R}^n$ and $\mathbb{R}$ the usual Euclidean metric and define $f : \mathbb{R}^n \to \mathbb{R}$ by

$$f(\mathbf{x}) = \|\mathbf{x}\|_*,$$

then, using the triangle inequality for $\| \ \|_*$ and the result of the previous paragraph,

$$|f(\mathbf{x}) - f(\mathbf{y})| = |\|\mathbf{x}\|_* - \|\mathbf{y}\|_*| \leq \|\mathbf{x} - \mathbf{y}\|_* \leq K\|\mathbf{x} - \mathbf{y}\|,$$

and so f is continuous. Now the unit sphere

$$S(\mathbf{0}, 1) = \{\mathbf{x} \in \mathbb{R}^n : \|\mathbf{x}\| = 1\}$$

is closed and bounded for the usual Euclidean metric, and so, by Theorem 4.44, f is bounded on $S(\mathbf{0}, 1)$. In particular, we can find a $\mathbf{k} \in S(\mathbf{0}, 1)$ where f attains its minimum, that is to say,

$$f(\mathbf{k}) \leq f(\mathbf{x})$$

for all $\mathbf{x}$ with $\|\mathbf{x}\| = 1$.

Since $\|\mathbf{k}\| = 1$, we have $\mathbf{k} \neq \mathbf{0}$ and thus $\|\mathbf{k}\|_* > 0$. Let us set $L = \|\mathbf{k}\|_*$. If $\mathbf{x} \neq \mathbf{0}$, then $\|\mathbf{x}\|^{-1}\mathbf{x} \in S(\mathbf{0}, 1)$, so

$$\|\mathbf{x}\|^{-1}\|\mathbf{x}\|_* = \|\|\mathbf{x}\|^{-1}\mathbf{x}\|_* = f(\|\mathbf{x}\|^{-1}\mathbf{x}) \geq f(\mathbf{k}) = \|\mathbf{k}\|_* = L,$$

and so $\|\mathbf{x}\|_* \geq L\|\mathbf{x}\|$. The case $\mathbf{x} = \mathbf{0}$ is trivial, so we have established

$$K\|\mathbf{x}\| \geq \|\mathbf{x}\|_* \geq L\|\mathbf{x}\|$$

for all $\mathbf{x} \in \mathbb{R}^n$, as required. ■

Exercise 10.46. *(i) Let $a_j \in \mathbb{R}$ for $1 \leq j \leq n$ and write*

$$\|\mathbf{x}\|_w = \sum_{j=1}^n a_j|x_j|.$$

State and prove necessary and sufficient conditions for $\| \ \|_w$ to be a norm on $\mathbb{R}^n$.

---

[4]Experts may be interested in Exercises K.203 and K.204. Non-experts should ignore this footnote.

(ii) Let $\| \ \|$ be the usual Euclidean norm on $\mathbb{R}^n$. Show that, if $n \geq 2$, then given any K we can find a norm $\| \ \|_*$ and points $\mathbf{y}_1$ and $\mathbf{y}_2$ such that

$$\|\mathbf{y}_1\|_* > K\|\mathbf{y}_1\| \text{ and } \|\mathbf{y}_2\| > K\|\mathbf{y}_2\|_*.$$

Does this result remain true if $n = 1$? Give reasons.

The fact that all norms on a finite-dimensional space are, essentially, the same meant that, in Chapters 4 and 6 and elsewhere, we did not really need to worry about which norm we used. By the same token, it obscured the importance of the appropriate choice of norm in approaching problems in analysis. However, once we consider infinite-dimensional spaces, the situation is entirely different.

Exercise 10.47. *Consider s_{00}, the space of real sequences $\mathbf{a} = (a_n)_{n=1}^{\infty}$ such that all but finitely many of the a_n are zero.*

(i) Show that if we use the natural definitions of addition and scalar multiplication

$$(a_n) + (b_n) = (a_n + b_n), \ \lambda(a_n) = (\lambda a_n),$$

then s_{00} is a vector space.

(ii) Show that the following definitions all give norms on s_{00}:

$$\begin{aligned}
\|\mathbf{a}\|_\infty &= \max_{n\geq 1} |a_n|, \\
\|\mathbf{a}\|_w &= \max_{n\geq 1} |na_n|, \\
\|\mathbf{a}\|_1 &= \sum_{n=1}^{\infty} |a_n|, \\
\|\mathbf{a}\|_2 &= \left(\sum_{n=1}^{\infty} |a_n|^2\right)^{1/2}, \\
\|\mathbf{a}\|_u &= \sum_{n=1}^{\infty} n|a_n|.
\end{aligned}$$

(iii) For each of the twenty-five possible pairs of norms $\| \ \|_A$ and $\| \ \|_B$ from part (ii), establish whether or not there exists a K such that $K\|\mathbf{a}\|_A \geq \|\mathbf{a}\|_B$ for all $\mathbf{a} \in s_{00}$. Show that none of the norms are Lipschitz equivalent.

(iv) Find a family of norms $\| \ \|_\alpha$ on s_{00} $[\alpha > 0]$ such that $\| \ \|_\alpha$ and $\| \ \|_\beta$ are not Lipschitz equivalent if $\alpha \neq \beta$.

When we study an algebraic structure we also study those maps which preserve algebraic structure. Thus, if we have a map $\theta : G \to H$ between groups, we want θ to preserve group multiplication, that is, we want $\theta(xy) = \theta(x)\theta(y)$. Such maps are called homomorphisms. Similarly, when we study an analytic structure, we also study those maps which preserve analytic structure. Consider for example a map f between metric spaces which preserves 'sequences tending to limits'. We then have

$$x_n \to x_0 \text{ implies } f(x_n) \to f(x_0)$$

and study continuous functions.

If we study objects which have linked analytic and algebraic structures, we will thus wish to study maps between them which preserve both algebraic and analytic structures. For normed vector spaces this means that we wish to study continuous linear maps. This fact is obscured by the fact that linear maps between finite-dimensional vector spaces are automatically continuous.

Exercise 10.48. *Suppose that U and V are finite-dimensional vector spaces with norms $\| \|_U$ and $\| \|_V$. Show that any linear map $T : U \to V$ is automatically continuous.*

However, as we shall see in Exercise 10.53, when we deal with infinite-dimensional vector spaces, not all linear maps need be continuous.

Here is a simple but important characterisation of those linear maps which are continuous.

Lemma 10.49. *Let $(U, \| \|_U)$ and $(V, \| \|_V)$ be normed vector spaces. Then a linear map $T : U \to V$ is continuous if and only if there exists a K such that $\|T\mathbf{u}\|_V \leq K\|\mathbf{u}\|_U$ for all $\mathbf{u} \in U$.*

Exercise 10.50. *Prove Lemma 10.49. [One possible proof follows that of Lemma 10.41 very closely.]*

Exercise 10.51. *By observing that the space of polynomials of degree n or less has finite dimension, or otherwise, prove the following result. There exists a constant C_n such that*

$$\sup_{t\in[0,1]} |P'(t)| \leq C_n \sup_{t\in[0,1]} |P(t)|$$

for all real polynomials P of degree n or less.

State, with proof, whether we can find a C independent of n such that

$$\sup_{t\in[0,1]} |P'(t)| \leq C \sup_{t\in[0,1]} |P(t)|$$

for all real polynomials P.

State, with proof, whether we can find a constant A_n depending on n such that

$$\sup_{t\in[0,1]} |P'(t)| \leq A_n \sup_{t\in[0,1]} |P(t)|$$

for all real polynomials P of degree n or less.

The close link between Lemma 10.49 and Lemma 10.41 is highlighted in the next exercise.

Exercise 10.52. *Let* $\| \ \|_1$ *and* $\| \ \|_2$ *be two norms on a vector space* U.

(i) Show that the following statements are equivalent.

(a) If $\|\mathbf{x}_n - \mathbf{x}_0\|_1 \to 0$, *then* $\|\mathbf{x}_n - \mathbf{x}_0\|_2 \to 0$.

(b) The identity map $I : (U, \| \ \|_1) \to (U, \| \ \|_2)$ *from* U *with norm* $\| \ \|_1$ *to* U *with norm* $\| \ \|_2$ *is continuous.*

(c) There exists a K *such that* $K\|\mathbf{u}\|_1 \geq \|\mathbf{u}\|_2$ *for all* $\mathbf{u} \in U$.

(ii) Write down, and prove equivalent, three similar statements (a)′, (b)′ and (c)′, where (c)′ is the statement

(c)′ There exist $K > L > 0$ *such that* $K\|\mathbf{u}\|_1 \geq \|\mathbf{u}\|_2 \geq L\|\mathbf{u}\|_1$ *for all* $\mathbf{u} \in U$.

Here are some examples of continuous and discontinuous linear maps[5].

Exercise 10.53. *Consider the vector space* s_{00} *defined in Exercise 10.47 and the norms*

$$\begin{aligned}
\|\mathbf{a}\|_\infty &= \max_{n\geq 1} |a_n|, \\
\|\mathbf{a}\|_w &= \max_{n\geq 1} |na_n|, \\
\|\mathbf{a}\|_1 &= \sum_{n=1}^{\infty} |a_n|, \\
\|\mathbf{a}\|_2 &= \left(\sum_{n=1}^{\infty} |a_n|^2\right)^{1/2}, \\
\|\mathbf{a}\|_u &= \sum_{n=1}^{\infty} n|a_n|
\end{aligned}$$

given there.

(i) For each of the twenty-five possible pairs of norms $\| \ \|_A$ *and* $\| \ \|_B$ *listed above, state, with reasons, whether the identity map* $I : (s_{00}, \| \ \|_A) \to (s_{00}, \| \ \|_B)$ *from* s_{00} *with norm* $\| \ \|_A$ *to* s_{00} *with norm* $\| \ \|_B$ *is continuous.*

(ii) Show that the map $T : s_{00} \to \mathbb{R}$ *defined by* $T\mathbf{a} = \sum_{j=1}^{\infty} a_j$ *is linear. If we give* $\mathbb{R}$ *the usual Euclidean norm and* s_{00} *one of the five norms listed above, state, with reasons, whether* T *is continuous.*

[5]If the reader returns to this example after studying the chapter on completeness, she may note that the normed spaces given here are not complete. The question of the existence of discontinuous linear maps between complete normed vector spaces involves more advanced ideas, notably the axiom of choice. However, the consensus is that it is *unreasonable* to expect all linear maps between complete normed vector spaces to be continuous, in the same way, and for much the same reasons, as it is *unreasonable* to expect all sets in $\mathbb{R}^3$ to have volume (see page 164).

(iii) Show that the map $S : s_{00} \to s_{00}$ defined by $S\mathbf{a} = \mathbf{b}$ with $b_j = ja_j$ is linear. For each of the twenty-five possible pairs of norms $\| \ \|_A$ and $\| \ \|_B$ listed above, state, with reasons, whether S, considered as a map from s_{00} with norm $\| \ \|_A$ to s_{00} with norm $\| \ \|_B$, is continuous.

Once we have Lemma 10.49, the way is open to an extension of the idea of an operator norm investigated in Section 6.2.

Definition 10.54. *Let U and V be vector spaces with norms $\| \ \|_U$ and $\| \ \|_V$. If $\alpha : U \to V$ is a continuous linear map, then we set*

$$\|\alpha\| = \sup_{\|\mathbf{x}\|_U \leq 1} \|\alpha\mathbf{x}\|_V.$$

Exercise 10.55. *(i) Let $U = V = s_{00}$ and let $\|\mathbf{a}\|_U = \|\mathbf{a}\|_V = \sum_{n=1}^{\infty} |a_n|$. Show that if $T : U \to V$ is defined by $T\mathbf{a} = \mathbf{b}$ with $b_j = (1 - j^{-1})a_j$, then T is a continuous linear map. However, there does not exist an $\mathbf{a} \in U$ with $\mathbf{a} \neq \mathbf{0}$ such that $\|T\mathbf{a}\|_V = \|T\|\|\mathbf{a}\|_U$. [See also Exercise 11.16.]*

(ii) If U and V are two finite-dimensional normed vector spaces and $T : U \to V$ is linear, can we always find an $\mathbf{a} \in U$ with $\mathbf{a} \neq \mathbf{0}$ such that $\|T\mathbf{a}\|_V = \|T\|\|\mathbf{a}\|_U$? Give reasons.

In exactly the way we proved Lemma 6.15, we can prove the following results.

Exercise 10.56. *Let U and V be vector spaces with norms $\| \ \|_U$ and $\| \ \|_V$ and let $\alpha, \beta : U \to V$ be continuous linear maps.*

(i) If $\mathbf{x} \in U$, then $\|\alpha\mathbf{x}\|_V \leq \|\alpha\| \, \|\mathbf{x}\|_U$.

(ii) $\|\alpha\| \geq 0$.

(iv) If $\|\alpha\| = 0$, then $\alpha = 0$.

(v) If $\lambda \in \mathbb{R}$, then $\lambda\alpha$ is a continuous linear map and $\|\lambda\alpha\| = |\lambda|\|\alpha\|$.

(vi) (The triangle inequality) $\alpha + \beta$ is a continuous linear map, and

$$\|\alpha + \beta\| \leq \|\alpha\| + \|\beta\|.$$

(vii) If W is a vector space with norm $\| \ \|_W$ and $\gamma : V \to W$ is a continuous linear map, then $\gamma\alpha$ is a continuous linear map and $\|\gamma\alpha\| \leq \|\gamma\| \, \|\alpha\|$.

We restate part of Exercise 10.56 in the language of this chapter.

Lemma 10.57. *Let U and V be vector spaces with norms $\| \ \|_U$ and $\| \ \|_V$. The space $\mathcal{L}(U, V)$ of continuous linear maps is a vector space, and the operator norm is a norm on $\mathcal{L}(U, V)$.*

Although we shall not develop the theme further, we note that we now have an appropriate definition for differentiation of functions between general normed vector spaces.

Definition 10.58. *Let U and V be vector spaces with norms $\| \|_U$ and $\| \|_V$. Suppose that E is a subset of of U and $\mathbf{x}$ a point such that there exists a $\delta > 0$ with $B(\mathbf{x}, \delta) \subseteq E$. We say that $\mathbf{f} : E \to V$ is differentiable at $\mathbf{x}$, if we can find a continuous linear map $\alpha : U \to V$ such that, when $\mathbf{h} \in B(\mathbf{x}, \delta)$,*

$$\bigstar \qquad \mathbf{f}(\mathbf{x}+\mathbf{h}) = \mathbf{f}(\mathbf{x}) + \alpha\mathbf{h} + \boldsymbol{\epsilon}(\mathbf{x},\mathbf{h})\|\mathbf{h}\|_U$$

where $\|\boldsymbol{\epsilon}(\mathbf{x},\mathbf{h})\|_V \to 0$ as $\|\mathbf{h}\|_U \to 0$. We write $\alpha = D\mathbf{f}(\mathbf{x})$ or $\alpha = \mathbf{f}'(\mathbf{x})$.

If E is open and $\mathbf{f}$ is differentiable at each point of E, we say that $\mathbf{f}$ is differentiable on E.

Notice that the only important change from Definition 6.4 is that we specifically demand that $\alpha = D\mathbf{f}(\mathbf{x})$ is a *continuous* linear function.

The best way to understand what is going on is probably to do the next exercise.

Exercise 10.59. *State and prove the appropriate extension of the chain rule given as Lemma 6.19.*

There is no difficulty in extending all our work on differentiation from finite-dimensional to general normed spaces. The details are set out in Dieudonné's book ([**14**], Chapter VIII).

Exercise 10.60. *The title of this section was 'Norms and the interaction of algebra and analysis', but much of it was really pure algebra in the sense that it did not use the fundamental axiom of analysis. Go back through the section and note which results are genuine results of analysis and which are just algebra disguised as analysis.*

10.5. Geodesics ♡

This section introduces an idea which is important in more advanced work but which will not be used elsewhere in this book. The discussion will be informal.

Suppose we wish to build a road joining two points of the plane. The cost per unit length of road will depend on the nature of the terrain. If the cost of building a short stretch of length δs near a point (x, y) is (to first order) $g(x, y)\delta s$, then, subject to everything being well behaved, the cost of road Γ will be

$$\int_\Gamma g(x, y)\, ds.$$

The reader may use whatever definition of line integral she is comfortable with. In particular, she does not need to have read Section 9.5. It is tempting to define the distance $d(A,B)$ between two points A and B in the plane as

$$d(A,B) = \inf\left\{\int_\Gamma g(x,y)\,ds : \Gamma \text{ joining } A \text{ and } B\right\}.$$

If we include amongst our conditions that g is continuous and $g(x,y) > 0$ everywhere, we see that d is a metric.

Plausible Statement 10.61. *(i) $d(A,B) \geq 0$ for all A and B.*

(ii) $d(A,B) = 0$ implies $A = B$.

(iii) $d(A,B) = d(B,A)$ for all A and B.

(iv) $d(A,C) \leq d(A,B) + d(B,C)$ for all A, B and C.

I have carefully used 'inf' rather than 'min' in the definition of $d(A,B)$. One problem is that I have not specified the kind of Γ that is permissible. (Notice that the argument needed to obtain part (iv) of Statement 10.61 means that we cannot just use 'smooth'.) However, it can be shown that, if we choose a suitable class for the possible Γ and a suitably well behaved g, then the minimum is attained. If Γ_0 is a path from A to B such that

$$\int_{\Gamma_0} g(x,y)\,ds = d(A,B),$$

we call Γ_0 a geodesic. (The geodesic need not be unique: consider road building for two towns at diametrically opposite points of a circular marsh.) If any two points are joined by a geodesic, then $d(A,B)$ is the 'length of a geodesic path joining A and B', where length refers to 'cost', and not to Euclidean distance.

Let us try to use these ideas to find a metric on the upper half-plane

$$H = \{z \in \mathbb{C} : \Im z > 0\}$$

which is invariant under Möbius transformation. (If you have not met Möbius transformations, skip to after Exercise 10.64, where I restate the problem.) More precisely, we want a metric d such that, if T is a Möbius map mapping H bijectively to itself, then $d(Tz_1, Tz_2) = d(z_1, z_2)$ for all $z_1, z_2 \in H$. Of course, it is possible that no such metric exists, but we shall see where the question leads.

Lemma 10.62. *The set $\mathcal{H}$ of Möbius maps T such that $T|_H : H \to H$ is bijective is a subgroup of the group $\mathcal{M}$ of all Möbius transformations. The subgroup $\mathcal{H}$ is generated by the transformations T_a with $a \in \mathbb{R}$, D_λ with λ real and $\lambda > 0$, and J, where*

$$\begin{aligned} T_a(z) &= a + z, \\ D_\lambda(z) &= \lambda z, \\ J(z) &= -z^{-1}. \end{aligned}$$

Sketch proof. The fact that $\mathcal{H}$ is a subgroup of $\mathcal{M}$ follows from easy general principles (see Exercise 10.63). We check directly that (if a is real and $\lambda > 0$) T_a, D_λ and J lie in $\mathcal{H}$. Thus the composition of elements of the stated type will also lie in $\mathcal{H}$.

We now wish to identify $\mathcal{H}$. If $T \in \mathcal{H}$, then T is a well behaved bijective map on $\mathbb{C}^* = \mathbb{C} \cup \{\infty\}$ and must take the boundary of H to the boundary of H. Thus, writing R for the real axis, we know that T takes $R \cup \{\infty\}$ to $R \cup \{\infty\}$. Thus, if $T(\infty) \neq \infty$, then $T(\infty) = a$ with a real, and we set $M_1 = JT_{-a}$. If $T(\infty) = \infty$, we take M_1 to be the identity map. In either case, $M_1T \in \mathcal{H}$ and $M_1T(\infty) = \infty$. We now observe that $M_1T(0) \in R \cup \{\infty\}$, by our previous argument, and that $M_1T(0) \neq \infty$, since Möbius maps are bijections. Thus $M_1T(0) = b$ with b real. Set $M_2 = T_{-b}$. We have $M_2M_1T \in \mathcal{H}$, $M_2M_1T(0) = 0$ and $M_2M_1T(\infty) = \infty$. But any Möbius map M which fixes 0 and ∞ must have the form $M(z) = \mu z$ for some $\mu \neq 0$. If M takes H to H, then μ is real and positive. Thus $M_2M_1T = D_\lambda$ for some $\lambda > 0$, and $T = M_1^{-1}M_2^{-1}D_\lambda$. We have shown that any element $\mathcal{H}$ is the composition of maps of the stated type, and the lemma follows. ▲

(The remark on page 222 applies here too. We know so much about Möbius transformations that the argument above is more a calculation than a proof. However, for more complicated conformal maps than are generally found in undergraduate courses, arguments involving 'boundaries' may be misleading.)

Exercise 10.63. *(Most readers will already know the content of this exercise.) Let X be a set and $S(X)$ the collection of bijections of X. Show that $S(X)$ is a group under the operation of composition. If G is a subgroup of $S(X)$, Y a subset of X and we define*

$$K = \{f \in G : f(Y) = Y\},$$

show that K is a subgroup of G.

Exercise 10.64. *If $|a| < 1$ and M_a is the Möbius map M_a given by*

$$M_a z = \frac{z - a}{a^* z - 1},$$

show that $|M_a e^{i\theta}| = 1$ for all real θ. Deduce, stating any properties of Möbius transforms that you need, that M_a maps the unit disc $D = \{z : |z| < 1\}$ to itself and interchanges 0 and a. Show that the set $\mathcal{H}'$ of Möbius maps T such that $T|_H : D \to D$ is bijective is a subgroup of the group $\mathcal{M}$ of all Möbius transformations generated by the transformations M_a with $|a| < 1$ and the rotations.

Lemma 10.62 reduces our problem to one of finding a metric d on the half plane $H = \{z : \Im z > 0\}$ such that $d(Tz_1, Tz_2) = d(z_1, z_2)$ for all

$z_1, z_2 \in H$, whenever T is one of the transformations T_a with $a \in \mathbb{R}$, D_λ with $\lambda > 0$, and J defined by

$$\begin{aligned} T_a(z) &= a + z, \\ D_\lambda(z) &= \lambda z, \\ J(z) &= -\, z^{-1}. \end{aligned}$$

We try using the ideas of this section and defining

$$d(z_1, z_2) = \inf \left\{ \int_\Gamma g(z)\, ds \; : \; \Gamma \text{ joining } z_1 \text{ and } z_2 \right\},$$

for some appropriate strictly positive function $g : H \to \mathbb{R}$. (Throughout, we write $z = x + iy$ and identify the plane $\mathbb{R}^2$ with $\mathbb{C}$ in the usual manner.)

Arguing informally, this suggests that, to first order,

$$d(z + \delta z, z) = g(z)\delta s = g(z)|(z + \delta z) - z|.$$

If T is one of the transformations we are considering, then we must have, to first order,

$$\begin{aligned} g(z)|(z + \delta z) - z| &= d(z + \delta z, z) = d(T(z + \delta z), Tz) \\ &= g(Tz)|T(z + \delta z) - T(z)| = g(Tz)|T'(z)||\delta z|, \end{aligned}$$

and so

★ $$g(z) = g(Tz)|T'(z)|.$$

Taking $T = T_a$, we obtain $g(z) = g(z+a)$ for all real a, so $g(x+iy) = h(y)$ for some well behaved function $h : (0, \infty) \to (0, \infty)$.

Taking $T = D_\lambda(z)$, we obtain $g(z) = \lambda g(\lambda z)$ for all real strictly positive λ. Thus $h(y) = \lambda h(\lambda y)$ for all λ, $y > 0$. Taking $\lambda = 1/y$, we obtain $h(y) = Ay$ for some $A > 0$. The factor A merely scales everything, so we take $A = 1$ and decide to experiment with $g(x + iy) = 1/y$.

Exercise 10.65. *Verify that, if $g(x + iy) = 1/y$ and $T = J$, then equation ★ holds.*

Exercise 10.66. *Suppose that $\gamma : [0, 1] \to H$ is well behaved and $T \in \mathcal{H}$. Let $\tilde{\gamma} = T \circ \gamma$ (that is, $\tilde{\gamma}(z) = T(\gamma(z))$). If Γ is the path described by γ and $\tilde{\Gamma}$ is the path described by $\tilde{\gamma}$, show, using whatever definition of the line integral you wish, that*

$$\inf \left\{ \int_\Gamma g(z)\, ds \right\} = \inf \left\{ \int_{\tilde{\Gamma}} g(z)\, ds \right\}$$

for $T = T_a$, $T = D_\lambda$ and $T = J$. Conclude that the equality holds for all $T \in \mathcal{H}$.

Exercise 10.66 shows that, if we set

$$d(z_1, z_2) = \inf\left\{\int_\Gamma \frac{1}{y}\,ds \;:\; \Gamma \text{ joining } z_1 \text{ and } z_2\right\},$$

then we do, indeed, get an invariant metric (that is to say, a metric with $d(Tz_1, Tz_2) = d(z_1, z_2)$ for all $z_1, z_2 \in H$ whenever $T \in \mathcal{H}$).

To find out more about this metric we need to find the geodesics, and to do this we use the methods of the calculus of variation described in Section 8.4. We shall use the methods in a *purely exploratory* manner with no attempt at rigour. (Notice that, even if we did things rigorously, we would only get necessary conditions.) Suppose that we wish to find the path Γ_0 which minimises $\int_\Gamma \frac{1}{y}\,ds$ among all paths Γ from $z_1 = x_1 + iy_1$ to $z_2 = x_2 + iy_2$ with $x_1 < x_2$ and y_1, $y_2 > 0$. It seems reasonable to look for a path given by $z = x + iy(x)$, where $y : [x_1, x_2] \to (0, \infty)$ is well behaved. We thus wish to minimise

$$\int_{x_1}^{x_2} \frac{1}{y(x)}(1 + y'(x)^2)^{1/2}\,dx = \int_{x_1}^{x_2} G(y(x), y'(x))\,dx,$$

where $G(v, w) = v^{-1}(1 + w^2)^{1/2}$. The Euler–Lagrange equation gives, via Exercise 8.63 (i),

$$G(y(x), y'(x)) - y'(x)G_{,2}(y(x), y'(x)) = c,$$

where c is a constant. Writing the equation more explicitly, we get

$$\frac{(1 + y'^2)^{1/2}}{y} - \frac{y'^2}{y(1 + y'^2)^{1/2}} = c,$$

so that

$$1 = cy(1 + y'^2)^{1/2}.$$

Setting $a = c^{-1}$, we obtain

$$a^2 = y^2(1 + y'^2),$$

so that, solving formally,

$$\frac{y\,dy}{(a^2 - y^2)^{1/2}} = dx$$

and

$$-(a^2 - y^2)^{1/2} = x - b,$$

for some constant $b \in \mathbb{R}$. Thus

$$(x - b)^2 + y^2 = a^2.$$

We expect the geodesic to be the arc of a circle with its centre on the real axis and passing through the two points z_1 and z_2.

Exercise 10.67. *Suppose (x_1, y_1), $(x_2, y_2) \in \mathbb{R}^2$ and $x_1 \neq x_2$. Show that there is one and only one circle with its centre on the real axis passing through the two points (x_1, y_1) and (x_2, y_2).*

What happens if $\Re z_1 = \Re z_2$? One way of guessing is to consider the geodesic path between z_1 and $z_2 + \delta$, where δ is real and non-zero. If we let $\delta \to 0$, the appropriate circle arcs approach a straight line joining z_1 and z_2, so we would expect this to be the solution.

Exercise 10.68. *Attack the geodesic problem by considering paths given by $z = x(y) + iy$, where $y : [y_1, y_2] \to \mathbb{R}$ is well behaved. (The difficulties are mainly notational, the formulae of the variational calculus assume that y is a function of x, and it requires a certain amount of clear headedness to deal with the case when x is a function of y. You should be able to make use of Exercise 8.63 (ii).) Check, by choosing appropriate constants, that your solutions include straight lines perpendicular to the x-axis.*

Now that we know (or at least guess) what the geodesics are, we can see (at least if we know a little about Möbius maps) a different way of showing this.

Exercise 10.69. *We work in $\mathbb{C}^* = \mathbb{C} \cup \{\infty\}$.*

(i) Show that, if Γ is a circle with centre on the real axis, there is a $T \in \mathcal{H}$ such that $T(\Gamma)$ is a circle with centre on the real axis passing through the origin.

(ii) Show that if Γ is a circle with centre on the real axis passing through the origin, there is a $T \in \mathcal{H}$ such that $T(\Gamma)$ is a line perpendicular to the real axis.

(iii) Show that if Γ is a circle with centre on the real axis or a line perpendicular to the real axis, there is a $T \in \mathcal{H}$ such that $T(\Gamma)$ is the imaginary axis.

Exercise 10.66 and Exercise 10.69 show that the following two theorems are equivalent.

Theorem 10.70. *The geodesic path between iy_1 and iy_2 [y_1, y_2 real, unequal and strictly positive] is a straight line.*

Theorem 10.71. *If $z_1, z_2 \in H$ with $z_1 \neq z_2$, the geodesic path between them is an arc of the circle through z_1 and z_2, unless $\Re z_1 = \Re z_2$, in which case, it is the straight line between the two points.*

Exercise 10.72. *Explain why Theorem 10.70 implies Theorem 10.71.*

We now turn to the proof of Theorem 10.70.

Sketch proof of Theorem 10.70. Let $Z : [0,1] \to H$ be a well behaved function such that $Z(0) = y_1$ and $Z(1) = y_2$ with $y_2 > y_1$. We write $Z(t) = X(t) + iY(t)$ with $X(t)$ and $Y(t)$ real, and take Γ to be the path described by Z. We observe that

$$\begin{aligned}\int_\Gamma \frac{1}{y}\,ds &= \int_0^1 \frac{1}{Y(t)}(X'(t)^2 + Y'(t)^2)^{1/2}\,dt \geq \int_0^1 \frac{(Y'(t)^2)^{1/2}}{Y(t)}\,dt\\ &= \int_0^1 \frac{|Y'(t)|}{Y(t)}\,dt \geq \int_0^1 \frac{Y'(t)}{Y(t)}\,dt = [\log Y(t)]_0^1\\ &= \log y_2 - \log y_1 = \int_{y_1}^{y_2} \frac{1}{t}\,dt = \int_{\Gamma_0} \frac{1}{y}\,ds,\end{aligned}$$

where Γ_0 is the straight line path from y_1 to y_2. We observe that the argument above shows that

$$\int_\Gamma \frac{1}{y}\,ds = \int_{\Gamma_0} \frac{1}{y}\,ds$$

only if $X'(t)^2 = 0$ and $Y'(t) \geq 0$ for all $t \in [0,1]$ and so, by simple arguments, $X(t) = 0$ and $Y'(t) \geq 0$ for all $t \in [0,1]$. Thus Γ_0 is the unique geodesic. ▲

Remark: This is only a sketch of a proof, because we have not really decided which curves will be eligible for paths. The proof strategy of 'project the path onto the y-axis and compare' ought to work for any reasonable definition of line integral and any eligible path.

Exercise 10.73. *We work in $\mathbb{R}^2$. Suppose that we want a metric defined by*

$$d(A,B) = \inf\left\{\int_\Gamma g(x,y)\,ds : \Gamma \textit{ joining } A \textit{ and } B\right\},$$

which is invariant under translation and rotation. Copy the investigation above going through the following steps.

(i) By considering points which are close and working to first order, show that we should try g constant. Without loss of generality take $g = 1$.

(ii) Use the calculus of variations to suggest the form of the geodesics.

(iii) Prove that your guess is correct.

(iv) Show that our metric is the usual Euclidean metric.

Exercise 10.74. *In this section we defined a metric d on H invariant under $\mathcal{H}$, but did not calculate it. We could find d by using the strategy of Exercise 10.69, but, in this exercise, we follow the easier path of allowing someone else to find the answer and then verifying it.*

If $z, w \in H$, we set

$$\rho(z,w) = \log \frac{|z-w^*| + |z-w|}{|z-w^*| - |z-w|}.$$

(i) Show that $|z-w^*| - |z-w| > 0$ for all $z, w \in H$, so ρ is well defined.
(ii) Show that $\rho(Tz, Tw) = \rho(z,w)$ for all $T \in \mathcal{H}$.
(iii) Show that $d(iy_1, iy_2) = \rho(iy_1, iy_2)$ for all y_1, $y_2 > 0$.
(iv) Deduce that $d = \rho$.

Chapter 11

Complete Metric Spaces

11.1. Completeness

If we examine the arguments of Section 10.3 we see that they are all mere algebra. What must we introduce to do genuine analysis on metric spaces? We cannot use a variant of the fundamental axiom because there is no order on our spaces[1]. Instead, we use a generalisation of the general principle of convergence.

Definition 11.1. *If (X, d) is a metric space, we say that a sequence of points $x_n \in X$ is Cauchy if, given any $\epsilon > 0$, we can find $n_0(\epsilon)$ such that $d(x_p, x_q) < \epsilon$ for all p, $q \geq n_0(\epsilon)$.*

Definition 11.2. *A metric space (X, d) is complete if every Cauchy sequence converges.*

In this context, Theorem 4.68 states that $\mathbb{R}^n$ with the Euclidean metric is complete.

The contraction mapping theorem (Theorem 12.3) and its applications will provide a striking example of the utility of this concept. However, this section is devoted to providing background examples of spaces which are and are not complete.

If you want to see completeness in action immediately, you should do the next example.

[1]There is an appropriate theory for objects with order (lattices) hinted at in Appendix D, but we shall not pursue this idea further.

Exercise 11.3. *We say that a metric space (X, d) has no isolated points if, given $y \in X$ and $\epsilon > 0$, we can find an $x \in X$ such that $0 < d(x, y) < \epsilon$. Show by the methods of Exercise 1.41 that a complete non-empty metric space with no isolated points is uncountable.*

Give an example of a countable metric space. Give an example of an uncountable metric space all of whose points are isolated.

The next lemma gives a good supply of metric spaces which are complete and of metric spaces which are not complete.

Lemma 11.4. *Let (X, d) be a complete metric space. If E is a subset of X and we define $d_E : E^2 \to \mathbb{R}$ by $d_E(u, v) = d(u, v)$ whenever $u, v \in E$, then (E, d_E) is complete if and only if E is closed in (X, d).*

Proof. This is just a matter of definition chasing.

Observe that any Cauchy sequence x_n in E is a Cauchy sequence in X and so converges to a point x in X. If E is closed, then $x \in E$ and $d_E(x_n, x) = d(x_n, x) \to 0$ as $n \to \infty$. Thus (E, d_E) is complete whenever E is closed.

Suppose now that (E, d_E) is complete. If $x_n \in E$ and $d(x_n, x) \to 0$ for some $x \in X$, we know (by the argument of Lemma 4.67 if you need it) that x_n is a Cauchy sequence in X and so a Cauchy sequence in E. Since (E, d_E) is complete, we can find a $y \in E$ such that $d_E(x_n, y) \to 0$. Now $d(x_n, y) = d_E(x_n, y) \to 0$, so, by the uniqueness of limits, $y = x$ and $x \in E$. Thus E is closed in (X, d). ■

Thus, for example, the closed interval $[a, b]$ is complete for the usual metric, but the open interval (a, b) is not.

Exercise 11.5. *Let (X, d) be a metric space and E a subset of X. Define $d_E : E^2 \to \mathbb{R}$ as in Lemma 11.4.*

(i) Show that, if E is not closed in (X, d), then (E, d_E) is not complete.

(ii) Give an example where E is closed in (X, d) but (E, d_E) is not complete.

(iii) Give an example where (X, d) is not complete but (E, d_E) is.

The reader is warned that, at least in my opinion, it is harder than it looks to prove that a metric space is or is not complete, and it is easy to produce plausible but unsatisfactory arguments in this context.

If we wish to show that a metric space (X, d) is incomplete, the natural way to proceed is to find a Cauchy sequence x_n and show that it does not converge. However, we must show that x_n does not converge to any point in X, and not that 'x_n does not converge to the point that it looks as though it ought to converge to'. Among the methods available for a correct proof are the following.

(1) Embed (X, d) in a larger metric space $(\tilde{X}, \tilde{d})$ (that is, find $(\tilde{X}, \tilde{d})$ such that $\tilde{X} \supseteq X$ and $\tilde{d}(x, y) = d(x, y)$ for $x, y \in X$) and show that there is an $x \in \tilde{X} \setminus X$ such that $\tilde{d}(x_n, x) \to 0$. (See Exercise 11.5.)

(2) For each fixed $x \in X$ show that there is a $\delta(x) > 0$ and an $N(x)$ (both depending on x) such that $d(x_n, x) > \delta(x)$ for $n > N(x)$.

(3) Show that the assumption $d(x_n, x) \to 0$ for some $x \in X$ leads to a contradiction.

Of course, no list of this sort can be exhaustive. None the less, it is good practice to ask yourself not simply whether your proof is correct but also what strategies it employs.

Here are a couple of examples.

Example 11.6. *Consider s_{00}, the space of real sequences $\mathbf{a} = (a_n)_{n=1}^{\infty}$ such that all but finitely many of the a_n are zero, introduced in Exercise 10.47. The norm defined by*

$$\|\mathbf{a}\|_1 = \sum_{n=1}^{\infty} |a_n|$$

is not complete.

Proof. Set

$$\mathbf{a}(n) = (1, 2^{-1}, 2^{-2}, \ldots, 2^{-n}, 0, 0, \ldots).$$

We observe that, if $m \geq n$,

$$d(\mathbf{a}(n), \mathbf{a}(m)) = \sum_{j=n+1}^{m} 2^{-j} \leq 2^{-n} \to 0$$

as $n \to \infty$, and so the sequence $\mathbf{a}(n)$ is Cauchy.

However, if $\mathbf{a} \in s_{00}$, we know that there is an N such that $a_j = 0$ for all $j \geq N$. It follows that

$$d(\mathbf{a}(n), \mathbf{a}) \geq 2^{-N}$$

whenever $n \geq N$, and so the sequence $\mathbf{a}(n)$ has no limit. ∎

[The proof above used method 2. For an alternative proof using method 1, see page 255. For a generalisation of the result see Exercise K.207.]

The next example needs a result which is left to the reader to prove. (You will need Lemma 8.33.)

Exercise 11.7. *Let $b > a$. Consider $C([a,b])$, the set of continuous functions $f : [a,b] \to \mathbb{R}$. If we set*

$$\|f - g\|_1 = \int_a^b |f(x) - g(x)|\, dx,$$

show that $\| \ \|_1$ is a norm.

Lemma 11.8. *The normed space of Exercise 11.7 is not complete.*

Proof. This proof uses method 3. With no real loss of generality, we take $[a,b] = [-1,1]$. Let

$$\begin{aligned} f_n(x) &= -1 && \text{for } -1 \leq x \leq -1/n, \\ f_n(x) &= nx && \text{for } -1/n \leq x \leq 1/n, \\ f_n(x) &= 1 && \text{for } 1/n \leq x \leq 1. \end{aligned}$$

If $m \geq n$, then

$$\|f_n - f_m\|_1 = \int_{-1}^{1} |f_n(x) - f_m(x)|\, dx \leq \int_{-1/n}^{1/n} 1\, dx = \frac{2}{n} \to 0$$

as $n \to \infty$, and so the sequence f_n is Cauchy.

Suppose, if possible, that there exists an $f \in C([-1,1])$ such that $\|f - f_n\|_1 \to 0$ as $n \to \infty$. Observe that, if $n \geq N$, we have

$$\int_{1/N}^{1} |f(x) - 1|\, dx = \int_{1/N}^{1} |f(x) - f_n(x)|\, dx \leq \int_{-1}^{1} |f(x) - f_n(x)|\, dx \to 0$$

as $n \to \infty$. Thus

$$\int_{1/N}^{1} |f(x) - 1|\, dx = 0$$

and, by Lemma 8.33, $f(x) = 1$ for $x \in [1/N, 1]$. Since N is arbitrary, $f(x) = 1$ for all $0 < x \leq 1$. A similar argument shows that $f(x) = -1$ for all $-1 \leq x < 0$. Thus f fails to be continuous at 0, and we have a contradiction. By reductio ad absurdum, the Cauchy sequence f_n has no limit. ∎

Lemma 11.8 is important as an indication of the unsatisfactory results of using too narrow a class of integrable functions.

The next exercise goes over very similar ground but introduces an interesting set of ideas.

Exercise 11.9. *Let $b > a$. Consider $C([a,b])$, the set of continuous functions $f : [a,b] \to \mathbb{R}$. If $f,\ g \in C([a,b])$ and we define*

$$\langle f, g \rangle = \int_a^b f(t)g(t)\, dt,$$

show that $(C([a,b]), \langle\ ,\ \rangle)$ *is an inner product space. More formally, show that*

(i) $\langle f,f\rangle \geq 0$ *with equality if and only if* $f=0$,

(ii) $\langle f,g\rangle = \langle g,f\rangle$,

(iii) $\langle \lambda f,g\rangle = \lambda\langle f,g\rangle$,

(iv) $\langle f,g+h\rangle = \langle f,g\rangle + \langle f,h\rangle$

for all f, g, $h \in C([a,b])$ *and all* $\lambda \in \mathbb{R}$.

Use the arguments of Lemmas 4.2 and 4.4 to show that setting

$$\|f\|_2 = \langle f,f\rangle^{1/2} = \left(\int_a^b f(t)^2\,dt\right)^{1/2}$$

gives a norm on $C([a,b])$.

Show, however, that $(C([a,b]), \|\ \|_2)$ *is not a complete normed space.*

If we wish to show that a metric space (X,d) is complete, the natural way to proceed is to take a Cauchy sequence x_n and show that it must converge. However, we must show that x_n actually converges to a point in X and not that 'the sequence x_n looks as though it ought to converge'. In many cases the proof proceeds through the following steps.

(A) The sequence x_n converges in some sense (but not the sense we want) to an object x.

(B) The object x actually lies in X.

(C) The sequence x_n actually converges to x in the sense we want.

Here is an example.

Example 11.10. *The set* l^1 *of real sequences* $\mathbf{a}$ *with* $\sum_{j=1}^{\infty}|a_j|$ *convergent forms a vector space if we use the natural definitions of addition and scalar multiplication*

$$(a_n)+(b_n)=(a_n+b_n),\ \ \lambda(a_n)=(\lambda a_n).$$

If we set

$$\|\mathbf{a}\|_1 = \sum_{j=1}^{\infty}|a_j|,$$

then $(l^1, \|\ \|_1)$ *is a complete normed space.*

Proof. We know that the space of all sequences forms a vector space, so we only have to show that l^1 is a subspace. Clearly $\mathbf{0}\in l^1$, and, since $\sum_{j=1}^{N}|\lambda a_j| = |\lambda|\sum_{j=1}^{N}|a_j|$, we have $\lambda\mathbf{a}\in l^1$ whenever $\mathbf{a}\in l^1$ and $\lambda\in\mathbb{R}$.

Suppose $\mathbf{a}$, $\mathbf{b} \in l^1$. We have

$$\sum_{j=1}^{N} |a_j + b_j| \leq \sum_{j=1}^{N} |a_j| + \sum_{j=1}^{N} |b_j| \leq \sum_{j=1}^{\infty} |a_j| + \sum_{j=1}^{\infty} |b_j|,$$

so, since an increasing sequence bounded above tends to a limit, it follows that $\sum_{j=1}^{\infty} |a_j + b_j|$ converges and $\mathbf{a}+\mathbf{b} \in l^1$. Hence l^1 is, indeed, a subspace of the space of all sequences. It is easy to check that $\| \; \|_1$ is a norm.

I shall label the next three paragraphs in accordance with the discussion just before this example.

Step A: Suppose $\mathbf{a}(n)$ is a Cauchy sequence in $(l^1, \| \; \|_1)$. For each fixed j,

$$|a_j(n) - a_j(m)| \leq \|\mathbf{a}(n) - \mathbf{a}(m)\|_1,$$

so $a_j(n)$ is a Cauchy sequence in $\mathbb{R}$. The general principle of convergence tells us that $a_j(n)$ tends to a limit a_j as $n \to \infty$.

Step B: Since any Cauchy sequence is bounded, we can find a K such that $\|\mathbf{a}(n)\|_1 \leq K$ for all n. We observe that

$$\begin{aligned}\sum_{j=1}^{N} |a_j| &\leq \sum_{j=1}^{N} |a_j - a_j(n)| + \sum_{j=1}^{N} |a_j(n)| \leq \sum_{j=1}^{N} |a_j - a_j(n)| + \|\mathbf{a}(n)\|_1 \\ &\leq \sum_{j=1}^{N} |a_j - a_j(n)| + K \to K\end{aligned}$$

as $n \to \infty$. Thus $\sum_{j=1}^{N} |a_j| \leq K$ for all N, and so $\sum_{j=1}^{\infty} |a_j|$ converges. We have shown that $\mathbf{a} \in l^1$.

Step C: We now observe that, if $n, m \geq M$,

$$\begin{aligned}\sum_{j=1}^{N} |a_j - a_j(n)| &\leq \sum_{j=1}^{N} |a_j - a_j(m)| + \sum_{j=1}^{N} |a_j(m) - a_j(n)| \\ &\leq \sum_{j=1}^{N} |a_j - a_j(m)| + \|\mathbf{a}(m) - \mathbf{a}(n)\|_1 \\ &\leq \sum_{j=1}^{N} |a_j - a_j(m)| + \sup_{p,q \geq M} \|\mathbf{a}(p) - \mathbf{a}(q)\|_1 \\ &\to \sup_{p,q \geq M} \|\mathbf{a}(p) - \mathbf{a}(q)\|_1\end{aligned}$$

as $m \to \infty$. Thus $\sum_{j=1}^{N} |a_j - a_j(n)| \leq \sup_{p,q \geq M} \|\mathbf{a}(p) - \mathbf{a}(q)\|_1$ for all N, and so

$$\|\mathbf{a} - \mathbf{a}(n)\|_1 \leq \sup_{p,q \geq M} \|\mathbf{a}(p) - \mathbf{a}(q)\|_1$$

for all $n \geq M$. Recalling that the sequence $\mathbf{a}(m)$ is Cauchy, we see that $\|\mathbf{a} - \mathbf{a}(n)\|_1 \to 0$ as $n \to \infty$, and we are done. ■

The method of proof of Step C, and, in particular, the introduction of the 'irrelevant m' in the first set of inequalities, is very useful but requires some thought to master.

Exercise 11.11. *In the proof above we stated that 'any Cauchy sequence is bounded'. Give one line proofs of the more precise statements that follow.*

(i) If (X, d) is a metric space, x_n is a Cauchy sequence in X and $a \in X$, then we can find a K such that $d(a, x_n) < K$ for all $n \geq 1$.

(ii) If $(V, \|\ \|)$ is a normed vector space and $\mathbf{x}_n$ is a Cauchy sequence in V, then we can find a K such that $\|\mathbf{x}_n\| < K$ for all $n \geq 1$.

Alternative proof of Example 11.6. We wish to show that s_{00}, with norm

$$\|\mathbf{a}\|_1 = \sum_{j=1}^{\infty} |a_j|,$$

is not complete. Observe that we can consider s_{00} as a subspace of l^1 and that the norm on l^1 agrees with our norm on s_{00}. Now set

$$\mathbf{a}(n) = (1,\ 2^{-1},\ 2^{-2},\ \dots,\ 2^{-n},\ 0,\ 0,\ \dots),$$

and

$$\mathbf{a} = (1,\ 2^{-1},\ 2^{-2},\ \dots,\ 2^{-n},\ 2^{-n-1},\ 2^{-n-2},\ \dots).$$

Since $\mathbf{a}(n) \in s_{00}$ for all n and $\|\mathbf{a}(n) - \mathbf{a}\|_1 \to 0$ as $n \to \infty$, we see that s_{00} is not closed in l^1, and so, by Exercise 11.5, $(s_{00}, \|\ \|_1)$ is not complete. ■

Here are two exercises using the method of proof of Example 11.10.

Exercise 11.12. *Let U be a complete vector space with norm $\|\ \|$. Show that the set $l^1(U)$ of sequences*

$$\underline{u} = (\mathbf{u}_1,\ \mathbf{u}_2,\ \mathbf{u}_3,\ \dots)$$

with $\sum_{j=1}^{\infty} \|\mathbf{u}_j\|$ convergent forms a vector space if we use the natural definitions of addition and scalar multiplication

$$(\mathbf{u}_n) + (\mathbf{v}_n) = (\mathbf{u}_n + \mathbf{v}_n),\ \ \lambda(\mathbf{u}_n) = (\lambda\mathbf{u}_n).$$

Show that, if we set

$$\|\underline{u}\|_1 = \sum_{j=1}^{\infty} \|\mathbf{u}_j\|,$$

then $(l^1, \|\ \|_1)$ is a complete normed space.

Exercise 11.13. *Show that the set l^∞ of bounded real sequences forms a vector space if we use the usual definitions of addition and scalar multiplication.*

Show further that, if we set

$$\|\mathbf{a}\|_\infty = \sup_{j\geq 1} |a_j|,$$

then $(l^\infty, \| \ \|_\infty)$ is a complete normed space.

[We shall prove a slight generalisation as Theorem 11.35, so the reader may wish to work through this exercise before she meets the extension.]

Exercise 11.14. *Consider the set c_0 of real sequences $\mathbf{a} = (a_1, a_2, a_3, \dots)$ such that $a_n \to 0$. Show that c_0 is a subspace of l^∞ and a closed subset of $(l^\infty, \| \ \|_\infty)$. Deduce that $(c_0, \| \ \|_\infty)$ is a complete normed space.*

Exercise K.208 contains another example of an interesting and important infinite-dimensional complete normed space.

The final result of this section helps show why the operator norm is so useful in more advanced work.

Lemma 11.15. *Let U and V be vector spaces with norms $\| \ \|_U$ and $\| \ \|_V$. Suppose further that $\| \ \|_V$ is complete. Then the operator norm is a complete norm on the space $\mathcal{L}(U,V)$ of continuous linear maps from U to V.*

Proof. Once again our argument falls into three steps.

Step A: Suppose that T_n is a Cauchy sequence in $(\mathcal{L}(U,V), \| \ \|)$. For each fixed $\mathbf{u} \in U$,

$$\|T_n\mathbf{u} - T_m\mathbf{u}\|_V = \|(T_n - T_m)\mathbf{u}\|_V \leq \|T_n - T_m\|\|\mathbf{u}\|_V,$$

so $T_n\mathbf{u}$ is a Cauchy sequence in V. Since V is complete, it follows that $T_n\mathbf{u}$ tends to a limit $T\mathbf{u}$, say.

Step B: In Step A we produced a map $T : U \to V$. We want to show that, in fact, $T \in \mathcal{L}(U,V)$. Observe first that, since T_n is linear

$$\begin{aligned}
&\|T(\lambda_1\mathbf{u}_1 + \lambda_2\mathbf{u}_2) - \lambda_1 T\mathbf{u}_1 - \lambda_2 T\mathbf{u}_2\|_V \\
&= \|(T(\lambda_1\mathbf{u}_1 + \lambda_2\mathbf{u}_2) - \lambda_1 T\mathbf{u}_1 - \lambda_2 T\mathbf{u}_2) - (T_n(\lambda_1\mathbf{u}_1 + \lambda_2\mathbf{u}_2) - \lambda_1 T_n\mathbf{u}_1 - \lambda_2 T_n\mathbf{u}_2)\|_V \\
&= \|(T(\lambda_1\mathbf{u}_1 + \lambda_2\mathbf{u}_2) - T_n(\lambda_1\mathbf{u}_1 + \lambda_2\mathbf{u}_2)) - \lambda_1(T\mathbf{u}_1 - T_n\mathbf{u}_1) - \lambda_2(T\mathbf{u}_2 - T_n\mathbf{u}_2)\|_V \\
&\leq \|T(\lambda_1\mathbf{u}_1 + \lambda_2\mathbf{u}_2) - T_n(\lambda_1\mathbf{u}_1 + \lambda_2\mathbf{u}_2)\|_V + |\lambda_1|\|T\mathbf{u}_1 - T_n\mathbf{u}_1\|_V + |\lambda_2|\|T\mathbf{u}_2 - T_n\mathbf{u}_2\|_V \\
&\to 0
\end{aligned}$$

as $n \to \infty$. Thus

$$\|T(\lambda_1\mathbf{u}_1 + \lambda_2\mathbf{u}_2) - \lambda_1 T\mathbf{u}_1 - \lambda_2 T\mathbf{u}_2\|_V = 0,$$

so

$$T(\lambda_1\mathbf{u}_1 + \lambda_2\mathbf{u}_2) - \lambda_1 T\mathbf{u}_1 - \lambda_2 T\mathbf{u}_2 = \mathbf{0},$$

and T is linear.

Next, observe that, since every Cauchy sequence is bounded, we can find a K such that $\|T_n\| \leq K$ for all n. It follows that $\|T_n\mathbf{u}\|_V \leq K\|\mathbf{u}\|_U$ for each n. Thus

$$\|T\mathbf{u}\|_V \leq \|T\mathbf{u} - T_n\mathbf{u}\|_V + \|T_n\mathbf{u}\|_V \leq \|T\mathbf{u} - T_n\mathbf{u}\|_V + K\|\mathbf{u}\|_U \to K\|\mathbf{u}\|_U$$

as $n \to \infty$, and so $\|T\mathbf{u}\|_V \leq K\|\mathbf{u}\|_U$ for all $\mathbf{u} \in U$. Thus T is continuous.

Step C: Finally we need to show that $\|T - T_n\| \to 0$ as $n \to \infty$. To do this we use the trick of 'the irrelevant m' introduced in the proof of Example 11.10:

$$\begin{aligned}
\|T\mathbf{u} - T_n\mathbf{u}\|_V &\leq \|T\mathbf{u} - T_m\mathbf{u}\|_V + \|T_m\mathbf{u} - T_n\mathbf{u}\|_V \\
&\leq \|T\mathbf{u} - T_m\mathbf{u}\|_V + \|(T_m - T_n)\mathbf{u}\|_V \\
&\leq \|T\mathbf{u} - T_m\mathbf{u}\|_V + \|T_m - T_n\|\|\mathbf{u}\|_U \\
&\leq \|T\mathbf{u} - T_m\mathbf{u}\|_V + \sup_{p,q \geq M} \|T_p - T_q\|\|\mathbf{u}\|_U \\
&\to \sup_{p,q \geq M} \|T_p - T_q\|\|\mathbf{u}\|_U
\end{aligned}$$

as $m \to \infty$. Thus $\|T\mathbf{u} - T_n\mathbf{u}\|_V \leq \sup_{p,q \geq M} \|T_p - T_q\|\|\mathbf{u}\|_U$ for all $\mathbf{u} \in U$, and so

$$\|T - T_n\| \leq \sup_{p,q \geq M} \|T_p - T_q\|$$

for all $n \geq M$. Recalling that the sequence T_m is Cauchy, we see that $\|T - T_n\| \to 0$ as $n \to \infty$, and we are done. ■

Remark: In this book we are mainly concerned with the case when U and V are finite-dimensional. In this special case, $\mathcal{L}(U, V)$ is finite-dimensional and, since all norms on a finite-dimensional space are equivalent (Theorem 10.45), the operator norm is automatically complete.

Exercise 11.16. *In Exercise 10.55, U and V are not complete. Give an example along the same lines involving complete normed spaces.*

11.2. The Bolzano–Weierstrass property

In the previous section we introduced the notion of a complete metric space as a generalisation of the general principle of convergence. The reader may ask why we did not choose to try for some generalisation of the Bolzano–Weierstrass theorem instead. One answer is that it is generally agreed that the correct generalisation of the Bolzano–Weierstrass property is via the notion of compactness and that compactness is best studied in the context of topological spaces (a concept more general than metric spaces). A second answer, which the reader may find more satisfactory, is given in the discussion below which concludes in Theorem 11.23.

We make the following definition.

Definition 11.17. *We say that a metric space (X, d) has the Bolzano–Weierstrass property*[2] *if every sequence $x_n \in X$ has a convergent subsequence.*

Lemma 11.18. *A metric space (X, d) with the Bolzano–Weierstrass property is complete.*

Proof. Suppose that x_n is a Cauchy sequence. By definition, given any $\epsilon > 0$, we can find $n_0(\epsilon)$ such that $d(x_p, x_q) < \epsilon$ for all $p, q \geq n_0(\epsilon)$. By the Bolzano–Weierstrass property, we can find $n(j) \to \infty$ and $x \in X$ such that $n(j) \to \infty$ and $x_{n(j)} \to x$ as $j \to \infty$.

Thus, given any $\epsilon > 0$, we can find a J such that $n(J) \geq n_0(\epsilon/2)$ and $d(x, x_{n(J)}) < \epsilon/2$. Since $n(J) \geq n_0(\epsilon/2)$, we know that, whenever $m \geq n_0(\epsilon/2)$, we have $d(x_{n(J)}, x_m) < \epsilon/2$, and so

$$d(x, x_m) \leq d(x, x_{n(J)}) + d(x_{n(J)}, x_m) < \epsilon/2 + \epsilon/2 = \epsilon.$$

Thus $x_n \to x$ as $n \to \infty$. ∎

Exercise 11.19. *We work in a metric space (X, d).*

(i) Show that if $x_n \to x$ as $n \to \infty$, then the sequence x_n is Cauchy. (Any convergent sequence is Cauchy.)

(ii) If x_n is a Cauchy sequence and we can find $n(j) \to \infty$ and $x \in X$ such that $n(j) \to \infty$ and $x_{n(j)} \to x$ as $j \to \infty$, then $x_n \to x$. (Any Cauchy sequence with a convergent subsequence is convergent.)

To show that the converse of Lemma 11.18 is false, it suffices to consider $\mathbb{R}$ with the usual topology. To give another counterexample (in which, additionally, the metric is bounded) we introduce a dull but useful metric space.

Exercise 11.20. *Let X be any set. We define $d : X^2 \to \mathbb{R}$ by*

$$\begin{aligned} d(x, y) &= 1 \qquad \text{if } x \neq y, \\ d(x, x) &= 0. \end{aligned}$$

(i) Show that d is a metric.

(ii) Show that $d(x_n, x) \to 0$ as $n \to \infty$ if and only if there exists an N such that $x_n = x$ for all $n \geq N$.

(iii) Show that (X, d) is complete.

(iv) If $y \in X$, show that the closed unit ball $\bar{B}(y, 1) = \{x : d(x, y) \leq 1\} = X$.

[2] In more advanced work we say that (X, d) is sequentially compact.

(v) If X is infinite, show, using (ii) or otherwise, that X does not have the Bolzano–Weierstrass property.

(vi) Show also that every subset of X is both open and closed.

We call the metric d of the previous lemma the *discrete metric.*

In order to characterise metric spaces having the Bolzano–Weierstrass property, we must introduce a further definition.

Definition 11.21. *We say that (X, d) is totally bounded if, given any $\epsilon > 0$, we can find $y_1, y_2, \dots, y_N \in X$ such that $\bigcup_{j=1}^{N} B(y_j, \epsilon) = X$.*

In other words, (X, d) is totally bounded if, given any $\epsilon > 0$, we can find a finite set of open balls of radius ϵ covering X.

Lemma 11.22. *If (X, d) is a metric space with the Bolzano–Weierstrass property, then it is totally bounded.*

Proof. If (X, d) is not totally bounded, then we can find an $\epsilon > 0$ such that no finite set of open balls of radius ϵ covers X. Choose any $x_1 \in X$. We obtain $x_2, x_3, \dots$ inductively as follows. Once $x_1, x_2, \dots, x_n$ have been fixed, we observe that $\bigcup_{j=1}^{n} B(x_j, \epsilon) \neq X$, so we can choose $x_{n+1} \notin \bigcup_{j=1}^{n} B(x_j, \epsilon)$.

Now consider the sequence x_j. By construction, $d(x_i, x_j) \geq \epsilon$ for all $i \neq j$, and so, if $x \in X$, we have $\max(d(x, x_i), d(x, x_j)) \geq \epsilon/2$ for all $i \neq j$. Thus the sequence x_j has no convergent subsequence, and (X, d) does not have the Bolzano–Weierstrass property. ∎

Theorem 11.23. *A metric space (X, d) has the Bolzano–Weierstrass property if and only if it is complete and totally bounded.*

Proof. Necessity follows from Lemmas 11.18 and 11.22.

To prove sufficiency, we first suppose that (X, d) is complete and totally bounded. Let x_n be a sequence in X. We wish to show that it has a convergent subsequence.

The key observation is contained in this paragraph. Suppose that A is a subset of X such that $x_n \in A$ for infinitely many values of n, and suppose $\epsilon > 0$. Since X is totally bounded, we can find a finite set of open balls B_1, $B_2, \dots, B_M$, each of radius ϵ, such that $\bigcup_{m=1}^{M} B_m = X$. It follows that $\bigcup_{m=1}^{M} A \cap B_m = A$, and, for at least one of the balls B_m, it must be true that $x_n \in A \cap B_m$ for infinitely many values of n. Thus we have shown that there is an open ball of radius ϵ such that $x_n \in A \cap B$ for infinitely many values of n.

It follows that we can construct inductively a sequence of open balls B_1, $B_2, \dots$ such that B_r has radius 2^{-r} and $x_n \in \bigcap_{s=1}^{r} B_s$ for infinitely many values of n $[r = 1, 2, \ldots]$. Pick $n(1) < n(2) < n(3) < \ldots$ such that $x_{n(r)} \in \bigcap_{s=1}^{r} B_s$. If $p, q > r$, then $x_{n(p)}, x_{n(q)} \in B_r$, and so $d(x_{n(p)}, x_{n(q)}) < 2^{-r+1}$. Thus the sequence $x_{n(r)}$ is Cauchy and, since X is complete, it converges. ∎

Exercise 11.24. *Show that the open interval* $(0, 1)$ *with the usual Euclidean metric is totally bounded but does not have the Bolzano–Weierstrass property.*

Exercise 11.25. *Use the completeness of the Euclidean norm on* $\mathbb{R}^m$ *and Theorem 11.23 to show that a closed bounded subset of* $\mathbb{R}^m$ *with the usual Euclidean norm has the Bolzano–Weierstrass property. (Thus we have an alternative proof of Theorem 4.16, provided we do not use Bolzano–Weierstrass to prove the completeness of* $\mathbb{R}^m$*.)*

Exercise 11.26. *Show that a metric space* (X, d) *is totally bounded if and only if every sequence in* X *has a Cauchy subsequence.*

We shall not make a great deal of use of the concept of the Bolzano–Weierstrass property in the remainder of this book. Thus, although the results that follow are quite important, the reader should treat them merely as a revision exercise for some of the material of Section 4.3.

Our first result is a generalisation of Theorem 4.41.

Definition 11.27. *If* (X, d) *is a metric space, we say that a subset* A *has the Bolzano–Weierstrass property*[3] *if the metric subspace* (A, d_A) *(where* d_A *is the restriction of the metric* d *to* A*) has the Bolzano–Weierstrass property.*

Exercise 11.28. *(i) Let* (X, d) *and* (Z, ρ) *be metric spaces. Show that, if* K *is a subset of* X *with the Bolzano–Weierstrass property and* $f : X \to Z$ *is continuous, then* $f(K)$ *has the Bolzano–Weierstrass property.*

(ii) Let (X, d) *be a metric space with the Bolzano–Weierstrass property and let* $\mathbb{R}^p$ *have the usual Euclidean norm. Show, by using part (i), or otherwise, that, if* $\mathbf{f} : X \to \mathbb{R}^p$ *is a continuous function, then* $\mathbf{f}(K)$ *is closed and bounded.*

Exercise 11.29. *State and prove the appropriate generalisation of Theorem 4.44.*

Exercise 11.30. *(This generalises Exercise 4.48.) Let* (X, d) *be a metric space with the Bolzano–Weierstrass property. Show that if* $K_1, K_2, \ldots$ *are closed sets such that* $K_1 \supseteq K_2 \supseteq \ldots$*, then* $\bigcap_{j=1}^{\infty} K_j \neq \varnothing$.

The following is a natural generalisation of Definition 4.60.

[3]It is more usual to say that A is compact. However, although the statement 'A has the Bolzano–Weierstrass property' turns out to be equivalent to 'A is compact' for metric spaces, this is not true in more general contexts.

Definition 11.31. *Let (X, d) and (Z, ρ) be metric spaces. We say that a function $f : X \to Z$ is uniformly continuous if, given $\epsilon > 0$, we can find a $\delta(\epsilon) > 0$ such that, if $x, y \in X$ and $d(x, y) < \delta(\epsilon)$, we have*

$$\rho(f(x), f(y)) < \epsilon.$$

The next exercise generalises Theorem 4.63.

Exercise 11.32. *Let (X, d) and (Z, ρ) be metric spaces. If (X, d) has the Bolzano–Weierstrass property, then any continuous function $f : X \to Z$ is uniformly continuous.*

11.3. The uniform norm

This section is devoted to one the most important norms on functions. We shall write $\mathbb{F}$ to mean either $\mathbb{R}$ or $\mathbb{C}$.

Definition 11.33. *If E is a non-empty set, we write $\mathcal{B}(E)$ (or, more precisely, $\mathcal{B}_{\mathbb{F}}(E)$) for the set of bounded functions $f : E \to \mathbb{F}$. The uniform norm $\| \ \|_\infty$ on $\mathcal{B}(E)$ is defined by $\|f\|_\infty = \sup_{x \in E} |f(x)|$.*

Exercise 11.34. *If we use the standard operations, show that $\mathcal{B}(E)$ is vector space over $\mathbb{F}$ and $\| \ \|_\infty$ is a norm.*

Show also that, if $f, g \in \mathcal{B}(E)$ and we write $f \times g(x) = f(x)g(x)$, then $f \times g \in \mathcal{B}(E)$ and $\|f \times g\|_\infty \leq \|f\|_\infty \|g\|_\infty$.

The norm $\| \ \|_\infty$, just defined, is called the uniform norm, sup norm or ∞ norm.

Theorem 11.35. *The uniform norm on $\mathcal{B}(E)$ is complete.*

Proof. This follows the model set up in Section 11.1. If reader has done Exercise 11.13, the argument will be completely familiar.

Suppose f_n is a Cauchy sequence in $(\mathcal{B}(E), \| \ \|_\infty)$. For each fixed $x \in E$

$$|f_n(x) - f_m(x)| \leq \|f_n - f_m\|_\infty,$$

so $f_n(x)$ is a Cauchy sequence in $\mathbb{F}$. The general principle of convergence tells us that $f_n(x)$ tends to a limit $f(x)$ as $n \to \infty$.

Since any Cauchy sequence is bounded, we can find a K such that $\|f_n\|_\infty \leq K$ for all n. We observe that

$$\begin{aligned}|f(x)| &\leq |f(x) - f_n(x)| + |f_n(x)| \leq |f(x) - f_n(x)| + \|f_n\|_\infty \\ &\leq |f(x) - f_n(x)| + K \to K\end{aligned}$$

as $n \to \infty$. Thus $|f(x)| \leq K$ for all $x \in E$, and so $f \in \mathcal{B}(E)$.

Finally we need to show that $\|f - f_n\|_\infty \to 0$ as $n \to \infty$. To do this we use the usual trick of the 'irrelevant m', observing that

$$\begin{aligned}|f(x) - f_n(x)| &\le |f(x) - f_m(x)| + |f_m(x) - f_n(x)| \\ &\le |f(x) - f_m(x)| + \|f_m - f_n\|_\infty \\ &\le |f(x) - f_m(x)| + \sup_{p,q\ge M} \|f_p - f_q\|_\infty \to \sup_{p,q\ge M} \|f_p - f_q\|_\infty\end{aligned}$$

as $m \to \infty$. Thus $|f(x) - f_n(x)| \le \sup_{p,q\ge M} \|f_p - f_q\|_\infty$ for all $x \in E$, and so

$$\|f - f_n\|_\infty \le \sup_{p,q\ge M} \|f_p - f_q\|_\infty$$

for all $n \ge M$. Recalling that the sequence f_m is Cauchy, we see that $\|f - f_n\|_\infty \to 0$ as $n \to \infty$, and we are done. ■

The space $\mathcal{B}(E)$ is not very interesting in itself, but, if E is a metric space, it has a very interesting subspace.

Definition 11.36. *If (E, d) is a non-empty metric space, we write $\mathcal{C}(E)$ (or, more precisely, $\mathcal{C}_{\mathbb{F}}(E)$) for the set of bounded continuous functions $f : E \to \mathbb{F}$.*

The next remark merely restates what we already know.

Lemma 11.37. *If (E, d) is a non-empty metric space, then $\mathcal{C}(E)$ is a vector subspace of $\mathcal{B}(E)$. Further, if f, $g \in \mathcal{C}(E)$, the pointwise product $f \times g \in \mathcal{C}(E)$.*

However, the next result is new and crucial.

Theorem 11.38. *If (E, d) is a non-empty metric space, then $\mathcal{C}(E)$ is a closed subset of $\mathcal{B}(E)$ under the uniform metric.*

This has the famous '$\epsilon/3$ proof'[4].

Proof. Let $f_n \in \mathcal{C}(E)$, $f \in \mathcal{B}(E)$ and $\|f_n - f\|_\infty \to 0$. We wish to show that f is continuous on E, and to do this it suffices to show that f is continuous at any specified point $x \in E$.

Let $\epsilon > 0$ and $x \in E$ be given. Since $\|f_n - f\|_\infty \to 0$ as $n \to \infty$, it follows, in particular, that there exists an N with

$$\|f_N - f\|_\infty < \epsilon/3.$$

[4]Or, according to a rival school of thought, the '3ϵ proof'.

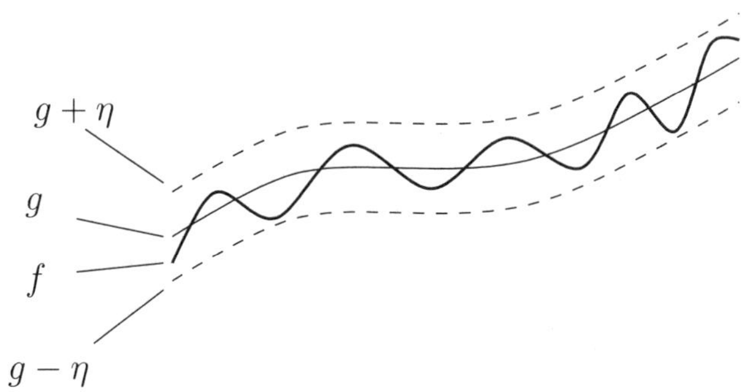

Figure 11.1. The uniform norm snake

Since f_N is continuous at x, we can find a $\delta > 0$ such that $|f_N(x) - f_N(t)| \leq \epsilon/3$ for all $t \in E$ with $d(x,t) < \delta$. It follows that

$$\begin{aligned}|f(x) - f(t)| &= |(f(x) - f_N(x)) + (f_N(x) - f_N(t)) + (f_N(t) - f(t))| \\ &\leq |f(x) - f_N(x)| + |f_N(x) - f_N(t)| + |f_N(t) - f(t))| \\ &\leq \|f - f_N\|_\infty + |f_N(x) - f_N(t)| + \|f_N - f\|_\infty \\ &< \epsilon/3 + \epsilon/3 + \epsilon/3 = \epsilon\end{aligned}$$

for all $t \in E$ with $d(x,t) < \delta$. ■

The key to the argument above is illustrated in Figure 11.1. Suppose that $\|f - g\|_\infty < \eta$. Then f is trapped in a snake of radius η with g as its backbone. In particular, if g is continuous, f cannot be 'terribly discontinuous'.

Since $\mathcal{C}(E)$ is a closed subset of $\mathcal{B}(E)$, Theorem 11.35 and Lemma 11.4 gives us another important theorem.

Theorem 11.39. *If (E,d) is a non-empty metric space, then the uniform metric on $\mathcal{C}(E)$ is complete.*

If E is a closed bounded subset of $\mathbb{R}^m$ with the Euclidean metric (or, more generally, if (E,d) is a metric space with the Bolzano–Weierstrass property), then, by Theorem 4.44 (or its easy generalisation to metric spaces with the Bolzano–Weierstrass property), all continuous functions $f : E \to \mathbb{F}$ are bounded. In these circumstances, we shall write $C(E) = \mathcal{C}(E)$ (or, more precisely, $C_\mathbb{F}(E) = \mathcal{C}_\mathbb{F}(E)$) and refer to the space $C(E)$ of continuous functions on E equipped with the uniform norm.

Exercise 11.40. *The question of which subsets of $C(E)$ have the Bolzano–Weierstrass property is quite hard and will not be tackled here. To get some understanding of the problem, show by considering $f_n(x) = \sin n\pi x$, or* *otherwise, that* $\{f \in C([0,1]) : \|f\|_\infty \leq 1\}$ *does not have the Bolzano–Weierstrass property.*

The reader has now met three different norms on $C([a,b])$ (two of them introduced in Lemma 11.8 and Exercise 11.9):

$$\|f\|_\infty = \sup_{t\in[a,b]} |f(t)|,$$

$$\|f\|_1 = \int_a^b |f(t)|\,dt,$$

$$\|f\|_2 = \left(\int_a^b |f(t)|^2\,dt\right)^{1/2}.$$

They represent different answers to the question 'when do two continuous functions f and g resemble each other'. If we say that f and g are close only if $\|f-g\|_\infty$ is small, then, however small $|f(x)-g(x)|$ is over 'most of the range', if $|f(x)-g(x)|$ is large anywhere, we say that f and g are far apart. For many purposes this is too restrictive, and we would like to say that f and g are close if 'on average' $|f(x)-g(x)|$ is small, that is to say, $\|f-g\|_1$ is small. For a communications engineer, to whom

$$\|f\|_2^2 = \int_a^b |f(t)|^2\,dt$$

is a measure of the power of a signal, the obvious measure of the closeness of f and g is $\|f-g\|_2$.

The problem of finding an appropriate measure of dissimilarity crops up in many different fields. Here are a few examples.

(a) In weather forecasting, how do we measure how close the forecast turns out to be to the true weather?

(b) In archaeology, ancient graves contain various 'grave goods'. Presumably graves which have many types of grave goods in common are close in time and those with few in common are distant in time. What is the correct measure of similarity between graves?

(c) In high definition TV and elsewhere, pictures are compressed for transmission and then reconstituted from the reduced information. How do we measure how close the final picture is to the initial one?

(d) Machines find it hard to read handwriting. People find it easy. How can we measure the difference between curves so that the same words written by different people give curves which are close but different words are far apart?

Since there are many different problems, there will be many measures of closeness. We should not be surprised that mathematicians use many different metrics and norms.

We conclude the section with a generalisation of Theorem 11.39. The proof provides an opportunity to review the contents of this section. In Exercise K.252 we use the result with $V = \mathbb{R}^2$.

Exercise 11.41. *Let (E, d) be a non-empty metric space and $(V, \| \ \|)$ a complete normed space. We write $\mathcal{C}_V(E)$ for the set of bounded continuous functions $\mathbf{f} : E \to V$, and set*

$$\|\mathbf{f}\|_\infty = \sup_{x \in E} \|\mathbf{f}(x)\|_V$$

whenever $\mathbf{f} \in \mathcal{C}_V(E)$. Show that $(\mathcal{C}_V(E), \| \ \|_\infty)$ is a complete normed vector space.

11.4. Uniform convergence

Traditionally, the material of the previous section has been presented in a different but essentially equivalent manner.

Definition 11.42. (Uniform convergence.) *If E is a non-empty set and $f_n : E \to \mathbb{F}$ and $f : E \to \mathbb{F}$ are functions, we say that f_n* converges uniformly *to f as $n \to \infty$ if, given any $\epsilon > 0$, we can find an $n_0(\epsilon)$ with the property that $|f_n(x) - f(x)| < \epsilon$ for all $x \in E$ and all $n \geq n_0(\epsilon)$.*

Remark 1: For a generalisation see Exercise K.226.

Remark 2: On page 64 we placed the definition of uniform continuity in parallel with the definition of continuity. In the same way, the reader should compare the definition of uniform convergence with the notion of convergence that we have used so far and which we shall now call pointwise convergence.

Definition 11.43. (Pointwise convergence.) *If E is a non-empty set and $f_n : E \to \mathbb{F}$ and $f : E \to \mathbb{F}$ are functions, we say that f_n* converges pointwise *to f as $n \to \infty$ if, given any $\epsilon > 0$ and any $x \in E$, we can find an $n_0(\epsilon, x)$ with the property that $|f_n(x) - f(x)| < \epsilon$ for all $n \geq n_0(\epsilon, x)$.*

Once again, 'uniform' means independent of the choice of x.

Theorem 11.38 now takes the following form.

Theorem 11.44. *If (E, d) is a non-empty metric space and $f_n : E \to \mathbb{F}$ forms a sequence of continuous functions tending uniformly to f, then f is continuous.*

More briefly, the uniform limit of continuous functions is continuous.

Proof of Theorem 11.44 from Theorem 11.38. The problem we must face is that the f_n need not be bounded.

To get round this, choose N such that $|f_N(x) - f(x)| < 1$ for all $x \in E$ and all $n \geq N$. If we set $g_n = f_n - f_N$, then

$$|g_n(x)| \leq |f_n(x) - f(x)| + |f_N(x) - f(x)| < 2,$$

and so $g_n \in \mathcal{C}(E)$ for all $n \geq N$. If we set $g = f - f_N$, then $g \in \mathcal{B}(E)$, and

$$\|g_n - g\|_\infty = \sup_{x \in E} |g_n(x) - g(x)| = \sup_{x \in E} |f_n(x) - f(x)| \to 0.$$

By Theorem 11.38, $g \in \mathcal{C}(E)$, and so $f = g + f_N$ is continuous. ∎

The same kind of simple argument applied to Theorem 11.39 gives the so-called 'general principle of uniform convergence'.

Theorem 11.45. (General principle of uniform convergence.) *Suppose that (E, d) is a non-empty metric space and $f_n : E \to \mathbb{F}$ is a continuous function $[n \geq 1]$. The sequence f_n converges uniformly to a continuous function f if and only if, given any $\epsilon > 0$, we can find an $n_0(\epsilon)$ such that $|f_n(x) - f_m(x)| < \epsilon$ for all n, $m \geq n_0(\epsilon)$ and all $x \in E$.*

This theorem is also known as the GPUC by those who do not object to theorems which sound as though they were a branch of the secret police.

Exercise 11.46. *Prove Theorem 11.45 from Theorem 11.39.*

Exercise 11.47. *(i) Prove Theorem 11.44 and Theorem 11.45 directly. (Of course, the proofs will be very similar to those of Theorem 11.38 and Theorem 11.35. If you prefer simply to look things up, proofs are given in practically every analysis textbook.)*

(ii) Prove Theorem 11.38 from Theorem 11.44 and Theorem 11.39 from Theorem 11.45.

The next two examples are very important for understanding the difference between pointwise and uniform convergence.

Example 11.48. (The witch's hat.) *Define $f_n : [0, 2] \to \mathbb{R}$ by*

$$\begin{aligned} f_n(x) &= 1 - n|x - n^{-1}| && \text{for } |x - n^{-1}| \leq n^{-1}, \\ f_n(x) &= 0 && \text{otherwise.} \end{aligned}$$

Then the f_n are continuous functions such that $f_n(x) \to 0$ as $n \to \infty$ for each x but $f_n \not\to 0$ uniformly.

More briefly, pointwise convergence does not imply uniform convergence. We sketch the witch's hat in Figure 11.2.

Proof. Observe that, if $x \neq 0$, then $x \geq 2N^{-1}$ for some positive integer N, and so, provided $n \geq N$,

$$f_n(x) = 0 \to 0$$

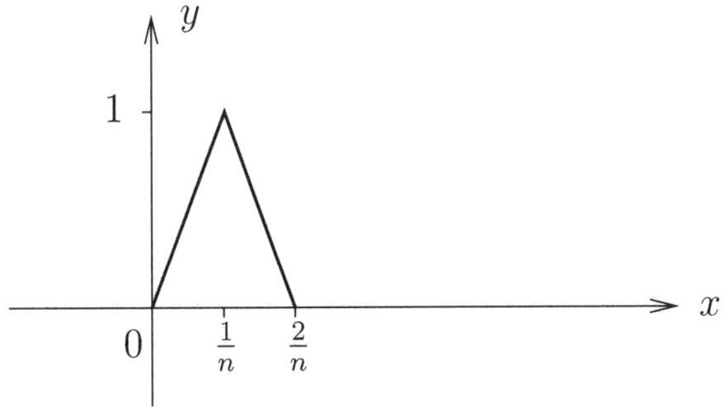

Figure 11.2. The witch's hat

as $n \to \infty$. On the other hand, if $x = 0$,

$$f_n(0) = 0 \to 0$$

as $n \to \infty$. Thus $f_n(x) \to 0$ for each $x \in [0, 2]$.

We observe that $\|f_n\|_\infty \geq f_n(n^{-1}) = 1 \nrightarrow 0$, so $f_n \nrightarrow 0$ uniformly as $n \to 0$. ■

Example 11.49. *Define $f_n : [0, 1] \to \mathbb{R}$ by $f_n(x) = x^n$. Then $f_n(x) \to f(x)$ as $n \to \infty$, where $f(x) = 0$ for $0 \leq x < 1$ but $f(1) = 1$.*

Thus the pointwise limit of continuous functions need not be continuous. We leave the verification to the reader.

Exercise 11.50. *Draw a diagram to illustrate Example 11.49 and prove the result.*

Uniform convergence is a very useful tool when dealing with integration.

Theorem 11.51. *Let $f_n \in C([a, b])$. If $f_n \to f$ uniformly, then $f \in C([a, b])$ and*

$$\int_a^b f_n(x)\, dx \to \int_a^b f(x)\, dx.$$

Students often miss the full force of this theorem because it is so easy to prove. Note, however, that we need to prove that the second integral actually exists. (You should look briefly at Exercise 9.1 before proceeding. If you want an example of a sequence of continuous functions whose pointwise limit is not Riemann integrable, consult the harder Exercise K.175.)

Proof. Since f is the uniform limit of continuous functions, it is itself continuous and therefore Riemann integrable. By the sup $\times$ length inequality,

$$\begin{aligned}\left|\int_a^b f(t)\,dt - \int_a^b f_n(t)\,dt\right| &= \left|\int_a^b (f(t)-f_n(t))\,dt\right| \leq \int_a^b |f(t)-f_n(t)|\,dt \\ &\leq (b-a)\|f-f_n\|_\infty \to 0\end{aligned}$$

as $n \to \infty$. ∎

Theorem 11.51 should be considered in the context of the following two examples.

Example 11.52. (The tall witch's hat.) *Define $f_n : [0,2] \to \mathbb{R}$ by*

$$\begin{aligned} f_n(x) &= n(1-n|x-n^{-1}|) && \text{for } |x-n^{-1}| \leq n^{-1}, \\ f_n(x) &= 0 && \text{otherwise.}\end{aligned}$$

Then the f_n are continuous functions such that $f_n(x) \to 0$ as $n \to \infty$ but

$$\int_0^2 f_n(x)\,dx \nrightarrow 0$$

as $n \to \infty$.

Example 11.53. (Escape to infinity.) *Define $f_n : \mathbb{R} \to \mathbb{R}$ by*

$$\begin{aligned} f_n(x) &= n^{-1}(1-n^{-1}|x|) && \text{for } |x| \leq n, \\ f_n(x) &= 0 && \text{otherwise.}\end{aligned}$$

Then the f_n are continuous functions such that $f_n(x) \to 0$ uniformly as $n \to \infty$ but

$$\int_{-\infty}^{\infty} f_n(x)\,dx \nrightarrow 0$$

as $n \to \infty$.

[We gave a similar example of escape to infinity in Exercise 5.24.]

Exercise 11.54. *(i) Draw a diagram to illustrate Example 11.52 and prove the result.*

(ii) Draw a diagram to illustrate Example 11.53 and prove the result.

One way of contrasting Theorem 11.51 with Example 11.53 is to think of pushing a piston down a cylinder with water at the bottom. Eventually we must stop because the water has nowhere to escape to. However, if we place a glass sheet on top of another glass sheet, any water between them escapes outwards.

The following example, which requires some knowledge of probability theory, illustrates the importance of the phenomenon exhibited in Example 11.53.

Exercise 11.55. *Let $X_1, X_2, \ldots$ be independent identically distributed random variables. Suppose further that X_1 has continuous density distribution f. We know that $X_1 + X_2 + \cdots + X_n$ then has a continuous distribution f_n.*

(i) Explain why $\int_{-\infty}^{\infty} f_n(x)\,dx = 1$.

(ii) In the particular case $f(x) = (2\pi)^{-1/2}\exp(-x^2/2)$, state the value of $f_n(x)$ and show that $f_n(x) \to 0$ everywhere. Thus

$$\lim_{n\to\infty} \int_{-\infty}^{\infty} f_n(x)\,dx \neq \int_{-\infty}^{\infty} \lim_{n\to\infty} f_n(x)\,dx.$$

(iii) If Y is a real-valued random variable with continuous density distribution g_Y and $a > 0$, show that aY has continuous density distribution g_{aY} given by $g_{aY}(x) = a^{-1}g(a^{-1}x)$. What is the density for $-aY$?

(iv) In the particular case investigated in part (ii), show that

(a) If $1/2 > \alpha \geq 0$, then $n^\alpha f_n(n^\alpha x) \to 0$ uniformly on $\mathbb{R}$ as $n \to \infty$ and

$$\lim_{n\to\infty} \int_{-\infty}^{\infty} n^\alpha f_n(n^\alpha x)\,dx \neq \int_{-\infty}^{\infty} \lim_{n\to\infty} n^\alpha f_n(n^\alpha x)\,dx.$$

(b) If $\alpha = 1/2$, then $n^\alpha f_n(n^\alpha x) = f(x)$ and

$$\lim_{n\to\infty} \int_{-\infty}^{\infty} n^\alpha f_n(n^\alpha x)\,dx = \int_{-\infty}^{\infty} \lim_{n\to\infty} n^\alpha f_n(n^\alpha x)\,dx.$$

(c) If $\alpha > 1/2$, then $n^\alpha f_n(n^\alpha x) \to 0$ for each $x \neq 0$ as $n \to \infty$, but $n^\alpha f_n(n^\alpha 0) \to \infty$.

(v) Draw diagrams illustrating the three cases in (iv) and give a probabilistic interpretation in each case.

(vi) How much further you can go, for general f, depends on your knowledge of probability. If you know any of the terms Tchebychev inequality, central limit theorem or Cauchy distribution, discuss how they apply here.

In any case, I hope I have demonstrated that when talking about things like

$$\frac{X_1 + X_2 + \cdots + X_n}{n^\alpha}$$

we expect the interchange of limits only to work in exceptional (but therefore profoundly interesting) cases.

Exercise 11.56. *As I have already noted, we gave a similar example to Example 11.53 in Exercise 5.24. We followed that by a dominated convergence theorem for sums (Lemma 5.25). Can you formulate a similar dominated convergence theorem for integrals? A possible version is given as Exercise K.245.*

Traditionally, Theorem 11.51 is always paired with the following result which is proved by showing that it is really a disguised theorem on integration!

Theorem 11.57. *Suppose that $f_n : [a,b] \to \mathbb{R}$ is differentiable on $[a,b]$ with continuous derivative f_n' (we take the one-sided derivative at end points as usual). Suppose that $f_n(x) \to f(x)$ as $n \to \infty$ for each $x \in [a,b]$ and suppose that f_n' converges uniformly to a limit F on $[a,b]$. Then f is differentiable with derivative F.*

First proof. Since f_n' is continuous and $f_n' \to F$ uniformly, F is continuous and Theorem 11.51 tells us that

$$\int_c^t f_n'(x)\,dx \to \int_c^t F(x)\,dx$$

as $n \to \infty$ for all $t,\ c \in [a,b]$. By the fundamental theorem of the calculus (in the form of Theorem 8.42), we know that $\int_c^t f_n'(x)\,dx = f_n(t) - f_n(c)$, and so

$$f(t) - f(c) = \int_c^t F(x)\,dx.$$

Since F is continuous, another application of the fundamental theorem of the calculus (this time in the form of Theorem 8.37) tells us that f is differentiable with

$$f'(t) = F(t),$$

as required. ■

This proof is easy but rather roundabout. We present a second proof which is harder but much more direct.

Second proof of Theorem 11.57. Our object in this proof is to show that $|f(x+h) - f(x) - F(x)h|$ decreases faster than linearly as $h \to 0$ [when $x,\ x+h \in [a,b]$]. We start in an obvious way by observing that

(1)
$$|f(x+h) - f(x) - F(x)h| \leq |f_n(x+h) - f_n(x) - f_n'(x)h| \\ + \big|\big(f(x+h) - f(x) - F(x)h\big) - \big(f_n(x+h) - f_n(x) - f_n'(x)h\big)\big|.$$

The first term in the inequality can be estimated by the mean value inequality:

(2) $$|f_n(x+h) - f_n(x) - f_n'(x)h| \leq |h| \sup_{0<\theta<1} |f_n'(x+\theta h) - f_n'(x)|.$$

To estimate $\sup_{0<\theta<1} |f_n'(x+\theta h) - f_n'(x)|$, we reverse the argument of the '$\epsilon/3$ theorem' (Theorem 11.38). We know that F is continuous because it is the uniform limit of continuous functions. Thus, given $\epsilon > 0$, we can find a

$\delta(\epsilon) > 0$ such that $|F(y) - F(x)| < \epsilon/3$ whenever $y \in [a,b]$ and $|x-y| < \delta(\epsilon)$. Choosing an $N(\epsilon)$ such that $\|f'_n - F\|_\infty < \epsilon/3$ for all $n \geq N(\epsilon)$, we have

$$(3) \quad \begin{aligned} |f'_n(y) - f'_n(x)| &\leq |f'_n(y) - F(y)| + |F(y) - F(x)| + |F(x) - f'_n(x)| \\ &\leq 2\|f'_n - F\|_\infty + |F(y) - F(x)| < 2\epsilon/3 + \epsilon/3 = \epsilon \end{aligned}$$

for all $y \in [a,b]$ with $|x - y| < \delta(\epsilon)$ and all $n \geq N(\epsilon)$. Using this result, we see that inequality (2) gives

$$(2') \qquad |f_n(x+h) - f_n(x) - f'_n(x)h| \leq \epsilon|h|,$$

so that inequality (1) gives, in turn,

$$\begin{aligned} (1') \ & |f(x+h) - f(x) - F(x)h| \\ & \leq \big|\big(f(x+h) - f(x) - F(x)h\big) - \big(f_n(x+h) - f_n(x) - f'_n(x)h\big)\big| + \epsilon|h|, \end{aligned}$$

for all $x + h \in [a,b]$ with $|h| < \delta(\epsilon)$, and all $n \geq N(\epsilon)$. Allowing $n \to \infty$, we obtain

$$(1'') \qquad |f(x+h) - f(x) - F(x)h| \leq \epsilon|h|,$$

for all $x + h \in [a,b]$ with $|h| < \delta(\epsilon)$, and this is what we want. ■

One major advantage of the second proof is that it generalises easily to many dimensions.

Exercise 11.58. *Let Ω be an open set in $\mathbb{R}^m$. Suppose that $\mathbf{f}_n : \Omega \to \mathbb{R}^p$ is a differentiable function on Ω with continuous derivative $D\mathbf{f}_n$. Suppose that $\mathbf{f}_n(\mathbf{x}) \to \mathbf{f}(\mathbf{x})$ as $n \to \infty$ for each $\mathbf{x} \in \Omega$ and suppose that there is a function $\Theta : \Omega \to \mathcal{L}(\mathbb{R}^m, \mathbb{R}^p)$ (that is, Θ is a function on Ω whose values are linear maps from $\mathbb{R}^m$ to $\mathbb{R}^p$) such that*

$$\sup_{\mathbf{x}\in\Omega} \|\Theta(\mathbf{x}) - D\mathbf{f}_n(\mathbf{x})\| \to 0$$

(that is, $D\mathbf{f}_n$ converges uniformly to Θ in the operator norm) as $n \to \infty$. Then $\mathbf{f}$ is differentiable with derivative Θ.

Although Exercise 11.58 is not hard, it provides a useful exercise in understanding notation. We note, for example, that $D\mathbf{f}_n$ is a function on Ω whose values are linear maps from $\mathbb{R}^m$ to $\mathbb{R}^p$. The statement that $D\mathbf{f}_n$ is continuous must therefore be interpreted as

$$\|D\mathbf{f}_n(\mathbf{x} + \mathbf{h}) - D\mathbf{f}_n(\mathbf{x})\| \to 0$$

as $\|\mathbf{h}\|_2 \to 0$ for all $\mathbf{x} \in \Omega$, where $\| \ \|_2$ is the Euclidean norm on $\mathbb{R}^m$ and $\| \ \|$ is the operator norm on $\mathcal{L}(\mathbb{R}^m, \mathbb{R}^p)$.

Exercise 11.59. *(Easy but optional.) Rewrite Exercise 11.58 as a theorem on differentiation of functions between general normed vector spaces in accordance with Definition 10.58. Check that essentially the same proof works in this more general case.*

The reader knows how to turn results on the limits of sequences into results on infinite sums and vice versa (see Exercise 4.72 if necessary). Applied to Theorems 11.51 and 11.57, the technique produces the following results.

Theorem 11.60. (Term by term integration.) *Let $g_j : [a, b] \to \mathbb{R}$ be continuous. If $\sum_{j=1}^{n} g_j$ converges uniformly as $n \to \infty$, then*

$$\int_a^b \sum_{j=1}^{\infty} g_j(x)\, dx = \sum_{j=1}^{\infty} \int_a^b g_j(x)\, dx.$$

Theorem 11.61. (Term by term differentiation.) *Let $g_j : [a, b] \to \mathbb{R}$ be differentiable with continuous derivative. If $\sum_{j=1}^{n} g_j(x)$ converges for each x and $\sum_{j=1}^{n} g_j'$ converges uniformly as $n \to \infty$, then $\sum_{j=1}^{\infty} g_j$ is differentiable and*

$$\frac{d}{dx}\left(\sum_{j=1}^{\infty} g_j(x)\right) = \sum_{j=1}^{\infty} g_j'(x).$$

As a typical example of the use of Theorem 11.57, we use it to extend Theorem 8.57 on differentiation under the integral to a much more useful result.

Theorem 11.62. (Differentiation under an infinite integral.) *Let $(c', d') \supseteq [c, d]$. Suppose $g : [0, \infty) \times (c', d') \to \mathbb{R}$ is continuous and that the partial derivative $g_{,2}$ exists and is continuous. Suppose further, that there exists a continuous function $h : [0, \infty) \times (c, d) \to \mathbb{R}$ with $|g_{,2}(x, y)| \leq h(x)$ for all (x, y) and such that $\int_0^{\infty} h(x)\, dx$ exists and is finite. Then, if $G(y) = \int_0^{\infty} g(x, y)\, dx$ exists for all $y \in (c, d)$, we have G differentiable on (c, d) with*

$$G'(y) = \int_0^{\infty} g_{,2}(x, y)\, dx.$$

Proof. Note that $H(y) = \int_0^{\infty} g_{,2}(x, y)\, dx$ exists by comparison (Lemma 9.6). Set $G_n(y) = \int_0^{n} g(x, y)\, dx$. By Theorem 8.57, G_n is differentiable with

$$G_n'(y) = \int_0^{n} g_{,2}(x, y)\, dx.$$

Since

$$|G_n'(y) - H(y)| = \left|\int_n^{\infty} g_{,2}(x, y)\, dx\right| \leq \int_n^{\infty} h(x)\, dx \to 0,$$

we see that G_n' converges uniformly to H on (c, d). By hypothesis, $G_n(y) \to G(y)$ on (c, d), so, by Theorem 11.57, G is differentiable with derivative H on (c, d). This is the required result. ∎

You should be careful when using this theorem to check that the hypotheses actually apply. Exercise K.243 illustrates what can go wrong if we do not prevent 'escape to infinity'.

11.5. Power series

The object of this section and the next is to show how the notion of uniform convergence is used in two topics of practical importance.

We make use of a result which is too trivial to constitute a theorem but too useful to leave unnamed.

Lemma 11.63. (The Weierstrass M-test.) *Consider functions $f_n : E \to \mathbb{C}$. Suppose that we can find positive real numbers M_n such that $|f_n(x)| \leq M_n$ for all $x \in E$ and all $n \geq 1$. If $\sum_1^\infty M_n$ converges, then $\sum_1^\infty f_n$ converges uniformly on E.*

Proof. Let $\epsilon > 0$. By the easy part of the general principle of convergence we can find an $N_0(\epsilon)$ such that $\sum_{r=n}^m M_n \leq \epsilon$ for all $m \geq n \geq N_0(\epsilon)$. It follows that

$$\left|\sum_{r=n}^m f_n(x)\right| \leq \sum_{r=n}^m |f_n(x)| \leq \sum_{r=n}^m M_n \leq \epsilon$$

for all $m \geq n \geq N_0(\epsilon)$ and all $x \in E$. By the general principle of uniform convergence (Theorem 11.45), $\sum_1^\infty f_n$ converges uniformly on E. ∎

This book has been mainly about real analysis. When we talked about functions from $\mathbb{R}^n$ to $\mathbb{R}^m$ we made a point of the fact that we cannot divide a vector by a vector. There are, however, two exceptions to this rule. The first is that $\mathbb{R}$, itself, is a vector space where division is permitted. The second is that, if we give $\mathbb{R}^2$ the algebraic structure of $\mathbb{C}$, we again obtain a system in which division is permitted. This enables us to develop a theory of differentiation for functions $f : \mathbb{C} \to \mathbb{C}$ running in parallel with the theory of *one-dimensional* real differentiation. Note that we use the usual metric $d(z_1, z_2) = |z_1 - z_2|$ throughout.

Definition 11.64. *Let Ω be an open set in $\mathbb{C}$ and let $z \in \Omega$. We say that $f : \Omega \to \mathbb{C}$ is differentiable at z with derivative $f'(z)$ if*

$$\left|\frac{f(z+h) - f(z)}{h} - f'(z)\right| \to 0$$

as $h \to 0$.

Exercise 11.65. *Check that the definition is equivalent to the statement*

$$f(z+h) = f(z) + f'(z)h + \epsilon(h)|h|$$

for $z + h \in \Omega$, where $\epsilon(h) \to 0$ as $h \to 0$.

Exercise 11.66. *Convince yourself that the elementary theory of complex differentiation is the same as that of real differentiation. For example, you could run through the general results leading to the following result:*

Let $a_r \in \mathbb{C}$ $[0 \le r \le N]$, $b_s \in \mathbb{C}$ $[0 \le s \le M-1]$, $b_M = 1$, *and let* $P(z) = \sum_{r=0}^{N} a_r z^r$, $Q(z) = \sum_{s=0}^{M} b_s z^s$. *Then* $\Omega = \{z \in \mathbb{C} : Q(z) \ne 0\}$ *is open and the quotient* $P/Q : \mathbb{C} \to \mathbb{C}$ *is everywhere differentiable.*

Exercise 11.66 is routine and not meant to be taken very seriously. The next two results are important in their own right and provide an opportunity to review the content and proof of two important earlier results.

Exercise 11.67. (A mean value inequality.) *Suppose that* Ω *is an open set in* $\mathbb{C}$ *and that* $f : \Omega \to \mathbb{C}$ *is differentiable at all points of* Ω. *Suppose, further, that the straight line segment*

$$L = \{(1-t)z_1 + tz_2 : 0 \le t \le 1\}$$

joining z_1 *and* z_2 *lies in* Ω *and that*

$$|f'(z)| \le K$$

for all $z \in L$.

Explain why we can find a real θ *such that* $e^{i\theta}(f(z_2) - f(z_1))$ *is real and positive. Show that the function* $F : [0,1] \to \mathbb{R}$ *given by*

$$F(t) = \Re\big(e^{i\theta}(f((1-t)z_1 + tz_2) - f(z_2))\big)$$

is differentiable on $(0,1)$, *and find its derivative. By applying some form of mean value theorem or mean value inequality (many versions will work) to* F, *show that*

$$|f(z_2) - f(z_1)| \le K|z_2 - z_1|.$$

Exercise 11.68. *By using the ideas of the second proof of Theorem 11.57, prove the following result.*

Let Ω *be an open set in* $\mathbb{C}$. *Suppose that* $f_n : \Omega \to \mathbb{C}$ *is differentiable at all points of* Ω *with continuous derivative* f_n'. *Suppose that* $f_n(z) \to f(z)$ *as* $n \to \infty$ *for each* $z \in \Omega$, *and suppose that there is a function* $g : \Omega \to \mathbb{C}$ *such that*

$$\sup_{z \in \Omega} |g(z) - f_n'(z)| \to 0$$

(that is, f_n' *converges uniformly to* g*). Then* f *is differentiable at all points of* Ω *with derivative* g.

To use this result we need to recall the definition of the radius of convergence of a power series and the proof that makes the definition possible. The reader can look these things up on page 70, but it would, I think, be helpful to redo the work as an exercise.

Exercise 11.69. *(i) Suppose that* $a_n \in \mathbb{C}$. *Show that if* $\sum_{n=0}^{\infty} a_n z_0^n$ *converges for some* $z_0 \in \mathbb{C}$, *then* $\sum_{n=0}^{\infty} a_n z^n$ *converges for all* $z \in \mathbb{C}$ *with* $|z| < |z_0|$.

(ii) Use (i) to show that either $\sum_{n=0}^{\infty} a_n z^n$ *converges for all* z *(in which case we say the series has infinite radius of convergence) or there exists an* $R \geq 0$ *such that* $\sum_{n=0}^{\infty} a_n z^n$ *converges for* $|z| < R$ *and diverges for* $|z| > R$ *(in which case we say that the series has radius of convergence* R*).*

We can now improve this result slightly.

Lemma 11.70. *Suppose that* $a_n \in \mathbb{C}$ *and* $\sum_{n=0}^{\infty} a_n z_0^n$ *has radius of convergence* R. *If* $0 \leq \rho < R$, *then* $\sum_{n=0}^{\infty} a_n z^n$ *converges uniformly for all* $|z| \leq \rho$.

Proof. Choose z_0 with $\rho < |z_0| < R$. We know that $\sum_{n=0}^{\infty} a_n z_0^n$ converges, and so $a_n z_0^n \to 0$ as $n \to \infty$. It follows that there exists an M such that $|a_n z_0^n| \leq M$ for all $n \geq 1$. Thus, if $|z| \leq \rho$, we have

$$|a_n z^n| = |a_n z_0^n| \frac{|z|^n}{|z_0|^n} \leq M \left(\frac{\rho}{|z_0|} \right)^n,$$

and so, by the Weierstrass M-test, $\sum_{n=0}^{\infty} a_n z^n$ converges uniformly for all $|z| \leq \rho$. ■

It is very important to bear in mind the following easy example.

Exercise 11.71. *Show that* $\sum_{n=0}^{\infty} z^n$ *has radius of convergence* 1 *but that* $\sum_{n=0}^{\infty} z^n$ *does not converge uniformly for* $|z| < 1$.

Thus a power series converges uniformly in any disc with centre 0 of *strictly smaller* radius than the radius of convergence, but, in general, the condition *strictly smaller* cannot be dropped.

The next result is also easy to prove.

Lemma 11.72. *Suppose that* $a_n \in \mathbb{C}$ *and* $\sum_{n=0}^{\infty} a_n z^n$ *has radius of convergence* R. *Then* $\sum_{n=1}^{\infty} n a_n z^{n-1}$ *also has radius of convergence* R.

Proof. Let $|w| < R$. Choose $w_0 \in \mathbb{C}$ with $|w| < |w_0| < R$ and $\rho \in \mathbb{R}$ with $|w| < \rho < |w_0|$. We know that $\sum_{n=0}^{\infty} a_n w_0^n$ converges, so, arguing as before, there exists an M such that $|a_n w_0^n| \leq M$ for all $n \geq 0$. Thus

$$|n a_n \rho^{n-1}| = |w_0|^{-1} |a_n w_0^n| \frac{n\rho^{n-1}}{|w_0|^{n-1}} \leq |w_0|^{-1} M n \left(\frac{\rho}{|w_0|} \right)^{n-1} \to 0.$$

(We use the fact that, if $|x| < 1$, then $nx^n \to 0$ as $n \to \infty$. There are many proofs of this, including Exercise K.9.) Thus we can find an M' such that $|n a_n \rho^{n-1}| \leq M'$ for all $n \geq 1$. Our usual argument now shows that $\sum_{n=1}^{\infty} n a_n w^{n-1}$ converges.

We have shown that the radius of convergence of $\sum_{n=1}^{\infty} na_n z^{n-1}$ is at least R. An easier version of the same argument shows that if $\sum_{n=1}^{\infty} na_n z^{n-1}$ has radius of convergence S, then the radius of convergence of $\sum_{n=0}^{\infty} a_n z^n$ is at least S. Thus the radius of convergence of $\sum_{n=1}^{\infty} na_n z^{n-1}$ is exactly R. ∎

We can now combine Exercise 11.68, Lemma 11.70 and Lemma 11.72 to obtain our main result on power series.

Theorem 11.73. *Suppose that $a_n \in \mathbb{C}$ and $\sum_{n=0}^{\infty} a_n z^n$ has radius of convergence $R > 0$. Set $\Omega = \{z : |z| < R\}$ (if $R = \infty$, then $\Omega = \mathbb{C}$) and define $f : \Omega \to \mathbb{C}$ by*

$$f(z) = \sum_{n=0}^{\infty} a_n z^n.$$

Then f is everywhere differentiable on Ω, and

$$f'(z) = \sum_{n=1}^{\infty} na_n z^{n-1}.$$

More briefly, a power series can be differentiated term by term within its circle of convergence.

Proof. We wish to show that, if $|w| < R$, then f is differentiable at w with the appropriate derivative. To this end, choose a ρ with $|w| < \rho < R$. By Lemma 11.70, we know that

$$\sum_{n=0}^{N} a_n z^n \to f(z)$$

uniformly for $|z| \leq \rho$. By Lemma 11.70, we know that there exists a function $g : \Omega \to \mathbb{C}$ such that $\sum_{n=1}^{N} na_n z^{n-1} \to g(z)$ as $N \to \infty$ for all $z \in \Omega$. Using Lemma 11.70 again, we have

$$\sum_{n=1}^{N} na_n z^{n-1} \to g(z)$$

uniformly for $|z| \leq \rho$. Since

$$\frac{d}{dz}\left(\sum_{n=0}^{N} a_n z^n\right) = \sum_{n=1}^{N} na_n z^{n-1},$$

Exercise 11.68 now tells us that f is differentiable in $\{z : |z| < \rho\}$ with

$$f'(z) = \sum_{n=1}^{\infty} na_n z^{n-1}.$$

Since $|w| < \rho$, we are done. ∎

Remark: The proof above is more cunning than is at first apparent. Roughly speaking, it is often hard to prove directly that $\sum_{n=0}^{\infty} a_n z^n$ has a certain property for all $|z| < R$, the radius of convergence, but relatively easy to show that $\sum_{n=0}^{N} a_n z^n$ has a certain property for all $|z| < R'$, whenever $R' < R$. However, if we choose $R_1 < R_2 < \dots$ with $R_N \to R$, we then know that $\sum_{n=0}^{\infty} a_n z^n$ will have the property for all

$$z \in \bigcup_{r=1}^{\infty} \{z : |z| < R_N\} = \{z : |z| < R\},$$

and we are done. (We give two alternative proofs of Theorem 11.73 in Exercise K.258 and Exercise K.260.)

Here are two useful corollaries.

Exercise 11.74. *Suppose that $a_n \in \mathbb{C}$ and $\sum_{n=0}^{\infty} a_n z^n$ has radius of convergence $R > 0$. Set $\Omega = \{z : |z| < R\}$ and define $f : \Omega \to \mathbb{C}$ by*

$$f(z) = \sum_{n=0}^{\infty} a_n z^n.$$

Show that f is infinitely differentiable on Ω and $a_n = f^{(n)}(0)/n!$.

In other words, if f can be expanded in a power series about 0, then that power series must be the Taylor series.

Exercise 11.75. (Uniqueness of power series.) *Suppose that $a_n \in \mathbb{C}$, and $\sum_{n=0}^{\infty} a_n z^n$ has radius of convergence $R > 0$. Set $\Omega = \{z : |z| < R\}$ and define $f : \Omega \to \mathbb{C}$ by*

$$f(z) = \sum_{n=0}^{\infty} a_n z^n.$$

If there exists a δ with $0 < \delta \leq R$ such that $f(z) = 0$ for all $|z| < \delta$, show, by using the preceding exercise, or otherwise, that $a_n = 0$ for all $n \geq 0$. [In Exercise K.268 we give a stronger result with a more direct proof.]

By restricting our attention to the real axis, we can obtain versions of all these results for real power series.

Lemma 11.76. *Suppose that $a_n \in \mathbb{R}$.*

(i) Either $\sum_{n=0}^{\infty} a_n x^n$ converges for all $x \in \mathbb{R}$ (in which case we say the series has infinite radius of convergence), or there exists an $R \geq 0$ such that $\sum_{n=0}^{\infty} a_n x^n$ converges for $|x| < R$ and diverges for $|x| > R$ (in which case we say the series has radius of convergence R).

(ii) If $0 \leq \rho < R$, then $\sum_{n=0}^{\infty} a_n x^n$ converges uniformly on $[-\rho, \rho]$.

(iii) The sum $f(x) = \sum_{n=0}^{\infty} a_n x^n$ is differentiable, term by term, on $(-R, R)$.

(iv) If $R > 0$, then f is infinitely differentiable and $a_n = f^{(n)}(0)/n!$.

(v) If f vanishes on $(-\delta, \delta)$ where $0 < \delta \leq R$, then $a_n = 0$ for all n.

Part (iv) should be read in conjunction with Cauchy's example of a well behaved function with no power series expansion round 0 (Example 7.5).

The fact that we can differentiate a power series term by term is important for two reasons. The first is that there is a very beautiful and useful theory of differentiable functions from $\mathbb{C}$ to $\mathbb{C}$ (called 'complex variable theory' or 'the theory of analytic functions'). In the initial development of the theory it is not entirely clear that there are any interesting functions for the theory to talk about. Power series provide such interesting functions.

The second reason is that it provides a rigorous justification for the use of power series in the solution of differential equations by methods of the type employed on page 89.

Exercise 11.77. *(i) Show that the sum $\sum_{n=0}^{\infty} \frac{z^n}{n!}$ has infinite radius of convergence.*

(ii) Let us set

$$e(z) = \sum_{n=0}^{\infty} \frac{z^n}{n!}$$

for all $z \in \mathbb{C}$. Show that e is everywhere differentiable and $e'(z) = e(z)$.

(iii) Use the mean value theorem of Exercise 11.67 to show that the function f defined by $f(z) = e(a - z)e(z)$ is constant. Deduce also that $e(a - z)e(z) = e(a)$ for all $z \in \mathbb{C}$ and $a \in \mathbb{C}$ and conclude that

$$e(z)e(w) = e(z + w)$$

for all $z, w \in \mathbb{C}$.

Here is another example.

Example 11.78. *Let $\alpha \in \mathbb{C}$. Solve the differential equation*

$$(1 + z)f'(z) = \alpha f(z)$$

subject to $f(0) = 1$.

Solution. We look for a solution of the form

$$f(z) = \sum_{n=0}^{\infty} a_n z^n$$

with radius of convergence $R > 0$. We differentiate term by term within the radius of convergence to get

$$(1 + z) \sum_{n=1}^{\infty} n a_n z^{n-1} = \alpha \sum_{n=0}^{\infty} a_n z^n,$$

whence

$$\sum_{n=0}^{\infty}((\alpha - n)a_n - (n+1)a_{n+1})z^n = 0$$

for all $|z| < R$. By the uniqueness result of Exercise 11.75, this gives

$$(\alpha - n)a_n - (n+1)a_{n+1} = 0,$$

so

$$a_{n+1} = \frac{\alpha - n}{n+1}a_n,$$

and, by induction,

$$a_n = A\frac{1}{n!}\prod_{j=0}^{n-1}(\alpha - j),$$

for some constant A. Since $f(0) = 1$, we have $A = 1$ and

$$f(z) = \sum_{n=0}^{\infty}\frac{1}{n!}\prod_{j=0}^{n-1}(\alpha - j)z^n.$$

If α is a positive integer N, say, then $a_j = 0$ for $j \geq N+1$ and we get the unsurprising result

$$f(z) = \sum_{n=0}^{N}\binom{N}{n}z^n = (1+z)^N.$$

From now on we assume that α is not a positive integer. If $z \neq 0$,

$$\frac{|a_{n+1}z^{n+1}|}{|a_n z^n|} = \frac{|\alpha - n|}{n+1}|z| = \frac{|1 - \alpha n^{-1}|}{1 + n^{-1}}|z| \to |z|$$

as $n \to \infty$, so, using the ratio test, $\sum_{n=0}^{\infty} a_n z^n$ has radius of convergence 1.

We have shown that, if there is a power series solution, it must be

$$f(z) = \sum_{n=0}^{\infty}\frac{1}{n!}\prod_{j=0}^{n-1}(\alpha - j)z^n.$$

Differentiating term by term, we see that, indeed, the f given is a solution valid for $|z| < 1$. ■

We have left open the possibility that the differential equation of Example 11.78 might have other solutions (such solutions would not have Taylor expansions). The uniqueness of the solution follows from general results developed later in this book (see Section 12.2). However, there is a simple proof of uniqueness in this case.

Example 11.79. *(i) Write $D = \{z : |z| < 1\}$. Let $\alpha \in \mathbb{C}$. Suppose that $f_\alpha : D \to \mathbb{C}$ satisfies*

$$(1+z)f'_\alpha(z) = \alpha f_\alpha(z)$$

and $f_\alpha(0) = 1$, whilst $g_{-\alpha} : D \to \mathbb{C}$ satisfies

$$(1+z)g'_{-\alpha}(z) = -\alpha g_{-\alpha}(z)$$

and $g_{-\alpha}(0) = 1$. Use the mean value theorem of Exercise 11.67 to show that $f_\alpha(z)g_{-\alpha}(z) = 1$ for all $z \in D$, and deduce that the differential equation

$$(1+z)f'(z) = \alpha f(z),$$

subject to $f(0) = 1$, has exactly one solution on D.

(ii) If α, $\beta \in \mathbb{C}$, show, using the notation of part (i), that

$$f_{\alpha+\beta}(z) = f_\alpha(z)f_\beta(z)$$

for all $z \in D$. State and prove a similar result for $f_\alpha(f_\beta(z))$.

Restricting to the real axis, we obtain the following version of our results.

Lemma 11.80. *Let α be a real number. Then the differential equation*

$$(1+x)f'(x) = \alpha f(x),$$

subject to $f(0) = 1$, has exactly one solution $f : (-1,1) \to \mathbb{R}$, which is given by

$$f(x) = \sum_{n=0}^{\infty} \frac{1}{n!} \prod_{j=0}^{n-1} (\alpha - j)x^n.$$

In Section 5.7 we developed the theory of the function $r_\alpha(x) = x^\alpha$ for $x > 0$ and α real. One of these properties is that

$$xr'_\alpha(x) = \alpha r_\alpha(x)$$

for all $x > 0$. We also have $r_\alpha(0) = 1$. Thus, if $g_\alpha(x) = r_\alpha(1+x)$, we have

$$(1+x)g'_\alpha(x) = \alpha g_\alpha(x)$$

for all $x \in (-1,1)$ and $g_\alpha(0) = 1$. Lemma 11.80 thus gives the following well known binomial expansion.

Lemma 11.81. *If $x \in (-1,1)$, then*

$$(1+x)^\alpha = \sum_{n=0}^{\infty} \frac{1}{n!} \prod_{j=0}^{n-1} (\alpha - j)x^n.$$

Exercise 11.82. *Use the same ideas to show that*

$$\log(1-x) = -\sum_{n=1}^{\infty} \frac{x^n}{n}$$

for $x \in (-1, 1)$.

Exercise 11.83. *(i) If you are unfamiliar with the general binomial expansion described in Lemma 11.81, write out the first few terms explicitly in the cases* $\alpha = -1$, $\alpha = -2$, $\alpha = -3$, $\alpha = 1/2$ *and* $\alpha = -1/2$. *Otherwise, go directly to part (ii).*

(ii) Show that

$$1 + \frac{1}{2}\left(\frac{2x}{1+x^2}\right) + \frac{1}{2} \times \frac{3}{4}\left(\frac{2x}{1+x^2}\right)^2 + \frac{1}{2} \times \frac{3}{4} \times \frac{5}{6}\left(\frac{2x}{1+x^2}\right)^3 + \dots$$

converges to $(1+x^2)/(1-x^2)$ *if* $|x| < 1$ *but converges to* $(1+x^2)/(x^2-1)$ *if* $|x| > 1$. *In* [**25**], *Hardy quotes this example to show the difficulties that arise if we believe that equalities which are true in one domain must be true in all domains.*

Exercise 11.84. *(i) If* $x > 0$ *and we set* $g(t) = (x-t)/(1-t)$ *for* $0 \le t \le x$, *show that* g *is decreasing on the interval* $[0, x]$ *and deduce that*

$$0 \le \frac{x-t}{1-t} \le x \text{ for all } 0 \le t \le x < 1.$$

Show similarly that

$$0 \le \left|\frac{x-t}{1-t}\right| \le |x| \text{ for all } -1 < x \le t \le 0.$$

(ii) Use Taylor's theorem with integral remainder (Theorem 8.51) to obtain the expansions for $(1+x)^\alpha$ *and* $\log(1-x)$.

[This is a slightly unfair question, since the forms of the Taylor remainder given in this book are not particularly well suited to the problem. If the reader consults other texts she will find forms of the remainder which will work more easily. She should then ask herself what the point of these forms of remainder is, apart from obtaining Taylor series which are much more easily obtained by finding the power series solution of an appropriate differential equation.]

Many textbooks on mathematical methods devote some time to the process of solving differential equations by power series. The results of this section justify the process.

Slogan: *The formal process of solving a differential equation by power series yields a correct result within the radius of convergence of the power series produced.*

The slogan becomes a theorem once we specify the type of differential equation to be solved.

Note however, that, contrary to the implied promise of some textbooks on mathematical methods, power series solutions are not always as useful as they look.

Exercise 11.85. *We know that* $\sum_{n=0}^{\infty}(-1)^n x^{2n}/(2n)!$ *converges everywhere to* $\cos x$. *Try and use this formula, together with a hand calculator to compute* $\cos 100$. *Good behaviour in the sense of the pure mathematician merely means 'good behaviour in the long run', and the 'long run' may be too long for any practical use.*

Can you suggest and implement a sensible method[5] *to compute* $\cos 100$.

Exercise 11.86. *By considering the relations that the coefficients must satisfy, show that there is no power series solution for the equation*

$$x^3 y'(x) = -2y(x)$$

with $y(0) = 0$ *valid in some neighbourhood of* 0.

*Show, however, that the system does have a well behaved solution. [*Hint: *Example 7.5.]*

If the reader is prepared to work quite hard, Exercise K.272 gives a good condition for the existence of a power series solution for certain typical differential equations.

We end this section with a look in another direction.

Exercise 11.87. *If* $z \in \mathbb{C}$ *and* n *is a positive integer, we can define* $n^{-z} = e^{-z \log n}$. *By using the Weierstrass M-test, or otherwise, show that, if* $\epsilon > 0$,

$$\sum_{n=1}^{\infty} n^{-z} \text{ converges uniformly for } \Re z > 1 + \epsilon.$$

We call the limit $\zeta(z)$. *Show further that* ζ *is differentiable on the range considered. Deduce that* ζ *is well defined and differentiable on the half plane* $\{z \in \mathbb{C} : \Re z > 1\}$. *(*$\zeta$ *is the famous Riemann zeta function.)*

11.6. Fourier series ♡

In this section we shall integrate complex-valued functions. The definition used is essentially that of Definition 8.70.

Definition 11.88. *If* $f : [a, b] \to \mathbb{C}$ *is such that* $\Re f : [a, b] \to \mathbb{R}$ *and* $\Im f : [a, b] \to \mathbb{R}$ *are Riemann integrable, then we say that* f *is Riemann integrable*

[5]Pressing the cos button is sensible, but not very instructive.

and

$$\int_a^b f(x)\,dx = \int_a^b \Re f(x)\,dx + i\int_a^b \Im f(x)\,dx.$$

We leave it to the conscientious reader to check that the integral behaves as it ought to behave.

If the reader has attended a course on mathematical methods, she will probably be familiar with the notion of the Fourier series of a periodic function.

Definition 11.89. *If $f:\mathbb{R}\to\mathbb{C}$ is continuous and periodic with period 2π (that is, $f(t+2\pi)=f(t)$ for all t) and m is an integer, we set*

$$\hat{f}(m) = \frac{1}{2\pi}\int_{-\pi}^{\pi} f(t)\exp(-imt)\,dt.$$

Fourier claimed[6], in effect, that

$$f(t) = \sum_{n=-\infty}^{\infty} \hat{f}(n)\exp(int).$$

We now know that the statement is false in the sense that there exist continuous functions such that

$$\sum_{n=-N}^{N} \hat{f}(n)\exp(int_0) \nrightarrow f(t_0)$$

as $N\to\infty$ for some t_0, but true in many other and deeper senses.

The unraveling of the various ways in which Fourier's theorem holds took a century and a half[7] and was one of the major influences on the rigorisation of analysis. In this section we shall merely provide a simple condition on $\hat{f}$ which ensures that Fourier's statement holds in its original form for a given function f.

Our discussion hinges on the following theorem, which is very important in its own right.

Theorem 11.90. (Uniqueness of the Fourier series.) *If $f:\mathbb{R}\to\mathbb{C}$ is continuous and periodic with period 2π and $\hat{f}(n)=0$ for all n, then $f=0$.*

To prove this result it turns out to be sufficient to prove an apparently weaker result. (See Exercises 11.93 and 11.94.)

Lemma 11.91. *If $f:\mathbb{R}\to\mathbb{R}$ is continuous and periodic with period 2π and $\hat{f}(n)=0$ for all n, then $f(0)=0$.*

[6] Others had had the idea before, but Fourier 'bet the farm on it'.

[7] Supposing the process to have terminated.

Proof. Suppose $f(0) \neq 0$. Without loss of generality we may suppose that $f(0) > 0$ (otherwise, we can consider $-f$). By continuity, we can find an ϵ with $1 > \epsilon > 0$ such that $|f(t) - f(0)| < f(0)/2$, and so $f(t) > f(0)/2$ for all $|t| \leq \epsilon$. Now choose $\eta > 0$ such that $2\eta + \cos\epsilon < 1$ and set $P(t) = \eta + \cos t$.

Since $P(t) = (\eta + \frac{1}{2}e^{it} + \frac{1}{2}e^{-it})$, we have

$$P(t)^N = \sum_{k=-N}^{k=N} b_{N\,k} e^{ikt}$$

for some $b_{N\,k}$, and so

$$\int_{-\pi}^{\pi} f(t)P(t)^N\,dt = \sum_{k=-N}^{k=N} b_{N\,k} \int_{-\pi}^{\pi} f(t)e^{ikt}\,dt = \sum_{k=-N}^{k=N} b_{N\,k}\hat{f}(-k) = 0$$

for all N.

Since f is continuous on $[-\pi, \pi]$, it is bounded; so there exists a K such that $|f(t)| \leq K$ for all $t \in [-\pi, \pi]$. Since $P(0) = \eta + 1$, we can find an $\epsilon' > 0$ with $\epsilon > \epsilon'$ such that $P(t) \geq 1 + \eta/2$ for all $|t| \leq \epsilon'$. Finally we observe that $|P(t)| \leq 1 - \eta$ for $\epsilon \leq |t| \leq \pi$. Putting all our information together, we obtain

$$\begin{aligned} f(t)P(t)^n &\geq f(0)(1+\eta/2)^N/2 && \text{for all } |t| \leq \epsilon', \\ f(t)P(t)^n &\geq 0 && \text{for all } \epsilon' \leq |t| \leq \epsilon, \\ |f(t)P(t)^n| &\leq K(1-\eta)^N && \text{for all } \epsilon \leq |t| \leq \pi. \end{aligned}$$

Thus

$$\begin{aligned} 0 &= \int_{-\pi}^{\pi} f(t)P(t)^N\,dt \\ &= \int_{|t|\leq\epsilon'} f(t)P(t)^N\,dt + \int_{\epsilon'\leq|t|\leq\epsilon} f(t)P(t)^N\,dt + \int_{\epsilon\leq|t|\leq\pi} f(t)P(t)^N\,dt \\ &\geq \epsilon' f(0)(1+\eta/2)^N + 0 - 2\pi K(1-\eta)^N \to \infty. \end{aligned}$$

The assumption that $f(0) \neq 0$ has led to a contradiction, and the required result follows by reductio ad absurdum. ■

Exercise 11.92. *Draw sketches illustrating the proof just given.*

Exercise 11.93. *(i) If $g : \mathbb{R} \to \mathbb{C}$ is continuous and periodic with period 2π and $a \in \mathbb{R}$, we write $g_a(t) = g(t-a)$. Show that $\hat{g}_a(n) = \exp(ina)\hat{g}(n)$.*

(ii) By translation, or otherwise, prove the following result. If $f : \mathbb{R} \to \mathbb{R}$ is continuous and periodic with period 2π and $\hat{f}(n) = 0$ for all n, then $f = 0$.

Exercise 11.94. *(i) If $g : \mathbb{R} \to \mathbb{C}$ is continuous and periodic with period 2π, show that $\widehat{g^*}(n) = (\hat{g}(-n))^*$.*

(ii) By considering $f + f^*$ and $f - f^*$, or otherwise, prove Theorem 11.90.

We can now state and prove our promised result on Fourier sums.

Theorem 11.95. *If $f : \mathbb{R} \to \mathbb{C}$ is continuous and periodic with period 2π and $\sum_{n=-\infty}^{\infty} |\hat{f}(n)|$ converges, then*

$$\sum_{n=-N}^{N} \hat{f}(n)\exp(int) \to f(t)$$

uniformly as $N \to \infty$.

Proof. Since $|\hat{f}(n)\exp(int) + \hat{f}(-n)\exp(-int)| \leq |\hat{f}(n)| + |\hat{f}(-n)|$, the Weierstrass M-test tells us that $\sum_{n=-N}^{N} \hat{f}(n)\exp(int)$ converges uniformly to $g(t)$, say. Since the uniform limit of continuous functions is continuous, g is continuous. We wish to show that $g = f$.

Observe that, since $|\exp(-imt)| = 1$, we have

$$\begin{aligned}\sum_{n=-N}^{N} \hat{f}(n)\exp(i(n-m)t) &= \exp(-imt)\sum_{n=-N}^{N} \hat{f}(n)\exp(int)\\ &\to \exp(-imt)g(t)\end{aligned}$$

uniformly as $N \to \infty$, so by Theorem 11.51, we have

$$\begin{aligned}\sum_{n=-N}^{N} \hat{f}(n)\frac{1}{2\pi}\int_{-\pi}^{\pi} \exp(i(n-m)t)\,dt &= \frac{1}{2\pi}\int_{-\pi}^{\pi}\sum_{n=-N}^{N} \hat{f}(n)\exp(i(n-m)t)\,dt\\ &\to \frac{1}{2\pi}\int_{-\pi}^{\pi} \exp(-imt)g(t)\,dt = \hat{g}(m).\end{aligned}$$

Now $\frac{1}{2\pi}\int_{-\pi}^{\pi} \exp(irt)\,dt$ takes the value 1 if $r = 0$ and the value 0 otherwise, so we have shown that $\hat{f}(m) \to \hat{g}(m)$ as $N \to \infty$. Thus $\hat{f}(m) = \hat{g}(m)$ for all m and, by the uniqueness of Fourier series (Theorem 11.90), we have $f = g$, as required. ■

Exercise 11.96. *If $f : \mathbb{R} \to \mathbb{C}$ is periodic with period 2π and has continuous second derivative, show, by integrating by parts twice, that*

$$\hat{f}(n) = -\frac{1}{n^2}\widehat{f''}(n)$$

for all $n \neq 0$. Deduce that

$$|\hat{f}(n)| \leq \frac{1}{n^2}\sup_{t\in[-\pi,\pi]} |f''(t)|$$

for all $n \neq 0$, and that

$$\sum_{n=-N}^{N} \hat{f}(n)\exp(int) \to f(t)$$

uniformly as $N \to \infty$.

Exercise 11.97. *Suppose $f : \mathbb{R} \to \mathbb{R}$ is a 2π periodic function with $f(x) = -f(-x)$ for all x and $f(x) = x(\pi - x)$ for $0 \leq x \leq \pi$. Show that*

$$f(x) = \frac{8}{\pi} \sum_{m=0}^{\infty} \frac{\sin(2m+1)x}{(2m+1)^3}$$

for all x, and, by choosing a particular x, show that

$$\sum_{m=0}^{\infty} \frac{(-1)^m}{(2m+1)^3} = \frac{\pi^3}{32}.$$

Exercise 11.98. *Suppose $f : \mathbb{R} \to \mathbb{R}$ is a 2π periodic continuous function with $\sum_{n=-\infty}^{\infty} |n\hat{f}(n)|$ convergent. Show that f is differentiable and*

$$f'(t) = \sum_{n=-\infty}^{\infty} in\hat{f}(n) \exp(int).$$

Chapter 12

Contraction Mappings and Differential Equations

12.1. Banach's contraction mapping theorem

This chapter and the next depend on the famous contraction mapping theorem by which Banach transformed a 'folk-technique' into a theorem.

Definition 12.1. *Let (X, d) be a metric space and $T : X \to X$ a mapping. We say that $w \in X$ is a fixed point of T if $Tw = w$. We say that T is a contraction mapping if there exists a positive number $K < 1$ with $d(Tx, Ty) \leq Kd(x, y)$ for all $x, y \in X$.*

The next exercise is easy but helps suggest the proof of the theorem that follows.

Exercise 12.2. *Let (X, d) be a metric space and $T : X \to X$ a contraction mapping with a fixed point w. Suppose that $x_0 \in X$ and we define x_n inductively by $x_{n+1} = Tx_n$. Show that $d(x_n, w) \to 0$ as $n \to \infty$.*

Theorem 12.3. (The contraction mapping theorem.) *A contraction mapping on a non-empty complete metric space has a unique fixed point.*

Proof. Suppose $1 > K > 0$, (X, d) is a non-empty complete metric space and $T : X \to X$ has the property $d(Tx, Ty) \leq Kd(x, y)$ for all $x, y \in X$.

We show first that, if T has a fixed point, it is unique. For suppose $Tw = w$ and $Tz = z$. We have

$$d(z, w) = d(Tz, Tw) \leq Kd(z, w)$$

so, since $K < 1$, $d(z, w) = 0$ and $z = w$.

To prove that a fixed point exists, choose any $x_0 \in X$ and define x_n inductively by $x_{n+1} = Tx_n$. (The preceding exercise shows this is a good idea.) By induction,

$$d(x_{n+1}, x_n) = d(Tx_n, Tx_{n-1}) \leq Kd(x_n, x_{n-1}) \leq \cdots \leq K^n d(x_1, x_0),$$

and so, by the triangle inequality, we have, whenever $m > n$

$$d(x_m, x_n) \leq \sum_{j=n}^{m-1} d(x_{j+1}, x_j) \leq \sum_{j=n}^{m-1} K^j d(x_1, x_0) \leq \frac{K^n}{1-K} d(x_1, x_0) \to 0$$

as $n \to \infty$. Thus the sequence x_n is Cauchy. Since (X, d) is complete, we can find a w such that $d(x_n, w) \to 0$ as $n \to \infty$.

We now show that w is indeed a fixed point. To do this, we observe that

$$\begin{aligned} d(Tw, w) &\leq d(Tw, x_{n+1}) + d(x_{n+1}, w) = d(Tw, Tx_n) + d(x_{n+1}, w) \\ &\leq Kd(w, x_n) + d(x_{n+1}, w) \to 0 + 0 = 0 \end{aligned}$$

as $n \to \infty$. Thus $d(Tw, w) = 0$ and $Tw = w$. ■

Wide though the conditions are, the reader should exercise caution before attempting to widen them further.

Example 12.4. *(i) If $X = \{-1, 1\}$, d is ordinary distance and the map $T : X \to X$ is given by $Tx = -x$, then (X, d) is a complete metric space and $d(Tx, Ty) = d(x, y)$ for all x, $y \in X$, but T has no fixed point.*

(ii) If $X = [1, \infty)$, d is Euclidean distance and

$$Tx = 1 + x + \exp(-x),$$

then (X, d) is a complete metric space and $d(Tx, Ty) < d(x, y)$ for all x, $y \in X$, but T has no fixed point.

(iii) If $X = (0, \infty)$, d is ordinary distance and the map $T : X \to X$ is given by $Tx = x/2$, then (X, d) is a metric space and T is a contraction mapping, but T has no fixed point.

Exercise 12.5. *Verify the statements made in Example 12.4. In each case, state the hypothesis in Theorem 12.3 which is not satisfied. In each case, identify the point at which the proof of Theorem 12.3 fails.*

The contraction mapping theorem is not the only important fixed point theorem in mathematics. Exercise 1.39 gives another fixed point result which can be generalised substantially. (For example, if

$$B = \{\mathbf{x} : \|\mathbf{x}\| \leq 1\}$$

is the unit ball in $\mathbb{R}^n$, then any continuous map of B into itself has a fixed point.) However, the standard proofs involve algebraic topology and are beyond the scope of this book.

12.2. Existence of solutions of differential equations

We use the contraction mapping theorem to show that a wide class of differential equations actually have a solution.

We shall be looking at equations of the form

$$y' = f(t, y).$$

Our first, simple but important, result is that this problem on differential equations can be turned into a problem on integral equations. (We shall discuss why this may be expected to be useful after Exercise 12.7.)

Lemma 12.6. *If $f : \mathbb{R}^2 \to \mathbb{R}$ is continuous, t_0, $y_0 \in \mathbb{R}$ and $\delta > 0$, then the following two statements are equivalent.*

(A) The function $y : (t_0 - \delta, t_0 + \delta) \to \mathbb{R}$ is differentiable and satisfies the equation $y'(t) = f(t, y(t))$ for all $t \in (t_0 - \delta, t_0 + \delta)$ together with the boundary condition[1] *$y(t_0) = y_0$.*

(B) The function $y : (t_0 - \delta, t_0 + \delta) \to \mathbb{R}$ is continuous and satisfies the condition

$$y(t) = y_0 + \int_{t_0}^{t} f(u, y(u))\, du$$

for all $t \in (t_0 - \delta, t_0 + \delta)$.

Proof. We show that (A) implies (B). Suppose that y satisfies condition (A). Since y is differentiable, it is continuous. Thus, since f is continuous, y' is continuous and one of the standard forms of the fundamental theorem of the calculus (Theorem 8.42) gives

$$y(t) - y(t_0) = \int_{t_0}^{t} f(u, y(u))\, du;$$

so, since $y(t_0) = y_0$,

$$y(t) = y_0 + \int_{t_0}^{t} f(u, y(u))\, du$$

[1]The reader may prefer 'initial condition'.

for all $t \in (t_0 - \delta, t_0 + \delta)$, as required.

The fact that (B) implies (A) is an immediate consequence of the fundamental theorem of the calculus in the form of Theorem 8.37. ■

Exercise 12.7. *If $f : \mathbb{R}^2 \to \mathbb{R}$ is n times differentiable, then any solution of $y'(t) = f(t, y(t))$ is $n+1$ times differentiable.*

Remark: Most mathematicians carry in their minds a list of operations which are or are not likely to to be troublesome. Such a list will probably contain the following entries.

less troublesome	more troublesome
multiplication	division
interpolation	extrapolation
averaging	differencing
integration	differentiation
direct calculation	finding inverses

Integration produces a better behaved function, differentiation may well produce a worse behaved function. The integral of an integrable function is an integrable function, the derivative of a differentiable function need not be differentiable. The contraction mapping theorem concerns a map $T : X \to X$, so to apply it we must be sure that our operation T does not take us out of our initial space. This is much easier to ensure if T involves integration rather than differentiation.

Theorem 12.8. *Suppose $f : \mathbb{R}^2 \to \mathbb{R}$ is continuous, t_0, $y_0 \in \mathbb{R}$ and $\delta > 0$. Suppose further that there exists a $K > 0$ such that $K\delta < 1$ and*

$$\bigstar \qquad |f(t,u) - f(t,v)| \leq K|u - v|$$

for all $t \in [t_0 - \delta, t_0 + \delta]$ and all u and v. Then there exists a unique $y : [t_0 - \delta, t_0 + \delta] \to \mathbb{R}$ which is continuous and satisfies the condition

$$y(t) = y_0 + \int_{t_0}^{t} f(u, y(u))\, du$$

for all $t \in [t_0 - \delta, t_0 + \delta]$.

Proof. We know that $C([t_0 - \delta, t_0 + \delta])$, the space of continuous functions on $[t_0 - \delta, t_0 + \delta]$ with the uniform norm $\| \ \|_\infty$, is complete. Now consider the map $T : C([t_0 - \delta, t_0 + \delta]) \to C([t_0 - \delta, t_0 + \delta])$ given by

$$(Tg)(t) = y_0 + \int_{t_0}^{t} f(u, g(u))\, du.$$

If $t_0 + \delta \geq t \geq t_0$, we have

$$\begin{aligned}
|(Tg)(t) - (Th)(t)| &= \left| \int_{t_0}^{t} f(u, g(u)) - f(u, h(u))\, du \right| \\
&\leq \int_{t_0}^{t} |f(u, g(u)) - f(u, h(u))|\, du \\
&\leq \int_{t_0}^{t} K|g(u) - h(u)|\, du \\
&\leq (t - t_0)K\|g - h\|_\infty \leq K\delta\|g - h\|_\infty,
\end{aligned}$$

and a similar argument gives

$$|(Tg)(t) - (Th)(t)| \leq K\delta\|g - h\|_\infty$$

for $t_0 \geq t \geq t_0 - \delta$. Thus

$$\|Tg - Th\|_\infty \leq K\delta\|g - h\|_\infty,$$

and T is a contraction mapping.

The contraction mapping theorem tells us that T has a unique fixed point; that is, there exists a unique $y \in C([t_0 - \delta, t_0 + \delta])$ such that

$$y(t) = y_0 + \int_{t_0}^{t} f(u, y(u))\, du$$

for all $t \in [t_0 - \delta, t_0 + \delta]$, and this is the required result. ■

Exercise 12.9. *Restate Theorem 12.8 in terms of differential equations.*

Condition ★ is called a Lipschitz condition.

Exercise 12.10. *(i) Show that, if $f : \mathbb{R}^2 \to \mathbb{R}$ has continuous partial derivative $f_{,2}$, then, given any $[a, b]$ and $[c, d]$, we can find a K such that*

$$|f(t, u) - f(t, v)| \leq K|u - v|$$

for all $t \in [a, b]$ and u, $v \in [c, d]$.

(ii) If $f : \mathbb{R}^2 \to \mathbb{R}$ is given by $f(t, y) = |y|$, show that

$$|f(t, u) - f(t, v)| \leq K|u - v|$$

for all t, u and v, but f does not have a partial derivative $f_{,2}$ everywhere.

In the absence of a condition like ★, differential equations can have unexpected properties.

Exercise 12.11. *Consider the differential equation*

$$y' = 3y^{2/3}$$

with $y(0) = 0$. Show that it has the solution

$$\begin{aligned} y(t) &= (t-a)^3 && \text{for } t < a, \\ y(t) &= 0 && \text{for } a \le t \le b, \\ y(t) &= (t-b)^3 && \text{for } b < t \end{aligned}$$

whenever $a \le b$.

Exercise 12.11 is worth remembering whenever you are tempted to convert the useful rule of thumb 'first order differential equations involve one choice of constant' into a theorem.

Remark: It is easy to write down differential equations with no solution. For example, there is no real-valued solution to

$$(y')^2 + y^2 + 1 = 0.$$

However, it can be shown that the existence part of Theorem 12.8 continues to hold, even if we drop condition ★, provided merely that f is continuous. The reader may wish to ponder on the utility of an existence theorem in the absence of a guarantee of uniqueness.

There is no difficulty in extending the proof of Theorem 12.8 to higher dimensions. In the exercise that follows the norm is the usual Euclidean norm and $\mathbf{y}'(t) = (y_1'(t), y_2'(t), \dots, y_n'(t))$.

Exercise 12.12. *(i) Suppose $\mathbf{f} : \mathbb{R}^{n+1} \to \mathbb{R}^n$ is continuous, $t_0 \in \mathbb{R}$, $\mathbf{y}_0 \in \mathbb{R}^n$ and $\delta > 0$. Suppose, further, that there exists a $K > 0$ such that $K\delta < 1$ and*

$$\bigstar \qquad \|\mathbf{f}(t, \mathbf{u}) - \mathbf{f}(t, \mathbf{v})\| \le K\|\mathbf{u} - \mathbf{v}\|$$

for all $t \in [t_0 - \delta, t_0 + \delta]$. Then there exists a unique $\mathbf{y} : [t_0 - \delta, t_0 + \delta] \to \mathbb{R}^n$ which is continuous and satisfies the condition

$$\mathbf{y}(t) = \mathbf{y}_0 + \int_{t_0}^{t} \mathbf{f}(u, \mathbf{y}(u))\, du$$

for all $t \in [t_0 - \delta, t_0 + \delta]$.

(ii) With the notation and conditions of (i), $\mathbf{y}$ is the unique solution of

$$\mathbf{y}'(t) = \mathbf{f}(t, \mathbf{y}(t)),\ \mathbf{y}(t_0) = \mathbf{y}_0$$

on $(t_0 - \delta, t_0 + \delta)$.

Exercise 12.12 is particularly useful because it enables us to deal with higher order differential equations. To see how the proof below works, observe that the second order differential equation

$$y'' + y = 0$$

can be written as two first order differential equations

$$y' = w,\ w' = -y$$

or, vectorially, as a single first order differential equation

$$(y, w)' = (w, -y).$$

Lemma 12.13. *Suppose* $g : \mathbb{R}^{n+1} \to \mathbb{R}$ *is continuous,* $t_0 \in \mathbb{R}$, $y_j \in \mathbb{R}^n$ *for* $0 \leq j \leq n-1$, *and* $\delta > 0$. *Suppose, further, that there exists a* $K > 0$ *such that* $(K+1)\delta < 1$ *and*

$$\bigstar \qquad |g(t, \mathbf{u}) - g(t, \mathbf{v})| \leq K\|\mathbf{u} - \mathbf{v}\|$$

for all $t \in [t_0 - \delta, t_0 + \delta]$. *Then there exists a unique, n times differentiable, function* $y : (t_0 - \delta, t_0 + \delta) \to \mathbb{R}$ *with*

$$y^{(n)}(t) = g(t, y(t), y'(t), \ldots, y^{(n-1)}(t)) \text{ and } y^{(j)}(t_0) = y_j \text{ for } 0 \leq j \leq n-1.$$

Proof. This uses the trick described above. We define

$$\mathbf{f}(t, u_1, u_2, \ldots, u_n) = (u_1, u_2, \ldots, u_n, g(t, u_1, u_2, \ldots, u_{n-1})).$$

The differential equation $\mathbf{y}'(t) = \mathbf{f}(t, \mathbf{y}(t))$ is equivalent to the system of equations

$$y_j'(t) = f_j(t, \mathbf{y}(t)) \qquad [1 \leq j \leq n],$$

which for our choice of $\mathbf{f}$ becomes

$$\begin{aligned} y_j'(t) &= y_j(t) && [1 \leq j \leq n-1], \\ y_n'(t) &= g(t, y(t), y'(t), \ldots, y^{(n-1)}(t)). \end{aligned}$$

Taking $y(t) = y_1(t)$, this gives us $y_j(t) = y^{(j-1)}(t)$ and

$$y^{(n)}(t) = g(t, y(t), y'(t), \ldots, y^{(n-1)}(t)),$$

which is precisely the differential equation we wish to solve. Our boundary conditions

$$y^{(j)}(t_0) = y_j \text{ for } 0 \leq j \leq n-1$$

now take the form $\mathbf{y}(t_0) = \mathbf{y}_0$ with

$$\mathbf{y}_0 = (y_0, y_1, \ldots, y_{n-1}),$$

and we have reduced our problem to that studied in Exercise 12.12.

To prove existence and uniqueness we need only verify that $\mathbf{f}$ satisfies the appropriate Lipschitz condition. But

$$\begin{aligned} &\|\mathbf{f}(t, \mathbf{u}) - \mathbf{f}(t, \mathbf{v})\| \\ &\quad = \|(u_1 - v_1, u_2 - v_2, \ldots, u_{n-1} - v_{n-1}, g(t, u_1, u_2, \ldots, u_n) - g(t, v_1, v_2, \ldots, v_n))\| \\ &\quad \leq \|\mathbf{u} - \mathbf{v}\| + |g(t, u_1, u_2, \ldots, u_n) - g(t, v_1, v_2, \ldots, v_n)| \leq (K+1)\|\mathbf{u} - \mathbf{v}\|, \end{aligned}$$

so we are done. ∎

12.3. Local to global ♡

We proved Theorem 12.8 for functions f with

$$\bigstar \qquad |f(t,u) - f(t,v)| \leq K|u-v|$$

for all $t \in [t_0 - \delta, t_0 + \delta]$ and all u and v. However, this condition is more restrictive than is necessary.

Theorem 12.14. *Suppose $\eta > 0$ and $f : [t_0-\eta, t_0+\eta] \times [y_0-\eta, y_0+\eta] \to \mathbb{R}$ is a continuous function satisfying the condition*

$$\bigstar \qquad |f(t,u) - f(t,v)| \leq K|u-v|$$

whenever $t \in [t_0 - \eta, t_0 + \eta]$ and $u, v \in [y_0 - \eta, y_0 + \eta]$. Then we can find a $\delta > 0$ with $\eta \geq \delta$ such that there exists a unique differentiable function $y : (t_0 - \delta, t_0 + \delta) \to \mathbb{R}$ which satisfies the equation $y'(t) = f(t, y(t))$ for all $t \in (t_0 - \delta, t_0 + \delta)$ together with the boundary condition $y(t_0) = y_0$.

Proof. This is an easy consequence of Theorem 12.8. Define a function $\tilde{f} : \mathbb{R}^2 \to \mathbb{R}$ as follows.

$$\begin{aligned}
\tilde{f}(t,y) &= f(t,y) && \text{if } |t-t_0| \leq \eta,\ |y-y_0| \leq \eta,\\
\tilde{f}(t,y) &= f(t_0+\eta,y) && \text{if } t > t_0+\eta,\ |y-y_0| \leq \eta,\\
\tilde{f}(t,y) &= f(t_0-\eta,y) && \text{if } t < t_0-\eta,\ |y-y_0| \leq \eta,\\
\tilde{f}(t,y) &= \tilde{f}(t,y_0+\eta) && \text{if } y > y_0+\eta,\\
\tilde{f}(t,y) &= \tilde{f}(t,y_0-\eta) && \text{if } y < y_0-\eta.
\end{aligned}$$

We observe that $\tilde{f}$ is continuous and

$$|\tilde{f}(t,u) - \tilde{f}(t,v)| \leq K|u-v|$$

for all t, u and v.

If we choose $\tilde{\delta} > 0$ with $K\tilde{\delta} < 1$, then Theorem 12.8 tells us that there exists a unique differentiable function $\tilde{y} : (t_0 - \tilde{\delta}, t_0 + \tilde{\delta}) \to \mathbb{R}$ which satisfies the equation $\tilde{y}'(t) = \tilde{f}(t, \tilde{y}(t))$ for all $t \in (t_0 - \tilde{\delta}, t_0 + \tilde{\delta})$ together with the boundary condition $\tilde{y}(t_0) = y_0$. Since $\tilde{y}$ is continuous, we can find a $\delta > 0$ with $\eta \geq \delta$, $\tilde{\delta} \geq \delta$ and

$$|\tilde{y}(t) - y_0| < \eta$$

for all $t \in (t_0 - \delta, t_0 + \delta)$. If we set $y = y|_{(t_0-\delta, t_0+\delta)}$ (the restriction of y to $(t_0 - \delta, t_0 + \delta)$), then

$$(t, y(t)) \in [t_0 - \eta, t_0 + \eta] \times [y_0 - \eta, y_0 + \eta],$$

and so

$$f(t, y(t)) = \tilde{f}(t, y(t))$$

for all $t \in (t_0 - \delta, t_0 + \delta)$, so y is the unique solution of

$$y'(t) = f(t, y(t)),$$

as required. ■

Exercise 12.15. *(i) Describe $\tilde{f}$ in words.*

(ii) It is, I think, clear that $\tilde{f}$ is continuous and

$$|\tilde{f}(t,u) - \tilde{f}(t,v)| \leq K|u - v|$$

for all t, u and v. Carry out some of the detailed checking which would be required if someone demanded a complete proof.

Theorem 12.14 tells us, that under very wide conditions, the differential equation has a *local solution* through each (t_0, y_0). Does it have a *global solution*, that is, if $f : \mathbb{R}^2 \to \mathbb{R}$ is well behaved, can we find a solution for the equation $y'(t) = f(t, y(t))$ which is defined for all $t \in \mathbb{R}$? Our first result in this direction is positive.

Theorem 12.16. *Suppose $f : \mathbb{R}^2 \to \mathbb{R}$ is a continuous function satisfying the following condition. There exists a $K : [0, \infty) \to [0, \infty)$ such that*

$$|f(t,u) - f(t,v)| \leq K(R)|u - v|$$

whenever $|t| \leq R$. Then, given any $(t_0, y_0) \in \mathbb{R}^2$, there exists a unique differentiable function $y : \mathbb{R} \to \mathbb{R}$ which satisfies the equation $y'(t) = f(t, y(t))$ for all $t \in \mathbb{R}$ together with the boundary condition $y(t_0) = y_0$.

Note that it makes no difference how fast $K(R)$ increases.

Proof. This proof is worth studying since it is of a type which occurs in several places in more advanced work. We refer to the equation $y'(t) = f(t, y(t))$ for all $t \in \mathbb{R}$ together with the boundary condition $y(t_0) = y_0$ as 'the system'.

Our result will follow if we can show that the system has a unique solution on $[t_0, \infty)$ and on $(-\infty, t_0]$. The proof is essentially the same for the two cases, so we show that the system has a unique solution on $[t_0, \infty)$. Observe that, if we can show that the system has a unique solution on $[t_0, T)$ for all $T > t_0$, we shall have shown that the system has a unique solution on $[t_0, \infty)$. (Write $y_T : [t_0, T) \to \mathbb{R}$ for the solution on $[t_0, T)$. If $S \geq T$, then $y_S(t) = y_T(t)$ for all $t \in [t_0, T)$ by uniqueness. Thus we can define $y : [t_0, \infty) \to \mathbb{R}$ by $y(t) = y_T(t)$ for all $t \in [t_0, T)$. By construction y is a solution of the system on $[t_0, \infty)$. If $w : [t_0, \infty) \to \mathbb{R}$ is a solution of the system on $[t_0, \infty)$, then, by uniqueness on $[t_0, T)$, $w(t) = y_T(t) = y(t)$ for all $t_0 \leq t \leq T$ and all $T > t_0$. Since T was arbitrary, $w(t) = y(t)$ for all $t \in [t_0, \infty)$.) We can thus concentrate our efforts on showing that the system has a unique solution on $[t_0, T)$ for all $T > t_0$.

Existence: Let

$$E = \{\tau > t_0 \; : \; \text{the system has a solution on } [t_0, \tau)\}.$$

By Theorem 12.14, E is non-empty. If E is bounded, it has a supremum, T_0, say. Choose $R_0 > |T_0| + 2$ and set $K_0 = K(R_0)$. By hypothesis,

$$|f(t,u) - f(t,v)| \leq K_0|u - v|$$

whenever $|t - T_0| < 2$. Choose $\delta_0 > 0$ such that $1 > \delta_0$, $T_0 - t_0 > 2\delta_0$ and $K_0\delta_0 < 1$. Since T_0 is the supremum of E, we can find $T_1 \in E$ such that $T_1 > T_0 - \delta_0/3$. Let $y : [t_0, T_1), \to \mathbb{R}$ be a solution of the system and let $T_2 = T_1 - \delta_0/3$. By Theorem 12.14, there exists a unique $w : (T_2 - \delta_0, T_2 + \delta_0) \to \mathbb{R}$ such that

$$w'(t) = f(t, w(t)), \;\; w(T_2) = y(T_2).$$

The uniqueness of w means that $w(t) = y(t)$ for all t where both y and w are defined (that is, on $(T_2 - \delta_0, T_1)$). Setting

$$\tilde{y}(t) = y(t) \qquad \text{for } t < T_1,$$
$$\tilde{y}(t) = w(t) \qquad \text{for } t < T_2 + \delta_0,$$

we see that $\tilde{y} : [t_0, T_2 + \delta_0) \to \mathbb{R}$ is a solution of the system. Since $T_2 + \delta_0 > T_1$, we have a contradiction. Thus, by reductio ad absurdum, E is unbounded and the system has a solution on $[t_0, T)$ for all $T > t_0$.

Uniqueness: We need to show that if $T > t_0$ and y and w are solutions of the system on $[t_0, T)$, then $y(t) = w(t)$ for all $t \in [t_0, T)$. The proof is similar to, but simpler than, the existence proof just given. Let

$$E = \{T > \tau \geq t_0 \; : \; y(t) = w(t) \text{ for all } t \in [t_0, \tau]\}.$$

Since $t_0 \in E$, we know that E is non-empty. By definition, E is bounded and so has a supremum T_0. If $T_0 = T$, we are done. If not, $T_0 < T$. By continuity, $y(T_0) = w(T_0)$. As before, choose $R_0 > |T_0| + 2$ and set $K_0 = K(R_0)$. By hypothesis,

$$|f(t,u) - f(t,v)| \leq K_0|u - v|$$

whenever $|t - T_0| < 2$. Choose $\delta_0 > 0$ such that $1 > \delta_0$, $T_0 - t_0 > 2\delta_0$, $\tau - T_0 > 2\delta_0$, and $K_0\delta_0 < 1$. By Theorem 12.14, there exists a unique $z : (T_0 - \delta_0, T_0 + \delta_0) \to \mathbb{R}$ such that

$$z'(t) = f(t, z(t)), \;\; z(T_0) = y(T_0).$$

By uniqueness $y(t) = z(t) = w(t)$ for all $t \in (T_0 - \delta_0, T_0 + \delta_0)$. It follows that $y(t) = w(t)$ for all $t \in [t_0, T_0 + \delta_0)$ and so, by continuity, for all $t \in [t_0, T_0 + \delta]$. Thus $T_0 + \delta_0 \in E$, contradicting the definition of T_0. The desired result follows by contradiction. ■

Exercise 12.17. *Suppose $\eta > 0$ and $f : (a,b) \to \mathbb{R}$ is a continuous function such that, given any $t_1 \in (a,b)$ and any $y_1 \in \mathbb{R}$, we can find an $\eta(t_1, y_1) > 0$ and a $K(t_1, y_1)$ such that*

$$\bigstar \qquad |f(t,u) - f(t,v)| \leq K(t_1, y_1)|u - v|$$

whenever

$$t \in [t_1 - \eta(t_1, y_1), t_1 + \eta(t_1, y_1)] \text{ and } u,\ v \in [y_1 - \eta(t_1, y_1), y_1 + \eta(t_1, y_1)].$$

Show that, if y, $w : (a,b) \to \mathbb{R}$ are differentiable functions such that

$$y'(t) = f(t, y(t)),\ w'(t) = f(t, w(t)) \text{ for all } t \in (a,b)$$

and $y(t_0) = w(t_0)$ for some $t_0 \in (a,b)$, then $y(t) = w(t)$ for all $t \in (a,b)$.

Exercise 12.18. *Use Example 1.3 to show that, in the absence of the fundamental axiom, we cannot expect even very well behaved differential equations to have unique solutions.*

Looking at Theorem 12.16, we may ask if we can replace the condition

$$|f(t,u) - f(t,v)| \leq K(R)|u - v| \text{ whenever } |t| \leq R$$

by the condition

$$|f(t,u) - f(t,v)| \leq K(R)|u - v| \text{ whenever } |t|, |u|, |v| \leq R.$$

Unless the reader is very alert, the answer comes as a surprise followed almost at once by surprise that the answer came as a surprise.

Example 12.19. *Let $f(t,y) = 1 + y^2$. Then*

$$|f(t,u) - f(t,v)| \leq 2R|u - v|$$

whenever $|t|, |u|, |v| \leq R$. However, given t_0, $y_0 \in \mathbb{R}$, there does not exist a differentiable function $y : \mathbb{R} \to \mathbb{R}$ such that $y'(t) = f(t, y(t))$ for all $t \in \mathbb{R}$.

Proof. Observe first that

$$|f(t,u) - f(t,v)| = |u^2 - v^2| = |u + v||u - v| \leq (|u| + |v|)|u - v| \leq 2R|u - v|$$

whenever $|t|$, $|u|$, $|v| \leq R$.

We can solve the equation

$$y' = 1 + y^2$$

formally by considering

$$\frac{dy}{1 + y^2} = dt$$

and obtaining

$$\tan^{-1} y = t + a,$$

so $y(t) = \tan(t + a)$ for some constant a. We choose $\alpha \in [t_0 - \pi/2, t_0 + \pi/2]$ so that $y_0 = \tan(t_0 - \alpha)$ satisfies the initial condition and thus obtain

$$y(t) = \tan(t - \alpha)$$

for $\alpha - \pi/2 < t < \alpha + \pi/2$. We check that we have a solution by direct differentiation. Exercise 12.17 tells us that this is the only solution. Since $\tan(t - \alpha) \to \infty$ as $t \to \alpha + \pi/2$ through values of $t < \alpha + \pi/2$, the required result follows. ■

(An alternative proof is outlined in Exercise K.294.)

Exercise 12.20. *(i) Sketch, on the same diagram, various solutions of $y'(t) = 1 + y(t)^2$ with different initial conditions.*

(ii) Identify the point in our proof of Theorem 12.16 where the argument fails for the function $f(t, y) = 1 + y^2$.

We may think of local solutions as a lot of jigsaw pieces. Just looking at the pieces does not tell us whether they fit together to form a complete jigsaw.

Here is another example which brings together ideas from various parts of the book. Although the result is extremely important, I suggest that the reader not bother too much with the details of the proof.

Lemma 12.21. *If $z_0 \in \mathbb{C} \setminus \{0\}$ and $w_0 \in \mathbb{C}$, then the differential equation*

$$f'(z) = \frac{1}{z}$$

has a solution with $f(z_0) = w_0$ in the set

$$B(z_0, |z_0|) = \{z \in \mathbb{C} : |z - z_0| < |z_0|\}.$$

However, the same differential equation

$$f'(z) = \frac{1}{z}$$

has no solution valid in $\mathbb{C} \setminus \{0\}$.

Proof. Our first steps reflect the knowledge gained in results like Example 11.78 and Exercise 11.82. The power series $\sum_{j=1}^{\infty} \frac{(-1)^{j+1} z^j}{j}$ has radius of convergence 1. We define $h : B(0, 1) \to \mathbb{C}$ by

$$h(z) = \sum_{j=1}^{\infty} \frac{(-1)^{j+1} z^j}{j}.$$

Since we can differentiate term by term within the radius of convergence, we have

$$h'(z) = \sum_{j=1}^{\infty}(-1)^{j+1}z^{j-1} = \sum_{j=0}^{\infty}(-1)^j z^j = \frac{1}{1+z}$$

for all $|z| < 1$. Thus, if we set $f(z) = w_0 + h(1 + (z - z_0)z_0^{-1})$ for $z \in B(z_0, |z_0|)$, the chain rule gives

$$f'(z) = \frac{1}{z_0} \times \frac{1}{1 + (z - z_0)z_0^{-1}} = \frac{1}{z},$$

as desired. Simple calculation gives $f(z_0) = w_0$.

The second part of the proof is, as one might expect, closely linked to Example 5.71. Suppose, if possible, that there exists an $f : \mathbb{C} \setminus \{0\} \to \mathbb{C}$ satisfying the differential equation

$$f'(z) = \frac{1}{z}.$$

By replacing f by $f - f(1)$, we may suppose that $f(1) = 0$. Define $A : \mathbb{R} \to \mathbb{C}$ by $A(t) = -if(e^{it})$. Writing $A(t) = a(t) + ib(t)$ with $a(t)$ and $b(t)$ real, we see that $A'(t) = a'(t) + ib'(t)$ exists with value

$$\begin{aligned} A'(t) &= \lim_{\delta t \to 0} \frac{-if(e^{i(t+\delta t)}) + if(e^{it})}{\delta t} \\ &= \lim_{\delta t \to 0} \frac{f(e^{i(t+\delta t)}) - f(e^{it})}{e^{i(t+\delta t)} - e^{it}} \times \frac{e^{i\delta t} - 1}{\delta t} \times (-ie^{it}) \\ &= f'(e^{it})i(-ie^{it}) = \frac{e^{it}}{e^{it}} = 1. \end{aligned}$$

Thus $A(t) = t + A(0) = t$. In particular,

$$0 = A(0) = -if(1) = -if(e^{2\pi i}) = A(2\pi) = 2\pi,$$

which is absurd. Thus no function of the type desired can exist. ■

Exercise 12.22. *The proof above is one of the kind where the principal characters wear masks. Go through the above proof using locutions like 'the thing that ought to behave like* $\log z$ *if* $\log z$ *existed and behaved as we think it ought.'*

Exercise 12.23. *Write*

$$B_j = \{z \in \mathbb{C} : |z - e^{\pi i j/3}| < 1\}.$$

Show that there exist a function $f_1 : \bigcup_{j=0}^{3} B_j \to \mathbb{C}$ *with*

$$f_1'(z) = \frac{1}{z}, \; f_1(1) = 0$$

and a function $f_2 : \bigcup_{j=3}^{6} B_j \to \mathbb{C}$ *with*

$$f_2'(z) = \frac{1}{z}, \ f_2(1) = 0.$$

Find $f_1 - f_2$ *on* B_0 *and on* B_3.

We have a lot of beautifully fitting jigsaw pieces, but when we put too many together they overlap, instead of forming a complete picture. Much of complex variable theory can be considered as an extended meditation on Lemma 12.21.

If the reader is prepared to allow a certain amount of hand waving, here is a another example of this kind of problem. Consider the circle $\mathbb{T}$ obtained by 'rolling up the real real line like a carpet' so that the point θ is identified with the point $\theta + 2\pi$. If we seek a solution of the equation

$$f''(\theta) + \lambda^2 f(\theta) = 0,$$

where λ is real and positive, then we can always obtain 'local solutions' $f(\theta) = \sin(\lambda\theta + \theta_0)$ valid on any small part of the circle we choose, but only if λ is an integer can we extend it to the whole circle. When we start doing analysis on spheres, cylinders, tori and more complicated objects, the problem of whether we can combine 'local solutions' to form consistent 'global solutions' becomes more and more central.

The next exercise is straightforward and worthwhile, but long.

Exercise 12.24. *(i) State and prove the appropriate generalisation of Theorem 12.16 to deal with a vectorial differential equation*

$$\mathbf{y}'(t) = \mathbf{f}(t, \mathbf{y}(t)).$$

(ii) Use (i) to obtain the following generalisation of Lemma 12.13. Suppose $g : \mathbb{R}^{n+1} \to \mathbb{R}$ *is a continuous function satisfying the following condition. There exists a* $K : [0, \infty) \to \mathbb{R}$ *such that*

$$|g(t, \mathbf{u}) - g(t, \mathbf{v})| \leq K(R)\|\mathbf{u} - \mathbf{v}\|$$

whenever $|t| \leq R$. *Then, given any* $(t_0, y_0, y_1, \ldots, y_{n-1}) \in \mathbb{R}^{n+1}$, *there exists a unique* n *times differentiable function* $y : \mathbb{R} \to \mathbb{R}$ *with*

$$y^{(n)}(t) = g(t, y(t), y'(t), \ldots, y^{n-1}(t)) \text{ and } y^{(j)}(t_0) = y_j \text{ for } 0 \leq j \leq n-1.$$

Exercise 12.25. *In this book we have given various approaches to the exponential and trigonometric functions. Using the material of this section, we can give a particularly neat treatment which avoids the use of infinite sums.*

(i) Explain why there exists a unique differentiable function $e : \mathbb{R} \to \mathbb{R}$ *such that*

$$e'(x) = e(x) \text{ for all } x \in \mathbb{R}, \ e(0) = 0.$$

By differentiating the function f defined by $f(x) = e(a-x)e(x)$, show that $e(a-x)e(x) = e(a)$ for all x, $a \in \mathbb{R}$ and deduce that $e(x+y) = e(x)e(y)$ for all x, $y \in \mathbb{R}$. List all the properties of the exponential function that you consider important, and prove them.

(ii) Explain why there exist unique differentiable functions $s, c : \mathbb{R} \to \mathbb{R}$ such that

$$s'(x) = c(x),\ c'(x) = -s(x) \text{ for all } x \in \mathbb{R},\ s(0) = 0,\ c(0) = 1.$$

By differentiating the function f defined by $f(x) = s(a-x)c(x) + c(a-x)s(x)$, obtain an important addition formula for trigonometric functions. Obtain at least one other such addition formula in a similar manner. List all the properties of sin *and* cos *that you consider important, and prove them.*

(iii) Write down a differential equation for $T(x) = \tan x$ of the form

$$T'(x) = g(T(x)).$$

Explain why, without using properties of tan*, we know there exists a function T with $T(0) = 0$ satisfying this differential equation on some interval $(-a, a)$ with $a > 0$. State and prove, using a method similar to those used in parts (i) and (ii), a formula for $T(x+y)$ when x, y, $x+y \in (-a, a)$.*

12.4. Green's function solutions ♡

In this section we discuss how to solve the differential equation for real-valued functions on $[0, 1]$ given as

$$\bigstar \qquad y''(t) + a(t)y'(t) + b(t)y(t) = f(t)$$

subject to the conditions $y(0) = y(1) = 0$ by using the Green's function. We assume that a and b are continuous. Notice that we are dealing with a linear differential equation, so that, if y_1 and y_2 are solutions and $\lambda_1 + \lambda_2 = 1$, then $\lambda_1 y_1 + \lambda_2 y_2$ is also a solution. Notice also that the boundary conditions are different from those we have dealt with so far. Instead of specifying y and y' at one point, we specify y at two points.

Exercise 12.26. *(i) Check the statement about the solutions.*

(ii) Explain why there is no loss in generality in considering the interval $[0, 1]$ rather than the interval $[u, v]$.

Most of this section will be taken up with an informal discussion leading to a solution (given in Theorem 12.31) that can be verified in a couple of lines. However, the informal heuristics can be generalised to deal with many interesting problems, and the verification cannot.

When a ball hits a bat its velocity changes very rapidly because the bat exerts a very large force for a very short time. However, the position of the ball hardly changes at all during the short time the bat and ball are in contact. We try to model this by considering the system

$$y''(t) + a(t)y'(t) + b(t)y(t) = h_\eta(t),\ y(0) = y(1) = 0,$$

where

$$\begin{aligned} h_\eta(t) &\geq 0 \qquad &&\text{for all } t,\\ h_\eta(t) &= 0 &&\text{for all } t \notin [s-\eta, s+\eta],\\ \int_{s-\eta}^{s+\eta} h_\eta(t)\,dt &= 1, \end{aligned}$$

and $\eta > 0$, $[s-\eta, s+\eta] \subseteq [0,1]$. We have

$$\begin{aligned} y''(t) + a(t)y'(t) + b(t)y(t) &= 0 \text{ for } t \leq s-\eta,\ y(0) = 0,\\ y''(t) + a(t)y'(t) + b(t)y(t) &= 0 \text{ for } t \geq s+\eta,\ y(1) = 0, \end{aligned}$$

and

$$\begin{aligned} y'(s+\eta) - y'(s-\eta) &= \int_{s-\eta}^{s+\eta} y''(t)\,dt\\ &= \int_{s-\eta}^{s+\eta} h_\eta(t) - a(t)y'(t) - b(t)y(t)\,dt\\ &= 1 - \int_{s-\eta}^{s+\eta} a(t)y'(t)\,dt - \int_{s-\eta}^{s+\eta} b(t)y(t)\,dt. \end{aligned}$$

What happens as we make η small? Although y' changes very rapidly, we would expect its value to remain bounded (the velocity of the ball changes but remains bounded), so we would expect $\int_{s-\eta}^{s+\eta} a(t)y'(t)\,dt$ to become very small. We expect the value of y to change very little, so we certainly expect $\int_{s-\eta}^{s+\eta} b(t)y(t)\,dt$ to become very small.

If we now allow η to tend to zero, we are led to look at the system of equations

$$\begin{aligned} y''(t) + a(t)y'(t) + b(t)y(t) &= 0 \text{ for } t < s,\ y(0) = 0,\\ \bigstar\bigstar \qquad y''(t) + a(t)y'(t) + b(t)y(t) &= 0 \text{ for } t > s,\ y(1) = 0,\\ y(s+) = y(s-) = y(s),\ y'(s+) &- y'(s-) = 1. \end{aligned}$$

Here, as usual, $y(s+) = \lim_{t\to s,\ t>s} y(s)$ and $y(s-) = \lim_{t\to s,\ t<s} y(s)$. The statement $y(s+) = y(s-) = y(s)$ thus means that y is continuous at s. We write the system ★★ more briefly as

$$y''(t) + a(t)y'(t) + b(t)y(t) = \delta_s(t),\ y(0) = y(1) = 0,$$

where δ_c may be considered as 'a unit impulse at c' or 'the idealisation of $h_\eta(t)$ for small η' or a 'delta function at s' or a 'Dirac point mass at s' (this links up with Exercise 9.44 on Riemann-Stieljes integration).

By the previous section, we know that there exists a unique, twice differentiable function $y_1 : [0,1] \to \mathbb{R}$ such that

$$y_1''(t) + a(t)y_1'(t) + b(t)y_1(t) = 0,\ y_1(0) = 0,\ y_1'(0) = 1,$$

and a unique, twice differentiable, $y_2 : [0,1] \to \mathbb{R}$ such that

$$y_2''(t) + a(t)y_2'(t) + b(t)y_2(t) = 0,\ y_2(1) = 0,\ y_2'(1) = 1.$$

We make the following

key assumption: $\quad y_1(1) \neq 0$

(so that y_2 cannot be a scalar multiple of y_1).

If y is a solution of ★★, the uniqueness results of the previous section tell us that

$$y(t) = Ay_1(t) \text{ for } 0 \leq t < s,\ y(t) = By_2(t) \text{ for } s < t \leq 1$$

for appropriate constants A and B. Since $y(s+) = y(s-) = y(s)$, we can find a constant C such that $A = Cy_2(s)$, $B = Cy_1(s)$, and so

$$y(t) = Cy_1(t)y_2(s) \text{ for } 0 \leq t < s,\ y(t) = Cy_2(t)y_1(s) \text{ for } s < t \leq 1.$$

The condition $y'(s+) - y'(s-) = 1$ gives us

$$C\big(y_1'(s)y_2(s) - y_1(s)y_2'(s)\big) = 1,$$

and so, setting $W(s) = y_1'(s)y_2(s) - y_1(s)y_2'(s)$, and assuming, without proof for the moment, that $W(s) \neq 0$, we have

$$y(t) = y_1(t)y_2(s)W(s)^{-1} \text{ for } 0 \leq t \leq s,$$
$$y(t) = y_2(t)y_1(s)W(s)^{-1} \text{ for } s \leq t \leq 1.$$

Although we shall continue with our informal argument afterwards, we take time out to establish that W is never zero.

Definition 12.27. *If u_1 and u_2 are two solutions of*

$$y''(t) + a(t)y'(t) + b(t)y(t) = 0,$$

we define the associated Wronskian *W by $W(t) = u_1'(t)u_2(t) - u_1(t)u_2'(t)$.*

Lemma 12.28. *(i) If W is as in the preceding definition, then $W'(t) = -a(t)W(t)$ for all $t \in [0,1]$.*

(ii) If W is as in the preceding definition, then

$$W(t) = Ae^{-\int_0^t a(x)\,dx}$$

for all $t \in [0,1]$ and some constant A.

(iii) If y_1 and y_2 are as in the discussion above and the **key assumption** *holds, then*

$$W(s) = y_1'(s)y_2(s) - y_1(s)y_2'(s) \neq 0 \textit{ for all } s \in [0,1].$$

Proof. (i) Just observe that

$$\begin{aligned}W'(t) &= u_1''(t)u_2(t) + u_1'(t)u_2'(t) - u_1'(t)u_2'(t) - u_1(t)u_2''(t)\\ &= u_1''(t)u_2(t) - u_1(t)u_2''(t)\\ &= u_2(t)(-a(t)u_1'(t) - b(t)u_1(t)) - u_1(t)(-a(t)u_2'(t) - b(t)u_2(t))\\ &= -a(t)W(t).\end{aligned}$$

(ii) We solve the differential equation *formally*, obtaining

$$\frac{W'(t)}{W(t)} = -a(t),$$

whence

$$\log W(t) = -\int_0^t a(x)\,dx + \log A,$$

and so $W(t) = Ae^{-\int_0^t a(x)\,dx}$ for some constant A.

We *verify* directly that this is indeed a solution. The uniqueness results of the previous section (note that $W(0) = A$) show that it is the unique solution.

(iii) Observe that $W(1) = y_1'(1)y_2(1) - y_1(1)y_2'(1) = -y_1(1) \neq 0$ by the **key assumption**. Since W does not vanish at 1, part (ii) shows that it vanishes nowhere. ■

Exercise 12.29. *Prove part (ii) of Lemma 12.28 by considering the derivative of the function f given by $f(t) = W(t)\exp(\int_0^t a(x)\,dx)$.*

A more general view of the Wronskian is given by Exercise K.301.

We write $G(s,t) = y(s)$, where y is the solution we obtained to ★★, that is, we set

$$\begin{aligned}G(s,t) &= y_1(t)y_2(s)W(s)^{-1} \text{ for } 0 \le t \le s,\\ G(s,t) &= y_2(t)y_1(s)W(s)^{-1} \text{ for } s < t \le 1.\end{aligned}$$

The function $G : [0,1]^2 \to \mathbb{R}$ is called a Green's function.

We return to our informal argument. Since $G(s,t)$ is the solution of

$$y''(t) + a(t)y'(t) + b(t)y(t) = \delta_s(t),\ y(0) = y(1) = 0,$$

it follows, by linearity, that $y(t) = \sum_{j=1}^m \lambda_j G(s_j,t)$ is the solution of

$$y''(t) + a(t)y'(t) + b(t)y(t) = \sum_{j=1}^m \lambda_j \delta_{s_j}(t),\ y(0) = y(1) = 0.$$

In particular, if $f : [0,1] \to \mathbb{R}$, then $y_N(t) = N^{-1}\sum_{j=1}^N f(j/N)G(j/N,t)$ is the solution of

$$y''(t) + a(t)y'(t) + b(t)y(t) = N^{-1}\sum_{j=1}^{N} f(j/N)\delta_{j/N}(t),\ y(0) = y(1) = 0.$$

Now imagine yourself pushing a large object. You could either give a continuous push, applying a force of magnitude $f(t)$, or give a sequence of sharp taps $N^{-1}\sum_{j=1}^N f(j/N)\delta_{j/N}(t)$. As you make the interval between the taps ever smaller (reducing the magnitude of each individual tap proportionally) the two ways of pushing the object become more and more alike, and

$$N^{-1}\sum_{j=1}^{N} f(j/N)\delta_{j/N} \to f \text{ in some way which we cannot precisely define.}$$

It is therefore plausible that, as $N \to \infty$,

$$y_N \to y_* \text{ in some way to be precisely determined later,}$$

where y_* is the solution of

$$y_*''(t) + a(t)y_*'(t) + b(t)y_*(t) = f(t),\ y_*(0) = y_*(1) = 0.$$

It also seems very likely that

$$y_N(t) = N^{-1}\sum_{j=1}^{N} f(j/N)G(j/N,t) \to \int_0^1 f(s)G(s,t)\,dt$$

(We could prove this rigorously, but a chain is as strong as its weakest link and the real difficulties lie elsewhere.)

We are therefore led to the following plausible statement.

Plausible Statement 12.30. *The equation*

$$y''(t) + a(t)y'(t) + b(t)y(t) = f(t),\ y(0) = y(1) = 0$$

has the solution

$$y(t) = \int_0^1 f(s)G(s,t)\,ds.$$

We summarise the results of our informal argument in the next theorem which we then prove formally.

Theorem 12.31. *Suppose that f, a, $b : [0,1] \to \mathbb{R}$ are continuous. By Exercise 12.24, there exist a unique twice differentiable $y_1 : [0,1] \to \mathbb{R}$ such that*

$$y_1''(t) + a(t)y_1'(t) + b(t)y_1(t) = 0,\ y_1(0) = 0,\ y_1'(0) = 1$$

and a unique twice differentiable $y_2 : [0,1] \to \mathbb{R}$ *such that*

$$y_2''(t) + a(t)y_2'(t) + b(t)y_2(t) = 0, \ y_2(1) = 0, \ y_2'(1) = 1.$$

We make the following

$$\textbf{key assumption:} \qquad y_1(1) \neq 0.$$

If we set $W(t) = y_2'(t)y_1(t) - y_2(t)y_1'(t)$*, then* W *is never zero and we may define* $G : [0,1]^2 \to \mathbb{R}$ *by*

$$G(s,t) = y_1(t)y_2(s)W(s)^{-1} \text{ for } 0 \leq t \leq s,$$
$$G(s,t) = y_2(t)y_1(s)W(s)^{-1} \text{ for } s < t \leq 1.$$

With this notation, G *is continuous, so we can define*

$$y(t) = \int_0^1 G(s,t)f(s)\,ds.$$

The function y*, so defined, is twice differentiable and satisfies*

$$\bigstar \qquad y''(t) + a(t)y'(t) + b(t)y(t) = f(t)$$

together with the conditions $y(0) = y(1) = 0$*. Moreover, this solution is unique.*

Proof. We have already established the contents of the first paragraph. The proof of the continuity of G is left to the reader (see Exercise 12.32 (iii) for a general result from which this follows). To show that y is differentiable and satisfies $\bigstar$, we observe that

$$\begin{aligned} y(t) &= \int_0^t G(s,t)f(s)\,ds + \int_t^1 G(s,t)f(s)\,ds \\ &= \int_0^t y_2(t)y_1(s)W(s)^{-1}f(s)\,ds + \int_t^1 y_1(t)y_2(s)W(s)^{-1}f(s)\,ds \\ &= y_2(t)\int_0^t y_1(s)W(s)^{-1}f(s)\,ds + y_1(t)\int_t^1 y_2(s)W(s)^{-1}f(s)\,ds \end{aligned}$$

[Experience shows that people get into frightful muddles over this calculation. At each stage you must ask yourself 'is s bigger than t or vice versa and what does this mean for the definition of $G(s,t)$?']

We now use the product rule and the fundamental theorem of the calculus to show that y is twice differentiable with

$$\begin{aligned} y'(t) &= y_2'(t)\int_0^t y_1(s)W(s)^{-1}f(s)\,ds + y_2(t)y_1(t)W(t)^{-1}f(t)\\ &\quad + y_1'(t)\int_t^1 y_2(s)W(s)^{-1}f(s)\,ds - y_1(t)y_2(t)W(t)^{-1}f(t)\\ &= y_2'(t)\int_0^t y_1(s)W(s)^{-1}f(s)\,ds + y_1'(t)\int_t^1 y_2(s)W(s)^{-1}f(s)\,ds \end{aligned}$$

and

$$\begin{aligned} y''(t) &= y_2''(t)\int_0^t y_1(s)W(s)^{-1}f(s)\,ds + y_2'(t)y_1(t)W(t)^{-1}f(t)\\ &\quad + y_1''(t)\int_t^1 y_2(s)W(s)^{-1}f(s)\,ds - y_1'(t)y_2(t)W(t)^{-1}f(t)\\ &= y_2''(t)\int_0^t y_1(s)W(s)^{-1}f(s)\,ds + y_1''(t)\int_t^1 y_2(s)W(s)^{-1}f(s)\,ds\\ &\quad + (y_2'(t)y_1(t) - y_1'(t)y_2(t))W(t)^{-1}f(t)\\ &= y_2''(t)\int_0^t y_1(s)W(s)^{-1}f(s)\,ds + y_1''(t)\int_t^1 y_2(s)W(s)^{-1}f(s)\,ds + f(t), \end{aligned}$$

so that

$$\begin{aligned} &y''(t) + a(t)y'(t) + b(t)y(t)\\ &= (y_1''(t) + a(t)y_1'(t) + b(t)y_1(t))\int_t^1 y_2(s)W(s)^{-1}f(s)\,ds\\ &\quad + (y_2''(t) + a(t)y_2'(t) + b(t)y_2(t))\int_0^t y_1(s)W(s)^{-1}f(s)\,ds + f(t)\\ &= 0\times\int_t^1 y_2(s)W(s)^{-1}f(s)\,ds + 0\times\int_0^t y_1(s)W(s)^{-1}f(s)\,ds + f(t) = f(t), \end{aligned}$$

as required.

To prove uniqueness, suppose that u_1 and u_2 are solutions of ★ satisfying $u_1(0) = u_2(0) = u_1(1) = u_2(1) = 0$. If we set $u = u_1 - u_2$, then, by linearity,

$$u''(t) + a(t)u'(t) + b(t)u(t) = 0,\ u(0) = 0,\ u(1) = 0.$$

Suppose $u'(0) = \lambda$. Then u and λy_1 both satisfy

$$w''(t) + a(t)w'(t) + b(t)w(t) = 0,\ w(0) = 0,\ w'(0) = \lambda,$$

so, by the uniqueness results of the previous section, $u = \lambda y_1$. But $y_1(1) \neq 0$ and $u(1) = 0$, so $\lambda = 0$ and $u = 0$. Thus $u_1 = u_2$, as required. ■

Exercise 12.32. *Let A and B be subsets of some metric space and consider a function $f : A \cup B \to \mathbb{R}$.*

(i) Show that, if f is not continuous at x, then at least one of the following two statements must be true:

(α) We can find $a_n \in A$ with $a_n \to x$ and $f(a_n) \nrightarrow f(x)$.

(β) We can find $b_n \in B$ with $b_n \to x$ and $f(b_n) \nrightarrow f(x)$.

(ii) Give an example where $f|_A$ and $f|_B$ are continuous, but f is not.

(iii) Show that, if A and B are closed, then the continuity of $f|_A$ and $f|_B$ implies the continuity of f.

[For further remarks along these lines, see Exercise K.197.]

The proof of Theorem 12.31 is rather disappointing, since it uses only rather elementary results and gives no hint as to how to proceed in more general circumstances. (In Exercise K.307, I outline a proof which involves slightly less calculation and slightly harder theorems, but it still does not get to the heart of the matter.) However the general idea of using 'delta functions' or 'impulses' to study ordinary and, particularly, partial differential equations is very important in both pure and applied mathematics. (The acoustics of concert halls are tested by letting off starting pistols and recording the results.)

From the point of view of the pure mathematician, one of the chief advantages of the Green's function method is that it converts problems on differentiation to problems on integration, with the benefits pointed out on page 290.

In the terminology of more advanced work, we have shown how differential operators like $y \mapsto Sy$, where

$$Sy(t) = y''(t) + a(t)y'(t) + b(t)y(t),$$

can be linked with better behaved integral operators like

$$f \mapsto Tf \text{ with } Tf(t) = \int_0^1 G(s,t)f(s)\,ds.$$

Note that we have shown $STf = f$ for f continuous, but note also that, if f is merely continuous, Sf need not be defined. The Green's function G is an example of an integral kernel[2]. More formally, if we write

$$Jf(s) = \int_0^1 K(s,t)f(s)\,ds,$$

then J is called an integral operator with kernel K.

We end with an example to show that things really do go awry if our **key assumption** fails.

Example 12.33. *The equation*

$$y''(t) + \pi^2 y(t) = t$$

has no solution satisfying $y(0) = y(1) = 0$.

Proof. Suppose that y satisfies the equation

$$y''(t) + \pi^2 y(t) = t$$

and satisfies $y(0) = 0$. If we set

$$w(t) = \pi^{-2}t + \frac{y'(0) - \pi^{-2}}{\pi}\sin \pi t,$$

then, by direct calculation,

$$w''(t) + \pi^2 w(t) = t$$

and $w(0) = 0 = y(0)$, $w'(0) = y'(0)$, so, by the uniqueness results of the previous section, $y = w$. In particular, $y(1) = w(1) \neq 0$. ■

[2]We shall not discuss these matters much further, but most of the new words in this last paragraph are well worth dropping.

> You must lie among the daisies and discourse in novel phrases of your complicated
> state of mind,
> The meaning doesn't matter if it's only idle chatter of a transcendental kind.
> And everyone will say,
> As you walk your mystic way,
> 'If this young man expresses himself in terms too deep for me,
> Why, what a very singularly deep young man this deep young man must be!'

(Gilbert and Sullivan, *Patience*)

Chapter 13

Inverse and Implicit Functions

13.1. The inverse function theorem

We start with an exercise giving a very simple example of a technique that mathematicians have used for centuries.

Exercise 13.1. *We work in* $\mathbb{R}$.

(i) Suppose x *and* y *are close to 1. If* $(x+\eta)^2 = y$, *show that*

$$\eta \approx (y - x^2)/2.$$

(ii) This suggests the following method for finding the positive square root of y. *Take* x_0 *close to the square root of* y *(for example* $x_0 = 1$*) and define* x_n *inductively by* $x_n = x_{n-1} + (y - x_{n-1}^2)/2$. *We expect* x_n *to converge to the positive square root of* y.

(a) Try this for $y = 1.21$ *and for* $y = 0.81$.

(b) Try this for $y = 100$.

(iii) Sketch $x - x^2/2$ *for* $1/2 \leq x \leq 3/2$. *Show that if* $|y - 1| \leq 1/2$ *and we define* Tx *by*

$$Tx = x + (y - x^2)/2,$$

then $|Tx - 1| \leq 1/2$ *whenever* $|x - 1| \leq 1/2$. *Show that* $T : [1/2, 3/2] \to [1/2, 3/2]$ *is a contraction mapping, and deduce that the proposed method for finding square roots works if* $|y - 1| \leq 1/2$.

(iv) Find a reasonable approximation to the positive square root of 150 *by observing that*

$$(150)^{1/2} = 12(150/144)^{1/2}$$

and using the method proposed in part (ii).

The discussion that follows may be considered as a natural extension of the ideas above. We work in $\mathbb{R}^n$ with the usual Euclidean norm.

Lemma 13.2. *Consider a function* $\mathbf{f} : \mathbb{R}^m \to \mathbb{R}^m$ *such that* $\mathbf{f}(\mathbf{0}) = \mathbf{0}$. *Suppose there exist a* $\delta > 0$ *and an* η *with* $1 > \eta > 0$ *such that*

$$\|(\mathbf{f}(\mathbf{x}) - \mathbf{f}(\mathbf{y})) - (\mathbf{x} - \mathbf{y})\| \leq \eta \|(\mathbf{x} - \mathbf{y})\|$$

for all $\|\mathbf{x}\|, \|\mathbf{y}\| \leq \delta$. *Then, if* $\|\mathbf{y}\| \leq (1 - \eta)\delta$, *there exists one and only one solution of the equation*

$$\bigstar \qquad \mathbf{f}(\mathbf{x}) = \mathbf{y}$$

with $\|\mathbf{x}\| < \delta$. *Further, if we denote this solution by* $\mathbf{g}(\mathbf{y})$, *we have*

$$\|\mathbf{g}(\mathbf{y}) - \mathbf{y}\| \leq \eta(1 - \eta)^{-1} \|\mathbf{y}\|.$$

Since Lemma 13.2 is most important to us when η is small, there is no harm in concentrating our attention on this case. Lemma 13.2 is then an instance of the

Slogan: Anything which is close to the identity has an inverse.

(This is a slogan, not a theorem. See, for example, Exercise 13.7.) The inequality

$$\|(\mathbf{f}(\mathbf{x}) - \mathbf{f}(\mathbf{y})) - (\mathbf{x} - \mathbf{y})\| \leq \eta \|(\mathbf{x} - \mathbf{y})\|$$

is then seen as a perturbation of the trivial equality

$$\|(I(\mathbf{x}) - I(\mathbf{y})) - (\mathbf{x} - \mathbf{y})\| = \|\mathbf{0}\| = 0\|(\mathbf{x} - \mathbf{y})\|.$$

If we try to solve equation $\bigstar$ by the method of Exercise 13.1, we are led to the observation that, if such a solution exists, with value $\mathbf{u}$, say, then, if $\mathbf{x}$ is close to $\mathbf{u}$,

$$\mathbf{u} - \mathbf{x} \approx \mathbf{f}(\mathbf{u}) - \mathbf{f}(\mathbf{x}) = \mathbf{y} - \mathbf{f}(\mathbf{x}),$$

and so

$$\mathbf{u} \approx \mathbf{x} + (\mathbf{y} - \mathbf{f}(\mathbf{x})).$$

This suggests that we should start with an $\mathbf{x}_0$ close to the solution of equation $\bigstar$ (for example, $\mathbf{x}_0 = \mathbf{y}$) and define $\mathbf{x}_n$ inductively by $\mathbf{x}_n = T\mathbf{x}_{n-1}$, where

$$T\mathbf{x} = \mathbf{x} + (\mathbf{y} - \mathbf{f}(\mathbf{x})).$$

If we add the contraction mapping as a further ingredient, we arrive at the following proof of Lemma 13.2.

Proof of Lemma 13.2. We set

$$X = \bar{B}(\mathbf{0}, \delta) = \{\mathbf{x} \in \mathbb{R}^m : \|\mathbf{x}\| \leq \delta\}.$$

Since X is closed in $\mathbb{R}^m$, we know that $(X, \| \ \|)$ is complete.

Let $\|y\| \leq (1-\eta)\delta$. If we set

$$T\mathbf{x} = \mathbf{x} + (\mathbf{y} - \mathbf{f}(\mathbf{x})),$$

then (since $\mathbf{f}(\mathbf{0}) = \mathbf{0}$)

$$\begin{aligned} \|T\mathbf{x}\| &\leq \|\mathbf{y}\| + \|\mathbf{x} - \mathbf{f}(\mathbf{x})\| \\ &\leq \|\mathbf{y}\| + \|(\mathbf{f}(\mathbf{x}) - \mathbf{f}(\mathbf{0})) - (\mathbf{x} - \mathbf{0})\| \\ &\leq \|\mathbf{y}\| + \eta\|\mathbf{x} - \mathbf{0}\| \\ &\leq (1-\eta)\delta + \eta\|\mathbf{x}\| \leq \delta, \end{aligned}$$

whenever $\mathbf{x} \in X$. Thus T is a well defined function $T : X \to X$.

If $\mathbf{x}$, $\mathbf{x}' \in X$, then

$$\|T\mathbf{x} - T\mathbf{x}'\| = \|(\mathbf{x} - \mathbf{x}') - (\mathbf{f}(\mathbf{x}) - \mathbf{f}(\mathbf{x}'))\| \leq \eta\|\mathbf{x} - \mathbf{x}'\|,$$

by the hypotheses of the lemma. Thus T is a contraction mapping and has a unique fixed point $\mathbf{u} \in X$ such that

$$\mathbf{u} = T\mathbf{u} = \mathbf{u} + (\mathbf{y} - \mathbf{f}(\mathbf{u})),$$

or, equivalently,

$$\mathbf{f}(\mathbf{u}) = \mathbf{y},$$

as required. To obtain the final inequality, we return to the original proof of the contraction mapping theorem (Theorem 12.3). Observe, as we did there, that, if $\mathbf{x} \in X$,

$$\|T^n\mathbf{x} - T^{n-1}\mathbf{x}\| \leq \eta\|T^{n-1}\mathbf{x} - T^{n-2}\mathbf{x}\| \leq \eta^{n-1}\|T\mathbf{x} - \mathbf{x}\|,$$

and so

$$\|T^n\mathbf{x} - \mathbf{x}\| \leq \sum_{j=1}^{n} \|T^j\mathbf{x} - T^{j-1}\mathbf{x}\| \leq \sum_{j=1}^{n} \eta^{j-1}\|T\mathbf{x} - \mathbf{x}\| \leq (1-\eta)^{-1}\|T\mathbf{x} - \mathbf{x}\|.$$

Since $T^n\mathbf{x} \to \mathbf{u}$,

$$\begin{aligned} \|\mathbf{u} - \mathbf{x}\| &\leq \|\mathbf{u} - T^n\mathbf{x}\| + \|T^n\mathbf{x} - \mathbf{x}\| \\ &\leq \|\mathbf{u} - T^n\mathbf{x}\| + (1-\eta)^{-1}\|T\mathbf{x} - \mathbf{x}\| \to (1-\eta)^{-1}\|T\mathbf{x} - \mathbf{x}\| \end{aligned}$$

as $n \to \infty$, it follows that

$$\|\mathbf{u} - \mathbf{x}\| \leq (1-\eta)^{-1}\|T\mathbf{x} - \mathbf{x}\|.$$

If we take $\mathbf{x} = \mathbf{y}$, the last inequality takes the form

$$\begin{aligned}\|\mathbf{u} - \mathbf{y}\| &\leq (1-\eta)^{-1}\|T\mathbf{y} - \mathbf{y}\| = (1-\eta)^{-1}\|\mathbf{y} - f(\mathbf{y})\| \\ &= (1-\eta)^{-1}\|(\mathbf{y} - \mathbf{0}) - (f(\mathbf{y}) - f(\mathbf{0}))\| \\ &\leq \eta(1-\eta)^{-1}\|\mathbf{y} - \mathbf{0}\| = \eta(1-\eta)^{-1}\|\mathbf{y}\|,\end{aligned}$$

as required. ■

Exercise 13.3. *Let (X, d) be a metric space (not necessarily complete) and $T : X \to X$ a mapping such that $d(Tx, Ty) \leq Kd(x, y)$ for all x, $y \in X$ and some $K < 1$. Suppose that T has a fixed point w. If $x_0 \in X$ and we define x_n inductively by $x_{n+1} = Tx_n$, show that $d(x_0, w) \leq (1-K)^{-1}d(x_0, x_1)$.*

Lemma 13.2 provides the core of the proof of our next result. Here, and for the rest of the chapter, we use, in addition to the Euclidean norm on $\mathbb{R}^m$, the operator norm on the space of linear maps $\mathcal{L}(\mathbb{R}^m, \mathbb{R}^m)$.

Lemma 13.4. *Consider a function $\mathbf{f} : \mathbb{R}^m \to \mathbb{R}^m$ such that $\mathbf{f}(\mathbf{0}) = \mathbf{0}$ and there exists a $\delta_0 > 0$ such that $\mathbf{f}$ is differentiable in the open ball $B(\mathbf{0}, \delta_0)$. If $D\mathbf{f}$ is continuous at $\mathbf{0}$ and $D\mathbf{f}(\mathbf{0}) = I$ (the identity map), then we can find a δ_1 with $\delta_0 \geq \delta_1 > 0$ and a $\rho > 0$ such that, if $\|y\| \leq \rho$, there exists one and only one solution of the equation*

$$\bigstar \qquad \mathbf{f}(\mathbf{x}) = \mathbf{y}$$

with $\|\mathbf{x}\| < \delta_1$. Further, if we denote this solution by $\mathbf{g}(\mathbf{y})$, the function $\mathbf{g}$ is differentiable at $\mathbf{0}$ with $D\mathbf{g}(\mathbf{0}) = I$.

Proof. Since $D\mathbf{f}$ is continuous at $\mathbf{0}$ and $D\mathbf{f} = I$, we can find a $\delta_1 > 0$ such that $\delta_0 > \delta_1 > 0$ and

$$\|D\mathbf{f}(\mathbf{w}) - I\| < 1/2$$

for $\|\mathbf{w}\| \leq \delta_1$. Applying the mean value inequality to the function $\mathbf{h} = \mathbf{f} - I$, we obtain

$$\begin{aligned}\|(\mathbf{f}(\mathbf{x}) - \mathbf{f}(\mathbf{y})) - (\mathbf{x} - \mathbf{y})\| = \|\mathbf{h}(\mathbf{x}) - \mathbf{h}(\mathbf{y})\| &\leq \|\mathbf{x} - \mathbf{y}\| \sup_{\|\mathbf{w}\| \leq \delta_1} \|D\mathbf{h}(\mathbf{w})\| \\ &= \|\mathbf{x} - \mathbf{y}\| \sup_{\|\mathbf{w}\| \leq \delta_1} \|D\mathbf{f}(\mathbf{w}) - I\| \leq \|\mathbf{x} - \mathbf{y}\|/2\end{aligned}$$

for all $\|\mathbf{x}\|$, $\|\mathbf{y}\| \leq \delta_1$. Setting $\rho = \delta_1/2$, we see, by Lemma 13.2, that, if $\|\mathbf{y}\| \leq \rho$, there exists one and only one solution of the equation

$$\bigstar \qquad \mathbf{f}(\mathbf{x}) = \mathbf{y}$$

with $\|\mathbf{x}\| < \delta_1$. For the rest of the proof we denote this solution by $\mathbf{g}(\mathbf{y})$.

To discuss the behaviour of $\mathbf{g}$ near $\mathbf{0}$, we echo the discussion of the first paragraph. Let $\eta > 0$ be given. Since $D\mathbf{f}$ is continuous at $\mathbf{0}$ and $D\mathbf{f} = I$, we can find a $\delta(\eta) > 0$ such that $\delta_1 > \delta(\eta) > 0$ and

$$\|D\mathbf{f}(\mathbf{w}) - I\| < \eta$$

for $\|\mathbf{w}\| \leq \delta(\eta)$. By exactly the same reasoning as in the first paragraph,

$$\|\mathbf{f}(\mathbf{x}) - (\mathbf{f}(\mathbf{y}) - (\mathbf{x} - \mathbf{y}))\| \leq \eta\|\mathbf{x} - \mathbf{y}\|$$

for all $\|\mathbf{x}\|, \|\mathbf{y}\| \leq \delta(\eta)$. The last sentence of Lemma 13.2 now tells us that, if $\|\mathbf{y}\| \leq (1 - \eta)\delta(\eta)$, the unique solution of the equation

$$\bigstar \qquad \mathbf{f}(\mathbf{x}) = \mathbf{y}$$

with $\|\mathbf{x}\| < \delta(\eta)$, which we have already agreed at the end of the previous paragraph to denote by $\mathbf{g}(\mathbf{y})$, satisfies

$$\|\mathbf{g}(\mathbf{y}) - \mathbf{y}\| \leq \eta(1 - \eta)^{-1}\|\mathbf{y}\|.$$

In less roundabout language,

$$\|\mathbf{g}(\mathbf{y}) - \mathbf{y}\| \leq \eta(1 - \eta)^{-1}\|\mathbf{y}\|$$

whenever $\|\mathbf{y}\| \leq (1 - \eta)\delta(\eta)$. Since $\mathbf{f}(\mathbf{0}) = \mathbf{0}$, we have $\mathbf{g}(\mathbf{0}) = \mathbf{0}$ and

$$\mathbf{g}(\mathbf{y}) - \mathbf{g}(\mathbf{0}) = I\mathbf{y} + \epsilon(\mathbf{y})\|\mathbf{y}\|$$

with $\|\epsilon(\mathbf{y})\| \to 0$ as $\|\mathbf{y}\| \to 0$. Thus $\mathbf{g}$ is differentiable at $\mathbf{0}$ with derivative I. ■

Exercise 13.5. *In the second paragraph of the proof just given it says 'By exactly the same reasoning as in the first paragraph'. Fill in the details.*

Exercise 13.6. *The only point of proving Lemma 13.2 was to expose the inner workings of the proof of Lemma 13.4. Prove Lemma 13.4 directly. (The proof is essentially the same, but combining the proofs of the two lemma is more economical.)*

The next exercise shows that simply knowing that $\mathbf{f}$ is differentiable at $\mathbf{0}$ with derivative I is not enough to give the existence of a local inverse.

Exercise 13.7. *(i) Let $f : \mathbb{R} \to \mathbb{R}$ be given by $f(x) = x + x^2\sin(1/x)$ for $x \neq 0$, $f(0) = 0$. Show, by using the definition of differentiability, or otherwise, that f is differentiable at 0 with $f'(0) = 1$. By considering the maxima and minima of f, or otherwise, show that there exists a sequence of non-zero $y_n \to 0$ such that the equation $f(x) = y_n$ has more than one solution.*

(ii) Let $f : \mathbb{R} \to \mathbb{R}$ be given by

$$
\begin{aligned}
f(x) &= 1 && \text{for } x > 1,\\
f(x) &= 1/n && \text{for } 1/n \geq x > 1/(n+1),\ n \text{ a strictly positive integer},\\
f(0) &= 0 && \\
f(x) &= -f(-x) && \text{for } x < 0.
\end{aligned}
$$

Show that f is differentiable at 0 with $f'(0) = 1$ but that there exists a sequence of non-zero $y_n \to 0$ such that the equation $f(x) = y_n$ has no solution.

It might be thought that Lemma 13.4 refers to a very special situation, but we can now apply another

Slogan: Anything which works for the identity will work, in a suitable form, for invertible elements. Moreover, the general case can be obtained from the special case of the identity.

Lemma 13.8. *Consider a function $\mathbf{f} : \mathbb{R}^m \to \mathbb{R}^m$ such that $\mathbf{f}(\mathbf{0}) = \mathbf{0}$ and there exists a $\delta_0 > 0$ such that $\mathbf{f}$ is differentiable in the open ball $B(\mathbf{0}, \delta_0)$. If $D\mathbf{f}$ is continuous at $\mathbf{0}$ and $D\mathbf{f}(\mathbf{0})$ is invertible, then we can find a δ_1 with $\delta_0 \geq \delta_1 > 0$ and a $\rho > 0$ such that, if $\|\mathbf{y}\| \leq \rho$, there exists one and only one solution of the equation*

$$\bigstar \qquad \mathbf{f}(\mathbf{x}) = \mathbf{y}$$

with $\|\mathbf{x}\| < \delta_1$. Further, if we denote this solution by $\mathbf{g}(\mathbf{y})$, the function $\mathbf{g}$ is differentiable at $\mathbf{0}$ with $D\mathbf{g}(\mathbf{0}) = D\mathbf{f}(\mathbf{0})^{-1}$.

Proof. Set $\alpha = D\mathbf{f}(\mathbf{0})$ and

$$\mathbf{F}(\mathbf{x}) = \alpha^{-1}\mathbf{f}(\mathbf{x}).$$

The proof now runs along totally predictable and easy lines.

By using standard chain rules (or simple direct calculations), we see that the function $\mathbf{F} : \mathbb{R}^m \to \mathbb{R}^m$ satisfies $\mathbf{F}(\mathbf{0}) = \mathbf{0}$ and $\mathbf{F}$ is differentiable in the open ball $B(\mathbf{0}, \delta_0)$ with derivative $D\mathbf{F}$ given by

$$D\mathbf{F}(\mathbf{x}) = \alpha^{-1}D\mathbf{f}(\mathbf{x}).$$

Thus $D\mathbf{F}$ is continuous at $\mathbf{0}$ and $D\mathbf{F}(\mathbf{0}) = I$. It follows, by Lemma 13.4, that there exist a δ_1 with $\delta_0 \geq \delta_1 > 0$ and a $\tilde{\rho} > 0$ such that if $\|\tilde{\mathbf{y}}\| \leq \delta_1$, there exists one and only one solution of the equation

$$\bigstar\bigstar \qquad \mathbf{F}(\mathbf{x}) = \tilde{\mathbf{y}}$$

with $\|\mathbf{x}\| < \tilde{\rho}$. Further, if we denote this solution by $\mathbf{G}(\tilde{\mathbf{y}})$, the function $\mathbf{G}$ is differentiable at $\mathbf{0}$ with $D\mathbf{G}(\mathbf{0}) = I$.

Equation ★★ can be rewritten as

$$\alpha^{-1}\mathbf{f}(\mathbf{x}) = \tilde{\mathbf{y}}.$$

Taking $\mathbf{y} = \alpha\tilde{\mathbf{y}}$, and writing

$$\mathbf{g}(\mathbf{t}) = \mathbf{G}(\alpha^{-1}\mathbf{t}),$$

we can rewrite the conclusion of the last paragraph as follows. If $\|\alpha^{-1}\mathbf{y}\| \leq \tilde{\rho}$ there exists one and only one solution of the equation

★ $$\mathbf{f}(\mathbf{x}) = \mathbf{y}$$

with $\|\mathbf{x}\| < \delta_1$. Further, if we denote this solution by $\mathbf{g}(\mathbf{y})$, the function $\mathbf{g}$ is differentiable at $\mathbf{0}$ with $D\mathbf{g}(\mathbf{0}) = \alpha^{-1} = D\mathbf{f}(\mathbf{0})^{-1}$.

If we set $\rho = \|\alpha^{-1}\|^{-1}\tilde{\rho}$, then, whenever $\|\mathbf{y}\| \leq \rho$, we have

$$\|\alpha^{-1}\mathbf{y}\| \leq \|\alpha^{-1}\|\|\mathbf{y}\| \leq \tilde{\rho}$$

and we have obtained all the conclusions of the lemma. ■

A simple translation argument transforms Lemma 13.8 to an apparently more general form.

Lemma 13.9. *Consider a function $\mathbf{f} : \mathbb{R}^m \to \mathbb{R}^m$ and a $\mathbf{w} \in \mathbb{R}^m$ such that there exists a $\delta_0 > 0$ such that $\mathbf{f}$ is differentiable in the open ball $B(\mathbf{w}, \delta_0)$. If $D\mathbf{f}$ is continuous at $\mathbf{w}$ and $D\mathbf{f}(\mathbf{w})$ is invertible, then we can find a δ_1 with $\delta_0 \geq \delta_1 > 0$ and a $\rho > 0$ such that, if $\|\mathbf{y} - \mathbf{f}(\mathbf{w})\| \leq \rho$, there exists one and only one solution of the equation*

★ $$\mathbf{f}(\mathbf{x}) = \mathbf{y}$$

with $\|\mathbf{x} - \mathbf{w}\| < \delta_1$. Further, if we denote this solution by $\mathbf{g}(\mathbf{y})$, the function $\mathbf{g}$ is differentiable at $\mathbf{f}(\mathbf{w})$ with $D\mathbf{g}(\mathbf{f}(\mathbf{w})) = D\mathbf{f}(\mathbf{w})^{-1}$.

Exercise 13.10. *Prove Lemma 13.9 from Lemma 13.8.*

We have now done most of the hard work of this section, but Lemma 13.9 can be made much more useful if we strengthen its hypotheses.

Slogan: Analysis is done not at points but on open sets.

Our first two results are preliminary.

Lemma 13.11. *(i) We write ι for the identity map $\iota : \mathbb{R}^n \to \mathbb{R}^n$. If $\alpha : \mathbb{R}^n \to \mathbb{R}^n$ is linear and $\|\iota - \alpha\| < 1$, then α is invertible.*

(ii) Suppose that $\beta, \gamma : \mathbb{R}^n \to \mathbb{R}^n$ are linear and β is invertible. If $\|\beta - \gamma\| < \|\beta^{-1}\|^{-1}$, then γ is invertible.

(iii) Consider a function $\mathbf{f} : \mathbb{R}^m \to \mathbb{R}^m$ which is differentiable on an open set U. If $D\mathbf{f}$ is continuous and invertible at a point $\mathbf{u}_0 \in U$, then we can find an open set $B \subseteq U$ with $\mathbf{u}_0 \in B$ such that $D\mathbf{f}$ is invertible at every point of B.

Proof. (i) Since we are working with finite-dimensional vector spaces, α is invertible if and only if it is injective, and α is injective if and only if $\ker\alpha = \{\mathbf{0}\}$. If $\mathbf{x} \neq 0$, then

$$\|\alpha\mathbf{x}\| = \|\mathbf{x} - (\iota - \alpha)\mathbf{x}\| \geq \|\mathbf{x}\| - \|(\iota - \alpha)\mathbf{x}\| \geq \|\mathbf{x}\| - \|\iota - \alpha\|\|\mathbf{x}\| > 0,$$

and so $\alpha\mathbf{x} \neq \mathbf{0}$. The result follows.

(ii) Observe that

$$\|\iota - \beta^{-1}\gamma\| = \|\beta^{-1}(\beta - \gamma)\| \leq \|\beta^{-1}\|\|(\beta - \gamma)\| < 1,$$

and so $\beta^{-1}\gamma$ is invertible. Write $\theta = (\beta^{-1}\gamma)^{-1}$ and observe that $(\theta\beta^{-1})\gamma = \theta(\beta^{-1}\gamma) = \iota$. Thus, since we are dealing with finite-dimensional spaces, γ is invertible with inverse $\theta\beta^{-1}$.

(iii) By the continuity of $D\mathbf{f}$ we can find an open set $B \subseteq U$ with $\mathbf{u}_0 \in B$ such that

$$\|D\mathbf{f}(\mathbf{u}_0) - D\mathbf{f}(\mathbf{u})\| < \|D\mathbf{f}(\mathbf{u}_0)^{-1}\|^{-1}.$$

The stated result follows from part (ii). ■

(Both Lemma 13.11 and its proof may be classified as 'quick and dirty'. A more leisurely approach, which reveals much more of what is actually going on, is given in Exercise K.310.)

Lemma 13.12. *Consider a function $\mathbf{f} : \mathbb{R}^m \to \mathbb{R}^m$ which is differentiable on an open set U. If $D\mathbf{f}$ is continuous and invertible at every point of U, then $\mathbf{f}(U)$ is open.*

Proof. Suppose $\mathbf{w} \in U$. Since U is open, we can find a $\delta_0 > 0$ such that the open ball $B(\mathbf{w}, \delta_0)$ is a subset of U. By hypothesis, $D\mathbf{f}$ is continuous at $\mathbf{w}$ and $D\mathbf{f}(\mathbf{w})$ is invertible. Thus, by Lemma 13.9, we can find a δ_1 with $\delta_0 \geq \delta_1 > 0$ and a $\rho > 0$ such that, if $\|\mathbf{y} - \mathbf{f}(\mathbf{w})\| \leq \rho$, there exists a solution of the equation

$$\bigstar \qquad \mathbf{f}(\mathbf{x}) = \mathbf{y}$$

with $\|\mathbf{x} - \mathbf{w}\| < \delta_1$. It follows that

$$B(\mathbf{f}(\mathbf{w}), \rho) \subseteq \mathbf{f}(B(\mathbf{w}, \delta_0)) \subseteq \mathbf{f}(U).$$

We have shown that $\mathbf{f}(U)$ is open ■

Theorem 13.13. (Inverse function theorem.) *Consider a function $\mathbf{f} : \mathbb{R}^m \to \mathbb{R}^m$ which is differentiable on an open set U. Suppose, further, that $D\mathbf{f}$ is continuous at every point of U, that $\mathbf{w} \in U$, and that $D\mathbf{f}(\mathbf{w})$ is invertible. Then we can find an open set $B \subseteq U$ with $\mathbf{w} \in B$ and an open set V such that*

(i) $\mathbf{f}|_B : B \to V$ is bijective, and

(ii) $\mathbf{f}|_B^{-1} : V \to B$ *is differentiable with*

$$D\mathbf{f}|_B^{-1}(\mathbf{f}(\mathbf{u})) = (D\mathbf{f}(\mathbf{u}))^{-1}$$

for all $\mathbf{u} \in B$.

Proof. Suppose $\mathbf{w} \in U$. By Lemma 13.11, we can find a $\delta_0 > 0$ such that the open ball $B(\mathbf{w}, \delta_0)$ is a subset of U and $D\mathbf{f}(\mathbf{u})$ is invertible at every point $\mathbf{u} \in B(\mathbf{w}, \delta_0)$.

We now use the same argument which we used in Lemma 13.12. We know that $D\mathbf{f}$ is continuous at $\mathbf{w}$ and $D\mathbf{f}(\mathbf{w})$ is invertible. Thus, by Lemma 13.9, we can find a δ_1 with $\delta_0 \geq \delta_1 > 0$ and a $\rho > 0$ such that if $\|\mathbf{y} - \mathbf{f}(\mathbf{w})\| \leq \rho$ there exists one and only one solution of the equation

$$\bigstar \qquad \mathbf{f}(\mathbf{x}) = \mathbf{y}$$

with $\|\mathbf{x} - \mathbf{w}\| < \delta_1$. Set $B = B(\mathbf{w}, \delta_1)$ and apply Lemma 13.12 and Lemma 13.9. ∎

A slight strengthening of Theorem 13.13 is given in Exercise K.317. The following cluster of easy exercises is intended to illuminate various aspects of the inverse function theorem.

Exercise 13.14. *Suppose* U *and* V *are open subsets of* $\mathbb{R}^m$ *and* $\mathbf{f} : U \to V$, $\mathbf{g} : V \to U$ *are such that* $\mathbf{g} \circ \mathbf{f}$ *is the identity map on* U. *Show that if* $\mathbf{f}$ *is differentiable at* $\mathbf{u} \in U$ *and* $\mathbf{g}$ *is differentiable at* $\mathbf{f}(\mathbf{u})$, *then* $(D\mathbf{f})(\mathbf{u})$ *and* $(D\mathbf{g})(\mathbf{f}(\mathbf{u}))$ *are invertible and*

$$(D\mathbf{g})(\mathbf{f}(\mathbf{u})) = \big((D\mathbf{f})(\mathbf{u})\big)^{-1}.$$

Exercise 13.15. *(A traditional examination question.) Let* $f : \mathbb{R} \to \mathbb{R}$ *be given by* $f(x) = x^3$. *Show that* f *is bijective but* $f'(0) = 0$ *(so the derivative of* f *at* 0 *is not invertible).*

Exercise 13.16. *Consider the open interval* $(-4, 4)$. *Find* $f((-4, 4))$ *for the functions* $f : \mathbb{R} \to \mathbb{R}$ *given by*

(i) $f(x) = \sin x$,

(ii) $f(x) = \sin 10^{-2}x$,

(iii) $f(x) = x^2$,

(iv) $f(x) = x^3 - x$.

In each case comment briefly on the relation of your result to Lemma 13.12.

Exercise 13.17. *Let*

$$U = \{(x, y) \in \mathbb{R}^2 : 1 < x^2 + y^2 < 2\}$$

and define $\mathbf{f} : U \to \mathbb{R}$ *by*

$$\mathbf{f}(x, y) = \left(\frac{x^2 - y^2}{(x^2 + y^2)^{1/2}}, \frac{2xy}{(x^2 + y^2)^{1/2}} \right).$$

(i) Show that U *is open.*

(ii) Show that $\mathbf{f}$ *is differentiable on* U *and that* $D\mathbf{f}$ *is continuous and invertible at every point of* U.

(iii) By using polar coordinates, discover why $\mathbf{f}$ *is defined as it is. Show that* $\mathbf{f}(U) = U$ *but* $\mathbf{f}$ *is not injective.*

Once the ideas behind the proof of the inverse function theorem are understood, it can be condensed into a very short argument. In Dieudonné's account the essential content of both this section and the next is stated and proved in more general form in about two pages ([**14**] Chapter X, Theorem 10.2.1). We leave it to the reader to undertake this condensation[1]. The usual approach to the inverse function theorem uses the contraction mapping theorem in a rather more subtle way than the approach adopted here. I outline the alternative approach in Appendix F.

13.2. The implicit function theorem ♡

The contents of this section really belong to a first course in differential geometry. However, there is an old mathematical tradition called 'pass the parcel' by which lecturers assume that all the hard but necessary preliminary work has been done 'in a previous course'[2]. In accordance with this tradition, lecturers in differential geometry frequently leave the proof of the implicit function theorem in the hands of an, often mythical, earlier lecturer in analysis. Even when the earlier lecturer actually exists, this has the effect of first exposing the students to a proof of a result whose use they do not understand, and then making them use a result whose proof they have forgotten.

My advice to the reader, as often in this book, is not to take this section too seriously. *It is far more important to make sure that you are confident of the meaning and proof of the inverse function theorem than that you worry about the details of this section.*

---

[1]There is a Cambridge story about an eminent algebraic geometer who presented his subject entirely without diagrams. However, from time to time, when things got difficult, he would hide part of the blackboard, engage in some rapid but hidden chalk work, rub out the result and continue.

[2]Particular topics handled in this way include determinants, the Jordan normal form, uniqueness of prime factorisation and various important inequalities.

Consider the function $h : \mathbb{R}^2 \to \mathbb{R}$ given by $h(x, y) = x^2 + y^2$. We know that the contour (or level) line

$$x^2 + y^2 = 1$$

can be represented by a graph

$$x = (1 - y^2)^{1/2}$$

close to the point $(x, y) = (0, 1)$. We also know that this representation fails close to the point $(x, y) = (1, 0)$, but that near that point we can use the representation

$$y = (1 - x^2)^{1/2}.$$

Leaping rapidly to a conclusion, we obtain the following slogan.

Slogan: If $f : \mathbb{R}^2 \to \mathbb{R}$ behaves well in a neighbourhood of a point (x_0, y_0), then at least one of the following two statements must be true.

(a) There exist a $\delta > 0$ and a well behaved bijective function $g : (-\delta, \delta) \to \mathbb{R}$ such that $g(0) = y_0$ and $f(x + x_0, g(x)) = f(x_0, y_0)$ for all $x \in (-\delta, \delta)$.

(b) There exist a $\delta > 0$ and a well behaved bijective function $g : (-\delta, \delta) \to \mathbb{R}$ such that $g(0) = x_0$ and $f(g(y), y + y_0) = f(x_0, y_0)$ for all $y \in (-\delta, \delta)$.

Exercise 13.18. *Consider the contour $x^3 - y^2 = 0$. Show that we can find $g_1 : \mathbb{R} \to \mathbb{R}$ and $g_2 : \mathbb{R} \to \mathbb{R}$ such that*

$$x^3 - g_1(x)^2 = 0 \qquad \text{for all } x \in \mathbb{R},$$
$$g_2(y)^3 - y^2 = 0 \qquad \text{for all } y \in \mathbb{R},$$

but g_1 is differentiable everywhere and g_2 is not. Explain in simple terms why this is the case.

One way of looking at our slogan is to consider a walker on a hill whose height is given by $f(x, y)$ at a point (x, y) on a map. The walker seeks to walk along a path of constant height. She is clearly going to have problems if she starts at a strict maximum since a step in any direction takes her downward. A similar difficulty occurs at a strict minimum. A different problem occurs at a saddle point (see Figure 13.1). It appears from the picture that, if f is well behaved, there are not one but two paths of constant height passing through a saddle point, and this will create substantial difficulties[3]. However, these are the only points which present problems.

[3]Remember Buridan's ass, which, placed between two equally attractive bundles of hay, starved to death because it was unable to find a reason for starting on one bundle rather than the other.

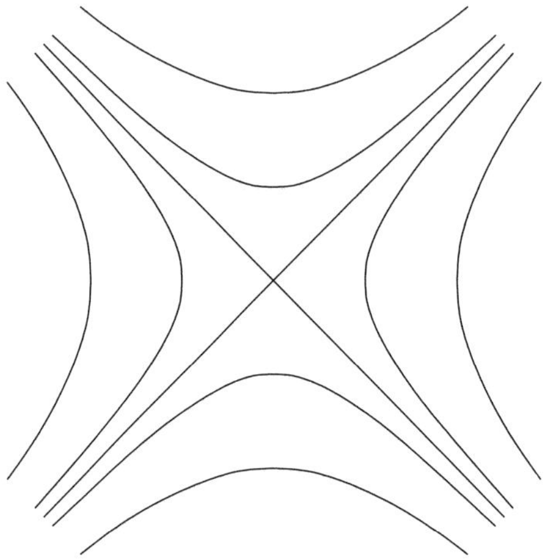

Figure 13.1. Problems at a saddle

Another way to look at our slogan is to treat it as a problem in the calculus (our arguments will, however, continue to be informal). Suppose that

$$f(x, g(x)) = f(x_0, y_0).$$

Assuming that everything is well behaved, we can differentiate with respect to x, obtaining

$$f_{,1}(x, g(x)) + g'(x) f_{,2}(x, g(x)) = 0,$$

and so, provided that $f_{,2}(x, y) \neq 0$,

$$g'(x) = -\frac{f_{,1}(x, g(x))}{f_{,2}(x, g(x))}.$$

Thus, provided that f is sufficiently well behaved and

$$f_{,2}(x_0, y_0) \neq 0,$$

our earlier work on differential equations tells us that there exists a local solution for g.

The two previous paragraphs tend to confirm the truth of our slogan and show that there exist *local* contour lines in the neighbourhood of any point (x_0, y_0) where at least one of $f_{,1}(x_0, y_0)$ and $f_{,2}(x_0, y_0)$ does not vanish. We shall not seek to establish the global existence of contour lines[4]. Instead we seek to extend the ideas of our slogan to higher dimensions.

We argue informally, assuming good behaviour as required. Consider a function $\mathbf{f} : \mathbb{R}^m \times \mathbb{R}^n \to \mathbb{R}^n$. Without loss of generality, we may suppose that $\mathbf{f}(\mathbf{0}, \mathbf{0}) = \mathbf{0}$. An appropriate generalisation of the questions considered above is to ask about solutions to the equation

$$\mathbf{f}(\mathbf{x}, \mathbf{g}_{\mathbf{h}}(\mathbf{x})) = \mathbf{h}, \tag{1}$$

where $\mathbf{h}$ is fixed. (Note that $\mathbf{x} \in \mathbb{R}^m$, $\mathbf{g}_{\mathbf{h}}(\mathbf{x}) \in \mathbb{R}^n$ and $\mathbf{h} \in \mathbb{R}^n$. Since everything is local, we suppose $\|\mathbf{h}\|$ is small and we only consider $(\mathbf{x}, \mathbf{g}_{\mathbf{h}}(\mathbf{x}))$ close to $(\mathbf{0}, \mathbf{0})$.) Before proceeding to the next paragraph the reader should convince herself that the question we have asked is a natural one.

The key step in resolving this problem is to rewrite it. Define $\tilde{\mathbf{f}}$, $\tilde{\mathbf{g}} : \mathbb{R}^m \times \mathbb{R}^n \to \mathbb{R}^m \times \mathbb{R}^n$ by

$$\tilde{\mathbf{f}}(\mathbf{x}, \mathbf{y}) = (\mathbf{x}, \mathbf{f}(\mathbf{x}, \mathbf{y}))$$
$$\tilde{\mathbf{g}}(\mathbf{x}, \mathbf{y}), = (\mathbf{x}, \mathbf{g}_{\mathbf{y}}(\mathbf{x})).$$

If equation (1) holds, then

$$\tilde{\mathbf{f}}(\tilde{\mathbf{g}}(\mathbf{x}, \mathbf{h})) = \tilde{\mathbf{f}}(\mathbf{x}, \mathbf{g}_{\mathbf{h}}(\mathbf{x})) = \big(\mathbf{x}, \mathbf{f}(\mathbf{g}_{\mathbf{h}}(\mathbf{x}), \mathbf{x})\big) = (\mathbf{x}, \mathbf{h}),$$

and so

$$\tilde{\mathbf{f}}(\tilde{\mathbf{g}}(\mathbf{x}, \mathbf{h})) = (\mathbf{x}, \mathbf{h}). \tag{2}$$

Conversely, if equation (2) holds, then equation (1) follows.

To solve equation (2) we need to use the inverse mapping results of the previous section, and, to use those results, we need to know if $D\tilde{\mathbf{f}}(\mathbf{x}, \mathbf{y})$ is invertible at a given point $(\mathbf{x}, \mathbf{y})$. Now

$$D\tilde{\mathbf{f}}(\mathbf{x}, \mathbf{y})(\mathbf{u}, \mathbf{v}) = \big(\mathbf{u}, D\mathbf{f}(\mathbf{x}, \mathbf{y})(\mathbf{u}, \mathbf{v})\big),$$

and so $D\tilde{\mathbf{f}}(\mathbf{x}, \mathbf{y})$ is invertible if and only if the linear map $\alpha : \mathbb{R}^n \to \mathbb{R}^n$, given by

$$\alpha\mathbf{v} = D\mathbf{f}(\mathbf{x}, \mathbf{y})(\mathbf{0}, \mathbf{v}),$$

is invertible. (We present two proofs of this last statement in the next two exercises. Both exercises take some time to state and hardly any time to do.)

Exercise 13.19. *In this exercise we use* **column** *vectors. Thus we write*

$$\tilde{\mathbf{f}}\begin{pmatrix}\mathbf{x}\\ \mathbf{y}\end{pmatrix} = \begin{pmatrix}\mathbf{x}\\ \mathbf{f}\binom{\mathbf{x}}{\mathbf{y}}\end{pmatrix}, \quad \tilde{\mathbf{g}}\begin{pmatrix}\mathbf{x}\\ \mathbf{y}\end{pmatrix} = \begin{pmatrix}\mathbf{x}\\ \mathbf{g}_{\mathbf{y}}(\mathbf{x})\end{pmatrix}.$$

[4]Obviously the kind of ideas considered in Section 12.3 will play an important role in such an investigation. One obvious problem is that, when we look at a contour in one location, there is no way of telling if it will not pass through a saddle point at another.

Suppose that $D\mathbf{f}\begin{pmatrix}\mathbf{x}\\ \mathbf{y}\end{pmatrix}$ *has matrix* C *with respect to the standard basis. Explain why* C *is an* $n \times (n+m)$ *matrix which we can therefore write as*

$$C = (B\ A),$$

with A *an* $n \times n$ *matrix and* B *an* $n \times m$ *matrix.*

Show that $D\tilde{\mathbf{f}}\begin{pmatrix}\mathbf{x}\\ \mathbf{y}\end{pmatrix}$ *has matrix*

$$E = \begin{pmatrix} I & 0 \\ B & A \end{pmatrix},$$

where 0 *is the* $n \times m$ *matrix consisting of entirely of zeros and* I *is the* $m \times m$ *identity matrix. By considering* $\det E$*, or otherwise, show that* E *is invertible if and only if* A *is. Show that* A *is the matrix of the linear map* $\alpha : \mathbb{R}^n \to \mathbb{R}^n$ *given by*

$$\alpha\mathbf{v} = D\mathbf{f}\begin{pmatrix}\mathbf{x}\\ \mathbf{y}\end{pmatrix}\begin{pmatrix}\mathbf{0}\\ \mathbf{v}\end{pmatrix}.$$

Exercise 13.20. *Let* W *be the direct sum of two subspaces* U *and* V*. If* $\gamma : W \to V$ *is a linear map, show that the map* $\alpha : V \to V$*, given by* $\alpha\mathbf{v} = \gamma\mathbf{v}$ *for all* $\mathbf{v} \in V$*, is linear.*

Now define $\tilde{\gamma} : W \to W$ *by* $\tilde{\gamma}(\mathbf{u}+\mathbf{v}) = \gamma(\mathbf{u}+\mathbf{v}) + \mathbf{u}$ *for* $\mathbf{u} \in U$*,* $\mathbf{v} \in V$*. Show that* $\tilde{\gamma}$ *is a well defined linear map, that* $\ker(\tilde{\gamma}) = \ker(\alpha)$ *and that* $\tilde{\gamma}(W) = U + \alpha(V)$*. Conclude that* $\tilde{\gamma}$ *is invertible if and only if* α *is.*

We can now pull the strands together.

Theorem 13.21. (Implicit function theorem.) *Consider a function* $\mathbf{f} : \mathbb{R}^m \times \mathbb{R}^n \to \mathbb{R}^n$ *which is differentiable on an open set* U*. Suppose further that* $D\mathbf{f}$ *is continuous at every point of* U*, that* $(\mathbf{x}_0, \mathbf{y}_0) \in U$ *and that the linear map* $\alpha : \mathbb{R}^n \to \mathbb{R}^n$ *given by*

$$\alpha\mathbf{t} = D\mathbf{f}(\mathbf{x}_0, \mathbf{y}_0)(\mathbf{0}, \mathbf{t})$$

is invertible. Then we can find an open set B_1 *in* $\mathbb{R}^m$*, with* $\mathbf{x} \in B_1$*, and an open set* B_2 *in* $\mathbb{R}^m$*, with* $\mathbf{f}(\mathbf{x}_0, \mathbf{y}_0) \in B_2$*, such that, if* $\mathbf{z} \in B_2$*, there exists a differentiable map* $\mathbf{g}_{\mathbf{z}} : B_1 \to \mathbb{R}^m$ *with* $(\mathbf{x}, \mathbf{g}_{\mathbf{z}}(\mathbf{x})) \in U$ *and*

$$\mathbf{f}(\mathbf{x}, \mathbf{g}_{\mathbf{z}}(\mathbf{x})) = \mathbf{z}$$

for all $\mathbf{x} \in B_1$*.*

(We give a slight improvement on this result in Theorem 13.26, but Theorem 13.21 contains the essential result.)

Proof. Define $\tilde{\mathbf{f}} : \mathbb{R}^m \times \mathbb{R}^n \to \mathbb{R}^m \times \mathbb{R}^n$ by

$$\tilde{\mathbf{f}}(\mathbf{x}, \mathbf{y}) = (\mathbf{x}, \mathbf{f}(\mathbf{x}, \mathbf{y})).$$

If $(\mathbf{x}, \mathbf{y}) \in U$, then $D\tilde{\mathbf{f}}(\mathbf{x}, \mathbf{y})$ exists and

$$D\tilde{\mathbf{f}}(\mathbf{x}, \mathbf{y})(\mathbf{s}, \mathbf{t}) = \big(\mathbf{s}, D\mathbf{f}(\mathbf{x}, \mathbf{y})(\mathbf{s}, \mathbf{t})\big).$$

In particular, since α is invertible, $D\tilde{\mathbf{f}}(\mathbf{x}_0, \mathbf{y}_0)$ is invertible. It follows, by the inverse function theorem (Theorem 13.13), that we can find an open set $B \subseteq U$ with $(\mathbf{x}_0, \mathbf{y}_0) \in B$ and an open set V such that

(i) $\tilde{\mathbf{f}}|_B : B \to V$ is bijective, and

(ii) $\tilde{\mathbf{f}}|_B^{-1} : V \to B$ is differentiable.

Let us define $\mathbf{G} : V \to \mathbb{R}^m$ and $\mathbf{g} : V \to \mathbb{R}^n$ by

$$\tilde{\mathbf{f}}|_B^{-1}(\mathbf{v}) = (\mathbf{G}(\mathbf{v}), \mathbf{g}(\mathbf{v}))$$

for all $\mathbf{v} \in V$. We observe that $\mathbf{g}$ is everywhere differentiable on V. By definition,

$$\begin{aligned}(\mathbf{x}, \mathbf{z}) &= \tilde{\mathbf{f}}|_B(\tilde{\mathbf{f}}|_B^{-1}(\mathbf{x}, \mathbf{z})) \\ &= \tilde{\mathbf{f}}|_B(\mathbf{G}(\mathbf{x}, \mathbf{z}), \mathbf{g}(\mathbf{x}, \mathbf{z})) \\ &= (\mathbf{G}(\mathbf{x}, \mathbf{z}), \mathbf{f}(\mathbf{G}(\mathbf{x}, \mathbf{z}), \mathbf{g}(\mathbf{x}, \mathbf{z}))),\end{aligned}$$

and so

$$\mathbf{G}(\mathbf{x}, \mathbf{z}) = \mathbf{x}$$

and

$$\mathbf{z} = \mathbf{f}(\mathbf{x}, \mathbf{g}(\mathbf{x}, \mathbf{z}))$$

for all $(\mathbf{x}, \mathbf{z}) \in V$.

We know that V is open and $(\mathbf{x}_0, \mathbf{f}(\mathbf{x}_0, \mathbf{y}_0)) \in V$, so we can find an open set B_1 in $\mathbb{R}^m$, with $\mathbf{x}_0 \in B_1$, and an open set B_2 in $\mathbb{R}^n$, with $\mathbf{f}(\mathbf{x}_0, \mathbf{y}_0) \in B_2$, such that $B_1 \times B_2 \subseteq V$. Setting

$$\mathbf{g}_{\mathbf{z}}(\mathbf{x}) = \mathbf{g}(\mathbf{x}, \mathbf{z})$$

for all $\mathbf{x} \in B_1$ and $\mathbf{z} \in B_2$, we have the required result. ■

Remark 1: We obtained the implicit function theorem from the inverse function theorem by introducing new functions $\tilde{\mathbf{f}}$ and $\tilde{\mathbf{g}}$. There is, however, no reason why we should not obtain the implicit function theorem directly by following the same kind of method as we used to prove the inverse function theorem. Here is the analogue of Lemma 13.2

Lemma 13.22. *Consider a function* $\mathbf{f} : \mathbb{R}^{m+n} \to \mathbb{R}^n$ *such that* $\mathbf{f}(\mathbf{0}, \mathbf{0}) = \mathbf{0}$. *Suppose that there exist a* $\delta > 0$ *and an* η *with* $1 > \eta > 0$ *such that*

$$\|(\mathbf{f}(\mathbf{x}, \mathbf{t}) - \mathbf{f}(\mathbf{x}, \mathbf{s})) - (\mathbf{s} - \mathbf{t})\| \leq \eta \|(\mathbf{s} - \mathbf{t})\|$$

for all $\|\mathbf{x}\|$, $\|\mathbf{s}\|$, $\|\mathbf{t}\| \leq \delta$ $[\mathbf{x} \in \mathbb{R}^m,\ \mathbf{s}, \mathbf{t} \in \mathbb{R}^n]$. *Then, if* $\|\mathbf{h}\| \leq (1-\eta)\delta$, *there exists one and only one solution of the equation*

$$\bigstar \qquad \mathbf{f}(\mathbf{x}, \mathbf{u}) = \mathbf{h}$$

with $\|\mathbf{u}\| < \delta$. *Further, if we denote this solution by* $\mathbf{g_h}(\mathbf{x})$, *we have*

$$\|\mathbf{g_h}(\mathbf{x}) - \mathbf{h}\| \leq \eta(1-\eta)^{-1}\|\mathbf{h}\|.$$

Exercise 13.23. *Prove Lemma 13.22. Sketch, giving as much or as little detail as you wish, the steps from Lemma 13.22 to Theorem 13.21. You should convince yourself that the inverse function theorem and the implicit function theorem are fingers of the same hand.*

Remark 2: In the introduction to this section we obtained contours as the solutions of an ordinary differential equation of the form

$$g'(x) = -\frac{f_{,1}(x, g(x))}{f_{,2}(x, g(x))}.$$

The situation for the general implicit function theorem is genuinely more complicated. Suppose, for example that we have a function $\mathbf{f} : \mathbb{R}^4 \to \mathbb{R}^2$ and we wish to find $(u, v) : \mathbb{R}^2 \to \mathbb{R}^2$ so that (at least locally)

$$\mathbf{f}(x, y, u(x, y), v(x, y)) = \mathbf{h},$$

that is,

$$f_1(x, y, u(x, y), v(x, y)) = h_1,$$
$$f_2(x, y, u(x, y), v(x, y)) = h_2.$$

On differentiating, we obtain

$$f_{1,1}(x, y, u(x, y), v(x, y)) + f_{1,3}(x, y, u(x, y), v(x, y))\frac{\partial u}{\partial x} + f_{1,4}(x, y, u(x, y), v(x, y))\frac{\partial v}{\partial x} = 0,$$
$$f_{1,1}(x, y, u(x, y), v(x, y)) + f_{1,3}(x, y, u(x, y), v(x, y))\frac{\partial u}{\partial y} + f_{1,4}(x, y, u(x, y), v(x, y))\frac{\partial v}{\partial y} = 0,$$
$$f_{2,1}(x, y, u(x, y), v(x, y)) + f_{2,3}(x, y, u(x, y), v(x, y))\frac{\partial u}{\partial x} + f_{2,4}(x, y, u(x, y), v(x, y))\frac{\partial v}{\partial x} = 0,$$
$$f_{2,1}(x, y, u(x, y), v(x, y)) + f_{2,3}(x, y, u(x, y), v(x, y))\frac{\partial u}{\partial y} + f_{2,4}(x, y, u(x, y), v(x, y))\frac{\partial v}{\partial y} = 0,$$

so we are faced with a system of simultaneous partial differential equations. (For a very slightly different view of the matter, see Exercise E.2.)

Remark 3: We proved Theorem 13.21 under the assumption that the linear map $\alpha : \mathbb{R}^n \to \mathbb{R}^n$ given by

$$\alpha\mathbf{t} = D\mathbf{f}(\mathbf{x}_0, \mathbf{y}_0)(\mathbf{0}, \mathbf{t})$$

is invertible. This condition gives unnecessary prominence to the last n coordinates. To get round this, we introduce a new definition.

Definition 13.24. *We say that a linear map $\beta : \mathbb{R}^{n+m} \to \mathbb{R}^n$ has* full rank *if β is surjective.*

The next, very easy, lemma gives some context.

Lemma 13.25. *Suppose that a linear map $\beta : \mathbb{R}^{n+m} \to \mathbb{R}^n$ has matrix B with respect to standard coordinates. (We use column vectors.) Then the following conditions are equivalent.*

(i) β has full rank.

(ii) B has n linearly independent columns.

(iii) By permuting the coordinate axes, we can ensure that the last n columns of B are linearly independent.

(iv) By permuting the coordinate axes, we can ensure that the linear map $\alpha : \mathbb{R}^n \to \mathbb{R}^n$ given by

$$\alpha\mathbf{t} = \beta(\mathbf{t}, \mathbf{0})$$

is invertible.

Proof. To see that (i) implies (ii), observe that

$$\operatorname{rank}\beta = \dim \beta\mathbb{R}^{n+m} = \dim \operatorname{span}\{\mathbf{b}_1,\, \mathbf{b}_2,\, ,\, \dots,\, \mathbf{b}_{n+m}\},$$

where $\mathbf{b}_j$ is the jth column of B. If β has full rank, then, since any spanning set contains a basis, n of the $\mathbf{b}_j$ must be linearly independent.

It is clear that (ii) implies (iii). To see that (iii) implies (iv), observe that, if α has matrix A with respect to standard coordinates after our permutation, then A is an $n \times n$ matrix whose columns are the first n columns of B and so linearly independent. Thus A is invertible, and so α is.

To see that (iv) implies (i), observe that

$$\{\mathbf{b}_1,\, \mathbf{b}_2,\, ,\, \dots,\, \mathbf{b}_{n+m}\} \supseteq \{\mathbf{a}_1,\, \mathbf{a}_2,\, ,\, \dots,\, \mathbf{a}_n\},$$

where $\mathbf{a}_k$ is the kth column of A, the matrix of α with respect to standard coordinates. Thus

$$\mathbb{R}^n \supseteq \operatorname{span}\{\mathbf{b}_1,\, \mathbf{b}_2,\, ,\, \dots,\, \mathbf{b}_{n+m}\} \supseteq \operatorname{span}\{\mathbf{a}_1,\, \mathbf{a}_2,\, ,\, \dots,\, \mathbf{a}_n\} = \mathbb{R}^n,$$

so $\beta\mathbb{R}^{n+m} = \mathbb{R}^n$ and β has full rank. ∎

The appropriate modification of Theorem 13.21 is now obvious.

Theorem 13.26. *Consider a function $\mathbf{f} : \mathbb{R}^{m+n} \to \mathbb{R}^n$ which is differentiable on an open set U. Suppose further that $D\mathbf{f}$ is continuous at every point of U, that $\mathbf{w}_0 \in U$ and that $D\mathbf{f}(\mathbf{w}_0)$ has full rank. Then we can permute the coordinate axes in such a way that, if we write $\mathbf{w}_0 = (\mathbf{x}_0, \mathbf{y}_0) \in \mathbb{R}^m \times \mathbb{R}^n$, we can find an open set B_1 in $\mathbb{R}^m$, with $\mathbf{x}_0 \in B_1$, and an open set B_2 in*

$\mathbb{R}^n$ with $\mathbf{f}(\mathbf{w}_0) \in B_2$ such that, if $\mathbf{z} \in B_2$, there exists a differentiable map $\mathbf{g}_\mathbf{z} : B_1 \to \mathbb{R}^m$ with $(\mathbf{x}, \mathbf{g}_\mathbf{z}(\mathbf{x})) \in U$ and

$$\mathbf{f}(\mathbf{x}, \mathbf{g}_\mathbf{z}(\mathbf{x})) = \mathbf{z}$$

for all $\mathbf{x} \in B_1$.

Remark 4: In Appendix C, I emphasise the importance of a large stock of examples, but I talk mainly about examples of badly behaved functions. It is also true that mathematicians find it hard to lay their hands on a sufficiently diverse collection of well behaved functions. In this book we started with the polynomials. To these we added the function $x \mapsto 1/x$ and got the rational functions $x \mapsto P(x)/Q(x)$ with P and Q polynomials. We then added power series and solutions of differential equations. (Exercise K.95 is relevant here.) The inverse function theorem gives us a new collection of interesting functions, and the implicit function theorem greatly extends this collection. The implicit function theorem is particularly important in giving us interesting examples of well behaved functions $f : \mathbb{R}^n \to \mathbb{R}$ when $n \geq 2$.

13.3. Lagrange multipliers ♡

Suppose that $t : \mathbb{R}^4 \to \mathbb{R}$ and $\mathbf{f} : \mathbb{R}^4 \to \mathbb{R}^2$ are well behaved functions, and we wish to maximise $t(\mathbf{x})$ subject to the condition $\mathbf{f}(\mathbf{x}) = \mathbf{h}$.

This will involve a fair amount of calculation, so it seems reasonable to simplify our task as much as possible. One obvious simplification is to translate so that $\mathbf{h} = (0, 0)$ and $\mathbf{f}(0, 0, 0, 0) = (0, 0)$. We now wish to investigate whether $\mathbf{x} = (0, 0, 0, 0)$ maximises $t(\mathbf{x})$ subject to the condition $\mathbf{f}(\mathbf{x}) = (0, 0)$. A more interesting simplification is to observe that, since we are assuming good behaviour, $D\mathbf{f}(\mathbf{0})$ will have full rank. We can therefore find invertible linear maps $\alpha : \mathbb{R}^4 \to \mathbb{R}^4$ and $\beta : \mathbb{R}^2 \to \mathbb{R}^2$ such that $\beta D\mathbf{f}(\mathbf{0})\alpha$ has matrix

$$J = \begin{pmatrix} 1 & 0 & 0 & 0 \\ 0 & 1 & 0 & 0 \end{pmatrix}$$

with respect to the standard basis (and using the column vector representation).

If we set

$$\mathbf{F}(\mathbf{x}) = \beta(\mathbf{f}(\alpha\mathbf{x})), \ T(\mathbf{x}) = t(\alpha\mathbf{x}),$$

then our problem becomes that of investigating whether $\mathbf{x} = (0, 0, 0, 0)$ maximises $T(\mathbf{x})$, subject to the condition $\mathbf{F}(\mathbf{x}) = (0, 0)$. By the implicit function theorem, we can find well behaved $u, v : \mathbb{R}^2 \to \mathbb{R}^2$ such that $u(0, 0) = v(0, 0) = (0, 0)$ and (at least if z and w are small)

$$\mathbf{F}(u(z, w), v(z, w), z, w) = (0, 0).$$

By the chain rule, $D\mathbf{f}(\mathbf{0})$ has matrix J with respect to the standard basis (using the column vector representation), and so

$$u_{,1}(0,0) = u_{,2}(0,0) = v_{,1}(0,0) = v_{,2}(0,0) = 0.$$

Before proceeding further, the reader should do the next exercise.

Exercise 13.27. *Explain why our conditions on* $\mathbf{F}$ *tell us that near* $(0,0,0,0)$

$$\mathbf{F}(x,y,z,w) \approx (x,y).$$

Suppose that, in fact, $\mathbf{F}(x,y,z,w) = (x,y)$ *exactly. Find* u *and* v*. What conditions on* T *ensure that* $\mathbf{x} = (0,0,0,0)$ *maximises* $T(\mathbf{x})$ *subject to the condition* $\mathbf{F}(\mathbf{x}) = (0,0)$*?*

Since (at least locally) $\mathbf{F}(u(z,w), v(z,w), z, w) = (0,0)$, our problem reduces to studying whether $(z,w) = (0,0)$ maximises

$$\tau(z,w) = T(u(z,w), v(z,w), z, w).$$

Since every maximum is a stationary point, we first ask if $(z,w) = (0,0)$ is a stationary point, that is, if

$$\frac{\partial \tau}{\partial z}(0,0) = \frac{\partial \tau}{\partial w}(0,0) = 0.$$

By the chain rule,

$$\begin{aligned}\frac{\partial \tau}{\partial z}(z,w) = T_{,1}(u(z,w), v(z,w), z, w)\frac{\partial u}{\partial z}(z,w) + T_{,2}(u(z,w), v(z,w), z, w)\frac{\partial v}{\partial z}(z,w)\\ + T_{,3}(u(z,w), v(z,w), z, w),\end{aligned}$$

and so

$$\frac{\partial \tau}{\partial z}(0,0) = T_{,3}(0,0,0,0).$$

A similar calculation gives

$$\frac{\partial \tau}{\partial w}(0,0) = T_{,4}(0,0,0,0),$$

so $(0,0)$ is a stationary point of τ if and only if

$$T_{,3}(0,0,0,0) = T_{,4}(0,0,0,0) = 0.$$

We have found a necessary condition for $(0,0,0,0)$ to maximise $T(\mathbf{x})$ subject to $\mathbf{F}(\mathbf{x}) = \mathbf{0}$, and so for $(0,0,0,0)$ to maximise $t(\mathbf{x})$ subject to $\mathbf{f}(\mathbf{x}) = \mathbf{0}$.

In order to make this condition usable, we must translate it back into the terms of our original problem, and this is most easily done by expressing it in coordinate-free notation. Observe that, if we write

$$l(\mathbf{x}) = T(\mathbf{x}) - T_{,1}(\mathbf{0})F_1(\mathbf{x}) - T_{,2}(\mathbf{0})F_2(\mathbf{x}),$$

then

$$\begin{aligned}
l_{,1}(\mathbf{0}) &= T_{,1}(\mathbf{0}) - T_{,1}(\mathbf{0})F_{1,1}(\mathbf{x}) - T_{,2}(\mathbf{0})F_{2,1}(\mathbf{0}) = T_{,1}(\mathbf{0}) - T_{,1}(\mathbf{0}) = 0,\\
l_{,2}(\mathbf{0}) &= T_{,2}(\mathbf{0}) - T_{,1}(\mathbf{0})F_{1,2}(\mathbf{x}) - T_{,2}(\mathbf{0})F_{2,2}(\mathbf{0}) = T_{,2}(\mathbf{0}) - T_{,2}(\mathbf{0}) = 0,\\
l_{,3}(\mathbf{0}) &= T_{,3}(\mathbf{0}) - T_{,1}(\mathbf{0})F_{1,3}(\mathbf{x}) - T_{,2}(\mathbf{0})F_{2,3}(\mathbf{0}) = T_{,3}(\mathbf{0}),\\
l_{,4}(\mathbf{0}) &= T_{,4}(\mathbf{0}) - T_{,1}(\mathbf{0})F_{1,4}(\mathbf{x}) - T_{,2}(\mathbf{0})F_{2,4}(\mathbf{0}) = T_{,4}(\mathbf{0}),
\end{aligned}$$

so, if $(0,0,0,0)$ is a maximum,

$$D(T - T_{,1}(\mathbf{0})F_1 - T_{,2}(\mathbf{0})F_2)(\mathbf{0}) = \mathbf{0}.$$

Thus, if $(0,0,0,0)$ gives a maximum and $\theta : \mathbb{R}^2 \to \mathbb{R}$ is the linear map given by $\theta(x,y) = T_{,1}(\mathbf{0})x + T_{,2}(\mathbf{0})y$, we have

$$D(T - \theta\mathbf{F})(\mathbf{0}) = \mathbf{0}.$$

Using the definitions of T and $\mathbf{F}$, this gives

$$D(t\alpha - \theta\beta\mathbf{f}(\alpha))(\mathbf{0}) = \mathbf{0},$$

so by the chain rule

$$(Dt(\mathbf{0}) - \theta\beta D\mathbf{f}(\mathbf{0}))\alpha = \mathbf{0}.$$

Since α is invertible, this is equivalent to saying

$$(Dt(\mathbf{0}) - \theta\beta D\mathbf{f}(\mathbf{0})) = \mathbf{0},$$

and, since β is invertible, this, in turn, is equivalent to saying that there exists a linear map $\phi : \mathbb{R}^2 \to \mathbb{R}$ such that

$$(Dt(\mathbf{0}) - \phi D\mathbf{f}(\mathbf{0})) = \mathbf{0},$$

so, using the chain rule once again,

$$D(t - \phi\mathbf{f})(\mathbf{0}) = \mathbf{0}.$$

Writing $\phi(x,y) = \lambda_1 x + \lambda_2 y$, we see that that $\mathbf{x} = (0,0,0,0)$ maximises $T(\mathbf{x})$, subject to the condition $\mathbf{F}(\mathbf{x}) = (0,0)$, only if we can find λ_1 and λ_2 such that $t - \lambda_1 F_1 - \lambda_2 F_2$ has $(0,0,0,0)$ as a stationary point.

The argument generalises easily to any number of dimensions.

Exercise 13.28. *Prove Lemma 13.29.*

Lemma 13.29. (Lagrangian necessity.) *Consider functions $\mathbf{f} : \mathbb{R}^{m+n} \to \mathbb{R}^n$ and $t : \mathbb{R}^{m+n} \to \mathbb{R}$ which are differentiable on an open set U. Suppose further that $D\mathbf{f}$ is continuous and of full rank at every point of U. If $\mathbf{h} \in \mathbb{R}^n$, then, if $\mathbf{z} \in U$ is such that*

(i) $\mathbf{f}(\mathbf{z}) = \mathbf{h}$, and

(ii) $t(\mathbf{z}) \geq t(\mathbf{x})$ for all $\mathbf{x} \in U$ such that $\mathbf{f}(\mathbf{x}) = \mathbf{h}$,

then it follows that there exists a $\boldsymbol{\lambda} \in \mathbb{R}^n$ such that the function $t - \boldsymbol{\lambda} \cdot \mathbf{f} : \mathbb{R}^{m+n} \to \mathbb{R}$ has $\mathbf{z}$ as a stationary point.

Exercise 13.30. *Suppose that $f : \mathbb{R}^2 \to \mathbb{R}$ and $t : \mathbb{R}^2 \to \mathbb{R}$ are very well behaved. Convince yourself that, in this case, Lemma 13.29 reads as follows. A man takes the path $f(x,y) = h$ over a hill of height $t(x,y)$. He reaches the highest point of the path at a point (x_0, y_0) where the path and the contour line $t(x,y) = t(x_0, y_0)$ have a common tangent. Generalise to higher dimensions.*

Lemma 13.29 gives us a recipe for finding the maximum of $t(\mathbf{x})$, subject to the conditions $\mathbf{x} \in U$ and $\mathbf{f}(\mathbf{x}) = \mathbf{h}$, when $t : \mathbb{R}^{n+m} \to \mathbb{R}$ and $\mathbf{f} : \mathbb{R}^{n+m} \to \mathbb{R}^n$ are well behaved functions and U is open.

(1) Set

$$L(\boldsymbol{\lambda}, \mathbf{x}) = t(\mathbf{x}) - \boldsymbol{\lambda} \cdot \mathbf{f}(\mathbf{x})$$

with $\boldsymbol{\lambda} \in \mathbb{R}^n$. ($L$ is called *the Lagrangian.*)

(2) For each fixed $\boldsymbol{\lambda}$, find the set $E(\boldsymbol{\lambda})$ of stationary points of $L(\boldsymbol{\lambda},\ )$ lying within U. (The previous sentence uses the notational convention described in the paragraph labled *Abuse of Language* on Page 396[5].)

(3) Now vary $\boldsymbol{\lambda}$ and find all $\boldsymbol{\lambda}^*$ and $\mathbf{x}^* \in E(\boldsymbol{\lambda}^*)$ such that $\mathbf{f}(\mathbf{x}^*) = \mathbf{h}$. The maximum, if it exists, will lie among these $\mathbf{x}^*$.

Put like this, the whole procedure has the same unreal quality that instructions on how to ride a bicycle would have for someone who had only seen a photograph of a bicycle. I suspect that the only way to learn how to use Lagrange's method is to use it. I expect that most of my readers will indeed be familiar with Lagrange's method, and in any case this is a book on the theory of analysis. However, I include one simple example of the method in practice.

Example 13.31. *Find the circular cylinder of greatest volume with given surface area. (That is, find the optimum dimensions for a tin of standard shape.)*

Calculation. Let the height of the cylinder be h and the radius be r. We wish to maximise the volume

$$V(r,h) = \pi r^2 h,$$

subject to keeping the surface area

$$A(r,h) = 2\pi r^2 + 2\pi r h$$

fixed, with $A(r,h) = A$ $[r,\ h > 0]$, say.

---

[5]This convention was thoroughly disliked by several of the readers of my manuscript but, as one of them remarked, 'No really satisfactory notation exists'. The reader should feel free to use her own notation. In particular she may prefer to use a placeholder and write $L(\boldsymbol{\lambda}, \cdot)$

The instructions for step (1) tell us to form the Lagrangian

$$L(\lambda, r, h) = V(r, h) - \lambda A(r, h) = \pi r^2 h - \lambda(2\pi r^2 + 2\pi r h).$$

The instructions for step (2) tell us to seek stationary values for $L(\lambda, \ , \ )$ when λ is fixed. We thus seek to solve

$$\frac{\partial L}{\partial r}(\lambda, r, h) = 0, \ \frac{\partial L}{\partial h}(\lambda, r, h) = 0,$$

that is,

$$2\pi r h - \lambda(4\pi r - 2\pi h) = 0, \ \pi r^2 - 2\lambda\pi r = 0,$$

or, simplifying,

★ $$rh - \lambda(2r + h) = 0, \ r - 2\lambda = 0.$$

Thus, if λ is fixed and positive, $L(\lambda, \ , \ )$ has a unique stationary point given by $r = r_\lambda = 2\lambda$, $h = h_\lambda = 4\lambda$. (If $\lambda \leq 0$, there are no stationary points consistent with our restrictions $r > 0$, $h > 0$.)

We now proceed to step (3), which requires us to find λ such that

$$A = A(r_\lambda, h_\lambda),$$

that is,

$$A = A(2\lambda, 4\lambda) = 24\pi\lambda^2.$$

This has a unique solution with $\lambda > 0$. We know, on general grounds (see Exercise 13.32), that a maximum exists; so, since Lagrange's method must produce the maximum point, and produces only one point in this case, this point must be the maximum.

The neatest way of writing our solution (and one that we could have obtained directly by eliminating λ from the two equations of ★) is that the cylinder of maximum volume with given surface area has height twice its radius. ■

It must be admitted that the standard tin does not follow our prescription very closely. (Presumably the cost of materials is relatively small compared to other costs. We must also remember that, near a maximum, small changes produce very small effects, so the penalties for deviating are not high[6].)

Exercise 13.32. *We work with the notation and hypotheses introduced in the calculations in Example 13.31. Choose r_0 and h_0 such that $r_0, h_0 > 0$*

[6]The matter is taken about as far as it can be without actually consulting a tin can manufacturer in an enjoyable article [**47**] by Roe. He points out several factors that we have neglected and suggests that very large tins are closer to our suggested shape.

and $A(r_0, h_0) = A$. *Show that there exists an* $R > 1$ *such that, if* $r > R$ *or* $R^{-1} > r > 0$ *and* $A(r, h) = A$, *then* $V(r, h) \leq V(r_0, h_0)/2$. *Show that*

$$K = \{(r, h) : A(r, h) = A,\ R \geq r \geq R^{-1},\ h > 0\}$$

is a closed bounded set in $\mathbb{R}^2$. *Deduce that* $V(r, h)$ *attains a maximum on* K, *and conclude that* $V(r, h)$ *attains a maximum on*

$$\{(r, h) : A(r, h) = A,\ r > 0,\ h > 0\}.$$

Exercise 13.33. *Suppose that* $f : \mathbb{R}^2 \to \mathbb{R}$ *and* $t : \mathbb{R}^2 \to \mathbb{R}$ *are given by* $f(x, y) = y$ *and* $t(x, y) = y^2 - x^2$. *We seek the maximum value of* $t(x, y)$ *subject to* $f(x, y) = a$. *Show that* $L(\lambda, \ ) = t - \lambda f$ *is stationary at* $(0, \lambda/2)$, *and that the only value of* λ *which gives* $f(0, \lambda/2) = a$ *is* $\lambda = 2a$. *[We again use the notation described on Page 396.]*

(i) Show that $(0, a)$ *does indeed maximise* $t(x, y)$ *subject to* $f(x, y) = a$.

(ii) Show, however, that $L(2a, \ ) = t - 2af$ *does not have a maximum at* $(0, a)$.

(iii) Draw a diagram in the manner of Exercise 13.30 to illustrate what is happening.

Exercise 13.34. *Suppose that* $f : \mathbb{R}^2 \to \mathbb{R}$ *and* $t : \mathbb{R}^2 \to \mathbb{R}$ *are given by* $f(x, y) = x - \alpha y^2$ *and* $t(x, y) = (x - 1)^2 + y^2$. *We seek maxima and minima of* t *subject to* $f(x, y) = 0$. *Show that* $L(\lambda, \ ) = t - \lambda f$ *is stationary at* $(1 + \lambda/2, 0)$, *and that the only value of* λ *which gives* $f(0, 0) = 0$ *is* $\lambda = -2$.

(i) Show that, if $\alpha > 1/2$, $(0, 0)$ *minimises* $t(x, y)$ *subject to* $f(x, y) = 0$. *Show, moreover, that* $L(-2, \ ) = t - 2f$ *has a strict minimum at* 0.

(ii) Show that, if $\alpha = 1/2$, $(0, 0)$ *minimises* $t(x, y)$ *subject to* $f(x, y) = 0$. *Show that* $L(-2, \ )$ *has a minimum at* 0 *but this is not a strict minimum.*

(iii) Show that, if $\alpha < 1/2$, $(0, 0)$ *maximises* $t(x, y)$ *subject to* $f(x, y) = 0$. *Show, however, that* $L(-2, \ )$ *does not have a maximum at* 0.

(iv) Draw a diagram in the manner of Exercise 13.30 to illustrate (as far as you can, do not overdo it) what is happening.

The following remark is entirely trivial but extremely useful.

Lemma 13.35. (Lagrangian sufficiency.) *Consider a set* X *and functions* $\mathbf{f} : X \to \mathbb{R}^n$ *and* $t : X \to \mathbb{R}$. *Set*

$$L(\boldsymbol{\lambda}, x) = t(x) - \boldsymbol{\lambda} \cdot \mathbf{f}(x)$$

for all $\boldsymbol{\lambda} \in \mathbb{R}^n$ *and* $x \in X$.

Suppose $\mathbf{h} \in \mathbb{R}^n$, $\boldsymbol{\lambda}^* \in \mathbb{R}^n$, *and* $x^* \in X$ *are such that* $\mathbf{f}(x^*) = \mathbf{h}$ *and*

$$L(\boldsymbol{\lambda}^*, x^*) \geq L(\boldsymbol{\lambda}^*, x)$$

for all $x \in X$*. Then*

$$t(x^*) \geq t(x)$$

for all $x \in X$ *such that* $\mathbf{f}(x) = \mathbf{h}$.

Proof. The proof is shorter than the statement. Observe that, under the given hypotheses,

$$\begin{aligned} t(x^*) &= t(x^*) - \boldsymbol{\lambda}^* \cdot \mathbf{f}(x^*) + \boldsymbol{\lambda}^* \cdot \mathbf{h} = L(\boldsymbol{\lambda}^*, x^*) + \boldsymbol{\lambda}^* \cdot \mathbf{h} \\ &\geq L(\boldsymbol{\lambda}^*, x) + \boldsymbol{\lambda}^* \cdot \mathbf{h} = t(x) - \boldsymbol{\lambda}^* \cdot \mathbf{f}(x) + \boldsymbol{\lambda}^* \cdot \mathbf{h} = t(x) \end{aligned}$$

for all $x \in X$ such that $\mathbf{f}(x) = \mathbf{h}$, as required. ■

It is worth noting that Lemma 13.35 (our Lagrangian sufficiency condition) does not require f or t to be well behaved or make any demands on the underlying space X. This is particularly useful since, if the reader notes where Lagrange's method is used, she will find that it is often used in circumstances much more general than those envisaged in Lemma 13.29 (our Lagrangian necessity condition).

However, the very simple examples given in Exercises 13.33 and 13.34 show that there is a very big gap between our Lagrangian necessity and sufficiency conditions even when we restrict ourselves to well behaved functions on finite-dimensional spaces. In the first three chapters of his elegant text *Optimization Under Constraints* [**51**] Whittle considers how this gap can be closed. It turns out that in certain very important practical cases (in particular those occurring in linear programming) the sufficient condition is also necessary, but that, from the point of view of the present book, these cases have very special features.

In general, the Lagrangian method is very effective in suggesting the correct answer, but when that answer has been found it is often easier to prove correctness by some other method (compare the treatment of geodesics in Section 10.5, where we found Theorem 10.71 by one method but proved it by another).

The alternative strategy, once the Lagrangian method has produced a possible solution, is to claim that 'it is obvious from physical considerations that this is indeed the correct solution'. This works well provided at least one of the following conditions applies:

(1) you are answering an examination question, or

(2) you have genuine physical insight, or

(3) you are reasonably lucky.

It is, of course, characteristic of bad luck that it strikes at the most inconvenient time.

Chapter 14

Completion

14.1. What is the correct question?

We cannot do interesting analysis on the rationals, but we can on the larger space of real numbers. On the whole, we cannot do interesting analysis on metric spaces which are not complete. Given such an incomplete metric space, can we find a larger complete metric space which contains it? Our first version of this question might run as follows.

Question A: If (X, d) is a metric space, can we find a complete metric space (Z, δ) such that $Z \supseteq X$ and $d(u, v) = \delta(u, v)$ for all u, $v \in X$?

Most mathematicians prefer to ask this question in a slightly different way.

Question A′: If (X, d) is a metric space, can we find a complete metric space $(Y, \tilde{d})$ and a map $\theta : X \to Y$ such that $\tilde{d}(\theta u, \theta v) = d(u, v)$?

Question A asks us to build a complete metric space containing (X, d). Question A′ asks us to build a complete metric space containing a perfect copy of (X, d). Since mathematicians cannot distinguish between the original space (X, d) and a perfect copy, they consider the two questions to be equivalent. However, both from a philosophical and a technical point of view, it is easier to handle Question A′.

Exercise 14.1. *Convince yourself* **either** *that Question A is equivalent to Question A′* **or** *that Question A′ is more appropriate than Question A.*

For the moment, we stick to Question A. A little thought shows that it does not have a unique answer.

Exercise 14.2. *Consider the open interval $X = (0,1)$ on $\mathbb{R}$ with the usual metric d.*

(i) Show that (X, d) is not complete.

(ii) Show that, if we take $Z = [0,1]$ with the usual metric δ, then (Z, δ) answers Question A.

(iii) Show that, if we take $Z = \mathbb{R}$ with the usual metric δ, then (Z, δ) answers Question A.

We therefore reformulate Question A as follows.

Question B: If (X, d) is a metric space, can we find a most economical complete metric space (Z, ρ) such that $Z \supseteq X$ and $d(u, v) = \rho(u, v)$ for all $u, v \in X$?

We can formulate an appropriate meaning for 'most economical' with the aid of the notion of density.

Definition 14.3. *If (X, d) is a metric space, we say that a subset E is* dense *in X if, given any $x \in X$, we can find $x_n \in E$ with $d(x_n, x) \to 0$ as $n \to \infty$.*

Exercise 14.4. *Show that the following statements about a subset E of a metric space (X, d) are equivalent.*

(i) E is dense in X.

(ii) Given $x \in X$ and $\epsilon > 0$, we can find a $y \in E$ such that $d(x, y) < \epsilon$.

(iii) If F is a closed subset of X with $F \supseteq E$, then $F = X$.

The notion of density is widely used in analysis, since, in some sense, a dense subset of a metric space acts as a 'skeleton' for the space. For the moment, let us see why it provides the appropriate definition of 'most economical' in the context of this chapter.

Lemma 14.5. *Suppose (X, d), (Y, ρ) and (Y', ρ') are metric spaces with the following properties.*

(i) (Y, ρ) and (Y', ρ') are complete.

(ii) There exist maps $\theta : X \to Y$ and $\theta' : X \to Y'$ such that $\rho(\theta u, \theta v) = \rho'(\theta' u, \theta' v) = d(u, v)$.

(iii) θX is dense in Y and $\theta' X$ is dense in Y'.

Then we can find a bijection $\phi : Y \to Y'$ such that $\phi\theta = \theta'$ and $\rho'(\phi w, \phi z) = \rho(w, z)$ for all $w, z \in Y$.

Exercise 14.6. *(i) Convince yourself that Lemma 14.5 can be roughly restated in terms of Question A as follows. Suppose (X, d) is a metric space and (Z, δ) and (Z', δ') are complete metric spaces such that $X \subseteq Z$, $X \subseteq Z'$, $\delta(u, v) = \delta'(u, v) = d(u, v)$ for all $u, v \in X$, and X is dense in (Z, δ) and in*

(Z', δ'). Then (Z, δ) and (Z', δ') have the same metric structure, and X sits in both metric spaces in the same way. [Note, you must be able to describe the purpose of the statement $\phi\theta = \theta'$ in this account.]

(ii) Convince yourself that the questions involved in Lemma 14.5 are better treated in the language of Question A' than in the language of Question A.

Proof of Lemma 14.5. If $y \in Y$, then, since θX is dense in (Y, ρ), we can find $x_n \in X$ with $\rho(\theta x_n, y) \to 0$. Since θx_n converges, it forms a Cauchy sequence in (Y, ρ). But

$$\rho'(\theta' x_n, \theta' x_m) = d(x_n, x_m) = \rho(\theta x_n, \theta x_m),$$

so $\theta' x_n$ is a Cauchy sequence in (Y', ρ') and so converges to a limit $\tilde{x}$, say.

Suppose $z_n \in X$ with $\rho(\theta z_n, y) \to 0$. By the argument of the previous paragraph, $\theta' z_n$ is a Cauchy sequence in (Y', ρ') and so converges to a limit $\tilde{z}$, say. We wish to show that $\tilde{x} = \tilde{z}$. To do this, observe that

$$\begin{aligned}
\rho'(\tilde{x}, \tilde{z}) &\leq \rho'(\tilde{x}, \theta' x_n) + \rho'(\tilde{z}, \theta' z_n) + \rho'(\theta' x_n, \theta' z_n) \\
&= \rho'(\tilde{x}, \theta' x_n) + \rho'(\tilde{z}, \theta' z_n) + \rho(\theta x_n, \theta z_n) \\
&\leq \rho'(\tilde{x}, \theta' x_n) + \rho'(\tilde{z}, \theta' z_n) + \rho(\theta x_n, y) + \rho(\theta z_n, y) \\
&\to 0 + 0 + 0 + 0 = 0,
\end{aligned}$$

as $n \to \infty$. Thus $\rho'(\tilde{x}, \tilde{z}) = 0$ and $\tilde{x} = \tilde{z}$. We have shown that we can define ϕy unambiguously by $\phi y = \tilde{x}$.

We have now defined $\phi : Y \to Y'$. To show that $\rho'(\phi w, \phi z) = \rho(w, z)$, choose $w_n \in X$ and $z_n \in X$ such that $\rho(\theta w_n, w)$, $\rho(\theta z_n, z) \to 0$ as $n \to \infty$. Then $\rho(\theta z_n, \theta w_n) \to \rho(z, w)$ as $n \to \infty$. (See Exercise 14.7 if necessary.) By definition, $\rho'(\theta' z_n, \phi z) \to 0$ and $\rho'(\theta' w_n, \phi w) \to 0$, so $\rho'(\theta' z_n, \theta' w_n) \to \rho'(\phi z, \phi w)$. Since

$$\rho'(\theta' z_n, \theta' w_n) = d(z_n, w_n) = \rho(\theta z_n, \theta w_n),$$

it follows that $\rho'(\phi w, \phi z) = \rho(w, z)$.

To show that $\phi\theta = \theta'$, choose any $x \in X$. If we set $x_n = x$ for each n, we see that $\rho(\theta x_n, \theta x) = 0 \to 0$ and $\rho'(\theta' x_n, \theta' x) = 0 \to 0$ as $n \to \infty$, so, by definition, $\phi\theta x = \theta' x$. Since x was arbitrary, it follows that $\phi\theta = \theta'$.

Finally, we must show that ϕ is a bijection. There are several ways of approaching this. We choose an indirect but natural approach. Observe that, interchanging the roles of Y and Y', the work done so far also shows that there is a map $\tilde{\phi} : Y' \to Y$ such that $\tilde{\phi}\theta' = \theta$ and $\rho(\tilde{\phi} w, \tilde{\phi} z) = \rho'(w, z)$ for all w, $z \in Y'$. Since

$$(\tilde{\phi}\phi)\theta = \tilde{\phi}(\phi\theta) = \tilde{\phi}\theta' = \theta,$$

we have $\tilde{\phi}\phi(y) = y$ for all $y \in \theta X$. Now θX is dense in Y, so, if $y \in Y$, we can find $y_n \in \theta X$ such that $\rho(y_n, y) \to 0$. Thus

$$\begin{aligned}\rho(\tilde{\phi}\phi(y), y) &\le \rho(\tilde{\phi}\phi(y), \tilde{\phi}\phi(y_n)) + \rho(\tilde{\phi}\phi(y_n), y_n) + \rho(y_n, y)\\ &= \rho(\tilde{\phi}\phi(y), \tilde{\phi}\phi(y_n)) + \rho(y_n, y) = \rho'(\phi(y), \phi(y_n)) + \rho(y_n, y)\\ &= \rho(y, y_n) + \rho(y_n, y) = 2\rho(y_n, y) \to 0\end{aligned}$$

as $n \to \infty$. Thus $\tilde{\phi}\phi(y) = y$ for all $y \in Y$. Similarly, $\phi\tilde{\phi}(y') = y'$ for all $y' \in Y'$. Thus $\tilde{\phi}$ is the inverse of ϕ, and ϕ must be bijective. ■

Exercise 14.7. *Let (X, d) be a metric space.*

(i) Show that

$$d(x, y) - d(u, v) \le d(x, u) + d(y, v),$$

and deduce that

$$|d(x, y) - d(u, v)| \le d(x, u) + d(y, v)$$

for all x, y, u, $v \in X$.

(ii) If $x_n \to x$ and $y_n \to y$, show that $d(x_n, y_n) \to d(x, y)$ as $n \to \infty$.

Exercise 14.8. *The proof of Lemma 14.5 looks quite complicated, but the difficulty lies in asking the right questions in the right order, rather than answering them. Outline the questions answered in the proof of Lemma 14.5, paying particular attention to the first two paragraphs.*

We can now pose Question B more precisely.

Question C: If (X, d) is a metric space, can we find a complete metric space (Z, ρ) such that $Z \supseteq X$, X is dense in (Z, ρ) and $d(u, v) = \rho(u, v)$ for all u, $v \in X$?

Question C′: If (X, d) is a metric space, can we find a complete metric space $(Y, \tilde{d})$ and a map $\theta : X \to Y$ such that θX is dense in $(Y, \tilde{d})$ and $\tilde{d}(\theta u, \theta v) = d(u, v)$ for all u, $v \in X$?

We shall leave these questions aside for the moment. Instead, we shall give some examples of how behaviour on a dense subset may force behaviour on the whole space. (Note, however that Exercise 5.72 and its preceeding discussion shows that this need not always be the case.)

Lemma 14.9. *Suppose (X, d) is a complete metric space with a dense subset E. Suppose that E is a vector space (over $\mathbb{F}$, where $\mathbb{F} = \mathbb{R}$ or $\mathbb{F} = \mathbb{C}$) with norm $\| \; \|_E$ such that $d(x, y) = \|x -_E y\|_E$ for all x, $y \in E$. Then X can be given the structure of a normed vector space with norm $\| \; \|$ such that E is a vector subspace of X and $d(x, y) = \|x - y\|$ for all x, $y \in X$.*

Proof. Suppose[1] that $(E, +_E, *_E, \| \|_E, \mathbb{F})$ is a normed vector space with addition denoted by $+_E$ and scalar multiplication denoted by $*_E$.

Suppose $x, y \in X$. We can find $x_n, y_n \in E$ such that $d(x_n, x), d(y_n, y) \to 0$. We know that x_n and y_n must be Cauchy sequences, and so

$$\begin{aligned}
d(x_n +_E y_n, x_m +_E y_m) &= \|(x_n +_E y_n) -_E (x_m +_E y_m)\|_E \\
&= \|(x_n -_E x_m) +_E (y_n -_E y_m)\|_E \le \|x_n -_E x_m\|_E + \|y_n -_E y_m\|_E \\
&= d(x_n, x_m) + d(y_n, y_m) \to 0
\end{aligned}$$

as $n, m \to \infty$. Thus $x_n +_E y_n$ is a Cauchy sequence in X and must have a limit z, say. Suppose now that we are given $x'_n, y'_n \in E$ such that $d(x'_n, x), d(y'_n, y) \to 0$. As before, $x'_n + y'_n$ must have a limit z', say. We want to show that $z = z'$. To do this, observe that, since

$$\begin{aligned}
&d(z, z') \le d(z, x_n +_E y_n) + d(x_n +_E y_n, x'_n +_E y'_n) + d(x'_n +_E y'_n, z) \\
&= d(z, x_n +_E y_n) + d(z', x'_n +_E y'_n) + \|(x_n +_E y_n) -_E (x'_n +_E y'_n)\| \\
&= d(z, x_n +_E y_n) + d(z', x'_n +_E y'_n) + \|(x_n -_E x'_n) +_E (y_n -_E y'_n)\|_E \\
&\le d(z, x_n +_E y_n) + d(z', x'_n +_E y'_n) + \|x_n -_E x'_n\|_E + \|y_n -_E y'_n\|_E \\
&= d(z, x_n +_E y_n) + d(z', x'_n +_E y'_n) + d(x_n, x'_n) + d(y_n, y'_n) \\
&\le d(z, x_n +_E y_n) + d(z', x'_n +_E y'_n) + d(x_n, x) + d(x'_n, x) + d(y_n, y) + d(y'_n, y) \\
&\to 0 + 0 + 0 + 0 + 0 + 0 = 0
\end{aligned}$$

as $n \to \infty$, it follows that $z = z'$. We can, therefore, define $x + y$ as the limit of $x_n +_E y_n$ when $x_n, y_n \in E$ and $d(x_n, x), d(y_n, y) \to 0$. Observe that, if $x, y \in E$, then, setting $x_n = x$, $y_n = y$, it follows that $x +_E y = x + y$.

We leave it as an exercise for the reader to show that if $\lambda \in \mathbb{F}$ and $x \in X$, we can define λx unambiguously as the limit of $\lambda *_E x_n$ when $x_n \in E$ and $d(x_n, x) \to 0$, and that, with this definition, $\lambda *_E x = \lambda x$ whenever $\lambda \in \mathbb{F}$ and $x \in E$.

Now we must check that X with the newly defined operations is indeed a vector space. This is routine. For example, if $x, y \in X$, and $\lambda \in \mathbb{F}$, we can find $x_n, y_n \in E$ such that $x_n \to x$ and $y_n \to y$. Since E is vector space,

$$\lambda *_E (x_n +_E y_n) = \lambda *_E x_n +_E \lambda *_E y_n.$$

But, by our definitions, $x_n +_E y_n \to x + y$, so $\lambda *_E (x_n +_E y_n) \to \lambda(x + y)$. Again, by definition, $\lambda *_E x_n \to \lambda x$ and $\lambda *_E y_n \to \lambda y$, so $\lambda *_E x_n +_E \lambda *_E y_n \to \lambda x + \lambda y$. The uniqueness of limits now gives us

$$\lambda(x + y) = \lambda x + \lambda y.$$

The remaining axioms are checked in the same way.

[1]This proof with its plethora of subscript E's may help explain why mathematicians are prepared to put up with some ambiguity in order to achieve notational simplicity.

Finally, we must check that there is a norm $\| \ \|$ on X such that $d(x,y) = \|x-y\|$. First, observe that, if x, $y \in X$, we can find x_n, $y_n \in E$ such that $x_n \to x$ and $y_n \to y$. We have

$$d(x_n, y_n) = \|x_n -_E y_n\|_E = d(x_n -_E y_n, 0),$$

where 0 is the zero vector in E (and so in X). Since

$$x_n -_E y_n = x_n +_E (-1) * y_n \to x + (-1)y = x - y,$$

we have $d(x_n -_E y_n, 0) \to d(x-y, 0)$. But $d(x_n, y_n) \to d(x,y)$ and the limit is unique, so we have

$$d(x,y) = d(x-y, 0).$$

We now set $\|a\| = d(a, 0)$ and check that $\| \ \|$ is indeed a norm. ■

Exercise 14.10. *Fill in the gaps in the proof of Lemma 14.9.*

(i) Show that, if $\lambda \in \mathbb{F}$ and $x \in X$, we can define λx unambiguously as the limit of $\lambda *_E x_n$ when $x_n \in E$ and $d(x_n, x) \to 0$, and that, with this definition, $\lambda *_E x = \lambda x$ whenever $\lambda \in \mathbb{F}$ and $x \in E$.

(ii) If you are of a careful disposition, check that all the axioms for a vector space hold for X with our operations. Otherwise, choose the two which seem to you hardest and check those.

(iii) Check that $\| \ \|$ defined at the end of the proof is indeed a norm.

Lemma 14.11. *Let (X,d) be a complete metric space with a dense subset E. Suppose that E is a vector space (over $\mathbb{F}$, where $\mathbb{F} = \mathbb{R}$ or $\mathbb{F} = \mathbb{C}$) with inner product $\langle \ , \ \rangle_E$ such that $d(x,y) = \|x -_E y\|_E$ for all x, $y \in E$, where $\| \ \|_E$ is the norm induced by the inner product. Then X can be given the structure of a vector space with inner product $\langle \ , \ \rangle$ such that E is a vector subspace of X and $\langle x,y\rangle_E = \langle x,y\rangle$ for all x, $y \in E$.*

Proof. By Lemma 14.9, X can be given the structure of a normed vector space with norm $\| \ \|$ such that E is a vector subspace of X and $d(x,y) = \|x-y\|$ for all x, $y \in X$. We need to show that it can be given an inner product consistent with this norm.

First, observe that, if x, $y \in X$, we can find x_n, $y_n \in E$ such that $x_n \to x$ and $y_n \to y$. Automatically, x_n and y_n form Cauchy sequences. Since any Cauchy sequence is bounded, we can find a K such that $\|x_n\|$, $\|y_n\| \leq K$ for all n. Thus, using the Cauchy–Schwarz inequality,

$$\begin{aligned}
|\langle x_n, y_n\rangle_E - \langle x_m, y_m\rangle_E| &= |\langle x_n, y_n - y_m\rangle_E + \langle x_n - x_m, y_m\rangle_E| \\
&\leq |\langle x_n, y_n - y_m\rangle_E| + |\langle x_n - x_m, y_m\rangle_E| \\
&\leq \|x_n\|\|y_n - y_m\| + \|x_n - x_m\|\|y_m\| \\
&\leq K\|y_n - y_m\| + K\|x_n - x_m\| \to 0 + 0 = 0,
\end{aligned}$$

and the $\langle x_n, y_n \rangle_E$ form a Cauchy sequence in $\mathbb{F}$ converging to t.

Suppose now that $x'_n, y'_n \in E$ are such that $x'_n \to x$ and $y'_n \to y$. As before, $\langle x'_n, y'_n \rangle_E$ must have a limit t', say. We want to show that $t = t'$. Note that, also as before, we can find a K' such that $\|x'_n\|, \|y'_n\| \le K'$ for all n. Thus, using the Cauchy–Schwarz inequality again,

$$\begin{aligned}|\langle x_n, y_n \rangle_E - \langle x'_n, y'_n \rangle_E| &= |\langle x_n, y_n - y'_n \rangle_E + \langle x_n - x'_n, y'_n \rangle_E| \\ &\le |\langle x_n, y_n - y'_n \rangle_E| + |\langle x_n - x'_n, y'_n \rangle_E| \\ &\le \|x_n\| \|y_n - y'_n\| + \|x_n - x'_n\| \|y'_n\| \\ &\le K\|y_n - y'_n\| + K'\|x_n - x'_n\| \to 0 + 0 = 0,\end{aligned}$$

so $t = t'$. We can thus define $\langle x, y \rangle$ unambiguously by choosing $x_n, y_n \in E$ such that $x_n \to x$ and $y_n \to y$ and taking $\langle x, y \rangle$ to be the limit of $\langle x_n, y_n \rangle_E$. Setting $x_n = x$, $y_n = y$ for all n, we see that $\langle x, y \rangle = \langle x, y \rangle_E$ whenever $x, y \in E$.

We observe that, if $x \in X$ and we choose $x_n \in E$ such that $x_n \to x$, then

$$\|x_n\|^2 = \langle x_n, x_n \rangle_E \to \langle x, x \rangle$$

and

$$\|x_n\|^2 = d(x_n, 0)^2 \to d(x, 0)^2 = \|x\|^2,$$

so, by the uniqueness of limits, $\|x\|^2 = \langle x, x \rangle$. The verification that $\langle\ ,\ \rangle$ is an inner product on X is left to the reader.

(An alternative proof which uses an important link between inner products and their derived norms is oulined in Exercises K.327 and K.328.) ■

Exercise 14.12. *Complete the proof of Lemma 14.11 by showing that $\langle\ ,\ \rangle$ is an inner product on X.*

By now the reader should have formed the opinion that what seemed to be a rather difficult proof technique when first used in the proof of Lemma 14.5 is, in fact, rather routine (though requiring continuous care). If she requires further convincing, she can do Exercise K.329.

14.2. The solution

In this section we show how, starting from a metric space (X, d), we can construct a complete metric space $(Y, \tilde{d})$ and a map $\theta : X \to Y$ such that θX is dense in $(Y, \tilde{d})$ and $\tilde{d}(\theta u, \theta v) = d(u, v)$ for all $u, v \in X$. We give a direct proof which can serve as a model in similar circumstances. (Exercise K.330 gives a quick and elegant proof which, however, only applies in this particular case.)

We start by considering the space $\mathbf{X}$ of Cauchy sequences $\mathbf{x} = (x_n)_{n=1}^{\infty}$. We write $\mathbf{x} \sim \mathbf{y}$ if $\mathbf{x}, \mathbf{y} \in \mathbf{X}$ and $d(x_n, y_n) \to 0$ as $n \to \infty$. The reader should verify the statements made in the next exercise.

Exercise 14.13. *Suppose* $\mathbf{x}$, $\mathbf{y}$, $\mathbf{z} \in \mathbf{X}$. *Then:*

(i) $\mathbf{x} \sim \mathbf{x}$.

(ii) If $\mathbf{x} \sim \mathbf{y}$, *then* $\mathbf{y} \sim \mathbf{x}$.

(ii) If $\mathbf{x} \sim \mathbf{y}$ *and* $\mathbf{y} \sim \mathbf{z}$, *then* $\mathbf{x} \sim \mathbf{z}$.

We write $[\mathbf{x}] = \{\mathbf{y} : \mathbf{y} \sim \mathbf{x}\}$ and call $[\mathbf{x}]$ the *equivalence class* of $\mathbf{x}$. We take Y to be the set of all such equivalence classes. The reader should verify the statement made in the next exercise.

Exercise 14.14. *Each* $\mathbf{x} \in \mathbf{X}$ *belongs to exactly one equivalence class in* Y.

We now want to define a metric on Y. The definition follows a pattern which is familiar from the previous section. Suppose a, $b \in Y$. If $a = [\mathbf{x}]$ and $b = [\mathbf{y}]$, then

$$\begin{aligned} |d(x_n, y_n) - d(x_m, y_m)| &\le |d(x_n, y_n) - d(x_n, y_m)| + |d(x_n, y_m) - d(x_m, y_m)| \\ &\le d(y_n, y_m) + d(x_n, x_m) \to 0 + 0 = 0 \end{aligned}$$

as n, $m \to \infty$. Thus $d(x_n, y_n)$ is a Cauchy sequence in $\mathbb{R}$ and tends to a limit t, say. Suppose now that $a = [\mathbf{x}']$ and $b = [\mathbf{y}']$. Since

$$\begin{aligned} |d(x_n, y_n) - d(x'_n, y'_n)| &\le |d(x_n, y_n) - d(x_n, y'_n)| + |d(x_n, y'_n) - d(x'_n, y'_n)| \\ &\le d(y_n, y'_n) + d(x_n, x'_n) \to 0 + 0 = 0, \end{aligned}$$

we have $d(x'_n, y'_n) \to t$ as $n \to \infty$. We can thus define $\tilde{d}([\mathbf{x}], [\mathbf{y}])$ unambiguously by

$$\tilde{d}([\mathbf{x}], [\mathbf{y}]) = \lim_{n\to\infty} d(x_n, y_n).$$

The reader should verify the statements made in the next exercise.

Exercise 14.15. *(i)* $(Y, \tilde{d})$ *is a metric space.*

(ii) If we set $\theta(x) = [(x, x, x, \dots)]$, *then* θ *is a map from* X *to* Y *such that* $\tilde{d}(\theta u, \theta v) = d(u, v)$.

(iii) If $a \in Y$, *then* $a = [\mathbf{x}]$ *for some* $\mathbf{x} \in \mathbf{X}$. *Show that* $\tilde{d}(\theta(x_n), a) \to 0$ *as* $n \to \infty$, *and so* $\theta(X)$ *is dense in* $(Y, \tilde{d})$.

Lemma 14.16. *If we adopt the hypotheses and notation introduced in this section, then* $(Y, \tilde{d})$ *is complete.*

Proof. Suppose $[\mathbf{y}(1)]$, $[\mathbf{y}(2)]$, $\dots$ is a Cauchy sequence in $(Y, \tilde{d})$. Since $\theta(X)$ is dense in $(Y, \tilde{d})$, we can find $x_j \in X$ such that $\tilde{d}(\theta(x_j), [\mathbf{y}(j)]) < j^{-1}$.

We observe that

$$\begin{aligned} d(x_j, x_k) &= \tilde{d}(\theta(x_j), \theta(x_k)) \\ &\leq \tilde{d}(\theta(x_j), [\mathbf{y}(j)]) + \tilde{d}(\theta(x_k), [\mathbf{y}(k)]) + \tilde{d}([\mathbf{y}(j)], [\mathbf{y}(k)]) \\ &< j^{-1} + k^{-1} + \tilde{d}([\mathbf{y}(j)], [\mathbf{y}(k)]), \end{aligned}$$

and so the x_j form a Cauchy sequence in (X, d).

We now wish to show that $\tilde{d}([\mathbf{y}(n)], [\mathbf{x}]) \to 0$ as $n \to \infty$. Let $\epsilon > 0$ be given. Since the sequence $[\mathbf{y}(n)]$ is Cauchy, we can find an M such that

$$\tilde{d}([\mathbf{y}(j)], [\mathbf{y}(k)]) < \epsilon/2 \text{ for all } j,\ k \geq M.$$

We now choose $N \geq M$ such that $N^{-1} < \epsilon/6$ and observe that the inequality proved in the last paragraph gives

$$d(x_j, x_k) < 5\epsilon/6 \text{ for all } j,\ k \geq N.$$

Thus

$$\tilde{d}([\mathbf{x}], \theta(x_k)) \leq 5\epsilon/6 \text{ for all } k \geq N,$$

and

$$\begin{aligned} \tilde{d}([\mathbf{x}], [\mathbf{y}(k)]) &\leq \tilde{d}([\mathbf{x}], \theta(x_k)) + \tilde{d}(\theta(x_k), [\mathbf{y}(k)]) \\ &< 5\epsilon/6 + k^{-1} \leq 5\epsilon/6 + N^{-1} < \epsilon \end{aligned}$$

for all $k \geq N$, so we are done. ■

Combining the results of this section with Lemma 14.5, we have the following theorem.

Theorem 14.17. *If (X, d) is a metric space, then there exists an essentially unique complete metric space $(Y, \tilde{d})$ such that X is dense in $(Y, \tilde{d})$ and $\tilde{d}$ restricted to X^2 is d.*

We call $(Y, \tilde{d})$ the *completion* of (X, d). Lemma 14.9 and Lemma 14.11 now give the following corollaries.

Lemma 14.18. *(i) The completion of a normed vector space is a normed vector space.*

(ii) The completion of an inner product vector space is an inner product vector space.

Exercise 14.19. *Readers of a certain disposition (with which the author sometimes sympathises and sometimes does not) will find the statements of Theorem 14.17 and Lemma 14.18 unsatisfactorily lax. Redraft them in the more rigorous style of Lemma 14.5 and Lemmas 14.9 and 14.11.*

Exercise 14.20. *If (X, d) is already complete, then the uniqueness of the completion means that (essentially) $(X, d) = (Y, \tilde{d})$. Go through the construction of $(Y, \tilde{d})$ in this section and explain in simple terms why the construction does indeed work as stated.*

14.3. Why do we construct the reals? ♡

Mathematicians know from experience that it is usually easier to ask questions than to answer them. However, in some very important cases, asking the correct question may represent a greater intellectual breakthrough than answering it.

From the time the study of Euclid's *Elements* was reintroduced into Europe until about 1750, few of those who studied it can have doubted that it described the geometry of the world, or, accepting that it did describe the geometry of the world, asked why it did so. Amongst those who did was the philosopher Kant, who proposed the following interesting answer. According to Kant, there exist certain *a priori* principles such that our minds are bound to interpret the world in terms of these a priori principles. The constraints imposed by the axioms of Euclidean geometry provided an example of such a priori principles.

Some mathematicians felt that the axioms stated by Euclid were not yet in the best form. In particular, they believed that the so-called parallel axiom[2] was unsatisfactory, and they sought to prove it from the other axioms.

Between 1750 and 1840 a number of independent workers saw that the questions.

(A) Why is the parallel axiom true?

(B) Why is Euclid's geometry true?

should be replaced by the new questions

(A′) Do there exist mathematical systems obeying all the axioms of Euclid except the parallel axiom and disobeying the parallel axiom?

(B′) Does Euclid's geometry describe the world?

Although Gauss did not publish his views, he asked both of these new questions and answered yes to the first and no to the second. He even tried to check the validity of Euclidean geometry by physical experiment[3]. In a

[2]This is now usually stated in the following form. Given a line l and a point P not on the line, there exists one and only one line l' through P which does not intersect l.

[3]Euclid uses the parallel axiom to prove that the sum of the angles of a triangle add up to 180 degrees. Gauss used surveying instruments to measure the angles of a triangle of mountain peaks. To within the limits of experimental error, the results confirmed Euclid's prediction.

private letter he stated that geometry should not be classed 'with arithmetic, which is purely a priori, but with mechanics'.

Gauss, Bolyai, Lobachevsky and Riemann developed the theory of various non-Euclidean geometries but did not show that they were consistent. Thus although they believed strongly that the answer to question (A′) was no, they did not prove it. (Certainly they did not prove it explicitly.) However, in the second half of the 19th century mathematicians (notably Beltrami) showed how to construct models for various non-Euclidean geometries using Euclidean geometry. In modern terms, they showed that, if Euclid's axioms (including the parallel axiom) form a consistent system, then the same set of axioms with the parallel axiom replaced by some appropriate alternative axiom remains consistent.

Finally, in 1917, Einstein's theory of general relativity confirmed the suspicion voiced from time to time by various farsighted mathematicians that question (B′) should be replaced by

(B″) Is Euclid's geometry the most convenient description of the world?

and that the answer to the new question was no.

In 1800, Euclid's geometry was the only rigorously developed part of mathematics. Initially, those, like Cauchy and Gauss, who sought to rigorise the calculus took the notion of magnitude (or as we would say real number) as implicit (or, as Kant would say, a priori), but it gradually became clear that the rigorous development of the calculus depended on (or was identical with) a close study of the real numbers. However, as I tried to show in Chapter 2, the properties of the real line that we want are by no means evident.

Faced with this problem, the natural way forward is to show that the complicated system constituted by the real numbers can be constructed from the simpler system of the rationals. Thus, if we believe that the rationals exist, we must believe that the reals exist. (Gauss and others had already shown how to construct $\mathbb{C}$ from $\mathbb{R}$, showing that, if the real numbers are acceptable, then so are the complex numbers.) Simpler constructions enabled mathematicians to obtain the rationals from the integers and the integers from the positive integers.

In the next sequence of exercises we outline these simpler constructions. The full verification of the statements made is long and tedious, and the reader should do as much or little as she wishes. Nothing in what follows depends on these exercises. I have felt free to use notions like equivalence relations, integral domains and so on which I have avoided in the main text.

Exercise 14.21. (Obtaining $\mathbb{Z}$ from $\mathbb{N}$.) *Here* $\mathbb{N} = \{0, 1, 2, \dots\}$. *We take* $(\mathbb{N}, +, \times)$ *with its various properties as given. Show that the relation*

$$(r, s) \sim (r', s') \text{ if } r + s' = r' + s$$

is an equivalence relation on $\mathbb{N}^2$.

We write $\mathbb{Z} = \mathbb{N}^2/\sim$ *for the set of equivalence classes*

$$[(r, s)] = \{(r', s') \in \mathbb{N}^2 \, : \, (r, s) \sim (r', s')\}.$$

Show that the following give well defined *operations:*

$$[(r, s)] + [(u, v)] = [(r + u, s + v)], \ [(r, s)] \times [(u, v)] = [(ru + sv, su + rv)].$$

Explain why you think we chose these definitions. Show that $(\mathbb{Z}, +, \times)$ *obeys the algebraic rules that it should (in more formal terms, that* $(\mathbb{Z}, +, \times)$ *is an integral domain). If you just wish to prove a selection of results, you might choose*

(a) There exists a unique $0 \in \mathbb{Z}$ *such that* $x + 0 = x$ *for all* $x \in \mathbb{Z}$.

(b) If $x \in \mathbb{Z}$, *then we can find a* $y \in \mathbb{Z}$ *such that* $x + y = 0$ *(we write* $y = -x$*).*

(c) If $x, y, z \in \mathbb{Z}$, *then* $x \times (y + z) = x \times y + x \times z$.

(d) There exists a unique $1 \in \mathbb{Z}$ *such that* $x \times 1 = x$ *for all* $x \in \mathbb{Z}$.

(e) If $z \neq 0$ *and* $x \times z = y \times z$, *then* $x = y$.

Show that the mapping $\theta : \mathbb{N} \to \mathbb{Z}$ *given by* $\theta(n) = [(n, 0)]$ *is injective and preserves addition and multiplication (that is,* $\theta(n + m) = \theta(n) + \theta(m)$ *and* $\theta(n \times m) = \theta(n) \times \theta(m)$*). Explain briefly why this means that we may consider* $\mathbb{N}$ *as a subset of* $\mathbb{Z}$.

Show that, if $\mathbb{P} = \theta(\mathbb{N})$, *then* $\mathbb{P}$ *obeys the appropriate version of axioms (P1) to (P3) in Appendix A. We write* $x > y$ *if* $x + (-y) \in \mathbb{P}$.

Exercise 14.22. (Obtaining $\mathbb{Q}$ from $\mathbb{Z}$.) *We take* $(\mathbb{Z}, +, \times)$ *with its various properties as given. Show that the relation*

$$(r, s) \sim (r', s') \text{ if } r \times s' = s \times r'$$

is an equivalence relation on $\mathbb{Z} \times (\mathbb{Z} \setminus \{0\})$.

We write $\mathbb{Q} = \mathbb{Z} \times (\mathbb{Z} \setminus \{0\})/\sim$ *for the set of equivalence classes*

$$[(r, s)] = \{(r', s') \in \mathbb{Z} \times (\mathbb{Z} \setminus \{0\}) \, : \, (r, s) \sim (r', s')\}.$$

Show that the following give well defined *operations:*

$$[(r, s)] + [(u, v)] = [(rv + su, sv)], \ [(r, s)] \times [(u, v)] = [(ru, sv)].$$

Explain why you think we chose these definitions. Show that $(\mathbb{Q}, +, \times)$ *obeys the algebraic rules (A1) to (A4), (M1) to (M4) and (D) set out in Appendix A (in algebraists' terms,* $(\mathbb{Q}, +, \times)$ *is a field). If you just wish to prove a selection of results, you might choose*

(a) There exists a unique $0 \in \mathbb{Q}$ *such that* $x + 0 = x$ *for all* $x \in \mathbb{Q}$.

(b) If $x \in \mathbb{Q}$, *then we can find a* $y \in \mathbb{Q}$ *such that* $x + y = 0$ *(we write* $y = -x$*).*

(c) If $x, y, z \in \mathbb{Q}$, *then* $x \times (y + z) = x \times y + x \times z$.

(d) There exists a unique $1 \in \mathbb{Z}$ *such that* $x \times 1 = x$ *for all* $x \in \mathbb{Q}$.

(e) If $x \neq 0$, *then we can find a* $y \in \mathbb{Q}$ *such that* $xy = 1$.

Define an appropriate $\mathbb{P}$ *which obeys the appropriate version of axioms (P1) to (P3) in Appendix A. [If you cannot do this, you do not understand what is going on.] We write* $x > y$ *if* $x + (-y) \in \mathbb{P}$. *In algebraists' terms,* $(\mathbb{Q}, +, \times, >)$ *is an ordered field.*

Define an injective map $\theta : \mathbb{Z} \to \mathbb{Q}$ *which preserves addition, multiplication and order (thus* $n > m$ *implies* $\theta(n) > \theta(m)$*) and show that it has the stated properties. Explain briefly why this means that we may consider* $\mathbb{Z}$ *as a subset (more strictly as a sub-ordered integral domain) of* $\mathbb{Q}$.

Exercise 14.23. (Obtaining $\mathbb{C}$ from $\mathbb{R}$.) *This is somewhat easier than the previous two exercises, since it does not involve equivalence classes and it is obvious that the definitions actually work. We take* $(\mathbb{R}, +, \times)$ *with its various properties as given.*

We write $\mathbb{C} = \mathbb{R}^2$ *and define the following operations on* $\mathbb{C}$:

$$(r, s) + (u, v) = (r + u, s + v), \quad (r, s) \times (u, v) = (ru - sv, rv + su).$$

Explain why you think we chose these definitions. Show that $(\mathbb{C}, +, \times)$ *obeys the algebraic rules (A1) to (A4), (M1) to (M4) and (D) set out in Appendix A (in algebraists' terms,* $(\mathbb{C}, +, \times)$ *is a field). If you just wish to prove a selection of results, you might choose*

(a) There exists a unique $0 \in \mathbb{C}$ *such that* $z + 0 = z$ *for all* $z \in \mathbb{C}$.

(b) If $z \in \mathbb{C}$, *then we can find a* $w \in \mathbb{C}$ *such that* $z + w = 0$ *(we write* $w = -z$*).*

(c) If $z, w, u \in \mathbb{C}$, *then* $z \times (w + u) = z \times w + z \times u$.

(d) There exists a unique $1 \in \mathbb{C}$ *such that* $z \times 1 = z$ *for all* $z \in \mathbb{C}$.

(e) If $z \neq 0$, *then we can find a* $w \in \mathbb{C}$ *such that* $zw = 1$.

Define an appropriate injective map $\theta : \mathbb{R} \to \mathbb{C}$ *which preserves addition and multiplication (thus* θ *is a field morphism), and show that it has the stated properties. Explain briefly why this means that we can consider* $\mathbb{R}$ *as a subfield of* $\mathbb{C}$.

Show that $(0, 1) \times (0, 1) = \theta(-1)$. *Show that, if we adopt the usual convention of considering* $\mathbb{R}$ *as a subfield of* $\mathbb{C}$ *and writing* $i = (0, 1)$, *then every element* z *of* $\mathbb{C}$ *can be written as* $z = a + bi$ *with* $a, b \in \mathbb{R}$ *in exactly one way.*

Show also that $(0, -1) \times (0, -1) = \theta(-1)$. *[This corresponds to the ambiguity introduced by referring to* i *as 'the square root of* -1*'. In* $\mathbb{C}$ *there are two square roots of* -1*, and there is no reason to prefer one square root rather than the other.]*

Mathematicians and philosophers have by now become so skilled in doubt that, whatever their private beliefs, few of them would care to maintain in public that the system of the positive integers (often called the 'natural numbers') is given a priori[4]. None the less, most mathematicians feel that, in some sense, the system of the positive integers is much more 'basic' or 'simple' or 'natural' than, say, the system of the complex numbers. If we can construct the complex numbers from the integers, we feel that much more confident in the system of the complex numbers.

Much of mathematics consists in solving problems and proving theorems, but part of its long-term progress is marked by changing and extending the nature of the problems posed. To a mathematician in 1800, the question 'Can you construct the reals from the rationals?' would be incomprehensible. One of the many obstacles to understanding the question would be the nature of the object constructed. Let us look at our construction of $\mathbb{Z}$ from $\mathbb{N}$ in Exercise 14.21. We say, in effect, that 'the integer 1 is the equivalence class of ordered pairs (n, m) of natural numbers such that their difference $n - m$ is the natural number 1'. The non-mathematician is entitled to object that, whatever the integer 1 may be, it is certainly not that. To this, the mathematician replies that the system constructed in Exercise 14.21 has exactly the properties that we wish the integers to have, and 'if it looks like a duck, swims like a duck and quacks like a duck, then it is a duck[5]'. To this, the philosopher objects that she can construct a clockwork toy that looks like a duck, swims like a duck and quacks like a duck. The mathematician replies that, if all that we want from a duck is that it should look like a duck, swim like a duck and quack like a duck, then, for our purposes, the clockwork toy is a duck.

14.4. How do we construct the reals? ♡

As the reader no doubt expects, we model the construction of the reals from the rationals on the completion arguments in Sections 14.1 and 14.2. As the reader no doubt also expects the construction will be long and relatively easy

[4]It is all very well to maintain that one plus one must always make two, but this fails when the one plus one are rabbits of different sexes or rain drops which run together. A stronger objection deals with statements like $n + m = m + n$. This statement might, perhaps, be a priori true if $n = 2$ and $m = 3$, but can it really be a priori true if n and m are integers so large that I could not write them down in my lifetime?

[5]This should be contrasted with the position in medical diagnosis. 'If it looks like a duck, swims like a duck and barks like a dog, then it is a duck. Every case presents atypical symptoms.'

with a great deal of routine verification left to her. However, it is important not to underestimate the subtlety of the task. Our earlier completion arguments used the existence and properties of the real numbers, our new arguments cannot. The reader should picture a street mime juggling non-existent balls. As the mime continues, the action of juggling slowly brings the balls into existence, at first in dim outline and then into solid reality.

We take the ordered field $(\mathbb{Q}, +, \times, >)$ as given. We recall that our definitions of convergence and Cauchy sequence apply in any ordered field. If $x_n \in \mathbb{Q}$ and $x \in \mathbb{Q}$, we say that $x_n \to x$ if, given any $\epsilon \in \mathbb{Q}$ with $\epsilon > 0$, we can find an $N(\epsilon) > 0$ such that

$$|x_n - x| < \epsilon \text{ for all } n \geq N(\epsilon) > 0.$$

We say that a sequence x_n in $\mathbb{Q}$ is Cauchy if, given any $\epsilon \in \mathbb{Q}$ with $\epsilon > 0$, we can find an $N(\epsilon) > 0$ such that

$$|x_n - x_m| < \epsilon \text{ for all } n,\, m \geq N(\epsilon) > 0.$$

We start by considering the space $\mathbf{X}$ of Cauchy sequences $\mathbf{x} = (x_n)_{n=1}^{\infty}$ in $\mathbb{Q}$. We write $\mathbf{x} \sim \mathbf{y}$ if $\mathbf{x}, \mathbf{y} \in \mathbf{X}$ and $x_n - y_n \to 0$ as $n \to \infty$. The reader should verify the statements made in the next exercise.

Exercise 14.24. *Suppose* $\mathbf{x}, \mathbf{y}, \mathbf{z} \in \mathbf{X}$. *Then:*

(i) $\mathbf{x} \sim \mathbf{x}$.

(ii) If $\mathbf{x} \sim \mathbf{y}$, *then* $\mathbf{y} \sim \mathbf{x}$.

(iii) If $\mathbf{x} \sim \mathbf{y}$ *and* $\mathbf{y} \sim \mathbf{z}$, *then* $\mathbf{x} \sim \mathbf{z}$.

We write $[\mathbf{x}] = \{\mathbf{y} : \mathbf{y} \sim \mathbf{x}\}$ and call $[\mathbf{x}]$ the *equivalence class* of $\mathbf{x}$. We take $\mathbb{R}$ to be the set of all such equivalence classes. The reader should verify the statements made in the next exercise.

Exercise 14.25. *Each* $\mathbf{x} \in \mathbf{X}$ *belongs to exactly one equivalence class in* $\mathbb{R}$.

Lemma 14.26. *(i) If* $\mathbf{x}, \mathbf{y} \in \mathbf{X}$, *then, taking* $\mathbf{x}+\mathbf{y}$ *to be the sequence whose nth term is* $x_n + y_n$, *we have* $\mathbf{x} + \mathbf{y} \in \mathbf{X}$.

(ii) If $\mathbf{x}, \mathbf{y}, \mathbf{x}', \mathbf{y}' \in \mathbb{R}$ *and* $\mathbf{x} \sim \mathbf{x}'$, $\mathbf{y} \sim \mathbf{y}'$, *then* $\mathbf{x} + \mathbf{y} \sim \mathbf{x}' + \mathbf{y}'$.

(iii) If $\mathbf{x}, \mathbf{y} \in \mathbf{X}$ *then, taking* $\mathbf{x} \times \mathbf{y}$ *to be the sequence whose nth term is* $x_n \times y_n$, *we have* $\mathbf{x} \times \mathbf{y} \in \mathbf{X}$.

(iv) If $\mathbf{x}, \mathbf{y}, \mathbf{x}', \mathbf{y}' \in \mathbb{R}$ *and* $\mathbf{x} \sim \mathbf{x}'$, $\mathbf{y} \sim \mathbf{y}'$, *then* $\mathbf{x} \times \mathbf{y} \sim \mathbf{x}' \times \mathbf{y}'$.

Proof. I leave parts (i) and (ii) to the reader.

(iii) Since $\mathbf{x}$ and $\mathbf{y}$ are Cauchy sequences, they are bounded; that is, we can find a K such that $|x_n|, |y_n| \leq K$ for all n. Thus (writing $ab = a \times b$)

$$\begin{aligned}|x_n y_n - x_m y_m| &\leq |x_n y_n - x_n y_m| + |x_n y_m - x_m y_m| \\ &= |x_n||y_n - y_m| + |y_m||x_n - x_m| \leq K(|y_n - y_m| + |x_n - x_m|),\end{aligned}$$

and so the $x_n y_n$ form a Cauchy sequence.

(iv) As in (iii), we can find a K such that $|x_n|$, $|y_n'| \leq K$ for all n, and so

$$\begin{aligned}|x_n y_n - x_n' y_n'| &\leq |x_n y_n - x_n y_n'| + |x_n y_n' - x_n' y_n'| \\ &= |x_n||y_n - y_n'| + |y_n'||x_n - x_n'| \leq K(|y_n - y_n'| + |x_n - x_n'|) \to 0\end{aligned}$$

as $n \to \infty$, as required. ■

Lemma 14.26 tells us that we can define addition and multiplication of real numbers by the formulae

$$[\mathbf{x}] + [\mathbf{y}] = [\mathbf{x} + \mathbf{y}], \ [\mathbf{x}] \times [\mathbf{y}] = [\mathbf{x} \times \mathbf{y}].$$

Lemma 14.27. *The system* $(\mathbb{R}, +, \times)$ *is a field (that is to say, it satisfies the rules (A1) to (A4), (M1) to (M4) and (D) set out in Appendix A).*

Proof. I claim, and the reader should check, that the verification is simple except possibly in the case of rule (M4). (We take $0 = [\mathbf{a}]$ with $a_j = 0$ for all j and $1 = [\mathbf{b}]$ with $b_j = 1$ for all j.)

Rule (M4) states that, if $[\mathbf{x}] \in \mathbb{R}$ and $[\mathbf{x}] \neq [\mathbf{a}]$ (where $a_j = 0$ for all j), then we can find a $[\mathbf{y}] \in \mathbb{R}$ such that $[\mathbf{x}] \times [\mathbf{y}] = [\mathbf{b}]$ (where $b_j = 1$ for all j).

To prove this, observe that, if $[\mathbf{x}] \neq [\mathbf{a}]$, then $x_n = x_n - a_n \nrightarrow 0$ as $n \to 0$. Thus we can find an $\epsilon \in \mathbb{Q}$ with $\epsilon > 0$ such that, given any N, we can find an $M \geq N$ such that $|x_M| > \epsilon$. Since the sequence x_n is Cauchy, we can find an $N(0)$ such that

$$|x_n - x_m| < \epsilon/2 \text{ for all } n, \ m \geq N(0) > 0.$$

Choose $N(1) \geq N(0)$ such that $|x_{N(1)}| > \epsilon$. We then have

$$|x_n| \geq |x_{N(1)}| - |x_n - x_{N(1)}| > \epsilon/2$$

for all $n \geq N(0)$.

Set $y_j = 1$ for $j < N(0)$ and $y_j = x_j^{-1}$ for $j \geq N(0)$. If n, $m \geq N(0)$, we have

$$|y_n - y_m| = |x_n - x_m||x_n|^{-1}|x_m|^{-1} \leq 4\epsilon^{-2}|x_n - x_m|,$$

so the sequence y_n is Cauchy. Since $x_j y_j = 1 = b_j$ for all $j \geq N(0)$, we have $[\mathbf{x}] \times [\mathbf{y}] = [\mathbf{b}]$, as required. ■

We now introduce an order on $\mathbb{R}$. This is an essential step before we can introduce distance and limits. We shall need the result of the following exercise.

Exercise 14.28. *(i) Let* $a_j = 0$ *for all* j. *By adapting the argument of the third paragraph of the proof of Lemma 14.27, show that, if* $[\mathbf{x}] \in \mathbb{R}$ *and* $[\mathbf{x}] \neq [\mathbf{a}]$, *then either*

(a) there exist an $N > 0$ *and an* $\epsilon \in \mathbb{Q}$ *with* $\epsilon > 0$ *such that* $x_n > \epsilon$ *for all* $n \geq N$*, or*

(b) there exist an $N > 0$ *and an* $\epsilon \in \mathbb{Q}$ *with* $\epsilon > 0$ *such that* $-x_n > \epsilon$ *for all* $n \geq N$.

(ii) Suppose that $[\mathbf{x}] \in \mathbb{R}$ *and there exist an* $N > 0$ *and an* $\epsilon \in \mathbb{Q}$ *with* $\epsilon > 0$ *such that* $x_n > \epsilon$ *for all* $n \geq N$*. Show that, if* $[\mathbf{y}] = [\mathbf{x}]$*, then there exists an* $N' > 0$ *such that* $y_n > \epsilon/2$ *for all* $n \geq N'$.

Lemma 14.29. *The set* $\mathbb{P}$ *of* $[\mathbf{x}] \in \mathbb{R}$ *such that there exist an* $N > 0$ *and an* $\epsilon \in \mathbb{Q}$ *with* $\epsilon > 0$ *such that* $x_n > \epsilon$ *for all* $n \geq N$ *is well defined. It has the following properties.*

(P1) If $[\mathbf{x}], [\mathbf{y}] \in \mathbb{P}$*, then* $[\mathbf{x}] + [\mathbf{y}] \in \mathbb{P}$.

(P2) If $[\mathbf{x}], [\mathbf{y}] \in \mathbb{P}$*, then* $[\mathbf{x}] \times [\mathbf{y}] \in \mathbb{P}$.

(P3) If $[\mathbf{x}] \in \mathbb{R}$*, then one and only one of the following three statements is true:* $[\mathbf{x}] \in \mathbb{P}$, $-[\mathbf{x}] \in \mathbb{P}$*, or* $[\mathbf{x}] = [\mathbf{a}]$*, where* $a_j = 0$ *for all* j.

Proof. The fact that $\mathbb{P}$ is well defined follows from part (ii) of Exercise 14.28. If $\epsilon_1 \in \mathbb{Q}$, $\epsilon_1 > 0$, $x_n \geq \epsilon_1$ for $n \geq N_1$ and $\epsilon_2 \in \mathbb{Q}$, $\epsilon_2 > 0$, $y_n \geq \epsilon_2$ for $n \geq N_2$, then $x_n + y_n \geq \epsilon_1 + \epsilon_2 > 0$ and $x_n \times y_n \geq \epsilon_1 \times \epsilon_2 > 0$ for all $n \geq \max(N_1, N_2)$, so conclusions (P1) and (P2) follow.

By part (i) of Exercise 14.28, we know that, if $[\mathbf{x}] \in \mathbb{R}$, then at least one of the three statements $[\mathbf{x}] \in \mathbb{P}$, $-[\mathbf{x}] \in \mathbb{P}$ or $[\mathbf{x}] = [\mathbf{a}]$ is true. Since the statements are exclusive, exactly one is true. ■

As usual, we write $[\mathbf{x}] > [\mathbf{y}]$ if $[\mathbf{x}] - [\mathbf{y}] \in \mathbb{P}$. If $[\mathbf{x}] \in \mathbb{R}$, then, as we have just shown, either $[\mathbf{x}] \in \mathbb{P}$, and we define $|[\mathbf{x}]| = [\mathbf{x}]$; or $-[\mathbf{x}] \in \mathbb{P}$, and we define $|[\mathbf{x}]| = -[\mathbf{x}]$; or $[\mathbf{x}] = [\mathbf{a}]$ (where $a_j = 0$ for all j), and we set $|[\mathbf{x}]| = [\mathbf{a}]$.

If $w \in \mathbb{Q}$, we set $w_n = w$ for all n and write $\theta(w) = [\mathbf{w}]$. We leave it to the reader to verify the next lemma.

Lemma 14.30. *The map* $\theta : \mathbb{Q} \to \mathbb{R}$ *is injective and preserves addition, multiplication and order.*

Note that $\theta(0) = [\mathbf{a}]$, where $a_j = 0$ for all j.

We have now shown that $(\mathbb{R}, +, \times, >)$ is an ordered field, and so our definitions of convergence and Cauchy sequence apply. If $[\mathbf{x}(n)] \in \mathbb{R}$ and $[\mathbf{x}] \in \mathbb{R}$, we say that $[\mathbf{x}(n)] \to [\mathbf{x}]$, if, given any $[\boldsymbol{\epsilon}] \in \mathbb{R}$ with $[\boldsymbol{\epsilon}] > \theta(0)$, we can find an $N([\boldsymbol{\epsilon}]) > 0$ such that

$$|[\mathbf{x}(n)] - [\mathbf{x}]| < [\boldsymbol{\epsilon}] \text{ for all } n \geq N([\boldsymbol{\epsilon}]).$$

We say that a sequence $[\mathbf{x}(n)]$ in $\mathbb{R}$ is Cauchy if, given any $[\epsilon] \in \mathbb{R}$ with $[\epsilon] > \theta(0)$, we can find an $N([\epsilon]) > 0$ such that

$$|[\mathbf{x}(n)] - [\mathbf{x}(m)]| < [\epsilon] \text{ for all } n,\, m \geq N([\epsilon]).$$

Lemma 14.31. *Given any $[\epsilon] \in \mathbb{R}$ with $[\epsilon] > \theta(0)$, we can find a $y \in \mathbb{Q}$ with $y > 0$ such that $[\epsilon] > \theta(y)$.*

Proof. This follows from our definition of $>$ via the set $\mathbb{P}$ and part (ii) of Exercise 14.28. ∎

Lemma 14.32. *Given any $[\mathbf{x}]$, $[\mathbf{y}] \in \mathbb{R}$ with $[\mathbf{x}] > [\mathbf{y}]$, we can find a $z \in \mathbb{Q}$ such that $[\mathbf{x}] > \theta(z) > [\mathbf{y}]$.*

Proof. By definition, $[\mathbf{x}] - [\mathbf{y}] > \theta(0)$; so, by Lemma 14.31, we can find a $v \in \mathbb{Q}$ with $v > 0$ such that $[\mathbf{x}] - [\mathbf{y}] > \theta(v) > \theta(0)$. Setting $u = v/5$, we obtain a $u \in \mathbb{Q}$ with $u > 0$ such that $[\mathbf{x}] - [\mathbf{y}] > \theta(5u) > \theta(0)$. It follows that we can find an M such that

$$x_j - y_j > 4u \text{ for all } j \geq M.$$

Now choose an $N \geq M$ such that $|y_j - y_k| < u$ and so, in particular,

$$y_j - y_k > -u \text{ for all } j,\, k \geq N.$$

Set $z = y_N + 2u$. By the displayed inequalities just obtained, we have

$$x_k > 4u + y_k = (y_N + 2u) + (y_k - y_N) + 2u > z + u$$

and

$$y_k < y_N + u \leq z - u$$

for all $k \geq N$. Thus $[\mathbf{x}] > \theta(z) > [\mathbf{y}]$, as required. ∎

Up to now, our arguments have only used the fact that $(\mathbb{Q}, +, \times, >)$ is an ordered field. We now use the fact, specific to $(\mathbb{Q}, +, \times, >)$, that every rational $x > 0$ can be written as $x = p/q$ with p and q strictly positive integers. Since $p/q \geq 1/q$, Lemma 14.31 implies the axiom of Archimedes.

Lemma 14.33. *We have $\theta(1/n) \to \theta(0)$ as $n \to \infty$.*

Exercise 14.34. *Give the details of the proof of Lemma 14.33.*

We now show that every Cauchy sequence in $(\mathbb{R}, +, \times, >)$ converges. The proof runs along the same lines as the proof of Lemma 14.16

Lemma 14.35. *If $[\mathbf{y}(1)]$, $[\mathbf{y}(2)]$, $[\mathbf{y}(3)]$, $\ldots$ is a Cauchy sequence in $\mathbb{R}$, then we can find an $[\mathbf{x}] \in \mathbb{R}$ such that $[\mathbf{y}(j)] \to [\mathbf{x}]$ as $j \to \infty$.*

Proof. Since $[\mathbf{y}(j)] + \theta(j^{-1}) > [\mathbf{y}(j)]$, it follows from Lemma 14.32 that we can find an $x_j \in \mathbb{Q}$ such that

$$[\mathbf{y}(j)] + \theta(j^{-1}) > \theta(x_j) > [\mathbf{y}(j)],$$

and so

$$|\theta(x_j) - [\mathbf{y}(j)]| < j^{-1}.$$

We observe that

$$\begin{aligned}\theta(|x_j - x_k|) &= |\theta(x_j) - \theta(x_k)| \\ &\leq |\theta(x_j) - [\mathbf{y}(j)]| + |\theta(x_k) - [\mathbf{y}(k)]| + |[\mathbf{y}(j)] - [\mathbf{y}(k)]| \\ &< \theta(j^{-1}) + \theta(k^{-1}) + |[\mathbf{y}(j)] - [\mathbf{y}(k)]|.\end{aligned}$$

Given any $\epsilon \in \mathbb{Q}$, we can find an N with the properties that $N^{-1} < \epsilon/3$ and $|[\mathbf{y}(j)] - [\mathbf{y}(k)]| < \theta(\epsilon/3)$ for all $j,\ k \geq N$. The results of this paragraph show that

$$|x_j - x_k| < \epsilon/3 + \epsilon/3 + \epsilon/3 = \epsilon$$

for all $j,\ k \geq N$, and so the x_j form a Cauchy sequence.

We now wish to show that $[\mathbf{y}(n)] \to [\mathbf{x}]$ as $n \to \infty$. Let $[\boldsymbol{\epsilon}] > \theta(0)$ be given. Since the sequence $[\mathbf{y}(n)]$ is Cauchy, we can find an M such that

$$|[\mathbf{y}(j)] - [\mathbf{y}(k)]| < \theta(1/2)[\boldsymbol{\epsilon}] \text{ for all } j,\ k \geq M.$$

We now use the axiom of Archimedes (Lemma 14.33) to show that we can pick an $N \geq M$ such that $\theta(N^{-1}) < \theta(1/6)[\boldsymbol{\epsilon}]$, and observe that the inequality proved in the last paragraph gives

$$|\theta(x_j) - \theta(x_k)| < \theta(1/2)[\boldsymbol{\epsilon}] \text{ for all } j,\ k \geq N.$$

Thus

$$|[\mathbf{x}] - \theta(x_k)| \leq \theta(1/2)[\boldsymbol{\epsilon}] \text{ for all } k \geq N,$$

and

$$|[\mathbf{x}] - [\mathbf{y}(k)]| \leq |[\mathbf{x}] - \theta(x_k)| + |\theta(x_k) - [\mathbf{y}(k)]| \leq \theta(1/2)[\boldsymbol{\epsilon}] + \theta(N^{-1}) < [\boldsymbol{\epsilon}]$$

for all $k \geq N$, so we are done. ■

Exercise 4.87 shows that the general principle of convergence and the axiom of Archimedes together imply the fundamental axiom of analysis. We have thus proved our desired theorem.

Theorem 14.36. $(\mathbb{R}, +, \times, >)$ *is an ordered field satisfying the fundamental axiom of analysis.*

As we remarked earlier, most of the construction given in this section applies to any ordered field. In Appendix G we push this idea a little further. As might be expected from Lemma 14.5 and the accompanying discussion, the real number system is unique (in an appropriately chosen sense). The details are given in Appendix A.

14.5. Paradise lost? ♡♡

Morris Kline once wrote that 'A proof tells us where to concentrate our doubts.' Our construction of the reals from the rationals is such a remarkable result that it merits further examination. We know that the rationals are countable and the reals are uncountable. Where in the proof does the switch from countable to uncountable occur? Inspection of the proof shows that the switch happens right at the beginning with the sentence 'We start by considering the space $\mathbf{X}$ of Cauchy sequences $\mathbf{x} = (x_n)_{n=1}^{\infty}$ in $\mathbb{Q}$.' If pressed to justify this step, I might split it into two:

(a) Consider the set $\mathbb{Q}^{\mathbb{N}}$ of all sequences $\mathbf{x} = (x_n)_{n=1}^{\infty}$ in $\mathbb{Q}$ (that is, all functions $\mathbf{x} : \mathbb{N} \to \mathbb{Q}$, where $\mathbb{N}$ is the set of strictly positive integers).

(b) Consider the subset $\mathbf{X}$ of $\mathbb{Q}^{\mathbb{N}}$ consisting of all sequences which satisfy the Cauchy condition.

Unfortunately my justification resembles the following pair of definitions.

(a′) Consider the set Ω of all sets.

(b′) Consider the subset Φ of Ω consisting of all $A \in \Omega$ such that $A \notin A$.

As the reader no doubt knows, the definition of Φ leads to a contradiction. If $\Phi \in \Phi$, then $\Phi \notin \Phi$, but if $\Phi \notin \Phi$, then $\Phi \in \Phi$.

There are two schools of thought regarding this paradox. The radical school wishes to ban all definitions of the form (a) and (b) from mathematics. At first sight this seems an easy task – a return to the simpler and healthier practices of our ancestors. However, a householder who investigates a trace of dry rot in one room may well discover that her whole house is so riddled with fungus that the whole structure must be torn down and rebuilt. It turns out that classical mathematics (that is, the kind of mathematics discussed in this book) depends so much on definitions and arguments which are unacceptable to the radicals that they have to rebuild mathematics from scratch.

The various schemes for rebuilding restrict mathematics to the study of 'constructible objects' and break decisively even at the most elementary level with the traditional practices of mathematicians. A striking example of this break is that, in the rebuilt structure, all functions are continuous. In Appendix H, I try to indicate why this is so.

The conservative school, to which most mathematicians belong, says that there are 'natural rules' which bar at least one of (a′) and (b′) whilst allowing both (a) and (b). Just as there are different radical theories with different criteria as to what is a 'constructible object', so there are different systems of conservative 'natural rules'. The standard system of rules (the Zermelo–Fraenkel system) is set out with great charm and clarity in Halmos's little masterpiece *Naive Set Theory* [**21**]. Halmos also shows how the system of positive integers can be constructed in a rather natural way using the Zermelo–Fraenkel axioms. The axioms for set theory must be supplemented by a system of rules setting out what kind of statements are allowed in mathematics[6] and what laws of inference we may use. Once you have read Halmos, then Johnstone's *Notes on Logic and Set Theory* [**27**] provides an excellent short[7] but rigorous account of all that the interested non-specialist needs to know.

All of the analysis in this book and almost all of mathematics that is presently known can be deduced from the small collection of Zermelo–Fraenkel axioms in the same way and with the same standard of rigour[8] as Euclid deduced his geometry from his axioms[9]. However, plausible as the Zermelo–Fraenkel axioms seem, their plausibility is based on our experience with finite sets, and to construct classical mathematics we must apply these axioms to infinite sets which lie outside our physical and mental experience. We cannot therefore be sure that the axioms may not lead to some subtle contradiction[10]. Kline quotes Poincaré, a witty, though not very profound, critic of set theory, as saying 'We have put a fence around the herd to protect it from the wolves, but we do not know whether some wolves were not already within the fence.'

Few mathematicians of the conservative school seem worried by this. Partly, this is because in nearly a century of continuous scrutiny no one has produced a contradiction. (Personally, I am less convinced that the human race will survive the next hundred years than that the Zermelo–Fraenkel system will survive the century.) Partly it is because, if a contradiction appeared in the Zermelo–Fraenkel system, we could simply use one of the

---

[6] Thus excluding 'This sentence is false or meaningless', 'The smallest positive integer not definable in less than fifty English words', 'When does the next train leave for London?' and so on.

[7] I emphasise the word short. The path from the Zermelo–Fraenkel axioms to the positive integers can be traversed easily in a 24 hour course.

[8] I choose my words carefully. Modern mathematicians have found gaps in Euclid's reasoning but none of these is large enough to invalidate the work.

[9] The massive Bourbaki volumes constitute a proof of this statement.

[10] Hilbert tried to avoid this problem by finding a model for the Zermelo–Fraenkel system inside some system like those proposed by the radicals which would be 'obviously' free of contradiction. Most mathematicians agree that Gödel's theorem makes it extremely unlikely that such a programme could be successful.

rival conservative systems of axioms or, if the worst came to the worst, we could admit that the radicals were right, after all, and study one of their reconstructed systems. (Modern mathematics thus appears as a ship towing a long line of lifeboats. If the ship sinks we move to the first lifeboat. If the first lifeboat sinks we move to the second, and so on.) Partly it is because the failure of the Zermelo–Fraenkel system would be one of the most interesting things that could happen to mathematics. And finally, we have come to accept that no really interesting part of mathematics can be proved free of contradiction[11].

So far in this section, I have tried to give a fair representation of the thoughts of others. For the short remainder of this section, I shall give my own views. Mathematics changes slowly. After ten years, some problems have been solved and some new problems have arisen, but the subject is recognisably the same. However, over a century it changes in ways which mathematicians at the beginning of that century could not conceive. The 20th century has, for example, seen special and general relativity, quantum mechanics, the electronic computer, the theorems of Gödel and Turing and Kolmogorov's axiomatisation of probability. The 19th century saw non-Euclidean geometry, complex variable theory, quaternions and Cantor's set theory. The 17th and 18th centuries saw the invention of the calculus.

Given this, it seems presumptuous to seek a permanent foundation for mathematics. It is a happy accident that almost all of mathematics, as we now know it, can be deduced from one set of axioms; but, just as history has taught us that there is not one geometry but many, so it is not unlikely that the future will present us with different mathematics deduced from radically different axiom schemes. The art of the mathematician will still consist in the rigorous proof of unexpected conclusions from simple premises, but those premises will change in ways that we cannot possibly imagine.

[11]A hungry fox tried to reach some clusters of grapes which he saw hanging from a vine trained on a tree, but they were too high. So he went off and comforted himself by saying 'They weren't ripe anyhow'. (Aesop)

Appendix A

Ordered Fields

Axioms for an ordered field. *An ordered field* $(\mathbb{F}, +, \times, >)$ *is a set* $\mathbb{F}$*, together with two binary operations* $+$ *and* $\times$ *and a binary relation* $>$*, obeying the following rules. (We adopt the notation* $ab = a \times b$*.)*

(A1) $a + (b + c) = (a + b) + c$.

(A2) $a + b = b + a$.

(A3) There is a unique element $0 \in \mathbb{F}$ *such that* $a + 0 = a$ *for all* $a \in \mathbb{F}$.

(A4) If $a \in \mathbb{F}$*, then we can find a* $-a \in \mathbb{F}$ *such that* $a + (-a) = 0$.

[Conditions (A1) to (A4) tell us that $(\mathbb{F}, +)$ *is a commutative group. We write* $a - b = a + (-b)$*.]*

(M1) $a(bc) = (ab)c$.

(M2) $ab = ba$.

(M3) There is a unique element $1 \in \mathbb{F}$ *such that* $a1 = a$ *for all* $a \in \mathbb{F}$.

(M4) If $a \in \mathbb{F}$ *and* $a \neq 0$*, then we can find an* $a^{-1} \in \mathbb{F}$ *such that* $aa^{-1} = 1$.

[Conditions (M1) to (M4) tell us, among other things, that $(\mathbb{F} \setminus \{0\}, \times)$ *is a commutative group.]*

(D) $a(b + c) = ab + ac$.

[Condition (D) is called the distributive law.]

If we write $\mathbb{P} = \{a \in \mathbb{F} : a > 0\}$*, then:*

(P1) If $a, b \in \mathbb{P}$*, then* $a + b \in \mathbb{P}$.

(P2) If $a, b \in \mathbb{P}$*, then* $ab \in \mathbb{P}$.

(P3) If $a \in \mathbb{F}$, then one and only one of the following three statements is true: $a \in \mathbb{P}$, $-a \in \mathbb{P}$ or $a = 0$.

We call $\mathbb{P}$ the set of strictly positive elements of $\mathbb{F}$, and write $a > b$ if and only if $a - b \in \mathbb{P}$.

Note that, although we state the axioms here, we shall not use them explicitly in this book. *Everything depending on these axioms is, so far as we are concerned, mere algebra.*

We write $a \geq 0$ if $a > 0$ or $a = 0$.

Here is an example of the kind of algebra that we would need to do to make our arguments complete.

Definition A.1. *We work in an ordered field $\mathbb{F}$ using the notation just introduced. If $a \in \mathbb{F}$, we write*

$$|a| = \begin{cases} a & \text{if } a \geq 0, \\ -a & \text{otherwise.} \end{cases}$$

We call $|a|$ the absolute value of a.

Lemma A.2. *We work in an ordered field $\mathbb{F}$. Suppose $a, b \in \mathbb{F}$. Then*

(i) $|a| \geq 0$,

(ii) $|(|a|)| = |a|$,

(iii) $|-a| = |a|$,

(iv) $|ab| = |a||b|$,

(v) $|a| + |b| \geq |a + b|$.

Proof. (This is just a sketch. We asume the consequence of all the axioms except the order axioms.) Parts (i) to (iii) can be proved by working through cases. To prove (iv), observe that part (iii) and (P3) allow us to assume without loss of generality that $a, b \geq 0$.

To prove (v), observe that (P3) allows us to assume without loss of generality that $|a| \geq |b|$. Part (iii) now allows us to assume without loss of generality that $a \geq 0$. If $b \geq 0$ then, by (P1), $a + b \geq 0$, so

$$|a| + |b| = a + b = |a + b|.$$

If $b < 0$, then $-b > 0$ so, using (P1),

$$(a - b) - (a + b) = (-b) + (-b) > 0,$$

and so

$$|a| + |b| = a - b > a + b = |a + b|.$$

■

The following series of exercises shows that, up to the appropriate isomorphism, the reals are the unique ordered field satisfying the fundamental axiom. Although the reader should probably go through such a proof at some time, it will become easier as she gets more experience (for example, by reading the first four sections of Chapter 14) and is not needed elsewhere in this book.

Exercise A.1. *We work in an ordered field $(\mathbb{F}, +, \times, >)$. You should cite explicitly each of the axioms from the list given on page 357 that you use. We write $1_{\mathbb{F}}$ for the unit of $\mathbb{F}$ (that is, for the unique element of $\mathbb{F}$ with $1_{\mathbb{F}}a = a$ for all $a \in \mathbb{F}$) and $0_{\mathbb{F}}$ for the zero of $\mathbb{F}$ (that is, for the unique element of $\mathbb{F}$ with $0_{\mathbb{F}} + a = a$ for all $a \in \mathbb{F}$).*

(i) By considering the equality

$$ab + \big(a(-b) + (-a)(-b)\big) = \big(ab + a(-b)\big) + (-a)(-b),$$

show that $(-a)(-b) = ab$ for all $a, b \in \mathbb{F}$.

(ii) By (i), we have, in particular, $a^2 = (-a)^2$. By applying axiom (P3), show that $a^2 > 0_{\mathbb{F}}$ whenever $a \neq 0_{\mathbb{F}}$.

(iii) Deduce that $1_{\mathbb{F}} > 0_{\mathbb{F}}$.

Exercise A.2. *We continue with the discussion of Exercise A.1. If n is a strictly positive integer, let us write*

$$n_{\mathbb{F}} = \overbrace{1_{\mathbb{F}} + 1_{\mathbb{F}} + \dots + 1_{\mathbb{F}}}^{n}.$$

Show, by induction, or otherwise, that $n_{\mathbb{F}} > 0_{\mathbb{F}}$, and so in particular $n_{\mathbb{F}} \neq 0_{\mathbb{F}}$.

In the language of modern algebra, we have shown that an ordered field has characteristic ∞. It is a standard result that any field of characteristic ∞ contains a copy of the rationals.

Exercise A.3. *We continue with the discussion of Exercise A.2. If n is a strictly negative integer, let us write $n_{\mathbb{F}} = -(-n)_{\mathbb{F}}$.*

(i) Show that if we set $\tau(n) = n_{\mathbb{F}}$, then $\tau : \mathbb{Z} \to \mathbb{F}$ is a well defined injective map with

$$\tau(n+m) = \tau(n) + \tau(m),$$
$$\tau(nm) = \tau(n)\tau(m),$$
$$\tau(n) > 0_{\mathbb{F}} \text{ whenever } n > 0,$$

for all $n, m \in \mathbb{Z}$.

(ii) If $n, m \in \mathbb{Z}$ and $m \neq 0$, let us set $\phi(n/m) = \tau(n)\tau(m)^{-1}$. Show that $\phi : \mathbb{Q} \to \mathbb{F}$ is a well defined (remember that $rn/rm = n/m$ for $r \neq 0$)

injective map with

$$\phi(x+y) = \phi(x) + \phi(y),$$
$$\phi(xy) = \phi(x)\phi(y),$$
$$\phi(x) > 0_{\mathbb{F}} \text{ whenever } x > 0,$$

for all $x,\ y \in \mathbb{Q}$.

So far we have merely been doing algebra.

Exercise A.4. *We continue with the discussion of Exercise A.3, but now we assume that* $\mathbb{F}$ *satisfies the fundamental axiom. Let us write* $\mathbb{K} = \phi(\mathbb{Q})$ *(informally,* $\mathbb{K}$ *is a copy of* $\mathbb{Q}$ *in* $\mathbb{F}$*). Prove the following slight improvement of Lemma 1.26. If* $x \in \mathbb{F}$*, we can find* $x_j \in \mathbb{K}$ *with* $x_1 \leq x_2 \leq \dots$ *such that* $x_j \to x$ *as* $j \to \infty$.

Exercise A.5. *We continue with the discussion of Exercise A.4, so, in particular, we assume that* $\mathbb{F}$ *satisfies the fundamental axiom.*

(i) If $x \in \mathbb{R}$*, then, by Exercise A.4, we can find* $x_j \in \mathbb{Q}$ *with* $x_1 \leq x_2 \leq \dots$ *such that* $x_j \to x$*. Show, by using the fundamental axiom, that there is an* $a \in \mathbb{F}$ *such that* $\phi(x_j) \to a$ *as* $j \to \infty$.

(ii) Show further that, if $x'_j \in \mathbb{Q}$ *and* $x'_j \to x$*, then* $\phi(x'_j) \to a$ *as* $j \to \infty$.

(iii) Conclude that we may define $\theta : \mathbb{R} \to \mathbb{F}$ *by adopting the procedure of part (i) and setting* $\theta(x) = a$.

(iv) Show that $\theta : \mathbb{R} \to \mathbb{F}$ *is a bijective map.*

(v) Show that, if $x,\ y \in \mathbb{R}$*, then*

$$\theta(x+y) = \theta(x) + \theta(y),$$
$$\theta(xy) = \theta(x)\theta(y),$$
$$\theta(x) > 0_{\mathbb{F}} \text{ whenever } x > 0.$$

Thus the real numbers are (for the purposes of analysis) unique.

Appendix B

Countability

Most of my readers will already have met countability and will not need to read this appendix. If you have not met countability, this appendix gives a 'quick and dirty' introduction[1].

Definition B.1. *A set E is called* countable *if $E = \varnothing$ or the elements of E can be written out as a sequence e_j (possibly with repetition), that is, we can find $e_1, e_2, \ldots$ such that*

$$E = \{e_1, e_2, e_3 \ldots\}.$$

Lemma B.2. *A set E is countable if and only if E is finite or the elements of E can be written out as a sequence e_j without repetition, that is, we can find $e_1, e_2, \ldots$, all unequal, such that*

$$E = \{e_1, e_2, e_3 \ldots\}.$$

Proof. If the elements of E can be written out as a sequence e_j (possibly with repetition), then either E is finite or we can obtain E as a sequence without repetition by striking out all terms equal to some previous one.

If E is finite, then either $E = \varnothing$, so E is countable by definition, or we can write $E = \{e_j : 1 \leq j \leq N\}$. Setting $e_n = e_N$ for $n \geq N$, we have

$$E = \{e_1, e_2, e_3 \ldots\}.$$

■

Exercise B.3. *Show that any subset of a countable set is countable.*

[1] The standard treatment via injectivity gives more flexible methods which are easily extended to more general situations

$$\begin{array}{lllll} b_1 & b_3 & b_6 & b_{10} & \ldots \\ b_2 & b_5 & b_9 & \ldots & \\ b_4 & b_8 & \ldots & & \\ b_7 & \ldots & & & \\ \ldots & & & & \end{array} \qquad \begin{array}{lllll} a_{1,1} & a_{1,2} & a_{1,3} & a_{1,4} & \ldots \\ a_{2,1} & a_{2,2} & a_{2,3} & \ldots & \\ a_{3,1} & a_{3,2} & \ldots & & \\ a_{4,1} & \ldots & & & \\ \ldots & & & & \end{array}$$

Figure B.1. A sequence of sequences set out as a sequence

The key result on countability, at least so far as we are concerned, is given in the next theorem.

Theorem B.4. *If A_1, A_2, ... are countable, then so is $\bigcup_{j=1}^{\infty} A_j$.*

In other words, the countable union of countable sets is countable[2].

Proof. We leave it to the reader to give the proof when all of the A_j are empty (easy) or not all the A_j are empty but all but finitely many are (either modify the proof below or adjoin infinitely many copies of one of the non-empty A_j).

In the remaining case, infinitely many of the A_j are non-empty and, by striking out all the empty sets, we may assume that all of the A_j are non-empty. Let us write

$$A_j = \{a_{j,1},\ a_{j,2},\ a_{j,3},\ \ldots\}.$$

If we write

$$N(r) = 1 + 2 + \cdots + (r-1) = r(r-1)/2$$

and set $b_{N(r)+k} = a_{r-k+1,k}$ $[1 \leq k \leq r,\ 1 \leq r]$ (so that we follow the pattern shown in Figure B.1), we see that

$$\bigcup_{j=1}^{\infty} A_j = \{b_1,\ b_2,\ b_3 \ldots\},$$

and so $\bigcup_{j=1}^{\infty} A_j$ is countable. ■

Exercise B.5. *Carry out the proofs asked for in the first sentence of the proof of Theorem B.4*

Theorem B.4 forms the basis for an argument[3] known as the 'hamburger argument'. (Consider the set A_j of hamburgers to be found within a sphere of radius j kilometres with centre the middle of Trafalgar square. Clearly A_j

[2] I would prefer to say that the counted union of counted sets can be counted, but tradition is against me.

[3] Marianna Csörnyei claims that this is taught in Hungarian primary schools, but she may exaggerate.

is finite and so countable. But the set A of all hamburgers in the universe is given by $A = \bigcup_{j=1}^{\infty} A_j$ so, since the countable union of countable sets is countable, A is countable. We can enumerate all the hamburgers in the universe as a sequence.)

Lemma B.6. *The set $\mathbb{Q}$ of rational numbers is countable.*

Proof. Let

$$A_j = \{r/s : |r| + |s| \leq j,\ s \neq 0,\ r,\ s \in \mathbb{Z}\}.$$

The set A_j is finite and so countable. But $\mathbb{Q} = \bigcup_{j=1}^{\infty} A_j$ so, since the countable union of countable sets is countable, $\mathbb{Q}$ is countable. ∎

In Exercise 1.41 we showed that the closed interval $[a, b]$ with $a < b$ is not a countable set, and so (by Exercise B.3) $\mathbb{R}$ is uncountable. Lemma B.6 thus provides an indirect but illuminating proof that $\mathbb{R} \neq \mathbb{Q}$.

Exercise B.7. (Existence of transcendentals.) *A real number x is called algebraic if*

$$\sum_{r=0}^{n} a_r x^r = 0$$

for some $a_r \in \mathbb{Z}$ $[0 \leq r \leq n]$, $a_n \neq 0$ and $n \geq 1$ (in other words, x is a root of a non-trivial polynomial with integer coefficients). The first proof that not all real numbers are algebraic was due to Liouville (see Exercise K.12). Cantor used the idea of countability to give a very different proof.

Let A_j be the set of all real roots of all polynomials $P(t) = \sum_{r=0}^{n} a_r t^r$ such that $a_r \in \mathbb{Z}$ $[0 \leq r \leq n]$, $a_n \neq 0$ and $n \geq 1$ and

$$n + \sum_{r=0}^{n} |a_r| \leq j.$$

By considering properties of the A_j, show that the set of algebraic numbers is countable. Deduce that not all real numbers are algebraic.

Exercise B.8. *(i) A set S of non-negative real numbers is such that there exists a positive constant K such that*

$$\sum_{x \in T} x < K$$

for every finite subset T of S. Prove that the set of non-zero elements of S is countable.

(ii) A set S of real numbers is such that, whenever $a_1, a_2, \ldots$ are distinct members of S,

$$\sum_{n=1}^{\infty} a_n \text{ converges.}$$

Show that we can find $b_1, b_2, \ldots$ *such that* $\sum_{n=1}^{\infty} b_n$ *is absolutely convergent and*

$$S \subseteq \{0\} \cup \{b_n : n \geq 1\}.$$

Exercise B.9. *Consider a function* $f : (a, b) \to \mathbb{R}$. *Recall that we say that* c *is a strict maximum (or strict local maximum) of* f *if we can find* α *and* β *such that* $a \leq \alpha < c < \beta \leq b$ *such that*

$$f(x) < f(c) \text{ for all } x \in (\alpha, \beta) \text{ with } x \neq c.$$

By associating each such c *with an interval with* rational *end points, or otherwise, show that the set* E *of strict maxima is countable.*

Give an example to show that E *can be infinite.*

Does the result remain true if we replace (a, b) *by the real line* $\mathbb{R}$*? Does an appropriate version of the result (to be stated) remain true if we replace* (a, b) *by* $\mathbb{R}^2$*? Does the result remain true if we replace 'strict maxima' by 'maxima' (that is, replace* $f(x) < f(c)$ *in the defining condition by '*$f(x) \leq f(c)$*')?*

It is worth remarking that Exercise B.7 reflects a general principle.

Plausible statement B.10. *The set of real numbers that we can describe in any way is countable.*

Plausible argument. I have only a finite number N of symbols on my typewriter. Thus the set A_j of real numbers that I can describe using j keystrokes has at most N^j members. Since A_j is finite and so countable, we see that the set $\bigcup_{j=1}^{\infty} A_j$ of real numbers that I can describe in any way is countable. ▲

The reader should be warned that there are severe philosophical and mathematical obstacles in the way of any attempt to make precise the statement and argument above. Never the less, most mathematicians are inclined to accept the general thrust of the argument. **When we talk about** $\mathbb{R}$**, we are talking about a set most of whose members we cannot describe.** Here is another reason to prefer rigour to intuition.

Appendix C

The Care and Treatment of Counterexamples

Mathematics students dislike questions beginning 'find a proof or counterexample'. They know that it is much harder to prove or disprove a result if you are not sure which of the two possible outcomes will occur.

In research we are faced with many uncertainties. We do not know whether our problem is easy, hard or beyond our capabilities (whereas in an exam we know that the problems will not exceed a certain level of difficulty). We often do not know if the solution of a particular problem will open a path to further progress or turn out to be a dead end. And, whatever our suspicions or hopes, we do not know whether the result we seek is true or false.

In these circumstances, mathematicians often adopt a two-pronged attack. First they try to prove the result. When their attack fails, they try to identify the point where it fails and try to construct a counterexample centred round the point of failure. If they cannot construct a counterexample they try to identify the reason why they cannot do so, and this may give them a hint as to how get round their previous difficulty in the proof. By alternating between determined attempts to seek a proof and determined attempts to find a counterexample they hope to arrive at a useful result.

If their attack ends in a counterexample, the counterexample may suggest how to strengthen hypotheses or weaken conclusions so as to produce a true theorem.

If the attack ends in a theorem, the matter does not finish there. Examination questions and exercises such as are found in this book are made easier by the fact that we know that all the information supplied is relevant. We know that, in real life, problems are made much harder by an excess of irrelevant information[1]. We expect, sometimes wrongly but often rightly, that a proof which depends on unnecessary hypotheses will be less clear than one which only uses the essential hypotheses. Thus, once a theorem is established, we usually embark on a systematic programme of weakening hypotheses and strengthening conclusions, not only in the hope of obtaining a stronger theorem but also in the hope of obtaining a better proof. A theorem is therefore frequently accompanied by a collection of counterexamples showing that hypotheses cannot be weakened or conclusions strengthened.

What I have said so far is probably obvious to the reader and to most students of mathematics. However, many students fail to draw the obvious moral. Since counterexamples are not ends in themselves but objects which we study in the hope of obtaining positive results,

counterexamples should be as simple as possible.

Let me illustrate this slogan by considering the notion of continuity. The first question we must ask after defining the notion of a continuous function is whether all functions are continuous. Here is the simplest counterexample I can think of.

Example C.1. *If we define* $f : \mathbb{R} \to \mathbb{R}$ *by*

$$\begin{aligned} f(t) &= 0 \qquad \textit{for } t \neq 0, \\ f(0) &= 1, \end{aligned}$$

then f *is not continuous at* 0.

The reader may feel that this is artificial since f can be redefined at one point (by setting $f(0) = 0$) in such a way as to make it continuous. We may therefore ask for a function which cannot be so defined.

Example C.2. *If we define* $H : \mathbb{R} \to \mathbb{R}$ *by*

$$\begin{aligned} H(t) &= 0 \qquad \textit{for } t < 0, \\ H(t) &= 1 \qquad \textit{for } t > 0, \\ H(0) &= a, \end{aligned}$$

then, whatever the value of the real number a, H *is not continuous at* 0.

[1]Remember the old riddle. A bus sets out from a depot with only the driver on board. At the first stop two people get on. At the next, three get on and one gets off. At the next, five get on and two get off. At the next, four get off and two get on. At the next three get off and two get on. At the next five get on and two get off. How many stops has the bus made?

Proof. If $f : \mathbb{R} \to \mathbb{R}$ is continuous at 0, then

$$f(1/n) - f(-1/n) \to f(0) - f(0) = 0$$

as $n \to \infty$. Since

$$H(1/n) - H(-1/n) = 1 \nrightarrow 0$$

as $n \to \infty$, H is not continuous at 0. ■

The discontinuity exhibited by H is of a particularly simple kind, since the right and left limits

$$H(0+) = \lim_{t \to 0,\ t>0} H(t) \text{ and } H(0-) = \lim_{t \to 0,\ t<0} H(t)$$

exist. We may, therefore, ask if there exist functions without such left and right limits.

Example C.3. *If we define* $f : \mathbb{R} \to \mathbb{R}$ *by*

$$\begin{aligned} f(t) &= 1/t \qquad \text{for } t \neq 0, \\ f(0) &= 0, \end{aligned}$$

then f *has neither a left limit nor a right limit at* 0.

We may feel that this is unsatisfactory and ask whether this phenomenon can occur for a bounded function.

Example C.4. *If we define* $f : \mathbb{R} \to \mathbb{R}$ *by*

$$\begin{aligned} f(t) &= \sin(1/t) \qquad \text{for } t \neq 0, \\ f(0) &= 0, \end{aligned}$$

then f *is bounded and has neither a left limit nor a right limit at* 0.

Proof. Observe that $f(1/(2n\pi)) = 0 \to 0$ but $f(1/((2n + \frac{1}{2})\pi) = 1 \to 1$ as $n \to \infty$. Thus $f(t)$ does not tend to a limit as $t \to 0$ through values of $t > 0$. Similarly, $f(t)$ does not tend to a limit as $t \to 0$ through values of $t < 0$. ■

Exercise C.5. *Find a bounded function* $f : \mathbb{R} \to \mathbb{R}$ *such that* $f(t) \to f(0)$ *as* $t \to 0$ *through values of* $t > 0$ *but* $f(t)$ *does not tend to a limit as* $t \to 0$ *through values of* $t < 0$.

The next exercise is a trivial complementary observation with a one line proof.

Exercise C.6. *Show that if* $f : \mathbb{R} \to \mathbb{R}$ *is such that* $f(t) \to f(0)$ *as* $t \to 0$ *through values of* $t > 0$ *and* $f(t) \to f(0)$ *as* $t \to 0$ *through values of* $t < 0$, *then* f *is continuous at* 0.

In view of Exercise C.6 we may ask if a function $F : \mathbb{R}^2 \to \mathbb{R}$ such that $F(\mathbf{u}t) \to F(\mathbf{0})$ as $t \to 0$ for all unit vectors $\mathbf{u}$ is automatically continuous at $\mathbf{0} = (0, 0)$. This is a harder question and, instead of sitting around trying to think of a function which might provide a counterexample, we actively construct one 'by hand'.

Exercise C.7. *(i) Let $r_n = 1/n$ and $\theta_n = \pi/(2n)$ and take*

$$\mathbf{x}_n = (r_n \cos\theta_n, r_n \sin\theta_n).$$

Show that we can find $\delta_n > 0$ such that $\|\mathbf{x}_n\|/2 > \delta_n > 0$ and, writing

$$B_n = \{\mathbf{y} : \|\mathbf{x}_n - \mathbf{y}\| \le \delta_n\},$$

we know that the following statement is true. If $\mathbf{u}$ is any unit vector, the line $\{t\mathbf{u} : t \in \mathbb{R}\}$ intersects at most one of the B_n. Give a sketch to illustrate your result.

(ii) Show that there exists a continuous function $g_n : \mathbb{R}^2 \to \mathbb{R}$ such that

$$g_n(\mathbf{x}_n) = 1,$$
$$g_n(\mathbf{y}) = 0 \qquad \text{for all } \mathbf{y} \notin B_n.$$

Explain why setting $F(\mathbf{x}) = \sum_{n=1}^{\infty} g_n(\mathbf{x})$ gives a well defined function. Show that $F(\mathbf{u}t) \to F(\mathbf{0})$ as $t \to 0$ for all unit vectors $\mathbf{u}$, but F is not continuous at $\mathbf{0}$.

Exercise C.8. *(This exercise requires material from Chapter 6.)*

(i) Show that we can find a twice continuously differentiable function $h : \mathbb{R} \to \mathbb{R}$ such that

$$h(0) = 1,$$
$$h(t) = 0 \qquad \text{for all } |t| > 1/2,$$
$$1 \ge h(t) \ge 0 \qquad \text{for all } t.$$

If $\delta > 0$ and $\mathbf{z} \in \mathbb{R}^2$, sketch the function $g : \mathbb{R}^2 \to \mathbb{R}$ given by $g(\mathbf{x}) = h(\delta^{-1}\|\mathbf{z} - \mathbf{x}\|)$.

(ii) By taking $F = \sum_{n=1}^{\infty} n^{-1} g_n$ with appropriately chosen g_n (you will need to be more careful in the choice than we were in the previous exercise), show that there exists a continuous function $F : \mathbb{R}^2 \to \mathbb{R}$ with directional derivative zero in all directions at $\mathbf{0}$ but which is not differentiable at $\mathbf{0}$.

(iii) Why is the result of Exercise C.8 stronger than that of Example 7.32?

[We give a slight strengthening of this result in Exercise K.345.]

Exercise C.9. *Consider a function $f : \mathbb{R}^2 \to \mathbb{R}$.*

(i) Show that f is continuous at $\mathbf{0} = (0, 0)$ if and only if, whenever we have a sequence $\mathbf{x}_n \to \mathbf{0}$, it follows that $f(\mathbf{x}_n) \to f(\mathbf{0})$ as $n \to \infty$.

(ii) Show that f is continuous at $\mathbf{0}$ if and only if, whenever we have a continuous function $\gamma : [0,1] \to \mathbb{R}^2$ with $\gamma(0) = \mathbf{0}$, it follows that $f(\gamma(t)) \to f(\mathbf{0})$ as $t \to 0$ through values of $t > 0$.

Thus, if $f(\mathbf{x})$ tends to $f(\mathbf{0})$ along every *path leading to $\mathbf{0}$, then f is continuous. However, as we have seen, the result fails if we replace 'every path' by 'every straight-line path'.*

If we wish to understand the ways in which a function can fail to be continuous at a point, it is surely more instructive to begin with the simple Example C.1 rather than rush to the complicated Exercise C.7. In the same way, if we wish to give an example of a continuous function which fails to be differentiable, it is surely best to look at the function $f : \mathbb{R} \to \mathbb{R}$ given by $f(x) = |x|$, rather than the fairly intricate construction of a continuous nowhere differentiable function given in Exercise K.251.

Proceeding from the simple to the complex as we did in studying discontinuity at a point has the further advantage that it gives us a number of examples of functions to think about rather than just one. To quote Halmos,

> If I had to describe my conclusions [on how to study mathematics] in one word, I'd say *examples.* They are to me of paramount importance. Every time I learn a new concept (free group, or pseudodifferential operator, or paracompact space) I look for examples — and, of course, non-examples. The examples should include, whenever possible, the typical ones and the extreme degenerate ones. Yes, to be sure, $\mathbb{R}^2$ is a real vector space, but so is the space of all real polynomials with real coefficients and the set of all analytic functions defined on the open disc — and what about the 0-dimensional case. Are there examples that satisfy all the conditions of the definition except $1\mathbf{x} = \mathbf{x}$? Is the set of all real-valued monotone increasing functions defined on, say, the unit interval a real vector space? How about all the monotone ones, both increasing and decreasing ...
>
> A good stock of examples, as large as possible, is indispensable for a thorough understanding of any concept and when I want to learn something new, I make it my first job to build one. Sometimes to be sure, that might require more of the theory than the mere definition reveals. When we first learn about transcendental numbers, for instance, it is not at all obvious that such things exist. Just the opposite happens when we are first told about measurable sets. There are plenty of them, there is no doubt about that; what is less

> easy is to find a single example of a set that is not like that[2]. [The quotation is from *I Want to be a Mathematician*[**22**], a book well worth reading in its entirety.]

This quotation brings me to my second point. Halmos talks about typical examples and illustrates his remarks with the idea of a vector space. In some sense typical vector spaces exist, since we have the following classification theorem from algebra.

Theorem C.10. *A finite-dimensional vector space over $\mathbb{R}$ is isomorphic to $\mathbb{R}^n$ for some positive integer n.*

[There is a similar but duller result for infinite-dimensional vector spaces.]

There is no such result for continuous functions, and I suspect that, to misquote Haldane, 'The typical continuous function is not only odder than we imagine, it is odder than we can possibly imagine'[3]. This is why mathematicians demand so much rigour in proof. When we showed, in Theorem 4.44, that every real-valued continuous function on a closed bounded set in $\mathbb{R}^n$ is bounded and attains its bounds, we proved a result which applied not only to all continuous functions that we know but to all continuous functions that the human race will ever know and to continuous functions which nobody will ever know and, if my belief is right, to continuous functions which are so wild as to be literally unimaginable[4].

Even if we restrict ourselves to well behaved functions, I would argue that it is much harder than it looks to obtain a sufficiently large library of typical functions. (See Remark 4 on page 328.)

On the other hand, when we talk about an example or a counterexample we are talking about a single object, and we do not need the obsessive rigour demanded by a proof which will apply to many objects. To show that a unicorn exists we need only exhibit a single unicorn. To show that no unicorn exists requires much more careful argument.

Demonstrations involving a single known object can be much simpler than those involving a multitude of unknown objects.

[2]The set considered in Exercise 8.4 is an example. (T.W.K.)

[3]In spite of Theorem C.10, this may be true of vector spaces as well. Conway used to say that one dimension is easy, two harder, three very hard, four still harder, ..., that somewhere about twenty eight things became virtually impossible but once you reached infinite dimensions things became easier again.

[4]Since the 1890's it has been known that there exist everywhere differentiable functions which take a strict local maximum value at a dense set of points. Until you read this footnote, did you imagine that such a thing was possible? Can you imagine it now? For a modern proof see [**28**].

Appendix D

A More General View of Limits

One of the more tedious parts of the standard treatment of analysis given in this book is the repeated definition of various kinds of limits. Here is a small selection.

Definition D.1. *If a_n is a sequence of real numbers and a is a real number, we say that $a_n \to a$ as $n \to \infty$ if, given any $\epsilon > 0$, we can find an $n_0(\epsilon)$ such that $|a - a_n| < \epsilon$ for all $n > n_0(\epsilon)$.*

Definition D.2. *If a_n is a sequence of real numbers and a is a real number, we say that $a_n \to \infty$ as $n \to \infty$ if, given any K, we can find an $n_0(K)$ such that $a_n > K$ for all $n > n_0(K)$.*

Definition D.3. *If $f : (a, b) \to \mathbb{R}$ is a function and $t \in (a, b)$, we say that $f(x) \to c$ as $x \to t$ if, given any $\epsilon > 0$, we can find a $\delta_0(\epsilon)$ such that $|f(x) - c| < \epsilon$ for all $x \in (a, b)$ with $0 \neq |t - x| < \delta_0(\epsilon)$.*

Definition D.4. *If $f : (a, b) \to \mathbb{R}$ is a function and $t \in (a, b)$, we say that $f(x) \to c$ as $x \to t$ through values of $x > t$ if, given any $\epsilon > 0$, we can find a $\delta_0(\epsilon)$ such that $|f(x) - c| < \epsilon$ for all $x \in (a, b)$ with $0 < x - t < \delta_0(\epsilon)$.*

In theory, each of these definitions should be accompanied by a collection of lemmas showing that each limit behaves in the way every other limit behaves. In practice, such lemmas are left to the reader. (In this book I have tended to look most carefully at definitions of the type of Definition D.1.) Experience seems to show that this procedure works quite well, both from the pedagogic and the mathematical point of view, but it is, to say the least, untidy.

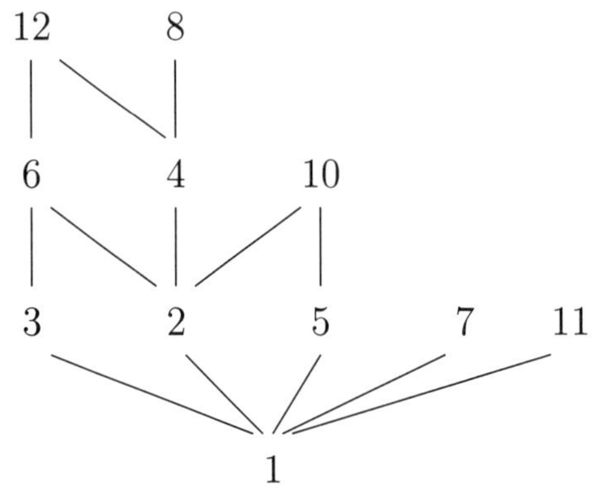

Figure D.1. A family tree for division

In his book *Limits, A New Approach to Real Analysis* [**2**], Beardon proposes an approach based on directed sets which avoids all these difficulties[1]. For a mixture of reasons, some good and some not so good, the teaching of elementary analysis is extremely conservative[2] and it remains to be seen if this innovation will be widely adopted.

Definition D.5. *A relation $\succ$ on a non-empty set X is called a direction if*

(i) whenever $x \succ y$ and $y \succ z$, then $x \succ z$, and

(ii) whenever $x, y \in X$, we can find $z \in X$ such that $z \succ x$ and $z \succ y$.

We call $(X, \succ)$ a directed set.

Notice that we do not demand that 'all points are comparable', that is, we do not demand that, if $x \neq y$, then either $x \succ y$ or $y \succ x$. (You should reread Definition D.5 bearing this in mind.) We will use this extra liberty in Exercises D.14 and D.16.

Exercise D.6. *Consider the set $\mathbb{N}^+$ of strictly positive integers. Write $n \succ m$ if m divides n. In Figure D.1, I draw a family tree of the integers 1 to 12 with n connected to m by a descending line if $n \succ m$. Extend the tree to cover the integers 1 to 25.*

Show that $\succ$ is a direction.

Give an example of two incomparable integers (that is, integers n and m such that it is not true that $n = m$ or $m \succ n$ or $n \succ m$).

[1]The notion of a directed set goes back to E. H. Moore in 1910, but Beardon's is the only elementary text I know which uses it. I am told that Heuser's *Lehrbuch der Analysis* also makes use of the same ideas, but my informant doubted 'if 1350 pages of analysis would be regarded as an elementary textbook outside a German university'.

[2]Interestingly, both [**11**] and [**5**], though radical in style, are entirely standard in content.

We can use the notion of direction to define a limit.

Definition D.7. *If $\succ$ is a direction on a non-empty set X and f is a function from X to $\mathbb{F}$ (where $\mathbb{F}$ is an ordered field such as $\mathbb{R}$ or $\mathbb{Q}$), we say that $f(x) \to a$, with respect to the direction $\succ$, if $a \in \mathbb{F}$ and, given any $\epsilon > 0$, we can find an $x_0(\epsilon) \in X$ such that $|f(x) - a| < \epsilon$ for all $x \succ x_0(\epsilon)$.*

The reader should have no difficulty in proving the following version of Lemma 1.6.

Exercise D.8. *Let $\succ$ be a direction on a non-empty set X, let f, g be functions from X to $\mathbb{F}$ and let a, b, c be elements of $\mathbb{F}$. Prove the following results.*

(i) The limit is unique. That is, if $f(x) \to a$ and $f(x) \to b$, with respect to the direction $\succ$, then $a = b$.

(ii) Suppose Y is a non-empty subset of X with the property that if whenever $x, y \in Y$, we can find $z \in Y$ such that $z \succ x$ and $z \succ y$. Then, if $\succ_Y$ is the restriction of the direction $\succ$ to Y, $\succ_Y$ is a direction on Y.

Suppose, in addition, that, given any $x \in X$, we can find a $y \in Y$ such that $y \succ x$. If $f(x) \to a$, with respect to the direction $\succ$, and f_Y is the restriction of f to Y, then $f_Y(x) \to a$, with respect to the direction $\succ_Y$.

(iii) If $f(x) = c$ for all $x \in X$, then $f(x) \to c$, with respect to the direction $\succ$.

(iv) If $f(x) \to a$ and $g(x) \to b$, with respect to the direction $\succ$, then, $f(x) + g(x) \to a + b$.

(v) If $f(x) \to a$ and $g(x) \to b$, with respect to the direction $\succ$, then $f(x)g(x) \to ab$.

(vi) Suppose that $f(x) \to a$, with respect to the direction $\succ$. If $f(x) \neq 0$ for each $x \in X$ and $a \neq 0$, then $f(x)^{-1} \to a^{-1}$.

(vii) If $f(x) \leq A$ for each $x \in X$ and $f(x) \to a$, with respect to the direction $\succ$, then $a \leq A$. If $g(x) \geq B$ for each $x \in X$ and $g(x) \to b$, with respect to the direction $\succ$, then $b \geq B$.

As one might expect, we can recover Lemma 1.6 from Exercise D.8.

Exercise D.9. *(i) If $\mathbb{N}^+$ is the set of strictly positive integers, show that $>$ (with its ordinary meaning) is a direction on $\mathbb{N}^+$. Show further that, if f is a function from $\mathbb{N}^+$ to $\mathbb{F}$ (an ordered field) and $a \in \mathbb{F}$, then $f(n) \to a$, with respect to the direction $>$, if and only if $f(n) \to a$ as $n \to \infty$ in the sense of Definition 1.5.*

(ii) Deduce Lemma 1.6 from Exercise D.8.

(iii) Show that (i) remains true if we replace $>$ by $\geq$. Show that (i) remains true if we replace $>$ by $\succ$ with $n \succ m$ if $n \geq 10m + 4$. Thus different succession relations can produce the same notion of limit.

The real economy of this approach appears when we extend it.

Exercise D.10. *(i) Let a, t and b be real with $a < t < b$. Show that, if we define the relation $\succ$ on $(a,b) \setminus \{t\}$ by $x \succ y$ if $|x-t| < |y-t|$, then $\succ$ is a direction. Suppose $f : (a,b) \to \mathbb{R}$ is a function and $c \in \mathbb{R}$. Show that $f(x) \to c$, with respect to the direction $\succ$, if and only if $f(x) \to c$ as $x \to t$, in the traditional sense of Definition D.3.*

(ii) Deduce the properties of the traditional limit of Definition D.3 from Exercise D.8.

(iii) Give a treatment of the classical 'limit from above' defined in Definition D.4 along the lines laid out in parts (i) and (ii).

Exercise D.11. *Obtain a multidimensional analogue of Definition D.7 following the lines of Definition 4.8, and prove a multidimensional version of Exercise D.8 along the lines of Lemma 4.9.*

A little thought shows how to bring Definition D.2 into this circle of ideas.

Definition D.12. *If $\succ_X$ is a direction on a non-empty set X, $\succ_Y$ is a direction on a non-empty set Y and f is a function from X to Y, we say that $f(x) \to *_Y$ as $x \to *_X$ if, given $y \in Y$, we can find $x_0(y) \in X$ such that $f(x) \succ_Y y$ for all $x \succ x_0(y)$.*

Exercise D.13. *(i) Show that if we take $X = \mathbb{N}^+$, $\succ_X$ to be the usual relation $>$ on X, $Y = \mathbb{R}$ and $\succ_Y$ to be the usual relation $>$ on Y, then saying that a function $f : X \to Y$ has the property $f(x) \to *_Y$ as $x \to *_X$ is equivalent to the classical statement $f(n) \to \infty$ as $n \to \infty$.*

(ii) Give a similar treatment for the classical statement that a function $f : \mathbb{R} \to \mathbb{R}$ satisfies $f(x) \to -\infty$ as $x \to \infty$.

(iii) Give a similar treatment for the classical statement that a function $f : \mathbb{R} \to \mathbb{R}$ satisfies $f(x) \to a$ as $x \to \infty$.

In all the examples so far, we have only used rather simple examples of direction. Here is a more complicated one.

Exercise D.14. *This exercise assumes a knowledge of Section 8.2, where we defined the Riemann integral. It uses the notation of that section. Consider the set X of ordered pairs $(\mathcal{D}, \mathcal{E})$, where $\mathcal{D}$ is the dissection*

$$\mathcal{D} = \{x_0, x_1, \ldots, x_n\} \text{ with } a = x_0 \leq x_1 \leq x_2 \leq \cdots \leq x_n = b,$$

and

$$\mathcal{E} = \{t_1, t_2, \ldots, t_n\} \text{ with } x_{j-1} \le t_j \le x_j.$$

If $f : [a, b] \to \mathbb{R}$ *is a bounded function, we write*

$$\sigma(f, \mathcal{D}, \mathcal{E}) = \sum_{j=1}^{n} f(t_j)(x_j - x_{j-1}).$$

Show that, if we write $(\mathcal{D}', \mathcal{E}') \succ (\mathcal{D}, \mathcal{E})$ *when* $\mathcal{D}' \supseteq \mathcal{D}$, *then* $\succ$ *is a direction on* X. *(Note that we place no conditions on* $\mathcal{E}$ *and* $\mathcal{E}'$.*) Show that not all dissections are comparable.*

Using the results of Section 8.2, show that f *is Riemann integrable, in the sense of Section 8.2, with integral* I *if and only if*

$$\sigma(f, \mathcal{D}, \mathcal{E}) \to I$$

with respect to the direction $\succ$.

Here is another example, this time depending on the discussion of metric spaces in Section 10.3. If we wish to define the notion of a limit for a function between two metric spaces, the natural classical procedure is to produce something along the lines of Definition 10.35.

Definition D.15. *Let* (X, d) *and* (Z, ρ) *be metric spaces, and* f *a map from* X *to* Z. *Suppose that* $x \in X$ *and* $z \in Z$. *We say that* $f(y) \to z$ *as* $y \to x$ *if, given* $\epsilon > 0$, *we can find a* $\delta(\epsilon, x) > 0$ *such that, if* $y \in X$ *and* $d(x, y) < \delta(\epsilon, x)$, *we have*

$$\rho(f(y), z) < \epsilon.$$

Here is an alternative treatment using direction.

Exercise D.16. *Let* (X, d) *and* (Z, ρ) *be metric spaces and* $x \in X$ *and* $z \in Z$. *We take* $\mathcal{X}$ *to be the collection of open sets in* X *which contain* x *and* $\mathcal{Z}$ *to be the collection of open sets in* Z *which contain* z.

Show that if we define a relation on $\mathcal{X}$ *by* $U \succ_{\mathcal{X}} V$ *if* $V \supseteq U$, *then* $\succ_{\mathcal{X}}$ *is a direction on* $\mathcal{X}$. *Is it always true that two elements of* $\mathcal{X}$ *are comparable?*

Let $\succ_{\mathcal{Z}}$ *be defined similarly. If* f *is a map from* X *to* Z, *show that* $f(y) \to z$ *as* $y \to x$ *in the sense of Definition D.15 if and only if* $f(y) \to *_{\mathcal{Z}}$ *as* $x \to *_{\mathcal{X}}$ *in the sense of Definition D.12.*

The advantage of the approach given in Exercise D.16 is that it makes no reference to the metric and raises the possibility of doing analysis on more general objects than metric spaces.

We close with a couple of interesting observations (it will be more convenient to use Definition D.7 than our more general Definition D.12).

Exercise D.17. *Suppose that* $\succ$ *is a direction on a non-empty set* X *and* f *is a function from* X *to* $\mathbb{R}$.

Suppose that there exists an $M \in \mathbb{R}$ *such that* $f(x) \leq M$ *for all* $x \in X$, *and suppose that* f *is 'increasing' in the sense that* $x \succ y$ *implies* $f(x) \leq f(y)$. *Show that there exists an* $a \in \mathbb{F}$ *such that that* $f(x) \to a$ *with respect to the direction* $\succ$.

[Hint: *Think about the supremum.*]

If the reader considers the matter, she may recall points in the book where such a result would have been useful.

Exercise D.18. *Prove Lemma 9.4 using Lemma D.17.*

Exercise D.19. *The result of Lemma D.17 is the generalisation of the statement that every bounded increasing sequence in* $\mathbb{R}$ *has a limit. Find a similar generalisation of the general principle of convergence. Use it to do Exercise 9.12.*

If the reader wishes to see more, I refer her to the elegant and efficient treatment of analysis in [**2**].

Appendix E

Traditional Partial Derivatives

One of the most troublesome culture clashes between pure mathematics and applied is that to an applied mathematician variables like x and t have meanings such as position and time, whereas to a pure mathematician all variables are 'dummy variables' or 'place-holders' to be interchanged at will. To a pure mathematician, v is an arbitrary function defined by its effect on a variable, so that $v(t) = At^3$ means precisely the same thing as $v(x) = Ax^3$ whereas, to an applied mathematician who thinks of v as a velocity, the statements $v = At^3$ and $v = Ax^3$ mean very different (indeed incompatible) things.

The applied mathematician thinks of $\frac{dv}{dt}$ as representing $\frac{\delta v}{\delta t}$ 'when everything is so small that second order quantities can be neglected'. Since

$$\frac{\delta v}{\delta t}\frac{\delta t}{\delta x} = \frac{\delta v}{\delta x},$$

it is obvious to the applied mathematician that

$$\text{(A)} \qquad \frac{dv}{dt} = \frac{dv}{dx}\frac{dx}{dt},$$

but the more rigid notational conventions of the pure mathematicians prevent them from thinking of v as two different functions (one of t and one of x) in the same formula. The closest a pure mathematician can get to equation (A) is the chain rule

$$\frac{d}{dt}v(x(t)) = v'(x(t))x'(t).$$

Now consider a particle moving along the x-axis so its position is x at time t. Since

$$\frac{\delta x}{\delta t}\frac{\delta t}{\delta x} = 1,$$

it is obvious to the applied mathematician that

$$\frac{dx}{dt}\frac{dt}{dx} = 1,$$

and so

$$\text{(B)} \qquad \frac{dt}{dx} = 1 \Big/ \frac{dx}{dt}$$

What does equation (B) mean? In the expression $\frac{dx}{dt}$ we treat x as a function of t, which corresponds to common sense, but in the expression $\frac{dt}{dx}$ we treat t as a function of x, which seems a little odd (how can the position of a particle influence time?). However, if the particle occupies each particular position at a unique time, we can read off time from position and so, in this sense, t is indeed a function of x. A pure mathematician would say that the function $x : \mathbb{R} \to \mathbb{R}$ is invertible and replace equation (B) by the inverse function formula

$$\frac{d}{dt}x^{-1}(t) = \frac{1}{x'(x^{-1}(t))}.$$

One of the reasons why this formula seems more complicated than equation (B) is that the information on where to evaluate $\frac{dx}{dt}$ and $\frac{dt}{dx}$ has been suppressed in equation (B), which ought to read something like

$$\frac{dt}{dx}(x_0) = 1 \Big/ \frac{dx}{dt}(t_0)\ ,$$

or

$$\left.\frac{dt}{dx}\right|_{x=x_0} = 1 \Big/ \left.\frac{dx}{dt}\right|_{t=t_0},$$

where the particle is at x_0 at time t_0.

The clash between the two cultures is still more marked when it comes to partial derivatives. Consider a gas at temperature T, held at a pressure P in a container of volume V and isolated from the outside world. The applied mathematician knows that P depends on T and V, and so writes $P = P(T, V)$ or $P = P(V, T)$. (To see the difference in conventions between pure and applied mathematics, observe that, to a pure mathematician, functions $f : \mathbb{R}^2 \to \mathbb{R}$ such that $f(x, y) = f(y, x)$ form a very restricted class!) Suppose that, initially, $T = T_0$, $P = P_0$, $V = V_0$. If we change T_0 to $T_0 + \delta T$ whilst keeping V fixed, then P changes to $P_0 + \delta P$ and the applied mathematician thinks of $\left.\dfrac{\partial P}{\partial T}\right|_{V=V_0,T=T_0}$ as representing $\dfrac{\delta P}{\delta T}$ 'when everything is so small that second order quantities can be neglected'. In other words, $\left.\dfrac{\partial P}{\partial T}\right|_{V=V_0,T=T_0}$ is the rate of change of P when T varies but V is kept fixed. Often applied mathematicians write

$$\frac{\partial P}{\partial T} = \left.\frac{\partial P}{\partial T}\right|_{V=V_0,T=T_0},$$

but you should note that this condensed notation suppresses the information that V (rather than, say, $V+T$) should be kept constant when T is varied.

There is good reason to suppose that there is a well behaved function $g : \mathbb{R}^3 \to \mathbb{R}$ such that, if a particular gas is at temperature T, held at pressure P in a container of volume V, then the temperature, pressure, volume triple $(T,\ P,\ V)$ satisfies

$$g(T,P,V) = 0$$

(at least for a wide range of values of T, P and V). We may link the pure mathematician's partial derivatives with those of the applied mathematician by observing that, if $(T_0,\ P_0,\ V_0)$ and $(T_0+\delta T,\ P_0+\delta P,\ V_0)$ are possible temperature, pressure, volume triples, then, to first order,

$$\begin{aligned} 0 &= g(T_0+\delta T, P_0+\delta P, V_0) \\ &= g(T_0,P_0,V_0) + g_{,1}(T_0,P_0,V_0)\delta T + g_{,2}(T_0,P_0,V_0)\delta P \\ &= g_{,1}(T_0,P_0,V_0)\delta T + g_{,2}(T_0,P_0,V_0)\delta P, \end{aligned}$$

and so, to first order,

$$\frac{\delta P}{\delta T} = -\frac{g_{,1}(T_0,P_0,V_0)}{g_{,2}(T_0,P_0,V_0)}.$$

It follows that

$$\left.\frac{\partial P}{\partial T}\right|_{V=V_0,T=T_0} = -\frac{g_{,1}(T_0,P_0,V_0)}{g_{,2}(T_0,P_0,V_0)}.$$

Essentially identical calculations show that

$$\left.\frac{\partial T}{\partial V}\right|_{P=P_0,V=V_0} - -\frac{g_{,3}(T_0,P_0,V_0)}{g_{,1}(T_0,P_0,V_0)}$$

and

$$\left.\frac{\partial V}{\partial P}\right|_{T=T_0,P=P_0} = -\frac{g_{,2}(T_0,P_0,V_0)}{g_{,3}(T_0,P_0,V_0)}.$$

Putting the last three equations together, we obtain

$$\left.\frac{\partial P}{\partial T}\right|_{V=V_0,T=T_0} \left.\frac{\partial T}{\partial V}\right|_{P=P_0,V=V_0} \left.\frac{\partial V}{\partial P}\right|_{T=T_0,P=P_0} = -1.$$

This is a very beautiful equation. It can be made much more mysterious by leaving implicit what we have made explicit, and writing

$$\text{(C)} \qquad \frac{\partial P}{\partial T}\frac{\partial T}{\partial V}\frac{\partial V}{\partial P} = -1.$$

If we further neglect to mention that T, P and V are restricted to the surface $g(T,P,V)=0$, we get 'an amazing result that common sense could not possibly have predicted ... It is perhaps the first time in our careers, for most of us, that we do not understand 3-dimensional space' ([**35**], page 65).

Exercise E.1. *Obtain the result corresponding to equation (C) for four variables. Without going into excessive details, indicate the generalisation to* n *variables with* $n \geq 2$. *(Check that if* $n = 2$ *you get a result corresponding to equation (B).)*

The difference between the 'pure' and 'applied' treatment is reflected in a difference in language. The pure mathematician speaks of 'differentiation of functions on many-dimensional spaces' and the applied mathematician speaks of 'differentiation of functions of many variables'. Which approach is better depends on the problem at hand. Although (T,P,V) is a triple of real numbers, it is not a vector, since pressure, volume and temperature are quantities of different types. Treating (T,P,V) as a vector is like adding apples to pears. The 'geometric' pure approach will yield no insight, since there is no geometry to consider. On the other hand, theories, like electromagnetism and relativity, with a strong geometric component will benefit from a treatment which does not disguise the geometry.

I have tried to show that the two approaches run in parallel but that, although statements in one language can be translated 'sentence by sentence' into the other, there is no word for word dictionary between them. A good mathematician can look at a problem in more than one way. In particular, a good mathematician will 'think like a pure mathematician when doing pure mathematics and like an applied mathematician when doing applied mathematics'. (Great mathematicians think like themselves when doing mathematics.)

Exercise E.2. *(In this exercise you should assume that everything is well behaved.) Rewrite the following statement in pure mathematics notation and prove it using whichever notation (pure, applied or mixed) you prefer.*

Suppose that the equations

$$f(x,y,z,w) = 0,$$
$$g(x,y,z,w) = 0$$

can be solved to give z and w as functions of (x, y). Then

$$\frac{\partial z}{\partial x} = -\left(\frac{\frac{\partial f}{\partial x}\frac{\partial g}{\partial w} - \frac{\partial f}{\partial w}\frac{\partial g}{\partial x}}{\frac{\partial f}{\partial z}\frac{\partial g}{\partial w} - \frac{\partial f}{\partial w}\frac{\partial g}{\partial z}}\right),$$

$$\frac{\partial z}{\partial y} = -\left(\frac{\frac{\partial f}{\partial y}\frac{\partial g}{\partial w} - \frac{\partial f}{\partial w}\frac{\partial g}{\partial y}}{\frac{\partial f}{\partial z}\frac{\partial g}{\partial w} - \frac{\partial f}{\partial w}\frac{\partial g}{\partial z}}\right).$$

Exercise E.3. *Each of the four variables p, V, T and S can be regarded as a well behaved function of any two of the others. In addition we have a variable U which may be regarded as a function of any two of the four variables above and satisfies*

$$\left.\frac{\partial U}{\partial S}\right|_V = T, \quad \left.\frac{\partial U}{\partial V}\right|_S = -p.$$

(i) Show that

$$\left.\frac{\partial V}{\partial S}\right|_p = \left.\frac{\partial T}{\partial p}\right|_S.$$

(ii) By finding two expressions for $\frac{\partial^2 U}{\partial p \partial V}$, or otherwise, show that

$$\left.\frac{\partial S}{\partial V}\right|_p \left.\frac{\partial T}{\partial p}\right|_V - \left.\frac{\partial S}{\partial p}\right|_V \left.\frac{\partial T}{\partial V}\right|_p = 1.$$

Exercise E.4. *You should only do this exercise if you have met the change of variable formula for multiple integrals. Recall that, if $(u, v) : \mathbb{R}^2 \to \mathbb{R}^2$ is a well behaved bijective map, we write*

$$J = \frac{\partial(u, v)}{\partial(x, y)} = \det \begin{pmatrix} \frac{\partial u}{\partial x} & \frac{\partial u}{\partial y} \\ \frac{\partial v}{\partial x} & \frac{\partial v}{\partial y} \end{pmatrix}$$

and call J the Jacobian determinant[1]*. Recall the the useful formula*

$$\frac{\partial(u, v)}{\partial(x, y)} \frac{\partial(x, y)}{\partial(u, v)} = 1.$$

Restate it in terms of $D\mathbf{f}$ and $D\mathbf{f}^{-1}$ evaluated at the correct points. (If you cannot see what is going on, look at the inverse function theorem (Theorem 13.13 (ii)) and at our discussion of equation (B) in this appendix.)

Restate and prove the formula

$$\frac{\partial(u, v)}{\partial(s, t)} \frac{\partial(s, t)}{\partial(x, y)} = \frac{\partial(u, v)}{\partial(x, y)}$$

in the same way.

[1] J is often just called the Jacobian, but we distinguish between the Jacobian matrix (see Exercise 6.9) and its determinant.

Another Approach to the Inverse Function Theorem

The object of this appendix is to outline a variant on the approach to the inverse function theorem given in Section 13.1.

We begin with a variation on Lemma 13.2.

Lemma F.1. *Consider a function* $\mathbf{f} : \mathbb{R}^m \to \mathbb{R}^m$ *such that* $\mathbf{f}(\mathbf{0}) = \mathbf{0}$. *Suppose that there exist a* $\delta > 0$ *and an* η *with* $1 > \eta > 0$ *such that*

$$\|(\mathbf{f}(\mathbf{x}) - \mathbf{f}(\mathbf{y})) - (\mathbf{x} - \mathbf{y})\| \leq \eta\|\mathbf{x} - \mathbf{y}\|$$

for all $\|\mathbf{x}\|, \|\mathbf{y}\| \leq \delta$. *Then there exists one and only one continuous function* $\mathbf{g} : \bar{B}(\mathbf{0}, (1-\eta)\delta) \to \bar{B}(\mathbf{0}, \delta)$ *such that* $\mathbf{g}(\mathbf{0}) = \mathbf{0}$ *and*

$$\bigstar \qquad \mathbf{f}(\mathbf{g}(\mathbf{y})) = \mathbf{y}$$

for all $\mathbf{y} \in \bar{B}(\mathbf{0}, (1-\eta)\delta)$.

Proof. This is a clever modification of the proof of Lemma 13.2. Let $X = \bar{B}(\mathbf{0}, (1-\eta)\delta)$. Consider the space C of continuous functions $\mathbf{h} : X \to \mathbb{R}^m$ with the uniform norm

$$\|\mathbf{h}\|_\infty = \sup_{\mathbf{t} \in X} \|\mathbf{h}(\mathbf{t})\|.$$

We know that $(C, \| \ \|_\infty)$ is complete. Let

$$E = \{\mathbf{h} : \|\mathbf{h}\|_\infty \leq \delta \text{ and } \mathbf{h}(\mathbf{0}) = \mathbf{0}\}.$$

Since E is closed in C, we know that E is complete under the uniform norm.

Let $\mathbf{h} \in E$. We define $T\mathbf{h}$ as a function on X by

$$T\mathbf{h}(\mathbf{y}) = \mathbf{h}(\mathbf{y}) + (\mathbf{y} - \mathbf{f}(\mathbf{h}(\mathbf{y}))),$$

so (following exactly the same calculations as in Lemma 13.2, but with $\mathbf{h}(\mathbf{y})$ in place of $\mathbf{x}$)

$$\begin{aligned}\|T\mathbf{h}(\mathbf{y})\| &\leq \|\mathbf{y}\| + \|\mathbf{h}(\mathbf{y}) - \mathbf{f}(\mathbf{h}(\mathbf{y}))\| \\ &= \|\mathbf{y}\| + \|(\mathbf{f}(\mathbf{h}(\mathbf{y})) - \mathbf{f}(\mathbf{0})) - (\mathbf{h}(\mathbf{y}) - \mathbf{0})\| \\ &\leq \|\mathbf{y}\| + \eta\|\mathbf{h}(\mathbf{y}) - \mathbf{0}\| \\ &\leq (1-\eta)\delta + \eta\|\mathbf{h}(\mathbf{y})\| < \delta\end{aligned}$$

whenever $\mathbf{y} \in X$. It is easy to check that, in addition, $T\mathbf{h}(\mathbf{0}) = \mathbf{0}$. Thus $T\mathbf{h} \in X$ and T is a well defined function $T : X \to X$.

If $\mathbf{h}$, $\mathbf{k} \in X$, then we see (by following exactly the same calculations as in Lemma 13.2, but with $\mathbf{h}(\mathbf{y})$ and $\mathbf{k}(\mathbf{y})$ in place of $\mathbf{x}$ and $\mathbf{x}'$) that

$$\begin{aligned}\|T\mathbf{h}(\mathbf{y}) - T\mathbf{k}(\mathbf{y})\| &= \|(\mathbf{h}(\mathbf{y}) - \mathbf{k}(\mathbf{y})) - (\mathbf{f}(\mathbf{h}(\mathbf{y})) - \mathbf{f}(\mathbf{k}(\mathbf{y})))\| \\ &\leq \eta\|\mathbf{h}(\mathbf{y}) - \mathbf{k}(\mathbf{y})\|\end{aligned}$$

for all $\mathbf{y} \in X$. Thus

$$\|T\mathbf{h} - T\mathbf{k}\|_\infty \leq \eta\|\mathbf{h} - \mathbf{k}\|_\infty$$

and T is a contraction mapping. If follows that T has a unique fixed point $\mathbf{g} \in E$ such that

$$\mathbf{g} = T\mathbf{g},$$

and so

$$\mathbf{g}(\mathbf{y}) = T\mathbf{g}(\mathbf{y}) = \mathbf{g}(\mathbf{y}) + (\mathbf{y} - \mathbf{f}(\mathbf{g}(\mathbf{y})),$$

that is,

$$\mathbf{f}(\mathbf{g}(\mathbf{y})) = \mathbf{y}$$

for all $\mathbf{y} \in X$, as required. ■

The new version of Lemma 13.2 gives rise to a new version of Lemma 13.4.

Lemma F.2. *Consider a function $\mathbf{f} : \mathbb{R}^m \to \mathbb{R}^m$ such that $\mathbf{f}(\mathbf{0}) = \mathbf{0}$ and there exists a $\delta_0 > 0$ such that $\mathbf{f}$ is differentiable in the open ball $B(\mathbf{0}, \delta_0)$. If $D\mathbf{f}$ is continuous at $\mathbf{0}$ and $D\mathbf{f}(\mathbf{0}) = I$ (the identity map), then we can find a δ_1 with $\delta_0 \geq \delta_1 > 0$ and a $\rho > 0$ such that there exists one and only one continuous function $\mathbf{g} : \bar{B}(\mathbf{0}, (1-\eta)\delta) \to \bar{B}(\mathbf{0}, \rho)$ with $\mathbf{g}(\mathbf{0}) = \mathbf{0}$ and*

$$\bigstar \qquad \mathbf{f}(\mathbf{g}(\mathbf{y})) = \mathbf{y}$$

for all $\mathbf{y} \in \bar{B}(\mathbf{0}, (1-\eta)\delta)$.

We leave the proof to the reader.

The reader will note that, though Lemma F.2 is stronger in some ways than Lemma 13.4, it is weaker in others. In particular, we have not shown that $\mathbf{g}$ is differentiable at $\mathbf{0}$. We deal with this by proving the following lemma.

Lemma F.3. *Consider a function* $\mathbf{f}:\mathbb{R}^m \to \mathbb{R}^m$ *such that* $\mathbf{f}(\mathbf{0}) = \mathbf{0}$ *and* $\mathbf{f}$ *is differentiable at* $\mathbf{0}$ *with* $D\mathbf{f}(\mathbf{0}) = I$. *Suppose that* U *is an open set containing* $\mathbf{0}$ *and* $\mathbf{g} : U \to \mathbb{R}^m$ *is a function such that* $\mathbf{g}(\mathbf{0}) = \mathbf{0}$, $\mathbf{f}(\mathbf{g}(\mathbf{y})) = \mathbf{y}$ *for all* $\mathbf{y} \in U$ *and* $\mathbf{g}$ *is continuous at* $\mathbf{0}$. *Then* $\mathbf{f}$ *is differentiable at* $\mathbf{0}$ *with* $D\mathbf{f}(\mathbf{0}) = I$.

Proof. Observe that

$$\mathbf{f}(\mathbf{h}) = \mathbf{h} + \boldsymbol{\epsilon}(\mathbf{h})\|\mathbf{h}\|,$$

where $\|\boldsymbol{\epsilon}(\mathbf{h})\| \to 0$ as $\|\mathbf{h}\| \to 0$. Thus, if $\mathbf{k} \in U$,

$$\mathbf{f}(\mathbf{g}(\mathbf{k})) = \mathbf{g}(\mathbf{k}) + \boldsymbol{\epsilon}(\mathbf{g}(\mathbf{k}))\|\mathbf{g}(\mathbf{k})\|,$$

and so

$$\mathbf{k} = \mathbf{g}(\mathbf{k}) + \boldsymbol{\epsilon}(\mathbf{g}(\mathbf{k}))\|\mathbf{g}(\mathbf{k})\|. \tag{†}$$

Since $\mathbf{g}$ is continuous at $\mathbf{0}$, we know that $\|\boldsymbol{\epsilon}(\mathbf{g}(\mathbf{k}))\| \to 0$ as $\|\mathbf{k}\| \to 0$. In particular, we can find an open set $V \subset U$ with $\mathbf{0} \in V$ such that

$$\|\boldsymbol{\epsilon}(\mathbf{g}(\mathbf{k}))\| < 1/2$$

whenever $\mathbf{k} \in V$, and so, using (†),

$$\|\mathbf{k}\| \geq \|\mathbf{g}(\mathbf{k})\| - \|\mathbf{g}(\mathbf{k})\|/2 = \|\mathbf{g}(\mathbf{k})\|/2$$

for all $\mathbf{k} \in V$.

Using (†) again, we see that

$$\|\mathbf{g}(\mathbf{k}) - \mathbf{k}\| \leq \|\boldsymbol{\epsilon}(\mathbf{g}(\mathbf{k}))\|\|\mathbf{g}(\mathbf{k})\| \leq 2\|\boldsymbol{\epsilon}(\mathbf{g}(\mathbf{k}))\|\|\mathbf{k}\|$$

for all $\mathbf{k} \in V$. Thus, since, as already noted above, $\|\boldsymbol{\epsilon}(\mathbf{g}(\mathbf{k}))\| \to 0$ as $\|\mathbf{k}\| \to 0$, it follows that

$$\mathbf{g}(\mathbf{k}) = \mathbf{k} + \boldsymbol{\eta}(\mathbf{k})\|\mathbf{k}\|,$$

where $\|\boldsymbol{\eta}(\mathbf{k})\| \to 0$ as $\|\mathbf{k}\| \to 0$. The result is proved. ∎

Combining this result with Lemma F.2 we obtain the following version of Lemma 13.4.

Lemma F.4. *Consider a function* $\mathbf{f}:\mathbb{R}^m \to \mathbb{R}^m$ *such that* $\mathbf{f}(\mathbf{0}) = \mathbf{0}$ *and there exists a* $\delta_0 > 0$ *such that* $\mathbf{f}$ *is differentiable in the open ball* $B(\mathbf{0}, \delta_0)$. *If* $D\mathbf{f}$ *is continuous at* $\mathbf{0}$ *and* $D\mathbf{f}(\mathbf{0}) = I$ *(the identity map), then we can find a*

δ_1 with $\delta_0 \geq \delta_1 > 0$ and a $\rho > 0$ such that there exists a unique continuous function $\mathbf{g} : \bar{B}(\mathbf{0}, (1-\eta)\delta) \to \bar{B}(\mathbf{0}, \rho)$ with $\mathbf{g}(\mathbf{0}) = \mathbf{0}$ and

$$\bigstar \qquad \mathbf{f}(\mathbf{g}(\mathbf{y})) = \mathbf{y}$$

for all $\mathbf{y} \in \bar{B}(\mathbf{0}, (1-\eta)\delta)$. Moreover, $\mathbf{g}$ is differentiable at $\mathbf{0}$ with $D\mathbf{g}(\mathbf{0}) = I$.

We can now rejoin the path taken in Section 13.1 to obtain the inverse function theorem (Theorem 13.13).

Completing Ordered Fields

In Section 14.4 we constructed the reals from the rationals. However, our arguments, up to and including the proof of Lemma 14.32, made no use of specific properties of the rationals. In fact, we can push our arguments a little further and prove the general principle of convergence (Lemma 14.35) without using specific properties of the rationals.

If we examine our proof of Lemma 14.35, we see that it makes use of the sequence $\theta(j^{-1})$ and the fact that $\theta(j^{-1}) > \theta(0)$ and $\theta(j^{-1}) \to \theta(0)$. The fact that $\theta(j^{-1}) \to \theta(0)$ is the axiom of Archimedes and required specific properties of the rationals in its proof. To provide a proof of Lemma 14.35 independent of the properties of the rationals, all we need is a sequence $[\boldsymbol{\eta}(j)] > \theta(0)$ with $[\boldsymbol{\eta}(j)] \to \theta(0)$ as $j \to \infty$. The situation is clarified by the following observation.

Lemma G.1. *Let $(\mathbb{F}, +, \times, >)$ be an ordered field. Then at least one of the two following statements is true.*

(A) If x_n is a Cauchy sequence in $\mathbb{F}$, then there exists an N such that $x_n = x_N$ for all $n \geq N$.

(B) We can find a sequence η_n in $\mathbb{F}$ with $\eta_n \neq 0$ for all n but with $\eta_n \to 0$ as $n \to \infty$.

Proof. If (A) is false, we can find a Cauchy sequence x_n such that, given any N we can find an $M > N$, with $x_M \neq x_N$. We can thus find $n(1) < n(2) < n(3) < \dots$ such that $x_{n(j+1)} \neq x_{n(j)}$. Setting $\eta_j = |x_{n(j+1)} - x_{n(j)}|$, we obtain a sequence with the properties required by (B). ∎

Exercise G.2. *(i) Explain why any $(\mathbb{F}, +, \times, >)$ satisfying alternative (A) in Lemma G.1 automatically satisfies the general principle of convergence.*

(ii) Show that we can replace the words 'at least one' by 'exactly one' in the statement of Lemma G.1.

Proof of Lemma 14.35 without using the axiom of Archimedes. If alternative (A) of Lemma G.1 applies, the result is automatic. If not, then alternative (B) applies and we can find $[\boldsymbol{\eta}(j)] \in \mathbb{R}$ such that $[\boldsymbol{\eta}(j)] > \theta(0)$ and $[\boldsymbol{\eta}(j)] \to \theta(0)$ as $j \to \infty$. By Lemma 14.32, it follows that we can find an $x_j \in \mathbb{Q}$ such that

$$[\mathbf{y}(j)] + [\boldsymbol{\eta}(j)] > \theta(x_j) > [\mathbf{y}(j)].$$

From now on the proof follows the same path as our original proof. More specifically, we show first that x_j is Cauchy in $\mathbb{Q}$ and then that $[\mathbf{y}(j)] \to [\mathbf{x}]$. ■

Exercise G.3. *Fill in the details of the proof just given.*

Taken together, the parts of the construction of $\mathbb{R}$ which only used the fact that $\mathbb{Q}$ is an ordered field yield a more general construction.

Lemma G.4. *Given any ordered field $(\mathbb{F}, +, \times, >)$ we can find an ordered field $(\mathbb{H}, +, \times, >)$ and an injective map $\theta : \mathbb{F} \to \mathbb{H}$ which preserves addition, multiplication and order with the following properties.*

(i) The general principle of convergence holds for $\mathbb{H}$.

(ii) If $a, b \in \mathbb{H}$ and $b > a$, we can find an $x \in \mathbb{F}$ with $b > \theta(x) > a$.

In Exercise 1.31 we constructed an object whose properties we restate in the next lemma.

Lemma G.5. *There exist an ordered field $(\mathbb{F}, +, \times, >)$ and an injective map $\phi : \mathbb{Q} \to \mathbb{F}$ which preserves addition, multiplication and order, and is such that there exists an $\eta \in \mathbb{F}$ with $\eta > 0$ and $\phi(1/n) > \eta$ for all $n \geq 1$ $[n \in \mathbb{Z}]$.*

Applying Lemma G.4 to the field of Lemma G.5, we obtain the following result.

Lemma G.6. *There exists an ordered field $(\mathbb{H}, +, \times, >)$ obeying the general principle of convergence and an injective map $\psi : \mathbb{Q} \to \mathbb{H}$ which preserves addition, multiplication and order, and is such that there exists an $\eta \in \mathbb{F}$ with $\eta > 0$ and $\psi(1/n) > \eta$ for all $n \geq 1$ $[n \in \mathbb{Z}]$.*

Thus we cannot deduce the axiom of Archimedes from the general principle of convergence, and so we cannot deduce the fundamental axiom of analysis from the general principle of convergence.

The reader may ask whether there actually exist ordered fields satisfying alternative (A) of Lemma G.1. She may also remark that, since any convergent sequence is a Cauchy sequence, it follows that the only convergent sequences for such a field are constant from some point on. Since the treatment of analysis in this book is based on sequences, she may therefore enquire if it is possible to do analysis on a space where there are no interesting convergent sequences. The answers to her questions are that there do exist ordered fields satisfying alternative (A) and that it is possible to do analysis without using sequences. Analysis without sequences and without metrics is the subject of general topology. But that, as they say, is another story.

Appendix H

Constructive Analysis

The purpose of this appendix is to try and give the reader a vague idea of how a radical reconstruction of analysis might work. It is based on the version proposed by Bishop[1]. (A follower of Bishop might well call it a caricature. If so, it is not intended as a hostile caricature.)

In classical analysis we often talk about objects that we cannot program a computer to find. In our new analysis the objects that we consider must have a 'meaning as a calculation'. The old notion of a real number x has no such meaning and is replaced by the following version. A real number $\mathbf{x}$ is a computer program which, given a positive integer n, produces a rational number x_n subject to the consistency condition, that if we consider the effect of trying our computer program on two integers n and m we have

$$\bigstar \qquad |x_n - x_m| \leq 10^{-n} + 10^{-m}.$$

An unreconstructed classical analyst or unsophisticated computer scientist would say that 'given a positive integer n' our program provides a rational x_n correct to within 10^{-n} of the true answer', but to the radical analyst there is no such object as the 'true answer'[2].

Here are some examples.

(a) If p is a rational number, we define define $p^* = \mathbf{x}$ by $x_n = p$.

(b) We define $\mathbf{e}$ by $e_n = \sum_{j=0}^{n+10} 1/j!$.

---

[1]There is an excellent account of Bishop's 'Constructive Analysis' in the first few chapters of [**7**]. The first serious attempt to found a radical analysis was due to Brouwer, who called his version 'Intuitionism'.

[2]Of course we can imagine the true answer, but we can also imagine unicorns. Or we can say that the true answer exists but is unknowable, in which case we might be better off doing theology rather than mathematics.

Exercise H.1. *Verify that* 1^* *and* $\mathbf{e}$ *satisfy the consistency condition* ★.

We can add two real numbers as follows. Given the programs for $\mathbf{x}$ and $\mathbf{y}$, we construct a program for $\mathbf{x}+\mathbf{y}$ as follows. Given a positive number n, we use the old programs to give us x_{n+2} and y_{n+2}. We set $z_n = x_{n+2}+y_{n+2}$ and write $\mathbf{z}=\mathbf{x}+\mathbf{y}$.

Exercise H.2. *Verify that* $\mathbf{x}+\mathbf{y}$ *satisfies the consistency condition* ★.

Multiplication is slightly more complicated. Given the programs for $\mathbf{x}$ and $\mathbf{y}$, we construct a program for $\mathbf{x}\times\mathbf{y}$ as follows. Given a positive number n, we first use the old programs to give us x_0 and y_0. We find the smallest positive integers m_1 and m_2 such that $10^{m_1} \geq |x_0|$ and $10^{m_2} \geq |y_0|$, and take $N = m_1+m_2+n+8$. We now set $w_n = x_N \times y_N$ and write $\mathbf{w}=\mathbf{x}\times\mathbf{y}$.

Exercise H.3. *Verify that* $\mathbf{x}\times\mathbf{y}$ *satisfies the consistency condition* ★.

We say that $\mathbf{x}=\mathbf{y}$ if we can prove, by looking at the programs involved, that $|x_n - y_n| \leq 2\times 10^{-n}$ for all n. We say that $\mathbf{x}\neq\mathbf{y}$ if we can find an N such that $|x_N - y_N| > 2\times 10^{-N}$.

Exercise H.4. *(i) Show that* $1^* + 1^* = 2^*$.

(ii) Show that if $\mathbf{x}$ *and* $\mathbf{y}$ *are real numbers, then* $\mathbf{x}+\mathbf{y}=\mathbf{y}+\mathbf{x}$.

(iii) We define $\mathbf{e}'$ *by* $e'_n = \sum_{j=0}^{n+9} 1/j!$. *Show that* $\mathbf{e}'$ *is a real number and* $\mathbf{e}'=\mathbf{e}$.

(iv) Show that $\mathbf{e}\times\mathbf{e}\neq\mathbf{e}$.

Notice that, given two real numbers, the radical analyst will say that there are three possibilities:

(a) $\mathbf{x}=\mathbf{y}$,

(b) $\mathbf{x}\neq\mathbf{y}$,

(c) neither (a) nor (b) is true.

Notice that for a radical analyst the nature of equality changes with time since new knowledge may enable us to move a pair of real numbers from class (c) to one of class (a) or class (b). The conservative analyst, in contrast, believes that, given real numbers x and y, it is true that $x=y$ or $x\neq y$ *even though it may be true that nobody will ever know which*[3]. In general the radical analyst replaces the dichotomy

true/false

of the conservative analyst by the much more careful trichotomy

proved true/proved false/not proved true or false.

[3]Compare someone who believes that he is being spied on by invisible aliens using undetectable rays.

The physicist who feels that this is too cautious might care to ask the opinion of Schrödinger's cat.

It is, I hope, clear what we should consider a function to be in our new analysis. A function $\mathbf{f} : \mathbb{R} \to \mathbb{R}$ is a program such that, when we give it a real $\mathbf{x}$ as a callable sub-routine, we obtain a real number $\mathbf{f}(\mathbf{x})$.

Thus, for example, we can define a function $\mathbf{m}$ corresponding to the classical 'modulus' function m with $m(x) = |x|$ as follows. Given $\mathbf{x}$ and n, compute x_{n+2}. Set $z_n = |x_{n+2}|$. We take $\mathbf{m}(\mathbf{x}) = \mathbf{z}$.

Exercise H.5. *Define a function* ***exp*** *corresponding to the classical* exp. *[It would be a good idea for your program to look first at* x_0 *'to gain some idea of the size of* $\mathbf{x}$*'.]*

Suppose now that a conservative analyst, anxious to be helpful, tries to define a function $\mathbf{H}$ corresponding to the classical Heaviside function $H : \mathbb{R} \to \mathbb{R}$ given by $H(x) = 0$ for $x < 0$, $H(x) = 1$ for $x \geq 0$. We need a program such that, when we give it a real $\mathbf{x}$ as a callable sub-routine, we obtain a real number $\mathbf{z} = \mathbf{H}(\mathbf{x})$. Suppose we ask our program for the value of z_2. Our program cannot ask for the value of x_n for all n, so it must ask for some subset of $x_0, x_1, \dots, x_N$, where N may depend on $x_0, x_1, \dots, x_M$ but M itself must be fixed (otherwise the program might not terminate). Suppose $x_0 = x_1 = \dots = x_N = 0$. We know that $\mathbf{H}(\mathbf{x}) = 0^*$ or $\mathbf{H}(\mathbf{x}) = 1^*$, so, by the definition of equality, either $z_2 > 3/4$ and $\mathbf{H}(\mathbf{x}) = 1^*$ or $z_2 < 1/4$ and $\mathbf{H}(\mathbf{x}) = 0^*$. But the next term in our sequence might be $x_{N+1} = -10^{-(N+1)}/2$ (in which case $\mathbf{H}(\mathbf{x}) = 0^*$ and $z_2 < 1/4$) or $x_{N+1} = 10^{-(N+1)}/2$ (in which case $\mathbf{H}(\mathbf{x}) = 1^*$ and $z_2 > 3/4$), so our program does not produce the required answer[4].

The reader will now remember that our basic counterexample, given in Example 1.3, to which we returned over and over again, depended on being able to construct functions of Heaviside type. Someone who believes that mathematical intuition comes only from the physical world might rush to the conclusion that every function in our new analysis must satisfy the intermediate value theorem. However, the surest source of intuition for our new analysis is the world of computing (or, more specifically, numerical analysis), and anyone who has done numerical analysis knows that there is no universal 'zero-finding' program.

[4]The conservative analyst now tells the radical analyst that all functions in the new analysis must be continuous. (This is what the present author, a typical sloppy conservative analyst, said on page 354.) However, the conservative classical analyst's proof (which is a development of the argument just given) depends on classical analysis! The radical analyst's views on such a 'proof' are those of an anti-smoking campaigner on a statement by a tobacco company executive. In some radical reconstructions of analysis it is possible to prove that all functions are continuous, in others (such as the one given here) there is, at present, no proof that all functions are continuous and no example of a function which is not.

What would be the statement of the intermediate value theorem in our new analysis? Surely, it would run something like this. We have a program **P** which, when given a function **f** with $\mathbf{f}(0^*) = 1^*$ and $\mathbf{f}(1^*) = -1^*$ as a callable subroutine and a positive integer n, produces a rational number z_n with $0 \leq z_n \leq 1$ such that

(i) the sequence z_n satisfies our consistency condition ★ and

(ii) $\mathbf{f}(\mathbf{z}) = 0^*$.

Exercise H.6. *(i) Let p and q be rational numbers. Show that our new analysis contains functions $\mathbf{f}_{p,q}$ corresponding to the classical function $f_{p,q}$ which is the simplest continuous piecewise linear function with $f_{p,q}(t) = 1$ for $t \leq 0$, $f_{p,q}(1/3) = p$, $f_{p,q}(2/3) = q$ and $f_{p,q}(t) = -1$ for $t \geq 1$.*

(ii) Sketch $f_{p,q}$ when p and q are very small (note that p and q may be positive or negative). Explain why your favourite root finding program will have problems with this function.

*(iii) Using the same kind of argument as we used to show the non-existence of **H**, convince yourself that a program **P** of the type described in the paragraph before this exercise does not exist.*

Of course, just as numerical analysis contains zero-finding algorithms which will work on functions satisfying reasonable conditions, so our new analysis contains versions of the intermediate value theorem which will work on functions satisfying reasonable conditions. The radical analyst has no difficulty in proving that

$$\mathbf{exp}(\mathbf{x}) = \mathbf{y}$$

has one and only one solution (for $\mathbf{y} \neq 0^*$).

When English students first learn that nouns in other languages have gender they burst into incredulous laughter, and few English people entirely lose the view that the French speak French in public and English in private. Because I have tried to explain a version of radical analysis in terms of conservative analysis, it looks as though radical analysis is simply a curious version of conservative analysis[5]. However, someone trained in the radical tradition would see conservative analysis as a curious offshoot of radical analysis in which distorted versions of known objects obeyed twisted versions of familiar theorems. There is no primal language, or world view, or real line, but there are many languages, world views and real lines, each of which can be understood to a large extent in terms of another but can only be fully understood on its own terms.

[5]One problem faced by anyone trying to reconstruct analysis is given in the traditional reply of the book burner. 'If these writings of the Greeks agree with the book of God, they are useless and need not be preserved: if they disagree they are pernicious and ought to be destroyed.' (See Gibbon's *Decline and Fall*, Chapter LI, and note the accompanying commentary.)

Appendix I

Miscellany

This appendix consists of short notes on various topics which I feel should be mentioned but which did not fit well into the narrative structure of this book.

Compactness: When mathematicians generalised analysis from the study of metric spaces to the study of more general 'topological spaces', they needed a concept to replace the Bolzano–Weierstrass property discussed in this book. After some experiment, they settled on a property called 'compactness'.

It can be shown that a metric space has the property of compactness (more briefly, a metric space is compact) when considered as a topological space if and only if it has the Bolzano–Weierstrass property[1]. (See Exercises K.219 and K.220 if you would like to know more.) We showed in Theorem 4.16 that a subset of $\mathbb{R}^n$ with the standard metric has the Bolzano–Weierstrass property if and only if it is closed and bounded.

Thus, if you read of a 'compact subset E of $\mathbb{R}^n$', you may translate this as 'a closed and bounded subset of $\mathbb{R}^n$' and, if you read of a 'compact metric space', you may translate this as 'a metric space having the Bolzano–Weierstrass property'. However, you must remember that a closed bounded metric space need not have the Bolzano–Weierstrass property (see Exercise 11.20) and that, *for topological spaces, the Bolzano–Weierstrass property is not equivalent to compactness.*

[1]In more advanced work we say that a metric space with the Bolzano–Weierstrass property is sequentially compact.

Abuse of language: The language of mathematics is a product of history as well as logic, and sometimes the forces of history are stronger than those of logic. Logically, we need to talk about *the value* $\sin x$ that *the function* sin takes at the point x, but, traditionally, mathematicians have talked about the function '$\sin x$'. To avoid this problem mathematicians tend to use phrases like 'the function $f : \mathbb{R} \to \mathbb{R}$ given by $f(x) = x^2$' or 'the map $x \mapsto x^2$'. In Section 13.3, when we wish to talk about the map $\mathbf{x} \mapsto L(\lambda, \mathbf{x}) = t(\mathbf{x}) - \lambda f(\mathbf{x})$ we write '$L(\lambda,\ ) = t - \lambda f$'. (Many mathematicians dislike leaving a blank space in a formula and use a place holder '$\cdot$' instead. They write '$L(\lambda, \cdot) = t(\cdot) - \lambda f(\cdot)$'.)

To a 19th century mathematician and to most of my readers this may appear unnecessary, but the advantage of the extra care appears when we need to talk about the map $x \mapsto 1$. However, from time to time, mathematicians revert to their traditional habits.

Bourbaki calls such reversions '*abuses of language* without which any mathematical text runs the risk of pedantry, not to say unreadability.'

Perhaps the most blatant abuse of language in this book concerns sequences. Just as $f(x)$ is not a function f, but the value of f at x, so a_n is not the sequence

$$a_1,\ a_2,\ a_3,\ a_4 \ldots,$$

but the nth term in such a sequence. Wherever I refer to 'the sequence a_n', I should have used some phrase like 'the sequence $(a_n)_{n=1}^{\infty}$'. Perhaps future generations will write like this, but it seemed to me that the present generation would simply find it distracting.

Non-uniform notation: When Klein gave his lectures on *Elementary Mathematics from an Advanced Standpoint* [**29**], he complained,

> There are a great many symbols used for each of the vector operations and, so far, it has proved impossible to produce a generally accepted notation. A commission was set up for this purpose at a scientific meeting at Kassel (1903). However, its members were not even able to come to a complete agreement among themselves. None the less, since their intentions were good, each member was willing to meet the others part way and the result was to bring three new notations into existence[2]! My experience in such things inclines me to the belief that real agreement is only possible if there are powerful economic interests behind such a move. ... But

[2]In later editions Klein recorded the failure of similar committees set up at the Rome International Mathematical Congress (1908).

> there are no such interests involved in vector calculus and so we must agree, for better or worse, to let every mathematician cling to the notation which he finds most convenient or – if he is dogmatically inclined – the only correct one.

Most mathematics students believe that there should be a unique notation for each mathematical concept. This is not entirely reasonable. Consider the derivative of a function $f : \mathbb{R} \to \mathbb{R}$. If I am interested in the slope of the curve $y = f(x)$ it is natural to use the suggestive Leibniz notation $\frac{dy}{dx}$. If I am interested in the function which is the derivative of f it is is more natural to consider f', and if I am interested in the operation of differentiation rather than the functions it operates on I may well write Df. Different branches of mathematics may have genuine reasons for preferring different notations for the same thing.

Even if she disagrees, the student must accept that different notations exist and there is nothing that she can do about it. A quick visit to the library reveals the following notations for the same thing (here $f : \mathbb{R}^3 \to \mathbb{R}$ is a well behaved function):

$$\frac{\partial^3 f}{\partial x^2 \partial z},\ f_{xxz},\ f'''_{xxz},\ D_{xxz}f,\ f_{,113},\ f_{113}, f'''_{113},\ D_{113}f,\ D_{(2,0,1)}f, D^{(2,0,1)}f$$

and several others.

Here are three obvious pieces of advice.

(1) If you are writing a piece of mathematics and several notations exist, you must make it clear to the reader which one you are using.

(2) If you are reading a piece of mathematics which uses an unfamiliar notation, try to go along with it rather than translating it back into your favourite notation. The advantage of a familiar notation is that it carries you along without too much thought, the advantage of an unfamiliar notation is that it makes you think again — and, who knows, the new notation might turn out to better than the old.

(3) Do not invent a new notation when a reasonably satisfactory one already exists.

Left and right derivatives: In many ways, the natural subsets of $\mathbb{R}^m$ to do analysis on are the open sets U, since, given a function $f : U \to \mathbb{R}^n$ and a point $\mathbf{x} \in U$, we can then examine the behaviour of f 'in all directions round $\mathbf{x}$', on a ball

$$B(\mathbf{x}, \epsilon) = \{\mathbf{y} : \|\mathbf{y} - \mathbf{x}\| < \epsilon\},$$

for some, sufficiently small, $\epsilon > 0$.

However, as we saw in Theorem 4.41 and Theorem 4.63, continuous functions behave particularly well on closed bounded sets. This creates a certain tension, as the next exercise illustrates.

Exercise I.1. *(i) Show that the map $\mathbb{R}^m \to \mathbb{R}$ given by $\mathbf{x} \mapsto \|\mathbf{x}\|$ is continuous.*

(ii) If E is a closed bounded set, show that there exists an $\mathbf{e}_0 \in E$ such that $\|\mathbf{e}_0\| \geq \|\mathbf{x}\|$ for all $\mathbf{x} \in E$. Give an example to show that $\mathbf{e}_0$ need not be unique.

(iii) Show that the only subset of $\mathbb{R}^m$ which is both open and closed and bounded is the empty set.

This tension is often resolved by considering functions defined on an open set U, but working on closed bounded subsets of U. In the special case when we work in one dimension and deal with intervals, we can use a different trick.

Definition I.2. *Let $b > a$ and consider $f : [a, b] \to \mathbb{R}$. We say that f has right derivative $f'(a+)$ at a if*

$$\frac{f(a+h) - f(a)}{h} \to f'(a+)$$

as $h \to 0$ through values $h > 0$.

Exercise I.3. *Define the left derivative $f'(b-)$, if it exists.*

When mathematicians write that f is differentiable on $[a, b]$ with derivative f', they are using the following convention.

Definition I.4. *If $f : [a, b] \to \mathbb{R}$ is differentiable at each point of (a, b) and has right derivative $f'(a+)$ at a and left derivative $f'(b-)$ at b, we say that f is differentiable on $[a, b]$ and write $f'(a) = f'(a+)$, $f'(b) = f'(b-)$.*

Observe that this is not radically different from our other suggested approach.

Lemma I.5. *If $f : [a, b] \to \mathbb{R}$ is differentiable on $[a, b]$, we can find an everywhere differentiable function $\tilde{f} : \mathbb{R} \to \mathbb{R}$ with $\tilde{f}(t) = f(t)$ for $t \in [a, b]$.*

Proof. Set

$$\begin{aligned}
\tilde{f}(t) &= f(a) + f'(a)(t-a) && \text{for } t < a,\\
\tilde{f}(t) &= f(t) && \text{for } a \leq t \leq b,\\
\tilde{f}(t) &= f(b) + f'(b)(t-b) && \text{for } b < t.
\end{aligned}$$

It is easy to check that $\tilde{f}$ has the required properties. ∎

Exercise I.6. *If $f : [a,b] \to \mathbb{R}$ is differentiable on $[a,b]$ with continuous derivative, show that we can find an everywhere differentiable function $\tilde{f} : \mathbb{R} \to \mathbb{R}$ with continuous derivative such that $\tilde{f}(t) = f(t)$ for $t \in [a,b]$.*

If $f : [a,b] \to \mathbb{R}$ is differentiable on $[a,b]$ and $f' : [a,b] \to \mathbb{R}$ is differentiable on $[a,b]$, it is natural to say that f is twice differentiable on $[a,b]$ with derivative $f'' = (f')'$, and so on.

Exercise I.7. *If $f : [a,b] \to \mathbb{R}$ is twice differentiable on $[a,b]$ with continuous second derivative, show that we can find an everywhere twice differentiable function $\tilde{f} : \mathbb{R} \to \mathbb{R}$ with continuous second derivative such that $\tilde{f}(t) = f(t)$ for $t \in [a,b]$. Generalise this result.*

Piecewise definitions: Occasionally mathematicians define functions as piecewise continuous, piecewise continuously differentiable and so on. The underlying idea is that the graph of the function is made up of a finite number of well behaved pieces.

Definition I.8. *A function $f : [a,b] \to \mathbb{R}$ is piecewise continuous if we can find*

$$a = x_0 < x_1 < \cdots < x_n = b$$

and continuous functions $g_j : [x_{j-1}, x_j] \to \mathbb{R}$ such that $f(x) = g_j(x)$ for all $x \in (x_{j-1}, x_j)$ and all $1 \leq j \leq n$

Definition I.9. *A function $f : [a,b] \to \mathbb{R}$ is piecewise linear (respectively differentiable, continuously differentiable, infinitely differentiable, etc.) if it is continuous and we can find*

$$a = x_0 < x_1 < \cdots < x_n = b$$

such that $f|_{[x_{j-1},x_j]} : [x_{j-1}, x_j] \to \mathbb{R}$ is linear (respectively differentiable, continuously differentiable, infinitely differentiable, etc.) for all $1 \leq j \leq n$.

Notice that Definition I.8 does not follow the pattern of Definition I.9.

Exercise I.10. *(i) Show by means of an example that a piecewise continuous function need not be continuous.*

(ii) Show that a piecewise continuous function is bounded.

Exercise I.11. *(This is a commentary on Theorem 8.32.) (i) Show that any piecewise continuous function $f : [a,b] \to \mathbb{R}$ is Riemann integrable.*

(ii) Show that, if $f : [a,b] \to \mathbb{R}$ is continuous on (a,b) and bounded on $[a,b]$, then f is Riemann integrable.

The definitions just considered are clearly rather ad hoc. Numerical analysts use a more subtle approach via the notion of a *spline* (see, for example, Chapters 18 and onwards in [**44**]).

Appendix J

Executive Summary

The summary is mainly intended for experts but may be useful for revision. It may also be more useful than the index if you want to track down a particular idea. Material indicated by $\langle\heartsuit$ ... $\heartsuit\rangle$ or by $\langle\heartsuit\heartsuit$ ... $\heartsuit\heartsuit\rangle$ is not central to the main argument. Material indicated by [...] is in appendices or exercises; this material either is not central to the main argument or is such that most students will have met it in other courses.

Introduction to the real number system

Need for rigorous treatment of analysis (p. 1). Limits in $\mathbb{R}$, subsequences, sums and products (p. 3). Continuity of functions from $\mathbb{R}$ to $\mathbb{R}$ (p. 7). The real numbers $\mathbb{R}$ form an ordered field obeying the fundamental axiom that every increasing bounded sequence converges (p. 9). Axiom of Archimedes (p. 10). [Decimal expansion (Exercise 1.32, p. 13).] The intermediate value theorem, proof by lion hunting (p. 15). [Countability (Appendix B, p. 361). The real numbers are uncountable (Exercise 1.41, p. 17). Cantor's proof of the existence of transcendentals (Exercise B.7, p. 363). Explicit construction of a transcendental number (Exercise K.12, p. 409).] Differentiation and the mean value inequality (one-dimensional case) (p. 18). Intermediate value theorem equivalent to fundamental axiom (p. 22). $\langle\heartsuit\heartsuit$Further informal discussion of the status of the fundamental axiom (p. 25).$\heartsuit\heartsuit\rangle$

Equivalents of the fundamental axiom

Supremum, existence for bounded non-empty sets equivalent to fundamental axiom, use as proof technique (p. 31). Theorem of Bolzano–Weierstrass, equivalent to fundamental axiom, use as proof technique (p. 37).

Higher dimensions

$\mathbb{R}^m$ as an inner product space, Cauchy–Schwarz and the Euclidean norm (p. 43). Limits in $\mathbb{R}^m$ (p. 46). Theorem of Bolzano–Weierstrass in $\mathbb{R}^m$ (p. 47). Open and closed sets (p. 48). Theorem of Bolzano–Weierstrass in the context of closed bounded subsets of $\mathbb{R}^m$ (p. 49). Continuity for many-dimensional spaces (p. 53). The image of a continuous function on a closed bounded subset of $\mathbb{R}^m$ is closed and bounded, a real-valued continuous function on a closed bounded subset of $\mathbb{R}^m$ is bounded and attains its bounds (p. 56). [The intersection of nested, non-empty, closed, bounded sets is non-empty (Exercise 4.48, p. 59).] Rolle's theorem and the one-dimensional mean value theorem (p. 59). Uniform continuity, a continuous function on a closed bounded subset of $\mathbb{R}^m$ is uniformly continuous (p. 64).

Sums

(This material is treated in $\mathbb{R}^m$.) General principle of convergence (p. 66). Absolute convergence implies convergence for sums (p. 68). Comparison test (p. 69). Complex power series and the radius of convergence (p. 70). ⟨♡Ratio test and Cauchy's condensation test (p. 69). Conditional convergence, alternating series test, Abel's test, rearrangement of conditionally convergent series (p. 75). Informal discussion of the problem of interchanging limits (p. 79). Dominated convergence theorem for sums, rearrangement of absolutely convergent series, Fubini's theorem for sums (p. 82). The exponential function mainly for $\mathbb{R}$ but with mention of $\mathbb{C}$, multiplication of power series (p. 88). The trigonometric functions, notion of angle (p. 95). The logarithm on $(0, \infty)$, problems in trying to define a complex logarithm (p. 99). Powers (p. 105). Fundamental theorem of algebra (p. 109).♡⟩

Differentiation from $\mathbb{R}^n$ to $\mathbb{R}^m$

Advantages of geometric approach, definition of derivative as a linear map, Jacobian matrix (p. 117). Operator norm, chain rule and other elementary properties of the derivative (p. 123). Mean value inequality (p. 130). Simple local and global Taylor theorems in one dimension, Cauchy's example of a function with no non-trivial Taylor expansion, Taylor theorems depend on the fundamental axiom (p. 135). Continuous partial derivatives imply differentiability, symmetry of continuous second order derivatives, informal treatment of higher order local Taylor theorems, informal treatment of higher order derivatives as symmetric multilinear maps (p. 139). Discussion, partly informal, of critical points, hill and dale theorem (p. 147).

Riemann integration

Need for precise definition of integral and area, Vitali's example (p. 161). Definition of the Riemann integral via upper and lower sums, elementary properties, integrability of monotonic functions (p. 164). Integrability of continuous functions, fundamental theorem of the calculus, Taylor's theorem with integral form of remainder (p. 173). ⟨♡ Differentiation under the integral for finite range, Euler–Lagrange equation in calculus of variations, use and limitations, Weierstrass type example (p. 181).♡⟩ Brief discussion of the Riemann integral of $\mathbb{R}^m$-valued functions, $\| \int f\| \leq \int \|f\|$ (p. 192).

⟨♡ Class of Riemann integrable functions not closed under pointwise convergence (p. 195). Informal discussion of improper Riemann integration (p. 197). Informal and elementary discussion of multiple integrals (no change of variable formula), Fubini for continuous functions on a rectangle (p. 201), Riemann–Stieltjes integration (p. 206). Rectifiable curves and line integrals, Schwarz's example showing the problems that arise for surfaces (p. 213).♡⟩

Metric spaces

⟨♡ Usefulness of generalising notion of distance illustrated by Shannon's theorem on the existence of good codes (p. 221).♡⟩ Metric spaces, norms, limits, continuity, open sets (p. 229). All norms on a finite-dimensional space are Lipschitz equivalent (p. 234). Continuity of functions between normed spaces (p. 238). ⟨♡ Informal discussion of geodesics illustrated by the Poincaré metric on the upper half plane (p. 241).♡⟩

Complete metric spaces

Definition of completeness, examples of complete and incomplete metric spaces including among incomplete ones the L^1 norm on $C([a,b])$ (p. 249). Completeness and total boundedness, equivalence of the conjunction of these properties with the Bolzano–Weierstrass property (p. 257).

Uniform metric is complete, restatement of result in classical terms (uniform limit of continuous functions is continuous, general principle of uniform convergence) (p. 261). Uniform convergence, integration and differentiation, restatement for infinite sums, differentiation under an infinite integral (p. 267). Local uniform convergence of power series, power series differentiable term by term, rigorous justification of power series solution of differential equations (p. 273). ⟨♡ An absolutely convergent Fourier series converges to the appropriate function (p. 282).♡⟩

Contraction mapping theorem

Banach's contraction mapping theorem (p. 287). Existence of solutions of differential equations by Picard's method (p. 289). ⟨♡ Informal discussion of existence and non-existence of global solutions of differential equations (p. 294). Green's function solutions for second order linear differential equations (p. 301).♡⟩

The inverse function theorem (p. 311). ⟨♡ The implicit function theorem (p. 320). Lagrange multipliers and Lagrangian necessary condition (p. 328). Lagrangian sufficient condition, problems in applying Lagrange multiplier methods (p. 333).♡⟩

Completion of metric spaces

Density, completion of metric space, inheritance of appropriate structures such as inner product (p. 335). Proof of existence of completion (p. 341). ⟨♡ Informal discussion of construction of $\mathbb{Z}$ from $\mathbb{N}$, $\mathbb{Q}$ from $\mathbb{Z}$, $\mathbb{C}$ from $\mathbb{R}$ (p. 344). Construction of $\mathbb{R}$ from $\mathbb{Q}$ (p. 348).♡⟩ ⟨♡♡ Rapid, optimistic and informal discussion of foundational issues (p. 354).♡♡⟩

Appendix K

Exercises

At an elementary level, textbooks consist of a little explanation (usually supplemented by a teacher) and a large number of exercises of a routine nature. At a higher level the number of exercises decreases and the exercises themselves become harder and more variable in difficulty. At the highest level there may be no exercises at all. We may say that such books consist of a single exercise: *read and understand the contents.* Because I would be happy if students treated my text in this manner, I have chosen to put most of the exercises in an appendix. Some answers, hints and so on can be found on an AMS page at `www.ams.org/bookpages/gsm-62` with a copy on my home page at `http://www.dpmms.cam.ac.uk/~twk/`.

I suspect that readers will gain most by tackling those problems which interest them. To help them make a choice, I have labeled them in the following manner [**2.1, P**]. The number **2.1** tells you that Section 2.1 may be relevant, and the letters and symbols that follow have the meanings given below. Like many similar labeling systems, it is not entirely satisfactory.

↑: Follows on from the preceding exercise.

↑↑: Follows on from an earlier exercise.

S: Rather shorter or easier than the general run of exercises.

M: Methods type exercise. Forget theoretical niceties and concentrate on getting an answer.

M!: Just try and get an answer by fair means or foul.

T: This exercise leads you through a standard piece of theory.

T!: This exercise leads you through a standard piece of theory but in a non-standard way.

P: Problem type exercise.

G: Uses general background rather than material in this book.

H: The result of this exercise is not standard.

H!: The result of this exercise is highly non-standard. Only do this exercise if you are really interested.

K.1. (Irrationality of $\sqrt{2}$.) [1.1, G, T] The reader presumably knows the classic proof that the equation $n^2 = 2m^2$ has no non-zero integer solutions (in other words, $x^2 = 2$ has no solution in $\mathbb{Q}$). Here are two others.

(i) Show that, if $n^2 = 2m^2$, then

$$(2m - n)^2 = 2(n - m)^2.$$

Deduce that, if n and m are strictly positive integers with $n^2 = 2m^2$, we can find strictly positive integers n' and m' with $n'^2 = 2m'^2$ and $n' < n$. Conclude that the equation $n^2 = 2m^2$ has no non-zero integer solutions.

(ii) Our second argument requires more thought but is also more powerful. We use it to show that, if N is a positive integer which is not a perfect square, then the equation $x^2 = N$ has no rational solution.

To this end we suppose that x is a positive rational with $x^2 = N$. Explain why we can find a least strictly positive integer m such that mx is an integer and why we can find an integer k with $k+1 > x > k$. Set $m' = mx - mk$ and show that m' is an integer, that $m'x$ is an integer and that $m > m' \geq 1$. The required result follows by contradiction. (This argument and its extensions are discussed in [**3**].)

(iii) Apostol gave the following beautiful geometric version of the argument of part (i) (see Figure K.1). It will appeal to all fans of Euclidean geometry and can be ignored by everybody else.

Suppose that n and m are strictly positive integers with $n^2 = 2m^2$, Draw a triangle ABC with AB and BC of length m units and CA of length n units. Explain why ABC is a right angled triangle. Let the circle Γ with centre A passing through B cut AC at D and let the tangent to Γ at D cut BC at E. Show that BE, ED and DC all have the same length. Deduce that ECD is a right angled triangle with sides ED and DC of integer length m' units and side CE of integer length n' units. Show that $n'^2 = 2m'^2$ and $n > n' \geq 1$. Deduce that the equation $n^2 = 2m^2$ has no non-zero integer solutions.

K.2. [1.2, P] If $d_n \geq 1$ for all n and $d_n + d_n^{-1}$ tends to a limit as $n \to \infty$, show that d_n tends to a limit.

Show, by giving an example, that the result is false if we replace the condition $d_n \geq 1$ by $d_n \geq k$ for some fixed k with $0 < k < 1$.

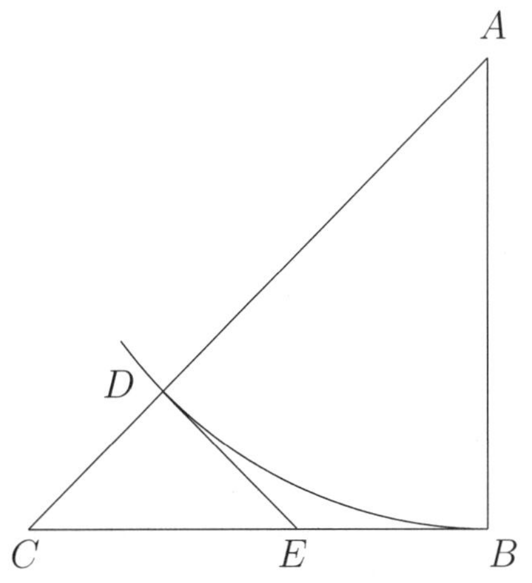

Figure K.1. Apostol's construction.

Can you find a result similar to that of the first paragraph involving $d_n \in \mathbb{C}$?

K.3. [1.4, P] Prove that, if

$$a_1 > b_1 > 0 \text{ and } a_{n+1} = \frac{a_n + b_n}{2}, \quad b_{n+1} = \frac{2a_n b_n}{a_n + b_n},$$

then $a_n > a_{n+1} > b_{n+1} > b_n$. Prove that, as $n \to \infty$, a_n and b_n both tend to the limit $\sqrt{(a_1 b_1)}$.

Use this result to give an example of an increasing sequence of rational numbers tending to a limit l which is not rational.

K.4. [1.3, P, S] Let $f, g : \mathbb{R} \to \mathbb{R}$ be functions such that the composition $g \circ f$ is continuous at x.

(i) Show, by means of examples, that the continuity of f at x is neither a necessary nor a sufficient condition for the continuity of g at $f(x)$.

(ii) Show, by means of examples, that the continuity of g at $f(x)$ is neither a necessary nor a sufficient condition for the continuity of f at x.

K.5. [1.3, P] The function $f : [0,1] \to [0,1]$ is defined by

$$f(x) = \begin{cases} 1 - x & \text{if } x \text{ is rational,} \\ x & \text{if } x \text{ is irrational.} \end{cases}$$

Consider the following statements, prove those which are true and demonstrate the falsity of the remainder.

(i) $f(f(x)) = x$ for all $x \in [0,1]$.

(ii) $f(x) + f(1-x) = 1$ for all $x \in [0,1]$.

(iii) f is bijective.

(iv) f is everywhere discontinuous in $[0, 1]$.

(v) $f(x + y) - f(x) - f(y)$ is rational for all x, $y \in [0, 1]$ such that $x + y \in [0, 1]$.

K.6. [1.3, P] Enumerate the rationals in $[0, 1]$ as a sequence q_1, q_2, ... and let H be the Heaviside function given by $H(t) = 0$ for $t \leq 0$ and $H(t) = 1$ for $t > 0$. Show that the equation

$$f(x) = \lim_{N \to \infty} \sum_{n=1}^{N} 2^{-n} H(x - q_n)$$

gives a well defined increasing function $f : [0, 1] \to \mathbb{R}$ which is discontinuous at rational points and continuous at irrational points.

K.7. [1.4, P] A bounded sequence of real numbers a_n satisfies the condition

$$a_n \leq \frac{a_{n-1} + a_{n+1}}{2}$$

for all $n \geq 1$. By showing that the sequence $a_{n+1} - a_n$ is decreasing, or otherwise, show that the sequence a_n converges.

What, if anything, can we deduce if the sequence is unbounded? What, if anything, can we deduce if the sequence a_n is bounded but we have the reverse inequality

$$a_n \geq \frac{a_{n-1} + a_{n+1}}{2}$$

for all $n \geq 1$? Prove your answers.

K.8. [1.4, P] The function $f : (0, 1) \to \mathbb{R}$ is continuous on the open interval $(0, 1)$ and satisfies $0 < f(x) < x$. The sequence of functions $f_n : (0, 1) \to \mathbb{R}$ is defined by

$$f_1(x) = f(x), \; f_n(x) = f(f_{n-1}(x)) \text{ for } n \geq 2.$$

Prove that $f_n(x) \to 0$ as $n \to \infty$ for all $x \in (0, 1)$.

Show, by means of a counterexample, that the result need not be true if f is not continuous everywhere in $(0, 1)$.

Enter any number on your calculator. Press the sine button repeatedly. What appears to happen? Prove your conjecture.

K.9. [1.4, T] (i) If $\delta > 0$, use the binomial theorem to show that

$$(1 + \delta)^n \geq \binom{n}{2} \delta^2.$$

Deduce that $n^{-1}(1 + \delta)^n \to \infty$ as $n \to \infty$. Show that, if $|x| < 1$, then $nx^n \to 0$ as $n \to \infty$.

(ii) Show, more generally, that, if $\delta > 0$, and k is any integer, then $n^{-k}(1+\delta)^n \to \infty$ as $n \to \infty$. Show that, if $|x| < 1$, then $n^k x^n \to 0$ as $n \to \infty$.

K.10. [1.4, P] Let $x_1, x_2, \ldots$ be a sequence of positive real numbers and suppose that

$$\frac{x_1 + x_2 + \cdots + x_n}{n} \to 1$$

as $n \to \infty$. Choose $m(n)$ so that $1 \le m(n) \le n$ and

$$x_{m(n)} = \max(x_1, x_2, \ldots, x_n).$$

Show that $x_n/n \to 0$ and deduce that $x_{m(n)}/n \to 0$ as $n \to \infty$.

By comparing x_r^α with $x_r x_{m(n)}^{\alpha-1}$, or otherwise, show that

$$\frac{x_1^\alpha + x_2^\alpha + \cdots + x_n^\alpha}{n^\alpha} \to \begin{cases} 0 & \text{if } \alpha > 1, \\ \infty & \text{if } 0 \le \alpha < 1, \end{cases}$$

as $n \to \infty$.

K.11. [1.5, P] Which of the following statements are true and which are false? Give a proof or counterexample.

(i) The sum of a rational number and an irrational number is irrational.

(ii) The product of a rational number and an irrational number is irrational.

(iii) If x is a real number and $\epsilon > 0$, we can find an irrational number y with $|x - y| < \epsilon$.

(iv) If each x_n is irrational and $x_n \to x$, then x is irrational.

(v) If x and y are irrational, then x^y is irrational.

[*Hint:* This is an old chestnut. Consider $a = 2^{1/2}$ and observe that $(a^a)^a = 2$.]

K.12. (Liouville's theorem.) [1.5, T] A real number x is called algebraic if

$$\sum_{r=0}^{n} a_r x^r = 0$$

for some $a_r \in \mathbb{Z}$ $[0 \le r \le n]$, $a_n \ne 0$ and $n \ge 1$ (in other words, x is a root of a non-trivial polynomial with integer coefficients). Liouville proved that not all real numbers are algebraic. Here is a version of his proof.

(i) If x and y are real and n is a strictly positive integer, show, by extracting the factor $x - y$ from $x^n - y^n$, that

$$|x^n - y^n| \le n(\max(|x|, |y|))^{n-1}|x - y|.$$

(ii) If P is a polynomial and $R > 0$, show, using (i), that there exists a K, depending on P and R, such that

$$|P(x) - P(y)| \leq K|x - y|$$

whenever $|x|, |y| \leq R$.

(iii) Suppose that

$$P(t) = \sum_{r=0}^{n} a_r t^r$$

for some $a_r \in \mathbb{Z}$ $[0 \leq r \leq n]$, $a_n \neq 0$ and $n \geq 1$. Show that $N^n P(M/N)$ is an integer whenever M is an integer and N is a strictly positive integer. Conclude that, if M and N are as stated, either $P(M/N) = 0$ or $|P(M/N)| \geq N^{-n}$.

(iv) Let P, R, K, M and N be as in parts (ii) and (iii). Suppose that η is a real number with $P(\eta) = 0$. If η and M/N lie in the interval $[-R, R]$, show that

$$N^{-n} \leq K|\eta - M/N|.$$

(v) Use (iv) to prove that, if η is an algebraic number which is not a rational number, then we can find n and L (depending on η) such that

$$|\eta - M/N| \geq LN^{-n}$$

whenever M is an integer and N is a strictly positive integer.

(vi) Explain why $\sum_{k=1}^{\infty} 10^{-k!}$ converges. Let us write $\lambda = \sum_{k=1}^{\infty} 10^{-k!}$. Show that

$$|\lambda - \sum_{k=1}^{J} 10^{-k!}| \leq \frac{10}{9} \times 10^{-(J+1)!}.$$

By considering $N = 10^{J!}$ and $M = \sum_{k=1}^{J} 10^{J!-k!}$, or otherwise, show that (a) λ is irrational and (b) λ is not an algebraic number.

Note that we have shown that λ is *too well approximable by rationals to be an algebraic number.* We discuss approximation problems again in Exercises K.13 and K.14.

(vii) In the first part of the exercise I deliberately avoided using the mean value inequality (Theorem 1.7). Prove (ii) directly, using the mean value inequality.

(viii) Show that no real number of the form

$$\sum_{k=1}^{\infty} m_k 10^{-k!}$$

with m_k taking the value 1 or 2 can be algebraic. If you know enough about countability, show that the collection of numbers of the form just given is uncountable.

K.13. (Continued fractions.) [1.5, T] In this exercise we take a first glance at continued fractions.

Consider the following process. If $x = x_0 \in [0,1)$, either $x_0 = 0$ and we stop, or we set

$$r_1 = \left[\frac{1}{x_0}\right]$$

(that is, r_1 is the integer part of $1/x_0$). We set

$$x_1 = \frac{1}{x_0} - r_1,$$

so that x_1 is the fractional part of $1/x_0$ and $x_1 \in [0,1)$. We now repeat the process so that (unless the process terminates) we have $x_n \in [0,1)$,

$$r_{n+1} = \left[\frac{1}{x_n}\right], \quad x_{n+1} = \frac{1}{x_n} - r_{n+1}.$$

Show that, if the process terminates, x must be rational.

For the rest of the exercise we shall consider the case in which the process does not terminate, unless we explicitly state otherwise. Show, by induction, or otherwise, that

★ $x = (r_1 + (r_2 + (r_3 + (r_4 + \ldots (r_n + (r_{n+1} + x_{n+1})^{-1})^{-1} \ldots)^{-1})^{-1})^{-1})^{-1}.$

This formula is usually written out as

$$x = \cfrac{1}{r_1 + \cfrac{1}{r_2 + \cfrac{1}{r_3 + \cfrac{1}{r_4 + \cdots + \cfrac{1}{r_n + \cfrac{1}{r_{n+1} + x_{n+1}}}}}}},$$

but this works better in handwriting than in print so we shall use the compact (but less suggestive) notation

$$x = [r_1, r_2, r_3, r_4, \ldots, r_n, r_{n+1} + x_{n+1}].$$

We are interested in what happens when we consider the 'truncation'

$$y_{n+1} = [r_1, r_2, r_3, r_4, \ldots, r_n, r_{n+1}].$$

Write out the formula for y_n in the style of ★. Show by induction, using the formula $x_{n+1} = \frac{1}{x_n} - r_{n+1}$, or otherwise, that

★★ $$x = \frac{p_{n+1} + p_n x_{n+1}}{q_{n+1} + q_n x_{n+1}},$$

where

$$p_{n+1} = r_{n+1}p_n + p_{n-1}, \; q_{n+1} = r_{n+1}q_n + q_{n-1},$$

and $p_0 = 0$, $q_0 = 1$, $p_1 = 1$, $q_1 = r_1$. If you have met Euclid's algorithm, you should note that the recurrence equations for p_n and q_n are precisely those which occur in that process. By considering the appropriate interpretation of the right hand side of equation ★★ when $x_{n+1} = 0$, show that

$$y_n = \frac{p_n}{q_n}$$

for $n \geq 1$.

By using the recurrence equations for p_n and q_n, show that

$$p_{n-1}q_n - p_n q_{n-1} = (-1)^n$$

for all $n \geq 1$. Use this fact together with equation ★★ to show that

$$x - \frac{p_{n+1}}{q_{n+1}} = \frac{(-1)^n x_{n+1}}{q_{n+1}(q_{n+1} + q_n x_{n+1})}$$

for all $n \geq 0$. Hence show that $x - p_n/q_n$ is alternately positive and negative, and that

★★★ $$\left| x - \frac{p_n}{q_n} \right| < \frac{1}{q_{n+1}q_n}$$

for all $n \geq 1$. By looking at the recurrence relations for q_n, show that $q_n \geq n$ and deduce that $p_n/q_n \to x$ as $n \to \infty$. It is thus natural to say that we have expressed x as an infinite continued fraction

$$x = \cfrac{1}{r_1 + \cfrac{1}{r_2 + \cfrac{1}{r_3 + \cfrac{1}{r_4 + \dots}}}}.$$

We know that the process of constructing successive x_n cannot terminate if x is irrational, but we did not discuss the case x rational. Show that, if r, s, p, q are integers with $s > r \geq 0$, $q > p \geq 0$, then either $r/s = p/q$ or

$$\left| \frac{r}{s} - \frac{p}{q} \right| \geq \frac{1}{qs}.$$

By considering equation ★★★, conclude that, if x is rational, the process of forming x_n will indeed terminate. It is easy to see (but I leave it up the reader to decide whether to check the details) that we can then express any rational x with $0 < x < 1$ as a continued fraction

$$x = [r_1, r_2, r_3, r_4, \ldots, r_N]$$

for appropriate r_j and N.

In this exercise we have shown how to express any real number x with $0 < x < 1$ as a continued fraction. (So any real number can be written in the form m or $m + y$, where m is an integer and y can be expressed as a continued fraction.) We have not discussed whether a general continued fraction

$$\cfrac{1}{s_1 + \cfrac{1}{s_2 + \cfrac{1}{s_3 + \cfrac{1}{s_4 + \ldots}}}},$$

with s_n a strictly positive integer, converges. (A fairly simple adaptation of the argument above shows that it does. Of course, we need to use the fundamental axiom of analysis at some point.)

K.14. [1.5, T, ↑] In this exercise we continue the discussion of Exercise K.13. We shall suppose, as before, that $0 < x < 1$ and x is irrational. One major advantage of continued fractions is that they give very good rational approximations, as was shown in ★★★, which gave us the estimate

$$(\dagger) \qquad \left| x - \frac{p_n}{q_n} \right| < \frac{1}{q_{n+1}q_n} \leq \frac{1}{q_n^2}.$$

Conclude that, if x is a real number, we can find integers p and q with $q \geq 1$ such that

$$\left| x - \frac{p}{q} \right| < \frac{1}{q^2}.$$

Since the larger q_n is (for fixed n), the better the approximation (†), it is natural to ask how small q_n can be.

Use induction and the recurrence relation for q_n obtained in Exercise K.13 to show that $q_n \geq F_n$, where

$$F_{n+1} = F_n + F_{n-1}$$

and $F_0 = 1$, $F_1 = 1$. By directly solving the recurrence relation, if you know how, or by inductive verification, if you do not, show that

$$F_n = \frac{\tau_1^{n+1} - \tau_2^{n+1}}{\sqrt{5}},$$

where $\tau_1 = (1+\sqrt{5})/2$ and $\tau_2 = (1-\sqrt{5})/2$. Explain why, if n is sufficiently large, F_n is the integer closest to $\tau_1^{n+1}/\sqrt{5}$. (Actually, the result is true for all $n \geq 0$, and you might wish to prove this stronger statement.)

It is interesting to look for an example of a 'worst behaved' continued fraction. If $q_n = F_n$, then the associated recurrence relation tells us that $r_n = 1$ and suggests we look at

$$x = \cfrac{1}{1+\cfrac{1}{1+\cfrac{1}{1+\cfrac{1}{1+\dots}}}}.$$

Since we have not proved the convergence of this continued fraction, our argument is now merely heuristic. Explain why

$$x = \frac{1}{1+x}$$

and deduce that, if $1 > x > 0$, we must have $x = (\sqrt{5}-1)/2$. Returning to full rigour, show that, if $x = (\sqrt{5}-1)/2$, then the process of Exercise K.13 does, indeed, give

$$x = \cfrac{1}{1+\cfrac{1}{1+\cfrac{1}{1+\cfrac{1}{1+\dots}}}},$$

as required.

Hardy and Wright's delightful *The Theory of Numbers* [**26**] contains a discussion of the problem of approximating real numbers by rationals, and shows that $(\sqrt{5}-1)/2$ and its relatives are indeed the 'worst approximable' numbers. (Chapter X of [**26**] deals with continued fractions and Chapter XI with problems of approximation by rationals.) We discussed a related problem in Exercise K.12.

K.15. (Lévy's universal chord theorem.) [1.6, P] Suppose $f : [0,1] \to \mathbb{R}$ is continuous and $f(0) = f(1)$. By considering the function $g : [0,1/2] \to \mathbb{R}$ defined by $g(x) = f(x+\frac{1}{2}) - f(x)$, show that there exist a, $b \in [0,1]$ such that $b - a = \frac{1}{2}$ and $f(b) = f(a)$. Show, more generally, that there exist α_n, $\beta_n \in [0,1]$ such that $\beta_n - \alpha_n = \frac{1}{n}$ and $f(\beta_n) = f(\alpha_n)$. (This is Lévy's universal chord theorem. For more discussion, including a proof that you cannot replace $\frac{1}{n}$ by a more general h, see [**8**], page 109.)

Show that we can find $a_n, b_n \in [0,1]$ such that $a_n \leq a_{n+1} < b_{n+1} \leq b_n$, $b_n - a_n = 2^{-n}$ and $f(b_n) = f(a_n)$.

Now suppose that $\delta > 0$, that f is defined on the larger interval $(-\delta, 1+\delta)$ and that $f : (-\delta, 1+\delta) \to \mathbb{R}$ is everywhere differentiable. Show that, if $f(0) = f(1)$, then there exists a $c \in [0,1]$ such that $f'(c) = 0$. (A little extra thought gives the full Rolle's theorem (Theorem 4.52).)

K.16. [1.6, P] Suppose that $f : [a,b] \to \mathbb{R}$ is increasing and $g : [a,b] \to \mathbb{R}$ is continuous. Show that, if $f(a) \geq g(a)$ and $f(b) \leq g(b)$, then there exists a $c \in [a,b]$ such that $f(c) = g(c)$.

Is the result of the first paragraph true if we replace '$f(a) \geq g(a)$ and $f(b) \leq g(b)$' by '$f(a) \leq g(a)$ and $f(b) \geq g(b)$'? Is the result of the first paragraph true if we replace 'g continuous' by 'g decreasing'? Is the result of the first paragraph true if we replace 'f increasing' by 'f continuous'? Give proofs or counterexamples as appropriate.

K.17. [3.1, P] Let $c \in (a,b)$. Suppose $f_n : (a,b) \to \mathbb{R}$ is continuous at c for each $n \geq 1$.

Show that, if $g : (a,b) \to \mathbb{R}$ is given by

$$g(x) = \max(f_1(x), f_2(x), \dots, f_N(x)),$$

then g is continuous at c, but show, by means of an example, that, even if there exists a K such that $|f_n(x)| \leq K$ for all n and all $x \in (a,b)$, if we define $G : (a,b) \to \mathbb{R}$ by

$$G(x) = \sup(f_1(x), f_2(x), \dots),$$

then G need not be continuous at c.

K.18. [3.1, P] (This requires good command of the notion of countability)

(i) Suppose that E is a bounded uncountable set of real numbers. Show that there exist real numbers α and β with $\alpha < \beta$ such that both of the sets

$$\{e \in E : e < \gamma\} \text{ and } \{e \in E : e > \gamma\}$$

are uncountable if and only if $\alpha < \gamma < \beta$.

(ii) State and prove the appropriate version of (ii) if we omit the condition that E be bounded.

(iii) Give an example of a bounded infinite set E of real numbers such that there does not exist a real γ with the property that both

$$\{e \in E : e < \gamma\} \text{ and } \{e \in E : e > \gamma\}$$

are infinite.

K.19. [3.2, P, S] Consider a sequence of real numbers x_n. Which of the following statements are always true and which are sometimes true and sometimes false? Give proofs or examples.

(i) There is a strictly increasing subsequence. (That is, we can find $n(j)$ with $n(j+1) > n(j)$ and $x_{n(j+1)} > x_{n(j)}$ for all $j \geq 1$.)

(ii) There is an increasing subsequence. (That is, we can find $n(j)$ with $n(j+1) > n(j)$ and $x_{n(j+1)} \geq x_{n(j)}$ for all $j \geq 1$.)

(iii) If there is no increasing subsequence, there must be a strictly decreasing subsequence.

(iv) If there is no strictly increasing subsequence and no strictly decreasing subsequence, then we can find an N such that $x_n = x_N$ for all $n \geq N$.

K.20. [3.2, P] (i) Suppose that a_n is a bounded sequence of real numbers. Show that, if $n(p) \to \infty$ as $p \to \infty$ and $a_{n(p)} \to a$, then

$$\limsup_{n\to\infty} a_n \geq a \geq \liminf_{n\to\infty} a_n.$$

(ii) Suppose that $\limsup_{n\to\infty} a_n = 1$ and $\liminf_{n\to\infty} a_n = 0$. Show, by means of examples, that it may happen that 1 and 0 are the only numbers which are the limits of convergent subsequences of the a_n, or it may happen that every a with $0 \leq a \leq 1$ is such a limit. [You may wish to investigate precisely what sets can occur, but you will need the notion of a closed set.]

(iii) Suppose that a_n and b_n are bounded sequences of real numbers. Show, by means of an example, that the equation

$$\limsup_{n\to\infty}(a_n + b_n) \stackrel{?}{=} \limsup_{n\to\infty} a_n + \limsup_{n\to\infty} b_n$$

need not hold. By giving proofs or counterexamples, establish whether the following relations always hold:

$$\limsup_{n\to\infty}(a_n + b_n) \stackrel{?}{\leq} \limsup_{n\to\infty} a_n + \limsup_{n\to\infty} b_n,$$

$$\limsup_{n\to\infty}(a_n + b_n) \stackrel{?}{\geq} \limsup_{n\to\infty} a_n + \liminf_{n\to\infty} b_n.$$

K.21. (Ptolomey's inequality.) [4.1, S] We work in a vector space with inner product. By squaring $\left\| \frac{\mathbf{x}}{\|\mathbf{x}\|^2} - \frac{\mathbf{y}}{\|\mathbf{y}\|^2} \right\|$ and rewriting the result, or otherwise, prove Ptolomey's inequality

$$\|\mathbf{x} - \mathbf{y}\|\,\|\mathbf{z}\| \leq \|\mathbf{y} - \mathbf{z}\|\,\|\mathbf{x}\| + \|\mathbf{z} - \mathbf{x}\|\,\|\mathbf{y}\|.$$

K.22. [4.2, P] We use the standard Euclidean norm on $\mathbb{R}^n$ and $\mathbb{R}^{n+1}$. Which of the following statements are true and which are false? Give proofs or counterexamples as appropriate.

(i) If E is a closed subset of $\mathbb{R}^{n+1}$, then

$$\{\mathbf{x} \in \mathbb{R}^n : (0, \mathbf{x}) \in E\}$$

is closed in $\mathbb{R}^n$.

(ii) If U is an open subset of $\mathbb{R}^{n+1}$, then

$$\{\mathbf{x} \in \mathbb{R}^n : (0, \mathbf{x}) \in U\}$$

is open in $\mathbb{R}^n$.

(iii) If E is a closed subset of $\mathbb{R}^n$, then

$$\{(0, \mathbf{x}) : \mathbf{x} \in E\}$$

is closed in $\mathbb{R}^{n+1}$.

(iv) If U is an open subset of $\mathbb{R}^n$, then

$$\{(0, \mathbf{x}) : \mathbf{x} \in U\}$$

is open in $\mathbb{R}^{n+1}$.

K.23. [4.2, T] We can extend Definition 4.34 as follows.

Let $E \subseteq \mathbb{R}^m$, $A \subseteq E$, $\mathbf{x} \in E$ and $\mathbf{l} \in \mathbb{R}^p$. Consider a function $\mathbf{f} : E \setminus \{\mathbf{x}\} \to \mathbb{R}^p$. We say that $\mathbf{f}(\mathbf{y}) \to \mathbf{l}$ as $\mathbf{y} \to \mathbf{x}$ through values of $\mathbf{y} \in A$ if, given $\epsilon > 0$, we can find a $\delta(\epsilon, \mathbf{x}) > 0$ such that, whenever $\mathbf{y} \in A$ and $0 < \|\mathbf{x} - \mathbf{y}\| < \delta(\epsilon, \mathbf{x})$, we have

$$\|\mathbf{f}(\mathbf{y}) - \mathbf{l}\| < \epsilon.$$

(i) Let $A \subseteq E \subseteq \mathbb{R}^m$, $\mathbf{x} \in E$ and $\mathbf{l} \in \mathbb{R}^p$. Consider a function $\mathbf{f} : E \setminus \{\mathbf{x}\} \to \mathbb{R}^p$. Show that, if A is finite (this includes the case $A = \varnothing$), then, automatically, $\mathbf{f}(\mathbf{y}) \to \mathbf{l}$ as $\mathbf{y} \to \mathbf{x}$ through values of $\mathbf{y} \in A$.

(ii) Let $E \subseteq \mathbb{R}^m$, $A \subseteq E$, $\mathbf{x} \in E$ and $\mathbf{l} \in \mathbb{R}^p$. Consider a function $\mathbf{f} : E \setminus \{\mathbf{x}\} \to \mathbb{R}^p$ such that $\mathbf{f}(\mathbf{y}) \to \mathbf{l}$ as $\mathbf{y} \to \mathbf{x}$ through values of $\mathbf{y} \in A$. Show that, if $B \subseteq A$, then $\mathbf{f}(\mathbf{y}) \to \mathbf{l}$ as $\mathbf{y} \to \mathbf{x}$ through values of $\mathbf{y} \in A$.

(iii) Let $E \subseteq \mathbb{R}^m$, $A \subseteq E$, $B \subseteq E$, $\mathbf{x} \in E$ and $\mathbf{l} \in \mathbb{R}^p$. Consider a function $\mathbf{f} : E \setminus \{\mathbf{x}\} \to \mathbb{R}^p$ such that $\mathbf{f}(\mathbf{y}) \to \mathbf{l}$ as $\mathbf{y} \to \mathbf{x}$ through values of $\mathbf{y} \in A$ and $\mathbf{f}(\mathbf{y}) \to \mathbf{l}$ as $\mathbf{y} \to \mathbf{x}$ through values of $\mathbf{y} \in B$. Show that $\mathbf{f}(\mathbf{y}) \to \mathbf{l}$ as $\mathbf{y} \to \mathbf{x}$ through values of $\mathbf{y} \in A \cup B$.

K.24. [4.2, P] We work in $\mathbb{R}^2$. The projections $\pi_1, \pi_2 : \mathbb{R}^2 \to \mathbb{R}$ are defined by $\pi_1(x, y) = x$ and $\pi_2(x, y) = y$.

(i) Show that $E = \{(x, 1/x) : x > 0\}$ is a closed set in $\mathbb{R}^2$ but that $\pi_1(E)$ is not closed.

(ii) Find a closed set E in $\mathbb{R}^2$ such that $\pi_2(E)$ is closed but $\pi_1(E)$ is not closed.

(iii) Show, however, that, if E is a closed set in $\mathbb{R}^2$ with $\pi_2(E)$ bounded, then $\pi_1(E)$ must be closed.

K.25. [4.2, P] (This echos Exercise K.24 but does not require it.) Let $E \subseteq \mathbb{R}^n$ and let $\mathbf{f} : E \to \mathbb{R}^m$ be a function. We call $G = \{(\mathbf{x}, \mathbf{f}(\mathbf{x})) : \mathbf{x} \in E\}$ the graph of $\mathbf{f}$.

(i) Show that, if G is bounded, then E is.

(ii) Show that we can have G closed and $\mathbf{f}$ continuous but E not closed.

(iii) Show that, if E is closed and $\mathbf{f}$ is continuous, then G is closed.

(iv) Show that G and E may be closed but $\mathbf{f}$ discontinuous.

(v) Show that, if G is closed and bounded, then E is closed and $\mathbf{f}$ is continuous.

K.26. [4.3, P] Let $E \subseteq \mathbb{R}^n$. A function $f : E \to \mathbb{R}$ is said to be upper semicontinuous on E if, given any $\mathbf{x} \in E$ and any $\epsilon > 0$, we can find a $\delta(\epsilon, \mathbf{x}) > 0$ such that $f(\mathbf{y}) < f(\mathbf{x}) + \epsilon$ for all $\mathbf{y} \in E$ with $\|\mathbf{y} - \mathbf{x}\| < \delta(\epsilon, \mathbf{x})$. Give an example of a function $f : [-1, 1] \to \mathbb{R}$ which is upper semicontinuous on $[-1, 1]$ but not continuous. If $E \subseteq \mathbb{R}^n$ and $f : E \to \mathbb{R}$ is such that both f and $-f$ are upper semicontinuous on E, show that f is continuous on E.

If $K \subseteq \mathbb{R}^n$ is closed and bounded and $f : K \to \mathbb{R}$ is upper semicontinuous, show that f is bounded above and attains its least upper bound. Is it necessarily true that f is bounded below? Is it necessarily true that, if f is bounded below, it attains its greatest lower bound? Give proofs or counterexamples as appropriate.

K.27. [4.3, P] (i) If $f : \mathbb{R} \to \mathbb{R}$ is continuous and periodic (that is to say, there exists an $a > 0$ with $f(x+a) = f(x)$ for all x), show that f is bounded. Does it attain its bounds?

(ii) Give an example of an unbounded periodic function.

(iii) Suppose that $g : \mathbb{R} \to \mathbb{R}$ is bounded on every interval $[0, a]$ and $g(x+1) - g(x) \to k$ as $x \to \infty$. Show that $g(x)/x \to k$ as $x \to \infty$.

(iv) Give an example of a function $G : \mathbb{R} \to \mathbb{R}$ such that $G(x+1) - G(x) \to 0$ as $x \to \infty$ but $G(x)/x \nrightarrow 0$ as $x \to \infty$.

(v) Suppose that $h(x)/x \to k$ as $x \to \infty$. Is it necessarily true that $h(x+1) - h(x) \to k$ as $x \to \infty$? Give a proof or a counterexample.

K.28. [4.3, H, S] Suppose that $f : (a, b) \to \mathbb{R}$ is a function with continuous derivative f'. The object of this exercise is to show, using the mean value inequality (Theorem 1.42), that, if $f'(t) > 0$ for all $t \in (a, b)$, then f is strictly increasing on (a, b). (We prove a stronger result in Lemma 4.50 using Theorem 4.49.)

To this end, suppose $a < x < y < b$.

(i) Using the fact that a continuous function on a closed bounded interval is bounded and attains its bounds, show that there exists an $m > 0$ such that $f'(t) \geq m$ for all $t \in [x, y]$.

(ii) Hence, using the mean value inequality or one of its corollaries, show that $f(y) - f(x) \geq m(y - x) > 0$.

K.29. [4.3, T!] Suppose x_1, x_2, ... are real numbers. Show that we can find closed intervals $K_n = [a_n, b_n]$ (where $b_n > a_n$) such that $x_n \notin K_n$ for all $n \geq 1$ but $K_n \subseteq K_{n-1}$ for $n \geq 2$. By applying the result of Exercise 4.48, show that there exists a real y with $y \neq x_n$ for all n (i.e., the reals are uncountable).

[The method of this exercise is quite close to that of Cantor's original proof.]

K.30. [4.3, T] Suppose $\alpha : \mathbb{R}^n \to \mathbb{R}^n$ is a linear map which is self-adjoint, that is to say, such that

$$\alpha(\mathbf{x}) \cdot \mathbf{y} = \mathbf{x} \cdot (\alpha \mathbf{y})$$

for all $\mathbf{x}$, $\mathbf{y} \in \mathbb{R}^n$. (We use the standard inner product notation.)

(i) Show that the map $f : \mathbb{R}^n \to \mathbb{R}$ given by $f(\mathbf{x}) = \mathbf{x} \cdot \alpha \mathbf{x}$ is continuous. Hence, stating carefully any theorem that you use, show that there is an $\mathbf{e} \in \mathbb{R}^n$ with $\|\mathbf{e}\| = 1$ such that

$$\mathbf{e} \cdot (\alpha \mathbf{e}) \geq \mathbf{x} \cdot (\alpha \mathbf{x}),$$

whenever $\|\mathbf{x}\| = 1$.

(ii) If $\mathbf{e}$ is as in (i), deduce that, if $\mathbf{h} \cdot \mathbf{e} = 0$ and $\|\mathbf{h}\| = 1$, then

$$(1 + \delta^2)\mathbf{e} \cdot (\alpha \mathbf{e}) \geq (\mathbf{e} + \delta \mathbf{h}) \cdot (\alpha(\mathbf{e} + \delta \mathbf{h}))$$

for all real δ. By considering what happens when δ is small, or otherwise, show that

$$\mathbf{e} \cdot (\alpha \mathbf{h}) + \mathbf{h} \cdot (\alpha \mathbf{e}) = 0,$$

and so $\mathbf{h} \cdot (\alpha \mathbf{e}) = 0$.

(iii) Part (ii) shows that $\mathbf{h} \cdot (\alpha \mathbf{e}) = 0$ whenever $\mathbf{h} \cdot \mathbf{e} = 0$. Explain why this tells us that $\alpha \mathbf{e} = \lambda \mathbf{e}$ for some real λ, and so the vector $\mathbf{e}$ found in (i) is an eigenvector.

(iv) Let $U = \{\mathbf{h} : \mathbf{e} \cdot \mathbf{h} = 0\}$. Explain why U is an $n-1$-dimensional subspace of V and $\alpha(U) \subseteq U$. If $\beta = \alpha|_U$, the restriction of α to U, show that β is selfadjoint. Hence use induction to show that $\mathbb{R}^n$ has an orthonormal basis $\mathbf{e}_1$, $\mathbf{e}_2$, ... , $\mathbf{e}_n$ of eigenvectors of α.

K.31. [4.3, T!, ↑] (i) Suppose that E is a subspace of $\mathbb{R}^n$ and $\mathbf{y} \notin E$. Show that the map $g : E \to \mathbb{R}$ given by

$$g(\mathbf{x}) = \|\mathbf{x} - \mathbf{y}\|$$

is continuous. Hence, show carefully (note that, except in the trivial case $E = \{\mathbf{0}\}$, E itself is not bounded) that there exists a $\mathbf{x}_0 \in E$ such that

$$\|\mathbf{x} - \mathbf{y}\| \geq \|\mathbf{x}_0 - \mathbf{y}\|$$

for all $\mathbf{x} \in E$.

(ii) Show that $\mathbf{x}_0 - \mathbf{y}$ is perpendicular to E (that is, $(\mathbf{x}_0 - \mathbf{y}) \cdot \mathbf{x} = 0$ for all $\mathbf{x} \in E$). (See Exercise K.30 (ii), if you need a hint.) Show that $\|\mathbf{x} - \mathbf{y}\| > \|\mathbf{x}_0 - \mathbf{y}\|$ for all $\mathbf{x} \in E$ with $\mathbf{x} \neq \mathbf{x}_0$ (in other words, $\mathbf{x}_0$ is unique).

(iii) Suppose that $\alpha : \mathbb{R}^m \to \mathbb{R}$ is linear and $\alpha \neq 0$. By setting

$$E = \{\mathbf{x} : \alpha\mathbf{x} = 0\}$$

and choosing an appropriate $\mathbf{y}$, show that there exists a unit vector $\mathbf{b}$, perpendicular to E. By considering the effect of α on the vector $\mathbf{x} - (\mathbf{b} \cdot \mathbf{x})\mathbf{b}$, show that there exists a vector $\mathbf{a}$ such that

$$\alpha\mathbf{x} = \mathbf{a} \cdot \mathbf{x}$$

for all $\mathbf{x} \in \mathbb{R}^m$.

K.32. (Hahn–Banach for $\mathbb{R}^n$.) [4.3, T!, ↑] This exercise extends some of the ideas of Exercise K.31. Recall that a set E in $\mathbb{R}^m$ is called convex if, whenever $\mathbf{e}_1, \mathbf{e}_2 \in E$ and $\lambda \in [0,1]$, we have $\lambda\mathbf{e}_1 + (1 - \lambda)\mathbf{e}_2 \in E$ (in other words, if two points lie in E, so does the line segment joining them).

(i) Suppose that E is a non-empty closed convex set in $\mathbb{R}^m$ and $\mathbf{y} \notin E$. Show that there exists an $\mathbf{x}_0 \in E$ such that

$$\|\mathbf{x} - \mathbf{y}\| \geq \|\mathbf{x}_0 - \mathbf{y}\|$$

for all $\mathbf{x} \in E$.

(ii) Suppose that $\mathbf{x}_0 = \mathbf{0}$. Show that $\mathbf{x} \cdot \mathbf{y} \leq 0$ for all $\mathbf{x} \in E$.

(iii) Returning to the general case, when all we know is that E is a non-empty closed convex set in $\mathbb{R}^m$ and $\mathbf{y} \notin E$, show that there exist an $\mathbf{a} \in \mathbb{R}^m$ and a real number c such that $\mathbf{x} \cdot \mathbf{a} \leq c$ for all $\mathbf{x} \in E$ but $\mathbf{a} \cdot \mathbf{y} > c$. (Results like this are called Hahn–Banach type theorems.)

K.33. (Extreme points.) [4.3, T!, ↑] Here is an application of Exercise K.32.

(i) Throughout this exercise, K is a non-empty closed bounded convex set in $\mathbb{R}^n$. We say that a point $\mathbf{z} \in K$ is an extreme point for K, if, whenever $\mathbf{x}, \mathbf{y} \in K$, $1 > \lambda > 0$ and $\lambda\mathbf{x} + (1 - \lambda)\mathbf{y} = \mathbf{z}$, then $\mathbf{x} = \mathbf{y} = \mathbf{z}$.

Show that the extreme points of the closed unit ball $\{\mathbf{x} \in \mathbb{R}^n : \|\mathbf{x}\| \leq 1\}$ are precisely those points $\mathbf{z} \in \mathbb{R}^n$ with $\|\mathbf{z}\| = 1$. Show that the extreme points of the closed cube $[-1, 1]^n$ are precisely the 2^n points $(z_1, z_2, \dots, z_n)$ with $|z_j| = 1$.

(ii) Suppose $\mathbf{w} \notin K$. By using Exercise K.32, show that there exist an $\mathbf{a} \in \mathbb{R}^n$ and a real number c such that $\mathbf{x} \cdot \mathbf{a} \leq c$ for all $\mathbf{x} \in K$ but $\mathbf{a} \cdot \mathbf{w} > c$. Quoting carefully any theorems that you use, show that there exists an $\mathbf{x}_0 \in K$ such that $\mathbf{x} \cdot \mathbf{a} \leq \mathbf{x}_0 \cdot \mathbf{a}$ for all $\mathbf{x} \in K$.

Show that

$$K' = \{\mathbf{u} : \mathbf{x}_0 + \mathbf{u} \in K \text{ and } \mathbf{u} \cdot \mathbf{a} = 0\}$$

is a non-empty closed convex subset of an $(n-1)$-dimensional subspace of $\mathbb{R}^n$. Show further that, if $\mathbf{u}_0$ is an extreme point of K', then $\mathbf{x}_0 + \mathbf{u}_0$ is an extreme point of K.

(iii) By using induction on the dimension n, or otherwise, show that every non-empty closed bounded convex set in $\mathbb{R}^n$ contains an extreme point.

(iv) Suppose that $T : \mathbb{R}^n \to \mathbb{R}$. Explain why there exists an $\mathbf{x}_0 \in K$ such that $T(\mathbf{x}_0) \geq T(\mathbf{x})$ for all $\mathbf{x} \in K$. By considering extreme points of the set

$$\{\mathbf{x} \in K : T(\mathbf{x}) = T(\mathbf{x}_0)\},$$

or otherwise, show that there exists an extreme point $\mathbf{e}$ of K with $T(\mathbf{e}) = T(\mathbf{x}_0)$. Thus every linear map $T : \mathbb{R}^n \to \mathbb{R}$ attains its maximum on a closed bounded convex set at some extreme point of that set. This observation is the foundation of the theory of linear optimisation.

(v) Suppose that L is a non-empty closed convex subset of K. If $L \neq K$ (so that there exists a $\mathbf{w} \in K$ with $\mathbf{w} \notin L$), use the ideas of (ii) to show that there exists an extreme point of K which does not lie in L.

K.34. (Krein–Milman for $\mathbb{R}^n$.) [4.3, T!, ↑] We continue with the ideas of Exercise K.33.

(i) If $\mathcal{L}$ is a collection of convex sets in $\mathbb{R}^n$, show that $\bigcap_{L\in\mathcal{L}} L$ is convex. If $\mathcal{M}$ is a collection of closed convex sets in $\mathbb{R}^n$, show that $\bigcap_{M\in\mathcal{M}} M$ is a closed convex set.

(ii) Explain why $\mathbb{R}^n$ is a closed convex set. If A is a subset of $\mathbb{R}^n$, we write

$$\operatorname{hull}(A) = \bigcap\{L \supseteq A : L \text{ is closed and convex}\}$$

and call hull(A) the closed convex hull of A. Show that hull(A) is indeed closed and convex, and that hull$(A) \supseteq A$. Show also that, if B is a closed convex set with $B \supseteq A$, then $B \supseteq$ hull(A).

[This argument parallels the standard approach to the closure and interior set out in Exercise K.198.]

(iii) Use (ii) and part (v) of Exercise K.33 to show that the closed convex hull of the set E of extreme points of a non-empty bounded closed convex subset K of $\mathbb{R}^n$ is the set itself. (More briefly, hull$(E) = K$.) Results of this kind are known as Krein–Milman theorems.

K.35. [4.3, T] This easy exercise introduces a concept that will be used in other exercises. We say that a non-empty collection $\mathcal{A}$ of subsets of some set X has *the finite intersection property* if, whenever $n \geq 1$ and $A_1, A_2, \dots, A_n \in \mathcal{A}$, we have $\bigcap_{j=1}^n A_j \neq \varnothing$.

(i) Fix N and let $\mathcal{A}(1)$ be the collection of subsets A of $\mathbb{Z}$ such that

$$A \cap \{1, 2, 3, \ldots, N, N+1\} \text{ has at least } N \text{ members.}$$

If $A_1, A_2, \ldots, A_N \in \mathcal{A}(1)$, show that $\bigcap_{j=1}^{N} A_j \neq \varnothing$. Show, however, that $\mathcal{A}(1)$ does not have the finite intersection property.

(ii) Let $\mathcal{A}(2)$ be the collection of subsets A of $\mathbb{Z}$ such that $\mathbb{N} \setminus A$ is finite. Show that $\mathcal{A}(2)$ has the finite intersection property but $\bigcap_{A \in \mathcal{A}(2)} A = \varnothing$.

(iii) (Requires countability.) Let $\mathcal{A}(3)$ be the collection of subsets A of $\mathbb{R}$ such that $\mathbb{R} \setminus A$ is finite. Show that if $A_1, A_2, \cdots \in \mathcal{A}(3)$ then $\bigcap_{j=1}^{\infty} A_j \neq \varnothing$ (so that, in particular, $\mathcal{A}(3)$ has the finite intersection property) but $\bigcap_{A \in \mathcal{A}(3)} A = \varnothing$.

(iv) Let $\mathcal{A}(4)$ be the collection of subsets A of $\mathbb{Z}$ of the form

$$A = \{kr : r \geq 1\}$$

for some $k \geq 1$. Show that $\mathcal{A}(4)$ has the finite intersection property but $\bigcap_{A \in \mathcal{A}(4)} A = \varnothing$.

K.36. [4.3, T, ↑] We work in $\mathbb{R}$. Let $\mathcal{K}$ be a collection of closed sets lying in some closed interval $[a, b]$ and suppose that $\mathcal{K}$ has the finite intersection property (see Exercise K.35).

(i) If $a \leq c \leq b$ and we write

$$\mathcal{K}_1 = \{K \cap [a, c] : K \in \mathcal{K}\} \text{ and } \mathcal{K}_2 = \{K \cap [c, b] : K \in \mathcal{K}\},$$

show that at least one of $\mathcal{K}_1$ and $\mathcal{K}_2$ must have the finite intersection property.

(ii) Use lion hunting to show that $\bigcap_{K \in \mathcal{K}} K \neq \varnothing$.

K.37. (The Heine–Borel theorem[1].) [4.3, T, ↑] We work in $\mathbb{R}^m$. Consider the following two statements.

(A) Let $R > 0$. If a collection $\mathcal{K}$ of closed sets lying in $[-R, R]^m$ has the finite intersection property, then $\bigcap_{K \in \mathcal{K}} K \neq \varnothing$.

(B) Let E be a bounded closed set. If a collection $\mathcal{U}$ of open sets is such that $\bigcup_{U \in \mathcal{U}} U \supseteq E$, then we can find an $n \geq 1$ and $U_1, U_2, \ldots, U_n \in \mathcal{U}$ such that $\bigcup_{j=1}^{n} U_n \supseteq E$.

(i) By considering sets of the form $K = E \setminus U$, deduce statement (B) from statement (A). Deduce statement (A) from statement (B).

[1]This theorem has been summarised by Conway as follows:

If E is closed and bounded, say Heine–Borel,
And also Euclidean, then we can tell
That, if it we smother
With a large open cover,
There's a finite refinement as well!

(ii) Prove statement (A) by a lion hunting argument along the lines of Exercises K.36 and 4.14. It follows from (i) that (B) holds. This result is known as the theorem of Heine–Borel.

(iii) Explain why Exercise 4.48 is a special case of statement (A).

(iv) Suppose that E is a set with the properties described in the second sentence of statement (B). Show that E is closed and bounded.

[*Hint:* To show E is closed, suppose $\mathbf{x} \notin E$ and consider sets U of the form

$$U = \{\mathbf{y} : r > \|\mathbf{x} - \mathbf{y}\| > 1/r\}$$

with $r > 1$.]

K.38. [4.3, H, ↑] (This complements Exercise K.36.) Suppose that $\mathbb{F}$ is an ordered field with the following property.

(A) Whenever $\mathcal{K}$ is a collection of closed sets lying in some closed interval $[a, b]$ and $\mathcal{K}$ has the finite intersection property, it follows that $\bigcap_{K \in \mathcal{K}} K \neq \varnothing$.

Suppose that E is a non-empty set in $\mathbb{F}$ which is bounded above. By considering

$$\mathcal{K} = \{[a, b] : a \in E \text{ and } b \geq e \text{ for all } e \in E\},$$

or otherwise, deduce that E has a supremum. Conclude that statement (A) is equivalent to the fundamental axiom of analysis.

K.39. [4.4, T!] In this exercise you may assume the standard properties of sin and cos, but not their power series expansion.

(i) By considering the sign of $f_1'(x)$, when $f_1(t) = t - \sin t$, show that

$$t \geq \sin t$$

for all $t \geq 0$.

(ii) By considering the sign of $f_2'(x)$, when $f_2(t) = \cos t - 1 + t^2/2!$, show that

$$\cos t \geq 1 - \frac{t^2}{2!}$$

for all $t \geq 0$.

(iii) By considering the sign of $f_3'(x)$, when $f_3(t) = \sin t - t + t^3/3!$, show that

$$\sin t \geq t - \frac{t^3}{3!}$$

for all $t \geq 0$.

(iv) State general results suggested by parts (i) to (iii), and prove them by induction. State and prove corresponding results for $t < 0$.

(v) Using (iv), show that

$$\sum_{n=0}^{N} \frac{(-1)^n t^{2n+1}}{(2n+1)!} \to \sin t$$

as $N \to \infty$ for all $t \in \mathbb{R}$. State and prove a corresponding result for cos. [This exercise could be usefully illustrated by computer graphics.]

K.40. (Jensen's inequality.) [4.4, T] Good inequalities are the diamonds of analysis, valuable but rare. Jensen earned his living in the telephone industry and pursued mathematics in his spare time. He has two marvelous inequalities named after him. This exercise discusses one of them.

We call a function $f : (a, b) \to \mathbb{R}$ *convex* if, whenever $x_1, x_2 \in (a, b)$ and $1 \geq \lambda \geq 0$, we have

$$\lambda f(x_1) + (1 - \lambda) f(x_2) \geq f(\lambda x_1 + (1 - \lambda) x_2).$$

(The centre of mass of a particle of mass λ at $(x_1, f(x_1))$ and a particle of mass $1 - \lambda$ at $(x_2, f(x_2))$ lies above the graph of the curve $y = f(x)$.)

(i) Suppose $g : (a, b) \to \mathbb{R}$ is twice differentiable with $g''(t) \geq 0$ for all $t \in (a, b)$. Suppose further that $a < x_1 < x_2 < b$ and $g(x_1) = g(x_2) = 0$. Sketch g. By using the mean value theorem twice, or otherwise, show that $g(t) \leq 0$ for all $t \in [x_1, x_2]$.

(ii) Suppose $f : (a, b) \to \mathbb{R}$ is twice differentiable with $f''(t) \geq 0$ for all $t \in (a, b)$. Sketch f. By using (i), or otherwise, show that f is convex.

We have now obtained a good supply of convex functions, and proceed to discuss a form of Jensen's inequality.

(iii) Suppose $f : (a, b) \to \mathbb{R}$ is convex. By writing

$$\sum_{j=1}^{n+1} \lambda_j x_j = \lambda_{n+1} x_{n+1} + (1 - \lambda_{n+1}) y_{n+1}$$

and using induction, or otherwise, show that if

$$x_1, x_2, \ldots, x_n \in (a, b), \ \lambda_1, \lambda_2, \ldots, \lambda_n \geq 0 \text{ and } \sum_{j=1}^{n} \lambda_j = 1,$$

then

$$\sum_{j=1}^{n} \lambda_j f(x_j) \geq f\left(\sum_{j=1}^{n} \lambda_j x_j\right).$$

This is Jensen's inequality.

(iv) The importance of Jensen's inequality lies in the fact that it has many classical inequalities as special cases. By taking $f(t) = -\log t$ and

$\lambda_j = 1/n$, obtain the arithmetic-geometric inequality

$$\frac{x_1 + x_2 + \cdots + x_n}{n} \geq (x_1 x_2 \ldots x_n)^{1/n}$$

whenever the x_j are strictly positive real numbers.

[We give a more sophisticated account of these matters in Exercise K.143.]

K.41. [4.4, P, ↑] (This uses Exercise K.40.) The four vertices A, B, C, D of a quadrilateral lie in anti-clockwise order on a circle of radius a and center O. We write $2\theta_1 = \angle AOB$, $2\theta_2 = \angle BOC$, $2\theta_3 = \angle COD$, $2\theta_4 = \angle DOA$. Find the area of the quadrilateral and state the relation that θ_1, θ_2, θ_3 and θ_4 must satisfy.

Use Jensen's inequality to find the largest possible area of a quadrilateral inscribed in a circle of radius a.

Use Jensen's inequality to find the largest possible area of an n-gon inscribed in a circle of radius a $[n \geq 3]$.

Use Jensen's inequality to find the smallest possible area of an n-gon circumscribing a circle of radius a $[n \geq 3]$.

[Compare Exercise K.325.]

K.42. [4.4, P] Suppose that $g : [0, 1] \to \mathbb{R}$ is a continuous function such that, for every $c \in (0, 1)$, we can find a k with

$$0 < c - k < c < c + k < 1 \text{ and } g(c) = \tfrac{1}{2}(g(c + k) + g(c - k)).$$

Show that, if $g(0) = g(1) = 0$, then $g(t) = 0$ for all $t \in [0, 1]$. What can you say if we drop the conditions on $g(0)$ and $g(1)$?

(In one dimension this is just a brain teaser, but it turns out that the generalisation to higher dimensions is a branch of mathematics in itself.)

K.43. [4.4, P] Define $f(x) = x^2 \sin(x^{-4})$ for $x \neq 0$, $f(0) = 0$. Show that $(f(h) - f(0))/h \to 0$ as $h \to 0$ and deduce that f is differentiable at 0. Use the chain rule to show that f is differentiable at all $x \neq 0$. Show that, although f is everywhere differentiable, its derivative is not continuous.

[This example is extended in Exercise K.183.]

K.44. (Darboux's theorem.) [4.4, T] Exercise K.43 shows that the derivative of a differentiable function need not be continuous. Oddly enough, it still satisfies a form of the intermediate value theorem. Suppose that $(\alpha, \beta) \supset [a, b]$, and that $f : (\alpha, \beta) \to \mathbb{R}$ is differentiable. Darboux's theorem asserts that, if k lies between $f'(a)$ and $f'(b)$, then there is a $c \in [a, b]$ with $f'(c) = k$.

Explain why there is no loss in generality in supposing that $f'(a) > k > f'(b)$. Set $g(x) = f(x) - kx$. By looking at $g'(a)$ and $g'(b)$, show that g cannot have a maximum at a or b. Use the idea of the proof of Rolle's theorem (Theorem 4.52) to show that there exists a $c \in (a, b)$ with $g'(c) = 0$, and deduce Darboux's theorem.

Use Darboux's theorem to show that, if $(\alpha, \beta) \supseteq [a, b]$ and $f : (\alpha, \beta) \to \mathbb{R}$ is twice differentiable with a local maximum at a and a local minimum at b, then $f''(c) = 0$ for some $c \in [a, b]$.

Darboux's theorem shows that not all functions can be derivatives. It is a very hard problem to characterise those functions $f : \mathbb{R} \to \mathbb{R}$ for which there exists an $F : \mathbb{R} \to \mathbb{R}$ with $F' = f$.

K.45. **[4.4, P]** (i) Suppose $f : \mathbb{R} \to \mathbb{R}$ is such that, for each $n \geq 1$, we can find x_n and y_n with $|x_n|, |y_n| \leq 1$, $|x_n - y_n| < n^{-1}$ and $|f(x_n) - f(y_n)| \geq 1$. Consider the following argument.

'Pick $z_n \in [-1, 1]$ lying between x_n and y_n. By the Bolzano–Weierstrass theorem, we can find a sequence $n(j) \to \infty$ and a $z \in [-1, 1]$ such that $z_{n(j)} \to z$ and so f must jump by at least 1 at z. Thus f cannot be continuous at z and in particular cannot be everywhere continuous.'

Which parts of the argument are correct and which incorrect? Is the conclusion correct? Give a proof or counterexample.

(ii) Suppose $f : \mathbb{R} \to \mathbb{R}$ is continuous and for each $n \geq 1$ we can find x_n and y_n with $|x_n|, |y_n| \leq 1$ such that $|x_n - y_n| < n^{-2}$ and $|f(x_n) - f(y_n)| \geq n^{-1}$. Consider the following argument.

'By the mean value theorem, we can find $z_n \in [-1, 1]$ with $|f'(z_n)| \geq n$. By the Bolzano–Weierstrass theorem, we can find a sequence $n(j) \to \infty$ and a $z \in [-1, 1]$ such that $z_{n(j)} \to z$, and so $f'(z)$ (if it exists) must be arbitrarily large. Thus f cannot be differentiable at z and in particular cannot be everywhere differentiable.'

Which parts of the argument are correct and which incorrect? Is the conclusion correct? Give a proof or counterexample.

K.46. **[4.4, H]** In this exercise we examine our proof of Rolle's theorem (Theorem 4.52) in detail.

(i) Obtain Rolle's theorem as the consequence of the following three statements.

(a) If $g : [a, b] \to \mathbb{R}$ is continuous, we can find $k_1, k_2 \in [a, b]$ such that $g(k_2) \leq g(x) \leq g(k_1)$ for all $x \in [a, b]$.

(b) If $g : [a, b] \to \mathbb{R}$ is a function with $g(a) = g(b)$ and we can find $k_1, k_2 \in [a, b]$ such that $g(k_2) \leq g(x) \leq g(k_1)$ for all $x \in [a, b]$, then at least one of the following three statements is true: $a < k_1 < b$, $a < k_2 < b$, or g is constant.

(c) If $g : [a, b] \to \mathbb{R}$ is differentiable at a point c with $a < c < b$ such that $g(x) \leq g(c)$ for all $x \in [a, b]$ (or such that $g(c) \leq g(x)$ for all $x \in [a, b]$), then $g'(c) = 0$.

(ii) Suppose we now work over $\mathbb{Q}$ as we did in Example 1.3. Prove that the following results still hold.

(b′) If $b > a$ and $f : \mathbb{Q} \to \mathbb{Q}$ is a function with $f(a) = f(b)$ with the property that we can find $a \leq k_1, k_2 \leq b$ such that $f(k_2) \leq f(x) \leq f(k_1)$ for all $a \leq x \leq b$, then at least one of the following three statements is true: $a < k_1 < b$, $a < k_2 < b$ or f is constant.

(c′) If $b > a$ and $f : \mathbb{Q} \to \mathbb{Q}$ is differentiable at a point c with $a < c < b$ such that $f(x) \leq f(c)$ for all $a \leq x \leq b$ (or such that $f(c) \leq f(x)$ for all $a \leq x \leq b$), then $f'(c) = 0$.

This shows that the key to our proof of the mean value theorem is the fact that, *if we work with the reals,* a continuous function on a closed bounded interval is bounded and attains its bounds. All the rest is mere algebra.

K.47. [4.4, H, S] (This is more of a comment than an exercise.)

Occasionally even a respected mathematician[2] may fall into the trap of believing that 'Theorem 4.52 is subtle only because we do not ask the derivative to be continuous and do not include the endpoints'. It is worth pointing out that, even in this case, there is no simple proof. If you ever find yourself believing that there is a simple proof, check it against the following example.

Consider the function $f : \mathbb{Q} \to \mathbb{Q}$ given in Example 1.3. Set

$$g(t) = 1 - t + f(t).$$

Show that the function $g : \mathbb{Q} \to \mathbb{Q}$ has continuous derivative g' and that g' satisfies the conclusions of the intermediate value theorem. (More formally, if $a, b \in \mathbb{Q}$ with $a < b$ and $\gamma \in \mathbb{Q}$ is such that either $g'(a) \leq \gamma \leq g'(b)$ or $g'(b) \leq \gamma \leq g'(a)$, then there exists a $c \in \mathbb{Q}$ with $a \leq c \leq b$ such that $g'(c) = \gamma$. The statement is complicated, but the verification is trivial.)

Show that $g(0) = g(2) = 0$, but that there does not exist a $c \in \mathbb{Q}$ with $g'(c) = 0$.

K.48. (Tchebychev polynomials.) [4.4, G, P] Prove that

$$(\cos\theta + i\sin\theta)^n = \cos n\theta + i\sin n\theta,$$

whenever n is a positive integer. (This is just to get you started, so, provided you make clear what you are assuming, you may make any assumptions you wish.)

[2]Name withheld.

By taking real and imaginary parts, show that there is a real polynomial of degree n such that

$$T_n(\cos\theta) = \cos n\theta \text{ for all real } \theta.$$

Write down $T_0(t)$, $T_1(t)$, $T_2(t)$, and $T_3(t)$ explicitly.

Prove that, if $n \geq 1$,

(a) $T_{n+1}(t) = 2tT_n(t) - T_{n-1}(t)$ for all $|t| \leq 1$,

(b) $T_{n+1}(t) = 2tT_n(t) - T_{n-1}(t)$ for all t,

(c) the coefficient of t^n in $T_n(t)$ is 2^{n-1},

(d) $|T_n(t)| \leq 1$ for all $|t| \leq 1$, and

(e) T_n has n distinct roots all in $(-1, 1)$.

Show also that, if n, $m \geq 1$, then

$$\int_{-1}^{1} \frac{T_n(x)T_m(x)}{(1-x^2)^{1/2}}\,dx = \frac{\pi}{2}\delta_{nm},$$

where $\delta_{nm} = 0$ if $n \neq m$ and $\delta_{nn} = 1$. What can you say if $m = 0$ and $n \geq 0$?

The T_n are called the Tchebychev[3] polynomials, after their discoverer.

K.49. [4.4, T, ↑] (This requires Exercises 4.58 and K.48.) Reread Exercise 4.58. Suppose now that $a = -1$, $b = 1$ and we choose the x_j of Exercise 4.58 to be the n roots of T_n (see part (d) of Exercise 4.58). Using part (b) of Exercise 4.58, show that $\prod_{j=1}^{n}(t - x_j) = 2^{-n+1}T_n(t)$. Conclude that, if $|f^{(n)}(x)| \leq A$ for all $x \in [-1, 1]$, then

$$|f(t) - P(t)| \leq \frac{A}{2^{n-1}n!}$$

for all $t \in [-1, 1]$.

Modify the ideas above to deal with a general interval $[a, b]$.

(Note that, whilst it is, indeed, true that, if you are going to interpolate a function by a polynomial, the zeros of the Tchebychev polynomial are good points of interpolation, it remains the case that it is rarely a good idea to interpolate a function by a polynomial of high degree. To see why the inequalities established in this question are less useful than they seem observe that $\sup_{x\in[a,b]} |f^{(n)}(x)|$ may grow very fast with n. For another example involving choosing appropriate points, see Exercise K.238.)

K.50. (The nth mean value theorem.) [4.4, T] The object of this exercise is to prove a version of Taylor's theorem corresponding to Theorem 4.49 (the one-dimensional mean value theorem).

[3] The modern view is that Tchebychev should have called himself Chebyshev. He seems to have preferred the transliteration Tchebicheff. Boas has the final word in his *Spelling Lesson* ([**8**], page 204).

(i) Reread Exercise 4.58.

(ii) Suppose that $f : (u, v) \to \mathbb{R}$ is $n + 1$ times differentiable and that $[a, b] \subseteq (u, v)$. Show that we can find a polynomial P of degree n such that $(f - P)^{(r)}(a) = 0$ for $r = 0, 1, \ldots, n$. Give P explicitly.

(iii) We are interested in the error

$$E(b) = f(b) - P(b).$$

To this end, we consider the function

$$F(x) = f(x) - P(x) - E(b)\left(\frac{x-a}{b-a}\right)^{n+1}.$$

Show that $F(a) = F'(a) = \cdots = F^{(n)}(a) = 0$ and $F(b) = 0$.

(iv) By using Rolle's theorem, show that $F'(c_1) = 0$ for some c_1 with $a < c_1 < b$. By repeating your argument n times, show that that $F^{(n+1)}(c) = 0$ for some c with $a < c < b$.

(v) Deduce that

$$0 = f^{(n+1)}(c) - \frac{(n+1)!}{(b-a)^{n+1}}E(b),$$

and so

$$f(b) - \sum_{j=0}^{n} \frac{f^{(j)}(a)}{j!}(b-a)^j = \frac{f^{(n+1)}(c)}{(n+1)!}(b-a)^{n+1}$$

for some c with $a < c < b$.

(vi) Show that

$$f(a) - \sum_{j=0}^{n} \frac{f^{(j)}(b)}{j!}(a-b)^j = \frac{f^{(n+1)}(c')}{(n+1)!}(a-b)^{n+1}$$

for some c' with $a < c' < b$.

(vii) Suppose that $f, g : (u, v) \to \mathbb{R}$ are $n + 1$ times differentiable and that $a \in (u, v)$. Show that, if

$$f(a) = f'(a) = \cdots = f^{(n)}(a) = g(a) = g'(a) = \cdots = g^{(n)}(a) = 0$$

but $g^{(n+1)}(a) \neq 0$, then, if $f^{(n+1)}$ and $g^{(n+1)}$ are continuous at a, it follows that

$$\frac{f(t)}{g(t)} \to \frac{f^{(n+1)}(a)}{g^{(n+1)}(a)}$$

as $t \to a$.

K.51. [4.4, **P**, ↑] Suppose that $f : \mathbb{R} \to \mathbb{R}$ is $2n+2$ times differentiable. By considering polynomials of the form $x^k(1-x)^l$, or otherwise, show that there is a unique polynomial P of degree $2n+1$ such that

$$P^{(r)}(0) = f^{(r)}(0) \text{ and } P^{(r)}(1) = f^{(r)}(1) \text{ for all } 0 \leq r \leq n.$$

Show that the error $E(y) = f(y) - P(y)$ at $y \in [0,1]$ is given by

$$E(y) = \frac{f^{(2n+2)}(\xi)}{(2n+2)!} y^{n+1}(y-1)^{n+1},$$

for some $\xi \in (0,1)$.

K.52. (Cauchy's mean value theorem.) [4.4, **P**] Suppose that $f, g : [a,b] \to \mathbb{R}$ are continuous functions with f and g differentiable on (a,b) and $g'(x) \neq 0$ for $x \in (a,b)$.

(i) Use the mean value theorem (Theorem 4.49) to show that $g(a) \neq g(b)$.

(ii) Show that we can find a real number A such that, setting

$$h(t) = f(t) - Ag(t),$$

the function h satisfies the conditions of Rolle's theorem (Theorem 4.52).

(iii) By applying Rolle's theorem to the function h in (ii), show that there exists a $c \in (a,b)$ such that

$$\frac{f(b)-f(a)}{g(b)-g(a)} = \frac{f'(c)}{g'(c)}.$$

(This is Cauchy's mean value theorem.)

(iv) Suppose that $f(a) = g(a) = 0$ and $f'(t)/g'(t) \to L$ as $t \to a$ through values of $t > a$. Prove the version of L'Hôpital's rule which states that, under these conditions, $f(t)/g(t) \to L$ as $t \to a$ through values of $t > a$.

(v) Suppose that $f, g : (u,v) \to \mathbb{R}$ are differentiable, that $a \in (u,v)$, $f(a) = g(a) = 0$ and $f'(t)/g'(t) \to L$ as $t \to a$. Show that $f(t)/g(t) \to L$ as $t \to a$.

(vi) Suppose that $f, g : (u,v) \to \mathbb{R}$ are differentiable, that $a \in (u,v)$,

$$f(a) = f'(a) = \cdots = f^{(n-1)}(a) = g(a) = g'(a) = g''(a) = \cdots = g^{(n-1)} = 0$$

and $f^{(n)}(t)/g^{(n)}(t) \to L$ as $t \to a$. Show that $f(t)/g(t) \to L$ as $t \to a$.

(vii) Suppose that $F : (u,v) \to \mathbb{R}$ is n times differentiable, that $a \in (u,v)$ and that $F^{(n)}$ is continuous at a. By setting

$$f(t) = F(t) - \sum_{j=0}^{n} \frac{F^{(j)}(a)}{j!}(t-a)^j$$

and $g(t) = (t-a)^n$ and applying (vi), show that

$$F(t) = \sum_{j=0}^{n} \frac{F^{(j)}(a)}{j!}(t-a)^j + \epsilon(t)|t-a|^n$$

where $\epsilon(t) \to 0$ as $t \to a$. (This is another version of the local Taylor's theorem, which we obtain by other arguments as Theorem 7.3.)

K.53. [4.4, P, ↑↑] (i) Write down the result of Exercise K.50 (v) in the specific case $n = 2$, $b - a = h$.

(ii) Throughout the rest of this exercise, $f : \mathbb{R} \to \mathbb{R}$ will be an everywhere twice differentiable function. Suppose that $|f(t)| \leq M_0$ and $|f''(t)| \leq M_2$ for all $t \in \mathbb{R}$. Show that

$$|f'(t)| \leq \frac{2M_0}{h} + \frac{hM_2}{h}$$

for all $h > 0$, and deduce, by choosing h carefully, that

$$|f'(t)| \leq (M_0 M_2)^{1/2}$$

for all $t \in \mathbb{R}$.

(iii) Which of the following statements are true and which false? Give a proof or counterexample as appropriate.

(a) Let $a < b$. If $|f(t)| \leq M_0$ and $|f''(t)| \leq M_2$ for all $t \in [a, b]$, then $|f'(a)| \leq (M_0 M_2)^{1/2}$.

(b) There exists an $L > 0$ such that, if $a + L \geq b$, $|f(t)| \leq M_0$ and $|f''(t)| \leq M_2$ for all $t \in [a, b]$, then $|f'(a)| \leq (M_0 M_2)^{1/2}$.

(c) Given M_0, $M_2 > 0$, we can find an $L > 0$ such that, if $a + L \geq b$, $|f(t)| \leq M_0$ and $|f''(t)| \leq M_2$ for all $t \in [a, b]$, then $|f'(a)| \leq (M_0 M_2)^{1/2}$.

(d) There exists an infinitely differentiable function $g : \mathbb{R} \to \mathbb{R}$ with $|g(t)| \leq 1$, $|g''(t)| \leq 1$ for all $t \in \mathbb{R}$ and $g'(0) = 1$.

(e) Given M_0, $M_2 > 0$, we can find an infinitely differentiable function $G : \mathbb{R} \to \mathbb{R}$ with $|G(t)| \leq M_0$ and $|G''(t)| \leq M_2$ for all $t \in \mathbb{R}$ and $G'(0) = (M_0 M_2)^{1/2}$.

(f) There exists a constant A such that, if $|f(t)| \leq M_0$ and $|f''(t)| \leq M_2$ for all $t \in \mathbb{R}$, then $|f'(t)| \leq A(M_0 M_2)^{1/4}$ for all $t \in \mathbb{R}$.

K.54. [4.5, P] We consider functions $f : (0, 1) \to \mathbb{R}$. Give a proof or a counterexample for each of these statements:

(i) If f is uniformly continuous, then f is bounded.

(ii) If f is uniformly continuous, then f is bounded and attains its bounds.

(iii) If f is continuous and bounded and attains its bounds, then f is uniformly continuous.

K.55. [4.5, P] Give a proof or a counterexample for each of these statements.

(i) If $f : \mathbb{R} \to \mathbb{R}$ is continuous and $|f(x)|$ tends to a limit as $|x| \to \infty$, then f is uniformly continuous.

(ii) If $f : \mathbb{R} \to \mathbb{R}$ is uniformly continuous, then $|f(x)|$ tends to a limit as $|x| \to \infty$.

(iii) If $f : \mathbb{R} \to \mathbb{R}$ is continuous and bounded, then f is uniformly continuous.

(iv) If $f : \mathbb{R} \to \mathbb{R}$ is uniformly continuous, then f is bounded.

(v) If $f : \mathbb{C} \to \mathbb{C}$ is continuous and $|f(z)|$ tends to a limit as $|z| \to \infty$, then f is uniformly continuous.

(vi) If $f : \mathbb{C} \to \mathbb{C}$ is uniformly continuous, then so is $|f| : \mathbb{C} \to \mathbb{R}$.

(vii) If $f : \mathbb{R} \to \mathbb{C}$ is continuous and $|f| : \mathbb{R} \to \mathbb{R}$ is uniformly continuous, then f is uniformly continuous.

(vii) If $f,\ g : \mathbb{R} \to \mathbb{R}$ are uniformly continuous, then so is their product $f \times g$.

(viii) If $f,\ g : \mathbb{R} \to \mathbb{R}$ are uniformly continuous, then so is their composition $f \circ g$.

K.56. [4.5, P] If E is a non-empty subset of $\mathbb{R}^n$, we define $f : \mathbb{R}^n \to \mathbb{R}$ by

$$f(\mathbf{x}) = \inf\{\|\mathbf{x} - \mathbf{y}\| \ :\ \mathbf{y} \in E\}.$$

Show that f is uniformly continuous.

Show also that E is closed if and only if, given $\mathbf{x} \in \mathbb{R}^n$, we can find $\mathbf{y} \in E$ such that $\|\mathbf{x} - \mathbf{y}\| = f(\mathbf{x})$.

K.57. [4.5, P] Suppose that $f : \mathbb{Q} \to \mathbb{R}$ is uniformly continuous on $\mathbb{Q}$. The object of this exercise is to show that f has a unique continuous extension to $\mathbb{R}$.

(i) By using the general principle of convergence, show that, if $x \in \mathbb{R}$, $x_n \in \mathbb{Q}$ $[n \geq 1]$ and $x_n \to x$ as $n \to \infty$, then $f(x_n)$ tends to a limit.

(ii) Show that, if $x \in \mathbb{R}$ and $x_n,\ y_n \in \mathbb{Q}$ $[n \geq 1]$ are such that $x_n \to x$ and $y_n \to x$, then $f(x_n)$ and $f(y_n)$ tend to the same limit.

(iii) Conclude that there is a unique $F : \mathbb{R} \to \mathbb{R}$ defined by $f(x_n) \to F(x)$ whenever $x_n \in \mathbb{Q}$ and $x_n \to x$.

(iv) Explain why $F(x) = f(x)$ whenever $x \in \mathbb{Q}$.

(v) Show that F is uniformly continuous.

[See also Exercise K.333.]

K.58. [4.6, P] Suppose that the power series $\sum_{n=0}^{\infty} a_n z^n$ has radius of convergence R and the power series $\sum_{n=0}^{\infty} b_n z^n$ has radius of convergence S.

(i) Show that, if $R \neq S$, then $\sum_{n=0}^{\infty}(a_n+b_n)z^n$ has radius of convergence $\min(R,S)$.

(ii) Show that, if $R = S$, then $\sum_{n=0}^{\infty}(a_n+b_n)z^n$ has radius of convergence $T \geq R$.

(iii) Continuing with the notation of (ii), show, by means of examples, that T can take any value with $T \geq R$.

(iv) If $\lambda \neq 0$, find the radius of convergence of $\sum_{n=0}^{\infty} \lambda a_n z^n$. What happens if $\lambda = 0$?

(v) If $\infty > R > 0$, find the radius of convergence of $\sum_{n=0}^{\infty} a_n z^{n^2}$. (It may help to start by considering the special case $\sum_{n=0}^{\infty} 2^n z^{n^2}$.) What, if anything, can be said about the cases $R = 0$ and $R = \infty$?

(vi) Investigate what, if anything, we can say about the radius of convergence of $\sum_{n=0}^{\infty} c_n z^n$ if $|c_n| = \max(|a_n|, |b_n|)$. Do the same if $|c_n| = \min(|a_n|, |b_n|)$.

K.59. **[4.6, P]** Let E be a subset of the positive integers and let $h(n)$ be the number of elements in the set $\{m \in E : m \leq n\}$. We define a sequence a_n by taking $a_0 = 1$ and setting $a_{n+1} = a_n/2$ if $n \in E$, $a_{n+1} = 2a_n$ if $n \notin E$.

If $h(n)/n \to \gamma$, show that the the radius of convergence of $\sum_{n=0}^{\infty} a_n z^n$ depends only on γ, and find it. Show, by means of examples, that we can choose E so that $\sum_{n=0}^{\infty} a_n z^n$ diverges everywhere on its circle of convergence and that we can choose E so that $\sum_{n=0}^{\infty} a_n z^n$ converges everywhere on its circle of convergence.

K.60. **[4.6, T]** Consider the power series $\sum_{n=0}^{\infty} a_n z^n$. Show that, if the sequence $|a_n|^{1/n}$ is bounded, then $\sum_{n=0}^{\infty} a_n z^n$ has infinite radius of convergence if $\limsup_{n\to\infty} |a_n|^{1/n} = 0$, and radius of convergence

$$R = (\limsup_{n\to\infty} |a_n|^{1/n})^{-1}$$

otherwise. What can we say if the sequence $|a_n|^{1/n}$ is unbounded? Prove your statement.

This formula for the radius of convergence is of considerable theoretical importance but is hard to handle as a practical tool. It is usually best to use the definition directly.

K.61. **[4.6, T]** Suppose that the sequence a_n of strictly positive real numbers has the property that there exists a $K \geq 1$ with $K \geq a_{n+1}/a_n \geq K^{-1}$ for all $n \geq 1$. Prove that

$$\liminf_{n\to\infty}(a_{n+1}/a_n) \leq \liminf_{n\to\infty} a_n^{1/n} \leq \limsup_{n\to\infty} a_n^{1/n} \leq \limsup_{n\to\infty}(a_{n+1}/a_n).$$

Conclude that, if $a_{n+1}/a_n \to l$ as $n \to \infty$, then $a_n^{1/n} \to l$.

Give examples of sequences a_n such that

(i) $a_n^{1/n} \to l$ for some l, but

$$\liminf_{n\to\infty}(a_{n+1}/a_n) < l < \limsup_{n\to\infty}(a_{n+1}/a_n).$$

(ii) $\liminf_{n\to\infty}(a_{n+1}/a_n) = \limsup_{n\to\infty} a_n^{1/n} < \limsup_{n\to\infty}(a_{n+1}/a_n)$.

In how many different ways can you replace the $\leq$ in the initial formula by combinations of $<$ and $=$? Can all these possibilities occur? Justify your answer.

Comment very briefly on the connection with formulae for the radius of convergence of a power series given in Exercise K.60.

K.62. (Summation methods.) [4.6, T] If $b_j \in \mathbb{R}$, let us write

$$\mathcal{C}_n = \frac{b_0 + b_1 + \cdots + b_n}{n+1}.$$

(i) Let $\epsilon > 0$. Show that, if $|b_j| \leq \epsilon$ for all $j \geq 0$, then $|\mathcal{C}_n| \leq \epsilon$ for all $n \geq 0$.

(ii) Show that, if $N \geq 0$ and $b_j = 0$ for all $n \geq N$, then $\mathcal{C}_n \to 0$ as $n \to \infty$.

(iii) Show that, if $b_j \to 0$ as $j \to \infty$, then $\mathcal{C}_n \to 0$ as $n \to \infty$.

(iv) By considering $b_j - b$, or otherwise, show that, if $b_j \to b$ as $j \to \infty$, then $\mathcal{C}_n \to b$ as $n \to \infty$.

(v) Let $b_j = (-1)^j$. Show that $\mathcal{C}_n$ tends to a limit (to be found) but b_j does not.

(vi) Let $b_{2^m+k} = (-1)^m$ for $0 \leq k \leq 2^m - 1$ and $m \geq 1$. Show that $\mathcal{C}_n$ does not tend to a limit as $n \to \infty$.

(vii) If $0 < r < 1$ and $b_j \in \mathbb{R}$, let us write

$$\mathcal{A}_r = \sum_{n=0}^{\infty}(1-r)r^n b_n.$$

Show that, if $b_j \to b$ as $j \to \infty$, then $\mathcal{A}_r \to b$ as $r \to 1$ through values of $r < 1$. Give an example where $\mathcal{A}_r$ tends to a limit but b_j does not. Show that there exists a sequence b_j with $|b_j| \leq 1$ where $\mathcal{A}_r$ does not tend to a limit as $r \to 1$ through values of $r < 1$.

(viii) Let $\mathbf{b}_j \in \mathbb{R}^m$. Show that, if $\mathbf{b}_j \to \mathbf{b}$ as $j \to \infty$, then

$$\frac{\mathbf{b}_0 + \mathbf{b}_1 + \cdots + \mathbf{b}_n}{n+1} \to \mathbf{b}$$

You should give **two** proofs.

(A) By looking at coordinates and using the result for $\mathbb{R}$.

(B) Directly, not using earlier results.

(ix) Explain briefly why (vii) can be generalised in the manner of (viii).

K.63. [4.6, T, ↑] We continue with the notation of Exercise K.62. We call the limit of $\mathcal{C}_n$, if it exists, the Cesàro limit of the sequence b_j. We call the limit of $\mathcal{A}_r$, if it exists, the Abel limit of the sequence b_j. Now suppose $a_j \in \mathbb{R}$ and we set $b_j = \sum_{k=0}^{j} a_k$.

(i) Show that

$$\mathcal{C}_n = \frac{1}{n+1}\sum_{k=0}^{n}(n+1-k)a_k$$

and

$$\mathcal{A}_r = \sum_{k=0}^{\infty} r^k a_k.$$

If $\mathcal{C}_n$ tends to a limit, we say that the sequence a_j has that limit as a Cesàro sum and that the sequence a_j is Cesàro summable. If $\mathcal{A}_r$ tends to a limit, we say that the sequence a_j has that limit as an Abel sum and that the sequence a_j is Abel summable.

(ii) Explain very briefly why the results of Exercise K.62 (viii) and (ix) show that we can extend the definitions of (i) to the case $a_j \in \mathbb{C}$.

(iii) Let $a_j = z^j$ with $z \in \mathbb{C}$.

(a) Show that $\sum_{j=0}^{\infty} z^j$ converges if and only if $|z| < 1$, and find the sum when it exists.

(b) Show that the sequence z^j is Cesàro summable if and only if $|z| \leq 1$ and $z \neq 1$. Find the Cesàro sum when it exists.

(c) Show that the sequence z^j is Abel summable if and only if $|z| \leq 1$ and $z \neq 1$. Find the Abel sum when it exists.

K.64. [4.6, T, ↑] (This generalises parts of Exercise K.62.) Suppose that $u_{jk} \in \mathbb{R}$ for j, $k \geq 0$. Suppose that:

(1) If k is fixed, $u_{jk} \to 0$ as $j \to \infty$.

(2) $\sum_{k=0}^{\infty} u_{jk}$ is absolutely convergent, and there exists an M such that $\sum_{k=0}^{\infty} |u_{jk}| \leq M$ for each $j \geq 0$.

(3) $\sum_{k=0}^{\infty} u_{jk} \to 1$ as $j \to \infty$.

(i) Show that, if $b_j \in \mathbb{R}$ and $b_j \to b$ as $j \to \infty$, then

$$\sum_{k=0}^{\infty} u_{jk} b_k \to b$$

as $j \to \infty$.

(ii) Explain why the result just proved gives part (iv) of Exercise K.62.

(iii) Let $0 < r_n < 1$ and $r_n \to 1$. Use (i) to show that, in the notation of Exercise K.62 (vii), if $b_j \to b$ as $j \to \infty$, then $\mathcal{A}_{r_n} \to b$ as $n \to \infty$. Deduce Exercise K.62 (vii).

(iv) State and prove an extension of part (i) along the lines of Exercise K.62 (viii).

(v) If $\lambda > 0$ and $b_j \in \mathbb{C}$, let us write

$$\mathcal{B}_\lambda = \sum_{n=0}^{\infty} \frac{\lambda^n}{n!} e^{-\lambda} b_n.$$

Show that, if $b_j \to b$ as $j \to \infty$, then $\mathcal{B}_\lambda \to b$ as $\lambda \to \infty$.

(vi) Much as in the introduction to Exercise K.63, we suppose $a_j \in \mathbb{C}$ and set $b_j = \sum_{r=0}^{j} a_j$. Using the notation of part (v), we say that, if $\mathcal{B}_\lambda$ tends to a limit, then the sequence a_j has that limit as a Borel sum, and that the sequence a_j is Borel summable.

Show that the sequence z^j is Borel summable if and only if $\Re z < 1$. Find the Borel sum when it exists.

K.65. [4.6, P, ↑] Let us write G for the set of real double sequences $U = (u_{jk})_{j,k \geq 0}$ which satisfy conditions (1), (2) and (3) of Exercise K.64. By providing a proof or counterexample as appropriate, establish which of the following statements are true and which are false.

(i) If b_j is a bounded real sequence, there exists a $U \in G$ such that $\sum_{k=0}^{\infty} u_{jk} b_k$ tends to a limit as $j \to \infty$. [*Hint:* Bolzano–Weierstrass.]

(ii) If b_j is a bounded real sequence which does not tend to a limit, there exist U, $V \in G$ such that $\sum_{k=0}^{\infty} u_{jk} b_k$ and $\sum_{k=0}^{\infty} v_{jk} b_k$ converge to different limits as $j \to \infty$.

(iii) If b_j is a bounded real sequence, we can find real numbers α and β such that there exists a $U \in G$ having the property that $\sum_{k=0}^{\infty} u_{jk} b_k$ tends to λ if and only if $\lambda \in [\alpha, \beta]$.

(iv) If $U \in G$, we can find a bounded real sequence b_j such that $\sum_{k=0}^{\infty} u_{jk} b_k$ does not tend to a limit as $j \to \infty$.

(v) If b_j is a bounded real sequence, we can find a $U \in G$ such that $\sum_{k=0}^{\infty} u_{jk} b_k$ does not tend to a limit as $j \to \infty$.

K.66. [4.6, T!, ↑] The object of this exercise is to prove the converse of part (i) of Exercise K.64. In other words, we want to show that, if $u_{jk} \in \mathbb{R}$ for j, $k \geq 0$ and

$$\sum_{k=0}^{\infty} u_{jk} b_k \to b$$

as $j \to \infty$ whenever $b_j \in \mathbb{R}$ and $b_j \to b$ as $j \to \infty$, then it follows that the u_{jk} satisfy conditions (1), (2) and (3) of Exercise K.64. (This is quite hard work, and simpler proofs exist using more advanced techniques.)

(i) By choosing particular sequences b_k, show that, if the u_{jk} satisfy the stated hypothesis, then they must satisfy conditions (1) and (3).

(ii) Show that, if $\sum_{k=0}^{\infty} u_{jk} b_k$ exists for each convergent sequence b_k, then $\sum_{k=0}^{\infty} u_{jk}$ is absolutely convergent. (Look at Exercise 5.11 if you need a hint.)

(iii) Suppose that b_k is given for $0 \le k \le N-1$ with $|b_k| \le L$. Suppose that $\eta > 0$, $\epsilon > 0$, $|u_k| \le \eta$ for $1 \le k \le N-1$ and $\sum_{k=1}^{M-1} |u_k| \ge K$ for some $M > N$. Show that we can find b_k $[N \le k \le M-1]$ such that $|b_k| \le \epsilon$ for $N \le k \le M-1$ and

$$\sum_{k=0}^{M-1} u_k b_k \ge (K - N\eta)\epsilon - NL\eta.$$

(iii) Suppose that the u_{jk} satisfy conditions (1) and (3) together with the conclusion of (ii), but do not satisfy condition (2). Show, by induction, or otherwise, that we can find a sequence of integers $0 = N(0) < N(1) < N(2) < \dots$, a sequence of integers $0 = j(0) < j(1) < (2) < \dots$, a sequence of real numbers $1 = \epsilon(0) > \epsilon(1) > \epsilon(2) > \dots$ with $2^{-r} \ge \epsilon(r) > 0$, and a sequence b_k of real numbers such that

$$|b_k| \le \epsilon(r) \qquad \text{for } N(r) \le k \le N(r+1) - 1,$$

$$\sum_{k=0}^{N(r+1)-1} u_{j(r)k} b_k \ge 2^r + 1 \qquad \text{and}$$

$$\left| \sum_{k=N(r+1)}^{\infty} u_{j(r)k} x_k \right| \le 1 \qquad \text{if } |x_k| \le \epsilon(r+1) \text{ for all } k \ge N(r+1).$$

Show that $|\sum_{k=0}^{\infty} u_{j(r)k} b_k| \ge 2^r$, and use contradiction to deduce the result stated at the beginning of the exercise.

K.67. [5.2, P] We work in $\mathbb{R}^m$. Show that the following statements are equivalent.

(i) $\sum_{n=1}^{\infty} \mathbf{x}_n$ converges absolutely.

(ii) Given any sequence ϵ_n with $\epsilon_n = \pm 1$, $\sum_{n=1}^{\infty} \epsilon_n \mathbf{x}_n$ converges.

K.68. [5.2, P] Let S_1 and S_2 be a partition of $\mathbb{Z}^+$. (That is to say, $S_1 \cup S_2 = \mathbb{Z}^+$ and $S_1 \cap S_2 = \varnothing$.) Show that, if $a_n \ge 0$ for all $n \ge 1$, then $\sum_{n=1}^{\infty} a_n$ converges if and only if $\sum_{n \in S_1} a_n$ converges (that is to say, $\sum_{n \in S_1,\, n \le N} a_n$ tends to a limit as $N \to \infty$) and $\sum_{n \in S_2} a_n$ converges.

Establish, using proofs or counterexamples, which parts of the first paragraph remain true and which become false if we drop the condition $a_n \geq 0$.

Suppose that a_n, $b_n \geq 0$ and both $\sum_{n=1}^{\infty} a_n$ and $\sum_{n=1}^{\infty} b_n$ converge. Show that, if $\max(a_n, b_n) \geq c_n \geq 0$, then $\sum_{n=1}^{\infty} c_n$ converges.

Hence show that, if $a_n \geq 0$ for all n and $\sum_{n=1}^{\infty} a_n$ converges, then

$$\sum_{n=1}^{\infty} \frac{a_n^{1/2}}{n} \text{ converges.}$$

Give an alternative proof of this result using the Cauchy–Schwarz inequality.

Suppose now that a_n, $b_n \geq 0$, $\sum_{n=1}^{\infty} a_n$ and $\sum_{n=1}^{\infty} b_n$ diverge, and that $\max(a_n, b_n) \leq c_n$. Does it follow that $\sum_{n=1}^{\infty} c_n$ diverges? Give a proof or a counterexample.

K.69. (Testing for convergence of sums.) [5.2, P] There is no certain way for finding out if a sum $\sum_{n=0}^{\infty} \mathbf{a}_n$ converges or not. However, there are a number of steps that you might run through.

(1) Is the matter obvious? If $\mathbf{a}_n \nrightarrow \mathbf{0}$ then the sum cannot converge. Does the ratio test work?

(2) Are the first few terms atypical? The convergence or otherwise of $\sum_{n=N}^{\infty} \mathbf{a}_n$ determines that of $\sum_{n=1}^{\infty} \mathbf{a}_n$.

(3) (Mainly applies to examination questions.) Is the sum $\sum_{n=0}^{\infty} \mathbf{a}_n$ obtained from a sum $\sum_{n=0}^{\infty} \mathbf{b}_n$ whose behaviour is known by adding or omitting irrelevant terms? (For examples, see (i) and (ii) below.)

(4) Check for absolute convergence before checking for convergence. It is generally easier to investigate absolute convergence than convergence, and, of course, absolute convergence implies convergence.

(5) To test for the convergence of a sum of positive terms (such as arises when we check for absolute convergence), try to find another sum whose behaviour is known and which will allow you to use the comparison test.

(6) To test for the convergence of a sum of decreasing positive terms, consider using the integral test (see Lemma 9.6) or the Cauchy condensation test.

(7) If (5) and (6) fail, try to combine them or use some of the ideas of parts (iv) and (v) below. Since all that is needed to find whether the sum $\sum_{n=1}^{\infty} a_n$ of positive terms converges is to discover whether the partial sums are bounded (that is, there exists a K with $\sum_{n=1}^{N} a_n \leq K$ for all N), it is rarely[4] hard to discover whether a naturally occurring sum of positive terms converges or not.

[4]But rarely does not mean never.

(8) If the series is not absolutely convergent, you may be able to show that it is convergent by using the alternating series test.

(9) If (8) fails, then the more general Abel's test (Lemma 5.15) may work.

(10) If (8) and (9) fail, then grouping of terms (see part (vii)) may help, but you must be careful (see parts (vi) and (viii)).

(11) If none of the above is helpful and you are dealing with a naturally occurring infinite sum which is not absolutely convergent but which you hope is convergent, remember that the only way that such a series can converge is by 'near perfect cancellation of terms'. Is there a natural reason why the terms should (almost) cancel one another out?

(12) If you reach step (12), console yourself with the thought that where routine ends, mathematics begins[5].

(i) State, with proof, whether $\displaystyle\sum_{n=1}^{\infty} \frac{1+(-1)^n}{n}$ converges.

(ii) Let p_n be the nth prime. State, with proof, whether $\displaystyle\sum_{n=1}^{\infty} \frac{1}{p_n^2}$, converges.

(iii) State, with proof, whether $\displaystyle\sum_{n=1}^{\infty} n^{-1-1/n}$ converges.

(iv) Suppose $a_n \geq 0$ for each $n \geq 1$. Show that $\sum_{n=1}^{\infty} a_n$ converges if and only if we can find a sequence $N(j) \to \infty$ such that $\sum_{n=1}^{N(j)} a_n$ tends to a limit as $j \to \infty$.

(v) [This is an variation on the integral test described in Lemma 9.6.] Suppose that we can find a positive continuous function $f : [1, \infty] \to \mathbb{R}$ and a constant K with $K \geq 1$ such that

$$Ka_n \geq f(t) \geq K^{-1}a_n \geq 0 \text{ for all } n \leq t \leq n+1.$$

Show that $\sum_{n=1}^{\infty} a_n$ converges if and only if $\int_1^{\infty} f(t)\,dt$ does.

(vi) Find a sequence $a_n \in \mathbb{R}$ such that $\sum_{n=1}^{2N} a_n$ tends to a limit as $N \to \infty$ but $\sum_{n=1}^{\infty} a_n$ does not converge.

(vii) Let $\mathbf{a}_n \in \mathbb{R}^m$. Suppose that there exists a strictly positive integer M such that $\sum_{n=1}^{MN} \mathbf{a}_n$ tends to a limit as $N \to \infty$, and suppose that $\mathbf{a}_n \to \mathbf{0}$ as $n \to \infty$. Show that $\sum_{n=1}^{\infty} \mathbf{a}_n$ converges.

[5]Minkowski was walking through Göttingen when he passed a young man lost in deep thought. 'Don't worry' said Minkowski, 'it is sure to converge.'

(viii) Find a sequence $a_n \in \mathbb{R}$ and a sequence $N(j) \to \infty$ as $j \to \infty$ such that $\sum_{n=1}^{N(j)} a_n$ tends to a limit as $j \to \infty$ and $a_n \to 0$ as $n \to \infty$, yet $\sum_{n=1}^{\infty} a_n$ diverges.

K.70. [5.2, P] Let a_n, b_n non-negative real numbers.

(i) Show that, if $\sum_{n=1}^{\infty} a_n$ and $\sum_{n=1}^{\infty} b_n$ converge, so does $\sum_{n=1}^{\infty}(a_n b_n)^{1/2}$.

(ii) Show that, if $\sum_{n=1}^{\infty} a_n$ converges, so does $\sum_{n=1}^{\infty}(a_n a_{n+1})^{1/2}$.

(iii) Suppose that the a_n form a decreasing sequence. Show that, if $\sum_{n=1}^{\infty}(a_n a_{n+1})^{1/2}$ converges, so does $\sum_{n=1}^{\infty} a_n$.

(iv) Give an example with $\sum_{n=1}^{\infty}(a_n a_{n+1})^{1/2}$ convergent and $\sum_{n=1}^{\infty} a_n$ divergent.

K.71. [5.2, P] We work in the real numbers. Are the following true or false? Give a proof or counterexample as appropriate.

(i) If $\sum_{n=1}^{\infty} a_n^4$ converges, then $\sum_{n=1}^{\infty} a_n^5$ converges.

(ii) If $\sum_{n=1}^{\infty} a_n^5$ converges, then $\sum_{n=1}^{\infty} a_n^6$ converges.

(iii) If $a_n \geq 0$ for all n and $\sum_{n=1}^{\infty} a_n$ converges, then $na_n \to 0$ as $n \to \infty$.

(iv) If $a_n \geq 0$ for all n and $\sum_{n=1}^{\infty} a_n$ converges, then $n(a_n - a_{n-1}) \to 0$ as $n \to \infty$.

(v) If a_n is a decreasing sequence of positive numbers and $\sum_{n=1}^{\infty} a_n$ converges, then $na_n \to 0$ as $n \to \infty$.

(vi) If a_n is a decreasing sequence of positive numbers and $na_n \to 0$ as $n \to \infty$, then $\sum_{n=1}^{\infty} a_n$ converges.

(vii) If $\sum_{n=1}^{\infty} a_n$ converges, then $\sum_{n=1}^{\infty} n^{-1/4} a_n$ converges.

(viii) If $\sum_{n=1}^{\infty} a_n^2$ converges, then $\sum_{n=1}^{\infty} n^{-3/4}|a_n|$ converges.

(ix) If $\sum_{n=1}^{\infty} a_n$ converges, then $\sum_{n=1}^{\infty} n^{-1}|a_n|$ converges.

(x) If $\sum_{n=1}^{\infty} a_n$ converges, then $\sum_{n=1}^{\infty} n^{-5/4}|a_n|$ converges.

K.72. [5.2, P, S] Show that, if $a_n \geq 0$ and k is a strictly positive integer, the convergence of $\sum_{n=1}^{\infty} a_n^k$ implies the convergence of $\sum_{n=1}^{\infty} a_n^{k+1}$.

Suppose now that we drop the condition that all the a_n be positive. By considering a series of real numbers a_n with

$$a_{3n+1} = 2f(n),\ a_{3n+2} = -f(n),\ a_{3n+3} = -f(n)$$

for a suitable $f(n)$, show that we may have

$$\sum_{n=1}^{\infty} a_n \text{ convergent but } \sum_{n=1}^{\infty} a_n^k \text{ divergent for all } k \geq 2.$$

K.73. (Kronecker's Lemma.) [5.2, T] The following variation on Abel's test (see Lemma 5.15 and Exercise 5.17) is useful in advanced probability theory.

(i) Suppose that μ_j is an increasing sequence of real strictly positive numbers, that $\mathbf{b}_j \in \mathbb{R}^m$ and that there exists a K such that $\|\sum_{j=1}^n \mu_j^{-1}\mathbf{b}_j\| \leq K$ for all $n \geq 1$. Show that $\|\mu_n^{-1}\sum_{j=1}^n \mathbf{b}_j\| \leq 2K$ for all $n \geq 1$.

(ii) Suppose that μ_j is an increasing sequence of real strictly positive numbers with $\mu_j \to \infty$ as $j \to \infty$, and that $\mathbf{b}_j \in \mathbb{R}^m$ is a sequence such that the sum $\sum_{j=1}^\infty \mu_j^{-1}\mathbf{b}_j$ converges. Prove Kronecker's lemma which states that $\|\mu_n^{-1}\sum_{j=1}^n \mathbf{b}_j\| \to 0$ as $n \to \infty$.

(iii) Suppose now that $m = 1$, $b_j > 0$ and $\mu_j > 0$. Kronecker's lemma tells us that, if (a) μ_j is an increasing sequence, (b) $\mu_j \to \infty$ and (c) $\sum_{j=1}^\infty \mu_j^{-1}b_j$ converges, then $\mu_n^{-1}\sum_{j=1}^n b_j \to 0$. Show, by means of examples, that this result may fail if any one of the conditions (a), (b) or (c) is false.

K.74. [5.2, P, S] (A traditional brain teaser[6]) Consider the sum

$$\frac{1}{1} + \frac{1}{2} + \cdots + \frac{1}{8} + \frac{1}{10} + \cdots \frac{1}{18} + \frac{1}{20} + \cdots \frac{1}{88} + \frac{1}{100} + \cdots$$

obtained by summing those $\frac{1}{n}$ where the decimal expression of n does not contain the digit 9. Show that the sum converges.

K.75. (Euler's γ.) [5.2, T] Explain why

$$\frac{1}{n+1} \leq \int_n^{n+1} \frac{1}{x}\,dx.$$

Hence, or otherwise, show that, if we write

$$T_n = \sum_{r=1}^n \frac{1}{r} - \log n,$$

we have $T_{n+1} \leq T_n$ for all $n \geq 1$. Show also that $1 \geq T_n \geq 0$. Deduce that T_n tends to a limit γ (Euler's constant) with $1 \geq \gamma \geq 0$. [It is an indication of how little we know about specific real numbers that, after three centuries, we still do not know whether γ is irrational. Hardy is said to have offered his chair to anyone who could decide this.]

(ii) By considering $T_{2n} - T_n$, show that

$$1 - \frac{1}{2} + \frac{1}{3} - \frac{1}{4} + \frac{1}{5} - \frac{1}{6} + \frac{1}{7} - \frac{1}{8} \cdots = \log 2.$$

(iii) By considering $T_{4n} - \frac{1}{2}T_{2n} - \frac{1}{2}T_n$, show that

$$1 + \frac{1}{3} - \frac{1}{2} + \frac{1}{5} + \frac{1}{7} - \frac{1}{4} + \cdots = \frac{3}{2}\log 2.$$

[6] Brian Blank has traced it back to an article by A. J. Kempner entitled *A Curious Convergent Series* in the American Mathematical Monthly of February 1914.

This famous example is due to Dirichlet. It gives a specific example where rearranging a non-absolutely convergent sum changes its value.

(iv) Show that

$$\frac{1}{2}+\frac{1}{4}+\cdots+\frac{1}{2n}-\frac{1}{2}\log n \to \frac{1}{2}\gamma,$$
$$1+\frac{1}{3}+\frac{1}{5}+\cdots+\frac{1}{2n+1}-\frac{1}{2}\log n \to \frac{1}{2}\gamma+\log 2$$

as $n \to \infty$.

Deduce that

$$1+\frac{1}{3}+\dots\frac{1}{2p-1}-\frac{1}{2}-\frac{1}{4}-\dots-\frac{1}{2q}+\frac{1}{2p}+\frac{1}{2p+2}+\dots+\frac{1}{4p-1}-\dots,$$

where p positive terms alternate with q negative terms, has sum $\log 2 + (1/2)\log(p/q)$.

(v) Use the ideas of (iv) to show how the series may be arranged to converge to any desired sum. (Be careful. Convergence of a subsequence of sums does not imply convergence of the whole sequence.)

K.76. [5.3, P] We work in $\mathbb{C}$.

(i) Review the proof of Abel's lemma in Exercise 5.17. Show that, if $|\sum_{j=0}^{N} a_j| \leq K$ for all N and λ_j is a decreasing sequence of positive terms with $\lambda_j \to 0$ as $j \to \infty$, then $\sum_{j=0}^{\infty} \lambda_j a_j$ converges and

$$\left|\sum_{j=0}^{\infty} \lambda_j a_j\right| \leq K\lambda_0.$$

(ii) Suppose that $\sum_{j=0}^{\infty} b_j$ converges. Show that, given $\epsilon > 0$, we can find an $N(\epsilon)$ such that

$$\left|\sum_{j=M}^{\infty} b_j x^j\right| \leq \epsilon$$

whenever $M \geq N(\epsilon)$ and x is a real number with $0 \leq x < 1$. (This result was also obtained in Exercise K.62 (vii), but the suggested proof ran along different lines. It worth mastering both proofs.)

(iii) By considering the equation

$$\sum_{j=0}^{\infty} b_j x^j = \sum_{j=0}^{M-1} b_j x^j + \sum_{j=M}^{\infty} b_j x^j,$$

or otherwise, show that, if $\sum_{j=0}^{\infty} b_j$ converges, then

$$\sum_{j=0}^{\infty} b_j x^j \to \sum_{j=0}^{\infty} b_j$$

when $x \to 1$ through real values of x with $x < 1$.

(iv) Use the result of Exercise 5.38 together with part (iii) to prove the following result. Let a_j and b_j be sequences of complex numbers and write

$$c_n = \sum_{j=0}^{n} a_{n-j} b_j.$$

Then, if all the three sums $\sum_{j=0}^{\infty} a_j$, $\sum_{j=0}^{\infty} b_j$ and $\sum_{j=0}^{\infty} c_j$ converge (not necessarily absolutely), we have

$$\sum_{j=0}^{\infty} a_j \sum_{j=0}^{\infty} b_j = \sum_{j=0}^{\infty} c_j.$$

(v) By choosing $a_j = b_j = (-1)^j j^{-1/2}$, or otherwise, show that, if c_n is defined as in (iv), the two sums $\sum_{j=0}^{\infty} a_j$ and $\sum_{j=0}^{\infty} b_j$ may converge and yet $\sum_{j=0}^{\infty} c_j$ diverge.

K.77. [5.3, P] This exercise is fairly easy but is included to show that the simple picture painted in Theorem 4.84 of a 'disc of convergence' for power series fails in more general contexts. More specifically, we shall deal with the absolute convergence of $\sum_{n=0}^{\infty} \sum_{m=0}^{\infty} c_{n,m} x^n y^m$ with $c_{n,m} \in \mathbb{R}$ $[n,\ m \geq 0]$ and $x,\ y \in \mathbb{R}$. [We shall deal only with absolute convergence because, as we saw in Section 5.3, this means that the order in which we sum terms does not matter.]

(a) Show that, if there exists a $\delta > 0$ such that $\sum_{n=0}^{\infty} \sum_{m=0}^{\infty} c_{n,m} x^n y^m$ converges absolutely for $|x|,\ |y| \leq \delta$, then there exist a $\rho > 0$ and an $K > 0$ such that

$$|c_{n,m}| \leq K \rho^{n+m} \text{ for all } n,\ m \geq 0.$$

(b) Identify the set

$$E = \{(x, y) \in \mathbb{R}^2 : \sum_{n=0}^{\infty} \sum_{m=0}^{\infty} |c_{n,m} x^n y^m| \text{ converges}\}$$

in the following cases.

(i) $c_{n,m} = 1$ for all $n,\ m \geq 0$.

(ii) $c_{n,m} = \binom{n+m}{n}$ for all $n,\ m \geq 0$.

(iii) $c_{2n,2m} = \binom{n+m}{n}$ for all $n,\ m \geq 0$, and $c_{n,m} = 0$ otherwise.

(iv) $c_{n,m} = (n!m!)^{-1}$ for all $n,\ m \geq 0$.

(v) $c_{n,m} = n!m!$ for all n, $m \geq 0$.

(vi) $c_{n,n} = 1$, for all $n \geq 0$ and $c_{n,m} = 0$ otherwise.

(c) Let E be defined as in (b). Show that, if $(x_0, y_0) \in E$, then $(x, y) \in E$ whenever $|x| \leq |x_0|$ and $|y| \leq |y_0|$. Conclude that E is the union of rectangles of the form $[-u, u] \times [-v, v]$.

K.78. [5.3, T!] Part (ii) of this fairly easy exercise proves a version of the so-called 'monotone convergence theorem'. Suppose that $a_{j,n} \in \mathbb{R}$ for all j, $n \geq 1$, that $a_{j,n+1} \geq a_{j,n} \geq 0$ for all j, $n \geq 1$, and $a_{j,n} \to a_j$ as $n \to \infty$ for all $j \geq 1$. Suppose further that $\sum_{j=1}^{\infty} a_{j,n}$ converges for each n.

(i) Show that, given any $\epsilon > 0$ and any $M \geq 1$, we can find an $N(\epsilon, M)$ such that

$$\sum_{j=1}^{\infty} a_{j,n} \geq \sum_{j=1}^{M} a_j - \epsilon$$

for all $n \geq N(\epsilon, M)$. Show also that, if $\sum_{j=1}^{\infty} a_j$ converges, we have $\sum_{j=1}^{\infty} a_j \geq \sum_{j=1}^{\infty} a_{j,n}$ for all n.

(ii) Deduce that, under the hypotheses of this exercise, $\sum_{j=1}^{\infty} a_{j,n}$ converges for each n and $\sum_{j=1}^{\infty} a_{j,n} \to A$ as $n \to \infty$ if and only if $\sum_{j=1}^{\infty} a_j$ converges with value A.

(iii) Can we drop the hypothesis $a_{j,n} \geq 0$? Give a proof or counterexample.

K.79. [5.3, T!, ↑] In this exercise we work in $\mathbb{C}$. If $\sum_{n=0}^{\infty} a_n z^n$ converges to $f(z)$, say, for all $|z| < \delta$ and $\sum_{n=0}^{\infty} b_n z^n$ converges to $g(z)$, say, for all $|z| < \delta'$ $[\delta, \delta' > 0]$, it is natural to ask if $f(g(z)) = f \circ g(z)$ can be written as a power series in z for some values of z.

(i) Show that formal manipulation suggests that

$$f \circ g(z) \stackrel{?}{=} \sum_{n=0}^{\infty} c_n z^n \text{ with } c_n = \sum_{r=0}^{\infty} a_r \sum_{m(1)+m(2)+\dots+m(r)=n} b_{m(1)} b_{m(2)} \dots b_{m(r)}.$$

The rest of this exercise is devoted to showing that this equality is indeed true provided that

$$\sum_{n=0}^{\infty} |b_n z^n| < \delta,$$

where δ is the radius of convergence of $\sum_{n=0}^{\infty} a_n z^n$. The exercise is only worth doing if you treat it as an exercise in rigour. Since the case $z = 0$ is fairly trivial, we shall assume that $z \neq 0$.

(ii) Let us define c_{Nr} and C_{Nr} by equating coefficients of powers of w in the equations

$$\sum_{r=0}^{\infty} c_{Nr} w^r = \sum_{n=0}^{N} a_n \left(\sum_{m=0}^{N} b_m w^m \right)^n \text{ and } \sum_{r=0}^{\infty} C_{Nr} w^r = \sum_{n=0}^{N} |a_n| \left(\sum_{m=0}^{N} |b_m| w^m \right)^n .$$

(Note that $c_{Nr} = C_{Nr} = 0$ if r is large.)

Show that $C_{Nq}|z|^r \leq \sum_{n=0}^{\infty} |a_n| \kappa^q$, where $\kappa = \sum_{n=0}^{\infty} |b_n z^n|$. Use the fact that an increasing sequence bounded above converges, to show that $C_{Nq} \to C_q$ for some C_r as $N \to \infty$. Use the monotone convergence theorem (Exercise K.78) to show that

$$\sum_{n=0}^{\infty} |a_n| \sum_{m(1)+m(2)+\cdots+m(r)=n} |b_{m(1)} b_{m(2)} \dots b_{m(r)}|$$

converges to C_n. Deduce that

$$\sum_{r=0}^{\infty} a_r \sum_{m(1)+m(2)+\cdots+m(r)=n} b_{m(1)} b_{m(2)} \dots b_{m(r)}$$

converges, and now use the dominated convergence theorem (Lemma 5.25) to show that

$$c_{Nn} \to \sum_{r=0}^{\infty} a_r \sum_{m(1)+m(2)+\cdots+m(r)=n} b_{m(1)} b_{m(2)} \dots b_{m(r)}.$$

(iii) By further applications of the monotone and dominated convergence theorems to both sides of the equations

$$\sum_{r=0}^{\infty} C_{Nr} |z|^r = \sum_{n=0}^{N} |a_n| \left(\sum_{m=0}^{N} |b_m z|^m \right)^n \text{ and } \sum_{r=0}^{\infty} c_{Nr} z^r = \sum_{n=0}^{N} a_n \left(\sum_{m=0}^{N} b_m z^m \right)^n ,$$

show that the two sides of the following equation converge and are equal:

$$\sum_{n=0}^{\infty} a_n \left(\sum_{m=0}^{N} b_m w^m \right)^n = \sum_{n=0}^{\infty} c_n z^n,$$

where $c_n = \sum_{r=0}^{\infty} a_r \sum_{m(1)+m(2)+\cdots+m(r)=n} b_{m(1)} b_{m(2)} \dots b_{m(r)}$.

(iv) Let us say that a function $G : \mathbb{C} \to \mathbb{C}$ is 'locally expandable in power series at z_0' if we can find an $\eta > 0$ and $A_j \in \mathbb{C}$ such that $\sum_{j=0}^{\infty} A_j z^j$ has radius of convergence at least η and $G(z) = \sum_{j=0}^{\infty} A_j (z - z_0)^j$ for all $z \in \mathbb{C}$ with $|z - z_0| < \eta$. Show that, if $G : \mathbb{C} \to \mathbb{C}$ is locally expandable in power series at z_0 and $F : \mathbb{C} \to \mathbb{C}$ is locally expandable in power series at $G(z_0)$, then $F \circ G$ is locally expandable in power series at z_0. [In more advanced work this result is usually proved by a much more indirect route which reduces it to a trivial consequence of other theorems.]

K.80. [5.4, P] We work in $\mathbb{C}$. Consider a sum of the form $\sum_{n=0}^{\infty} b_n e^{nz}$. Show that there exists an X, which may be ∞, $-\infty$ (with appropriate conventions) or any real number, such that the sum converges whenever $\Re z < X$ and diverges whenever $\Re z > X$. Show, by means of examples, that any such value of X may occur.

Find the value of X for the sum

$$\sum_{n=0}^{\infty} \frac{2^n e^{nz}}{(n+1)^2}.$$

By using results from the first paragraph, show that, if $\sum_{n=0}^{\infty} c_n$ converges, then there exists a Y, which may be ∞ (with an appropriate convention) or any positive real number, such that $\sum_{n=0}^{\infty} c_n \cos nz$ converges absolutely whenever $|\Im z| < Y$ and diverges whenever $|\Im z| > Y$.

K.81. (The Wallis formula.) [5.4, P] Set

$$I_n = \int_0^{\pi/2} \sin^n x \, dx.$$

(i) Show that $I_{n+1} = \dfrac{n}{n+1} I_{n-1}$ for $n \geq 1$.

(ii) Show that $I_{n+1} \leq I_n \leq I_{n-1}$, and deduce that $I_{n+1}/I_n \to 1$ as $n \to \infty$.

(iii) By computing I_0 and I_1 directly, find I_{2n+1} and I_{2n} for all $n \geq 0$.

(iv) By applying (ii) and (iii), show that

$$\prod_{k=1}^{n} \frac{4k^2}{4k^2-1} \to \frac{\pi}{2} \text{ as } n \to \infty.$$

K.82. (Infinite products.) [5.4, T]

(i) Prove that, if $x \geq 0$, then $\exp x \geq 1 + x$.

(ii) Suppose $a_1, a_2, \ldots, a_n \in \mathbb{C}$. Show that

$$\left| \prod_{j=1}^{n} (1 + a_j) \right| \leq \prod_{j=1}^{n} (1 + |a_j|)$$

and

$$\left| \prod_{j=1}^{n} (1 + a_j) - 1 \right| \leq \prod_{j=1}^{n} (1 + |a_j|) - 1.$$

(iii) Suppose $a_1, a_2, \ldots, a_n, \ldots \in \mathbb{C}$ and $1 \leq n \leq m$. Show that

$$\left| \prod_{j=1}^{n} (1 + a_j) - \prod_{j=1}^{m} (1 + a_j) \right| \leq \exp\left(\sum_{j=1}^{n} |a_j| \right) \left(\exp\left(\sum_{j=n+1}^{m} |a_j| \right) - 1 \right).$$

(iv) Show that, if $\sum_{j=1}^{\infty} a_j$ converges absolutely, then $\prod_{j=1}^{n}(1+a_j)$ tends to a limit A, say. We write

$$A = \prod_{j=1}^{\infty}(1 + a_j).$$

(v) Suppose that $\sum_{j=1}^{\infty} a_j$ converges absolutely, and, in addition, that $a_n \neq -1$ for all n. By considering b_n such that

$$(1 + b_n)(1 + a_n) = 1,$$

or otherwise, show that $\prod_{j=1}^{\infty}(1 + a_j) \neq 0$.

(vi) Suppose $a_j > 0$ and $\sum_{j=1}^{\infty} a_j$ converges. If we write $R_n = \sum_{j=n+1}^{\infty} a_j$, show, by considering an appropriate infinite product, that $\sum_{j=1}^{\infty} a_j/R_j$ diverges.

K.83. [5.4, T, ↑] (i) Suppose that $\alpha > 0$. By expanding into geometric series and multiplying those series, show that

$$(1 - 2^{-\alpha})^{-1}(1 - 3^{-\alpha})^{-1} = \sum_{n\in S(2,3)} n^{-\alpha},$$

where $S(2,3)$ is the set of strictly positive integers with only 2 and 3 as factors, and we sum over the set $S(2,3)$ in order of increasing size of its elements.

(ii) Find a similar expression for $\prod_{j=1}^{N}(1 - p_j^{\alpha})^{-1}$, where $p_1, p_2, \dots, p_N$ are the first N primes. If $\alpha > 1$, use dominated convergence (Lemma 5.25) to show that

$$\prod_{p\in P} \frac{1}{1 - p^{-\alpha}} = \sum_{n=1}^{\infty} \frac{1}{n^{\alpha}},$$

where P is the set of primes and we range over P in order of increasing size of its elements.

(iii) By considering carefully what happens as $\alpha \to 1$ from above, show that

$$\prod_{p\in P,\ p\leq N} \frac{1}{1 - p^{-1}} \to \infty$$

as $N \to \infty$. Deduce that

$$\sum_{p\in P} \frac{1}{p} \text{ diverges.}$$

(iv) Part (iii) and its proof are due to Euler. It shows that, not merely are there an infinity of primes, but that, in some sense, they are quite common. Show that if $M(n)$ is the number of primes between 2^n and 2^{n+1}, then the sequence $2^{-\beta n}M(n)$ is unbounded whenever $\beta < 1$.

K.84. [5.4, P, ↑] Suppose that a_n is real and $0 \leq a_n < 1$ for all $n \geq 1$. Suppose further that $a_n \to 0$ as $n \to \infty$. Let the numbers P_n be defined by taking $P_0 = 1$ and

$$P_n = \begin{cases} (1+a_n)P_{n-1} & \text{if } P_{n-1} \leq 1, \\ (1-a_n)P_{n-1} & \text{if } P_{n-1} > 1, \end{cases}$$

for all $n \geq 1$. Show that P_n tends to a limit l, say.

What, if anything, can you say about l if $\sum_{n=1}^{\infty} a_n$ diverges? What, if anything, can you say about l in general? Prove your answers.

K.85. [5.4, P, S] By considering the partial products $\prod_{j=1}^{n}(1+z^{2^j})$ and using the dominated convergence theorem, show that

$$\prod_{j=1}^{\infty}(1+z^{2^j}) = \sum_{k=0}^{\infty} z^k = (1-z)^{-1}$$

for all $z \in \mathbb{C}$ with $|z| < 1$.

K.86. [5.5, P, G, S] Show that, if θ is not an integral multiple of $\pi/2$, then

$$\sum_{n=1}^{N} \frac{1}{2^n} \tan \frac{\theta}{2^n} = \frac{1}{2^n} \cot \frac{\theta}{2^n} - \cot \theta.$$

Deduce that $\sum_{n=1}^{\infty} \frac{1}{2^n} \tan \frac{\theta}{2^n}$ converges and

$$\frac{1}{\theta} = \cot \theta + \sum_{n=1}^{\infty} \frac{1}{2^n} \tan \frac{\theta}{2^n}.$$

K.87. (Vieta's formula.) [5.5, T] (See Exercise K.82 for the definition of an infinite product if you really need it.) Show that, if $x \in \mathbb{R}$ and $x \neq 0$, then

$$\sin(2^{-n}x) \prod_{j=1}^{n} \cos(2^{-j}x) = 2^{-n} \sin x.$$

Deduce that

$$\prod_{j=1}^{\infty} \cos(2^{-j}x) = \frac{\sin x}{x}.$$

Does the result hold if $x \in \mathbb{C}$ and $x \neq 0$?

By setting $x = \pi/2$, obtain Vieta's formula (probably the first published infinite product)

$$\frac{2}{\pi} = \sqrt{\frac{1}{2}} \sqrt{\frac{1}{2} + \frac{1}{2}\sqrt{\frac{1}{2}}} \sqrt{\frac{1}{2} + \frac{1}{2}\sqrt{\frac{1}{2} + \frac{1}{2}\sqrt{\frac{1}{2}}}} \dots$$

or (in a more reasonable notation) $\frac{2}{\pi} = \prod_{n=1}^{\infty} u_n$, where $u_{n+1}^2 = (1+u_n)/2$ and $u_1 = 2^{-1/2}$.

K.88. (π is irrational.) [5.5, T] In this exercise we prove that π^2 is irrational. Why does this show that π is irrational?

Suppose, if possible, that $\pi^2 = a/b$ with a and b strictly positive integers. Set

$$f(x) = \frac{x^n(1-x)^n}{n!},$$
$$G(x) = b^n \sum_{r=0}^{n} (-1)^r \pi^{2n-2r} f^{(2r)}(x).$$

(i) Show that

$$\frac{d}{dx}\big(G'(x)\sin \pi x - \pi G(x)\cos \pi x\big) = \pi^2 a^n f(x) \sin \pi x,$$

and deduce, by using the mean value theorem, or otherwise, that

$$\bigstar \qquad G(0) + G(1) = \pi \int_0^1 a^n f(x) \sin \pi x \, dx.$$

(ii) Show that the left hand side of equation ★ is an integer.

(iii) Show that, provided that n is large enough, the right hand side of equation ★ lies strictly between 0 and 1. Deduce, by reductio ad absurdum, that π^2 is irrational.

K.89. [5.6, P, G, S] We say that a function $f : A \to B$ is injective if $f(a) = f(a')$ implies $a = a'$. We say that a function $f : A \to B$ is surjective if, given $b \in B$, we can find an $a \in A$ such that $f(a) = b$. If f is both injective and surjective, we say that f is bijective.

(a) Consider the functions $f : A \to B$, $g : B \to C$ and their composition $g \circ f : A \to C$ given by $g \circ f(a) = g(f(a))$. Prove the following results.

(i) If f and g are surjective, then so is $g \circ f$.

(ii) If f and g are injective, then so is $g \circ f$.

(iii) If $g \circ f$ is injective, then so is f.

(iv) If $g \circ f$ is surjective, then so is g.

(b) Give an example where $g \circ f$ is injective and surjective but f is not surjective and g is not injective.

(c) If any of your proofs of parts (i) to (iv) of (a) involved contradiction, reprove them without such arguments[7].

[7]Conway refers to arguments of the form 'Assume X is true but Y is false. Since X implies Y it follows that Y is true. This contradicts our original assumption. Thus X implies Y.' as 'absurd absurdums'.

(d) Have you given the simplest possible example in (b)? (If you feel that this is not a proper question, let us ask instead for the smallest possible sets A and B.)

K.90. [5.6, P, S] Let a, b, c, $d \in \mathbb{R}$ and $ad - bc > 0$. Set $E = \mathbb{R}$, if $c = 0$, $E = \mathbb{R} \setminus \{-d/c\}$, otherwise. If $f : E \to \mathbb{R}$ is given by

$$f(x) = \frac{ax + b}{cx + d},$$

show that f can be written as the composition of functions (over appropriate ranges) of the form $x \mapsto x + \alpha$, $x \mapsto \beta x$ (with $\beta > 0$) and $x \mapsto -1/x$, where α and β are real. Use the chain rule to show that f is differentiable on E and $f'(x) > 0$ for all $x \in E$. (Of course, the result also follows by direct calculation.)

What can you say about the derivative of f if we replace the condition '$ad - bc > 0$' by the condition '$ad - bc < 0$'? Prove your answer. What happens if $ad - bc = 0$?

K.91. [5.6, P, G] (i) Show that a bijective function $f : [0, 1] \to [0, 1]$ is continuous if and only if it is strictly monotonic (that is, strictly increasing or strictly decreasing).

(ii) Suppose that $g : [0, 1] \to [0, 1]$ is injective. If g is continuous, must it necessarily be strictly monotonic? If g is strictly monotonic, must it necessarily be continuous? Give proofs or counterexamples as appropriate.

(iii) Rework part (ii) with 'injective' replaced by 'surjective'.

K.92. [5.6, P, S] Sketch the graph of $x(t) = (\log t)/t$ in the range $t > 0$. Sketch the set

$$\{(x, y) \in \mathbb{R}^2 \,:\, x^y = y^x,\ x, y > 0\}.$$

Provide sufficient calculation to show that your sketches are roughly correct.

Find all the solutions of $n^m = m^n$ with m and n positive integers.

Use the ideas of this exercise to decide which of the numbers e^π or π^e is the larger. (A traditional brain teaser.)

K.93. (Multiplication before Napier.) [5.6, T, G] (i) The ancient Greek astronomers drew up tables of chords where

$$\operatorname{chord} \alpha = 2 \sin(\alpha/2).$$

Prove that

$$\operatorname{chord} \theta \operatorname{chord}(\pi - \phi) = \operatorname{chord}(\theta + \phi) + \operatorname{chord}(\theta - \phi).$$

If $0 < x < 2$ and $0 < y < 2$, show that the following procedure computes xy using only addition, subtraction and table look-up.

(a) Find θ and ϕ so that $x = \operatorname{chord} \theta$, $y = \operatorname{chord}(\pi - \phi)$.

(b) Compute $\theta + \phi$ and $\theta - \phi$.

(c) Find $\mathrm{chord}(\theta + \phi)$ and $\mathrm{chord}(\theta - \phi)$.

(d) Compute $\mathrm{chord}(\theta + \phi) + \mathrm{chord}(\theta - \phi)$.

(ii) In fact we can multiply positive numbers using only addition, subtraction, division by 4 and a table of squares. Why? (Do think about this for a little before looking at Exercise 1.10 (iii) for a solution.)

If we just want to multiply two numbers together, the methods of (i) and (ii) are not much longer than using logarithms; but, if we want to multiply several numbers together, logarithms are much more convenient. (Check this statement by considering how you would multiply 10 numbers together by the various methods.) Logarithms also provide an easy method of finding αth roots. (Describe it.) We may say that logarithms are more useful computational tools because they rely on isomorphism rather than clever tricks.

[This exercise was suggested by the beautiful discussion in [**41**]. If the reader wants to know more about the historical context of logarithms, Phillips' book is the place to start.]

K.94. [5.6, P, S] (i) By differentiation, or otherwise, show that if α is real and $\beta > 1$, then $x^\alpha \beta^x \to \infty$ as $x \to \infty$. (If we take α to be a negative integer, we see why this result is sometimes stated as 'powers beat polynomials'.)

(ii) Arrange the functions

$$\left(\frac{1}{x}\right)^{1/2}, \ \left(\log\frac{1}{x}\right)^3, \ \exp\left\{\left(\log\frac{1}{x}\right)^{1/2}\right\}, \ \exp\left(\frac{1}{x}\right)$$

as $f_1(x)$, $f_2(x)$, $f_3(x)$, $f_4(x)$ in such a way that $f_r(x)/f_{r+1}(x) \to 0$ as $x \to 0$ through positive values of x. Justify your answer.

K.95. [5.6, T] (i) In parts (i) and (ii) we work over $\mathbb{C}$. If P is a polynomial with a root α of order $n \geq 1$, show that α is a root of P' of order $n-1$. (Less formally, if $(z-\alpha)^n$ divides $P(z)$ but $(z-\alpha)^{n+1}$ does not divide $P(z)$, then $(z-\alpha)^{n-1}$ divides $P(z)$ but $(z-\alpha)^n$ does not divide $P(z)$. You may find it helpful to consider the case $\alpha = 0$ first.)

(ii) By using part (i), or otherwise, show that we cannot find non-zero polynomials P and Q with no common factors satisfying the equation $(P'(z)Q(z) - P(z)Q'(z))z = Q(z)^2$ for all z.

(iii) For the rest of the exercise we work over $\mathbb{R}$. If $a < b$, show that we cannot find polynomials P and Q such that

$$\frac{d}{dx}\frac{P(x)}{Q(x)} = \frac{1}{x} \text{ for all } x \in (a, b).$$

Thus, if all that we previously have are the rational functions, the logarithm is a genuinely new function.

(iv) The argument of (iii) is essentially algebraic. We could obtain a similar result by means of analysis. Let $a > 0$. By considering the behaviour of

$$\frac{P(x)}{Q(x)} \log x \text{ as } x \to \infty,$$

show that we cannot find polynomials P and Q with $\log x = P(x)/Q(x)$ for $x \geq a$.

(v) Use arguments in the spirit of those above to give two proofs that the exponential function is not the ratio of two polynomials.

[There is a famous theorem of Liouville which shows that the function Φ defined by

$$\Phi(x) = \frac{1}{(2\pi)^{1/2}} \int_{-\infty}^{x} \exp(-t^2/2)\, dt$$

cannot be obtained from the standard functions exp, log, the trigonometric functions and the polynomials by arithmetic operations and composition. (Thus, Liouville's theorem tells us, at once, that

$$\frac{1}{2} + \frac{x^{15}\cos(\exp(-x^2))}{2\big(1 + x^{30}\cos(\exp(-x^2))\big)^{1/2}} \left(1 - \frac{\exp(-x^2/2)}{(2\pi)^{1/2}} \left(1 - \frac{1}{x^2+2}\right)\right)$$

cannot take the value $\Phi(x)$ for all real x.) The proofs of theorems like these tend to be via algebra (Galois theory) in the spirit of (iii), rather than via analysis in the spirit of (iv).]

K.96. [5.6, S] Show that, if x is real and $|x| \leq 1/2$, then we have $x \geq \log(1+x) \geq x - x^2$.

Deduce that, if k is a strictly positive integer, then

$$\left(1 - \frac{1}{n^2}\right)\left(1 - \frac{2}{n^2}\right) \cdots \left(1 - \frac{kn}{n^2}\right) \to e^{-k^2/2}$$

as $n \to \infty$.

Use the same ideas to show that, if α is real,

$$\left(1 + \frac{\alpha}{n}\right)^n \to e^{\alpha}$$

as $n \to \infty$. (Note that the proof in Exercise 5.50 applies in the complex case as well.)

K.97. [5.6, P] We shall give a number of alternative treatments of the logarithm. This is one I like less than some of the others.

Let $x_0 = x > 0$. The sequence of real strictly positive numbers x_n is defined by the recurrence relation

$$x_{n+1}^2 = x_n.$$

Prove that the two expressions $2^n(x_n - 1)$ and $2^n(1 - 1/x_n)$ both tend to the same limit as $n \to \infty$. Taking this limit as the *definition* of $\log x$, show that

(i) $\log 1 = 0$,

(ii) $\log xy = \log x + \log y$ for all x, $y > 0$,

(iii) $\log : (0, \infty) \to \mathbb{R}$ is an increasing function,

(iv) log is everywhere continuous,

(v) log is differentiable at 1 with derivative 1,

(vi) log is everywhere differentiable and $\log' x = 1/x$,

(vii) $\log : (0, \infty) \to \mathbb{R}$ is a bijective function.

K.98. [5.7, P, S] Let $f : \mathbb{R} \to \mathbb{R}$ be a strictly positive function. Show that $\left(\dfrac{f(x + \eta)}{f(x)}\right)^{1/\eta}$ tends to a non-zero limit l, say, as $\eta \to 0$ if and only if f is differentiable at x. If l exists, give its value in terms of $f(x)$ and $f'(x)$.

K.99. [5.7, P, S] Let a and β be real numbers such that $a > 0$ and $\beta \neq 0$. Show that there exists a unique γ (possibly depending on a and β) such that

$$\frac{\gamma x^{\beta - 1}}{x^\beta - a^\beta} - \frac{1}{x - a}$$

tends to a limit as $x \to a$, and find that limit.

K.100. [5.7, S] (This is really just a remark.) I have been unable to trace the story told in Exercise 5.81 to a sure source. However, one of Pierce's papers[8] does refer to the 'mysterious formula'

$$i^{-i} = (\sqrt{e})^{2\pi}$$

(which he makes still more mysterious by using a notation of his own invention). By remarking that $e^{i\pi/2} = e^{-5i\pi/2}$, show that, *even if we interpret this equation formally,* the left hand side of his equation is ambiguous. How did we prevent this ambiguity in part (ii) of Exercise 5.81?

K.101. (Functional equations.) [5.7, T]

(i) Suppose $f : \mathbb{R} \to \mathbb{R}$ is a function such that

$$f(x + y) = f(x) + f(y)$$

for all x, $y \in \mathbb{R}$. Let $f(1) = a$.

[8] *Linear Associative Algebra*, Vol 4 of the American Journal of Mathematics, see page 101.

(a) Find $f(n)$ for n a strictly positive integer.

(b) Find $f(n)$ for n an integer.

(c) Find $f(1/n)$ for n a non-zero integer.

(d) Find $f(x)$ for x a rational.

Now suppose that f is continuous. Show that $f(x) = ax$ for all $x \in \mathbb{R}$.

(ii) Suppose $\alpha : \mathbb{R}^n \to \mathbb{R}^m$ is a continuous function such that

$$\alpha(\mathbf{x} + \mathbf{y}) = \alpha(\mathbf{x}) + \alpha(\mathbf{y})$$

for all $\mathbf{x}$, $\mathbf{y} \in \mathbb{R}^n$. Show that α is a linear map.

(iii) Suppose $g : (0, \infty) \to (0, \infty)$ is a continuous function such that

$$g(xy) = g(x)g(y)$$

for all x, $y \in (0, \infty)$. Find the general form of g.

(iv) Suppose $g : \mathbb{R} \setminus \{0\} \to \mathbb{R} \setminus \{0\}$ is a continuous function such that

$$g(xy) = g(x)g(y)$$

for all x, $y \in \mathbb{R} \setminus \{0\}$. Find the general form of g.

K.102. [5.7, T, ↑] In this exercise we seek to characterise the continuous homomorphisms χ from the real numbers $\mathbb{R}$ under addition to the group $S^1 = \{\lambda \in \mathbb{C} : |\lambda| = 1\}$ under multiplication. In other words, we want to find all continuous functions $\chi : \mathbb{R} \to S^1$ such that

$$\chi(x + y) = \chi(x)\chi(y)$$

for all x, $y \in \mathbb{R}$.

(i) For each $t \in \mathbb{T}$, let $\theta(t)$ be the unique solution of

$$\chi(t) = \exp i\theta(t)$$

with $-\pi < \theta(t) \leq \pi$. Explain why we cannot establish the equation

$$\theta(t) \stackrel{?}{=} 2\theta(t/2)$$

for all t but we can find a $\delta > 0$ such that

$$\theta(t) = 2\theta(t/2)$$

for all $|t| < \delta$.

(ii) If $\theta(\delta/2) = \gamma$ and $\lambda = 2\delta^{-1}\gamma$, establish carefully that $\chi(t) = \exp i\lambda t$ for all t. Conclude that the continuous homomorphisms χ from $\mathbb{R}$ to S^1 are precisely the functions of the form $\chi(t) = \exp i\lambda t$ for some real λ.

(iii) Find the continuous homomorphisms from the group S^1 to itself.

K.103. [5.7, T, ↑] (i) Suppose $f : \mathbb{R} \to \mathbb{R}$ is a function such that

$$f(2t) = 2f(t)$$

for all $t \in \mathbb{R}$. Show that $f(0) = 0$.

(ii) Let p_j be the jth prime. Show that we can find a function $F : \mathbb{R} \to \mathbb{R}$ such that

$$F(2t) = 2F(t)$$

for all $t \in \mathbb{R}$, and $F(p_j^{-1}) = j$. Show that F is not continuous at 0.

(iii) Suppose $f : \mathbb{R} \to \mathbb{R}$ is a function such that

$$f(2t) = 2f(t)$$

for all $t \in \mathbb{R}$ and, in addition, f is differentiable at 0. Show that $f(t) = At$ for all $t \in \mathbb{R}$ and some $A \in \mathbb{R}$.

(iv) Show that we can find a function $F : \mathbb{R} \to \mathbb{R}$ such that

$$F(2t) = 2F(t)$$

for all $t \in \mathbb{R}$, and F is continuous at 0 but not differentiable there.

(v) Suppose $u : \mathbb{R} \to \mathbb{R}$ is a function such that

$$u(2t) = u(t)^2$$

for all $t \in \mathbb{R}$ and, in addition, u is differentiable at 0. Show that either $u(t) = 0$ for all $t \in \mathbb{R}$, or there exists an $\alpha > 0$ such that $u(t) = \alpha^t$ for all $t \in \mathbb{R}$.

K.104. [6.1, P] We work in $\mathbb{R}^3$ with the usual inner product. Consider the function $\mathbf{f} : \mathbb{R}^3 \to \mathbb{R}^3$ given by

$$\mathbf{f}(\mathbf{x}) = \frac{\mathbf{x}}{\|\mathbf{x}\|} \text{ for } \mathbf{x} \neq \mathbf{0}$$

and $\mathbf{f}(\mathbf{0}) = \mathbf{0}$. Show that $\mathbf{f}$ is differentiable except at $\mathbf{0}$, and

$$D\mathbf{f}(\mathbf{x})\mathbf{h} = \frac{\mathbf{h}}{\|\mathbf{x}\|} - \langle \mathbf{x}, \mathbf{h} \rangle \frac{\mathbf{x}}{\|\mathbf{x}\|^3}.$$

Verify that $D\mathbf{f}(\mathbf{x})\mathbf{h}$ is orthogonal to $\mathbf{x}$, and explain geometrically why this is the case.

K.105. (The Cauchy–Riemann equations.) [6.1, T] (This exercise is really the first theorem in a course on complex variable theory, but it can do the reader no harm to think about it in advance.)

We say that a function $f : \mathbb{C} \to \mathbb{C}$ is complex differentiable at z_0 if there exists an $f'(z_0) \in \mathbb{C}$ such that

$$\frac{f(z_0 + h) - f(z_0)}{h} \to f'(z_0) \text{ as } |h| \to 0.$$

Observe that $\mathbb{C}$ can be considered as the vector space $\mathbb{R}^2$, and so we can write

$$f(x+iy) = u(x,y) + iv(x,y)$$

with x, y, u and v real, obtaining a function $F : \mathbb{R}^2 \to \mathbb{R}^2$ given by

$$F\begin{pmatrix} x \\ y \end{pmatrix} = \begin{pmatrix} u(x,y) \\ v(x,y) \end{pmatrix}.$$

Show that the following statements are equivalent:

(i) f is complex differentiable at z_0.

(ii) F is differentiable at $\begin{pmatrix} x_0 \\ y_0 \end{pmatrix}$, and its Jacobian matrix at this point is given by

$$\begin{pmatrix} \frac{\partial u}{\partial x} & \frac{\partial u}{\partial y} \\ \frac{\partial v}{\partial x} & \frac{\partial v}{\partial y} \end{pmatrix} = \lambda \begin{pmatrix} \cos\theta & -\sin\theta \\ \sin\theta & \cos\theta \end{pmatrix}$$

with λ real and $\lambda \geq 0$.

(iii) F is differentiable at $\begin{pmatrix} x_0 \\ y_0 \end{pmatrix}$, and its derivative at this point is either the composition of a dilation and a rotation or the zero map.

(iv) F is differentiable at $\begin{pmatrix} x_0 \\ y_0 \end{pmatrix}$ and its partial derivatives at this point satisfy the Cauchy–Riemann conditions

$$\frac{\partial u}{\partial x} = \frac{\partial v}{\partial y}, \ \frac{\partial v}{\partial x} = -\frac{\partial u}{\partial y}.$$

K.106. [6.2, P, S] Suppose that $f : \mathbb{R}^2 \to \mathbb{R}$ be differentiable and we set $g(x) = f(x, c-x)$, where c is a constant. Show that $g : \mathbb{R} \to \mathbb{R}$ is differentiable, and find its derivative

(i) directly from the definition of differentiability,

and also

(ii) by using the chain rule.

Deduce that, if $f_{,1} = f_{,2}$ throughout $\mathbb{R}^2$, then $f(x,y) = h(x+y)$ for some differentiable function h.

K.107. [6.2, T] (The results of this exercise are due to Euler.) Let α be a real number. We say that a function $f : \mathbb{R}^m \setminus \{\mathbf{0}\} \to \mathbb{R}$ is homogeneous of degree α if

$$(\dagger) \qquad f(\lambda \mathbf{x}) = \lambda^\alpha f(\mathbf{x})$$

for all $\lambda > 0$ and all $\mathbf{x} \neq \mathbf{0}$.

(i) By differentiating both sides of (†) and choosing a particular value of λ, show that, if $f : \mathbb{R}^m \setminus \{\mathbf{0}\} \to \mathbb{R}$ is a differentiable function which is homogeneous of degree α, then

$$(\dagger\dagger) \qquad \sum_{j=1}^{m} x_j f_{,j}(\mathbf{x}) = \alpha f(\mathbf{x})$$

for all $\mathbf{x} \neq \mathbf{0}$.

(ii) Suppose, conversely, that $f : \mathbb{R}^m \setminus \{\mathbf{0}\} \to \mathbb{R}$ is a differentiable function satisfying (††). By setting up a differential equation for the function v defined by $v(\lambda) = f(\lambda\mathbf{x})$, or otherwise, show that f is homogeneous of degree α.

K.108. [6.2, T] (This is intended to make you think rather than come to any particular conclusion.)

(i) Suppose that $\alpha : \mathbb{R}^2 \to \mathbb{R}^2$ is a linear map and that its matrix with respect to the standard basis is

$$\begin{pmatrix} a & b \\ c & d \end{pmatrix}.$$

Show how $\|\alpha\|$ may be calculated. (There is no particular point in carrying through the algebra in detail.)

(ii) Suppose that $\alpha : \mathbb{R}^m \to \mathbb{R}^p$ is a linear map and that its matrix with respect to the standard basis is $A = (a_{ij})$. Show how $\|\alpha\|$ might be calculated. (If you know about Lagrange multipliers, this gives you an opportunity to apply them.) How practical do you think your method actually is?

We discuss this problem further in Exercises K.109 to K.111.

K.109. [6.2, T] A linear map $\alpha : \mathbb{R}^m \to \mathbb{R}^m$ is called *symmetric* if

$$\langle \alpha(\mathbf{x}), \mathbf{y} \rangle = \langle \mathbf{x}, \alpha(\mathbf{y}) \rangle$$

for all $\mathbf{x}$, $\mathbf{y} \in \mathbb{R}^m$. Show that $\alpha : \mathbb{R}^m \to \mathbb{R}^m$ is symmetric if and only if its matrix (a_{ij}) with respect to the standard base is symmetric, that is to say, $a_{ij} = a_{ji}$ for all i and j.

In algebra courses (and, indeed, in Exercise K.30) it is shown that the linear map $\alpha : \mathbb{R}^m \to \mathbb{R}^m$ is symmetric if and only if it has m orthonormal eigenvectors. Show that a symmetric linear map $\alpha : \mathbb{R}^m \to \mathbb{R}^m$ has operator norm equal to the largest absolute value of its eigenvalues, that is to say,

$$(\dagger) \qquad \|\alpha\| = \max\{|\lambda| : \lambda \text{ an eigenvalue of } \alpha\}.$$

By considering the linear map $\alpha : \mathbb{R}^2 \to \mathbb{R}^2$ with matrix

$$A = \begin{pmatrix} 0 & 1 \\ 0 & 0 \end{pmatrix},$$

or otherwise, show that the equality (†) may fail if α is not symmetric.

K.110. [6.2, T, ↑] Exercise K.109 tells us that the operator norm of a symmetric linear map $\alpha : \mathbb{R}^m \to \mathbb{R}^m$ is the largest absolute value of its eigenvalues, but does not tell us how to find this value.

(i) Our first thought might be to look at the roots of the characteristic polynomial. Examine the kind of calculations involved when $m = 50$. (Actually, matters are even worse than they look at first sight.)

(ii) Suppose α has eigenvalues λ_j with associated orthonormal eigenvectors $\mathbf{e}_j$. Choose an $\mathbf{x}_0 \in V$ and define $\mathbf{y}_k = \|\mathbf{x}_k\|^{-1}\mathbf{x}_k$ (if $\mathbf{x}_k = \mathbf{0}$ the process terminates) and $\mathbf{x}_{k+1} = \alpha(\mathbf{y}_k)$. If $\lambda_1 > |\lambda_j|$ $[2 \leq j \leq m]$, show that, except in a special case to be specified,

$$\|\mathbf{x}_k\| \to \lambda_1, \ \|\mathbf{y}_k - \mathbf{e}_1\| \to 0$$

as $k \to \infty$.

How must your answer be modified if $-\lambda_1 > |\lambda_j|$ $[2 \leq j \leq m]$?

(iii) Now suppose simply that $\alpha \neq 0$ and $|\lambda_1| \geq |\lambda_j|$ $[2 \leq j \leq m]$. Show that it remains true that, except in a special case to be specified,

$$\|\mathbf{x}_k\| \to |\lambda_1|$$

as $k \to \infty$.

If $\alpha \neq 0$, all the λ_j are positive and $\lambda_1 \geq \lambda_j$ $[2 \leq j \leq m]$, show that, except in a special case to be specified, $\mathbf{x}_k$ tends to a limit. Show that this limit is an eigenvector but (unless $\lambda_1 > \lambda_j$ $[2 \leq j \leq m]$) this eigenvector need not be a multiple of $\mathbf{e}_1$.

(iv) We thus have an algorithm for computing the largest absolute value of the eigenvalues of α by applying the iterative procedure described in (ii). The reader may raise various objections.

(a) 'The randomly chosen $\mathbf{x}_0 \in V$ may be such that the process fails.' Suppose that $V = \mathbb{R}^{50}$ and you use your favourite computer and your favourite random number generator. Estimate the probability that a randomly chosen $\mathbf{x}_0$ will be such that the process fails. Recalling that computers use finite precision arithmetic, discuss the effect of errors on the process.

(b) 'We give an iterative procedure rather than an exact formula.' Remember that if $V = \mathbb{R}^{50}$ we are seeking a root of a polynomial of degree 50. How do you propose to find an exact formula for such an object?

(c) 'We may have to make many iterations.' Estimate the number of operations required for each iteration if the matrix A associated with α is given, and conclude that, unless m is very large, each iteration will be so fast that the number of iterations hardly matters.

(v) The method is (usually) quite effective for finding the largest (absolute value) of the eigenvalues. Sometimes it is less effective in finding the associated eigenvector. Why should this be?

(vi) Discuss how you might go about finding the second largest (absolute value) of the eigenvalues.

(vii) (In the remainder of this question we drop the assumption that α is symmetric.) Suppose that α is a linear map $\alpha : \mathbb{C}^m \to \mathbb{C}^m$ whose eigenvalues satisfy $|\lambda_1| > |\lambda_j|$ $[2 \leq j \leq m]$. Show that the method outlined above for finding $|\lambda_1|$ will continue to work.

(viii) Let $\mathbf{e}_1$ and $\mathbf{e}_2$ be linearly independent vectors in $\mathbb{R}^2$ with $\|\mathbf{e}_1 - \mathbf{e}_2\|$ small. Let $\alpha : \mathbb{R}^2 \to \mathbb{R}^2$ be the linear map such that $\alpha\mathbf{e}_1 = \mathbf{e}_1$, $\alpha\mathbf{e}_2 = -\mathbf{e}_2$. Set

$$\mathbf{x} = \|\mathbf{e}_1 - \mathbf{e}_2\|^{-1}(\mathbf{e}_1 - \mathbf{e}_2).$$

and examine the the behaviour of $\alpha^r\mathbf{x}$.

(ix) Explain why our algorithm may fail for an $\alpha : \mathbb{C}^m \to \mathbb{C}^m$ whose eigenvalues merely satisfy $|\lambda_1| \geq |\lambda_j|$ $[2 \leq j \leq m]$.

(x) Explain why the method of (vii) may prove very slow even if $|\lambda_1| > |\lambda_j|$ $[2 \leq j \leq m]$. Why did we not have the problem in the real symmetric case?

K.111. [6.2, T, ↑] We now consider general linear maps $\alpha : \mathbb{R}^m \to \mathbb{R}^m$. In this general case mathematicians remain very interested in finding the largest absolute value of the eigenvalues. (Some of the reasons for this are set out in Exercises K.311 to K.315.) However, they are much less interested in finding $\|\alpha\|$, which, as I have stressed, is mainly used as a convenient theoretical tool.

If you really need to obtain $\|\alpha\|$, this exercise justifies one method of going about it. Recall that α^* is the linear map $\alpha^* : \mathbb{R}^m \to \mathbb{R}^m$ whose matrix $A^* = (b_{ij})$ with respect to the standard bases is given by $b_{ij} = a_{ji}$ for all i and j (where $A = (a_{ij})$ is the matrix of α with respect to the same basis).

(i) Show that

$$\langle \alpha\mathbf{x}, \mathbf{y} \rangle = \langle \mathbf{x}, \alpha^*\mathbf{y} \rangle$$

for all $\mathbf{x}$, $\mathbf{y} \in \mathbb{R}^m$.

(ii) Explain why

$$\langle \mathbf{x}, \alpha^*\alpha\mathbf{x} \rangle = \|\alpha\mathbf{x}\|^2$$

for all $\mathbf{x} \in \mathbb{R}^m$. Deduce that $\|\alpha^*\alpha\| \geq \|\alpha\|^2$ and so $\|\alpha^*\| \geq \|\alpha\|$.

(iii) Show that, in fact, $\|\alpha^*\| = \|\alpha\|$ and

$$\|\alpha^*\alpha\| = \|\alpha\|^2.$$

Show that $\alpha^*\alpha$ is a symmetric linear map all of whose eigenvalues are positive.

(iv) Use the results of (iii) and Exercise K.110 to give a method of obtaining $\|\alpha\|$.

(v) Estimate the number of operations required if the matrix A associated with α is given. What step requires the bulk of the calculations?

(vi) Find A^*A in the special case

$$A = \begin{pmatrix} 1 & 1 \\ 0 & 2 \end{pmatrix}.$$

What are the eigenvalues of A and A^*A? What is $\|A\|$?

K.112. [6.3, P, S] We work in $\mathbb{R}^2$. Let

$$B = \{\mathbf{x} : \|\mathbf{x}\| < 1\},\ \bar{B} = \{\mathbf{x} : \|\mathbf{x}\| \leq 1\} \text{ and } \partial B = \{\mathbf{x} : \|\mathbf{x}\| = 1\}.$$

Suppose that $f : \bar{B} \to \mathbb{R}$ is a continuous function which is differentiable on B. Show that, if $f(\mathbf{x}) = 0$ for all $\mathbf{x} \in \partial B$, there exists a $\mathbf{c} \in B$ with $Df(\mathbf{c}) = \mathbf{0}$.

Construct a continuous function $g : \bar{B} \to \mathbb{R}$ which is differentiable on B but such that we cannot find a linear map $\alpha : \mathbb{R}^2 \to \mathbb{R}$ such that $f - \alpha$ is constant on ∂B. [*Hint:* This is easy.]

Discuss the foregoing results in the light of Rolle's theorem (Theorem 4.52) and the proof of the one-dimensional mean value theorem given in Exercise 4.53.

K.113. [7.1, P] Consider a function $f : \mathbb{R} \to \mathbb{R}$. State and prove a necessary and sufficient condition for there to be a continuous function $F : \mathbb{R}^2 \to \mathbb{R}$ with

$$F(x, y) = \frac{f(x) - f(y)}{x - y}$$

whenever $x \neq y$.

Prove that, if f is twice differentiable, then F is everywhere differentiable.

K.114. [7.1, P, S] Suppose that $f : \mathbb{R} \to \mathbb{R}$ is once differentiable. Show that, if $f(x)/x \to 0$ as $x \to \infty$ and $f'(x) \to k$ as $x \to \infty$, then $k = 0$.

Give an example of a differentiable function $g : \mathbb{R} \to \mathbb{R}$ such that $g(x) \to 0$ (and so, in particular, $g(x)/x \to 0$) but $g'(x)$ does not tend to a limit as $x \to \infty$. (If you really need a hint, look at Exercise K.183.)

K.115. **[7.1, P, S]** Suppose that $q_1, q_2, \ldots$ is a strictly increasing sequence of strictly positive integers such that $\sum_{n=1}^{\infty} \frac{1}{q_n}$ diverges. Show that

$$\sum_{n=1}^{\infty} \log\left(1 - \frac{1}{q_n}\right) \text{ diverges but } \sum_{n=1}^{\infty} \left(\log\left(1 - \frac{1}{q_n}\right) + \frac{1}{q_n}\right) \text{ converges.}$$

K.116. **[7.3, P]** (This exercise has low theoretical interest, but tests your power of clear organisation.) Find the maximum of

$$ax^2 + bx + c$$

on the interval $[0, 10]$, where a, b and c are real constants.

K.117. **(Routh's rule.)** **[7.3, T]** It is Routh's misfortune to be known chiefly as the man who beat Maxwell in the Cambridge mathematics examinations, but he was an able mathematician and teacher[9]. Part (v) of this exercise gives his method for determining if a real symmetric matrix $A = (a_{ij})_{1 \le i,j \le n}$ is positive definite.

(i) We said that A is positive definite if all its eigenvalues are strictly positive. It is more usual to say that A is positive definite if we have $\sum_{i=1}^{n} \sum_{j=1}^{n} x_i a_{ij} x_j > 0$ for all $\mathbf{x} \neq \mathbf{0}$. Show that the two definitions are equivalent. [Think about the map $\mathbf{x} \mapsto \mathbf{x}^T A \mathbf{x}$ (with $\mathbf{x}$ as a column vector).]

(ii) By choosing a particular $\mathbf{x}$, show that, if A is positive definite, then $a_{11} > 0$.

(iii) If $a_{11} \neq 0$, show that (whether A is positive definite or not)

$$\sum_{i=1}^{n} \sum_{j=1}^{n} x_i a_{ij} x_j = a_{11} y_1^2 + \sum_{i=2}^{n} \sum_{j=2}^{n} x_i b_{ij} x_j,$$

where $y_1 = x_1 + a_{11}^{-1} a_{12} x_2 + a_{11}^{-1} a_{13} x_3 + \cdots + a_{11}^{-1} a_{1n} x_n$, and

$$b_{ij} = a_{ij} - \frac{a_{1i} a_{1j}}{a_{11}} = \frac{a_{ij} a_{11} - a_{1i} a_{1j}}{a_{11}}.$$

Hence, or otherwise, show carefully that A is positive definite if and only if $a_{11} > 0$ and the matrix $B = (b_{ij})_{2 \le i,j \le n}$ is positive definite. (Of course, we must perform the very simple verification that B is a real symmetric matrix.) The matrix B is called the *Schur complement* of a_{11} in A.

(iv) By considering the effect of row operations of the form

$$[\text{new } i\text{th row}] = [\text{old } i\text{th row}] - a_{i1} a_{11}^{-1} [\text{old 1st row}]$$

on A, or otherwise, show that $\det A = a_{11} \det B$.

---

[9] He was one of the two greatest Cambridge coaches (people who tried to make students do slightly better in mathematics examinations than they should, an occupation which I have always felt to be fairly harmless, if not terribly useful).

(v) Use (iii), ideas based on (iv) and induction on n to prove Routh's rule: The real symmetric matrix A is positive definite if and only if the determinants of the n leading minors

$$a_{11} = \det(a_{11}),\ \det\begin{pmatrix} a_{11} & a_{12} \\ a_{21} & a_{22} \end{pmatrix},\ \det\begin{pmatrix} a_{11} & a_{12} & a_{13} \\ a_{21} & a_{22} & a_{23} \\ a_{31} & a_{32} & a_{33} \end{pmatrix},\ \ldots,\ \det A$$

are all strictly positive.

(vi) (This is just a remark.) Show that A is negative definite if and only if $-A$ is positive definite.

K.118. [7.3, P] (i) Let $A(t)$ be a 2×2 real symmetric matrix with

$$A(t) = \begin{pmatrix} a(t) & b(t) \\ b(t) & c(t) \end{pmatrix}.$$

Suppose that a, b, $c : \mathbb{R} \to \mathbb{R}$ are continuous. Show that, if $A(t)$ has eigenvalues $\lambda_1(t)$ and $\lambda_2(t)$ with $\lambda_1(t) \geq \lambda_2(t)$, then λ_1 and λ_2 are continuous. Show that, if $A(0)$ is positive definite and $A(1)$ is negative definite, then there exists an $\alpha \in (0, 1)$ with $A(\alpha)$ singular. Show, more strongly, that, either we can find α_1 and α_2 with $1 > \alpha_1 > \alpha_2 > 0$ and $A(\alpha_1)$, $A(\alpha_2)$ singular, or there exists an $\alpha \in (0, 1)$ with $A(\alpha)$ the zero matrix.

(ii) Prove the result of Exercise 7.37 (ii) with $m = 2$.

(iii) Find a 2×2 real matrix $B(t) = (b_{ij}(t))_{1 \leq i,j \leq 2}$ such that the entries $b_{ij} : \mathbb{R} \to \mathbb{R}$ are continuous, $B(0) = I$ and $B(\pi) = -I$, but $B(t)$ is nowhere singular. [Think rotations.]

K.119. [7.3, M!] Four towns lie on the vertices A, B, C, D of a square. Find the shortest total length of the system of roads shown in Figure K.2, where the diagram is symmetric about lines through the centres of opposite sides of the square. By formal or informal arguments, show that this arrangement gives the shortest total length of a system of roads joining all four towns. (You should, at least, convince yourself of this.) Is the arrangement unique?

The interesting point here is that the most highly symmetric road systems do not give the best answer.

K.120. [7.3, P] (A very traditional question.) The great, grey-green, greasy Limpopo River is 1 kilometre wide and flows with negligible speed between parallel banks. A young elephant wishes to reach a fever tree $h > 0$ kilometres upstream on the other side as quickly as possible. He gallops a distance $h - x$ kilometres upstream and then plunges in and swims directly to the tree. If he gallops at a speed v kilometres an hour and swims at a speed u kilometres an hour with $u > v > 0$, what value of x should he choose?

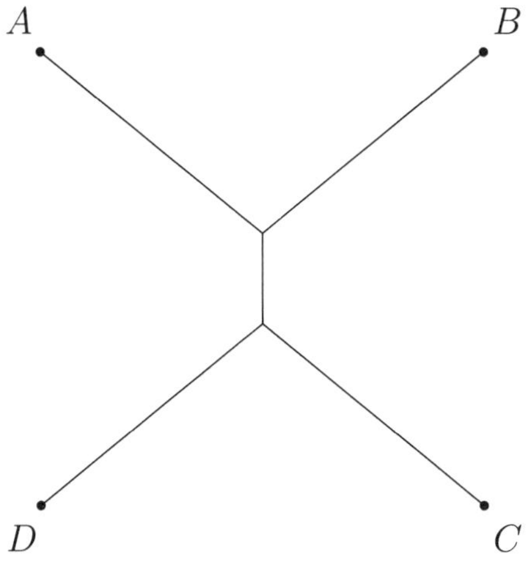

Figure K.2. Shortest road system for vertices of a square

Explain briefly why the character of the solution is different for large and small h.

K.121. [7.3, P] For each real value of a, find the points at which the function

$$y^2 - \tfrac{1}{3}x^3 + ax$$

attains its global maximum and global minimum in the closed disc $\{(x, y) : x^2 + y^2 \leq 1\}$.

K.122. [7.3, P] We continue with the notation of Exercise 7.30, but we do not assume $g(\theta) = -g(-\theta)$. Show that, if g is differentiable, then f is differentiable except possibly at $(0, 0)$ (there are various ways of doing this, but, whichever one you choose, be careful), For which g does f and have directional derivatives in all directions at $(0, 0)$? For which functions g is f differentiable at $(0, 0)$?

K.123. [7.3, P] (i) Let $a > b > 0$ and define $f : \mathbb{R}^2 \to \mathbb{R}$ by

$$f(x, y) = (y - ax^2)(y - bx^2).$$

Sketch the set $E = \{(x, y) : f(x, y) > 0\}$. Show that f has a minimum at $(0, 0)$ along each straight line through the origin. (Formally, if α and β are real numbers which are not both zero, the function $g : \mathbb{R} \to \mathbb{R}$ given by $g(t) = f(\alpha t, \beta t)$ has a strict local minimum at 0.) Show, however, that f has no minimum at $(0, 0)$. Does f have any minima?

(ii) Suppose $F : \mathbb{R}^2 \to \mathbb{R}$ is twice continuously differentiable with a non-singular Hessian at $(0, 0)$. Show that F has a minimum at $(0, 0)$ if and only if it has a minimum at $(0, 0)$ along each straight line through the origin. Why is this result consistent with part (i)?

[Like many examples in this subject, part (i) is due to Peano.]

K.124. [8.2, P] (i) Let F be the function given in Exercise 8.30. By considering $F - \frac{1}{2}$, or otherwise, show that if a bounded function $f : [a, b] \to \mathbb{R}$ is not Riemann integrable, it does not follow that $|f|$ is not Riemann integrable. If a bounded function $f : [a, b] \to \mathbb{R}$ is such that $|f|$ is not Riemann integrable, does it follow that f is not Riemann integrable? Give a proof or counterexample.

(ii) Let $f, g : [a, b] \to \mathbb{R}$ be bounded functions. If both f and g are not Riemann integrable, does it follow that $f + g$ is not Riemann integrable? If both f and g are not Riemann integrable, does it follow that $f+g$ is Riemann integrable? If f is Riemann integrable but g is not Riemann integrable, does it follow that $f+g$ is not Riemann integrable? If f is Riemann integrable but g is not Riemann integrable, does it follow that $f + g$ is Riemann integrable? (Give proofs or counterexamples.)

(iii) Compose and answer similar questions for the product fg and for λf, where λ is a non-zero real number.

K.125. [8.2, P] Define $f : [0, 1] \to \mathbb{R}$ by $f(p/q) = 1/q$ when p and q are coprime integers (that is, strictly positive integers with no common factor) with $1 \le p < q$, and $f(x) = 0$ otherwise.

(i) Show that f is Riemann integrable, and find $\int_0^1 f(x)\, dx$.

(ii) At which points, if any, is f continuous? Prove your answer.

(iii) At which points, if any, is f differentiable? Prove your answer.

Use the ideas of this exercise to define a function $g : \mathbb{R} \to \mathbb{R}$ which is unbounded on every interval $[a, b]$ with $b > a$.

K.126. [8.2, P] Suppose that $f : [0, 1] \to \mathbb{R}$ is Riemann integrable. Show that, if we write $g_n(x) = f(x^n)$, then $g_n : [0, 1] \to \mathbb{R}$ is Riemann integrable. State, with appropriate proof, which of the following conditions imply that

$$\int_0^1 f(x^n)\, dx \to f(0)$$

as $n \to \infty$.

(i) f is continuous on $[0, 1]$.

(ii) f is continuous at 0.

(iii) No further condition on f.

K.127. [8.2, P] Suppose $f : \mathbb{R} \to \mathbb{R}$ is differentiable, $b > a$ and $f(a) = f(b) = 0$. Show that there exists a $c \in [a, b]$ such that

$$|f'(c)| > \frac{4}{b-a} \int_a^b |f(t)|\, dt.$$

Show that we cannot replace 4 by some K with $K > 4$

K.128. [8.2, H] (i) Let

$$\mathcal{D} = \{x_0, x_1, \dots, x_n\} \text{ with } a = x_0 \le x_1 \le x_2 \le \cdots \le x_n = b$$

be a dissection of $[a, b]$ and let $\epsilon > 0$. Show that there exists a $\delta > 0$, depending on ϵ and $\mathcal{D}$, such that, if

$$\mathcal{D}' = \{y_0, y_1, \dots, y_m\} \text{ with } a = y_0 \le y_1 \le y_2 \le \cdots \le y_m = b$$

is a dissection of $[a, b]$ with $|y_j - y_{j-1}| < \delta$ for all j, then the total length of those intervals $[y_{k-1}, y_k]$ not completely contained in some $[x_{i-1}, x_i]$ is less than ϵ. More formally,

$$\sum_{i=1}^{n} \sum_{[y_{k-1}, y_k] \triangle [x_{i-1}, x_i] \neq \varnothing} |y_k - y_{k-1}| < \epsilon.$$

(ii) Show that, if $f : [a, b] \to \mathbb{R}$ is a function with $|f(t)| \le M$ for all t, $\mathcal{D}$ is a dissection of $[a, b]$ and $\epsilon > 0$, we can find an $\eta > 0$ (depending on M, ϵ and $\mathcal{D}$) such that if

$$\mathcal{D}' = \{y_0, y_1, \dots, y_m\} \text{ with } a = y_0 \le y_1 \le y_2 \le \cdots \le y_m = b$$

is a dissection of $[a, b]$ with $|y_j - y_{j-1}| < \eta$ for all j, then $S(f, \mathcal{D}') < S(f, \mathcal{D}) + \epsilon$ and $S(f, \mathcal{D}') > s(f, \mathcal{D}) - \epsilon$.

(iii) Show that a bounded function $f : [a, b] \to \mathbb{R}$ is Riemann integrable with integral I if and only if, given any $\epsilon > 0$, we can find an $\eta > 0$ such that, if

$$\mathcal{D} = \{x_0, x_1, \dots, x_n\} \text{ with } a = x_0 \le x_1 \le x_2 \le \cdots \le x_n = b$$

is a dissection of $[a, b]$ with $|x_j - x_{j-1}| < \eta$ for all j, then

$$S(f, \mathcal{D}) - s(f, \mathcal{D}) < \epsilon, \text{ and } |S(f, \mathcal{D}) - I| \le \epsilon.$$

(iv) Show that a bounded function $f : [a, b] \to \mathbb{R}$ is Riemann integrable with integral I if and only if, given any $\epsilon > 0$, we can find a positive integer N such that, if $n \ge N$ and

$$\mathcal{D} = \{x_0, x_1, \dots, x_n\} \text{ with } x_j = a + j(b-a)/N,$$

then

$$S(f, \mathcal{D}) - s(f, \mathcal{D}) < \epsilon, \text{ and } |S(f, \mathcal{D}) - I| \le \epsilon.$$

(v) Show that, if $f : [a,b] \to \mathbb{R}$ is Riemann integrable, then

$$\frac{b-a}{N}\sum_{j=0}^{N-1} f(a+j(b-a)/N) \to \int_a^b f(t)\,dt$$

as $N \to \infty$. Why is this result compatible with Dirichlet's counterexample (Exercise 8.30)?

K.129. [8.2, H] (This exercise is best treated fairly informally.) If we insist on considering the integral as the area under a curve, then our definition of the Riemann integral of a function which can be negative looks a bit odd. We could restrict ourselves initially to positive bounded functions and then extend to general functions in (at least) three ways.

(A) If $f : [a,b] \to \mathbb{R}$ is bounded and we can write $f = F - G$ with F and G positive and Riemann integrable, then we set

$$\int_a^b f(t)\,dt = \int_a^b F(t)\,dt - \int_a^b G(t)\,dt.$$

if the right hand side of the equation exists.

(B) If $f : [a,b] \to \mathbb{R}$ is bounded then we can write $f(t) = f_\kappa(t) - \kappa$ with f_κ positive and κ a real number. We set

$$\int_a^b f(t)\,dt = \int_a^b f_\kappa(t)\,dt - \kappa(b-a).$$

(C) If $f : [a,b] \to \mathbb{R}$ is bounded, we write $f_+(t)(t) = \max(f(t),0)$ and and $f_-(t) = \max(-f(t),0)$. We set

$$\int_a^b f(t)\,dt = \int_a^b f_+(t)\,dt - \int_a^b f_-(t)\,dt$$

if the right hand side of the equation exists.

Run through the checks that you must make to see that these definitions work as expected. For example, if we use (B), we must check that the value of the integral is independent of the value of κ chosen. We must also show that, if f and g are integrable, so are $f+g$ and fg. Case (C) is (I think) harder, and you may prefer simply to observe why it is harder and not proceed further.

Explain why we get the same integrable functions and the same integrals from the four definitions considered (the one we actually used, (A), (B) and (C)).

K.130. [8.2, H] No modern mathematician would expect any reasonable theory of integration on the rationals, for the following reason.

Since the rationals are countable, we can enumerate the elements of $[0,1] \cap \mathbb{Q}$. Enumerate them as $x_1, x_2, \dots$. Let J_r be an interval of length

2^{-r-1} containing x_r. We observe that the union of the J_r covers $[0,1] \cap \mathbb{Q}$ but that the total length of the intervals is only $\sum_{r=1}^{\infty} 2^{-r-1} = 1/2$.

(i) Show that, given any $\epsilon > 0$, we can find intervals J_1, J_2, ... of total length less than ϵ with $\bigcup_{r=1}^{\infty} J_r \supseteq \mathbb{Q}$.

(ii) Such a phenomenon cannot take place for the reals. Suppose that we are given intervals J_1, J_2, ... in $\mathbb{R}$ of total length $1-\epsilon$, where $1 > \epsilon > 0$. Show that we can find open intervals I_1, I_2, ... of total length less than $1-\epsilon/2$ such that $I_r \supseteq J_r$ for each r. Now set $K_j = [0,1] \setminus \bigcup_{r=1}^{j} I_r$. By applying Exercise 4.48, which tells that the intersection of bounded, closed, nested non-empty sets in $\mathbb{R}$ is itself non-empty, show that $\bigcap_{r=1}^{\infty} K_j \neq \varnothing$ and deduce that $\bigcup_{r=1}^{\infty} J_r \not\supseteq [0,1]$. We cannot put a quart into a pint pot[10].

(iii) It could be argued that the example above does not completely rule out a theory of integration for well behaved functions $f : \mathbb{Q} \to \mathbb{Q}$. To show that no such theory exists, consider the function $f : \mathbb{Q} \to \mathbb{Q}$ given by

$$\begin{aligned} f(x) &= 1 \qquad \text{if } x^2 < 2, \\ f(x) &= 0 \qquad \text{otherwise,} \end{aligned}$$

Examine how the procedure for defining an integral $\int_0^2 f(x)\,dx$ by means of upper and lower sums and integrals breaks down

K.131. (First integral mean value theorem.) [8.3, T]

(i) Suppose that $F : [a,b] \to \mathbb{R}$ is a continuous. function such that $\sup_{t\in[a,b]} F(t) \geq \lambda \geq \inf_{t\in[a,b]} F(t)$. Show that there exists a $c \in [a,b]$ with $F(c) = \lambda$.

(ii) Suppose that $w : [a,b] \to \mathbb{R}$ is Riemann integrable and non-negative on $[a,b]$. If $f : [a,b] \to \mathbb{R}$ is continuous, show that there exists a $c \in [a,b]$ with

$$\int_a^b f(t)w(t)\,dt = f(c)\int_a^b w(t)\,dt.$$

(This is a pretty result, but in the view of the present author it resembles the mean value theorem in being a mainly decorative extension of simpler inequalities.)

(iii) Show that (ii) may fail if we do not demand w positive. Show that it also holds if we demand w everywhere negative.

K.132. [8.3, T] (i) The real function $f : [a,b] \to \mathbb{R}$ is strictly increasing and continuous with inverse function g. Give a geometric interpretation of the equality

$$\int_a^b f(x)\,dx + \int_{f(a)}^{f(b)} g(y)\,dy = bf(b) - af(a),$$

[10]Or a litre into a half litre bottle.

and prove it by using upper and lower Riemann sums.

(ii) The real function $f : [0, T] \to \mathbb{R}$ is strictly increasing and continuous with inverse function g. If $f(0) = 0$, $T \geq X \geq 0$ and $f(T) \geq Y \geq 0$, give a geometric interpretation of the inequality

$$\int_0^X f(x)\,dx + \int_0^Y g(y)\,dy \geq XY$$

and prove it. For which values of X and Y does this inequality become an equality? Prove your answer.

(iii) Suppose p an q are strictly positive real numbers with $p^{-1}+q^{-1} = 1$. Show that

$$\frac{X^p}{p} + \frac{Y^q}{q} \geq XY.$$

For which values of X and Y does this inequality become an equality? (The importance of this inequality will be demonstrated in Exercise K.209 and the exercises which follow it.)

K.133. [8.3, H] Historically and pedagogically, the integrability of continuous functions is always linked with the proof that a continuous function on a closed set is uniformly bounded. The following exercise shows that this is not the only path.

(i) Reread Exercise 8.24 (i). Show that, under the assumptions of that exercise,

$$I^*_{[a,c]}(f|_{[a,c]}) + I^*_{[c,b]}(f|_{[c,b]}) = I^*_{[a,b]}(f|_{[a,b]}) \text{ and}$$
$$I_{*[a,c]}(f|_{[a,c]}) + I_{*[c,b]}(f|_{[c,b]}) = I_{*[a,b]}(f|_{[a,b]}),$$

where $I^*_{[a,c]}$ denotes the upper integral on $[a, c]$, and so on.

(ii) Let $f : [a, b] \to \mathbb{R}$ be continuous. Suppose, if possible, that f is not integrable. Explain why this means that there is a $\kappa > 0$ such that

$$I^*_{[a,b]}(f|_{[a,b]}) - I_{*[a,b]}(f|_{[a,b]}) \geq \kappa(b - a).$$

(iii) Suppose that $c_0 = (a+b)/2$. Show that at least one of the following two statements must be true:

$$I^*_{[a,c_0]}(f|_{[a,c_0]}) - I_{*[a,c_0]}(f|_{[a,c_0]}) \geq \kappa(c_0 - a),$$
$$I^*_{[c_0,b]}(f|_{[c_0,b]}) - I_{*[c_0,b]}(f|_{[c_0,b]}) \geq \kappa(b - c_0).$$

Use this remark as the basis for a lion hunting argument. Use the continuity of f to obtain a contradiction and deduce Theorem 8.32.

K.134. [8.3, H] For the reasons sketched in Section 9.3, it is hard to extend the change of variable formula given in Theorem 8.44 for one-dimensional integrals to many dimensions without a fundamental rethink about the nature of area. But, whether we face or evade this issue, the proof of Theorem 8.44 is essentially one-dimensional. In this exercise we give a lion hunting argument which could be extended to higher dimensions.

(i) We use the notation and hypotheses of Theorem 8.44. Suppose, if possible, that the theorem is false. Explain why this means that there is a $\kappa > 0$ such that

$$\left|\int_{g(c)}^{g(d)} f(s)\,ds - \int_c^d f(g(x))g'(x)\,dx\right| \geq \kappa(d-c).$$

(ii) Suppose that $e_0 = (c+d)/2$. Show that at least one of the following two statements must be true.

$$(a)\ \left|\int_{g(c)}^{g(e_0)} f(s)\,ds - \int_c^{e_0} f(g(x))g'(x)\,dx\right| \geq \kappa(e_0-c),$$

$$(b)\ \left|\int_{g(e_0)}^{g(d)} f(s)\,ds - \int_{e_0}^{d} f(g(x))g'(x)\,dx\right| \geq \kappa(d-e_0).$$

Use this remark as the basis for a lion hunting argument, and deduce Theorem 8.44 from the resulting contradiction.

K.135. [8.3, H, ↑] (i) Prove Lemma 8.49 (the formula for integration by parts) by direct calculation in the special case $G(x) = Ax + B$, $f(x) = Kx + L$.

(ii) Prove Lemma 8.49 by lion hunting along the lines of Exercise K.134.

K.136. [8.3, T, S] Let $f : [a, b] \to \mathbb{R}$ be differentiable (with one sided derivatives at the end points). Explain why, if $a \leq x_{j-1} < x_j \leq b$, then

$$\sup_{x\in[x_{j-1},x_j]} f'(t)(x_j - x_{j-1}) \geq f(x_j) - f(x_{j-1}) \geq \inf_{x\in[x_{j-1},x_j]} f'(t)(x_j - x_{j-1}),$$

and deduce the theorem of Darboux which states that, if f' is Riemann integrable, then

$$\int_a^b f'(t)\,dt = f(b) - f(a).$$

[This is a considerable generalisation of Exercise 8.43, but is not as final a result as it looks at first sight, since it need not be true that f' is Riemann integrable.]

K.137. [8.3, T] (i) Show directly, using uniform continuity and an argument along the lines of our proof of Theorem 8.32, that, if $f : [a, b] \to \mathbb{R}$ is continuous, then

$$\frac{b-a}{N}\sum_{j=0}^{N-1} f(a + j(b-a)/N) \to \int_a^b f(t)\,dt$$

as $N \to \infty$. (The general result, which holds for all Riemann integrable functions, was given in Exercise K.128 (v), but the special case is easier to prove.)

(ii) It is natural to ask whether we cannot define the integral of a continuous function directly in this manner, without going through upper and lower sums. Here we show one way in which it it can be done. We shall suppose that $f : [a, b] \to \mathbb{R}$ is a continuous function and write

$$S_N = \frac{b-a}{N} \sum_{j=0}^{N-1} f(a + j(b-a)/N).$$

As might be expected, our main tool will be uniform continuity.

Show that, given $\epsilon > 0$, we can find an $N_0(\epsilon)$ such that, if $N \geq N_0(\epsilon)$ and $P \geq 1$, then $|S_N - S_{NP}| < \epsilon$. Show that, if $M, N \geq N_0(\epsilon)$, then $|S_N - S_M| < 2\epsilon$. Deduce that S_N tends to a limit as $N \to \infty$. We can define this limit to be $\int_a^b f(t)\, dt$.

(iii) Explain briefly why the new definition gives the same value for the integral as the old.

The objections to this procedure are that it obscures the geometric idea of area, that many branches of pure and applied mathematics deal with functions like the Heaviside function which, though well behaved, are not continuous, and that it gives a special status to those points of the form $a + j(b-a)/N$. (These objections, although strong, are not, however, conclusive. The ideas sketched here are frequently used in obtaining various generalisations of our simple integral. An example of this is given in Exercise K.152.)

K.138. [8.3, T, S] Let $f : [-1, 1] \to \mathbb{R}$ be defined by $f(1/n) = 1$ for all integers $n \geq 1$, and $f(t) = 0$, otherwise. By finding appropriate dissections, show that f is Riemann integrable. If

$$F(t) = \int_0^t f(x)\, dx,$$

find F and show that F is everywhere differentiable. Observe that $F'(0) = f(0)$. Is f continuous at 0? Prove your answer.

K.139. [8.3, T] (i) Use the relation

$$r(r-1) = \frac{(r+1)r(r-1) - r(r-1)(r-2)}{3}$$

to find $\sum_{r=1}^{n} r(r-1)$. Use a similar relation to find $\sum_{r=1}^{n} r$, and deduce the value of $\sum_{r=1}^{n} r^2$.

(ii) By using the method of part (i), find $\sum_{r=1}^{n} r^3$. Show that

$$\sum_{r=1}^{n} r^m = (m+1)^{-1} n^{m+1} + P_m(n),$$

where P_m is a polynomial of degree at most m.

(iii) Use dissections of the form

$$\mathcal{D} = \{0,\, 1/n,\, 2/n,\, \dots,\, 1\}$$

to show that

$$\int_0^1 x^m\, dx = \frac{1}{m+1}$$

for any positive integer m.

(iv) Use the result of (iii)[11] to compute $\int_0^a x^m\, dx$ and $\int_a^b x^m\, dx$ for all values of a and b. Obtain the same result by using a version of the fundamental theorem of the calculus.

(v) Use dissections of the form

$$\mathcal{D} = \{0,\, 1/n,\, 2/n,\, \dots,\, 1\}$$

to show that

$$\int_0^1 e^x\, dx = e - 1.$$

Obtain the same result by using a version of the fundamental theorem of the calculus.

(vi) Use dissections of the form

$$\mathcal{D} = \{br^n,\, br^{n-1},\, br^{n-2},\, \dots,\, b\}$$

with $0 < r$ and $r^n = a/b$ to compute

$$\int_a^b x^m\, dx$$

for any positive integer m.

K.140. (Numerical integration.) [8.3, T] We saw in Exercise K.137 (i) that, if $f : [a,b] \to \mathbb{R}$ is continuous, then $\frac{b-a}{N} \sum_{j=0}^{N-1} f(a + j(b-a)/N)$ is a good approximation to $\int_a^b f(t)\, dt$. In this exercise we shall see that, if we know that f is better behaved, then we can get better approximations. The calculations give a good example of the use of global Taylor theorems such as Theorem 7.2.

[11]This method is due to Wallis, who built on earlier, more geometric, ideas and represents one of the high spots of analysis before the discovery of the fundamental theorem of the calculus united the theories of differentiation and integration. Wallis was appointed professor at the fiercely royalist university of Oxford as a reward for breaking codes for the parliamentary (that is, anti-royalist) side in the English civil war.

(i) Show that, if g is linear, then

$$\int_{-1}^{1} g(t)\,dt = g(-1) + g(1).$$

(ii) Suppose that $g : \mathbb{R} \to \mathbb{R}$ is twice differentiable with $|g''(t)| \leq K$ for all $t \in [-1, 1]$. Explain why $|g(t) - g(0) - g'(0)t| \leq Kt^2/2$ for all $t \in [-1, 1]$, and deduce that

$$\left| \int_{-1}^{1} g(t)\,dt - \big(g(-1) + g(1)\big) \right| \leq \frac{4K}{3}.$$

(iii) Suppose that $f : \mathbb{R} \to \mathbb{R}$ is twice differentiable with $|f''(t)| \leq K$ for all $t \in [a, b]$. If N is a strictly positive integer and $Nh = b - a$, show that

$$\left| \int_{a+(r-1)h}^{a+rh} f(t)\,dt - \frac{\big(f(a+(r-1)h) + f(a+rh)\big)h}{2} \right| \leq \frac{Kh^3}{6}$$

for all integers r with $1 \leq r \leq N$. Let

$$T_h f = \frac{h}{2}\big(f(a) + 2f(a+h) + 2f(a+2h) + \cdots + 2f(a+(N-1)h) + f(b)\big).$$

Show that

$$\bigstar \qquad \left| \int_{a}^{b} f(t)\,dt - T_h f \right| \leq \frac{K(b-a)h^2}{6}.$$

We call the approximation $\int_a^b f(t)\,dt \approx T_h f$ the *trapezium rule*. Speaking informally, we can say that 'the error in the trapezium rule decreases like the square of the step length h'.

(iv) Let $a = 0$, $b = \pi$, $Nh = \pi$ with N a strictly positive integer, and let $F(t) = \sin^2(Nt)$. Show that

$$\left| \int_{a}^{b} F(t)\,dt - T_h F \right| \geq A \sup_{t \in [a,b]} |F''(t)|(b-a)h^2,$$

where A is a strictly positive constant. Conclude that the bound in $\bigstar$ cannot be substantially improved. (In fact, it can be halved by careful thought about worst cases.)

(v) Show that, if g is constant, then

$$\int_{-1}^{1} g(t)\,dt = 2g(-1).$$

By imitating the earlier parts of this exercise, show that, if $f : \mathbb{R} \to \mathbb{R}$ is once differentiable with $|f'(t)| \leq K$ for all $t \in [a, b]$, N is a strictly positive integer, $Nh = b - a$ and we write

$$S_h f = h\big(f(a) + f(a+h) + f(a+2h) + \cdots + f(a+(N-1)h)\big),$$

then $\left|\int_a^b f(t)\,dt - S_h f\right| \leq K(b-a)h$. Speaking informally, we can say that 'the error when we use the approximation rule $\int_a^b f(t)\,dt \approx S_h f$ decreases like the step length h'. Show that we cannot improve substantially on the bound obtained.

(vi) Show that, if g is a cubic (that is to say, a polynomial of degree 3), then

$$\int_{-1}^{1} g(t)\,dt = \frac{g(-1)+4g(0)+g(1)}{3}.$$

(vii) Suppose that $f:\mathbb{R}\to\mathbb{R}$ is four times differentiable with $|f^{(4)}(t)| \leq K$ for all $t\in[a,b]$. If N is a strictly positive even integer and $Nh = b-a$, let us write

$$\begin{aligned}Q_h f = \frac{h}{3}\big(f(a) &+ 4f(a+h) + 2f(a+2h) + 4f(a+3h) + 2f(a+4h)\\ &+ \cdots + 2f(a+(N-2)h) + 4f(a+(N-1)h) + f(b)\big).\end{aligned}$$

Show that

$$\bigstar\bigstar \qquad \left|\int_a^b f(t)\,dt - Q_h f\right| \leq \frac{K(b-a)h^4}{90}.$$

We call the approximation $\int_a^b f(t)\,dt \approx Q_h f$ *Simpson's rule.* Speaking informally, we can say that 'the error in the Simpson's rule decreases like the fourth power of the step length h'. Show that the bound in $\bigstar\bigstar$ cannot be substantially improved. (Again, it can be halved by careful thought about worst cases.)

(The reader may be tempted to go on and consider more and more complicated rules along these lines. However, such rules involve assumptions about the behaviour of higher derivatives which are often unrealistic in practice.)

K.141. [8.3, T] The object of this exercise is to define the logarithm and the real exponential function, so no properties of those functions should be used. You should quote all theorems that you use, paying particular attention to those on integration.

We set $l(x) = \int_1^x \frac{1}{t}\,dt$.

(i) Explain why $l:(0,\infty)\to\mathbb{R}$ is a well defined function.

(ii) Use the change of variable formula for integrals to show that

$$\int_x^{xy} \frac{1}{t}\,dt = l(y)$$

whenever $x, y > 0$. Deduce that $l(xy) = l(x) + l(y)$.

(iii) Show that l is everywhere differentiable with $l'(x) = 1/x$.

(iv) Show that l is a strictly increasing function.

(v) Show that $l(x) \to \infty$ as $x \to \infty$.

(vi) Show that $l : (0, \infty) \to \mathbb{R}$ is a bijective function.

(vii) In Section 5.6 we defined the logarithm as the inverse of the exponential function $e : \mathbb{R} \to (0, \infty)$. Turn this procedure on its head by defining e as the inverse of l. Derive the main properties of e, taking particular care to quote those theorems on inverse functions that you require. When you have finished, glance through Section 5.4 to see if you have proved all the properties of the (real) exponential given there.

(viii) Check that

$$\frac{1}{1-t} = 1 + t + \cdots + t^n + \frac{t^{n+1}}{1-t}$$

for all $t \neq 1$, and show that

$$\left| \int_0^x \frac{t^{n+1}}{1-t}\, dt \right| \leq \frac{|x|^{n+2}}{1-|x|}$$

for all $|x| < 1$. Use these observations to obtain a Taylor series for $\log(1-x)$, giving the range over which your argument is valid.

(ix) Use (viii) to find the Taylor series of $\log\big((1+x)/(1-x)\big)$, giving the range over which your argument is valid. Show that

$$\log y = 2\left(\left(\frac{y-1}{y+1}\right) + \frac{1}{3}\left(\frac{y-1}{y+1}\right)^3 + \frac{1}{5}\left(\frac{y-1}{y+1}\right)^5 + \dots\right)$$

for all $y > 0$.

K.142. [8.3, P, S, ↑] Use the Taylor series for $\log(1+x)$ and some result on products like Exercise 5.38 to obtain

$$(\log(1+x))^2 = 2\sum_{n=2}^{\infty} \frac{(-1)^n}{n} S_{n-1} x^n$$

for all $|x| < 1$, where $S_m = \sum_{n=1}^{m} 1/n$.

K.143. (Convex functions.) [8.3, T] Recall from Exercise K.40 that we call a function $f : (a,b) \to \mathbb{R}$ *convex* if, whenever $x_1, x_2 \in (a,b)$ and $1 \geq \lambda \geq 0$, we have

$$\lambda f(x_1) + (1-\lambda) f(x_2) \geq f(\lambda x_1 + (1-\lambda)x_2).$$

(i) Suppose that $a < x_1 < x_2 < x_3 < b$. Show algebraically that

$$\frac{f(x_2) - f(x_1)}{x_2 - x_1} \leq \frac{f(x_3) - f(x_2)}{x_3 - x_2},$$

and illustrate the result graphically.

(ii) If $c \in (a, b)$, show that

$$\sigma_f(c+) = \inf\left\{\frac{f(c+h) - f(c)}{h} : c < c + h < b\right\}$$

exists and

$$\frac{f(c+h) - f(c)}{h} \to \sigma_f(c+)$$

as $h \to 0$ through values $h > 0$. State and prove the appropriate result for $\sigma_f(c-)$. Show also that $\sigma_f(c+) \geq \sigma_f(c-)$.

(iii) Using (ii), or otherwise, show that f is continuous at c. (Thus a convex function is automatically continuous.)

(iv) Using (ii), or otherwise, show that we can find a real B such that

$$f(x) - f(c) \geq B(x - c)$$

for all $x \in (a, b)$. (Notice that, if $\sigma_f(c+) = \sigma_f(c)$, then f is differentiable at c, $B = f'(c)$ and the line $y - f(c) = B(x - c)$ is the tangent, but that if $\sigma_f(c+) \neq \sigma_f(c)$, then B is not unique and f is not differentiable at c. Exercise K.144) is relevant here.)

(v) Suppose $a < \alpha < \beta < b$. Let $g : [\alpha, \beta] \to \mathbb{R}$ be a positive continuous function with $\int_\alpha^\beta g(t)\, dt = 1$. Show that $c = \int_\alpha^\beta tg(t)\, dt \in [\alpha, \beta]$ and prove, using (iv), or otherwise, that

$$\int_\alpha^\beta f(t)g(t)\, dt \geq f\left(\int_\alpha^\beta tg(t)\, dt\right).$$

(vi) Prove the result of Exercise K.40 (iii) by the method of (iv). If you know a little probability, use the method of (iv) to show that if X is a random variable taking values in $[\alpha, \beta]$ we have

$$\mathbb{E}f(X) \geq f(\mathbb{E}X).$$

(vii) If you have done Exercise K.32 (iii), try and obtain part (iv) as a consequence.

K.144. [8.3, T, ↑] (i) By considering the map $x \to |x|$, or otherwise, show that a convex function need not be differentiable everywhere.

(ii) We use the assumptions and notation of Exercise K.143 (ii). Show that, if $a < \alpha \leq c_1 < c_2 < \cdots < c_N \leq \beta < b$, then

$$\sum_{j=1}^{N} \big(\sigma_f(c_j+) - \sigma_f(c_j-)\big) \leq \sigma_f(\beta+) - \sigma_f(\alpha-).$$

Deduce, by using a 'hamburger argument' (see page 362), that the set of points in $[\alpha, \beta]$ where f is not differentiable is countable. Deduce that f is differentiable on (a, b) except at a countable set of points.

(iii) Show that, if $f, g : (a,b) \to \mathbb{R}$ are convex, then so are $f+g$ and μf when $\mu \geq 0$. If $f_n : (a,b) \to \mathbb{R}$ is convex for each n and $f : (a,b) \to \mathbb{R}$ is such that $f_n(x) \to f(x)$ as $n \to \infty$ for each $x \in (a,b)$, show that f is convex.

(iv) Construct an $f : (-1,1) \to \mathbb{R}$ which is convex but is not differentiable at any rational point.

K.145. [8.3, H, S] (i) Let $f : \mathbb{R} \to \mathbb{R}$ be continuous. Let $a_n < c < b_n$ and $a_n, b_n \to c$ as $n \to \infty$. Write $I_n = [a_n, b_n]$, $|I_n| = b_n - a_n$ and $\int_{I_n} f(t)\,dt = \int_{a_n}^{b_n} f(t)\,dt$. Show, by using the method of proof of Theorem 8.37, but not the theorem itself, that

$$\frac{1}{|I_n|}\int_{I_n} f(t)\,dt \to f(c).$$

(ii) How might this result generalise to higher dimensions (so we consider $f : \mathbb{R}^m \to \mathbb{R}$) and how might we prove the generalisation? (Without more work we cannot give a rigorous proof, but we can certainly see how the proof ought to run.)

K.146. [8.3, H!] We saw in Exercise K.130 that there is no hope of a successful theory of integration for functions $f : \mathbb{Q} \to \mathbb{Q}$. None the less, I think that looking at such functions can illuminate the role played by uniform continuity in the proof of Theorem 8.32.

(i) Suppose $f : [0,1] \cap \mathbb{Q} \to \mathbb{Q}$ is uniformly continuous. Show that, given any $\epsilon > 0$, we can find a dissection $\mathcal{D}$ of $[0,1] \cap \mathbb{Q}$ such that

$$S(f,\mathcal{D}) - s(f,\mathcal{D}) < \epsilon.$$

In what follows we shall construct a function $g : [0,1] \cap \mathbb{Q} \to \mathbb{Q}$ which is bounded and continuous but has the property that

$$S(g,\mathcal{D}) - s(g,\mathcal{D}) \geq 1$$

for every dissection $\mathcal{D}$ of $[0,1] \cap \mathbb{Q}$.

(ii) We start by working in $\mathbb{R}$. We say that an interval J is 'good' if

$$J = \{x \in [0,1] \,:\, (x-q)^2 \leq 2^{-m}\}$$

with q rational and m a strictly positive integer. We enumerate the elements of $[0,1] \cap \mathbb{Q}$ as a sequence $x_1, x_2, \ldots$ of distinct elements.

Show that, setting $\mathcal{S}_0 = \varnothing$, we can construct inductively finite collections $\mathcal{S}_n$ of disjoint good sets such that:

(a)$_n$ $\mathcal{S}_n = \mathcal{S}_{n-1} \cup \mathcal{K}_n \cup \mathcal{L}_n$ with $\mathcal{S}_{n-1}$, $\mathcal{K}_n$ and $\mathcal{L}_n$ disjoint.

(b)$_n$ The total lengths of the intervals in $\mathcal{K}_n \cup \mathcal{L}_n$ is less than 2^{-n-3}.

(c)$_n$ If J is a subinterval of $[0,1]$ of length at least 2^{-n} with the property that $J \cap \bigcup_{I \in \mathcal{S}_{n-1}} I = \varnothing$, then we can find a $K \in \mathcal{K}_n$ and an $L \in \mathcal{L}_n$ such that $J \supseteq K \cup L$.

(d)$_n$ $x_n \in \bigcup_{I\in\mathcal{S}_n} I$.

(iii) Show that the total length of the intervals making up $\mathcal{S}_n$ is always less than $1/4$.

Suppose that

$$\mathcal{D} = \{a_0, a_1, \dots, a_p\} \text{ with } 0 = a_0 \le a_1 \le a_2 \le \dots \le a_p = 1$$

is a dissection of $[0,1]$. Show that, provided that n is sufficiently large, the total length of those intervals $[a_{j-1}, a_j]$ with $1 \le j \le m$ such that we can find a $K \in \mathcal{K}_n$ and an $L \in \mathcal{L}_n$ with $[a_{j-1}, a_j] \supseteq K \cup L$ will be at least $1/2$.

(iv) Show that, if $q \in \mathbb{Q} \cap [0,1]$, then there is a unique $n \ge 1$ such that

$$q \in \bigcup_{I\in\mathcal{S}_n\setminus\mathcal{S}_{n-1}} I.$$

Explain why q lies in exactly one of $\bigcup_{K\in\mathcal{K}_n} K$ or $\bigcup_{L\in\mathcal{L}_n} L$. We define $g(q) = 1$ if $q \in \bigcup_{K\in\mathcal{K}_n} K$ and $g(q) = -1$ if $q \in \bigcup_{L\in\mathcal{L}_n} L$.

(v) We now work in $\mathbb{Q}$. Show that $g : [0,1] \cap \mathbb{Q} \to \mathbb{Q}$ is a bounded continuous function. By using (iii), or otherwise, show that

$$S(g, \mathcal{D}) - s(g, \mathcal{D}) \ge 1$$

for every dissection $\mathcal{D}$ of $[0,1] \cap \mathbb{Q}$.

K.147. [8.4, P] Suppose that $g : \mathbb{R}^2 \to \mathbb{R}$ is a differentiable function with continuous partial derivatives. Suppose that $g : \mathbb{R}^2 \to \mathbb{R}$ is a differentiable function with continuous partial derivatives. Examiners often ask you to show that

$$G(x) = \int_0^x g(x,t)\,dt$$

is differentiable and determine its derivatives.

(i) Why can you not just quote Theorem 8.57?

(ii) Set $F(x,y) = \int_0^y g(x,t)\,dt$. Show that F has partial derivatives. Show further that these partial derivatives are continuous.

(iii) Use the chain rule to show that G is differentiable and determine its derivatives.

(iv) Suppose that $h : \mathbb{R} \to \mathbb{R}$ is differentiable. Show that

$$H(x) = \int_0^{h(x)} g(x,t)\,dt$$

is differentiable, and determine its derivatives.

K.148. [8.4, P] The following exercise is included because it uses several of our theorems in a rather neat manner. (It is actually a conservation of energy result.) Suppose that $u : \mathbb{R}^2 \to \mathbb{R}$ has continuous partial second derivatives, that

$$\frac{\partial^2 u}{\partial x^2}(x,t) = \frac{\partial^2 u}{\partial t^2}(x,t) \text{ and that } \frac{\partial u}{\partial t}(0,t) = \frac{\partial u}{\partial t}(1,t) = 0.$$

Show that, if

$$E(t) = \int_0^1 \left(\frac{\partial u}{\partial t}(x,t)\right)^2 + \left(\frac{\partial u}{\partial x}(x,t)\right)^2 \, dx,$$

then E is a constant.

Identify explicitly the major theorems that you use. (The author required four major theorems to do this exercise, but you may have done it another way or have a different view about what constitutes a major theorem.)

K.149. [8.4, T] In Example 7.6 we constructed an infinitely differentiable function $E : \mathbb{R} \to \mathbb{R}$ with $E(t) = 0$ for $t \leq 0$ and $E(t) > 0$ for $t > 0$. If $\delta > 0$, sketch the function $H_\delta : \mathbb{R} \to \mathbb{R}$ given by

$$H_\delta(t) = \int_0^t E(\delta - x)E(x)\, dx.$$

By using functions constructed along these lines, or otherwise, prove the following mild improvement of Exercise 8.66.

If

$$I(f) = \int_0^1 \left(1 - (f'(x))^4\right)^2 + f(x)^2 \, dx$$

(as in Exercise 8.66), show that there exists a sequence of infinitely differentiable functions $f_n : [0,1] \to \mathbb{R}$ with $f_n(0) = f_n(1) = 0$ such that

$$I(f_n) \to 0.$$

K.150. [8.4, M] (This exercise should be treated as one in mathematical methods rather than in analysis.)

Show that the Euler–Lagrange equation for finding stationary values of an integral of the type

$$\int_a^b g(x,y)\, ds,$$

where the integral is an arc length integral (informally, 'ds is the element of arc length'), may be written

$$g_{,2}(x,y) - g_{,1}(x,y)y'(x) - \frac{y''(x)g(x,y)}{1 + y'(x)^2} = 0.$$

Hence, or otherwise, show (under the assumption that the problem admits a well behaved solution) that the curve joining 2 given points (on one side of the y axis) that minimises the surface area generated by rotating the curve about the y-axis is given by

$$x = c\cosh\frac{y+A}{C},$$

where A and C are constants.

K.151. [8.4, M] (This exercise should be treated as one in mathematical methods rather than in analysis.)

We wish to find a well behaved function $y : [a, b] \to \mathbb{R}$ which makes the integral

$$I = \int_a^b f(x, y, y', y'')\, dx$$

stationary subject to given values of y and y' at a and b. Derive an analogue of the Euler–Lagrange equation for y and solve the problem in the case where $a = 0$, $b = 1$, $y(0) = y'(0) = 0$, $y(1) = 0$, $y'(1) = 4$ and

$$f(x, y, y', y'') = \frac{1}{2}(y'')^2 - 24y.$$

Explain why your result cannot be a maximum.

K.152. [8.5, T] Use the argument of Exercise K.137 (ii) to show that, if $\mathbf{f} : [a, b] \to \mathbb{R}^m$ is a continuous function and we write

$$\mathbf{S}_N = \frac{b-a}{N}\sum_{j=0}^{N-1} \mathbf{f}(a + j(b-a)/N),$$

then $\mathbf{S}_N$ tends to a limit as $N \to \infty$. We can define this limit to be $\int_a^b \mathbf{f}(t)\, dt$.

(The advantage of the procedure is that we define $\int_a^b \mathbf{f}(t)\, dt$ directly without using components. It is possible to define $\int_a^b \mathbf{f}(t)\, dt$ directly for all Riemann integrable $\mathbf{f} : [a, b] \to \mathbb{R}^m$ by mixing the ideas of this exercise with the ideas we used to define the one-dimensional Riemann integral, but it is not clear that the work involved is worth it.)

K.153. [9.1, T] (i) Define $g_m : [0, 1] \to \mathbb{R}$ by $g_m(x) = 1 - m|\frac{1}{2} - x|$ for $|\frac{1}{2} - x| \leq m^{-1}$, $g_m(x) = 0$ otherwise. Show that there exists a $g : [0, 1] \to \mathbb{R}$, which you should define explicitly, such that $g_m(x) \to g(x)$ as $m \to \infty$ for each $x \in [0, 1]$.

(ii) If f_n is as in Exercise 9.1, show that there exists a sequence of continuous functions $g_{nm} : [0, 1] \to \mathbb{R}$ such that $g_{nm}(x) \to f_n(x)$ as $m \to \infty$ for each $x \in [0, 1]$.

(Thus repeated taking of limits may take us out of the class of Riemann integrable functions. In fact, the later and more difficult Exercise K.175 shows that it is possible to find a sequence of bounded continuous functions

$h_n : [0,1] \to \mathbb{R}$ and a function $h : [0,1] \to \mathbb{R}$ which is not Riemann integrable such that $h_n(x) \to h(x)$ as $n \to \infty$ for each $x \in [0,1]$.)

K.154. [9.2, P] Suppose that $f : [0,\infty) \to \mathbb{R}$ is a non-negative, non-increasing function. If $h > 0$, show that $\sum_{n=1}^{\infty} f(nh)$ converges if and only if $\int_0^\infty f(t)\,dt$ converges. If $\int_0^\infty f(t)\,dt$ converges, show that

$$h\sum_{n=1}^{\infty} f(nh) \to \int_0^\infty f(t)\,dt$$

as $h \to 0$ with $h > 0$.

Show that, given any integer $N \geq 1$ and any $\epsilon > 0$, there exists a continuous function $g : [0,1] \to \mathbb{R}$ such that $0 \leq g(x) \leq 1$ for all $x \in [0,1]$ and

$$\int_0^1 g(x)\,dx < \epsilon, \text{ but } \sum_{n=1}^{N} g(nN^{-1}) = 1.$$

Show that there exists a continuous function $G : [0,\infty) \to \mathbb{R}$ such that $G(x) \geq 0$ for all $x \geq 0$ and $G(x) \to 0$ as $x \to \infty$ but

$$h\sum_{n=1}^{\infty} G(nh) \not\to \int_0^\infty G(t)\,dt$$

as $h \to 0$ with $h > 0$.

K.155. [9.2, P] (i) Suppose that $g : (0,\infty) \to \mathbb{R}$ is everywhere differentiable with continuous derivative and $\int_1^\infty |g'(t)|\,dt$ converges. Show that $\sum_{n=1}^\infty g(n)$ converges if and only if $\int_1^\infty g(t)\,dt$ converges.

(ii) Deduce Lemma 9.6 for the case when f is differentiable.

(iii) Suppose g satisfies the hypotheses of the first sentence of part (i). By giving a proof or a counterexample, establish whether, if the sum $\sum_{n=1}^\infty g(n)$ is convergent, then it is automatically absolutely convergent.

K.156. (A weak Stirling's formula.) [9.2, T] Show that

$$\int_{n-1/2}^{n+1/2} \log x\,dx - \log n = \int_0^{1/2} \left(\log\left(1+\frac{t}{n}\right) + \log\left(1-\frac{t}{n}\right)\right)\,dt.$$

By using the mean value theorem, or otherwise, deduce that

$$\left|\int_{n-1/2}^{n+1/2} \log x\,dx - \log n\right| \leq \frac{4}{3n^2}.$$

(You may replace $4/(3n^2)$ by An^{-2} with A another constant, if you wish.) Deduce that $\int_{1/2}^{N+1/2} \log x\,dx - \log N!$ converges to a limit. Conclude that there is a constant C such that

$$\frac{n!}{(n+1/2)^{(n+1/2)}}e^{-n} \to C \text{ as } n \to \infty.$$

K.157. [9.2, P, ↑] By using Exercise K.156 and Abel's test, or otherwise, determine the values of $z \in \mathbb{C}$ for which $\sum_{n=0}^{\infty} \binom{2n}{n} z^n$ converges and prove your answer.

Answer the same question for $\sum_{n=1}^{\infty} n^{\alpha}\binom{2n}{n} z^n$, where α is real. (The answer will depend on the value of α.)

K.158. [9.2, T] (i) Show that the change of variable theorem works for infinite integrals of the form $\int_0^{\infty} f(x)\,dx$. More specifically, prove the following theorem.

Suppose that $f : [0,\infty) \to \mathbb{R}$ is continuous and $g : [0,\infty) \to \mathbb{R}$ is differentiable with continuous derivative. Suppose, further, that $g(t) \geq 0$ for all t, and that $g(0) = 0$ and $g(t) \to \infty$ as $t \to \infty$. Then if either side of the following equation exists, both do and they are equal:

$$\int_0^{\infty} f(s)\,ds = \int_0^{\infty} f(g(x))g'(x)\,dx.$$

Explain why this result is consistent with Example 9.18.

(ii) Explain why the function $f : [0,\infty) \to \mathbb{R}$ defined by $f(x) = (\sin x)/x$ for $x \neq 0$, $f(0) = 1$ is continuous at 0. It is traditional to write $f(x) = (\sin x)/x$ and ignore the fact that, strictly speaking, $(\sin 0)/0$ is meaningless. Sketch f.

If $I_n = \displaystyle\int_0^{n\pi} \frac{\sin x}{x}\,dx$, show, by using the alternating series test, that I_n tends to a strictly positive limit L, say. Deduce carefully that $\displaystyle\int_0^{\infty} \frac{\sin x}{x}\,dx$ exists with value L.

(iii) Let $I(t) = \displaystyle\int_0^{\infty} \frac{\sin tx}{x}\,dx$ for all $t \in \mathbb{R}$. Show using (i), or otherwise, that $I(t) = L$ for all $t > 0$, $I(0) = 0$, and $I(t) = -L$ for $t < 0$.

(iv) Find a continuous function $g : [0,\pi] \to \mathbb{R}$ such that $g(t) \geq 0$ for all $t \in [0,\pi]$, $g(\pi/2) > 0$ and

$$\left|\frac{\sin x}{x}\right| \geq \frac{g(x-n\pi)}{n}$$

for all $n\pi \leq x \leq (n+1)\pi$ and all integers $n \geq 1$. Hence, or otherwise, show that $\int_0^{\infty} |(\sin x)/x|\,dx$ fails to converge.

(v) For which values of real α does $\displaystyle\int_0^{\infty} \frac{\sin x}{x^{\alpha}}\,dx$ converge? Prove your answer.

[There are many methods for finding the constant L of part (ii). The best known uses complex analysis. I give a rather crude but direct method in the next exercise. Hardy found the matter sufficiently interesting to to justify writing two articles (pages 528–533 and 615–618 of [**23**]) comparing various methods.]

K.159. [9.2, T, ↑] Here is a way of evaluating

$$\int_0^\infty \frac{\sin x}{x}\,dx.$$

(We adopt the convention that, if we write $f(t) = \dfrac{\sin \lambda t}{t}$, then $f(0) = \lambda$.)

(i) Show, by using the formula for the sum of a geometric series, or otherwise, that

$$1 + 2\sum_{r=1}^{n} \cos rx = \frac{\sin\left((n+\frac{1}{2})x\right)}{\sin\frac{x}{2}}$$

for all $|x| \leq \pi$, and deduce that

$$2\pi = \int_{-\pi}^{\pi} \frac{\sin\left((n+\frac{1}{2})x\right)}{\sin\frac{x}{2}}\,dx.$$

(ii) If $\epsilon > 0$, show that

$$\int_{-\epsilon}^{\epsilon} \frac{\sin \lambda x}{x}\,dx \to \int_{-\infty}^{\infty} \frac{\sin x}{x}\,dx$$

as $\lambda \to \infty$.

(iii) If $\pi \geq \epsilon > 0$, show, by using the estimates from the alternating series test, or otherwise, that

$$\int_{-\epsilon}^{\epsilon} \frac{\sin\left((n+\frac{1}{2})x\right)}{\sin\frac{x}{2}}\,dx \to \int_{-\pi}^{\pi} \frac{\sin\left((n+\frac{1}{2})x\right)}{\sin\frac{x}{2}}\,dx = 2\pi$$

as $n \to \infty$.

(iv) Show that

$$\left|\frac{2}{x} - \frac{1}{\sin\frac{1}{2}x}\right| \to 0$$

as $x \to 0$.

(v) Combine the results above to show that

$$\int_0^\infty \frac{\sin x}{x}\,dx = \frac{\pi}{2}.$$

K.160. (Big Oh and little oh.) [**9.2, T**] The following notations are much used in branches of mathematics like number theory, algorithmics and combinatorics. Consider functions $f,\ g : \mathbb{N} \to \mathbb{R}$.

We say $f(n) = O(g(n))$ as $n \to \infty$ if there exist a constant $A > 0$ and an integer n_0 such that $|f(n)| \leq A|g(n)|$ for $n \geq n_0$.

We say $f(n) = \Omega(g(n))$ as $n \to \infty$ if there exist a constant $A > 0$ and an integer n_0 such that $|f(n)| \geq A|g(n)|$ for $n \geq n_0$.

We say $f(n) = o(g(n))$ if, given any $\epsilon > 0$, we can find an integer $n_1(\epsilon)$ such that $|f(n)| \leq \epsilon|g(n)|$ for $n \geq n_1(\epsilon)$.

We say $f(n) \sim g(n)$ as $n \to \infty$ if $f(n)/g(n) \to 1$ as $n \to \infty$.

(i) If $f(n) = O(g(n))$, does it follow that $f(n) = \Omega(g(n))$? Give a proof or counterexample. If $f(n) = O(g(n))$, must it necessarily be false that $f(n) = \Omega(g(n))$? Give a proof or counterexample.

(ii) Formulate the 11 other possible questions along the lines of (i) and resolve them. (This is not as tedious as it might seem, since they all have two-sentence answers.)

(iii) In more traditional terms, what does it mean to say that $f(n) = o(1)$ as $n \to \infty$? What does it mean to say that $f(n) = O(1)$ as $n \to \infty$?

(iv) Give an example of a pair of functions $f,\ g : \mathbb{N} \to \mathbb{R}$ such that $f(n),\ g(n) > 0$ for all n and neither of the two relations $f(n) = O(g(n))$, $g(n) = O(f(n))$ holds.

(v) Consider $f,\ g : \mathbb{R} \to \mathbb{R}$. Give definitions along the lines given above for the statement '$f(x) = O(g(x))$ as $x \to \infty$' and for the statement '$f(x) = o(g(x))$ as $x \to 0$'.

(vi) Recall or do exercise K.94, and write the result in the notation of this exercise.

Although the notations introduced in this exercise are very useful for stating results, they are sharp-edged tools which can do as much harm as good in the hands of the inexperienced. I strongly recommend translating any statements involving such notations back into classical form before trying to work with them.

K.161. [**9.2, P, ↑**] Are the following statements true or false? Give reasons.

(i) If $f_j,\ g_j : \mathbb{N} \to \mathbb{R}$ are positive functions such that $f_1(n) = O(g_1(n))$ and $f_2(n) = O(g_2(n))$, then $f_1(n) + f_2(n) = O(g_1(n) + g_2(n))$.

(ii) If $f_j,\ g_j : \mathbb{N} \to \mathbb{R}$ are functions such that $f_1(n) = O(g_1(n))$ and $f_2(n) = O(g_2(n))$, then $f_1(n) + f_2(n) = O(g_1(n) + g_2(n))$.

(iii) If $f_j,\ g_j : \mathbb{N} \to \mathbb{R}$ are positive functions such that $f_1(n) = O(g_1(n))$ and $f_2(n) = O(g_2(n))$, then $f_1(n) + f_2(n) = O\big(\max(g_1(n), g_2(n))\big)$.

(iv) $n! = O(2^{n^2})$ as $n \to \infty$.

(v) $\cos x - 1 + \frac{(\sin x)^2}{2!} = O(x^4)$ as $x \to 0$.

(vi) $\cos x - 1 + \frac{(\sin x)^2}{2!} - \frac{(\sin x)^4}{4!} = O(x^6)$ as $x \to 0$.

K.162. [9.2, P, ↑] (i) If $\alpha > -1$, show that

$$\sum_{r=1}^{n} r^\alpha \sim \frac{n^{\alpha+1}}{\alpha+1}.$$

(ii) State and prove a result similar to (i) for $\alpha = -1$.

(iii) If $\alpha < -1$, show that

$$\sum_{r=n}^{\infty} r^\alpha \sim \frac{n^{\alpha+1}}{(-\alpha)-1}.$$

(iv) Does there exist a modification of (iii) similar to (ii) for $\alpha = -1$? Give brief reasons.

K.163. [9.2, P, ↑] Suppose that $g : \mathbb{R} \to \mathbb{R}$ is twice differentiable and

$$g''(x) = O(x^{-\lambda}) \text{ as } x \to \infty$$

for some $\lambda > 0$. Prove that

$$\int_n^{n+1} (g(x) - g(n+\tfrac{1}{2}))\, dx = O(n^{-\lambda})$$

as $n \to \infty$. Deduce that, if $\lambda > 1$ and $|\int_1^X g(x)\, dx| \to \infty$ as $n \to \infty$, then

$$\sum_{r=1}^{n} g(r+\tfrac{1}{2}) \sim \int_1^{n+1} g(x)\, dx.$$

Deduce the result of Exercise K.156 as a special case.

K.164. [9.2, P] Show that $\int_0^{\pi/2} \log(\sin x)\, dx$ (defined to be the limit of $\int_\epsilon^{\pi/2} \log(\sin x)\, dx$ as $\epsilon \to 0$ through positive values of ϵ) exists. By considering $\int_0^{\pi/2} \log(\sin 2x)\, dx$ and exploiting various symmetries, show that

$$\int_0^{\pi/2} \log(\sin x)\, dx = -\frac{\pi}{2} \log 2.$$

K.165. [9.2, P] (i) Suppose that $f : (0, a] \to \mathbb{R}$ is a positive, monotonic decreasing, continuous function. Show that $\int_0^a f(x)\, dx$ (defined as the limit of $\int_\epsilon^a f(x)\, dx$ as $\epsilon \to 0$ through positive values of ϵ) exists if and only if $(a/N)\sum_{r=1}^N f(ra/N)$ tends to a limit. If these limits exist, show that

$$\frac{a}{N} \sum_{1}^{N} f\left(\frac{ra}{N}\right) \to \int_0^a f(x)\, dx \text{ as } N \to \infty.$$

(ii) Show that the result of part (i) is false if we omit the condition that f is monotonic decreasing.

(iii) Let $\omega = 2\pi/N$. By multiplying both sides of the equation by $z-1$, or otherwise, show that

$$\bigstar \qquad \sum_{r=0}^{N-1} z^r = \prod_{r=1}^{N-1} (z - \omega^r)$$

for all $z \neq 1$. Explain why $\bigstar$ continues to hold when $z = 1$, and deduce that

$$\prod_{r=1}^{N-1} \sin\frac{r\pi}{N} = \frac{N}{2^{N-1}}.$$

(iv) Use the result of (i) to show that $\int_0^{\pi/2} \log(\sin x)\,dx$ exists and has value $-\frac{\pi}{2}\log 2$.

K.166. [9.2, P] The function $f : [0,1] \to \mathbb{R} \cup \{\infty\}$ is defined as follows. Each $0 < x < 1$ is expressed as a decimal $x = .a_1a_2\ldots a_k\ldots$ with a_j an integer $0 \leq a_j \leq 9$. In ambiguous cases we use the non-terminating form. We put $f(x) = k$ if the kth digit a_k is the first digit 7 to occur in the expansion, if there is any digit 7. If there is none (and also at $x = 0$ and $x = 1$) we put $f(x) = \infty$. Prove that the function $f_X : [0,1] \to \mathbb{R}$ defined by $f_X(x) = \min(X, f(x))$ is integrable for all $X \geq 0$ and that

$$\int_0^1 f_X(x)\,dx \to 10$$

as $X \to \infty$.

Does it make any difference if we redefine f so at those points x where we previously had $f(x) = \infty$ we now have $f(x) = 0$?

K.167. [9.2, H] This exercise investigates another way of defining improper integrals.

(i) If $f : [a,b] \to \mathbb{R}$ and $R,\ S \geq 0$, we write

$$f_{R,S}(x) = \begin{cases} R & \text{if } f(x) \geq R, \\ f(x) & \text{if } R > f(x) > -S, \\ -S & \text{if } -S \geq f(x). \end{cases}$$

If $f|_{R,S} \in \mathcal{R}[a,b]$ for all $R,\ S \geq 0$ and there exists an L such that, given any $\epsilon > 0$, there exists an $R_0 > 0$ with

$$\left| \int_a^b f_{R,S}(x)\,dx - L \right| < \epsilon$$

for all $R,\ S \geq R_0$, then we say that $\mathcal{LV} \int_a^b f(x)\,dx$ exists with value L.

Show that, if $f|_{R,S} \in \mathcal{R}[a,b]$ for all R, $S \geq 0$, then $\mathcal{LV} \int_a^\infty f(x)\,dx$ exists if and only if $\mathcal{LV} \int_a^\infty |f(x)|\,dx$ exists.

(ii) Suppose that $f : [a,b] \to \mathbb{R}$ is continuous except at b and $f(x) \to \infty$ as $x \to b$. Show that the following two statements are equivalent.

(A) There exists an L such that, given any $\epsilon > 0$, there exists an $R_0 > 0$ with

$$\left| \int_a^b f_{R,S}(x)\,dx - L \right| < \epsilon$$

for all R, $S > R_0$.

(B) There exists an L' such that, given any $\epsilon > 0$, there exists a $\delta_0 > 0$ with

$$\left| \int_a^{b-\eta} f(x)\,dx - L' \right| < \epsilon$$

for all $0 < \eta < \delta_0$.

Show further that, if (A) and (B) hold, then $L = L'$.

(iii) Suppose that $f : [a,b] \to \mathbb{R}$ is continuous except at possibly at b. Show that statement (A) implies statement (B) but that the converse is false. (Think about Exercise K.158 and change of variable if you need a hint.)

[**Warning:** If you ever need to take the contents of this question seriously, you should switch to the Lebesgue integral.]

K.168. [9.3, P, S] By considering both orders of integration in

$$\int_0^X \int_a^b e^{-tx}\,dt\,dx,$$

where $b > a > 0$ and $X > 0$, show that

$$\int_0^X \frac{e^{-ax} - e^{-bx}}{x}\,dx = \int_a^b \frac{1 - e^{-tX}}{t}\,dt.$$

Show that

$$\int_a^b \frac{e^{-tX}}{t}\,dt \leq e^{-aX} \log\left(\frac{b}{a}\right),$$

and hence find

$$\int_0^\infty \frac{e^{-ax} - e^{-bx}}{x}\,dx.$$

K.169. [9.3, T] Suppose that $f : \mathbb{R}^2 \to \mathbb{R}$ is continuous. By using results on the differentiation of integrals, which should be quoted exactly, show that

$$\frac{d}{dt} \int_a^t \left(\int_c^d f(u,v)\,dv \right) du = \frac{d}{dt} \int_c^d \left(\int_a^t f(u,v)\,du \right) dv.$$

Deduce that

$$\int_a^b \left(\int_c^d f(u,v)\, dv \right) du = \int_c^d \left(\int_a^b f(u,v)\, du \right) dv.$$

In what way is this result weaker than that of Theorem 9.20?

K.170. [9.3, H] (i) Reread Exercise 5.32.

(ii) Let us define $f : [0,1] \times [0,1] \to \mathbb{R}$ by

$$\begin{aligned} f(x,y) &= 2^{2n} && \text{if } 2^{-n} < x < 2^{-n+1},\ 2^{-n} < y < 2^{-n+1} \text{ and } n \geq 1,\ n \in \mathbb{Z}, \\ f(x,y) &= -2^{2n+1} && \text{if } 2^{-n} < x < 2^{-n+1},\ 2^{-n-1} < y < 2^{-n} \text{ and } n \geq 1,\ n \in \mathbb{Z}, \\ f(x,y) &= 0 && \text{otherwise.} \end{aligned}$$

Sketch the sets where $f(x,y)$ takes various values. Show that the integral $\int_0^1 f(x,y)\, dx$ is well defined and calculate it for all values of y. Show that the integral $\int_0^1 f(x,y)\, dy$ is well defined, and calculate it for all values of x. Show that the two integrals

$$\int_0^1 \int_0^1 f(x,y)\, dx\, dy \text{ and } \int_0^1 \int_0^1 f(x,y)\, dy\, dx$$

are well defined, and calculate them. Show that the two integrals are not equal.

(iii) Suppose that $u : \mathbb{R} \to \mathbb{R}$ is continuous function with $u(t) = 0$ for $t < 1/4$ and $t > 3/4$, $u(t) \geq 0$ for all t and $\int_0^1 u(t)\, dt = 1$. Show that the function $g : [0,1]^2 \to \mathbb{R}$

$$g(x,y) = \sum_{n=1}^{\infty} \left(2^{2n} u(2^n x - 1) u(2^n y - 1) - 2^{2n+1} u(2^n x - 1) u(2^{n+1} y - 1) \right)$$

is well defined. Show that the integral $\int_0^1 f(x,y)\, dx$ is well defined, and calculate it for all values of y. Show that the integral $\int_0^1 f(x,y)\, dy$ is well defined, and calculate it for all values of x. Show that the two integrals

$$\int_0^1 \int_0^1 g(x,y)\, dx\, dy \text{ and } \int_0^1 \int_0^1 g(x,y)\, dy\, dx$$

are well defined but not equal. Show that g is continuous except at $(0,0)$, that the function $x \mapsto g(x,y)$ is an everywhere continuous function for all y, and that the function $y \mapsto g(x,y)$ is an everywhere continuous function for all x. Why does Theorem 9.20 fail?

(iv) The traditional counterexample is the following. Define the function $h : [0,1]^2 \to \mathbb{R}$ by $h(0,0) = 0$ and

$$h(x,y) = \frac{xy(x^2 - y^2)}{(x^2 + y^2)^3}$$

otherwise. Show that the integral $\int_0^1 h(x,y)\,dx$ is well defined, and calculate it for all values of y (you may find the substitution $w = x^2 + y^2$ useful). Show that the integral $\int_0^1 f(x,y)\,dy$ is well defined, and calculate it for all values of x. Show that the two integrals

$$\int_0^1 \int_0^1 h(x,y)\,dx\,dy \text{ and } \int_0^1 \int_0^1 h(x,y)\,dy\,dx$$

are well defined but not equal. Show that g is continuous except at $(0,0)$, that the function $x \mapsto g(x,y)$ is an everywhere continuous function for all y, and that the function $y \mapsto g(x,y)$ is an everywhere continuous function for all x.

(v) Which of examples (ii) and (iv) do you consider 'more natural'. Which do you feel you 'understand better'? Which is 'better'. Why? (Of course you can refuse to answer questions like this on the grounds that they are not mathematical, but I think you will lose something.)

K.171. (Interchange of infinite integrals.) **[9.3, T]** (i) Suppose that $g : [0,\infty) \to \mathbb{R}$. is continuous and $|g(x)| \leq A(1+x^2)^{-1}$ for all x. Show that $\int_0^\infty g(x)\,dx$ exists, that

$$\left|\int_0^\infty g(x)\,dx\right| \leq 2A,$$

and

$$\left|\int_0^R g(x)\,dx - \int_0^\infty g(x)\,dx\right| \leq AR^{-1},$$

for $R \geq 1$. [If you do not wish to use knowledge of $\tan^{-1}$, simply observe that $(1+x^2)^{-1} \leq 1$ for $0 \leq x \leq 1$ and $(1+x^2)^{-1} \leq x^{-2}$ for $x \geq 1$.]

(ii) Suppose that $f : [0,\infty)^2 \to \mathbb{R}$ is continuous and satisfies the condition $|f(x,y)| \leq A(1+x^2)^{-1}(1+y^2)^{-1}$ for all $(x,y) \in \mathbb{R}^2$. Show that $\int_0^\infty \int_0^\infty f(x,y)\,dx\,dy$ and $\int_0^\infty \int_0^\infty f(x,y)\,dy\,dx$ exist and that

$$\int_0^\infty \int_0^\infty f(x,y)\,dx\,dy = \int_0^\infty \int_0^\infty f(x,y)\,dy\,dx.$$

(iii) (Parts (iii) and (iv) run along similar lines to parts (i) and (ii) of Exercise K.170. You do not need to have done Exercise K.170 to do this exercise, but if you have, it is worth noting the similarities and the differences.) Reread Exercise 5.32. Let b_{rs} be as in part (ii) of that exercise. Define $F : [0,\infty)^2 \to \mathbb{R}$ by

$$F(x,y) = b_{rs} \text{ for } r-1 \leq x < r \text{ and } s-1 \leq y < s.$$

Show that $\int_0^\infty \int_0^\infty F(x,y)\,dx\,dy$ and $\int_0^\infty \int_0^\infty F(x,y)\,dy\,dx$ exist and that

$$\int_0^\infty \int_0^\infty F(x,y)\,dx\,dy \neq \int_0^\infty \int_0^\infty F(x,y)\,dy\,dx.$$

(iv) By modifying the construction in part (iii), or otherwise, find a continuous function $G : [0,\infty)^2 \to \mathbb{R}$ such that $\sup_{x^2+y^2\geq R} |G(x,y)| \to 0$ as $R \to \infty$ but

$$\int_0^\infty \int_0^\infty G(x,y)\,dx\,dy \neq \int_0^\infty \int_0^\infty G(x,y)\,dy\,dx.$$

K.172. [9.3, H, S] It is a weakness of the Riemann integral that there is no Fubini theorem for general Riemann integrable functions.

Let $A = [0,1] \times [0,1]$ and define $f : A \to \mathbb{R}$ by $f(x,0) = 1$ if x is rational, $f(x,y) = 0$ otherwise. By finding appropriate dissections, show that f is Riemann integrable. Explain why $F_2(y) = \int_0^1 f(t,y)\,dt$ is not defined for all y.

[See also the next exercise.]

K.173. [9.3, H] We say that a sequence $\mathbf{e}_1$, $\mathbf{e}_2$, ... of points in $A = [0,1] \times [0,1]$ is dense in A if, given $\mathbf{x} \in A$, we can find a $j \geq 1$ with $\|\mathbf{x} - \mathbf{e}_j\| < \epsilon$.

(i) By considering $\mathbb{Q} \times \mathbb{Q}$, or by direct construction, or otherwise, show that we can find a sequence $\mathbf{e}_1$, $\mathbf{e}_2$, ... of points which is dense in A.

(ii) By induction, or otherwise, show that we can find a sequence $\mathbf{w}_1 = (u_1, v_1)$, $\mathbf{w}_2 = (u_2, v_2)$, ... of points in A such that $\|\mathbf{w}_k - \mathbf{e}_k\| < 1/k$ and $u_i \neq u_j$, $v_i \neq v_j$ whenever $i \neq j$. Show that the sequence $\mathbf{w}_1$, $\mathbf{w}_2$, ... is dense in A.

(iii) Define $f : A \to \mathbb{R}$ by $f(\mathbf{x}) = 1$ if $\mathbf{x} = \mathbf{w}_j$ for some j, $f(\mathbf{x}) = 0$ otherwise. Show that, if x is fixed, $f(x,y) \neq 1$ for at most one value of y. Conclude that $F_1(x) = \int_0^1 f(x,s)\,ds$ exists and takes the value 0 for all y. Thus $\int_0^1 \left(\int_0^1 f(x,y)\,dy\right)\,dx$ exists. Similarly, $\int_0^1 \left(\int_0^1 f(x,y)\,dx\right)\,dy$ exists. Show, however, that f is not Riemann integrable.

K.174. [9.3, H] The object of this exercise is to construct a bounded open set in $\mathbb{R}$ which does not have Riemann length. Our construction will take place within the closed interval $[-2,2]$, and all references to dissections and upper sums and so forth will refer to this interval.

(i) Enumerate the rationals in $[0,1]$ as a sequence y_1, y_2, ... Let $U_j = \bigcup_{k=1}^j (y_k - 2^{-k-4}, y_k + 2^{-k-4})$ and $U = \bigcup_{k=1}^\infty (y_k - 2^{-k-4}, y_k + 2^{-k-4})$. Explain why U is open and $I^*(\mathbb{I}_U) \geq 1$.

(ii) We wish to show that $I_*(\mathbb{I}_U) < 1$, so that $\mathbb{I}_U$ is not Riemann integrable and U has no Riemann length. To this end, suppose, if possible, that $I_*(\mathbb{I}_U) \geq 1$. Explain why this means that we can find disjoint closed intervals $I_r = [a_r, b_r]$ $[1 \leq r \leq N]$ lying inside $[-2,2]$ such that

$$\sum_{r=1}^N (b_r - a_r) \geq 15/16 \text{ yet } \bigcup_{r=1}^N [a_r, b_r] \subset U.$$

(iii) Let $K_j = ([-2,2] \setminus U_j) \cap \bigcup_{r=1}^{N} [a_r, b_r]$. Explain why K_j is a closed bounded set. By considering the total length of the intervals making up U_j, show that $K_j \neq \varnothing$. Use Exercise 4.48 to show that

$$([-2,2] \setminus U) \cap \bigcup_{r=1}^{N} [a_r, b_r] \neq \varnothing.$$

Deduce, by reductio ad absurdum, that $I_*(\mathbb{I}_U) < 1$ and U has no Riemann length.

(iv) Show that there is a closed bounded set in $\mathbb{R}$ which does not have Riemann length. (This is a one-line argument.)

(v) Show that there are bounded closed sets and bounded open sets in $\mathbb{R}^2$ which do not have Riemann area.

[The first example of this type was found by Henry Smith. Smith was a major pure mathematician at a time and place (19th century Oxford) not particularly propitious for such a talent. He seems to have been valued more as a good College and University man than for anything else[12].]

K.175. [9.3, H, ↑] (This continues with the ideas of Exercise K.174 above.)

(i) If $(a,b) \subseteq [-2,2]$, find a sequence of continuous functions $f_n : [-2,2] \to \mathbb{R}$ such that $0 \leq f_n(x) \leq \mathbb{I}_{(a,b)}(x)$ for all n and $f_n(x) \to \mathbb{I}_{(a,b)}(x)$ as $n \to \infty$ for all $x \in [a,b]$.

(ii) If U_j and U are as in Exercise K.174, find a sequence of continuous functions $f_n : [-2,2] \to \mathbb{R}$ such that $0 \leq f_n(x) \leq \mathbb{I}_{U_n}(x)$ for all n and $f_n(x) \to \mathbb{I}_U(x)$ as $n \to \infty$ for all $x \in [a,b]$.

(iii) Show that it is possible to find a sequence of bounded continuous functions $h_n : [0,1] \to \mathbb{R}$ and a function $h : [0,1] \to \mathbb{R}$ which is not Riemann integrable such that $h_n(x) \to h(x)$ as $n \to \infty$, for each $x \in [0,1]$.

K.176. [9.4, T] Let $f : [a,b] \to \mathbb{R}$ be an increasing function.

(i) If $t_j \in [a,b]$ and $t_1 \leq t_2 \leq t_3 \leq \dots$, explain why $f(t_j)$ tends to a limit.

(ii) Suppose $x \in [a,b]$. If $t_j,\ s_j \in [a,b]$,

$$t_1 \leq t_2 \leq t_3 \leq \dots, \text{ and } s_1 \leq s_2 \leq s_3 \leq \dots,$$

$t_j \to x$, $s_j \to x$ and $t_j,\ s_j < x$ for all j, show that $\lim_{j\to\infty} f(t_j) = \lim_{j\to\infty} f(s_j)$. Show that the condition $t_j,\ s_j < x$ for all j cannot be omitted.

(iii) Show that, if $x \in (a,b]$ and $t \to x$ through values of t with $a \leq t < x$, then $f(t)$ tends to a 'left limit' $f(x-)$.

(iv) State and prove the appropriate result on 'right limits'.

[12] He even gave extra teaching on Sunday afternoon, telling his students that 'It was lawful on the Sabbath day to pull an ass out of the ditch'.

(v) If we write $J(x) = f(x+) - f(x-)$ for the 'jump' at x, show that $J(x) \geq 0$. Show that, if $a < x_1 < x_2 < \cdots < x_{N-1} < x_N < b$, then

$$\sum_{r=1}^{N} J(x_r) \leq f(b) - f(a).$$

(vi) Let us write $E_k = \{x \in (a, b) : J(x) \geq 1/k\}$. Use (v) to show that E_k is finite. Deduce that $E = \bigcup_{k=1}^{\infty} E_k$ is countable, and conclude that an increasing function is continuous except at a countable set of points.

(vii) If we define $\tilde{f}(x) = f(x+)$ when $x \in E \cup \{a\}$, $\tilde{f}(x) = f(x)$ otherwise, show that $\tilde{f} : [a, b] \to \mathbb{R}$ is a right continuous increasing function. If we define $\bar{f}(x) = f(x+)$ for all $x \in [a, b)$, $\bar{f}(b) = f(b)$, state and prove the relation between $\tilde{f}$ and $\bar{f}$.

(viii) If $g : \mathbb{R} \to \mathbb{R}$ is an increasing function, show that g is continuous except at a countable set of points E, say. If we define $\tilde{g}(x) = f(x+)$ when $x \in E$, $\tilde{g}(x) = g(x)$ otherwise, show that $\tilde{g} : \mathbb{R} \to \mathbb{R}$ is a right continuous increasing function.

(ix) Not all discontinuities are as well behaved as those of increasing functions. If $g : [-1, 1] \to \mathbb{R}$ is defined by $g(x) = \sin(1/x)$ for $x \neq 0$, show that there is no choice of value for $g(0)$ which will make g left or right continuous at 0.

K.177. [9.4, P] Define $H : \mathbb{R} \to \mathbb{R}$ by $H(t) = 0$ if $t < 0$, $H(t) = 1$ if $t \geq 0$.

(i) If $f(t) = \sin(1/t)$ for $t > 0$, and $f(t) = 0$ for $t < 0$, show that, whatever value we assign to $f(0)$, the function f is not Riemann–Stieltjes integrable with respect to H.

(ii) If $f(t) = \sin(1/t)$ for $t < 0$, and $f(t) = 0$ for $t > 0$, show that f is Riemann–Stieltjes integrable with respect to H if and only if $f(0) = 0$.

(iii) By reflecting on parts (i) and (ii), or otherwise, formulate and prove a necessary and sufficient condition for a bounded function $f : \mathbb{R} \to \mathbb{R}$ to be Riemann–Stieltjes integrable with respect to H.

K.178. [9.4, P, ↑] (i) Show that, if $u : \mathbb{R} \to \mathbb{R}$ is bounded, increasing and right continuous, then we can write

$$u(x) = U(x) + \sum_{j=1}^{\infty} \lambda_j H(x - x_j),$$

where $U : \mathbb{R} \to \mathbb{R}$ is bounded, increasing and continuous, H is the right continuous version of the Heaviside function, $\lambda_j \geq 0$ for all j and $\sum_{j=1}^{\infty} \lambda_j$ converges.

(ii) Show that, if $G : \mathbb{R} \to \mathbb{R}$ is bounded, increasing and right continuous and $f : \mathbb{R} \to \mathbb{R}$ is bounded, increasing and left continuous, then f is Riemann–Stieltjes integrable with respect to G.

(iii) Suppose $G : \mathbb{R} \to \mathbb{R}$ is bounded, increasing and right continuous and $f : \mathbb{R} \to \mathbb{R}$ is bounded, increasing and right continuous. Find necessary and sufficient conditions for f to be Riemann–Stieltjes integrable with respect to G, and prove that your statement is correct.

K.179. [9.4, P, ↑] (i) Let H be the Heaviside function discussed in the previous exercises. Give an example of a continuous function $f : \mathbb{R} \to \mathbb{R}$ such that $f(t) \geq 0$ for all t and f is not identically zero but

$$\int_{\mathbb{R}} f(x)\, dH(x) = 0.$$

(ii) For the remainder of the exercise $G : \mathbb{R} \to \mathbb{R}$ will always be a bounded, increasing and right continuous function. By reflecting on part (i) and the proof of Exercise 8.34, formulate a necessary and sufficient condition on G so that the following theorem holds.

If $f : \mathbb{R} \to \mathbb{R}$ is continuous and bounded, $f(t) \geq 0$ for all t and

$$\int_{\mathbb{R}} f(x)\, dG(x) = 0,$$

then $f(t) = 0$ for all t.

Prove that your condition is, indeed, necessary and sufficient.

(iii) By reflecting on part (ii), formulate a necessary and sufficient condition on G so that the following theorem holds.

If $f : \mathbb{R} \to \mathbb{R}$ is continuous and bounded and

$$\int_{\mathbb{R}} f(x)g(x)\, dG(x) = 0,$$

whenever $g : \mathbb{R} \to \mathbb{R}$ is continuous and bounded, it follows that $f(t) = 0$ for all t.

Prove that your condition is, indeed, necessary and sufficient.

K.180. (Bounded variation.) [9.4, T] We say that a function $F : [a, b] \to \mathbb{R}$ is of *bounded variation* if there exists a constant K such that, whenever $a = x_0 \leq x_1 \leq \cdots \leq x_n = b$, we have

$$\sum_{j=1}^{n} |f(x_j) - f(x_{j-1})| \leq K.$$

(i) Show that any bounded increasing function $G : [a, b] \to \mathbb{R}$ is of bounded variation.

(ii) Show that the sum of two functions of bounded variation is of bounded variation. Deduce, in particular, that the difference $F - G$ of two increasing functions $F, G : [a,b] \to \mathbb{R}$ is of bounded variation.

(iii) Show that any function of bounded variation is bounded.

(iv) Let $f : [-1,1] \to \mathbb{R}$ be defined by $f(x) = x\sin(1/x)$ for $x \neq 0$, $f(0) = 0$. Show that f is continuous everywhere on $[-1,1]$, but not of bounded variation. [For variations on this theme, consult Exercise K.183.]

K.181. [9.4, T, ↑] (This continues the previous exercise.) In this exercise it will be useful to remember that, to prove $A \leq B$, it is sufficient to prove that $A \leq B + \epsilon$ for all $\epsilon > 0$. Suppose $F : [a,b] \to \mathbb{R}$ is of bounded variation.

(i) Explain why we can define $V_F : [a,b] \to \mathbb{R}$ by taking $V_F(t)$ to be the supremum of all sums

$$\sum_{j=1}^{n} |F(x_j) - F(x_{j-1})|,$$

where $a = x_0 \leq x_1 \leq \cdots \leq x_n = t$ and $n \geq 1$.

Explain why we can define $F_+ : [a,b] \to \mathbb{R}$ by taking $F_+(t)$ to be the supremum of all sums

$$\sum_{j=0}^{n} (F(y_j) - F(x_j)),$$

where $a = x_0 \leq y_0 \leq x_1 \leq y_1 \leq x_2 \leq y_2 \leq \cdots \leq x_n \leq y_n = t$ and $n \geq 0$.

We define $F_- : [a,b] \to \mathbb{R}$ by taking $F_-(t)$ to be the supremum of all sums

$$\sum_{j=1}^{n} (F(y_{j-1}) - F(x_j)),$$

where $a = x_0 \leq y_0 \leq x_1 \leq y_1 \leq x_2 \leq y_2 \leq \cdots \leq x_n \leq y_n = t$ and $n \geq 1$.

(ii) Show that F_+ and F_- are increasing functions and

$$V_F = F_+ + F_-, \quad F = F(a) + F_+ - F_-.$$

In particular, we have shown that every function of bounded variation on $[a,b]$ is the difference of two increasing functions. We call the function $V_F(t)$ the *total variation* of F on $[a,t]$.

(iii) Suppose that we have two increasing functions $G_+, G_- : [a,b] \to \mathbb{R}$ such that $G_+(a) = G_-(a) = 0$ and $F = F(a) + G_+ - G_-$. Show that $G_+(t) \geq F_+(t)$ and $G_-(t) \geq F_-(t)$ for all $t \in [a,b]$.

Show, more precisely, that $G_+ - F_+$ and $G_- - F_-$ are increasing functions.

K.182. [9.4, T, ↑] We use the notation and hypotheses of the previous exercise. We need the results of Exercise K.176, which the reader should either do or reread before continuing.

(i) By Exercise K.176 there is a countable (possibly finite or empty) set E_+ such that

$$\lim_{t \to e,\ t>e} F_+(t) - \lim_{t \to e,\ t<e} F_+(t) > 0$$

for $e \in E_+$, and f is continuous at each $x \notin E_+$. The same result holds with F_+ and E_+ replaced by F_- and E_-. By using Exercise K.181 (iii), or otherwise, show that $E_+ \cap E_- = \varnothing$.

(ii) Use part (i) to show that F_+ and F_- are right continuous if F is. Show that F_+ and F_- are continuous if F is.

(iii) Suppose that F is continuously differentiable. Show that

$$F_+(t) = \int_a^t \max(F'(x), 0)\, dx, \ \ F_-(t) = -\int_a^t \min(F'(x), 0)\, dx.$$

[In setting out your proof, remember that a continuous function may change sign infinitely often in an interval.]

Conclude that F_+ and F_- are continuously differentiable if F is.

K.183. [9.4, P, ↑] (This exercise is not required for later parts of the sequence.) If α and β are real, let $f_{\alpha\beta} : [-1, 1] \to \mathbb{R}$ be defined by $f_{\alpha\beta}(x) = |x|^\alpha \sin(|x|^\beta)$ for $x \neq 0$, $f_{\alpha\beta}(0) = 0$.

(i) For which values of α and β is $f_{\alpha\beta}$ of bounded variation?

(ii) For which values of α and β is $f_{\alpha\beta}$ everywhere continuous?

(iii) For which values of α and β is $f_{\alpha\beta}$ everywhere differentiable?

(iv) For which values of α and β is $f_{\alpha\beta}$ everywhere differentiable with continuous derivative?

K.184. [9.4, T, ↑] (This exercise closely echos Exercise K.181.) We say that a function $G : \mathbb{R} \to \mathbb{R}$ is of bounded variation if there exists a constant K such that, whenever $x_0 < x_1 < \cdots < x_n$, we have

$$\sum_{j=1}^{n} |G(x_j) - G(x_{j-1})| \leq K.$$

(i) Show that a function $G : \mathbb{R} \to \mathbb{R}$ is of bounded variation if and only if it is the difference of two bounded increasing functions.

(ii) Consider a function $G : \mathbb{R} \to \mathbb{R}$. Show that its restriction $G|_{[a,b]}$ is of bounded variation for every closed interval $[a, b]$ if and only if it is the difference of two increasing functions.

(iii) Show that, if a function $G : \mathbb{R} \to \mathbb{R}$ is of bounded variation with $G(t) \to 0$ as $t \to \infty$, then there exist unique bounded increasing functions $G_+, G_- : \mathbb{R} \to \mathbb{R}$ such that $G_+(t)$, $G_-(t) \to 0$ as $t \to -\infty$ and $G = G_+ - G_-$ with the property that, if F_+, $F_- : \mathbb{R} \to \mathbb{R}$ are increasing functions with $G = F_+ - F_-$, then $F_+ - G_+$ and $F_- - G_-$ are increasing.

(iv) Suppose that G, G_+, G_- are as in (iii). Show that G_+ and G_- are right continuous if G is. Show that G_+ and G_- are continuous if G is. Show that G_+ and G_- are continuously differentiable if G is.

(v) We write $G_V = G_+ + G_-$. Identify G_V as the supremum of a certain set of sums, and prove that your identification is correct.

K.185. [9.4, T, ↑] (i) Suppose that $F, G : \mathbb{R} \to \mathbb{R}$ are bounded increasing right continuous functions. Show that a bounded function $f : \mathbb{R} \to \mathbb{R}$ is Riemann–Stieltjes integrable with respect to both F and G if and only if it is Riemann–Stieltjes integrable with respect to $F + G$. If f is Riemann–Stieltjes integrable with respect to both F and G, show that

$$\int_{\mathbb{R}} f(x)\, d(F+G)(x) = \int_{\mathbb{R}} f(x)\, dF(x) + \int_{\mathbb{R}} f(x)\, dG(x).$$

(ii) Let F_1, F_2, G_1, $G_2 : \mathbb{R} \to \mathbb{R}$ be bounded increasing right continuous functions with $F_1 - G_1 = F_2 - G_2$. If a bounded function $f : \mathbb{R} \to \mathbb{R}$ is Riemann–Stieltjes integrable with respect to F_1, F_2, G_1 and G_2, show that

$$\int_{\mathbb{R}} f(x)\, dF_1(x) - \int_{\mathbb{R}} f(x)\, dG_1(x) = \int_{\mathbb{R}} f(x)\, dF_2(x) - \int_{\mathbb{R}} f(x)\, dG_2(x).$$

K.186. [9.4, T, ↑] We use the notation and results of Exercise K.184. Although we do not use the results of Exercise K.185 directly, they show that the path chosen is a reasonable one.

If $G : \mathbb{R} \to \mathbb{R}$ is right continuous of bounded variation, then we know that G_+ and G_- are increasing bounded right continuous functions. We say that a bounded function $f : \mathbb{R} \to \mathbb{R}$ is Riemann–Stieltjes integrable with respect to G if it is Riemann–Stieltjes integrable with respect to both G_+ and G_-. We write

$$\int_{\mathbb{R}} f(x)\, dG(x) = \int_{\mathbb{R}} f(x)\, dG_+(x) - \int_{\mathbb{R}} f(x)\, dG_-(x).$$

Develop the theory of this extended integral along the lines of Section 9.4, starting at Exercise 9.38 (ii) and ending at Exercise 9.44. Note that, though this is easy, you must be careful to make the right adjustments. Thus, for example, the conclusion of the result corresponding to Exercise 9.38 (iii) is

$$\left| \int_{\mathbb{R}} f(x)\, dG(x) \right| \leq K \lim_{t \to \infty} G_V(t),$$

and, in Exercise 9.44 (ii), we can choose λ_j positive or negative.

K.187. (The trigonometric functions via arc length.) **[9.5, T]** In Section 5.5 we developed the theory of the trigonometric functions via their differential equations. In that development, the trigonometric functions come first and angle is defined in terms of those functions (see Exercise 5.56). Here we reverse the process.

(i) Consider the map $\gamma : [-1, 1] \to \mathbb{R}^2$ given by

$$\gamma(y) = ((1 - y^2)^{1/2}, y),$$

where we take the positive square root. Convince yourself that this represents an arc of the unit circle. If $1 > y \geq 0$, we define the angle $\theta(y)$ subtended at the origin by $\gamma(0)$ and $\gamma(y)$ to be length of the curve γ between those two points. If $0 \geq y > -1$, we define the angle $\theta(y)$ subtended at the origin by $\gamma(0)$ and $\gamma(y)$ to be minus the length of the curve γ between those two points.

Show that

$$\theta(y) = \int_0^y \frac{1}{(1 - t^2)^{1/2}}\, dt$$

for $-1 < y < 1$. Show that there is a real number ω such that $\theta(y) \to \omega$ as $y \to 1$ through values of $y < 1$, and $\theta(y) \to -\omega$ as $y \to -1$ through values of $y > -1$. We define $\theta(1) = \omega$ and $\theta(-1) = -\omega$.

(ii) Show that θ is a strictly increasing continuous function on $[-\omega, \omega]$. We may thus define a function $\sin : [-\omega, \omega] \to \mathbb{R}$ by

$$\sin t = \theta^{-1}(t).$$

Show that sin is once differentiable on $(-\omega, \omega)$ with $\sin' t = (1 - (\sin t)^2)^{1/2}$. Show that sin is twice differentiable on $(-\omega, \omega)$ with $\sin'' t = -\sin t$.

(iii) We now define $\sin t = \sin(2\omega - t)$ for $t \in [\omega, 3\omega]$. Show (paying particular attention to behaviour at ω) that $\sin : [-\omega, 3\omega] \to \mathbb{R}$ is a well defined continuous function and that sin is twice differentiable on $(-\omega, 3\omega)$ with $\sin'' t = -\sin t$.

(iv) Show that we can extend the definition of sin to obtain a twice differentiable function $\sin : \mathbb{R} \to \mathbb{R}$ with $\sin'' t = -\sin t$.

(v) Show that, if we define $\cos t = \sin(t - \omega)$, then $\cos^2 t + \sin^2 t = 1$ for all t and $0 \leq \cos s \leq 1$ for all $0 \leq s \leq \omega$. Explain why this gives the 'right geometrical meaning' to $\cos t$ for $0 \leq t \leq \omega$. Show also that, if we set $\pi = 2\omega$, this gives the right geometrical meaning for π in terms of the circumference formula for the circle.

K.188. (The trigonometric functions via area.) **[9.5, T, ↑]** We might argue that it is better to base our definition of angle on area rather than the more delicate concept of length. Consider the part of the unit circle shown

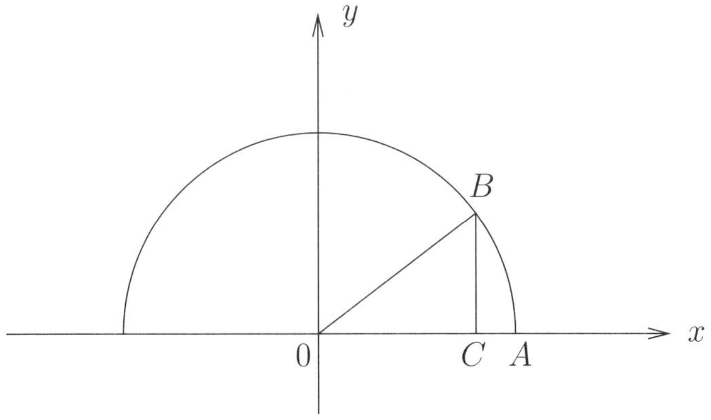

Figure K.3. The area of a sector

in Figure K.3 with A the point $(1, 0)$, B the point $(x, (1-x^2)^{1/2})$ and C the point $(x, 0)$. Explain why the area of the area of the sector OAB is

$$\theta(x) = \frac{x(1-x^2)^{1/2}}{2} + \int_x^1 (1-t^2)^{1/2}\,dt.$$

Use the definition of θ just given to obtain an appropriate definition of $\cos t$ in an appropriate range $[0, \omega]$. Show that cos has the properties we should expect. Extend your definition so that cos is defined on the whole real line and has the properties we expect.

K.189. [9.5, G, M!] (You should treat this exercise informally.) Here are some simple examples of a phenomenon which has to be accepted in any reasonably advanced account of length, area and volume.

(i) We work in $\mathbb{R}^2$. Let

$$D_n = \{(x, y) \,:\, (x-n)^2 + y^2 \leq (n+1)^{-2}\}.$$

Sketch $E = \bigcup_{n=1}^{\infty} D_n$ and convince yourself that the area of E ought to be $\sum_{n=1}^{\infty} \operatorname{area} D_n = \pi \sum_{n=1}^{\infty} (n+1)^{-2}$, which is finite, but that the length of the boundary of E ought to be $\sum_{n=1}^{\infty} \operatorname{circumference} D_n = 2\pi \sum_{n=1}^{\infty} (n+1)^{-1}$, which is infinite.

(ii) Construct a similar example in $\mathbb{R}^3$ involving volume and area.

(iii) We work in $\mathbb{C}$. Consider

$$E = \{0\} \cup \bigcup_{n=1}^{\infty} \bigcup_{r=1}^{n} \{z \,:\, |z - 2^{-n}\exp(2\pi i r/n)| \leq 2^{-n-3}\}.$$

Convince yourself that it is reasonable to say that E has finite area but that E has boundary of infinite length.

(iv) (Torricelli's trumpet, often called Gabriel's horn.) This example goes back historically to the very beginning of the calculus. Consider the volume of revolution E obtained by revolving the curve $y = 1/x$ for $x \geq 1$ around the x axis. Thus

$$E = \{(x, y, z) \in \mathbb{R}^3 : y^2 + z^2 \leq x^{-2},\ x \geq 1\}.$$

We write $f(x) = 1/x$. Convince yourself that the volume of E is given by $\pi \int_1^\infty f(x)^2\,dx$, which is finite[13], but the curved surface of E has area $2\pi \int_1^\infty f(x)(1 + f'(x)^2)^{1/2}\,dx$, which is infinite.

Thus, if we have a trumpet in the form of the curved surface of E, we would 'clearly' require an infinite amount of paint to paint its inside (since this has infinite area). Equally 'clearly', since E has finite volume, we can fill the trumpet up with a finite amount of paint and empty it again leaving the inside nicely painted!

Note that by 'winding our trumpet like a ball of knitting" we can keep it within a bounded set (compare parts (i) and (ii)).

(v) (Thor's drinking horn.) As a variation on these themes, consider the volumes of revolution K obtained by revolving the curve $y = x^{-1/2}$ for $x \geq 1$ around the x axis and K' obtained by revolving the curve $y = x^{-1/2} + x^{-1}$ for $x \geq 1$ around the x axis. Thus

$$K = \{(x, y, z) \in \mathbb{R}^3 : y^2 + z^2 \leq x^{-1},\ x \geq 1\} \text{ and}$$
$$K' = \{(x, y, z) \in \mathbb{R}^3 : y^2 + z^2 \leq (x^{-1/2} + x^{-1})^2,\ x \geq 1\}.$$

Show that K has infinite volume but that $K' \setminus K$ has finite volume. Sluse, who was the first to discover such an object, wrote to Huygens that he could design a 'glass of small weight which the hardest drinker could not empty'.

K.190. [9.5, T] In Section 9.5 we discussed how to define the line integral

$$\int_\gamma f(\mathbf{x})\,ds.$$

In this exercise we discuss the vector line integral

$$\int_\gamma f(\mathbf{x})\,d\mathbf{s}.$$

[13]This result is due to Torricelli. In modern terms, this was the first infinite integral ever to be considered. Torricelli first evaluated the integral by 'slicing' and then gave a 'proof by exhaustion' which met Greek (and thus modern) standards of rigour. The result created a sensation because it showed that a solid could have 'infinite extent' but finite volume. Thomas Hobbes, a political philosopher who fancied himself as mathematician, wrote of Torricelli's result: 'To understand this for sense, it is not required that a man should be a geometrician or a logician but that he should be mad' [**37**].

(i) In some sense, we should have

$$\int_\gamma f(\mathbf{x})\, d\mathbf{s} \approx \sum_{j=1}^{N} f(\mathbf{x}_j)(\mathbf{x}_j - \mathbf{x}_{j-1}).$$

Discuss informally how this leads to the following variation of the definition for $\int_\gamma f(\mathbf{x})\, ds$ found on page 217.

Let $x,\ y : [a,b] \to \mathbb{R}^2$ have continuous derivatives (with the usual conventions about left and right derivatives at end points) and consider the curve $\gamma : [a,b] \to \mathbb{R}^2$ given by $\gamma(t) = (x(t), y(t))$. If $f : \mathbb{R}^2 \to \mathbb{R}$ is continuous, then we define

$$\int_\gamma f(\mathbf{x})\, d\mathbf{s} = \int_a^b f(x(t), y(t))\gamma'(t)\, dt.$$

(Here, as usual, $\gamma'(t) = (x'(t), y'(t))$.)

(ii) Compute $\int_{\gamma_k} f(\mathbf{x})\, d\mathbf{s}$ for each of the six curves γ_k on page 213 and the function f given by $f(x,y) = x$.

(iii) (What follows requires Exercise K.186.) We can extend our definitions to general rectifiable curves as follows. Suppose that $\gamma : [a,b] \to \mathbb{R}^2$ is rectifiable. Show that, if we set $\gamma(t) = (X(t), Y(t))$, then the functions $X,\ Y : [a,b] \to \mathbb{R}$ are of bounded variation, and so we may define

$$\int_\gamma f(\mathbf{x})\, d\mathbf{s} = \left(\int_a^b f(X(t), Y(t))\, dX(t),\ \int_a^b f(X(t), Y(t))\, dY(t) \right).$$

Show that, if γ is sufficiently well behaved that the definition given in part (i) applies, then the definitions in (i) and (iii) give the same answer.

K.191. [10.2, H, G] Consider the system described in the first few paragraphs of Section 10.2, where n bits are transmitted and each bit has a probability p of being transmitted wrongly, independently of what happens to any other bit.

(i) Explain why the probability that there are k errors in all is given by

$$u_n(k) = \binom{n}{k} p^k (1-p)^{n-k}.$$

By looking at $u_n(k+1)/u_n(k)$, or otherwise, show that there is an integer k_0 with $u_n(k+1) \geq u_n(k)$ for $k \leq k_0 - 1$ and $u_n(k+1) \geq u_n(k)$ for $k_0 \leq k$. Show that $np - 1 < k_0 < np + 1$.

(ii) Let $\epsilon > 0$. Use the ideas of Lemma 10.5 to show that

$$\sum_{|k-np| \geq n\epsilon} u_n(k) \to 0$$

as $n \to \infty$. Deduce that, if we write N_n for the number of errors in a message of length n, then

$$(\dagger) \qquad \Pr\{|N_n - np| \geq \epsilon n\} \to 0,$$

and so, in particular,

$$\Pr\{N_n \geq (p+\epsilon)n\} \to 0,$$

as $n \to \infty$.

(iii) If we are prepared to use a little more probability theory, then, as the reader almost certainly knows, there is a general method for obtaining the result of (ii). (If the notation is unfamiliar, do not proceed further.) Write $X_j = 0$ if the jth bit is transmitted correctly and $X_j = 1$ if not. Show that $\mathbb{E}X_j = p$ and $\operatorname{var} X_j = p(1-p)$. Explain why $N_n = \sum_{j=1}^{n} X_j$, and deduce carefully that $\mathbb{E}N_n = pn$ and $\operatorname{var} N_n = np(1-p)$. Now use Tchebychev's inequality (Lemma 9.47 with $X = N_n - np$) to obtain equation ($\dagger$) in (ii).

K.192. [10.3, P, S] This exercise and the next deal with modifications to the axioms for a metric space. Recall that the standard axioms are

(A) $d(x,y) = 0$ if and only if $x = y$,

(B) $d(x,y) = d(y,x)$ for all $x, y \in X$,

(C) $d(x,y) + d(y,z) \geq d(x,z)$ for all $x, y, z \in X$.

(i) Suppose that $d : X^2 \to \mathbb{R}$ satisfies axiom (A) and

(C)′ $d(x,y) + d(z,y) \geq d(x,z)$ for all $x, y, z \in X$.

Show that (X,d) is a metric space.

(ii) Say everything that there is to say about 'anti-triangle' spaces (X,d) satisfying axioms (A) and (B) together with

(D) $d(x,y) + d(y,z) \leq d(x,z)$ for all $x, y, z \in X$.

What can you say if, in addition, we demand $d(x,y) \geq 0$ for all $x, y \in X$?

(iii) Suppose (X,d) satisfies axioms (A) and (C). Show that, if we set $\rho(x,y) = d(x,y) + d(y,x)$, then (X,ρ) is a metric space.

K.193. [10.3, T, S] This exercise is more interesting than the previous one because it deals with a fairly frequent situation. It requires knowledge of the notions of equivalence relation and equivalence class.

Suppose (X,d) satisfies axioms (B) and (C) together with

(A)′ $d(x,x) = 0$ for all $x \in X$.

Show that the relation $x \sim y$ if $d(x,y) = 0$ is an equivalence relation. If we write $[x]$ for the equivalence class of x and $\tilde{X} = X/\sim$ for the set of equivalence classes, show that $\tilde{d}([x],[y]) = d(x,y)$ gives a well defined function on $\tilde{X}^2$, and that $(\tilde{X},\tilde{d})$ is a metric space.

K.194. [10.3, T, G] This is easy, but requires enough group theory to understand the statement of the exercise.

Let G is a finite group with identity e generated by a subset X. We write $X^{-1} = \{x^{-1} : x \in X\}$. We set $\rho(e) = 0$ and, if $g \neq e$, we set

$$\rho(g) = \min\{N : \text{we can find } x_1, x_2, \ldots x_N \in X \cup X^{-1} \text{ such that } g = x_1 x_2 \ldots x_N\}.$$

Show that $d(g, h) = \rho(gh^{-1})$ defines a metric on G.

Show also that d is left invariant, that is to say, $d(gk, hk) = d(g, h)$ for all g, h, $k \in G$. Is it necessarily true that d is right invariant (i.e. $d(kg, kh) = d(g, h)$ for all g, h, $k \in G$)? [*Hint, if required:* Think about the dihedral group.]

We say that G has diameter $\max_{g \in G}(d(e, g))$ (with respect to the generating set X).

(i) Show that S_n, the permutation group on n elements, is generated by the set X_1 of elements of the form (ij). Show that S_n has diameter less that $A_1 n$ (for some constant A_1) with respect to X_1. Interpret this result in terms of card shuffling.

(ii) Show that S_n is generated by the set X_2 of elements of the form $(1j)$. Show that S_n has diameter less that $A_2 n$ (for some constant A_2) with respect to X_2.

(iii) Show that S_n is generated by the set X_3 consisting of the two elements (12) and $(123 \ldots n)$. Show that S_n has diameter less that $A_3 n^3$ (for some constant A_3) with respect to X_3.

(iv) Let $n \geq 2$. Consider the dihedral group D_n generated by a and b subject only to the relations $a^n = e$, $b^2 = e$ and $bab = a^{-1}$. Show that there exists an $A_4 > 0$ such that D_n has diameter at least $A_4 n$ with respect to $X_4 = \{a, b\}$.

(v) How many elements does S_n have? How many does D_n have? Comment very briefly on the relationship (or lack of relationship) between size and diameter shown by parts (i) to (iv).

K.195. [10.3, T, G] We work on $\mathbb{R}$.

(i) Suppose that $|x| > 4^{-k}$ for some integer k. Show, by induction on N, or otherwise, that if

$$x = \sum_{j=1}^{N} x_j \text{ with } |x_j| \leq 4^{-n(j)} \text{ for some integer } n(j) \geq 0 \ [1 \leq j \leq N],$$

then $\sum_{j=1}^{N} 2^{-n(j)} \geq 2^{-k}$.

(ii) Show that

$$d(x,y) = \inf\{\sum_{j=1}^{N} 2^{-n(j)} : x - y = \sum_{j=1}^{N} x_j \text{ with } |x_j| \leq 4^{-n(j)}$$
$$\text{for some integer } n(j) \geq 0 \ [1 \leq j \leq N]\}$$

defines a metric on $\mathbb{R}$ such that $d(x+a, y+a) = d(x,y)$ for all x, y, $a \in \mathbb{R}$.

(iii) Show that we can find a constant $K > 1$ such that

$$K|x-y|^{1/2} \geq d(x,y) \geq K^{-1}|x-y|^{1/2}.$$

for all x, y, $a \in \mathbb{R}$. Deduce that $d(x_n, y_n) \to 0$ if and only if $|x_n - y_n| \to 0$.

(iv) Sketch the graph of the function $f(x) = d(0,x)$. (This requires thought.)

K.196. [10.3, P] Let (X,d) and (Y,ρ) be metric spaces and let X_1 and X_2 be subsets of X such that $X_1 \cup X_2 = X$. Suppose $f : X \to Y$ is such that $f|_{X_j}$ is continuous as a function from X_j to Y. Which, if any, of the following statements are true and which may be false? (Give proofs or counterexamples as appropriate.)

(i) The function f is automatically continuous.

(ii) If X_1 and X_2 are closed, then f is continuous.

(iii) If X_1 and X_2 are open, then f is continuous.

(iv) If X_1 is closed and X_2 is open, then f is continuous.

Give $\mathbb{R}^n$ its usual Euclidean norm. Suppose $\mathbf{g} : X \to \mathbb{R}^n$ is continuous. Show that there is a continuous function $\mathbf{f} : X \to \mathbb{R}^n$ such that $\mathbf{f}(x) = \mathbf{g}(x)$ when $\|\mathbf{g}(x)\| < 1$ and $\|\mathbf{f}(x)\| = 1$ when $\|\mathbf{g}(x)\| \geq 1$.

K.197. [10.3, P] (i) Give an example of a metric space (X,d) which contains a set A which is both open and closed but $A \neq X$ and $A \neq \varnothing$.

(ii) Suppose (X,d) is a metric space, that A and B are open subsets of X with $A \cup B = X$, and that we are given a function $f : X \to \mathbb{R}$. Show that, if the restrictions $f|_A : A \to \mathbb{R}$ and $f|_B : B \to \mathbb{R}$ are continuous, then f is continuous. (This repeats part of Exercise K.196.)

(iii) Consider the interval $[0,1]$ with the usual metric. Suppose that A and B are open subsets of $[0,1]$ with $A \cup B = [0,1]$ and $A \cap B = \varnothing$. Define $f(x) = 1$ if $x \in A$ and $f(x) = 0$ if $x \in B$. By considering the result of (ii), show that $A = \varnothing$ or $B = \varnothing$. Deduce that the only subsets of $[0,1]$ which are both open and closed are $[0,1]$ and $\varnothing$.

(iv) Suppose that A is an open subset of $\mathbb{R}^n$ and $\mathbf{a} \in A$, $\mathbf{b} \notin A$. By considering $f : [0,1] \to \mathbb{R}$ given by $f(t) = 1$ if $(1-t)\mathbf{a} + t\mathbf{b} \in A$ and $f(t) = 0$ otherwise, show that A is not closed. Thus the only subsets of $\mathbb{R}^n$ which are both open and closed are $\mathbb{R}^n$ and $\varnothing$.

K.198. [10.3, T] Let (X, d) be a metric space. If A is a subset of X, show that the following definitions are equivalent.

(i) The point $x \in \bar{A}$ if and only if we can find $x_n \in A$ such that $d(x_n, x) \to 0$ as $n \to \infty$.

(ii) Let $\mathcal{F}$ be the collection of closed sets F such that $F \supseteq A$. Then $\bar{A} = \bigcap_{F \in \mathcal{F}} F$.

(iii) $\bar{A}$ is a closed set with $\bar{A} \supseteq A$ such that, if F is closed and $F \supseteq A$, then $F \supseteq \bar{A}$.

(It is worth noting that condition (iii) has the disadvantage that it is not clear that such a set always exists.) We call $\bar{A}$ the *closure* of A. We also write $\bar{A} = \operatorname{Cl} A$.

If B is a subset of X, show that the following definitions are equivalent.

(i)$'$ The point $x \in B^\circ$ if and only if we can find $\epsilon > 0$ such that $y \in B$ whenever $d(x, y) < \epsilon$.

(ii)$'$ Let $\mathcal{U}$ be the collection of open sets U such that $U \subseteq B$. Then $B^\circ = \bigcup_{U \in \mathcal{U}} U$.

State a condition (iii)$'$ along the lines of (iii), and show that it is equivalent to (i)$'$ and (ii)$'$.

We call B° the *interior* of B. We also write $B^\circ = \operatorname{Int} B$.

Show that $\operatorname{Cl} A = X \setminus \operatorname{Int}(X \setminus A)$ and $\operatorname{Int} A = X \setminus \operatorname{Cl}(X \setminus A)$. Show that $A = \operatorname{Cl} A$ if and only if A is closed and $A = \operatorname{Int} A$ if and only if A is open.

Suppose we work in $\mathbb{R}$ with the usual distance. Find the interior and closure of the sets $\mathbb{Q}$, $\mathbb{Z}$, $\{1/n : n \in \mathbb{Z},\ n \geq 1\}$, $[a, b]$, (a, b) and $[a, b)$ where $b > a$.

Let (X, d) and (Y, ρ) be metric spaces. Show that $f : X \to Y$ is continuous if and only if $f(\bar{A}) \subseteq \overline{f(A)}$ for all subsets A of X.

(An unimportant warning.) Find a metric space (X, d) and an $x \in X$ such that

$$\begin{gathered}\operatorname{Cl}\{y : d(x, y) < 1\} \neq \{y : d(x, y) \leq 1\} \text{ and}\\ \operatorname{Int}\{y : d(x, y) \leq 1\} \neq \{y : d(x, y) < 1\}.\end{gathered}$$

K.199. [10.3, P, ↑] We use the notation of Exercise K.198. Show that, if U is open,

$$\operatorname{Cl}(\operatorname{Int}(\operatorname{Cl} U)) = \operatorname{Cl} U.$$

Show that, starting from a given A, it is not possible to produce more than 7 distinct sets by repeated application of the operators Cl and Int.

Give an example of a set A in $\mathbb{R}$ with the usual metric which gives rise to 7 distinct sets in this way.

K.200. [10.4, T] The results of this exercise are simple but give some insight into the nature of the norm.

(i) Let $\| \ \|$ be a norm on a real vector space. Show that the closed unit ball

$$\Gamma = \bar{B}(\mathbf{0}, 1) = \{\mathbf{x} \in V : \|\mathbf{x}\| \leq 1\}$$

has the following properties.

(A) Γ is convex, that is to say, if $\mathbf{x}, \mathbf{y} \in \Gamma$ and $1 \geq \lambda \geq 0$, then $\lambda\mathbf{x} + (1-\lambda)\mathbf{y} \in \Gamma$.

(B) Γ is centrally symmetric, that is to say, if $\mathbf{x} \in \Gamma$, then $-\mathbf{x} \in \Gamma$.

(C) Γ is absorbing, that is to say, if $\mathbf{x} \in V$, then we can find a strictly positive real number λ with $\lambda\mathbf{x} \in \Gamma$.

(D) If $\mathbf{x} \neq \mathbf{0}$, then there exists a $t > 0$ such that (when $s \geq 0$) $s\mathbf{x} \in \Gamma$ if and only if $t \geq s$.

(ii) Show, conversely, that if Γ is a convex, centrally symmetric, absorbing set obeying condition (D), then Γ is the closed unit ball of a norm.

(iii) Suppose that Γ is a convex, centrally symmetric, absorbing set obeying the following condition.

(D)$'$ If $\mathbf{x} \neq \mathbf{0}$, then there exists a $\mu > 0$ such that $\mu\mathbf{x} \notin \Gamma$.

Show that there exists a norm $\| \ \|$ such that $\|\mathbf{x}\| < 1$ implies $\mathbf{x} \in \Gamma$ and $\|\mathbf{x}\| > 1$ implies $\mathbf{x} \notin \Gamma$.

(iv) If $\| \ \|_A$ and $\| \ \|_B$ are norms on V and $K > 0$, state and prove a necessary and sufficient condition in terms of the closed unit balls of the two norms for the inequality

$$\|\mathbf{x}\|_A \leq K\|\mathbf{x}\|_B$$

to hold for all $\mathbf{x} \in V$.

K.201. [10.4, P] Consider the real vector space l^∞ of real sequences $\mathbf{a} = (a_1, a_2, \dots)$ with norm $\|\mathbf{a}\|_\infty = \sup_{n \geq 1} |a_n|$. Show that the set

$$E = \{\mathbf{a} : \text{there exists an } N \text{ with } a_n = 0 \text{ for all } n \geq N\}$$

is a subspace of l^∞ but not a closed subset.

Show that, if we set $F = \{\mathbf{a} : a_{2n} = 0 \text{ for all } n \geq N\}$, then F is a closed infinite-dimensional subspace of l^∞.

Show that any finite-dimensional subspace of any normed vector space is closed.

K.202. [10.4, T] Let $(E, \| \ \|)$ be a real normed space and let $\mathbb{R}$ have its usual norm. Let $T : E \to \mathbb{R}$ be a linear map.

(i) Show that, if T is not the zero map, the null space

$$N = T^{-1}(0) = \{\mathbf{e} \in E : T\mathbf{e} = 0\}$$

is a subspace of E such that, if $\mathbf{y} \notin N$, then every $\mathbf{x} \in E$ can be written in exactly one way as $\mathbf{x} = \lambda\mathbf{y} + \mathbf{e}$ with $\lambda \in \mathbb{R}$ and $\mathbf{e} \in N$. Comment very briefly on what happens when T is the zero map.

(ii) Suppose that T is not the zero map and N is as in (i). Show that N is closed if and only if, whenever $\mathbf{y} \notin N$, there exists a $\delta(\mathbf{y}) > 0$ such that $\|\lambda\mathbf{y} + \mathbf{e}\| \geq |\lambda|\delta(\mathbf{y})$ for all $\lambda \in \mathbb{R}$ and $\mathbf{e} \in N$.

(iii) Show that T is continuous if and only if its null space $T^{-1}(0)$ is closed.

(iv) Find $T^{-1}(0)$, and check directly that the conclusions of (iii) are satisfied if we take $E = s_{00}$ and take T as in part (ii) of Exercise 10.53. (There are five norms to check.)

K.203. [10.4, H] This exercise and the one that follows are included more for the benefit of the author than the reader. I was worried whether the proof of Theorem 10.45 actually required the use of results from analysis. In this exercise I outline a simple example, supplied by Imre Leader, which should convince all but the most stubborn that the proof should indeed use analysis.

Consider the map $N : \mathbb{Q}^2 \to \mathbb{R}$ given by $N(x, y) = |x + y2^{1/2}|$. Show that:

(1) $N(\mathbf{x}) \geq 0$ for all $\mathbf{x} \in \mathbb{Q}^2$.

(2) If $N(\mathbf{x}) = 0$, then $\mathbf{x} = \mathbf{0}$.

(3) If $\lambda \in \mathbb{Q}$ and $\mathbf{x} \in \mathbb{Q}^2$, then $N(\lambda\mathbf{x}) = |\lambda|N(\mathbf{x})$.

(4) (The triangle inequality) If $\mathbf{x}, \mathbf{y} \in \mathbb{Q}^2$, then $N(\mathbf{x}+\mathbf{y}) \leq N(\mathbf{x})+N(\mathbf{y})$.

Let $\|(x, y)\| = \max(|x|, |y|)$. Show that, given any $\delta > 0$, we can find an $\mathbf{x} \in \mathbb{Q}^2$ such that

$$N(\mathbf{x}) < \delta\|\mathbf{x}\|.$$

Explain why, if $N : \mathbb{Q}^2 \to \mathbb{R}$ is any function satisfying conditions (1) to (4), there exists a $K > 0$ such that

$$N(\mathbf{x}) < K\|\mathbf{x}\|$$

for all $\mathbf{x} \in \mathbb{Q}^2$.

K.204. [10.4, H!, ↑↑] In this exercise we give another example of Imre Leader which proves conclusively that Theorem 10.45 is indeed a result of analysis. It is quite complicated, and I include it to provide pleasure for the expert[14] rather than enlightenment for the beginner. Our object is to prove the following result.

There is a map $N : \mathbb{Q}^3 \to \mathbb{Q}$ such that:

(1) $N(\mathbf{x}) \geq 0$ for all $\mathbf{x} \in \mathbb{Q}^3$.

(2) If $N(\mathbf{x}) = 0$, then $\mathbf{x} = \mathbf{0}$.

(3) If $\lambda \in \mathbb{Q}$ and $\mathbf{x} \in \mathbb{Q}^3$, then $N(\lambda\mathbf{x}) = |\lambda| N(\mathbf{x})$.

(4) (The triangle inequality) If $\mathbf{x}, \mathbf{y} \in \mathbb{Q}^3$, then $N(\mathbf{x}+\mathbf{y}) \leq N(\mathbf{x})+N(\mathbf{y})$.

(5) If we write $\|\mathbf{x}\|_\infty = \max_{1\leq j\leq 3} |x_j|$, then, given any $\delta > 0$, we can find a $\mathbf{x} \in \mathbb{Q}^3$ such that

$$N(\mathbf{x}) < \delta \|\mathbf{x}\|_\infty.$$

(i) Why does this result show that Theorem 10.45 is a result of analysis?

(ii) (The proof begins here. It will be helpful to recall the ideas and notation of Exercise K.200.) We work in $\mathbb{R}^2$ with the usual Euclidean norm $\| \ \|$. Suppose $\mathbf{u}_1, \mathbf{u}_2, \ldots$ are elements of $\mathbb{R}^2$ none of which are scalar multiples of any other and such that $1/2 \leq \|\mathbf{u}_j\| \leq 2$ for all $j \geq 1$. Show, by using induction, or otherwise, that we can find an $a_n \in \mathbb{Q}$ such that, writing $\mathbf{v}_n = a_n \mathbf{u}_n$, we have

(a) $1/2 < \|\mathbf{v}_n\| < 2$, and

(b) $\mathbf{v}_n \notin \Gamma_{n-1}$,

where Γ_0 is the closed disc $\bar{B}(\mathbf{0}, 1/2)$ of radius $1/2$ and centre $\mathbf{0}$ and, if $n \geq 1$,

$$\Gamma_n = \{\lambda\mathbf{x} + \mu\mathbf{v}_n \ : \ \mathbf{x} \in \Gamma_{n-1},\ |\lambda| + |\mu| = 1,\ \lambda,\ \mu \in \mathbb{R}\}.$$

(For those who know the jargon, Γ_n is the convex hull of $\Gamma_{n-1} \cup \{\mathbf{v}_n, -\mathbf{v}_n\}$.)

(iii) Continuing with the ideas of (ii), write Γ for the closure of $\bigcup_{n=0}^\infty \Gamma_n$. Show that Γ is a closed centrally symmetric convex set with

$$\bar{B}(\mathbf{0}, 1/2) \subseteq \Gamma \subseteq \bar{B}(\mathbf{0}, 2).$$

Show further that, if $\mu \in \mathbb{R}$, then $\mu\mathbf{v}_n \in \Gamma$ if and only if $|\mu| \leq 1$. Use the results of Exercise K.200 to deduce the existence of a norm $\| \ \|_*$ on $\mathbb{R}^2$ such that $\|\mathbf{u}_n\|_* \in \mathbb{Q}$ for all n and

$$2\|\mathbf{x}\| \geq \|\mathbf{x}\|_* \geq \|\mathbf{x}\|/2$$

for all $\mathbf{x} \in \mathbb{R}^2$.

(iv) Let

$$E = \{(a + b2^{1/2}, c + b3^{1/2}) \ : \ a,\ b,\ c \in \mathbb{Q}\}.$$

[14]Who can start by considering the question of whether we can replace $\mathbb{Q}^3$ by $\mathbb{Q}^2$ in the next paragraph.

By considering a suitable sequence $\mathbf{u}_n \in E$, show that there exists a norm $\| \ \|_*$ on $\mathbb{R}^2$ such that $\|\mathbf{e}\|_* \in \mathbb{Q}$ for all $e \in E$, and

$$2\|\mathbf{x}\| \geq \|\mathbf{x}\|_* \geq \|\mathbf{x}\|_*/2$$

for all $\mathbf{x} \in \mathbb{R}^2$.

(v) Define $N : \mathbb{Q}^3 \to \mathbb{Q}$ by $N(x,y,z) = \|(x + y2^{1/2}, z + y3^{1/2})\|_*$. Show that N has properties (1) to (5), as required.

K.205. [11.1, P] Suppose that (X,d) is a complete metric space. We say that a subset F of X has bounded diameter if there exists a K such that $d(x,y) < K$ for all x, $y \in F$, and we then say that F has diameter

$$\sup\{d(x,y) \,:\, x,\, y \in F\}.$$

Suppose that we have a sequence of closed subsets F_j of bounded diameter with $F_1 \supseteq F_2 \supseteq \dots$. Show that, if the diameter of F_n tends to 0, then $\bigcap_{j=1}^\infty F_j$ contains exactly one point.

Show that $\bigcap_{j=1}^\infty F_j$ may be empty if

(a) (X,d) is not complete, or

(b) the F_j have bounded diameters but the diameters do not tend to 0, or

(c) $X = \mathbb{R}$, d is the usual metric and the F_j do not have bounded diameters.

K.206. [11.1, P] (i) Find necessary and sufficient conditions on a function $f : \mathbb{R} \to \mathbb{R}$ for

$$\rho(x,y) = \rho(y,x) = f(y) - f(x) \text{ whenever } y \geq x,$$

to define a metric on $\mathbb{R}$. Prove your result.

(ii) Show that the metric ρ of (i) is complete if and only if f is continuous, $f(t) \to \infty$ as $t \to \infty$ and $f(t) \to -\infty$ as $t \to -\infty$.

(iii) Find necessary and sufficient conditions on a function $g : [0,\infty) \to \mathbb{R}$ for $d(x,y) = g(|x-y|)$ to define a metric on $\mathbb{R}$. Prove your result.

(iv) Suppose that g is increasing (some condition is required to exclude unpleasant behaviour). By considering $g(2^{-n})$, or otherwise, show that, if d is defined as in (iii), then there exists a $K > 0$ such that $d(x,y) \geq K|x-y|$ whenever $|x-y| \leq 1$. Does there always exist an $L > 0$ such that $L|x-y| \geq d(x,y)$ whenever $|x-y| \leq 1$? Give a proof or a counterexample.

(v) Suppose that g is increasing. Show that the metric d of (iii) is always complete. (It may be helpful to consider the case $g(t) \to 0$ as $t \to 0$ separately from the case $g(t) \not\to 0$.)

(vi) Let (X, θ) be a metric space. Show that the conditions you found in (iii) are sufficient to ensure that $\phi(x,y) = g(\theta(x,y))$ defines a metric. Are they always necessary? Give reasons.

K.207. [11.1, P] Consider the space s_{00} introduced in Exercise 10.47. Recall that s_{00} is the space of real sequences $\mathbf{a} = (a_n)_{n=1}^{\infty}$ such that all but finitely many of the a_n are zero. Our object in this exercise is to show that no norm on s_{00} can be complete. To this end, let us suppose that $\| \ \|$ is a norm on s_{00}.

(i) Write $E_n = \{\mathbf{a} \in s_{00} : a_j = 0 \text{ for all } j \geq n\}$. Show that E_n is closed.

(ii) If $\mathbf{b} \in E_{n+1} \setminus E_n$, show that there exists a $\delta > 0$ such that $\|\mathbf{b} - \mathbf{a}\| > \delta$ for all $\mathbf{a} \in E_n$.

(iii) If $\mathbf{h} \in E_{n+1} \setminus E_n$ and $\mathbf{a} \in E_n$, show that $\mathbf{a} + \lambda\mathbf{h} \in E_{n+1} \setminus E_n$ for all $\lambda \neq 0$.

(iv) Show that, given $\mathbf{x}(n) \in E_n$ and $\delta_n > 0$, we can find $\mathbf{x}(n+1) \in E_{n+1}$ such that $\|\mathbf{x}(n+1) - \mathbf{x}(n)\| < \delta_n/4$, and $0 < \delta_{n+1} < \delta_n/4$ such that $\|\mathbf{x}(n+1) - \mathbf{a}\| > \delta_{n+1}$ for all $\mathbf{a} \in E_n$.

(v) Starting with $\mathbf{x}(1) = \mathbf{0}$ and $\delta_1 = 1$, construct a sequence obeying the conclusions of part (iv) for all $n \geq 2$. Show that the sequence $\mathbf{x}(n)$ is Cauchy.

(vi) Continuing with the notation of (v), show that, if the sequence $\mathbf{x}(n)$ converges to some $\mathbf{y} \in s_{00}$, then $\|\mathbf{y} - \mathbf{x}(n+1)\| < \delta_{n+1}/2$ for each $n \geq 1$. Conclude that $\mathbf{y} \notin E_n$ for any n, and deduce the required result by reductio ad absurdum.

K.208. (The space l^2.) [11.1, T] (i) Let $a_j, b_j \in \mathbb{R}$. Use the triangle inequality for the Euclidean norm on $\mathbb{R}^m$ and *careful* handling of limits to show that, if $\sum_{j=1}^{\infty} a_j^2$ and $\sum_{j=1}^{\infty} b_j^2$ converge, so does $\sum_{j=1}^{\infty}(a_j + b_j)^2$. Show further that, in this case,

$$\left(\sum_{j=1}^{\infty}(a_j + b_j)^2\right)^{1/2} \leq \left(\sum_{j=1}^{\infty} a_j^2\right)^{1/2} + \left(\sum_{j=1}^{\infty} b_j^2\right)^{1/2}.$$

(ii) Show that the set l^2 of real sequences $\mathbf{a}$ with $\sum_{j=1}^{\infty} a_j^2$ convergent forms a vector space if we use the natural definitions of addition and scalar multiplication

$$(a_n) + (b_n) = (a_n + b_n), \ \lambda(a_n) = (\lambda a_n).$$

Show that, if we set $\|\mathbf{a}\|_2 = \left(\sum_{j=1}^{\infty} a_j^2\right)^{1/2}$, then $(l^2, \| \ \|_2)$ is a complete normed space.

(iii) The particular space $(l^2, \| \ \|_2)$ has a further remarkable property, to which we devote the next paragraph.

Show, using the Cauchy—Schwarz inequality for $\mathbb{R}^m$, or otherwise, that if $\mathbf{a}$, $\mathbf{b} \in l^2$, then $\sum_{j=1}^{\infty} |a_j b_j|$ converges. We may thus define

$$\mathbf{a} \cdot \mathbf{b} = \sum_{j=1}^{\infty} a_j b_j.$$

Show that this inner product satisfies all the conclusions of Lemma 4.1.

K.209. (Hölder's inequality.) [11.1, T] Throughout this exercise p and q are real numbers with $p > 1$ and $p^{-1} + q^{-1} = 1$.

(i) Show that $q > 1$.

(ii) Use the convexity of $-\log$ (see Exercise K.40) to show that, if x and y are strictly positive real numbers, then

$$xy \leq \frac{x^p}{p} + \frac{y^q}{q}.$$

Observe that this equality remains true if we merely assume that $x, y \geq 0$. (For an alternative proof, see Exercise K.131.)

(iii) Suppose that $f, g : [a, b] \to \mathbb{R}$ are continuous. Show that

$$\int_a^b |f(t)g(t)|\, dt \leq \frac{1}{p} \int_a^b |f(t)|^p\, dt + \frac{1}{q} \int_a^b |g(t)|^q\, dt.$$

Deduce that, if $F, G : [a, b] \to \mathbb{R}$ are continuous and

$$\int_a^b |F(t)|^p\, dt = \int_a^b |G(t)|^q\, dt = 1,$$

then

$$\int_a^b |F(t)G(t)|\, dt \leq 1.$$

(iv) By considering κF and μG with F and G as in (iii), or otherwise, show that, if $f, g : [a, b] \to \mathbb{R}$ are continuous, then

$$\int_a^b |f(t)g(t)|\, dt \leq \left(\int_a^b |f(t)|^p\, dt\right)^{1/p} \left(\int_a^b |g(t)|^q\, dt\right)^{1/q}.$$

This inequality is known as Hölder's inequality[15] for integrals.

(v) Show that the Cauchy–Schwarz inequality for integrals is a special case of Hölder's inequality.

---

[15] Mathematical nomenclature should not be confused with historic truth. Hölder's paper makes it clear that he is discussing an inequality previously proved by L. G. Rogers (1862–1933). Rogers was Professor of Mathematics at Yorkshire College (now the University of Leeds) and a fine mathematician 'with little ambition or desire for recognition'. In 1913, Ramanujan conjectured some remarkable identities which no-one could prove. In 1917, whilst looking through old journals, Ramanujan came across a 1894 paper of Rogers in which a more general form of the identities were proved. Rogers' election to the Royal Society followed.

K.210. [11.1, T, ↑] (i) Suppose that $f : [a, b] \to \mathbb{R}$ is a continuous function such that

$$\int_a^b |f(t)g(t)|\,dt \le A\left(\int_a^b |g(t)|^q\,dt\right)^{1/q}$$

for some constant A and all continuous functions $g : [a, b] \to \mathbb{R}$. By taking $g(t) = |f(t)|^\alpha$ for a suitable α, or otherwise, show that

$$\left(\int_a^b |f(t)|^p\,dt\right)^{1/p} \le A.$$

This result is known as the 'reverse Hölder inequality'.

(ii) By first applying Hölder's inequality to the right hand side of the inequality

$$\int_a^b |(f(t)+h(t))g(t)|\,dt \le \int_a^b |f(t)g(t)|\,dt + \int_a^b |h(t)g(t)|\,dt$$

and then using the reverse Hölder inequality, show that

$$\left(\int_a^b |f(t)+h(t)|^p\,dt\right)^{1/p} \le \left(\int_a^b |f(t)|^p\,dt\right)^{1/p} + \left(\int_a^b |h(t)|^p\,dt\right)^{1/p}$$

for all continuous functions $f, h : [a, b] \to \mathbb{R}$. (This inequality is due to Minkowski.)

(iii) Show that, if we set

$$\|f\|_p = \left(\int_a^b |f(t)|^p\,dt\right)^{1/p},$$

then $(C([a, b]), \| \ \|_p)$ is a normed space. Show, however, that $(C([a, b]), \| \ \|_p)$ is not complete.

(iv) Rewrite Hölder's inequality and the reverse Hölder inequality using the notation introduced in (iii).

K.211. (The l^p spaces.) [11.1, T, ↑] Except in part (v), p and q are real numbers with $p > 1$ and $p^{-1} + q^{-1} = 1$.

(i) By imitating the methods of Exercise K.209, or otherwise, show that, if $a_j, b_j \in \mathbb{R}$, then

$$\sum_{j=1}^n |a_j||b_j| \le \left(\sum_{j=1}^n |a_j|^p\right)^{1/p}\left(\sum_{j=1}^n |b_j|^q\right)^{1/q}.$$

Deduce carefully that, if $\sum_{j=1}^{\infty} |a_j|^p$ and $\sum_{j=1}^{\infty} |b_j|^q$ converge, then so does $\sum_{j=1}^{\infty} |a_j||b_j|$, and

$$\sum_{j=1}^{\infty} |a_j||b_j| \leq \left(\sum_{j=1}^{\infty} |a_j|^p\right)^{1/p} \left(\sum_{j=1}^{\infty} |b_j|^q\right)^{1/q}.$$

This inequality is known as Hölder's inequality for sums.

(ii) By imitating the methods of Exercise K.209, or otherwise, show that the collection l^p of real sequences $\mathbf{a} = (a_1, a_2, \dots)$ forms a normed vector space under the appropriate algebraic operations (to be specified) and norm

$$\|\mathbf{a}\|_p = \left(\sum_{j=1}^{\infty} |a_j|^p\right)^{1/p}.$$

(iii) By using the ideas of Example 11.10, or otherwise, show that $(l^p, \| \ \|_p)$ is complete.

(iv) Use the parallelogram law (see Exercise K.327 (i)) to show that the l^p norm is not derived from an inner product if $p \neq 2$. Do the same for $(C([a,b]), \| \ \|_p)$.

(v) We defined $(l^1, \| \ \|_1)$ in Exercise 11.10 and $(l^\infty, \| \ \|_\infty)$ in Exercise 11.13. Investigate Hölder's inequality and the reverse Hölder inequality for $p = 1$, $q = \infty$ and for $p = \infty$, $q = 1$. Carry out a similar investigation for $(C([a,b]), \| \ \|_1)$ (see Lemma 11.8) and $(C([a,b]), \| \ \|_\infty)$.

K.212. [11.1, T!, ↑] In this exercise we consider the l^p norm on $\mathbb{R}^n$. Thus

$$\|\mathbf{x}\|_p = \left(\sum_{j=1}^{n} |x_j|^p\right)^{1/p} \quad \text{for } 1 \leq p < \infty, \quad \|\mathbf{x}\|_\infty = \max_{1 \leq j \leq n} |x_j|.$$

Explain briefly why the $\| \ \|_p$ are norms.

(i) Let $n = 2$. Sketch the unit balls

$$\{\mathbf{x} \in \mathbb{R}^2 \ : \ \|\mathbf{x}\|_p \leq 1\}$$

for $p = 1$, $p = 2$ and $p = \infty$. Sketch unit balls for other values of p.

(ii) Let $n = 2$ and define $f_p : \mathbb{R}^2 \to \mathbb{R}$ by $f_p(\mathbf{x}) = \|\mathbf{x}\|_p$ $[1 \leq p \leq \infty]$. At which points is f_p differentiable? (The answer depends on p.) Prove your statements.

(iii) By first proving the result in the case when $\|\mathbf{x}\|_s = 1$, or otherwise, show that, if $1 \leq r \leq s \leq \infty$, then $\|\mathbf{x}\|_s \leq \|\mathbf{x}\|_r$ for all $\mathbf{x} \in \mathbb{R}^n$. Give an example of an $\mathbf{x} \in \mathbb{R}^n$ with $\|\mathbf{x}\|_s = \|\mathbf{x}\|_r = 1$.

(iv) By applying the Hölder's inequality of Exercise K.211 (i) with $a_j = |x_j|^r$ and $b_j = 1$, show that, if $1 \leq r \leq s < \infty$, then

$$\|\mathbf{x}\|_r \leq n^{\frac{1}{r}-\frac{1}{s}}\|\mathbf{x}\|_s$$

for all $\mathbf{x} \in \mathbb{R}^n$. Show that, with an appropriate interpretation, the result applies to all $1 \leq r \leq s \leq \infty$.

(v) Write

$$K_{r,s}(n) = \sup_{\mathbf{x}\in\mathbb{R}^n\setminus\{\mathbf{0}\}} \frac{\|\mathbf{x}\|_s}{\|\mathbf{x}\|_r}.$$

Show that $K_{r,s}(n) = 1$ if $1 \leq s \leq r \leq \infty$ and $K_{r,s}(n) = n^{\frac{1}{r}-\frac{1}{s}}$ if $1 \leq r \leq s \leq \infty$.

(vi) Consider the l^p spaces of Exercise K.211 (iii). Show that, if $1 \leq r < s \leq \infty$, then $l^r \subseteq l^s$, and $\|\mathbf{a}\|_r \leq \|\mathbf{a}\|_s$ for all $\mathbf{a} \in l^r \cap l^s$, but, given any K, we can find a $\mathbf{b} \in l^r \cap l^s$ with $\|\mathbf{b}\|_s \geq K\|\mathbf{b}\|_r$. Show that $l^r \neq l^s$.

(vii) Suppose that we define

$$\|\mathbf{x}\|_{*,p} = \left(n^{-1}\sum_{j=1}^{n}|x_j|^p\right)^{1/p} \quad \text{for } 1 \leq p < \infty, \ \|\mathbf{x}\|_{*,\infty} = \max_{1\leq j\leq n}|x_j| = \|\mathbf{x}\|_\infty$$

for all $\mathbf{x} \in \mathbb{R}^n$. Explain briefly why the $\| \ \|_{*,p}$ are norms. Show that, if $1 \leq r \leq s \leq \infty$,

$$\|\mathbf{x}\|_{*,r} \leq \|\mathbf{x}\|_{*,s} \leq n^{\frac{1}{r}-\frac{1}{s}}\|\mathbf{x}\|_{*,r}$$

and neither inequality can be improved.

(viii) Consider $C([0,1])$ with the norms defined in Exercise K.209. Show that, if $1 \leq r < s \leq \infty$, then $\|f\|_s \leq \|f\|_r$ for all $f \in C([0,1])$, but, given any K, we can find a $g \in C([0,1]$ with $\|g\|_r \geq K\|g\|_s$. What are the corresponding results if we replace $C([0,1])$ by $C([a,b])$? Give reasons.

K.213. [10.4, P, S] Let U and V be vector spaces with norms $\| \ \|_U$ and $\| \ \|_V$. Looking back on the special case discussed in Section 6.2 and Exercises K.108 to K.111, we would expect that it may be difficult (or, more usually, impossible) to calculate the operator norm $\|\alpha\|$ for a given element $\alpha \in \mathcal{L}(U,V)$.

(i) There are important exceptions to this rule when V is one-dimensional. Suppose that $b_j \in \mathbb{R}$ are such that

$$\sum_{j=1}^{\infty} b_j x_j$$

converges for all $\mathbf{x} \in l^p$. Use the results of Exercises K.209 and K.210 to show that, if $\mathbf{b} \in l^q$, then $T\mathbf{x} = \sum_{j=1}^{\infty} b_j x_j$ defines a continuous linear map from l^p to $\mathbb{R}$ with operator norm $\|T\| = \|\mathbf{b}\|_q$ (for the standard norms on l^p and $\mathbb{R}$).

(ii) There are other occasional (but unusual) exceptions. Suppose that $U = \mathbf{R}^n$ has one of the two norms $\| \ \|_1$ and $\| \ \|_\infty$ and $V = \mathbf{R}^m$ has one of the two norms $\| \ \|_1$ and $\| \ \|_\infty$ (see Exercise K.212). If $T : U \to V$ is a linear map represented by the matrix (a_{ij}) with respect to the standard basis, find $\|T\|$ in three of the four possible cases. In the fourth case, show that $\|T\|$ cannot be written in terms of the absolute values $|a_{ij}|$.

K.214. (The Hausdorff metric.) [11.1, T] Let (X, d) be a metric space and E a non-empty subset of X. For each $x \in X$, define

$$d(x, E) = \inf\{d(x, y) \ : \ y \in E\}.$$

Show that the map $f : X \to \mathbb{R}$ given by $f(x) = d(x, E)$ is continuous.

Now consider the set $\mathcal{E}$ of non-empty closed bounded sets in $\mathbb{R}^n$ with the usual Euclidean metric d. If $K,\ L \in \mathcal{E}$, define

$$\rho'(K, L) = \sup\{d(\mathbf{k}, L) \ : \ \mathbf{k} \in K\}$$

and

$$\rho(K, L) = \rho'(K, L) + \rho'(L, K).$$

Prove that, if $K,\ L,\ M \in \mathcal{E}$, $\mathbf{k} \in K$ and $\mathbf{l} \in L$, then

$$d(\mathbf{k}, M) \leq d(\mathbf{k}, \mathbf{l}) + \rho'(L, M),$$

and deduce, or prove otherwise, that

$$\rho'(K, M) \leq \rho'(K, L) + \rho'(L, M).$$

Hence show that $(\mathcal{E}, \rho)$ is a metric space.

Suppose now that (X, d) is $\mathbb{R}$ with the usual metric. Suppose K is closed and bounded in $\mathbb{R}^n$ (again with the usual Euclidean metric) and $f_n : K \to X$ and $f : K \to X$ are continuous functions. If f_n converges uniformly to f show that $\rho(f_n(K), f(K)) \to 0$ as $n \to \infty$. Is the converse true? (Give a proof or a counterexample.)

K.215. [11.1, T, ↑] The metric ρ defined in the previous exercise (Exercise K.214) is called the Hausdorff metric. It is clearly a good way of comparing the 'similarity' of two sets. Its utility is greatly increased by the observation that $(\mathcal{E}, \rho)$ is complete. The object of this exercise is to prove this result.

(i) If $K_n \in \mathcal{E}$ and $K_1 \supseteq K_2 \supseteq K_3 \supseteq \ldots$, explain why $K = \bigcap_{j=1}^{\infty} K_j \in \mathcal{E}$ and show that $\rho(K_n, K) \to 0$ as $n \to \infty$.

(ii) Suppose $L_n \in \mathcal{E}$ and $\rho(L_n, L_m) \leq 4^{-n}$ for all $1 \leq n \leq m$. If we set

$$K_j = L_j + \bar{B}(\mathbf{0}, 2^{-j}) = \{\mathbf{l} + \mathbf{x} \ : \ \mathbf{l} \in L_j,\ \|\mathbf{x}\| \leq 2^{-j}\},$$

show that $K_n \in \mathcal{E}$ and $K_1 \supseteq K_2 \supseteq K_3 \supseteq \ldots$. Show also that, if we write $K = \bigcap_{j=1}^{\infty} K_j$, we have $\rho(L_n, K) \to 0$ as $n \to \infty$.

(iii) Deduce that $(\mathcal{E}, \rho)$ is complete.

K.216. [11.2, T, ↑] (i) Consider $(\mathcal{E}, \rho)$, the space of closed bounded subsets of $\mathbb{R}^n$ with the Hausdorff metric. Show that it is not totally bounded.

(ii) Consider the subspace of $\mathcal{E}$ given by

$$\mathcal{F} = \{E \in \mathcal{E} : E \subseteq [-1,1] \times [-1,1]\}.$$

If $X_N = \{(r/N, s/N) : -N \leq r, s \leq N, \ r, s \in \mathbb{Z}\}$, show that, given any $E \in \mathcal{F}$, we can find a non-empty subset $Y \subseteq X_N$ with $d(E, Y) \leq 2N^{-1}$. Deduce that $\mathcal{F}$ with the restriction metric is totally bounded.

(iii) (Much of this part is repetition and need only be sketched.) Let (X, d) be a metric space. Write $\mathcal{H}$ for the collection of non-empty subsets of X with the Bolzano–Weierstrass property. Show that, if

$$h(E, F) = \sup_{e \in E} \inf_{f \in F} d(e, f) + \sup_{f \in F} \inf_{e \in E} d(e, f),$$

then $(\mathcal{H}, h)$ is a metric space. Show that $(\mathcal{H}, h)$ is complete if and only if (X, d) is. Show that $(\mathcal{H}, h)$ is totally bounded if and only if (X, d) is.

K.217. [11.2, T] Let $(V, \| \ \|)$ be a normed vector space and E a closed subspace (that is, a vector subspace which is also a closed subset). If $E \neq V$, there must exist a $\mathbf{z} \in V \setminus E$. By considering

$$\inf\{\|\mathbf{z} - \mathbf{y}\| : \mathbf{y} \in E\},$$

or otherwise, show that, given any $\epsilon > 0$, we can find an $\mathbf{x} \in V$ such that $\|\mathbf{x}\| = 1$ but

$$\|\mathbf{x} - \mathbf{e}\| > 1 - \epsilon \text{ for all } \mathbf{e} \in E.$$

By using Theorem 10.45, or otherwise, show that any finite-dimensional subspace of V is closed. (Omit this if you have already done Exercise K.201.)

Show that, if V is infinite-dimensional, we can find a sequence $\mathbf{x}_j \in V$ such that

$$\|\mathbf{x}_j\| = 1 \text{ for all } j, \text{ but } \|\mathbf{x}_i - \mathbf{x}_j\| > 1/2 \text{ for all } i \neq j.$$

Deduce that if $(V, \| \ \|)$ is a normed space, then the closed unit ball $\bar{B} = \{\mathbf{x} : \|\mathbf{x}\| \leq 1\}$ has the Bolzano–Weierstrass property if and only if V is finite-dimensional.

K.218. [11.2, P] (i) Consider l^∞, the space of bounded real sequences with norm $\| \ \|_\infty$ given by $\|\mathbf{x}\|_\infty = \sup_{n \geq 1} |x_n|$. Let κ_n be a positive real number $[n \geq 1]$. Show that the set

$$E = \{\mathbf{x} \in l^\infty : |x_n| \leq \kappa_n \text{ for all } n \geq 1\}$$

has the Bolzano–Weierstrass property if and only if $\kappa_n \to 0$ as $n \to \infty$.

(ii) Consider l^1, the space of real sequences whose sum is absolutely convergent, with norm $\| \|_1$ given by $\|\mathbf{x}\|_1 = \sum_{n=1}^{\infty} |x_n|$. Let κ_n be a positive real number $[n \geq 1]$. Show that the set

$$E = \{\mathbf{x} \in l^1 : 0 \leq x_n \leq \kappa_n \text{ for all } n \geq 1\}$$

has the Bolzano–Weierstrass property if and only if $\sum_{n=1}^{\infty} \kappa_n$ converges.

K.219. [11.2, T, ↑↑] This exercise takes up ideas which were discussed in Exercises K.35 to K.37.

We work in a metric space (X, d). Consider the following two possible properties of (X, d).

(A) If a collection $\mathcal{K}$ of closed sets has the finite intersection property, then $\bigcap_{K \in \mathcal{K}} K \neq \varnothing$.

(B) If a collection $\mathcal{U}$ of open sets is such that $\bigcup_{U \in \mathcal{U}} U = X$, then we can find an $n \geq 1$ and $U_1, U_2, \ldots, U_n \in \mathcal{U}$ such that $\bigcup_{j=1}^{n} U_n = X$.

(i) Show that (X, d) has property (A) if and only if it has property (B).

(ii) Suppose that (X, d) has the Bolzano–Weierstrass property and that $\mathcal{U}$ is a collection of open sets such that $\bigcup_{U \in \mathcal{U}} U = X$. Show that there exists a $\delta > 0$ such that, given any $x \in X$, we can find a $U \in \mathcal{U}$ with $B(x, \delta) \subseteq U$.

(iii) Use (ii) and Lemma 11.22 to show that, if (X, d) has the Bolzano–Weierstrass property, then it has property (B).

K.220. [11.2, T, ↑] This exercise continues Exercise K.219.

(i) Suppose that $x_n \in X$ for all n, but that the sequence x_n has no convergent subsequence. Show that, given any $x \in X$, we can find a $\delta_x > 0$ and an $N_x \geq 1$ such that $x_n \notin B(x, \delta_x)$ for all $n \geq N_x$.

(ii) By considering sets of the form $U_x = B(x, \delta_x)$, or otherwise, show that the space X described in (i) cannot have property (B).

(iii) Deduce that a metric space (X, d) has property (B) if and only if it has the Bolzano–Weierstrass property.

(iv) Use (iii) to prove parts (ii) and (iv) of Exercise K.37.

If (X, d) has property (B), we say that it is compact. We have shown that compactness is identical with the Bolzano–Weierstrass property for metric spaces. However, the definition of compactness makes no explicit use of the notions of 'distance' (that is, metric) or 'convergence', and can be generalised to situations where these concepts cease to be useful.

K.221. [11.3, P] If $f : [0, 1] \to \mathbb{R}$ is continuous, we write

$$\|f\|_\infty = \sup_{t \in [0,1]} |f(t)|, \text{ and } \|f\|_1 = \int_0^1 |f(t)| \, dt.$$

Consider the space $C^1([0,1])$ of continuously differentiable functions. We set

$$\|f\|_A = \|f\|_\infty + \|f\|_1,$$
$$\|f\|_B = \|f'\|_\infty,$$
$$\|f\|_C = \|f\|_\infty + \|f'\|_\infty,$$
$$\|f\|_D = |f(0)| + \|f'\|_1.$$

Which of these formulae define norms? Of those that are norms, which are complete? Consider those which are norms together with the norms $\| \ \|_\infty$ and $\| \ \|_1$. Which are Lipschitz equivalent and which not? Prove all your answers.

K.222. [11.3, P, S] (i) By differentiating, expanding in power series and then integrating term by term, find a power series expansion for $\sin^{-1} x$. State the theorems that you use and give (with reasons) the values of x for which you can show that the expansion is valid.

(ii) (For enthusiasts only.) Give (with reasons) the values of x for which the expansion is valid. [If you need hints, look at Exercises K.76 and K.96.]

K.223. [11.3, T,S] (Parts of this exercise repeat parts of Exercise K.221.) Weierstrass's example (see page 189) becomes very slightly less shocking if we realise that our discussion uses an inappropriate notion of when one function is close to another. In the next two exercises we discuss a more appropriate notion. It will become clear why much of the early work in what we now call the theory of metric spaces was done by mathematicians interested in the calculus of variations.

(i) Find a sequence of functions $f_n : [-1,1] \to \mathbb{R}$ which have continuous derivative (with the usual convention about derivative at the end points) such that $f_n(x) \to |x|$ uniformly on $[-1,1]$ as $n \to \infty$.

(ii) If $b > a$, let us write $C^1([a,b])$ for the space of functions on $[a,b]$ which have continuous derivative. Show that $C^1([a,b])$ is not closed in $(C([a,b]), \| \ \|_\infty)$, and deduce that $(C^1([a,b]), \| \ \|_\infty)$ is not complete.

(iii) If $f \in C^1([a,b])$, let us write

$$\|f\|_* = \|f\|_\infty + \|f'\|_\infty.$$

Show that $(C^1([a,b]), \| \ \|_*)$ is a complete normed space.

(iv) Consider

$$\mathcal{A} = \{f \in C^1([0,1]) \ : \ f(0) = f(1) = 0\}.$$

Show that $\mathcal{A}$ is a closed subset of $(C^1([0,1])$, and that $\mathcal{A}$ is a vector subspace of $C^1([0,1])$. Conclude that $(\mathcal{A}, \| \ \|_*)$ is a complete normed space. Show also that $(\mathcal{A}, \| \ \|_\infty)$ is not complete.

K.224. [11.3, T, ↑] We continue with the discussion and notation of Exercise K.223. The example we gave on page 189 involved studying the function $I : \mathcal{A} \to \mathbb{R}$ given by

$$I(f) = \int_0^1 (1 - (f'(x))^4)^2 + f(x)^2\,dx.$$

(i) Show that (if we give $\mathbb{R}$ the usual metric) the function $I : \mathcal{A} \to \mathbb{R}$ is not continuous if we give $\mathcal{A}$ the norm $\| \ \|_\infty$, but is continuous if we give $\mathcal{A}$ the norm $\| \ \|_*$.

(ii) Do (or recall) Exercise 8.35, which shows that

$$\inf\{I(f) \,:\, f \in \mathcal{A}\} = 0$$

but that $I(f) > 0$ for all $f \in \mathcal{A}$. This is the main point of the Weierstrass counterexample, and is unaffected by our discussion.

(iii) Write $f_0 = 0$. By using Exercise 8.66, or otherwise, show that we can find $f_n \in \mathcal{A}$ such that $\|f_n - f_0\|_\infty \to 0$ and $If_n \to 0$.

(iv) Show that, if $f_n \in \mathcal{A}$ and $If_n \to 0$, then the sequence f_n does not converge in $(\mathcal{A}, \| \ \|_*)$.

K.225. [11.3, T, ↑] Exercise K.224 shows that the question 'Is $f_0 = 0$ a local minimum for I?' should be asked using the norm $\| \ \|_*$ rather than the norm $\| \ \|_\infty$, but leaves the question unanswered.

(i) By considering functions of the form $f(x) = \epsilon \sin 2N\pi x$, or otherwise, show that, given any $\delta > 0$, we can find u, $v \in \mathcal{A}$ with $\|u\|_*$, $\|v\|_* < \delta$ and $Iu > 1 > Iv$. Conclude that $f_0 = 0$ is neither a local maximum nor a local minimum of J when we use the norm $\| \ \|_*$. (Thus the Euler–Lagrange search in Exercise 8.65 does not look at sufficiently many ways of approaching f_0.)

(ii) Find an infinitely differentiable function P such that $P(0) = 0$, $P(t) > 0$ for $t \neq 0$, $P'(1) = 0$ and $P''(1) > 0$ (so P has a strict local minimum at 1). [The simplest such function that I can think of is a polynomial.]

(iii) Define $J : \mathcal{A} \to \mathbb{R}$ by

$$J(f) = \int_0^1 P(1 - f'(x)^2) + f(x)^2\,dx.$$

Show that

$$\inf\{J(f) \,:\, f \in \mathcal{A}\} = 0$$

but that $J(f) > 0$ for all $f \in \mathcal{A}$. Show that J has a strict local minimum at $f_0 = 0$ when we use the norm $\| \ \|_*$ (that is to say, there exists a $\delta > 0$ such that $J(f) > J(f_0)$ whenever $\|f - f_0\|_* < \delta$). Show that we can find a sequence $f_n \in \mathcal{A}$ such that $f_n \to 0$ uniformly and $J(f_n) \to 0$.

(iv) Find an infinitely differentiable function $G : \mathbb{R}^2 \to \mathbb{R}$ such that, defining $K : \mathcal{A} \to \mathbb{R}$ by

$$K(f) = \int_0^1 G(f(x), f'(x))\, dx,$$

we have

$$\sup\{K(f) : f \in \mathcal{A}\} = 1, \ \inf\{K(f) : f \in \mathcal{A}\} = 0,$$

but $1 > K(f) > 0$ for all $f \in \mathcal{A}$.

K.226. [11.4, P] (i) If E is a non-empty set and (Y, ρ) a complete metric space, we write $\mathcal{B}(E)$ for the set of bounded functions $f : E \to Y$. The uniform distance d_∞ on $\mathcal{B}(E)$ is defined by $d_\infty(f, g) = \sup_{x \in E} \rho\big(f(x), g(x)\big)$. Show that $(\mathcal{B}(E), d_\infty)$ is a complete metric space.

(ii) Generalise and prove Theorem 11.38 and Theorem 11.39.

(iii) Let E be a non-empty set and (Y, ρ) a complete metric space. If $f_n : E \to Y$ and $f : E \to Y$ are functions, we say that f_n *converges uniformly* to f as $n \to \infty$ if, given any $\epsilon > 0$, we can find an $n_0(\epsilon)$ such that $\rho(f_n(x), f(x)) < \epsilon$ for all $x \in E$ and all $n \geq n_0(\epsilon)$.

Generalise and prove Theorem 11.44 and Theorem 11.45.

(iv) Suppose that we now drop the condition that (Y, ρ) is complete. Which of the results above continue to hold, and which fail?

[*Hint:* Take E to consist of one point.]

K.227. [11.4, P] (i) Show that the integral

$$\phi_\gamma(u) = \int_0^\infty e^{-ux} x^{\gamma - 1}\, dx$$

converges for all $u > 0$ and all $\gamma > 0$. By differentiating under the integral sign, integrating by parts and solving the resulting differential equation, show that

$$\phi_\gamma(u) = A_\gamma u^\gamma$$

where A_γ is a constant depending on γ. You should justify each step of your argument.

(ii) Obtain the result of (i) in the case when γ is an integer by integrating by parts.

(iii) Show that the integral

$$\psi(u) = \int_{-\infty}^\infty e^{-ux} e^{-x^2/2}\, dx$$

converges for all real u. By differentiating under the integral sign, integrating by parts and solving the resulting differential equation, show that

$$\psi(u) = A e^{-u^2/2}$$

for some constant A.

(iv) Obtain the results of (i) and (iii) by a change of variable. [However, the method of (iii) carries over to the case when u is complex, but the method of (iv) does not.]

K.228. [11.4, P, S] Suppose that $f : \mathbb{R} \to \mathbb{R}$ is infinitely differentiable everywhere and there exists a function $g : \mathbb{R} \to \mathbb{R}$ such that $f^{(n)}(x) \to g(x)$ uniformly on each bounded interval as $n \to \infty$. By considering the equation

$$\int_a^x f^{(n+1)}(t)\,dt = f^{(n)}(x) - f^{(n)}(a),$$

or otherwise, find a differential equation for g and show that $g(x) = Ce^x$ for some C. Must it be true that $f(x) = Ce^x$? Give reasons.

K.229. [11.4, P] Let $\alpha, \beta > 0$. We define $f_n : [0,1] \to \mathbb{R}$ by $f_n(x) = n^\alpha x^\beta (1-x)^n$. State, with proof, the values of α and β for which the following occur as $n \to \infty$.

(i) $f_n(x) \to 0$ for all $x \in [0,1]$.

(ii) $f_n \to 0$ uniformly on $[0,1]$.

(iii) $\int_0^1 f_n(x)\,dx \to 0$.

[*Hint:* Graph f_n. In (iii) it may be helpful to estimate $\int_{(r-1)/n}^{r/n} f_n(x)\,dx$.]

K.230. [11.4, P] Consider a sequence of functions $g_n : [a,b] \to \mathbb{R}$ and a continuous function $g : [a,b] \to \mathbb{R}$. Suppose that $g_n(x) \to g(x)$ as $n \to \infty$ for each $x \in [a,b]$.

(i) Show that $g_n \to g$ uniformly on $[a,b]$ if and only if, given any sequence $x_n \in [a,b]$ with $x_n \to x$, we have $g_n(x_n) \to g(x)$ as $n \to \infty$.

(ii) If we drop the condition that g is continuous, does the 'if' part remain true? Does the 'only if' part remain true?

(iii) If we replace $[a,b]$ by (a,b) (and restrict ourselves to sequences x_n with $x_n \to x \in (a,b)$), does the 'if' part remain true? Does the 'only if' part remain true?

(iv) If we replace $[a,b]$ by (a,b) but add the condition that the g_n must be continuous, does the 'if' part remain true? Does the 'only if' part remain true?

Give proofs or counterexamples as appropriate.

K.231. [11.4, P] (i) A continuous function $f : [0,1] \to \mathbb{R}$ has $f(1) = 1$ and satisfies

$$0 \leq f(x) < 1 \text{ for all } 0 \leq x < 1.$$

Show that, if $1 > \delta > 0$, then

$$n\int_0^{1-\delta}(f(x))^n\,dx \to 0$$

as $n \to \infty$. Deduce that, if the left derivative $f'(1)$ exists and is non-zero, then

$$n\int_0^1 (f(x))^n\,dx \to \frac{1}{f'(1)}$$

as $n \to \infty$.

(ii) A differentiable function $g : [0,1] \to \mathbb{R}$ satisfies

$$0 \le g(x) \le 1 \text{ for all } 0 \le x \le 1.$$

Show that, if $n\int_0^1 (g(x))^n\,dx$ remains bounded as $n \to \infty$, then $g(x) < 1$ for all $0 < x < 1$.

K.232. (Weierstrass approximation theorem.) [11.4, T] (i) If $1 > \eta > 0$ and m is a positive integer, define $S_{m,\eta} : [-1,1] \to \mathbb{R}$ by $S_{m,\eta}(t) = (1+\eta-t^2)^m$. By considering the behaviour of

$$T_{m,\eta}(t) = \left(\int_{-1}^1 S_{m,\eta}(x)\,dx\right)^{-1} S_{m,\eta}(t)$$

as $m \to \infty$, or otherwise, show that we can find a sequence of polynomials U_n such that

(a) $U_n(t) \ge 0$ for all $t \in [-1,1]$ and all $n \ge 1$,

(b) $U_n(t) \to 0$ uniformly on $[-1,-\epsilon] \cup [\epsilon,1]$ as $n \to \infty$ for any given $\epsilon > 0$, and

(c) $\displaystyle\int_{-1}^1 U_n(t)\,dt = 1$ for all $n \ge 1$.

(ii) Sketch the functions given by $V_n(t) = \displaystyle\int_{-1}^t U_n(s)\,ds$ and $W_n(t) = \displaystyle\int_{-1}^t V_n(s)\,ds$.

(iii) Show that, if $g : [-1,1] \to \mathbb{R}$ is given by $g(t) = 0$ for $t \in [-1,0]$ and $g(t) = t$ for $t \in [0,1]$, then there exists a sequence of real polynomials P_n with $P_n(t) \to g(t)$ uniformly on $[-1,1]$.

(iv) By considering functions of the form $cP_n(at+b)$, show that, if $\alpha \in [-1,1]$ and if $g_\alpha : [-1,1] \to \mathbb{R}$ is given by $g_\alpha(t) = 0$ for $t \in [-1,\alpha]$ and $f(t) = t - \alpha$ for $t \in [\alpha,1]$, then there exists a sequence of real polynomials Q_n with $Q_n(t) \to g_\alpha(t)$ uniformly on $[-1,1]$.

(v) If $h : [-1,1] \to \mathbb{R}$ is piecewise linear, show that there exists a sequence of real polynomials R_n with $R_n(t) \to h(t)$ uniformly on $[-1,1]$.

(vi) If $F : [-1, 1] \to \mathbb{R}$ is continuous, show that there exists a sequence of real polynomials S_n with $S_n(t) \to F(t)$ uniformly on $[-1, 1]$.

(vi) If $f : [a, b] \to \mathbb{R}$ is continuous, show that there exists a sequence of real polynomials P_n with $P_n(t) \to f(t)$ uniformly on $[a, b]$.

(vii) If $f : [a, b] \to \mathbb{C}$ is continuous, show that there exists a sequence of polynomials P_n with $P_n(t) \to f(t)$ uniformly on $[a, b]$.

K.233. [11.4, P, S, ↑] (This is easy if you see how to apply the result of Exercise K.232 appropriately.) Show that, given a continuous function $f : [0, 1] \to \mathbb{R}$, we can find a sequence of polynomials P_n such that $P_n(x^2) \to f(x)$ uniformly on $[0, 1]$ as $n \to \infty$.

Show, however, that there exists a continuous function $g : [-1, 1] \to \mathbb{R}$ such that we cannot find a sequence of polynomials Q_n with $Q_n(x^2) \to g(x)$ uniformly on $[-1, 1]$ as $n \to \infty$.

K.234. [11.4, P, T] This proof of Weierstrass's approximation theorem is due to Bernstein and requires some elementary probability theory.

The random variable $X_n(t)$ is the total number of successes in n independent trials in each of which the probability of success is t. Show, by using Tchebychev's inequality, or otherwise, that

$$\Pr\left(\left|\frac{X_n(t)}{n} - t\right| \geq \delta\right) \leq \frac{1}{n\delta^2}.$$

Suppose that $f : [0, 1] \to \mathbb{R}$ is continuous. Let $p_n(t) = \mathbb{E}f(X_n(t)/n)$ be the expectation of $f(X_n(t)/n)$. Show that $p_n \to f$ uniformly. (You will need to use the theorems which say that a continuous function on a closed interval is both uniformly continuous and bounded. You should indicate explicitly where you use these two facts.)

Show also that p_n is a polynomial of degree at most n (p_n is called a Bernstein polynomial). Deduce the result of Exercise K.232.

If you know about convex functions (for example, if you have done Exercise K.40), show that, if f is convex, then $p_n(x) \geq f(x)$ for all $x \in [0, 1]$.

K.235. [11.4, P] (i) (Dini's theorem) Let I be a bounded interval on the real line and let $f_n : I \to \mathbb{R}$ $[n \geq 1]$ and $f : I \to \mathbb{R}$ be functions such that $f_n(x) \to f(x)$ as $n \to \infty$ for each $x \in I$. Show that, if all four of the following conditions hold

(a) f_n is continuous on I for each n,

(b) f is continuous on I,

(c) for each x, the sequence $f_n(x)$ is decreasing,

(d) I is closed,

then $f_n \to f$ uniformly on I.

(ii) Exhibit counterexamples to show that the result is false if any one of the four conditions is omitted.

(iii) Let

$$p_0(t) = 0 \text{ and } p_n(t) = p_{n-1}(t) + \frac{1}{2}(t^2 - p_{n-1}(t)).$$

Show, by induction, or otherwise, that $0 \le p_{n-1}(t) \le p_n(t) \le |t|$ for all $t \in [-1,1]$. Use part (i) to show that $p_n(t) \to |t|$ uniformly on $[-1,1]$.

(iv) Sketch the function $h : \mathbb{R} \to \mathbb{R}$ given by $h(x) = |x| + |x-a|$ with $a > 0$.

(v) Use the result of (iii) to give an alternative proof of Exercise K.232 (iii).

(vi) Generalise the result of (i) as far as you can. (For example, you could take I to be a subset of $\mathbb{R}^n$ or a metric space.)

K.236. (Orthogonal polynomials.) [11.4, T, ↑↑] Let $a < b$ and let $w : [a,b] \to \mathbb{R}$ be a strictly positive continuous function.

(i) Show that the equation

$$\langle f, g\rangle = \int_a^b f(t)g(t)w(t)\,dt$$

defines an inner product on the space $C([a,b])$ of continuous functions from $[a,b]$ to $\mathbb{R}$. We write $\|f\|_2 = \langle f, f\rangle^{1/2}$. (This coincides with our usual notation if $w(t) = 1$ for all $t \in [a,b]$.)

(ii) Prove that there is a sequence $s_0, s_1, s_2, \ldots$ of polynomials with real coefficients such that

(a) s_n has exact degree n,

(b) each s_n has leading coefficient 1, and

(c) $\langle s_n, s_m\rangle = 0$ for $n \neq m$.

(iii) Show that, if P is a polynomial of degree at most n, then there exists one and only one solution to the equation

$$P = \sum_{j=0}^{n} a_j s_j$$

with $a_j \in \mathbb{R}$. Deduce, in particular, that $\langle s_n, Q\rangle = 0$ whenever Q is a polynomial of degree at most $n-1$.

(iv) Use the fact that the polynomials are uniformly dense in $C([a,b])$ (Weierstrass's theorem, Exercise K.232) to show that, if $f \in C([a,b])$,

$$\inf\left\{\left\|\sum_{j=0}^{n} a_j s_j - f\right\|_2 : a_j \in \mathbb{R} \text{ for } 0 \le j \le n\right\} \to 0$$

as $n \to \infty$.

(v) Show that,if $f \in C([a,b])$, then $\|\sum_{j=0}^{n} a_j s_j - f\|_2$ has a unique minimum, attained when

$$a_j = \hat{f}(j) = \frac{\langle f, s_j\rangle}{\langle s_j, s_j\rangle}.$$

Show that

$$\left\|\sum_{j=0}^{n} \hat{f}(j)s_j - f\right\|_2 \to 0$$

as $n \to \infty$.

(vi) By considering the expression

$$\int_a^b s_n(t)q(t)w(t)\,dt,$$

where

$$q(t) = (t-t_1)(t-t_2)\dots(t-t_k)$$

and $t_1, t_2, \dots, t_k$ are the points of (a,b) at which s_n changes sign, prove that s_n must have exactly n distinct zeros, all of which lie in the open interval (a,b).

K.237. (Legendre polynomials.) [11.4, T, ↑] Let $p_0(x) = 1$ and

$$p_n(x) = \frac{d^n}{dx^n}(x^2-1)^n$$

for $n \geq 1$. By integrating by parts, or otherwise, show that

$$\int_{-1}^{1} p_n(x)p_m(x)\,dx = \delta_{nm}\gamma_n,$$

where γ_n is to be found explicitly. (Recall that $\delta_{nm} = 0$ if $n \neq m$, $\delta_{nn} = 1$.)

Let $[a,b] = [-1,1]$, $w(x) = 1$. Show that there exists a c_n (to be found explicitly) such that, if we write $s_n = c_n p_n$, then the s_n obey the conditions of Exercise K.236 (ii). We call the p_n Legendre polynomials.

Show that

$$\int_{-1}^{1} t^k p_n(t)\,dt = 0 \text{ for } 0 \leq k \leq n-1, \text{ and } \int_{-1}^{1} t^n p_n(t)\,dt = \frac{2^{n+1}(n!)^2}{(2n+1)!}.$$

K.238. (Gaussian quadrature.) [11.4, T, ↑] (i) Suppose $x_1, x_2, \dots, x_n$ are distinct points in $[-1,1]$. Show that, if we write

$$e_j(x) = \prod_{i \neq j} \frac{x - x_i}{x_j - x_i},$$

then any polynomial P of degree at most $n-1$ satisfies

$$P(t)=\sum_{j=1}^{n} P(x_j)e_j(t)$$

for all $t\in[-1,1]$. Deduce that we can find real numbers $a_1, a_2, \dots, a_n$ such that

$$\bigstar \qquad \int_{-1}^{1} P(t)\,dt=\sum_{j=1}^{n} a_jP(x_j)$$

for all polynomials of degree $n-1$ or less.

(ii) Gauss realised that this idea can be made much more effective if we take the $x_1, x_2, \dots, x_n$ to be the zeros of the Legendre polynomial p_n (see Exercise K.237 and part (iv) of Exercise K.236), and this we shall do from now on. Show that, if P is a polynomial of degree $2n-1$ or less, we can write $P=p_nQ+R$, where Q and R are polynomials of degree $n-1$. Explain why $\int_{-1}^{1} p_n(t)Q(t)\,dt=0$. By using this result, and equation ★, show that

$$\bigstar\bigstar \qquad \int_{-1}^{1} P(t)\,dt=\sum_{j=1}^{n} a_jP(x_j)$$

for all polynomials P of degree $2n-1$ or less. (For another example of a choice involving zeros of an orthogonal polynomial, see Exercise K.49.)

(iii) By considering polynomials of the form $\prod_{i\neq j}(x-x_i)^2$, show that $a_j>0$ for each j. By considering the constant polynomial 1, show that $\sum_{j=1}^{n} a_j=2$.

Let us write

$$G_nf=\sum_{j=1}^{n} a_jf(x_j) \text{ and } If=\int_{-1}^{1} f(t)\,dt$$

whenever $f\in C([-1,1])$. Show that, if $f, g\in C([-1,1])$, then

$$|G_nf-G_ng|\leq 2\|f-g\|_\infty \text{ and } |If-Ig|\leq 2\|f-g\|_\infty.$$

Use the fact that the polynomials are uniformly dense in $C([-1,1])$ (Weierstrass's theorem, Exercise K.232) to show that

$$G_n(f)\rightarrow \int_{-1}^{1} f(t)\,dt$$

as $n\rightarrow\infty$ whenever $f\in C([-1,1])$.

(iv) Prove the following converse to the result of (ii). If b_j and y_j are such that

$$\int_{-1}^{1} P(t)\,dt=\sum_{j=1}^{n} b_jP(y_j),$$

then the y_j are the n zeros of the Legendre polynomial p_n.

K.239. [11.4, P, ↑↑] (i) Consider the orthogonal polynomials s_n of Exercise K.236. By imitating the arguments of that exercise, show that, if λ is real, then $s_n+\lambda s_{n-1}$ has at least $n-1$ distinct real roots and has no multiple roots. Deduce that $s_n + \lambda s_{n-1}$ has n distinct real roots.

(ii) If P and Q are real polynomials with a common real root at x, show that there exist real λ and μ, not both zero, such that $\mu P+\lambda Q$ has a multiple root at x. Deduce that s_n and s_{n-1} have no common root.

(iii) If P and Q are real polynomials such that P has real roots at x and y with $x < y$ and Q has no real root in $[x, y]$, show that there exists a real λ such that $P + \lambda Q$ has a multiple root at some point $w \in (x, y)$. Deduce that, between any two roots of s_n, there is a root of s_{n-1}.

K.240. [11.4, P] This exercise introduces ideas which are very useful in more advanced work. Let M be a strictly positive integer.

(i) Show, by using the Weierstrass M-test, or otherwise, that

$$\sum_{n=M}^{N} \frac{1}{(x-n)^2} \text{ and } \sum_{n=M}^{N} \frac{1}{(x+n)^2}$$

converge uniformly, and so

$$\sum_{n=M}^{N} \frac{1}{(x-n)^2} + \frac{1}{(x+n)^2}$$

converges uniformly to a limit $f_M(x)$, say, for $|x| \leq M/2$ as $N \to \infty$. Show that f_M is continuous on $[-M/2, M/2]$. Show that f_M is differentiable on $(-M/2, M/2)$.

(ii) By choosing appropriate values of M (depending on x), show that the function

$$F(x) = \sum_{n=-\infty}^{\infty} \frac{1}{(x-n)^2}$$

is well defined, continuous and differentiable on $\mathbb{R} \setminus \mathbb{Z}$. Show that, indeed, F is infinitely differentiable on $\mathbb{R} \setminus \mathbb{Z}$.

(iii) Show, reasonably carefully, that $F(x + 1) = F(x)$ for all $x \in \mathbb{R} \setminus \mathbb{Z}$. (In other words, F is periodic with period 1.)

K.241. [11.4, P, ↑] The following clever special argument allows us to identify the function F constructed in Exercise K.240. We retain the notation of that exercise.

(i) Suppose that $g : \mathbb{R} \to \mathbb{R}$ is a continuous function such that $g(x+1) = g(x)$ for all $x \in \mathbb{R}$. Show that there exists a K such that $|g(x)| \leq K$ for all

$x \in \mathbb{R}$. Give examples to show the conclusion may fail if we drop either the condition g continuous or the condition g periodic.

(ii) Suppose $g : \mathbb{R} \to \mathbb{R}$ is such that there exists a K with $|g(x)| \leq K$ and

$$g(x) = \frac{1}{4}\left(g\left(\frac{x}{2}\right) + g\left(\frac{x+1}{2}\right)\right)$$

for all x. Show that $g(x) = 0$ for all x.

(iii) Show that

$$\pi^2 \operatorname{cosec}^2(\pi x) - \frac{1}{x^2} \to \frac{\pi^2}{3}$$

as $x \to 0$. Deduce that, if we set

$$g(x) = F(x) - \pi^2 \operatorname{cosec}^2(\pi x) = f_1(x) - \left(\pi^2 \operatorname{cosec}^2(\pi x) - \frac{1}{x^2}\right)$$

for $x \in \mathbb{R} \setminus \mathbb{Z}$ and $g(n) = f_1(0) - \pi^2/3$ for $n \in \mathbb{Z}$, then g is a continuous periodic function.

(iv) Use part (ii) to show that, in fact, $g(x) = 0$, and so

$$\sum_{n=-\infty}^{\infty} \frac{1}{(x-n)^2} = \pi^2 \operatorname{cosec}^2(\pi x)$$

for all $x \in \mathbb{R} \setminus \mathbb{Z}$. Show also that $\sum_{n=1}^{\infty} n^{-2} = \pi^2/6$.

K.242. [11.4, T] (i) Define $f_n : [-1, 1] \to \mathbb{R}$ by $f_n(x) = (x^2 + n^{-2})^{1/2}$. Show that f_n' converges pointwise to a limit F, to be found, and f_n converges uniformly to a limit f, to be found. Show that f is not everywhere differentiable.

(ii) By considering the integral of appropriate witch's hats, or otherwise, find differentiable functions $f_n : [-1, 1] \to \mathbb{R}$ such that $f_n(x) = 0$ for $x \leq 0$ and $f_n(x) = 1$ for $x \geq 1/n$. Show that $f_n'(x) \to 0$ for each x and there is an f such that $f_n(x) \to f(x)$ for each x as $n \to \infty$, but that f is not continuous.

(iii) Why do these results not contradict Theorem 11.57?

K.243. [11.4, H] Suppose that $u, v : \mathbb{R} \to \mathbb{R}$ are infinitely differentiable, that $v(0) = 0$, that $u(t) \geq 0$ for all $t \in \mathbb{R}$ and that $\int_0^\infty u(x)\, dx = 1$. In this exercise we shall be interested in the function $g : \mathbb{R}^2 \to \mathbb{R}$ defined by

$$g(x, y) = yv(y)u(xy).$$

(i) Just to get an idea of what is going on, sketch g in the particular case when $v(y) = y$ and $u(x) = 0$ for $x < 1$ and $x > 2$.

(ii) Show that g has partial derivatives of all orders (thus g is infinitely differentiable). Show also that

$$\int_0^\infty g(x,y)\,dx = v(y)$$

for all y.

(iii) Show that, if $G(y) = \int_0^\infty g(x,y)\,dx$, then $G'(0) = v'(0)$, but

$$\int_0^\infty g_{,2}(x,0)\,dx = 0.$$

We can certainly choose v with $v'(0) \neq 0$ (for example, $v(y) = y$). Why does this not contradict Theorem 11.62?

K.244. [11.4, H, ↑] In this exercise we modify the ideas we used in Exercise K.243 to obtain an example relevant to Theorem 8.57.

Suppose that u, $v : \mathbb{R} \to \mathbb{R}$ have continuous derivatives, that $v(0) = 0$, that $u(t) \geq 0$ for all $t \in \mathbb{R}$, that $u(t) = 0$ if $t \leq 1$ or $t \geq 2$ and that $\int_0^\infty u(x)\,dx = 1$. Explain why $|t|^{-1/2}v(t) \to 0$ as $t \to 0$. In this exercise we shall be interested in the function $g : \mathbb{R}^2 \to \mathbb{R}$ defined by

$$g(x,y) = |y|^{-1/2}v(y)u(x|y|^{-1/2}) \text{ for } y \neq 0,\ g(x,0) = 0.$$

(i) Just to get an idea of what is going on, sketch g in the particular case when $v(y) = y$

(ii) Show that g is everywhere continuous.

(iii) Show that g has partial derivatives $g_{,1}$ and $g_{,2}$ everywhere and that these derivatives are continuous except, possibly, at $(0,0)$.

(iv) Show that if $G(y) = \int_0^1 g(x,y)\,dx$, then $G'(0) = v'(0)$, but

$$\int_0^1 g_{,2}(x,0)\,dx = 0.$$

We can certainly choose v with $v'(0) \neq 0$ (for example, $v(y) = y$). Why does this not contradict Theorem 8.57?

K.245. (A dominated convergence theorem.) [11.4, P] This result echos Lemma 5.25. Reread the proof of that lemma if you need a hint.

Suppose that $g : [0,\infty) \to \mathbb{R}$ is a positive continuous function such that $\int_0^\infty g(t)\,dt$ converges. If $f_n : [0,\infty) \to \mathbb{R}$ is a continuous function with $|f_n(t)| \leq g(t)$ for all $t \in [0,\infty)$ and $f_n \to f$ uniformly on $[0,\infty)$ as $n \to \infty$, show that all the integrals $\int_0^\infty f_n(t)\,dt$ $[n \geq 1]$ and $\int_0^\infty f(t)\,dt$ converge and

$$\int_0^\infty f_n(t)\,dt \to \int_0^\infty f(t)\,dt.$$

Can we replace the condition '$f_n \to f$ uniformly on $[0,\infty)$' by $f_n \to f$ uniformly on each interval $[0,X]$ with $X > 0$? Give reasons.

K.246. (A monotone convergence theorem.) [11.4, P, ↑] (i) Recall or do part (i) of Exercise K.235 (Dini's theorem). Deduce that, if $g_n : [a, b] \to \mathbb{R}$ $[n \geq 1]$ and $g : [a, b] \to \mathbb{R}$ are continuous functions such that $g_{n+1}(x) \geq g_n(x)$ and $g_n(x) \to g(x)$ for each $x \in [a, b]$, then

$$\int_a^b g_n(x)\, dx \to \int_a^b g(x)\, dx.$$

(ii) Suppose that $f_n : [0, \infty) \to \mathbb{R}$ $[n \geq 1]$ and $f : [0, \infty) \to \mathbb{R}$ are continuous functions such that $f_{n+1}(x) \geq f_n(x) \geq 0$ and $f_n(x) \to f(x)$ for each $x \in [0, \infty)$. Show that $\int_0^\infty f_n(t)\, dt$ converges for each n and $\int_0^\infty f_n(t)\, dt \to A$ as $n \to \infty$ if and only if $\int_0^\infty f_n(t)\, dt$ converges with value A. (Note that this result parallels part (ii) of Exercise K.78.

[The results of the last two exercises are quite useful, but suffer from the weakness that we must put conditions on the limit function f. The parallel results in Lebesgue theory tell us that, if the f_n are well behaved from the point of view of the theory, then, automatically, so is f.]

K.247. [11.4, P] (i) Suppose that d_1, d_2, ... are metrics on a space X. Show that

$$d(x, y) = \sum_{j=1}^{\infty} \frac{2^{-j} d_j(x, y)}{1 + d_j(x, y)}$$

is a metric on X.

(ii) Suppose that, in addition, $d_{n+1}(x, y) \geq d_n(x, y)$ for all n and all x, $y \in X$. Show that if (X, d_n) is complete for each n, then (X, d) is complete.

(iii) Consider the set

$$X = \{n^{-1} : n \in \mathbb{Z},\ n \geq 1\} \cup \{0_*, 0_{**}\}.$$

Set

$$\begin{aligned} d_*(n^{-1}, m^{-1}) = d_{**}(n^{-1}, m^{-1}) &= |n^{-1} - m^{-1}|, \\ d_*(0_*, m^{-1}) = d_{**}(0_{**}, m^{-1}) &= m^{-1}, \\ d_*(0_{**}, m^{-1}) = d_{**}(0_*, m^{-1}) &= d_*(0_{**}, 0_*) = d_{**}(0_*, 0_{**}) = 1 \end{aligned}$$

and $d_*(x, y) = d_*(y, x)$, $d_{**}(x, y) = d_{**}(y, x)$, $d_*(x, x) = d_{**}(x, x) = 0$. Show that (X, d_*) and (X, d_{**}) are complete metric spaces, but that, if we write $d(x, y) = d_*(x, y) + d_{**}(x, y)$, then (X, d) is a metric space which is not complete. Show that the second sentence of (ii) may be false if we omit the conditions of the first sentence.

(iv) Consider C^∞, the space of infinitely differentiable functions on $[0,1]$. Show that

$$d(f,g) = \sum_{j=0}^{\infty} \frac{2^{-j} \sup_{t\in[0,1]} |f^{(j)}(t) - g^{(j)}(t)|}{1 + \sup_{t\in[0,1]} |f^{(j)}(t) - g^{(j)}(t)|}$$

is a complete metric on C^∞. (Be careful, $\sup_{t\in[0,1]} |f^{(j)}(t) - g^{(j)}(t)|$ is not a metric on C^∞ if $j \neq 0$.)

Show that $d(f_n, f) \to 0$ as $n \to \infty$ if and only if $f_n^{(j)}(x) \to f(x)$ uniformly on $[0,1]$ for each j.

Let $g_n(x) = n^{-n} \sin(n+1)^2 \pi x$. Show that $d(g_n, 0) \to 0$ but

$$\sup_{j\geq 0} \sup_{x\in[0,1]} |g_n^{(j)}(x)| = \infty$$

(more formally, the set $\{|g_n^{(j)}(x)| \,:\, x \in [0,1],\ j \geq 0\}$ is unbounded) for all n.

(v) Consider the space $C(D)$ of continuous (but not necessarily bounded) functions $f : D \to \mathbb{C}$, where D is the open unit disc given by

$$D = \{z \in \mathbb{C} \,:\, |z| < 1\}.$$

Find a complete metric d such that $d(f_n, f) \to 0$ as $n \to \infty$ if and only if

$$\sup_{|z|\leq 1-n^{-1}} |f(z) - g(z)| \to 0$$

for all $n \geq 1$. Prove that you have, indeed, found such a metric.

K.248. (Pointwise convergence cannot be given by a metric.) [11.4, H] The object of this exercise is to show that the notion of a metric is not sufficient to cover all kinds of convergence. More specifically, we shall show that there does not exist a metric d on the space $C([0,1])$ of continuous functions such that $d(f_n, f) \to 0$ as $n \to \infty$ if and only if $f_n(x) \to f(x)$ for all $x \in [0,1]$. (A more transparent proof becomes available once you have studied analytic topology.)

The proof is by contradiction, so we assume that d is a metric such that $d(f_n, f) \to 0$ as $n \to \infty$ if and only if $f_n(x) \to f(x)$ for all $x \in [0,1]$.

(i) Let $g_n : [0,1] \to \mathbb{R}$ be a 'witch's hat' given by

$$\begin{aligned} g_n(x) &= 1 - n|x - n^{-1} - 2^{-1}| && \text{for } |x - n^{-1} - 2^{-1}| \leq n^{-1}, \\ g_n(x) &= 0 && \text{otherwise.} \end{aligned}$$

Show that $d(g_n, 0) \to 0$ as $n \to \infty$.

(ii) By using the ideas of (i), or otherwise, show that, given any closed interval $I \subseteq [0,1]$ and any $\epsilon > 0$, we can find a $g \in C([0,1])$ and a closed interval $J \subseteq I$ such that $d(g,0) \leq \epsilon$ but $g(x) \geq 1/2$ on J.

(iii) Show that we can find a sequence of closed intervals $I_1 \supseteq I_2 \supseteq I_3 \supseteq \dots$ and continuous functions f_n such that $d(f_n, 0) \leq 2^{-n}$ but $f_n(x) \geq 1/2$ on I_n.

(iv) Use Exercise 4.48 to derive a contradiction.

K.249. [11.4, T, S] (This sketches some background for the next exercise.) Let V be a vector space over $\mathbb{F}$. Suppose that $\| \ \|$ is a norm on V and d a metric on V such that $d(\mathbf{x}_n, \mathbf{x}) \to 0$ as $n \to \infty$ if and only if $\|\mathbf{x}_n - \mathbf{x}\| \to 0$. Show that, if $\mathbf{a} \in V$ and λ, $\lambda_n \in \mathbb{F}$, then:

(a) $d(\mathbf{a} + \mathbf{x}_n, \mathbf{a} + \mathbf{x}) \to 0$ if and only if $d(\mathbf{x}_n, \mathbf{x}) \to 0$.

(b) $d(\lambda \mathbf{x}_n, \mathbf{0}) \to 0$ whenever $d(\mathbf{x}_n, \mathbf{0}) \to 0$.

(c) $d(\lambda_n \mathbf{x}, \mathbf{0}) \to 0$ whenever $\lambda_n \to 0$.

These remarks make it easy to find metrics which do not have the convergence properties of norms. As an example, check which of properties (a), (b) and (c) hold for the discrete metric, and deduce that the discrete metric does not have the convergence properties of a norm. Do the same for the British Railway non-stop metric of Exercise 10.20.

Show, however, that the metric of part (ii) of Exercise K.247 has all the properties (a), (b) and (c). In the next exercise we show that, even so, it does not have the convergence properties of a norm.

K.250. (An interesting metric not derivable from a norm.) [11.4, H, ↑] In this exercise we shall show that there does not exist a norm $\| \ \|_D$ on the space $C^\infty([0,1])$ of continuous functions such that $\|f_n\|_D \to 0$ as $n \to \infty$ if and only if $f_n^{(j)}(x) \to 0$ uniformly on $[0,1]$ for each j. This shows that there does not exist a norm whose associated metric has the properties of the metric d defined in part (ii) of Exercise K.247.

The proof proceeds by contradiction. We assume that $\| \ \|_D$ has the properties given in the previous paragraph.

(i) By reductio ad absurdum, or otherwise, show that, for each integer $j \geq 0$, there exists an $\epsilon_j > 0$ such that we have $\|f\|_D \geq \epsilon_j$ whenever $\sup_{x \in [0,1]} |f^{(j)}(x)| \geq 1$.

(ii) Deduce that $\|g\|_D \geq \epsilon_j \sup_{x \in [0,1]} |g^{(j)}(x)|$ for all $g \in C^\infty([0,1])$ and all $j \geq 0$.

(iii) Show that, by choosing δ_k and N_k appropriately, and setting $f_k(x) = \delta_k \sin \pi N_k x$, or otherwise, that we can find $f_k \in C^\infty([0,1])$ such that

$$\sup_{x \in [0,1]} |f_k^{(j)}(x)| \leq 2^{-k} \qquad \text{when } 0 \leq j \leq k-1,$$

$$\sup_{x \in [0,1]} |f_k^{(k)}(x)| \geq 2^j \epsilon_j.$$

(iv) Show that $\|f_n\|_D \to \infty$ as $n \to \infty$, but $f_n^{(j)}(x) \to 0$ uniformly on $[0, 1]$ for each j. This contradiction gives the desired conclusion.

K.251. (A continuous nowhere differentiable function.) [11.4, T]

(i) Suppose that $f : [0, 1] \to \mathbb{R}$ is differentiable with continuous derivative. Explain why we can find a K, depending on f, such that

$$|f(x) - f(y)| \leq K|x - y|$$

for all $x, y \in [0, 1]$.

(ii) Suppose $f : [0, 1] \to \mathbb{R}$ is differentiable with continuous derivative. By considering

$$g(x) = f(x) + \epsilon \sin Nx$$

with N very large, show that, given any $\epsilon > 0$ and any $\eta > 0$, we can find a $g : [0, 1] \to \mathbb{R}$ with the following properties.

(a) g is differentiable with continuous derivative.

(b) $\|g - f\|_\infty \leq \epsilon$.

(c) Given any $x \in [0, 1]$, we can find a $y \in [0, 1]$ such that $|x - y| < \eta$ but $|g(x) - g(y)| > \epsilon/2$.

(iii) Set $g_0 = 0$. Show that we can find a sequence of functions $g_n : [0, 1] \to \mathbb{R}$ such that, whenever $n \geq 1$,

$(\mathrm{a})_n$ g_n is differentiable with continuous derivative.

$(\mathrm{b})_n$ $\|g_n - g_{n-1}\|_\infty \leq 2^{-3n}$.

$(\mathrm{c})_n$ Given any $x \in [0, 1]$, we can find a $y \in [0, 1]$ such that $|x - y| < 2^{-4n}$ but $|g_n(x) - g_n(y)| > 2^{-3n-1}$.

(iv) Show that g_n converges uniformly to a continuous function g. Show further that

$$\|g - g_n\|_\infty \leq \sum_{j=n+1}^{\infty} 2^{-3j} = 2^{-3n}/7.$$

Use this fact, together with $(\mathrm{c})_n$, to show that, given any $x \in [0, 1]$, we can find a $y \in [0, 1]$ such that $|x - y| < 2^{-4n}$ but

$$|g(x) - g(y)| > 2^{-3n-1} - 2^{-3n+1}/7 = 3 \times 2^{-3n-1}/7.$$

Deduce that

$$\sup_{y \in [0,1],\ y \neq x} \left| \frac{g(x) - g(y)}{x - y} \right| \geq \frac{3 \times 2^{n-1}}{7}$$

for all $n \geq 1$ and all x. Conclude that g is nowhere differentiable.

[At the beginning of the 20th century continuous, nowhere differentiable functions were considered to be playthings of pure mathematicians of uncertain taste. Classes of continuous nowhere differentiable functions now form the main subject of study in financial mathematics and parts of engineering.]

K.252. (A continuous space-filling curve.) [11.4, T] This exercise uses Exercise 11.41. We work in $\mathbb{R}^2$ and $\mathbb{R}$ with the usual Euclidean norms.

(i) Let (x_0, y_0), $(x_1, y_1) \in [0,1]^2$. Show, by specifying $\mathbf{f}$ piece by piece, or otherwise, that we can find a continuous function $\mathbf{f} : [0,1] \to [0,1]^2$ such that $\mathbf{f}(0) = (x_0, y_0)$, $\mathbf{f}(1) = (x_1, y_1)$ and $(rN^{-1}, sN^{-1}) \in \mathbf{f}([0,1])$ for each $0 \leq r, s \leq N$. (In other words, the curve described by $\mathbf{f}$ starts at (x_0, y_0), ends at (x_1, y_1) and passes through all points of the form (rN^{-1}, sN^{-1}).)

(ii) Suppose that $\mathbf{h} : [0,1] \to [0,1]^2$ is continuous and $0 = t_0 < t_1 < t_2 < \cdots < t_M = 1$. Show that, given any $\epsilon > 0$, we can find an $\eta > 0$ such that

$$t_0 + \eta < t_1 - \eta < t_1 + \eta < t_2 - \eta < t_2 + \eta < t_3 - \eta < t_3 + \eta < \cdots < t_M - \eta$$

and $\|\mathbf{h}(t) - \mathbf{h}(s)\| < \epsilon$ whenever t, $s \in [0,1]$ and $|t - s| \leq 2\eta$.

(iii) Let $\mathbf{g}_0 : [0,1] \to [0,1]^2$ be a continuous function such that $\mathbf{g}_0(0) = (0,0)$, $\mathbf{g}_0(1/3) = (0,1)$, $\mathbf{g}_0(2/3) = (1,0)$ and $\mathbf{g}_0(1) = (1,1)$, and let $E_0 = \{0,\ 1/3,\ 2/3,\ 1\}$. Show that we can find a sequence of functions $\mathbf{g}_n : [0,1] \to [0,1]^2$ and finite sets $E_n \subset [0,1]$ such that, if $n \geq 1$,

$(\text{a})_n$ $\|\mathbf{g}_n - \mathbf{g}_{n-1}\|_\infty \leq 2^{-n+4}$.

$(\text{b})_n$ If $a \in E_n$, then $\mathbf{g}_n(a) = (r2^{-n}, s2^{-n})$ for some integers r and s with $0 \leq r,\ s \leq 2^n$.

$(\text{c})_n$ If r and s are integers with $0 \leq r,\ s \leq 2^n$, then there exists an $a \in E_n$ with $\mathbf{g}_n(a) = (r2^{-n}, s2^{-n})$.

$(\text{d})_n$ If r and s are integers with $0 \leq r,\ s \leq 2^{n-1}$ and $a \in E_{n-1}$ satisfies $\mathbf{g}_{n-1}(a) = (r2^{-n+1}, s2^{-n+1})$, then, if u and v are integers with $0 \leq u,\ v \leq 2^n$ and

$$|u2^{-n} - r2^{n-1}|,\ |v2^{-n} - s2^{n-1}| \leq 2^{-n},$$

we can find $b \in E_n$ such that $|a - b| \leq 2^{-n}$ and $\mathbf{g}_n(b) = (r2^{-n}, s2^{-n})$.

[Conditions $(\text{a})_n$ and $(\text{d})_n$ are the important ones. The other two are included to help keep track of what is going on. Note that $\mathbf{g}_n$ will not be injective for $n \geq 1$ and that E_n will contain several points x with $\mathbf{g}_n(x) = (r2^{-n}, s2^{-n})$.]

(iv) Show that $\mathbf{g}_n$ converges uniformly to a continuous function $\mathbf{g} : [0,1] \to [0,1]^2$.

(v) Using the fact that every $x = [0,1]$ can be written $x = \sum_{j=1}^{\infty} \kappa_j 2^{-j}$ with $\kappa_j \in \{0,1\}$, or otherwise, use condition $(\text{d})_n$ to show that, if $(x,y) \in [0,1]$, we can find $t_n \in E_n$ $[n \geq 0]$ such that $|t_n - t_{n-1}| \leq 2^{-n}$ for $n \geq 1$ and $\|\mathbf{g}_n(t_n) - (x,y)\| \to 0$ as $n \to 0$. Explain why we can find a $t \in [0,1]$ with

$t_n \to t$ as $n \to \infty$ and use the inequality

$$\begin{aligned}\|\mathbf{g}(t) - (x,y)\| &\leq \|\mathbf{g}(t) - \mathbf{g}(t_n)\| + \|\mathbf{g}(t_n) - \mathbf{g}_n(t_n)\| + \|\mathbf{g}_n(t_n) - (x,y)\| \\ &\leq \|\mathbf{g}(t) - \mathbf{g}(t_n)\| + \|\mathbf{g} - \mathbf{g}_n\|_\infty + \|\mathbf{g}_n(t_n) - (x,y)\|\end{aligned}$$

to show that $\mathbf{g}(t) = (x,y)$. Conclude that $\mathbf{g} : [0,1] \to [0,1]^2$ is a continuous surjective map.

[I cannot claim any practical use for space filling curves, but the popularity of fractals means that mathematicians do come across continuous maps $\mathbf{h} : [0,1] \to [0,1]^2$ where the image $\mathbf{h}([0,1])$ is 'unexpectedly large'.]

K.253. (The devil's staircase.) [11.4, T] This result is less spectacular than the previous two and probably harder to explain. If you do not understand my presentation, the example can be found in most books on measure theory. (The set E defined in (iii) is called the Cantor set or the 'Cantor middle third set'.)

(i) We work on $[0,1]$. Write $E(0) = (3^{-1}, 2 \times 3^{-1})$ for the 'middle third' open interval of $[0,1]$. We have $[0,1] \setminus E(0) = [0,3^{-1}] \cup [2 \times 3^{-1}, 1]$, and we write $E_0 = (3^{-2}, 2 \times 3^{-2})$ for the middle third open interval of $[0, 3^{-1}]$ and $E_1 = (2 \times 3^{-1} + 3^{-2}, 2 \times 3^{-1} + 2 \times 3^{-2})$ for the middle third interval of $[2 \times 3^{-1}, 1]$. More generally, we set

$$E_{\epsilon(1),\epsilon(2),\dots,\epsilon(n)} = (\sum_{j=1}^{n} 2\epsilon(j)3^{-j} + 3^{-n-1}, \sum_{j=1}^{n} 2\epsilon(j)3^{-j} + 2 \times 3^{-n-1}),$$

where $\epsilon(j) = 0$ or $\epsilon(j) = 1$. Sketch

$$E(n) = \bigcup_{\epsilon(j)\in\{0,1\}} E_{\epsilon(1),\epsilon(2),\dots,\epsilon(n)}$$

for $n = 1, 2, 3$, and verify that $E(n)$ is the union of 2^n intervals each of length 3^{-n-1} and such that the intervals are the middle third open intervals of the closed intervals which make up $[0,1] \setminus \bigcup_{k=0}^{n-1} E_k$. (You should worry more about understanding the pattern than about rigour.)

(ii) We take $f_n : [0,1] \to \mathbb{R}$ to be the simplest piecewise linear function which satisfies $f_n(0) = 0$, $f_n(1) = 1$ and

$$f_n(x) = 2^{-1} \qquad \text{for } x \in E(0),$$

$$f_n(x) = 2^{-(m+1)} + \sum_{j=1}^{m} 2^{-j\epsilon(j)} \qquad \text{for } x \in E_{\epsilon(1),\epsilon(2),\dots,\epsilon(m)} \text{ and } 1 \leq m \leq n.$$

I sketch f_2 in Figure K.4. Sketch f_3 for yourself. Explain why f_n is an increasing function.

(iii) Show that $\|f_n - f_m\|_\infty \leq 2^{-n}$, and deduce that f_n converges uniformly to a function f which must therefore be continuous. Show that f is

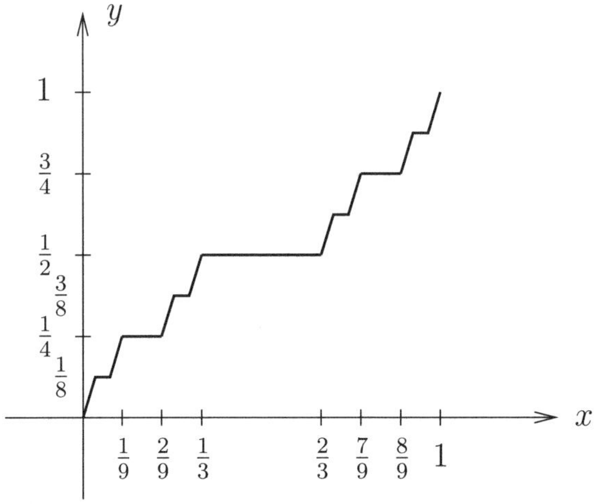

Figure K.4. The second approximation to the devil's staircase

increasing, that $f(0) = 0$, $f(1) = 1$ and

$$f(x) = 2^{-1} \qquad \text{for } x \in E(0),$$

$$f(x) = 2^{-(m+1)} + \sum_{j=1}^{m} 2^{-j\epsilon(j)} \qquad \text{for } x \in E_{\epsilon(1),\epsilon(2),\dots,\epsilon(m)} \text{ and all } m \geq 1.$$

Explain why f is differentiable on $E(m)$ with $f'(x) = 0$ for each $x \in E(m)$. Conclude that f is differentiable on $E = \bigcup_{m=0}^{\infty} E(m)$ with $f'(x) = 0$ for each $x \in E$.

(iv) Show that the total length of the intervals making up $\bigcup_{m=0}^{n} E(m)$ is $1 - (2/3)^{n+1}$. If $\mathbb{I}_E : [0,1] \to \mathbb{R}$ is given by $\mathbb{I}_E(x) = 1$ if $x \in E$ and $\mathbb{I}_E(x) = 0$ otherwise, show, by finding appropriate dissections, that $\mathbb{I}_E$ is Riemann integrable and

$$\int_0^1 \mathbb{I}_E(x)\,dx = 1.$$

Informally, we have shown that 'f has zero derivative on practically the whole of $[0,1]$, but still manages to increase from 0 to 1'! (Exercise K.338 continues this line of thought.)

(v) This part takes us back to Section 9.4, and, in particular, to the promise made on page 212 to exhibit a real-valued random variable which is not a simple mix of discrete and continuous. Our argument is informal (partly because we have not defined what we mean by a 'mix of discrete and continuous'). Define $F : \mathbb{R} \to \mathbb{R}$ by $F(x) = 0$ if $x < 0$, $F(x) = f(x)$ if $0 \leq x \leq 1$ and $F(x) = 1$ otherwise. Explain why F is continuous. Let X be the real-valued random variable with $\Pr\{X \leq x\} = F(x)$.

Suppose, if possible, that X is a 'mix of discrete and continuous'. Since F is continuous, X must, in fact, be a continuous random variable and so

$$F(x) = \Pr\{X \leq x\} = \int_{-\infty}^{x} g(t)\,dt$$

for some well behaved g. If g is Riemann integrable, then it is bounded on $[0,1]$ with $|g(x)| \leq K$ for all $x \in [0,1]$. Deduce that, under this assumption,

$$|F(x) - F(y)| \leq K(y-x)$$

for $0 \leq x \leq y \leq 1$. By showing that this last inequality is false for some x and y, obtain a contradiction.

Thus X must be a novel type of random variable.

[Graphs very similar to that of F turn up in the theory of differential equations, in particular for processes intended to mimic transition to turbulence in fluid flows.]

K.254. [11.4, T!] The object of this exercise and the next two is to define x^α and to obtain its properties directly rather than by the indirect definition used in Section 5.7. In this first exercise we deal with the case when α is rational. We assume the properties of x^n when n is an integer.

(i) Carefully quoting the results that you use, show that, if n is a strictly positive integer, then, given any $x > 0$, there exists a unique $r_{1/n}(x) > 0$ with $r_{1/n}(x)^n = x$. Show further (again carefully quoting the results that you use) that the function $r_{1/n} : (0,\infty) \to (0,\infty)$ is differentiable with

$$r'_{1/n}(x) = \frac{r_{1/n}(x)}{nx}.$$

(ii) Show that, if m, m', n and n' are strictly positive integers with $m/n = m'/n'$, then

$$(r_{1/n}(x))^m = (r_{1/n'}(x))^{m'}$$

for all $x > 0$. Explain why this means that we can define $r_{m/n}(x) = r_{1/n}(x)^m$ for all strictly positive integers m and n.

(iii) If α, $\beta \in \mathbb{Q}$ and α, $\beta > 0$, prove the following results.

(a) $r_{\alpha+\beta}(x) = r_\alpha(x) r_\beta(x)$ for all $x > 0$.

(b) $r_{\alpha\beta}(x) = r_\alpha(r_\beta(x))$ for all $x > 0$.

(c) $r_\alpha(xy) = r_\alpha(x) r_\alpha(y)$ for all $x,\ y > 0$.

(d) $r_\alpha(1) = 1$.

(e) The function $r_\alpha : (0,\infty) \to (0,\infty)$ is differentiable with

$$r'_\alpha(x) = \frac{\alpha r_\alpha(x)}{x}.$$

Show further that, if n is a strictly positive integer,

(f) $r_n(x) = x^n$ for all $x > 0$.

(iv) Show carefully how to define r_α for all $\alpha \in \mathbb{Q}$ and obtain results corresponding to those in part (iii).

K.255. [11.4, T!, ↑] We continue the arguments of Exercise K.254.

(i) Suppose that $x > 0$, $\alpha \in \mathbb{R}$ and $\alpha_n \in \mathbb{Q}$ with $\alpha_n \to \alpha$ as $n \to \infty$. Show that the sequence $r_{\alpha_n}(x)$ is Cauchy and so tends to a limit y, say.

Suppose that $\beta_n \in \mathbb{Q}$ with $\beta_n \to \alpha$ as $n \to \infty$. Show that $r_{\beta_n}(x) \to y$ as $n \to \infty$. We can thus write $r_\alpha(x) = y$ without risk of ambiguity.

Show also that $r_\alpha(x) > 0$.

(ii) Suppose that $\alpha \in \mathbb{R}$, $\alpha \neq 0$ and $\alpha_n \in \mathbb{Q}$ with $\alpha_n \to \alpha$ as $n \to \infty$. Show that, if $0 < a < b$, we have $r_{\alpha_n} \to r_\alpha$ uniformly on $[a, b]$. Deduce, quoting any results that you use, that the function $r_\alpha : (0, \infty) \to (0, \infty)$ is differentiable with

$$r'_\alpha(x) = \frac{\alpha r_\alpha(x)}{x}.$$

(iii) Prove the remaining results corresponding to to those in part (iii) of Exercise K.254.

K.256. [11.4, T!, ↑] Exercise K.255 does not give all the properties of x^α that we are interested in. In particular, it tells us little about the map $\alpha \mapsto x^\alpha$. If $a > 1$, let us write $P(t) = P_a(t) = a^t$. (The restriction to $a > 1$ is for convenience only and is removed in (viii).)

(i) Show that $P : \mathbb{R} \to (0, \infty)$ is an increasing function.

(ii) Show that $P(t) \to 1$ as $t \to 0$.

(iii) By using the relation $P(s+t) = P(s)P(t)$, show that P is everywhere continuous.

(iv) Show that, if P is differentiable at 0, then P is everywhere differentiable.

(v) Let $f(y) = n(y^{1/n} - 1) - (n+1)(y^{1/(n+1)} - 1)$. By considering f', or otherwise, show that $f(y) \geq 0$ for all $y \geq 1$. Hence, or otherwise, show that

$$\frac{P(1/n) - P(0)}{1/n} \text{ tends to a limit, } L \text{ say,}$$

as $n \to \infty$.

(vi) Show that, if $1/n \geq x \geq 1/(n+1)$, then

$$\frac{P(x) - P(0)}{x} \leq \frac{P(1/n) - P(0)}{1/n} \times \frac{n+1}{n}.$$

By using this and similar estimates, or otherwise, show that

$$\frac{P(x)-P(0)}{x} \to L$$

as $x \to 0$ through values of $x > 0$.

(vii) Show that, if $x \neq 0$

$$\frac{P(x)-P(0)}{x} = P(x)\frac{P(-x)-P(0)}{-x},$$

and deduce that

$$\frac{P(x)-P(0)}{x} \to L$$

as $x \to 0$. Conclude that P is differentiable at 0 and so everywhere with $P'(t) = LP(t)$.

(viii) If $a > 0$ and we set $P_a(t) = a^t$, show that P_a is differentiable with $P_a'(t) = L_a P'(t)$ for some constant L_a. Show that $L_a > 0$ if $a > 1$. What can you say about L_a for other values of a?

(ix) If $a, \lambda > 0$, show that $L_{a^\lambda} = \lambda L_a$. Deduce that there is a real number $e > 1$ such that, if we write

$$e(t) = e^t,$$

then e is a differentiable function with $e'(t) = e(t)$.

(x) Tear off the disguise of L_a.

We have now completed a direct attack on the problem of defining x^α and obtaining its main properties. It should be clear why the indirect definition $x^\alpha = \exp(\alpha \log x)$ is preferred.

K.257. [11.5, P] Show that

$$\cosh x \cos x = \sum_{n=0}^{\infty} \frac{4^n x^{4n}}{(4n)!}$$

for all x.

(a) By writing the left hand side as the sum of exponentials.

(b) By considering the product of the Taylor series of cos and cosh.

(c) By showing that both sides are the solution of a certain 4-th order differential equation with initial conditions.

(d) By using Taylor's theorem with the integral form of the remainder.

K.258. [11.5, P] Suppose that $\sum_{n=0}^{\infty} a_n z^n$ has radius of convergence greater than or equal to $R > 0$. Write $g(z) = \sum_{n=0}^{\infty} a_n z^n$ for $|z| < R$. Show that, if $|z_0| < R$, then we can find $b_0, b_1, \dots \in \mathbb{C}$ such that $\sum_{n=0}^{\infty} b_n z^n$ has radius of convergence greater than or equal to $R - |z_0|$ and $g(z) = \sum_{n=0}^{\infty} b_n (z - z_0)^n$ for $|z_0 - z| < R - |z_0|$. [This is quite hard work. If you have no idea how to attack it, first work out formally what the values of the b_n must be. Now try and use Lemma 5.26.]

Show that g is complex differentiable at z_0 with

$$g'(z_0) = b_1 = \sum_{n=1}^{\infty} n a_n z_0^{n-1}.$$

Deduce that a power series is differentiable and that its derivative is that obtained by term by term differentiation within its radius of convergence. (We thus have an alternative proof of Theorem 11.73.)

K.259. [11.4, H, ↑] (i) By considering $f_n(x) = A_n \sin\big(B_n x + C_n\big)$ for suitable A_N, B_N and C_N, show that we can find an infinitely differentiable function $f_n : \mathbb{R} \to \mathbb{R}$ such that (writing $g^{(0)}(x) = g(x)$)

$$|f_n^{(r)}(x)| \leq 2^{-n-1} \qquad \text{for all } n > r \geq 0 \text{ and all } x \in \mathbb{R},$$

yet

$$f_n^{(n)}(0) \geq (n!)^2 + 1 + \sum_{r=0}^{n-1} |f_r^{(n)}(0)|.$$

Explain why f_n tends uniformly to an infinitely differentiable function $f : \mathbb{R} \to \mathbb{R}$ with $f^{(n)}(0) \geq (n!)^2$. Show that the Taylor series

$$\sum_{n=0}^{\infty} \frac{f^{(n)}(0)}{n!} x^n$$

diverges for all $x \neq 0$.

(ii) The result of (i) can be extended as follows. Show that we can find a sequence y_1, y_2 ... of rational numbers such that each rational number occurs infinitely often in the sequence. (In other words, given a rational number q and an integer N, we can find an $n \geq N$ with $y_n = q$.)

By considering $f_n(x) = A_n \sin\big(B_n x + C_n\big)$ for suitable A_N, B_N and C_N, show that we can find an infinitely differentiable function $f_n : \mathbb{R} \to \mathbb{R}$ such that

$$|f_n^{(r)}(x)| \leq 2^{-n-1} \qquad \text{for all } n > r \geq 0 \text{ and all } x \in \mathbb{R}$$

yet

$$f_n^{(n)}(y_n) \geq (n!)^2 + 1 + \sum_{r=0}^{n-1} |f_r^{(n)}(y_n)|.$$

Deduce the existence of an infinitely differentiable function $f : \mathbb{R} \to \mathbb{R}$ whose Taylor series

$$\sum_{n=0}^{\infty} \frac{f^{(n)}(q)}{n!} (q+h)^n$$

diverges for all $h \neq 0$ and all rational numbers q.

(iii) The result of (ii) is even stronger than at first appears. Use the ideas of Exercise K.258 to show that, if a function $F : \mathbb{R} \to \mathbb{R}$ has derivatives of all orders at x and

$$F(x+h) = \sum_{n=0}^{\infty} \frac{F^{(n)}(x)}{n!} h^n$$

for all $|h| < R$, then F is infinitely differentiable at t for all $|t - x| < R$ and

$$F(t+k) = \sum_{n=0}^{\infty} \frac{F^{(n)}(t)}{n!} k^n$$

whenever $|k| < R - |t - x|$.

Conclude that the function f obtained in (ii) cannot have a valid Taylor expansion about any point.

K.260. [11.5, P] Here is another proof of Theorem 11.73. Suppose that $\sum_{n=0}^{\infty} a_n z^n$ has radius of convergence greater than or equal to $R > 0$.

(i) Show that, given any $\epsilon > 0$, we can find a $K(\epsilon)$ with $\sum_{n=0}^{\infty} |a_n z^n| < K(\epsilon)$ for all $|z| \leq R - \epsilon$.

(ii) Show that $\sum_{n=1}^{\infty} n a_n z^{n-1}$ and $\sum_{n=2}^{\infty} n(n-1) a_n z^{n-2}$ also have radius of convergence R.

(iii) Show that

$$\binom{n}{j} \leq n(n-1)\binom{n-2}{j}$$

for all $n - 2 \geq j \geq 0$, and deduce that

$$|(z+h)^n - z^n - nz^{n-1}h| \leq n(n-1)(|z| + |h|)^{n-2}$$

for all $z,\ h \in \mathbb{C}$.

(iv) Use (i), (ii) and (iii) to show that, if $0 < \delta$ and $|z| + |h| < R - \delta$, then

$$\left| \sum_{n=0}^{\infty} a_n (z+h)^n - \sum_{n=0}^{\infty} a_n z^n - h \sum_{n=1}^{\infty} n a_n z^{n-1} \right| \leq A(\delta)|h|^2,$$

where $A(\delta)$ depends only on δ (and not on h or z).

(v) Deduce that a power series is differentiable and that its derivative is that obtained by term by term differentiation within its radius of convergence.

K.261. [11.5, P] Consider the Legendre differential equation

$$(1 - x^2)y'' - 2xy' + l(l+1)y = 0,$$

where l is a real number. Find the general power series solution. Show that, unless l is a non-negative integer, every non-trivial power series solution has radius of convergence 1.

Show, however, that, if $l = n$, a non-negative integer, the general series solution can be written

$$y(x) = AP_n(x) + BQ_n(x),$$

where Q_n is a power series with radius of convergence 1, P_n is a polynomial of degree n, and A and B are arbitrary constants.

Verify that, when n is a non-negative integer, the function given by $v(x) = (x^2 - 1)^n$ satisfies the equation

$$(1 - x^2)v'(x) + 2nxv(x) = 0.$$

Deduce that $v^{(n)}(x)$ satisfies the Legendre differential equation with $l = n$. [Consult Exercise K.299 if you need a hint.] Conclude that P_n is a constant multiple of the function p_n defined in Exercise K.237.

K.262. [11.5, P] Obtain the general power series solution of

$$z^2\frac{d^2w}{dz^2} + z\frac{dw}{dz} + (z^2 - 1)w = 0.$$

For what values of z is your solution valid, and why? How many constants can you choose freely in your solution? Suggest (you are not asked to do more than suggest) a reason for this.

Answer the same questions for

$$z^2\frac{d^2w}{dz^2} + z\frac{dw}{dz} + (z^2 - 1)w = z^2.$$

Is it true that the equation

$$z^2\frac{d^2w}{dz^2} + z\frac{dw}{dz} + (z^2 - 1)w = P(z)$$

always has a power series solution when P is a polynomial? Give reasons.

K.263. [11.5, P] Using Lemma 11.81, or otherwise, find the power series expansion $\sum_{j=0}^{\infty} a_j x^j$ of $(1 + x)^{1/2}$ for x real and $|x| < 1$.

Show that the complex power series $\sum_{j=0}^{\infty} a_j z^j$ has radius of convergence 1. What is the power series for the product $\sum_{j=0}^{\infty} a_j z^j \sum_{j=0}^{\infty} a_j z^j$, and why? What is its radius of convergence?

By considering

$$(1 - z)^{1/2}\big((1 - z)^{-1/2}(1 + K^{-1}z)\big),$$

or otherwise, show that, if two power series of the same radius of convergence $R > 0$ are multiplied, the resulting power series may have any radius of convergence with value R or greater. (The case $R = 0$ is not really a question of analysis, and is left to enthusiasts.)

Show, more generally, that, if two power series with radius of convergence $R > 0$ and $S > 0$ are multiplied, the resulting power series may have any radius of convergence with value $\min(R, S)$ or greater.

K.264. [11.5, P] By modifying the proof of Abel's test (Lemma 5.15), or otherwise, prove the following result. Let E be a set and $\mathbf{a}_j : E \to \mathbb{R}^m$ a sequence of functions. Suppose that there exists a K such that

$$\left\| \sum_{j=1}^{n} \mathbf{a}_j(t) \right\| \leq K$$

for all $n \geq 1$ and all $t \in E$. If λ_j is a decreasing sequence of real positive numbers with $\lambda_j \to 0$ as $j \to \infty$, then $\sum_{j=1}^{\infty} \lambda_j \mathbf{a}_j$ converges uniformly on E.

Show, using similar ideas (see Exercise K.76 if necessary), that, if $b_n \in \mathbb{R}$ and $\sum_{n=0}^{\infty} b_n$ converges, then $\sum_{n=0}^{\infty} b_n x^n$ converges uniformly on $[0, 1]$. Explain why this implies that

$$\bigstar \qquad \int_0^1 \left(\sum_{n=0}^{\infty} b_n x^n \right) dx = \sum_{n=0}^{\infty} \frac{b_n}{n+1}.$$

Show that, provided we interpret the left hand integral as an improper integral, $\lim_{\epsilon \to 0,\ \epsilon > 0} \int_0^{1-\epsilon}$, equation $\bigstar$ remains true under the assumption that $\sum_{n=0}^{\infty} b_n/(n+1)$ converges.

Show that

$$1 - \frac{1}{2} + \frac{1}{3} - \frac{1}{4} + \cdots = \log 2$$

and

$$1 - \frac{1}{3} + \frac{1}{5} - \frac{1}{7} + \cdots = \frac{\pi}{4}.$$

Do these provide good methods for computing $\log 2$ and π? Why?

Recall that, if $a, b > 0$, then $a^{\log_a b} = b$. Show that

$$\log_2 e - \log_4 e + \log_8 e - \log_{16} e + \ldots$$

converges to a value to be found.

K.265. [11.5, G, S] (Only if you have done sufficient probability, but easy if you have.) Let X be a random variable taking positive integral values. Show that

$$\phi_X(t) = \mathbb{E} t^X = \sum_{n=0}^{\infty} \Pr(X = n) t^n$$

is well defined for all $t \in [-1, 1]$.

If $\mathbb{E}X$ is bounded, show that

$$\frac{\phi_X(1) - \phi_X(t)}{1 - t} \to \mathbb{E}X$$

as $t \to 1$ through values $t < 1$. Give an example of an X with $\mathbb{E}X$ infinite.

K.266. [11.5, T, S, ↑↑] Use the ideas of Exercise K.82 to prove the following extension. Suppose that we have a sequence of functions $a_n : \Omega \to \mathbb{C}$ and a sequence of positive real numbers M_n such that $\sum_{j=1}^{\infty} M_j$ converges absolutely and $|a_n(z)| \leq M_n$ for all $z \in \Omega$ and all n. Then

$$\prod_{j=1}^{N}(1 + a_j(z)) \to \prod_{j=1}^{\infty}(1 + a_j(z))$$

uniformly on Ω.

K.267. [11.5, P, S, ↑] We do not have the apparatus to exploit Exercise K.266 fully, but the following exercise is suggestive.

(i) Show that $S_N(z) = z\prod_{n=1}^{N}(1 - z^2 n^{-2})$ converges uniformly to a limit $S(z)$ on any disc $\{z : |z| < R\}$. Show that S is a well defined continuous function on $\mathbb{C}$.

(ii) Observe that $(1 - z^2 n^{-2}) = n^{-2}(n - z)(n + z)$. By considering $S_N(z+1)/S_N(z)$, or otherwise, show that $S(z+1) = -S(z)$. Show also that $S(z) = 0$ if and only if z is an integer.

[The product $S(z) = z\prod_{n=1}^{\infty}(1 - z^2 n^{-2})$ goes back to Euler, who identified it with a well known function which the reader should also be able to guess. An advantage of infinite products over infinite sums is that we can find zeros of the function much more easily.]

K.268. [11.5, T] (i) Suppose that $a_n \in \mathbb{C}$ and $\sum_{n=0}^{\infty} a_n z^n$ has radius of convergence $R > 0$. If $a_N \neq 0$ and $a_n = 0$ for $n < N$, use the inequality

$$\left|\sum_{n=0}^{q} a_n z^n\right| \geq |z|^N \left(|a_N| - \sum_{n=N+1}^{q} |a_n z^{n-N}|\right)$$

(for $q > N$), together with standard bounds on $|a_n z^n|$, to show that there exists a $\delta > 0$ such that

$$\left|\sum_{n=0}^{\infty} a_n z^n\right| \geq \frac{|a_N||z|^N}{2}$$

for all $|z| < \delta$.

(ii) Suppose that $a_n \in \mathbb{C}$ and $\sum_{n=0}^{\infty} a_n z^n$ has radius of convergence $R > 0$. Set $f(z) = \sum_{n=0}^{\infty} a_n z^n$ for $|z| < R$. Show, using (i), or otherwise, that, if there exist $w_m \to 0$ with $f(w_m) = 0$ and $w_m \neq 0$, then $a_n = 0$ for all n, and so $f = 0$.

(iii) Use Exercise K.258 to deduce that, if f is as in (ii), then either $f(z) = 0$ for all $|z| < R$, or, whenever $|z_0| < R$ and $f(z_0) = 0$, we can find a $\delta > 0$ with $|z_0| + \delta < R$ such that $f(z) \neq 0$ for all z with $|z - z_0| < \delta$. (We say that the zeros of a non-trivial power series must be isolated.)

K.269. [11.5, P, ↑] This is a commentary on the previous exercise. We shall consider functions from $\mathbb{R}$ to $\mathbb{R}$.

(i) Suppose that $a_n \in \mathbb{R}$ and $\sum_{n=0}^{\infty} a_n x^n$ has radius of convergence $R > 0$. Set $f(x) = \sum_{n=0}^{\infty} a_n x^n$ for $|x| < R$. Show that, if there exist $x_m \to 0$ with $f(x_m) = 0$ and $x_m \neq 0$, then $a_n = 0$ for all n, and so $f = 0$.

(ii) Explain why a polynomial of degree at most n with $n+1$ zeros must be zero.

(iii) Give an example of a power series with radius of convergence ∞ which has infinitely many zeros but is not identically zero.

(iv) Let F be as in Example 7.5. Set $G(x) = F(x) \sin x^{-1}$ for $x \neq 0$ and $G(0) = 0$. Show that G is infinitely differentiable everywhere (this will probably require you to go through much the same argument as we used for F). Show that there exist $x_m \to 0$ with $G(x_m) = 0$ and $x_m \neq 0$, but G is not identically zero.

K.270. [11.5, P] This exercise prepares the way for Exercise K.271. The first two parts will probably be familiar, but are intended to provide a hint for the third part.

(i) If $n \geq r \geq 1$, show that $\binom{n}{r-1} + \binom{n}{r} = \binom{n+1}{r}$.

(ii) Use induction to show that

$$(x+y)^n = \sum_{r=0}^{n} \binom{n}{r} x^r y^{n-r}$$

whenever n is a positive integer and x and y are real.

(iii) Suppose that n is a positive integer and x and y are real. Verify the next formula directly for $n = 0$, 1, 2, 3 (we take $\prod_{j=0}^{-1}(x-j) = 1$), and then prove it for all n.

$$\sum_{r=0}^{n} \binom{n}{r} \prod_{j=0}^{r-1}(x-j) \prod_{k=0}^{n-r-1} (y-k) = \prod_{q=0}^{n-1}(x+y-q).$$

K.271. [11.5, T, ↑] The following is a modification of Cauchy's proof of the binomial theorem. We shall work in $\mathbb{R}$. We take w as a fixed real number with $|w| < 1$.

(i) Use the ratio test to show that $\sum_{n=1}^{\infty} \frac{1}{n!} \prod_{j=0}^{n-1}(x-j) w^n$ converges. By thinking about your proof and the Weierstrass M-test, improve this result to show that $\sum_{n=1}^{\infty}\frac{1}{n!}\prod_{j=0}^{n-1}(x-j)w^n$ converges uniformly for $|x| \leq M$ whenever M is fixed. We write

$$f(x) = 1 + \sum_{n=1}^{\infty}\frac{1}{n!}\prod_{j=0}^{n-1}(x-j)w^n.$$

Explain why f is continuous.

(ii) Use Exercise 5.38 (first proved by Cauchy) and Exercise K.270 to show that

$$\bigstar \qquad f(x)f(y) = f(x+y)$$

for all x and y.

(iii) Show that $f(0) = 1$ and deduce that $f(x) \neq 0$ for all x. Deduce, quoting any theorem you need, that $f(x) > 0$ for all x.

(iv) Explain why we can define $g : \mathbb{R} \to \mathbb{R}$ by $g(x) = \log f(x)$ and why g is continuous. Show that

$$\bigstar\bigstar \qquad g(x) + g(y) = g(x+y)$$

for all x and y. Use Exercise K.101 (i) (first proved by Cauchy) to show that $g(x) = ax$, and so $f(x) = b^x$ for some real positive number b.

(v) Find b by considering $f(1)$. Deduce that

$$(1+w)^x = 1 + \sum_{n=1}^{\infty}\frac{1}{n!}\prod_{j=0}^{n-1}(x-j)w^n.$$

[One problem faced by anyone teaching elementary analysis in France is that every theorem seems to be called Cauchy's theorem. A glance at the exercise above shows why[16].]

K.272. [11.5, T, ↑↑] (Part (iii) of this exercise makes strong use of Exercise K.79.) Exercise 11.86 raises the question of when we can tell whether a differential equation of the form $u'(t) = f(t, u(t))$ has a power series solution. The following theorem of Cauchy (see the concluding remark of the previous exercise) gives a rather natural condition.

Suppose that $c_{n,m} \in \mathbb{R}$ $[n,\, m \geq 0]$ and that there exist a $\rho > 0$ and a $K > 0$ such that

$$|c_{n,m}| \leq K\rho^{n+m} \text{ for all } n,\, m \geq 0.$$

[16] Moreover, Cauchy's contributions to the rigorising of analysis form only a fragment of his contribution to pure and applied mathematics. The following quotation from a materials engineer can be echoed in many fields. '[Cauchy's paper of 1822] was perhaps the most important event in the history of elasticity since Hooke. After this, that science showed promise of becoming a practical tool for engineers rather than a happy hunting-ground for a few somewhat eccentric philosophers.' ([**19**], Chapter 3)

(Exercise K.77 (c) shows why we use this condition.) Then we can find a $\delta > 0$ with $\rho > \delta$ and $a_n \in \mathbb{R}$ such that $\sum_{n=0}^{\infty} a_n t^n$ converges to $u(t)$, say, for all $|t| < \delta$ and such that, writing

$$f(x,y) = \sum_{n=0}^{\infty} \sum_{m=0}^{\infty} c_{n,m} x^n y^m$$

for $|x|, |y| < \rho^{-1}$, we have $u(0) = 0$, $|u(t)| < \rho$ and

$$u'(t) = f(t, u(t))$$

for all $|t| \le \delta$.

(i) The object of this exercise is to prove Cauchy's theorem, but in the first two parts of the exercise we are merely interested in seeing what is going on and so we work in a non-rigorous manner. Assuming that a power series solution $u(t) = \sum_{n=0}^{\infty} a_n t^n$ with the required properties actually exists, show, by formal manipulation, that $a_0 = 0$ and $k a_k$ should be the coefficient of x^{k-1} in

$$\sum_{n=0}^{\infty} \sum_{m=0}^{\infty} c_{n,m} x^n \left(\sum_{r=0}^{\infty} a_r x^r \right)^m .$$

Explain why that coefficient is a finite sum only depending on a_0, a_1, ..., a_{k-1}. Show that, if we define $A_0 = 0$, and define A_1, A_2, ... formally by means of the equation

$$\bigstar \qquad \frac{d}{dt}\left(\sum_{n=0}^{\infty} A_n t^n \right) = \sum_{n=0}^{\infty} \sum_{m=0}^{\infty} K \rho^{n+m} t^n \left(\sum_{r=0}^{\infty} A_r t^r \right)^m ,$$

then $|a_k| \le A_k$.

(ii) We continue with the ideas of (i). Show that $\bigstar$ can be rewritten (at least, formally) as

$$\bigstar\bigstar \qquad \frac{dw}{dt} = \frac{K}{(1-\rho t)(1-\rho w)},$$

where $w(t) = \sum_{n=0}^{\infty} A_n t^n$. Solve equation $\bigstar\bigstar$ for $w(0) = 0$.

(iii) We now reverse the non-rigorous argument, but this time proceed rigorously. Show that equation $\bigstar\bigstar$ has a solution $w(t)$ with $w(0) = 0$ and the property that we can find A_j and $\eta > 0$ such that $\sum_{n=0}^{\infty} A_n t^n$ has radius of convergence at least η and $w(t) = \sum_{n=0}^{\infty} A_n t^n$ for all $|t| < \eta$. Show that the A_n satisfy equation $\bigstar$ for $|t| < \eta$. Deduce that, if a_k is given by equating coefficients in the equation

$$\sum_{n=0}^{\infty} \sum_{n=0}^{\infty} a_n x^n = \sum_{m=0}^{\infty} c_{n,m} x^n \left(\sum_{r=0}^{\infty} a_r x^r \right)^m ,$$

then $|a_k| \leq A_k$ and $\sum_{n=0}^{\infty} a_n t^n$ has radius of convergence at least η. Show that, if we set $u(t) = \sum_{n=0}^{\infty} a_n t^n$ for $|t| < \eta$, then $u(0) = 0$ and there exists a δ with $0 < \delta \leq \eta$ such that $|u(t)| < \rho$ for $|t| < \delta$. Show that δ and u satisfy the conclusions of Cauchy's theorem.

K.273. (Convolution.) [11.6, T] Let us write $C_P(\mathbb{R})$ for the set of continuous functions $f : \mathbb{R} \to \mathbb{R}$ which are periodic with period 2π.

(i) Show that, if $f, g \in C_P(\mathbb{R})$, then the equation

$$f * g(t) = \frac{1}{2\pi} \int_{-\pi}^{\pi} f(t-s)g(s)\, ds$$

gives a well defined, continuous, 2π periodic function $f * g : \mathbb{R} \to \mathbb{R}$. (Thus $f * g \in C_P(\mathbb{R})$.)

(ii) Show that, if $f, g \in C_P(\mathbb{R})$, then

$$f * g(t) = \frac{1}{2\pi} \int_{a}^{2\pi+a} f(t-s)g(s)\, ds$$

for all $t \in \mathbb{R}$ and $a \in \mathbb{R}$.

(iii) Show that, if $f, g, h \in C_P(\mathbb{R})$ and $\lambda \in \mathbb{R}$, then

$$(\lambda f) * g = \lambda(f * g),\ f * (g + h) = f * g + f * h,$$
$$f * g = g * f \text{ and } f * (g * h) = (f * g) * h.$$

(iv) Suppose that $f, g \in C_P(\mathbb{R})$ and f is differentiable with $f' \in C_P(\mathbb{R})$. Show that $f * g$ is differentiable and

$$(f * g)' = f' * g.$$

(v) Suppose that $f, u_n \in C_P(\mathbb{R})$, that $u_n(t) \geq 0$ for all $t \in \mathbb{R}$, that $u_n(t) = 0$ for $\pi/n \leq |t| \leq \pi$, and that

$$\frac{1}{2\pi} \int_{-\pi}^{\pi} u_n(t)\, dt = 1.$$

Show that $u_n * f \to f$ uniformly as $n \to \infty$.

(vii) Suppose that $f \in C_P(\mathbb{R})$ and $\epsilon > 0$. Show, by using parts (iv) and (v), that there exists a twice differentiable function g with $g, g', g'' \in C_P(\mathbb{R})$ such that $\|f - g\|_\infty \leq \epsilon$.

(viii) Let us write $C_P(\mathbb{C})$ for the set of continuous functions $f : \mathbb{R} \to \mathbb{C}$ which are periodic with period 2π. Show. in as much detail as you consider desirable, that parts (i) to (vi) hold with $C_P(\mathbb{R})$ replaced by $C_P(\mathbb{C})$

K.274. [11.6, T, ↑] We continue with the notation of the previous exercise.

(i) Suppose that $f, g \in C_P(\mathbb{C})$. Show that

$$\widehat{f * g}(n) = \hat{f}(n)\hat{g}(n)$$

for all integers n. (We say that 'taking Fourier coefficients converts convolution to multiplication'.)

(ii) Show that, if $f \in C_P(\mathbb{C})$, then

$$|\hat{f}(n)| \leq \|f\|_\infty.$$

(iii) Show that, if $f,\ g \in C_P(\mathbb{C})$, then

$$|\hat{f}(n)| \leq |\hat{g}(n)| + \|f - g\|_\infty.$$

(iv) Use part (iii) of this exercise, part (vii) of Exercise K.273 and Exercise 11.97 to show that, if $f \in C_P(\mathbb{C})$, then $\hat{f}(n) \to 0$ as $|n| \to \infty$. (This is a version of the important Riemann–Lebesgue lemma.)

(v) Suppose, if possible, that there exists an $e \in C_P(\mathbb{C})$ such that $e * f = f$ for all $f \in C_P(\mathbb{C})$. Find $\hat{e}$ and use (iv) to show that no such e can exist. (Informally, convolution on $C_P(\mathbb{C})$ has no unit.)

K.275. [11.6, T, ↑] This exercise uses the version of the Riemann–Lebesgue lemma proved in Exercise K.274 (iv). If $f \in C_P(\mathbb{R})$, we write

$$S_n(f,t) = \sum_{j=-n}^{n} \hat{f}(n) \exp(int).$$

(i) Suppose that $f_1,\ g_1 \in C_P(\mathbb{R})$ and $f_1(t) = g_1(t)\sin t$ for all t. Show that $\hat{f}_1(j) = \big(\hat{g}_1(j-1) - \hat{g}_1(j+1)\big)/2i$, and deduce that $S_n(f_1, 0) \to 0$ as $n \to \infty$.

(ii) Suppose that $f_2 \in C_P(\mathbb{R})$, $f_2(n\pi) = 0$ and f_2 is differentiable at $n\pi$ for all integers n. Show that there exists a $g_2 \in C_P(\mathbb{R})$ such that $f_2(t) = g_2(t)\sin t$ for all t, and deduce that $S_n(f_2, 0) \to 0$ as $n \to \infty$.

(iii) Suppose that $f_3 \in C_P(\mathbb{R})$, $f_3(0) = 0$ and f_3 is differentiable at 0. Write $f_4(t) = f_3(2t)$. Compute $\hat{f}_4(j)$ in terms of the Fourier coefficients of f_3. Show that $S_n(f_4, 0) \to 0$, and deduce that $S_n(f_3, 0) \to 0$ as $n \to \infty$.

(iv) Suppose that $f \in C_P(\mathbb{R})$, and f is differentiable at some point x. Show that $S_n(f, x) \to f(x)$ as $n \to \infty$.

K.276. [12.1, H] The result of this exercise is a special case of part of Lemma 13.4. However, precisely because it is a special case, some readers may find it a useful introduction to the ideas of Section 13.1.

(i) Let $\delta > 0$ and let $f : \mathbb{C} \to \mathbb{C}$ be an everywhere differentiable function such that $f(0) = 0$ and $|f'(z) - 1| \leq 1/2$ for all $|z| \leq \delta$. Set $X = \{z \in \mathbb{C} : |z| \leq \delta\}$. If $|w| \leq \delta/2$ and we define $Tz = z - f(z) + w$ for $z \in X$, show, by using the mean value inequality (Exercise 11.67), that $Tz \in X$ whenever $z \in X$. Thus T is a function $T : X \to X$.

(ii) We continue with the hypotheses and notation of part (i). Show that T is a contraction mapping, and deduce that the equation $f(z) = w$ has exactly one solution with $|z| \leq \delta$.

(iii) Let $F : \mathbb{C} \to \mathbb{C}$ be an everywhere differentiable function with continuous derivative. Suppose further that $F(0) = 0$ and $F'(0) = 1$. Show that there exists a $\delta > 0$ such that, if $|w| \leq \delta/2$, the equation $F(z) = w$ has exactly one solution with $|z| \leq \delta$.

(iv) Let $g : \mathbb{C} \to \mathbb{C}$ be an everywhere differentiable function with continuous derivative. Show that, if $g'(0) \neq 0$, there exist $\eta_1, \eta_2 > 0$ such that, if $|\omega - g(0)| \leq \eta_1$, the equation $g(z) = w$ has exactly one solution with $|z| \leq \eta_2$.

(v) If we omit the condition $g'(0) \neq 0$ in (iv), does it remain true that there exists an $\eta > 0$ such that, if $|\omega - g(0)| \leq \eta$, the equation $g(z) = w$ has a solution? Give a proof or counterexample.

(vi) If we omit the condition $g'(0) \neq 0$ in (iv) but add the condition $g''(0) \neq 0$, does it remain true that there exist $\eta_1, \eta_2 > 0$ such that, if $|\omega - g(0)| \leq \eta_1$, the equation $g(z) = w$ has at most one solution with $|z| \leq \eta_2$? Give a proof or counterexample.

K.277. [12.1, P, ↑] We work in $\mathbb{C}$. Consider the following statement.

There exists a $\delta > 0$ such that, if $|w| \leq \delta$, the equation $f_j(z) = w$ has a solution.

Give a proof or counterexample in each of the following cases.

(i) $f_1(z) = z^*$.

(ii) $f_2(z) = z + z^*$.

(iii) $f_3(z) = z + |z|$.

(iv) $f_4(z) = z + |z|^2$.

Is our statement true for all F in the following cases? Give a proof or counterexample.

(v) $f_5(z) = F(z) + |z|$, where $F : \mathbb{C} \to \mathbb{C}$ is an everywhere differentiable function with continuous derivative, $F(0) = 0$ and $F'(0) = 1$.

(vi) $f_6(z) = F(z) + |z|^2$, where $F : \mathbb{C} \to \mathbb{C}$ is an everywhere differentiable function with continuous derivative, $F(0) = 0$ and $F'(0) = 1$.

K.278. (Newton–Raphson.) [12.1, T] (i) Suppose that $f : \mathbb{R} \to \mathbb{R}$ is a twice differentiable function such that

$$\left| \frac{f(x)f''(x)}{f'(x)^2} \right| \leq \lambda$$

(thus, implicitly, $f(x) \neq 0$) for all x and some $|\lambda| < 1$. Show that the mapping

$$Tx = x - \frac{f(x)}{f'(x)}$$

is a contraction mapping, and deduce that f has a unique root y.

(ii) Suppose that $F : \mathbb{R} \to \mathbb{R}$ is a twice differentiable function such that

$$\left| \frac{F(x)F''(x)}{F'(x)^2} \right| \leq \lambda$$

for all $|x| \leq a$ and some $|\lambda| < 1$, and that $F(0) = 0$. Consider the mapping

$$Tx = x - \frac{F(x)}{F'(x)}.$$

Show that $T^n x \to 0$ whenever $|x| \leq a$.

Suppose that F is three times differentiable. By using a local Taylor theorem, or otherwise, show that there exists an $\eta > 0$ such that

$$|Tx| \leq \left(\left| \frac{F''(0)}{2F'(0)} \right| + 1 \right) x^2$$

whenever $|x| \leq \eta$.

(iii) If you know what the Newton–Raphson method is, comment on the relevance of the results of (i) and (ii) to that method. Comment in particular on the speed of convergence.

K.279. [12.1, G, P, S] (This is a short exercise and involves no analysis.) Suppose $f,\ g : X \to X$ are such that $fg = gf$. Show that, if f has a unique fixed point, then g has a fixed point. Can g have more than one fixed point? (Give a proof or counterexample.)

Show that, if we merely know that f has fixed points, it does not follow that g has any.

K.280. [12.1, P] If (X, d) is a complete metric space and $T : X \to X$ is a surjective map such that

$$d(Tx, Ty) \geq Kd(x, y)$$

for all $x, y \in X$ and some $K > 1$, show that T has a unique fixed point.

By considering the map $T : \mathbb{R} \to \mathbb{R}$ defined by $T(n + x) = 1 + 4n + 2x$ for $0 \leq x < 1$ and n an integer, or otherwise, show that the condition T surjective cannot be dropped.

K.281. [12.1, P] We work in $\mathbb{R}^m$ with the usual distance. Let E be a closed non-empty subset of $\mathbb{R}^m$ and let T be a map $T : E \to E$.

(i) Suppose $\|T(\mathbf{a}) - T(\mathbf{b})\| < \|\mathbf{a} - \mathbf{b}\|$ for all $\mathbf{a}, \mathbf{b} \in E$ with $\mathbf{a} \neq \mathbf{b}$. We saw in Example 12.4 that T need not have a fixed point. Show that, if T has a fixed point, it is unique.

(ii) Suppose $\|T(\mathbf{a}) - T(\mathbf{b})\| > \|\mathbf{a} - \mathbf{b}\|$ for all $\mathbf{a}, \mathbf{b} \in E$ with $\mathbf{a} \neq \mathbf{b}$. Show that T need not have a fixed point but, if T has a fixed point, it is unique.

(iii) Suppose $\|T(\mathbf{a}) - T(\mathbf{b})\| = \|\mathbf{a} - \mathbf{b}\|$ for all $\mathbf{a}, \mathbf{b} \in E$. Show that T need not have a fixed point and that, if T has a fixed point, it need not be unique.

(iv) Suppose now that E is non-empty, closed and bounded and

$$\|T(\mathbf{a}) - T(\mathbf{b})\| < \|\mathbf{a} - \mathbf{b}\|$$

for all $\mathbf{a}, \mathbf{b} \in E$ with $\mathbf{a} \neq \mathbf{b}$. By considering $\inf_{\mathbf{x} \in E} \|\mathbf{x} - T(\mathbf{x})\|$, or otherwise, show that T has a fixed point.

(v) Suppose that $E = \mathbb{R}^m$ and

$$\|T(\mathbf{a}) - T(\mathbf{b})\| < \|\mathbf{a} - \mathbf{b}\|$$

for all $\mathbf{a}, \mathbf{b} \in \mathbb{R}^m$ with $\mathbf{a} \neq \mathbf{b}$. Show that, if there exist a point $\mathbf{c} \in \mathbb{R}^m$ and a $K > 0$ such that $\|T^n \mathbf{c}\| < K$ for all positive integers n, then T has a fixed point.

K.282. [12.1, G, P, S] (This is easy if you see what is going on.) Show by means of an example that an isometry (distance preserving map) $f : X \to X$ on a metric space (X, d) need not be bijective.

Let (X, d) be a metric space. Show that the set of bijective isometries $f : X \to X$ forms a group $G(X)$ under composition.

By choosing X to be an appropriate closed bounded subset of $\mathbb{R}^2$ with the usual metric, or otherwise, prove the following statements about $G(X)$ when X has the Bolzano–Weierstrass property.

(i) $G(X)$ need not be Abelian.

(ii) There exists an X such that $G(X)$ has only one element.

(iii) There exists an X such that $G(X)$ has infinitely many elements.

If $n \geq 2$, give an example of an (X, d) with the Bolzano–Weierstrass property and a $T : X \to X$ such that T^m has no fixed points for $1 \leq m \leq n - 1$ but T^n does.

K.283. [12.1, P] (i) Let (X, d) be a metric space which has the Bolzano–Weierstrass property. Suppose $f : X \to X$ is *expansive* in the sense that

$$d(f(x), f(y)) \geq d(x, y)$$

for all $x, y \in X$. (Note that we do not assume that f is continuous.) If $a, b \in X$, we write $a_0 = a$ and $a_{n+1} = f(a_n)$ and define a sequence b_n similarly. Show that we can find a sequence of integers $n(1) < n(2) < \dots$

such that $d(a_{n(k)}, a_0) \to 0$ and $d(b_{n(k)}, b_0) \to 0$ as $n \to \infty$. Deduce that $d(a, b) = d(f(a), f(b))$ for all $a, b \in X$ (that is, f is an isometry).

Find a complete metric space (Y, ρ) and maps $g, h : Y \to Y$ such that

$$\rho(g(x), g(y)) \geq 2\rho(x, y),\ \rho(h(x), h(y)) \geq 2\rho(x, y)$$

and g is continuous but h is not.

(ii) Let (X, d) be a metric space with the Bolzano–Weierstrass property. Suppose $f : X \to X$ is an *isometry* in the sense that

$$d(f(x), f(y)) = d(x, y)$$

for all $x, y \in X$. If $a \in X$, we write $a_0 = a$ and $a_{n+1} = f(a_n)$. Show (as in (i)) that we can find a sequence of integers $n(1) < n(2) < \dots$ such that $d(a_{n(k)}, a_0) \to 0$ as $n \to \infty$, and deduce that f is bijective.

Find a complete metric space (Y, ρ) and a map $g : Y \to Y$ such that

$$\rho(g(x), g(y)) = \rho(x, y)$$

for all $x, y \in Y$, but g is not surjective.

K.284. [12.1, P, ↑] Let (X, d) be a metric space which has the Bolzano–Weierstrass property, and (Y, ρ) a metric space. Suppose $f : X \to Y$ and $g : Y \to X$ are such that $\rho(f(x), f(x')) \geq d(x, x')$ for all $x, x' \in X$ and $d(g(y), g(y')) \geq \rho(y, y')$ for all $y, y' \in Y$. By considering $g \circ f$ and using Exercise K.283, show that $\rho(f(x), f(x')) = d(x, x')$ for all $x, x' \in X$, $d(g(y), g(y')) = \rho(y, y')$ for all $y, y' \in Y$, and f and g are bijective. Is it necessarily true that f and g are inverse functions?

Give an example of two complete metric spaces (X, d) and (Y, ρ) having the Bolzano–Weierstrass property and functions $f : X \to Y$ and $g : Y \to X$ such that f is bijective, f and f^{-1} are continuous but $\rho(f(x), f(x')) \geq 2d(x, x')$ and g is an isometry but not bijective.

K.285. [12.1, P] Let $(V, \| \ \|)$ be a complete normed space. We say that a subset $A \subseteq V$ is *convex* if whenever $\mathbf{a}, \mathbf{b} \in A$ and $0 \leq \lambda \leq 1$ we have

$$\lambda\mathbf{a} + (1 - \lambda)\mathbf{b} \in A.$$

Suppose that A is a closed bounded convex subset of V and $\mathbf{f} : A \to A$ satisfies

$$\|\mathbf{f}(\mathbf{a}) - \mathbf{f}(\mathbf{b})\| \leq \|\mathbf{a} - \mathbf{b}\|$$

for all $\mathbf{a}, \mathbf{b} \in A$. By considering

$$\mathbf{g}(\mathbf{x}) = (1 - \epsilon)\mathbf{f}(\mathbf{x}) + \epsilon\mathbf{a}_0$$

for some $\mathbf{a}_0 \in A$ and $\epsilon > 0$, show that

$$\inf\{\|\mathbf{f}(\mathbf{a}) - \mathbf{a}\| \ : \ \mathbf{a} \in A\} = 0,$$

and deduce that $\mathbf{f}$ has a fixed point in A.

Give an example to show that the conclusion may be false if any one of the three conditions: convex, closed or bounded is omitted.

The following result is useful in the theory of Markov chains (see Exercise K.287). Suppose that $p_{ij} \geq 0$ for $1 \leq i, j \leq n$ and $\sum_{j=1}^{n} p_{ij} = 1$ for $1 \leq i \leq n$. Give $\mathbb{R}^n$ the norm $\| \ \|_1$, where $\|\mathbf{x}\|_1 = \sum_{j=1}^{n} |x_j|$. By choosing an appropriate A and $\mathbf{f}$, show that we can find $\pi_i \geq 0$ for $1 \leq i \leq n$ with $\sum_{i=1}^{n} \pi_i = 1$ such that

$$\sum_{i=1}^{n} \pi_i p_{ij} = \pi_j.$$

K.286. [12.1, P] Throughout this exercise (X, d) is a non-empty metric space with the Bolzano–Weierstrass property, and $T : X \to X$ is a map such that $d(Tx, Ty) \leq d(x, y)$ and such that, given any $x, y \in X$ with $x \neq y$, we can find a strictly positive integer $N(x, y)$ such that $d(T^{N(x,y)}x, T^{N(x,y)}y) < d(x, y)$.

(i) Show that the map $S : X \to \mathbb{R}$ given by $Sx = d(x, Tx)$ is continuous. Explain why this means that there is a $z \in X$ such that $d(z, Tz) \leq d(Tx, x)$ for all $x \in X$. Show that, in fact, $z = Tz$. Show that z is the unique fixed point of T.

(ii) Let $\epsilon > 0$ and write

$$U_n = \{x \in X \, : \, d(T^n x, z) < \epsilon\}.$$

Show that $U_n \subseteq U_{n+1}$ and that U_n is open for all $n \geq 1$. Write $U = \bigcup_{n=1}^{\infty} U_n$ and $F = X \setminus U$. Why does F have the Bolzano–Weierstrass property? Show that $Tx \in F$ whenever $x \in F$. Use part (i) to show that, if F were non-empty, there would exist a $w \in F$ with $Tw = w$. Deduce that F is indeed empty.

(iii) Show that, if $x \in X$, then $d(T^n x, z) \to 0$ as $n \to \infty$.

K.287. (Finite Markov chains.) [12.1, T!, ↑] In a finite Markov chain with states 1, 2, ... , m, if the system is in state i at time n it will be in state j at time $n + 1$ with probability p_{ij} irrespective of any earlier history. Thus, if it is in state i with probability q_i at time n and in state j with probability q'_j at time $n + 1$, we have

$$q'_j = \sum_{i=1}^{m} q_i p_{ij}.$$

The reader may ignore the previous paragraph, but it explains why we are interested in the space

$$X = \{\mathbf{q} \in \mathbb{R}^m \, : \, \sum_{i=1}^{m} q_i = 1, \ q_s \geq 0 \ [1 \leq s \leq m]\}$$

and the map $T : X \to \mathbb{R}^m$ given by $T\mathbf{q} = \mathbf{q}'$, where $q'_j = \sum_{i=1}^m q_i p_{ij}$ for all $1 \le j \le m$. Verify that, if, as we shall do from now on, we take $p_{ij} \ge 0$ for all $1 \le i,\, j \le m$ and $\sum_{j=1}^m p_{ij} = 1$ for all $1 \le i \le m$, then $T\mathbf{q} \in X$ for all $\mathbf{q} \in X$, and so we may consider T as a map from X to X.

Finally we have to choose a metric d. In most of this book the particular choice of metric has not been important, but here it is. We take

$$d(\mathbf{u}, \mathbf{v}) = \|\mathbf{u} - \mathbf{v}\|_1 = \sum_{i=1}^m |u_i - v_i|$$

for all $\mathbf{u},\, \mathbf{v} \in X$.

(i) Explain why (X, d) has the Bolzano–Weierstrass property.

(ii) Show that $d(T\mathbf{u}, T\mathbf{v}) \le d(\mathbf{u}, \mathbf{v})$ for all $\mathbf{u},\, \mathbf{v} \in X$.

(iii) Show that, if $n \ge 1$, then

$$(T^n \mathbf{q})_j = \sum_{i=1}^m q_i p_{ij}^{(n)}$$

for all $1 \le j \le m$, where $p_{ij}^{(n)} \ge 0$ for all $1 \le i,\, j \le m$ and $\sum_{j=1}^m p_{ij}^{(n)} = 1$ for all $1 \le i \le m$. If you know some probability, prove this both by an *algebraic* and a *probabilistic* argument.

(iv) Show that, if

$$\bigstar \qquad p_{ij}^{(N)} > 0 \text{ for all } 1 \le i,\, j \le m$$

for some $N \ge 1$, then $d(T^N\mathbf{u}, T^N\mathbf{v}) < d(\mathbf{u}, \mathbf{v})$ for all $\mathbf{u},\, \mathbf{v} \in X$ with $\mathbf{u} \ne \mathbf{v}$.

(v) Conclude, using Exercise K.286, that, provided only that $\bigstar$ holds for some $N \ge 1$, there is a unique $\boldsymbol{\pi} \in X$ such that $T\boldsymbol{\pi} = \boldsymbol{\pi}$ and that, if $\mathbf{q} \in X$, then $d(T^n\mathbf{q}, \boldsymbol{\pi}) \to 0$ as $n \to \infty$. If the associated Markov chain has been running for a long time, what can you say about the probability that it is in state i?

(vi) Suppose that $m = 2$, $p_{12} = p_{21} = 1$ and $p_{11} = p_{22} = 0$. Show that there is a unique $\boldsymbol{\pi} \in X$ such that $T\boldsymbol{\pi} = \boldsymbol{\pi}$. For which $\mathbf{q} \in X$ is it true that $d(T^n\mathbf{q}, \boldsymbol{\pi}) \to 0$ as $n \to \infty$? Why does your result not contradict (v)?

(vii) Suppose that $m = 2$, $p_{12} = p_{21} = 0$ and $p_{11} = p_{22} = 1$. Find all the $\boldsymbol{\pi} \in X$ such that $T\boldsymbol{\pi} = \boldsymbol{\pi}$. What can you say about the behaviour of $T^n\mathbf{q}$ as $n \to \infty$ for each $\mathbf{q} \in X$? Why does your result not contradict (v)?

(viii) Suppose that $m = 4$ and that $p_{ij} = 1/2$ if $1 \le i,\, j \le 2$ or $3 \le i,\, j \le 4$ and that $p_{ij} = 0$ otherwise. Find all the $\boldsymbol{\pi} \in X$ such that $T\boldsymbol{\pi} = \boldsymbol{\pi}$. What can you say about the behaviour of $T^n\mathbf{q}$ as $n \to \infty$ for each $\mathbf{q} \in X$?

(ix) Note that, by Exercise K.285, the equation $T\boldsymbol{\pi} = \boldsymbol{\pi}$ always has a solution in X even if $\bigstar$ does not hold.

K.288. (Countable Markov chains.) [12.1, T!] In this exercise and the next we consider Markov chains with an infinite number of states 1, 2, ... In algebraic terms we consider a collection of positive numbers p_{ij} $[1 \le i, j]$ such that $\sum_{j=1}^{\infty} p_{ij} = 1$ for each $i \ge 1$.

(i) Consider l^1, the space of real sequences whose sum is absolutely convergent, with norm $\| \ \|_1$ given by $\|\mathbf{x}\|_\infty = \sum_{n=1}^{\infty} |x_n|$. Show that, if we write $(T\mathbf{x})_j = \sum_{j=1}^{\infty} x_i p_{ij}$, then, T is a well defined continuous linear map from l^1 to l^1 with operator norm $\|T\| = 1$. Show further that, if we write

$$X = \{\mathbf{q} \in l^1 \ : \ \|\mathbf{q}\|_1 = 1 \text{ and } q_i \ge 0 \text{ for all } i \ge 1\},$$

then $T\mathbf{q} \in X$ whenever $\mathbf{q} \in X$. (Note that you are manipulating infinite sums and you must justify each step carefully. Results like Lemma 5.31 may be useful.) We write $(T^n\mathbf{x}) = \sum_{j=1}^{\infty} x_i p_{ij}^{(n)}$.

(ii) We shall only be interested in systems satisfying the following extension of ★. If $k \ge 1$, there exists an $N(k)$ such that

$$\bigstar\bigstar \qquad p_{ij}^{(N(k))} > 0 \text{ for all } 1 \le i, j \le k.$$

Show that, under this condition, given any $\mathbf{u}, \mathbf{v} \in X$ with $\mathbf{u} \ne \mathbf{v}$, we can find an $N \ge 1$ such that $d(T^N\mathbf{u}, T^N\mathbf{v}) < d(\mathbf{u}, \mathbf{v})$.

(iii) Show, under assumption ★★, that, if the equation $T\boldsymbol{\pi} = \boldsymbol{\pi}$ has a solution $\boldsymbol{\pi} \in X$, then it is unique. Let

$$Y = \{\mathbf{h} \in l^1 \ : \ h_i \ge 0 \text{ for all } i \ge 1\}.$$

Find all the solutions of the equation $T\mathbf{h} = \mathbf{h}$ with $\mathbf{h} \in Y$, first under the assumption that the equation $T\boldsymbol{\pi} = \boldsymbol{\pi}$ has a solution $\boldsymbol{\pi} \in X$, and second under the assumption that it does not.

(iv) Show, under assumption ★★, that, if the equation $T\boldsymbol{\pi} = \boldsymbol{\pi}$ has a solution $\boldsymbol{\pi} \in X$, then $\pi_i > 0$ for all i.

(v) Consider the following systems. In each case show we have a Markov chain satisfying ★★. By solving the appropriate difference equations, or otherwise, show that in case (a) the equation $T\boldsymbol{\pi} = \boldsymbol{\pi}$ has a solution $\boldsymbol{\pi} \in X$, but in cases (b) and (c) it does not.

(a) $p_{12} = 1$; $p_{i\,i-1} = 1/2$, $p_{i\,i} = p_{i\,i+1} = 1/4$ for $i \ge 2$; $p_{ij} = 0$ otherwise.

(b) $p_{12} = 1$; $p_{i\,i+1} = 1/2$, $p_{i\,i} = p_{i\,i-1} = 1/4$ for $i \ge 2$; $p_{ij} = 0$ otherwise.

(c) $p_{12} = 1$; $p_{i\,i-1} = p_{i\,i} = p_{i\,i+1} = 1/3$ for $i \ge 2$; $p_{ij} = 0$ otherwise.

K.289. [12.1, T!, ↑↑] This exercise continues the previous one. We assume that condition ★★ holds and that the equation $T\boldsymbol{\pi} = \boldsymbol{\pi}$ has a solution $\boldsymbol{\pi} \in X$. We will need part (ii) of Exercise K.218.

(i) Let $J \ge 1$ be fixed. Show, using part (iv) of question K.288, that we can find an $\mathbf{h} \in Y$ such that $T\mathbf{h} = \mathbf{h}$ and $h_J = 1$.

(ii) We continue with the notation of (i). Explain why X does not have the Bolzano–Weierstrass property but

$$X_J = \{\mathbf{q} \in X : h_i \geq q_i \text{ for all } i \geq 1\}$$

does. Explain why X_J is non-empty and show, carefully, that $T\mathbf{q} \in X_J$ whenever $\mathbf{q} \in X_J$. Deduce that, if $\mathbf{q} \in X_J$, then $\|T^n\mathbf{q} - \boldsymbol{\pi}\|_1 \to 0$ as $n \to \infty$.

(iii) Let $\mathbf{e}(p) \in l^1$ be given by $\mathbf{e}(p)_p = 1$, $\mathbf{e}(p)_i = 0$ otherwise. Explain why $\|T^n\mathbf{e}(p) - \boldsymbol{\pi}\|_1 \to 0$ as $n \to \infty$. Deduce further that, if $\mathbf{q} \in X$, then $\|T^n\mathbf{q} - \boldsymbol{\pi}\|_1 \to 0$. Interpret this result probabilistically.

K.290. [12.2, P] Let (X, d) be a complete metric space and $T : X \to X$ a mapping such that T^n is a contraction mapping. Show that T has a unique fixed point.

If you have done Exercise K.282, explain why the example asked for in the last paragraph of that exercise does not contradict the result of the first paragraph of this exercise.

Consider the space $C([a, b])$ with the uniform norm. Let ϕ be a continuous real-valued function on $\mathbb{R} \times [a, b]$ which satisfies the Lipschitz condition

$$|\phi(x, t) - \phi(y, t)| \leq M|x - y| \text{ for } x,\ y \in \mathbb{R} \text{ and } t \in [a, b],$$

and let $g \in C([a, b])$. Define $T : C([a, b]) \to C([a, b])$ by

$$(Th)(t) = g(t) + \int_a^t \phi(h(s), s)\, ds.$$

Show, by induction or otherwise, that

$$|(T^n h)(t) - (T^n k)(t)|_\infty \leq \frac{1}{n!} M^n (t - a)^n \|h - k\|_\infty$$

for all $h,\ k \in C[a, b]$ and $t \in [a, b]$.

Show that T has a fixed point. If g is constant, comment briefly on the associated theorem on the existence of solutions of a certain differential equation.

K.291. [12.2, P, S] Suppose that $f : \mathbb{R} \to \mathbb{R}$ is differentiable with $|f'(x)| < M$ for all $x \in \mathbb{R}$. Show that the system of equations

$$x_i = \sum_{j=1}^n a_{ij} f(x_j) + b_i \quad [1 \leq i \leq n]$$

with a_{ij} and b_i real has a unique solution provided that

$$\sum_{i=1}^n \sum_{j=1}^n a_{ij}^2 < \frac{1}{M^2}.$$

K.292. [12.2, P, S] Consider the differential equation for a function $y : \mathbb{R} \to \mathbb{R}$

$$\bigstar \qquad y'(t) = |y(t)|^\alpha, \ y(0) = 0,$$

where α is real and strictly positive. For which values of α does $\bigstar$ have only one solution and for which values of α does it have more than one? If $\bigstar$ has only one solution, prove this. If $\bigstar$ has more than one solution, give at least two solutions explicitly.

K.293. [12.2, P] Dieudonné is associated in most mathematicians' minds with the abstract approach of Bourbaki and his own text *Foundations of Modern Analysis* [**14**]. He may have had this in mind when he wrote the determinedly concrete *Infinitesimal Calculus* [**15**], from which this exercise is taken.

(i) Suppose $f : \mathbb{R}^2 \to \mathbb{R}$ is a continuous function with $f(t, u) < 0$ when $tu > 0$ and $f(t, u) > 0$ when $tu < 0$. Explain why $f(0, t) = f(u, 0) = 0$ for all t and u. If $y : \mathbb{R} \to \mathbb{R}$ is a differentiable function with

$$\bigstar \qquad y'(t) = f(t, y(t)) \text{ for all } t \in \mathbb{R}, \ y(0) = 0,$$

show, by considering the behaviour of y at a maximum and a minimum in each interval $[0, c]$ with $c > 0$, or otherwise, that $y(t) = 0$ for all $t \geq 0$. Show that $y(t) = 0$ for all $t \in \mathbb{R}$, and so $\bigstar$ has a unique solution.

(ii) Now define

$$f(t, u) = \begin{cases} -2t & \text{if } x \geq t^2, \\ -2x/t & \text{if } t^2 > x > -t^2, \\ 2t & \text{if } -t^2 \geq x. \end{cases}$$

Verify that f is well defined and continuous everywhere and satisfies the hypotheses of (i). Now set $y_0(t) = t^2$ and define

$$y_{n+1}(t) = \int_0^t f(s, y_n(s))\, ds.$$

Show that $y_n(t)$ fails to converge as $n \to \infty$ for any $t \neq 0$ although, as we know from (i), $\bigstar$ has a unique solution.

K.294. [12.3, P] (This exercise makes repeated use of the mean value inequality.)

(i) Suppose that $a > 0$, $y : [0, a) \to \mathbb{R}$ is differentiable, $y(0) \geq 1$ and

$$y'(t) \geq |y(t)|^\beta$$

with $\beta > 1$. Show that $y(t)$ is strictly increasing for $t \geq 0$ and $y(t) \geq t+1$ for $a > t \geq 0$. Let t_j be defined by taking $t_j = 0$ if $y(0) \geq 2^j$, $t_j = a$ if $y(t) < 2^j$ for all $0 \leq t < a$ and $y(t_j) = 2^j$ otherwise. Show that $t_{j+1} - t_j \leq 2^{(1-\beta)j}$, and deduce that $a \leq (1 - 2^{(1-\beta)})^{-1}$.

(ii) Deduce the result of Example 12.19.

(iii) Show that there exists a differentiable function $y : [0, \infty) \to \mathbb{R}$ such that $y(0) = e$ and

$$y'(t) = y(t) \log y(t).$$

K.295. (Euler's method.) [12.3, T] The next few exercises deal with Euler's method for finding approximate solutions of a differential equation. They were suggested by the typically neat treatment in [**46**]. You will find the discussion more interesting if you have used Euler's method before or if you try one of the excellent demonstrations which now exist on the Internet.

Suppose that we wish to solve the differential equation $y'(t) = f(t, y(t))$, $y(0) = y_0$ numerically on a computer. (We shall look at $y(t)$ for $t \geq 0$, but the same ideas apply when $t \leq 0$.) A natural approach is to seek (approximate) solutions at the points $x_r = rh$, where the strictly positive number h is called the 'step length' and r is a positive integer. Explain briefly why we expect

$$y\big((r+1)h\big) \approx y(rh) + f(rh, y(rh))h.$$

This suggests that our approximation y_r to $y(rh)$ could be obtained by solving the system of equations

$$\text{(A)} \qquad y_{r+1} = y_r + f(x_r, y_r)h.$$

This is called Euler's method.

Euler's method seems reasonable and works on suitable test examples. In this exercise we investigate its behaviour in general. We shall assume stronger conditions on f than we did in Theorem 12.8 and elsewhere in Sections 12.2 and 12.3. Specifically, we demand that $f : \mathbb{R}^2 \to \mathbb{R}$ is once differentiable and there exist constants K and L such that $|f_{,2}(t, u)| \leq K$ and

$$|f_{,1}(t, u)| + |f_{,2}(t, u) f(t, u)| \leq L$$

for all $(t, u) \in \mathbb{R}^2$. Explain briefly why these conditions imply the existence of a unique differentiable $y : \mathbb{R} \to \mathbb{R}$ such that

$$\text{(B)} \qquad y'(t) = f(t, y(t)), \ y(0) = y_0.$$

Explain why y is twice differentiable, and express $y''(t)$ in terms of $f(t, y(t))$, $f_{,1}(t, y(t))$ and $f_{,2}(t, y(t))$. By using the 2nd mean value theorem (Exercise K.50), or otherwise, show that

$$|y((n+1)h) - y(nh) - f(nh, y(nh))h| \leq \frac{Lh^2}{2}.$$

Use this result to show that

$$|y\big((n+1)h\big) - y_{n+1}| \leq (1 + Kh)|y(nh) - y_n| + \frac{Lh^2}{2}.$$

Deduce that

$$|y(Nh) - y_N| \leq \frac{Lh}{2K}\big((1+Kh)^{N+1} - 1\big).$$

In order to emphasise the dependence on h, let us write $y_n^{[h]} = y_n$. Show that, if the positive number a is fixed and we take $h_N = a/N$, then

$$|y(a) - y_N^{[h_N]}| \leq \frac{Lh_N}{2K}\big((1+Kh_N)^{N+1} - 1\big).$$

Observe (or recall) that, if $x \geq 0$, then

$$e^x - \left(1 + \frac{x}{m}\right)^m \geq \sum_{r=0}^{m}\left(\frac{1}{r!} - \binom{m}{r}\frac{1}{m^r}\right)x^r \geq 0,$$

and deduce that

(C) $$|y(a) - y_N^{[h_N]}| \leq \frac{Lh_N}{2K}(e^{aK} - 1)$$

and, in particular, $y_N^{[h_N]} \to y(a)$ as $N \to \infty$.

K.296. [12.3, T, ↑] We continue with the discussion of Exercise K.295. Informally, we may interpret the inequality (C) in Exercise K.295 as saying that 'our error estimate for Euler's method is roughly proportional to the step length'. Inequality (C) is, however, only an upper estimate for the error. Can we get a substantially better estimate? Consider the special case $f(t,u) = u$, $y_0 = 0$ and $a = 1$. Show that $y(t) = e^t$, $y_n = (1+y)^n$ and that

$$\begin{aligned} y(1) - y_N^{[h_N]} &= e - (1+N^{-1})^N \\ &\geq \sum_{r=0}^{N}\left(\frac{1}{r!} - \binom{N}{r}\frac{1}{N^r}\right) \geq \frac{1}{2N} = \frac{h_N}{2}. \end{aligned}$$

Informally, we may say 'in this case, the error for Euler's method decreases no faster than the step length'.

We see that (roughly speaking) halving the step length in Euler's method will double the amount of calculation and only halve the error. Thus Euler's method is a reasonable one if we want to solve *one* differential equation to a *reasonable* degree of accuracy, but unattractive if we want to solve *many* differential equations to a *high* degree of accuracy.

K.297. [12.3, T, ↑] We continue with the discussion of Exercises K.295 and K.296. So far we have ignored the fact that computers only work to a certain degree of accuracy. As a simple model to take account of this, we replace equation (A) of Exercise K.295 by

(D) $$\tilde{y}_{r+1} = \tilde{y}_r + f(x_r, \tilde{y}_r)h + \epsilon_r,$$

where the 'error' ϵ_r satisfies the condition $|\epsilon_r| \le \epsilon$ for some fixed $\epsilon > 0$. As before, we impose the initial condition $\tilde{y}_0 = 0$.

(i) As in Exercise K.295, we take $y_n^{[h]} = y_n$, fix a positive number a, and take $h_N = a/N$. Show that

$$\text{(E)} \qquad |y(a) - y_N^{[h_N]}| \le \left(\frac{Lh_N}{2K} + \frac{\epsilon}{h_N K}\right)(e^{aK} - 1).$$

What happens to the error estimate (that is, the right hand side of (E)) as $N \to \infty$ (that is, as the step length approaches 0)? Explain in simple terms why we should expect this.

(ii) Which positive value of h minimises $\dfrac{Lh}{2K} + \dfrac{\epsilon}{hK}$?

(iii) Show that the best bound on the error that we can guarantee using (E) is proportional to $\epsilon^{1/2}$.

(iv) Just as in Exercise K.296, we ask whether we can improve substantially on (iii). Once again, we consider the special case $f(t,u) = u$, $y_0 = 0$ and $a = 1$. A little experimentation shows that we can solve (D) if we choose $\epsilon_r = -(1 - rh)\epsilon$, giving

$$\text{(F)} \qquad \tilde{y}_{r+1} = (1 + N^{-1})\tilde{y}_r - (1 - rN^{-1})\epsilon.$$

Show that, for this choice of ϵ_r, we obtain

$$\tilde{y}_r = (1 - N^{-1})^r - r\epsilon,$$

and so

$$y(1) - \tilde{y}_N^{[h_N]} = e - \left((1 + N^{-1})^N - N\epsilon\right) \ge \frac{h_N}{2} + \frac{\epsilon}{h_N} \ge 2^{1/2}\epsilon^{1/2}.$$

Thus the answer to our question is no.

There exist several different ways of improving on Euler's method, but a good understanding of Euler's method is very helpful in understanding how anybody came to think of these more advanced techniques.

K.298. [12.3, H, ↑] (i) Let $a > 0$. By considering the differential equation $y'(x) = g(x)$ and using Exercise K.295, show that there exists a constant κ_a such that, if $g : \mathbb{R} \to \mathbb{R}$ is differentiable, $|g'(t)| \le M$ for all $t \in [0, a]$, N is a strictly positive integer and $Nh = a$, then

$$\left| \int_0^a g(t)\,dt - h \sum_{n=0}^{N-1} g(rh) \right| \le M\kappa_a h.$$

(It is easy to obtain this result directly with better values for κ_a. See, for example, Exercise K.140 (v), or just think a bit.)

(ii) Suppose that $a = \pi$, $Nh = a$ with N a strictly positive integer, and $G(t) = \sin^2(Nt)$. Show that

$$\left| \int_0^a G(t)\,dt - h\sum_{n=0}^{N-1} G(rh) \right| \geq \sup_{t\in[0,a]} |G(t)|.$$

(iii) Consider the situation described in Exercise K.295. Suppose that we replace the condition $|f_{,1}(t,u)| + |f_{,2}(t,u)f(t,u)| \leq L$ by the weaker condition $|f(t,u)| \leq L$ for all $(t,u) \in \mathbb{R}^2$. If $a > 0$ and $\epsilon > 0$ are fixed, is it possible to find an N_0 (depending only on K, L, a and ϵ) such that

$$|y(a) - y_N^{[h_N]}| \leq \epsilon$$

for all $N \geq N_0$?

K.299. (Leibniz rule.) [12.3, G] (i) If u, $v : \mathbb{R} \to \mathbb{R}$ are n times differentiable at x, show, by induction or otherwise, that the product uv is n times differentiable at x with

$$(uv)^{(n)}(x) = \sum_{r=0}^{n} \binom{n}{r} u^{(n-r)}(x)v^{(r)}(x).$$

(This is called the Leibniz rule.)

(ii) If $y : \mathbb{R} \to \mathbb{R}$ is defined by

$$(1+x^2)^{1/2}y(x) = \log(x + (1+x^2)^{1/2}),$$

show that y is differentiable on $(0,\infty)$ with

$$(1+x^2)y'(x) + xy(x) = 1.$$

Show, by using induction and the Leibniz rule, that y is infinitely differentiable, and find the Taylor series

$$\sum_{n=0}^{\infty} \frac{y^{(n)}(0)}{n!} x^n.$$

(iii) Find the radius of convergence of the Taylor series and, by using results on the differentiation of power series and the uniqueness of the solution of differential equations, show that, within the radius of convergence,

$$y(x) = \sum_{n=0}^{\infty} \frac{y^{(n)}(0)}{n!} x^n.$$

K.300. [12.4, P, S] Let the two functions $K : [0,1]^2 \to \mathbb{R}$ and $g : [0,1] \to \mathbb{R}$ be continuous. Explain why there exists an $M > 0$ such that $|K(s,t)| \leq M$ for all $(s,t) \in [0,1]^2$. Suppose that $|\lambda| < M^{-1}$. By finding an appropriate contraction mapping $T : C([0,1]) \to C([0,1])$, show that there exists a unique $f \in C([0,1])$ such that

$$f(t) = g(t) + \lambda \int_0^1 K(s,t)f(s)\,ds.$$

K.301. (The Wronskian.) [12.4, M] This should be treated as an exercise in calculus rather than analysis. We write

$$\begin{vmatrix} a_{11} & a_{12} & a_{13} \\ a_{21} & a_{22} & a_{23} \\ a_{31} & a_{32} & a_{33} \end{vmatrix} = \det \begin{pmatrix} a_{11} & a_{12} & a_{13} \\ a_{21} & a_{22} & a_{23} \\ a_{31} & a_{32} & a_{33} \end{pmatrix}.$$

(i) If $u_1, u_2, \dots, w_3$ are all differentiable, show that

$$\frac{d}{dt}\begin{vmatrix} u_1(t) & u_2(t) & u_3(t) \\ v_1(t) & v_2(t) & v_3(t) \\ w_1(t) & w_2(t) & w_3(t) \end{vmatrix}$$
$$= \begin{vmatrix} u_1'(t) & u_2'(t) & u_3'(t) \\ v_1(t) & v_2(t) & v_3(t) \\ w_1(t) & w_2(t) & w_3(t) \end{vmatrix} + \begin{vmatrix} u_1(t) & u_2(t) & u_3(t) \\ v_1'(t) & v_2'(t) & v_3'(t) \\ w_1(t) & w_2(t) & w_3(t) \end{vmatrix} + \begin{vmatrix} u_1(t) & u_2(t) & u_3(t) \\ v_1(t) & v_2(t) & v_3(t) \\ w_1'(t) & w_2'(t) & w_3'(t) \end{vmatrix}.$$

(ii) If u_1, u_2 and u_3 are three solutions of

$$y'''(t) + a(t)y''(t) + b(t)y'(t) + c(t)y(t) = 0,$$

we define their Wronskian W by

$$W(t) = \begin{vmatrix} u_1(t) & u_2(t) & u_3(t) \\ u_1'(t) & u_2'(t) & u_3'(t) \\ u_1''(t) & u_2''(t) & u_3''(t) \end{vmatrix}.$$

Use part (i) and results about determinants to show that

$$W'(t) = -a(t)W(t).$$

(iii) Generalise parts (i) and (ii). Reread the proof of of Lemma 12.28 (i) in the light of this exercise.

K.302. [12.4, T, S] The functions $f, g : \mathbb{R} \to \mathbb{R}$ are everywhere differentiable and their Wronskian $f'g - g'f$ never vanishes. By applying Rolle's theorem to f/g, or otherwise, show that if f has zeros at a and b with $a < b$, then g must have a zero strictly between a and b. Deduce that 'the zeros of f and g are interlaced'.

K.303. [12.4, P] (i) If $f_1, f_2 : [a,b] \to \mathbb{R}$ are once differentiable, we define the Wronskian $W(f_1, f_2)$ by

$$W(f_1, f_2) = \det \begin{pmatrix} f_1 & f_2 \\ f_1' & f_2' \end{pmatrix}$$

Show that, if f_1 and f_2 are linearly dependent (that is, we can find $\lambda_1, \lambda_2 \in \mathbb{R}$, not both zero such that $\lambda_1 f_1(t) + \lambda_2 f_2(t) = 0$ for all $t \in [a,b]$), then the Wronskian $W(f_1, f_2)$ is identically zero. By considering functions of the type g with $g(x) = 0$ for $x \leq c$, $g(x) = (x-c)^2$ for $x \geq c$, or otherwise, show that the converse is false.

(ii) Suppose now that f_1, f_2 are twice continuously differentiable. Show, by considering the Wronskian

$$W(f, f_1, f_2) = \det \begin{pmatrix} f & f_1 & f_2 \\ f' & f_1' & f_2' \\ f'' & f_1'' & f_2'' \end{pmatrix}$$

and using results on the existence and uniqueness of solutions of differential equations with given initial conditions, that the following result holds.

There exists a differential equation of the form

$$y''(x) + a_1(x)y'(x) + a_2(x)y(x) = 0$$

whose solutions are exactly the functions of the form $\lambda_1 f_1 + \lambda_2 f_2$ with $\lambda_1, \lambda_2 \in \mathbb{R}$ (in more sophisticated language, having f_1 and f_2 as a basis for the space of solutions) if and only if $W(f_1, f_2)$ is non-zero everywhere on $[a,b]$.

Generalise the results of this exercise to n functions $f_1, f_2, \dots, f_n$.

K.304. (Method of reduction of order.) [12.4, M]

(This exercise should be treated as a mathematical methods one.)

(i) Consider a differential equation

$$\bigstar \qquad y''(x) + p(x)y'(x) + q(x)y(x) = 0.$$

Suppose that u is a solution of $\bigstar$. Show that, if we write $y(x) = u(x)v(x)$ and $w(x) = u'(x)$, then $\bigstar$ takes the form of a *first order* differential equation in w. Explain why we would expect $y(x) = (A + B\int_0^x w(t)\,dt)u(x)$ with w a solution of the first order equation to give the general solution for $\bigstar$.

(ii) Show that $y(x) = \exp x$ gives a solution of

$$y''(x) - (2x+1)y'(x) + 2(x+1)y(x) = 0,$$

and use the method of (i) to find the general solution.

(iii) We know that $y(x) = \exp x$ gives a solution of

$$y''(x) - 2y'(x) + y(x) = 0.$$

Find the general solution by the method of (i).

(iv) Show how, if we know one solution of

$$y'''(x) + p_1(x)y''(x) + p_2(x)y'(x) + p_3(x)y(x) = 0,$$

we can reduce the problem of finding a general solution to that of solving a second order linear differential equation. Generalise.

K.305. (Method of variation of parameters.) [12.4, M, ↑]

(Like Exercise K.304, this exercise should be treated as a mathematical methods one.)

Consider a differential equation

★ $$y''(x) + p(x)y'(x) + q(x)y(x) = 0.$$

Let y_1 and y_2 be solutions of ★ whose Wronskian $W(x) = y_1'(x)y_2(x) - y_1(x)y_2'(x)$ never vanishes. Suppose that we wish to solve the differential equation

★★ $$y''(x) + p(x)y'(x) + q(x)y(x) = f(x).$$

(i) In view of the success of the method of Exercise K.304 we might be tempted to look for a solution of the form

$$y(x) = u_1(x)y_1(x) + u_2(x)y_2(x).$$

Make this substitution in ★★.

(ii) At the end of (i), you obtained a differential equation (†) for u_1 and u_2. It is a useful heuristic guide (though not always a reliable one) that two unknowns require two equations, so we add the further equation

(††) $$u_1'(x)y_1(x) + u_2'(x)y_2(x) = 0.$$

Use (††) to simplify (†), and then use the pair of equations to find u_1' and u_2'. Now find the general solution of ★★.

(iii) Discuss the relation of your result to the Green's function result given in Theorem 12.31.

(iv) Use the method of this exercise to find the general solution of

$$y''(x) - y(x) = \frac{2}{1 + e^x}.$$

(v) Try and extend the ideas of this exercise to the solution of

$$y'''(x) + p(x)y''(x) + q(x)y'(x) + r(x)y = f(x)$$

when three solutions y_1, y_2, y_3 with non-zero Wronskian (see Exercise K.301) are known.

[It would be hard to think of an example more opposed to the modern view that 'understanding precedes doing'. But even the most convinced proponent of modern ways must admit that it has a certain charm.]

K.306. [12.4, P] Let $C(S)$ be the space of continuous functions on the unit square

$$S = \{(x,y) : 0 \le x \le 1,\ 0 \le y \le 1\}$$

equipped with the uniform norm $\|K\|_\infty = \sup_{(x,y)\in S} |K(x,y)|$, and let $C(I)$ be the space of continuous functions on the unit interval $I = [0,1]$ equipped with the uniform norm. Let $\mathcal{L}$ be the space of continuous linear maps $L : C(I) \to C(I)$ equipped with the operator norm. If $K \in C(S)$ we set

$$(T_K(f))(x) = \int_0^1 K(x,y) f(y)\, dy.$$

Show that $T_K \in \mathcal{L}$.

Prove or disprove each of the following statements.

(i) If $K_n \to K$ in $C(S)$, then $T_{K_n} \to T_K$ in $\mathcal{L}$.

(ii) If $T_{K_n} \to T_K$ in $\mathcal{L}$, then $K_n \to K$ in $C(S)$.

(iii) The mapping $K \mapsto T_K$ is an injective map from $C(S)$ to $\mathcal{L}$.

(iv) The mapping $K \mapsto T_K$ is a surjective map from $C(S)$ to $\mathcal{L}$.

(v) If there exists a $\delta > 0$ such that $K(x,y) \ge \delta$ for all $(x,y) \in S$, then $K \mapsto T_K$ is injective.

(vi) If there exists a $\delta > 0$ such that $K(x,y) \ge \delta$ for all $(x,y) \in S$, then $K \mapsto T_K$ is surjective.

K.307. [12.4, T] (i) In Exercise K.147, you showed that if $g : \mathbb{R}^2 \to \mathbb{R}$ is a differentiable function with continuous partial derivatives, then

$$G(x) = \int_0^x g(x,t)\, dt$$

is differentiable with

$$G'(x) = g(x,x) + \int_0^x g_{,1}(x,t)\, dt.$$

Review this exercise.

(ii) Suppose that a, b, $f : \mathbb{R} \to \mathbb{R}$ are continuous. Explain why there exists a unique twice differentiable $y_1 : \mathbb{R} \to \mathbb{R}$ such that

$$y_1''(t) + a(t)y_1'(t) + b(t)y_1(t) = 0,\ y_1(0) = 0,\ y_1'(0) = 1,$$

and a unique twice differentiable $y_2 : \mathbb{R} \to \mathbb{R}$ such that

$$y_2''(t) + a(t)y_2'(t) + b(t)y_2(t) = 0,\ y_2(1) = 0,\ y_2'(1) = 1.$$

We make the following

key assumption: $y_1(1) \ne 0.$

Show that, if we define H, $\tilde{H} : \mathbb{R}^2 \to \mathbb{R}$ by $W = y_1y_2' - y_1'y_2$ and

$$H(s,t) = y_1(t)y_2(s)W(s)^{-1}, \; \tilde{H}(s,t) = y_2(t)y_1(s)W(s)^{-1},$$

then H and $\tilde{H}$ are twice differentiable functions with continuous second partial derivatives and

$$H_{,22}(s,t) + a(t)H_{,2}(s,t) + b(t)H(s,t) = 0, \; H(s,0) = 0,$$
$$\tilde{H}_{,22}(s,t) + a(t)\tilde{H}_{,2}(s,t) + b(t)\tilde{H}(s,t) = 0, \; \tilde{H}(s,1) = 0.$$

Check that $H(t,t) = \tilde{H}(t,t)$, $\tilde{H}_{,2}(t,t) - H_{,2}(t,t) = 1$.

(iii) We define $G : \mathbb{R}^2 \to \mathbb{R}$ by $G(s,t) = H(s,t)$ for $t \le s$ and $G(s,t) = \tilde{H}(s,t)$ for $s < t$. If we set

$$y(t) = \int_0^1 G(s,t)f(s)\,ds = \int_0^t H(s,t)f(s)\,ds + \int_t^1 \tilde{H}(s,t)f(s)\,ds,$$

show, using part (i) and the properties of H and $\tilde{H}$ established in part (ii) (but not using the definitions of H and $\tilde{H}$), that y is twice differentiable and satisfies

★ $$y''(t) + a(t)y' + b(t)y = f(t)$$

together with the conditions $y(0) = y(1) = 0$.

K.308. [12.4, M] Consider the differential equation

$$y^{(4)} - k^4y = f$$

for $y : [0,1] \to \mathbb{R}$ subject to the boundary conditions $y(0) = y'(0) = 0$, $y(1) = y'(1) = 0$, where $f : [0,1] \to \mathbb{R}$ is continuous and k is real.

By extending the discussion of the Green's function in Section 12.4, show that, provided that k does not take certain exceptional values to be identified, the system has the solution

$$y(x) = \int_0^1 G(x,t)f(t)\,dt,$$

where

$G(x,t) =$

$$\begin{cases} A(\sinh kx - \sin kx) + B(\cosh kx - \cos kx) & \text{if } 0 \le x \le t, \\ C(\sinh k(1-x) - \sin k(1-x)) + D(\cosh k(1-x) - \cos k(1-x)) & \text{if } t < x \le 1, \end{cases}$$

and A, B C, D are given by

$$\begin{pmatrix} A \\ B \\ C \\ D \end{pmatrix} = M^{-1} \begin{pmatrix} 0 \\ 0 \\ 0 \\ 1 \end{pmatrix},$$

where M is a 4×4 matrix to be given explicitly.

K.309. [12.4, M] Consider the equation for a damped harmonic oscillator

$$y''(t) + 2\beta y'(t) + \omega^2 y(t) = f(t),$$

where β and ω are strictly positive real numbers and $y : [0, \infty) \to \mathbb{R}$ satisfies $y(0) = y'(0) = 0$. Use Green's function methods to show that, if $\beta > \omega$,

$$y(t) = \alpha^{-1} \int_0^t f(s) e^{-\beta(t-s)} \sinh(\alpha(t-s))\, ds,$$

where α is the positive square root of $\beta^2 - \omega^2$, and obtain a similar result in the cases when $\omega > \beta$ and $\omega = \beta$.

Suppose that the driving force f is non-zero only when t is very small. Sketch the behaviour of $y(t)$ for large t. If y represents the position of a pointer, we would like $y(t)$ to remain fairly close to zero after a fairly short time. Explain why it is not a good idea to take β very small or very large. What rough value of β would you choose? (You may feel this is an improper question, but it is one engineers have to answer every day.)

K.310. [13.1, T] (Although this sequence of exercises seems to find a natural place here, it could have been placed earlier and linked with Section 11.1.) Let $(U, \| \ \|_U)$ be a complete normed vector space. By Exercise 11.15, the space $\mathcal{L}(U, U)$ of continuous linear maps with the operator norm $\| \ \|$ is complete.

(i) If $T \in \mathcal{L}(U, U)$ and $\|T\| < 1$, show that the sequence $S_n = \sum_{j=0}^n T^j$ is Cauchy. Deduce that S_n converges in the operator norm to a limit $S = \sum_{j=0}^\infty T^j$. Show also that $\|S\| \leq (1 - \|T\|)^{-1}$ and $\|I - S\| \leq \|T\|(1 - \|T\|)^{-1}$.

(ii) We continue with the notation and assumptions of (i). By looking at $S_n(I - T)$, show that $S(I - T) = I$. Show also that $(I - T)S = I$. Conclude that, if $A \in \mathcal{L}(U, U)$ and $\|I - A\| < 1$, then A is invertible and

$$\|A^{-1}\| \leq (1 - \|I - A\|)^{-1} \text{ and } \|I - A^{-1}\| \leq \|I - A\|(1 - \|I - A\|)^{-1}.$$

(iii) Suppose that $B \in \mathcal{L}(U, U)$ is invertible and $\|B - C\| < \|B^{-1}\|^{-1}$. Show that $A = B^{-1}C$ is invertible. Show that $A^{-1}B^{-1}C = I$ and that $CA^{-1}B^{-1} = BB^{-1}BA^{-1}B^{-1} = I$, so C is invertible with inverse $A^{-1}B^{-1}$. Show, further, that

$$\|C^{-1}\| \leq \|B^{-1}\|(1 - \|B^{-1}\|\|B - C\|)^{-1},$$
$$\|B^{-1} - C^{-1}\| \leq \|B^{-1}\|\|B - C\|(1 - \|B^{-1}\|\|B - C\|)^{-1}.$$

(iv) Let E be the set of invertible $C \in \mathcal{L}(U, U)$. Show that E is open in $(\mathcal{L}(U, U), \| \ \|)$. Show further that, if we define $\Theta : E \to E$ by $\Theta(C) = C^{-1}$, then Θ is a continuous function.

(v) Returning to the discussion in (i) and (ii), show that if $\|T\| < 1$, then

$$\|(I-T)^{-1} - I - T\| \leq \|T\|^2(1-\|T\|)^{-1}$$

and

$$\|(I+T)^{-1} - I + T\| \leq \|T\|^2(1-\|T\|)^{-1}.$$

Conclude that

$$\Theta(I+T) = I - T + \epsilon(T)\|T\|,$$

where $\epsilon : \mathcal{L}(U,U) \to \mathcal{L}(U,U)$ is such that $\|\epsilon(T)\| \to 0$ as $\|T\| \to 0$. Conclude that Θ is differentiable at I. [If you wish to confine yourself to finite-dimensional spaces, take U finite-dimensional, but there is no need.]

(vi) Show that Θ is everywhere differentiable on E with $D\Theta(B) = \Phi_B$, where $\Phi_B : \mathcal{L}(U,U) \to \mathcal{L}(U,U)$ is the linear map given by

$$\Phi_B(S) = -B^{-1}SB^{-1}.$$

Check that this reduces to a known result when $\dim U = 1$.

K.311. (Spectral radius.) [13.1, T, ↑] We continue with the ideas and notation of Exercise K.310.

(i) Suppose that $A \in \mathcal{L}(U,U)$. Show that

$$\|A^{jk+r}\| \leq \|A^j\|^k \|A\|^r.$$

(ii) Continuing with the hypotheses of (i), show that $\triangle = \inf_{n\geq 1} \|A^n\|^{1/n}$ exists and, by using the result of (i), or otherwise, that

$$\|A^n\|^{1/n} \to \triangle.$$

We call $\triangle$ the spectral radius of A, and write $\rho(A) = \triangle$.

(iii) Show that $\sum_{n=0}^{\infty} A^n$ converges in the operator norm if $\rho(A) < 1$ and diverges if $\rho(A) > 1$.

(iv) If $\rho(A) < 1$, show that $I - A$ is invertible and $(I-A)^{-1} = \sum_{n=0}^{\infty} A^n$.

K.312. [13.1, T, ↑] We continue with the ideas of Exercise K.311, but change our alphabet from Latin to Greek.

(i) If $\alpha \in \mathcal{L}(U,U)$ and λ is a scalar, show that $\rho(\lambda\alpha) = |\lambda|\rho(\alpha)$.

(ii) Write down a linear map $\beta : \mathbb{C}^m \to \mathbb{C}^m$ such that $\beta^{m-1} \neq 0$ but $\beta^m = 0$. What is the value of $\rho(\beta)$?

(iii) If the linear map $\alpha : \mathbb{C}^m \to \mathbb{C}^m$ has m distinct eigenvalues explain briefly why we can find an invertible linear map $\theta : \mathbb{C}^m \to \mathbb{C}^m$ such that $\theta\alpha\theta^{-1}$ has a diagonal matrix D with respect to the standard basis. By considering the behaviour of D^n, or otherwise, show that

$$\rho(\alpha) = \max\{|\lambda| : \lambda \text{ an eigenvalue of } \alpha\}.$$

(For an improvement of this result, see Exercise K.314.)

(iv) Give two linear maps $\alpha, \beta : \mathbb{C}^2 \to \mathbb{C}^2$ such that $\rho(\alpha) = \rho(\beta) = 0$ but $\rho(\alpha + \beta) = 1$.

K.313. (Cayley–Hamilton.) [13.1, T, ↑] This exercise requires you to know that any linear map $\alpha : \mathbb{C}^m \to \mathbb{C}^m$ has an upper triangular matrix with respect to some basis. Recall also that, if $\alpha : \mathbb{C}^m \to \mathbb{C}^m$, we can define its characteristic polynomial χ_α by the condition $\det(t\iota - \alpha) = \chi_\alpha(t)$ for all $t \in \mathbb{C}$. (Here and elsewhere, $\iota : \mathbb{C}^m \to \mathbb{C}^m$ is the identity map.)

(i) We work with the standard basis on $\mathbb{C}^m$. Explain why, if $\alpha : \mathbb{C}^m \to \mathbb{C}^m$ is linear, we can find an invertible linear map $\theta : \mathbb{C}^m \to \mathbb{C}^m$ such that $\beta = \theta\alpha\theta^{-1}$ has an upper triangular matrix with respect to the standard basis.

(ii) Show that, if $\beta : \mathbb{C}^m \to \mathbb{C}^m$ has an upper triangular matrix with respect to the standard basis and $\epsilon > 0$, then we can find $\gamma : \mathbb{C}^m \to \mathbb{C}^m$ with m distinct eigenvalues such that $\|\gamma - \beta\| < \epsilon$.

(iii) Deduce that, if $\alpha : \mathbb{C}^m \to \mathbb{C}^m$ is linear, we can find linear $\alpha_k : \mathbb{C}^m \to \mathbb{C}^m$ with m distinct eigenvalues and

$$\|\alpha_k - \alpha\| \to 0 \text{ as } k \to \infty.$$

(iv) (Pure algebra) If $\alpha : \mathbb{C}^m \to \mathbb{C}^m$ has n distinct eigenvalues, show, by considering the matrix of α with respect to a basis of eigenvectors, that

$$\chi_\alpha(\alpha) = 0.$$

(v) Show carefully, using (iii) and limiting arguments, that, if $\alpha : \mathbb{C}^m \to \mathbb{C}^m$ is linear, then

$$\chi_\alpha(\alpha) = 0.$$

This is the Cayley–Hamilton theorem.

[The argument given here is substantially longer than that used in algebra courses and is repugnant to the soul of any pure-minded algebraist. But it has its points of interest.]

(vi) Suppose that $\alpha, \beta : \mathbb{C}^m \to \mathbb{C}^m$ are linear maps. Observe that $\det(t\beta - \beta t) = 0$ for all $t \in \mathbb{C}$, but show that, if $m \geq 2$, we can choose α and β such that

$$\det(\alpha\beta - \beta\alpha) \neq 0.$$

Thus arguments of the form 'replace t by α' cannot be use to prove results like the Cayley–Hamilton theorem.

K.314. [13.1, T, ↑] We continue with the ideas of Exercise K.312.

(i) If $\alpha, \beta \in \mathcal{L}(U, U)$ and α and β commute, show that

$$\rho(\alpha+\beta) \leq \rho(\alpha)+\rho(\beta).$$

(Compare Exercise K.312 (iv).)

(ii) By using the ideas of Exercise K.313, or otherwise, show that, if $\alpha : \mathbb{C}^m \to \mathbb{C}^m$ is linear and $\epsilon > 0$, we can find a $\beta : \mathbb{C}^m \to \mathbb{C}^m$ such that $\|\beta\| < \epsilon$, α and β commute, $\alpha+\beta$ has m distinct eigenvalues, and every eigenvalue of $\alpha+\beta$ lies within a distance ϵ of some eigenvalue of α and vice versa.

(iii) By using (ii), or otherwise, show that, if $\alpha : \mathbb{C}^m \to \mathbb{C}^m$ is linear, then

$$\rho(\alpha) = \max\{|\lambda| : \ \lambda \text{ an eigenvalue of } \alpha\}.$$

[You may wish to read or reread Exercises K.109 to K.111 at this point.]

(iv) Explain why we can define the spectral radius $\rho(A)$ of an $n \times n$ matrix unambiguously as the spectral radius $\rho(\alpha)$ of a linear map represented by A with respect to some basis.

K.315. [13.1, T, ↑] In previous exercises we have shown that, if $\alpha : \mathbb{C}^m \to \mathbb{C}^m$ is linear, and we write

$$\rho(\alpha) = \max\{|\lambda| : \ \lambda \text{ an eigenvalue of } \alpha\},$$

then $\|\alpha^n\|^{1/n} \to \rho(\alpha)$.

Suppose that $\mathbf{b} \in \mathbb{C}^m$, we choose an arbitrary $\mathbf{x}_0 \in \mathbb{C}^m$ and we consider the sequence

$$\mathbf{x}_{n+1} = \mathbf{b} + \alpha \mathbf{x}_n.$$

If $\rho(\alpha) < 1$, give **two** proofs that

$$\|\mathbf{x}_n - (\iota - \alpha)^{-1}\mathbf{b}\| \to 0 \text{ as } n \to \infty :$$

(i) By finding $\mathbf{x}_n$ explicitly and using Exercise K.311.

(ii) By using the first paragraph of Exercise K.290.

By considering an eigenvector corresponding to an eigenvalue of largest modulus, show that the sequence will diverge if $\rho(\alpha) > 1$ and we choose $\mathbf{x}_0$ suitably. Show that, if $\rho(\alpha) = 1$, either we can find an $\mathbf{x}_0$ such the sequence diverges or, if the sequence always converges, we can find two starting points with different limits.

K.316. [13.1, T, ↑] Consider the problem of solving the equation

$$A\mathbf{x} = \mathbf{b},$$

where A is an $m \times m$ matrix and $\mathbf{x}$ and $\mathbf{b}$ are column vectors of length m. If m is small, then there is no reason to go beyond standard computational methods, and if m is large and A is a general matrix, we have no choice but to use standard methods. These involve storing all m^2 coefficients and, in the case of Gaussian elimination, require of the order of m^3 operations.

Suppose we have to deal with a matrix A such that A is close to I, in some sense to be determined later in the exercise, and there are only of the order of n non-zero coefficients in $I - A$ in a well organized pattern. (Such problems arise in the numerical solution of important partial differential equations.) The following method can then be employed. Choose $\mathbf{x}_0$ and define a sequence

$$\mathbf{x}_{n+1} = \mathbf{b} + (I - A)\mathbf{x}_n.$$

Using the ideas of earlier exercises, show that, under certain conditions, to be stated, $\mathbf{x}_n$ will tend to a unique solution of $A\mathbf{x} = \mathbf{b}$. Discuss the rapidity of convergence, and show that, under certain conditions, to be stated, only a few iterations will be required to get the answer to any reasonable degree of accuracy. Since each iteration requires, at worst, of the order of m^2 operations, and, in many cases, only of the order of m operations, this method is much more efficient than any known general purpose method.

The rest of the exercise consists of elaboration of this idea. We require Exercise K.315. Suppose that A is an $m \times m$ matrix and $A = D - U - L$ where L is strictly lower triangular, U is strictly upper triangular and D is diagonal with all diagonal terms non-zero. We seek to solve $A\mathbf{x} = \mathbf{b}$. The following iterative schemes have been proposed:

$$\text{Jacobi} \quad \mathbf{x}_{n+1} = D^{-1}(\mathbf{b} + (U + L)\mathbf{x}_n),$$
$$\text{Gauss–Seidel} \quad \mathbf{x}_{n+1} = (D - L)^{-1}(\mathbf{b} + U\mathbf{x}_n).$$

(Note that it is easy to invert a triangular matrix.) For each of these two schemes, give a necessary and sufficient condition in terms of the spectral radius of an appropriate matrix for the method to work.

A variation on the Gauss–Seidel method uses

$$\mathbf{x}_{n+1} = (D - \omega L)^{-1}(\omega\mathbf{b} + ((1 - \omega)D + \omega U)\mathbf{x}_n),$$

where ω is some fixed real number. Give a necessary and sufficient condition, in the form $\rho(H) < 1$, where H is an appropriate matrix, for the method to work. By showing that

$$\det H = (1 - \omega)^m,$$

or otherwise, show that the scheme must fail if $\omega < 0$ or $\omega > 2$.

K.317. [13.1, S, ↑↑] (This is a short exercise, but requires part (iv) of Exercise K.310.) Show that Theorem 13.13 can be strengthened by adding the following sentence at the end. 'Moreover, $D\mathbf{f}|_B^{-1}$ is continuous on B.'

K.318. [13.1, T, ↑↑] This is another exercise in the ideas of Exercise K.310. We work in a complete normed space $(U, \| \|_U)$ as before.

(i) Show that, if $\alpha \in \mathcal{L}(U, U)$, we can find $\exp \alpha \in \mathcal{L}(U, U)$ such that

$$\left\| \sum_{r=0}^{n} \frac{\alpha^r}{r!} - \exp \alpha \right\| \to 0$$

as $n \to \infty$.

(ii) Show carefully that, if α and β commute,

$$\exp \alpha \exp \beta = \exp(\alpha + \beta).$$

(iii) Show that, if α and β are general (not necessarily commuting) elements of $\mathcal{L}(U, U)$, then

$$\left\| h^{-2}(\exp(h\alpha) \exp(h\beta) - \exp(h\beta) \exp(h\alpha)) - (\alpha\beta - \beta\alpha) \right\| \to 0$$

as the real number $h \to 0$.

Conclude that, in general, $\exp(\alpha) \exp(\beta)$ and $\exp(\beta) \exp(\alpha)$ need not be equal. Deduce also that $\exp(\alpha + \beta)$ and $\exp(\alpha) \exp(\beta)$ need not be equal.

(iv) Show carefully (you must bear part (iii) in mind) that the map $\exp : \mathcal{L}(U, U) \to \mathcal{L}(U, U)$ is everywhere continuous.

K.319. [13.1, P, ↑] (i) Consider the map $\Theta_3 : \mathcal{L}(U, U) \to \mathcal{L}(U, U)$ given by $\Theta_3(\alpha) = \alpha^3$. Show that Θ is everywhere differentiable with

$$D\Theta_3(\alpha)\beta = \beta\alpha^2 + \alpha\beta\alpha + \alpha^2\beta.$$

(ii) State and prove the appropriate generalisation to the map $\alpha \mapsto \alpha^m$ with m a positive integer.

(iii) Show that exp, defined in Exercise K.318, is everywhere differentiable. [This requires care.]

K.320. [13.1, P, S, ↑↑] Suppose U is a finite-dimensional vector space over $\mathbb{C}$. Let $\alpha : U \to U$ be a linear map. If α has matrix representation A with respect to some basis, explain why, as $N \to \infty$, the entries of the matrix $\sum_{n=0}^{N} A^n/n!$ converge to the entries of a matrix $\exp A$ which represents $\exp \alpha$ with respect to the given basis. (See Exercise K.318.)

It is a theorem that any $\alpha \in \mathcal{L}(U, U)$ has an upper triangular matrix with respect to some basis. By using this fact, or otherwise, show that

$$\det(\exp \alpha) = e^{\operatorname{Trace} \alpha}.$$

K.321. [13.1, P] We work in the space $M_2(\mathbb{R})$ of 2×2 real matrices. We give $M_2(\mathbb{R})$ the associated operator norm.

(i) Show that the map $S : M_2(\mathbb{R}) \to M_2(\mathbb{R})$ given by $S(A) = A^2$ is everywhere differentiable with $DS(A)B = AB + BA$. (If you have done Exercise K.319, you may just quote it.)

(ii) Show that the matrix equation

$$A^2 = \begin{pmatrix} 0 & 1 \\ 0 & 0 \end{pmatrix}$$

has no solution.

(iii) Calculate explicitly all the solutions of

$$\begin{pmatrix} a & b \\ c & d \end{pmatrix}^2 = I.$$

What are the possible values of $\det A$? Describe geometrically the linear maps associated with the matrices A such that $A^2 = I$ and $\det A = 1$. Describe geometrically the linear maps associated with the matrices A such that $A^2 = I$ and A is diagonal.

(iv) Show that there are open sets G and H containing 0 (the zero matrix) such that the equation

$$A^2 = I + X$$

has exactly one solution of the form $A = I + Y$ with $Y \in H$ for each $X \in G$.

(v) Show that we cannot find open sets G and H containing 0 (the zero matrix) such that the equation

$$A^2 = I + X$$

has exactly one solution of the form

$$A = \begin{pmatrix} 1 & 0 \\ 0 & -1 \end{pmatrix} + Y$$

with $Y \in G$ for each $X \in H$. Identify which hypothesis of the inverse function theorem (Theorem 13.13) fails to hold and show, by direct calculation, that it does indeed fail.

(vi) For which B is it true that $B^2 = I$ and we can find open sets H and G containing 0 (the zero matrix) such that the equation

$$A^2 = I + X$$

has exactly one solution of the form $A = B + Y$ with $Y \in G$ for each $X \in H$? Give reasons for your answer.

K.322. [13.1, P, G] This exercise requires some knowledge of eigenvectors of symmetric linear maps. We work in $\mathbb{R}^3$ with the usual inner product. Suppose $\alpha : \mathbb{R}^3 \to \mathbb{R}^3$ is an antisymmetric linear map (that is, $\alpha^T = -\alpha$). Show that $\mathbf{x}$ and $\alpha\mathbf{x}$ are orthogonal and that the eigenvalues of α^2 must be non-positive real numbers. By considering eigenvectors of α and α^2, show that we can always find $\mu \in \mathbb{R}$ and three orthonormal vectors $\mathbf{e}_1$, $\mathbf{e}_2$ and $\mathbf{e}_3$ such that

$$\alpha\mathbf{e}_1 = \mu\mathbf{e}_2, \ \alpha\mathbf{e}_2 = -\mu\mathbf{e}_1, \ \alpha\mathbf{e}_3 = \mathbf{0}.$$

By choosing appropriate axes and using matrix representations, show that $\exp\alpha$ is a rotation.

K.323. [13.1, P, G, S] Show that every linear map $\alpha : \mathbb{R}^m \to \mathbb{R}^m$ is the sum of a symmetric and an antisymmetric linear map. Suppose α is an orthogonal map (that is, $\alpha\alpha^T = \iota$) with $\|\iota - \alpha\| < \epsilon$ and ϵ very small. Show that

$$\alpha = \iota + \epsilon\beta + \epsilon^2\gamma,$$

where β is antisymmetric, γ is antisymmetric and $\|\beta\|$, $\|\gamma\| \leq 2$.

K.324. [13.1, P, G, ↑↑] In this exercise we work with real $n \times n$ matrices. Although we use the exponential

$$\exp A = \sum_{j=0}^{\infty} \frac{A^n}{n!}$$

obtained in Exercise K.320, the main requirement is some elementary matrix theory. Our object is to find all 'isotropic matrices', that is to say, all matrices A such that $UAU^T = A$ whenever $UU^T = I$ and $\det U = 1$.

(i) Show that, if $UU^T = I$, then $\det U = \pm 1$.

(ii) If $B^T = -B$, show that $(\exp B)(\exp B)^T = 1$.

(iii) Show that the function $\theta : \mathbb{R} \to \mathbb{R}$ given by $\theta(h) = \det(\exp hB)$ is continuous. Deduce, using part (i), that, if $B^T = -B$, then $\det(\exp B) = 1$.

(iv) Let A be isotropic and $B^T = -B$. Explain why $(\exp hB)A = A(\exp hB)$ for all $h \in \mathbb{R}$ and, by considering what happens when $h \to 0$, show that $BA = AB$.

(v) By considering particular simple choices of B, or otherwise, show that, if $n \geq 3$, then the only isotropic matrices have the form λI with $\lambda \in \mathbb{R}$.

(vi) Find all the isotropic matrices when $n = 2$. (Note that you must check that the matrices you find are indeed isotropic.) Give a simple geometric characterisation of those isotropic matrices with determinant 1, and use it to characterise all isotropic matrices.

(vii) What happens when $n = 1$?

(viii) How would the results of (v), (vi) and (vii) change if we omitted the condition $\det U = 1$ in the definition of isotropic? Prove your statements.

K.325. [13.3, M] (Treat this as a 'methods question'.) The four vertices A, B, C, D of a quadrilateral lie in anti-clockwise order on a circle of radius a and center O. We write $2\theta_1 = \angle AOB$, $2\theta_2 = \angle BOC$, $2\theta_3 = \angle COD$, $2\theta_4 = \angle DOA$. Find the area of the quadrilateral and state the relation that θ_1, θ_2, θ_3 and θ_4 must satisfy.

Use Lagrange's method to find the form of a quadrilateral of greatest area inscribed in a circle of radius a.

Use Lagrange's method to find the form of an n-gon of greatest area inscribed in a circle $[n \geq 3]$.

Use Lagrange's method to find the form of an n-gon of least area circumscribing a circle $[n \geq 3]$.

[Compare Exercise K.41.]

K.326. [13.3, T] Let p and q be strictly positive real numbers with $p^{-1} + q^{-1} = 1$. Suppose that $y_1, y_2, \ldots, y_n, c > 0$. Explain why there must exist $x_1, x_2, \ldots, x_n \geq 0$ with $\sum_{j=1}^{n} x_j^p = c$ and

$$\sum_{j=1}^{n} x_j y_j \geq \sum_{j=1}^{n} t_j y_j \text{ whenever } t_1, t_2, \ldots, t_n \geq 0 \text{ with } \sum_{j=1}^{n} t_j^p = c.$$

Use the Lagrange multiplier method to find the x_j. Deduce from your answer that

$$\sum_{j=1}^{n} |a_j b_j| \leq \left(\sum_{j=1}^{n} |a_j|^p\right)^{1/p} \left(\sum_{j=1}^{n} |b_j|^q\right)^{1/q}$$

whenever a_j, $b_j \in \mathbb{C}$. Under what conditions does equality hold?

This gives an alternative proof of the first result in Exercise K.211 (i).

K.327. (The parallelogram law.) [14.1, T] Except in the last part of this exercise, we work in a real normed vector space $(V, \| \ \|)$.

(i) Suppose that V has a real inner product $\langle \ , \ \rangle$ such that $\langle \mathbf{x}, \mathbf{x} \rangle = \|\mathbf{x}\|^2$ for all $\mathbf{x} \in V$. Show that

$$\|\mathbf{x} + \mathbf{y}\|^2 + \|\mathbf{x} - \mathbf{y}\|^2 = 2\|\mathbf{x}\|^2 + 2\|\mathbf{y}\|^2$$

for all $\mathbf{x}$, $\mathbf{y} \in V$. (This is called the parallelogram law.)

(ii) Show also that

$$4\langle \mathbf{x}, \mathbf{y} \rangle = \|\mathbf{x} + \mathbf{y}\|^2 - \|\mathbf{x} - \mathbf{y}\|^2$$

for all $\mathbf{x}$, $\mathbf{y} \in V$. (This is called the polarisation identity.)

(iii) Use the parallelogram law to obtain a relation between the lengths of the sides and the diagonals of a parallelogram in Euclidean space.

(iv) Use the triangle inequality to show that $|\|\mathbf{x}\| - \|\mathbf{y}\|| \leq \|\mathbf{x} - \mathbf{y}\|$, and use this result together with the polarisation identity and the parallelogram law to give another proof of the Cauchy–Schwarz inequality.

(v) Suppose now that $(V, \langle\ ,\ \rangle)$ is a complex inner product space with norm $\|\ \|$ derived from the inner product. Show that the parallelogram law holds in the same form as before, and obtain the new polarisation identity

$$4\langle \mathbf{x}, \mathbf{y}\rangle = \|\mathbf{x} + \mathbf{y}\|^2 - \|\mathbf{x} - \mathbf{y}\|^2 + i\|\mathbf{x} + i\mathbf{y}\|^2 - i\|\mathbf{x} - i\mathbf{y}\|^2.$$

(vi) Show that the uniform norm on $C([0,1])$ is not derived from an inner product. (That is to say, there does not exist an inner product $\langle\ ,\ \rangle$ with $\|f\|_\infty^2 = \langle f, f\rangle$ for all $f \in C([0,1])$.)

K.328. [14.1, T, ↑] The parallelogram law of Exercise K.327 actually characterises norms derived from an inner product, although the proof is slightly trickier than one might expect.

(i) Let $(V, \|\ \|)$ be a real normed space such that

$$\|\mathbf{x} + \mathbf{y}\|^2 + \|\mathbf{x} - \mathbf{y}\|^2 = 2\|\mathbf{x}\|^2 + 2\|\mathbf{y}\|^2$$

for all $\mathbf{x}$, $\mathbf{y} \in V$. It is natural to try setting

$$\langle \mathbf{x}, \mathbf{y}\rangle = 4^{-1}\big(\|\mathbf{x} + \mathbf{y}\|^2 - \|\mathbf{x} - \mathbf{y}\|^2\big).$$

Show that $\langle \mathbf{x}, \mathbf{y}\rangle = \langle \mathbf{y}, \mathbf{x}\rangle$ for all $\mathbf{x}$, $\mathbf{y} \in V$ and that $\langle \mathbf{x}, \mathbf{x}\rangle = \|\mathbf{x}\|^2$ (so that, automatically, $\langle \mathbf{x}, \mathbf{x}\rangle \geq 0$ with equality if and only if $\mathbf{x} = \mathbf{0}$).

(ii) The remaining inner product rules are harder to prove. Show that

$$\|\mathbf{u} + \mathbf{v} + \mathbf{w}\|^2 + \|\mathbf{u} + \mathbf{v} - \mathbf{w}\|^2 = 2\|\mathbf{u} + \mathbf{v}\|^2 + 2\|\mathbf{w}\|^2$$

and use this to establish that

$$\langle \mathbf{u} + \mathbf{w}, \mathbf{v}\rangle + \langle \mathbf{u} - \mathbf{w}, \mathbf{v}\rangle = 2\langle \mathbf{u}, \mathbf{v}\rangle \tag{1}$$

for all $\mathbf{u}$, $\mathbf{v}$, $\mathbf{w} \in V$. Use equation (1) to establish that

$$\langle 2\mathbf{u}, \mathbf{v}\rangle = 2\langle \mathbf{u}, \mathbf{v}\rangle \tag{2}$$

and then use equations (1) and (2) to show that

$$\langle \mathbf{x}, \mathbf{v}\rangle + \langle \mathbf{y}, \mathbf{v}\rangle = \langle \mathbf{x} + \mathbf{y}, \mathbf{v}\rangle$$

for all $\mathbf{x}$, $\mathbf{y}$, $\mathbf{v} \in V$.

(iii) Establish the equation

$$\langle \lambda\mathbf{x}, \mathbf{y}\rangle = \lambda\langle \mathbf{x}, \mathbf{y}\rangle$$

for all positive integer values of λ, then for all integer values, for all rational values and then for all real values of λ.

(iv) Use the fact that the parallelogram law characterises norms derived from an inner product to give an alternative proof of Lemma 14.11 in the real case.

(v) Extend the results of this exercise to complex vector spaces.

K.329. [14.1, P] Suppose (X, d) is a complete metric space with a dense subset E. Suppose that E is a vector space (over $\mathbb{F}$, where $\mathbb{F} = \mathbb{R}$ or $\mathbb{F} = \mathbb{C}$) with norm $\| \ \|_E$ such that $d(x, y) = \|x -_E y\|_E$ for all $x, y \in E$. Suppose further that there is map $M_E : E^2 \to E$ such that, writing $M_E(x, y) = xy$, we have

(i) $x(yz) = (xy)z$,

(ii) $(x + y)z = xz + yz$, $z(x + y) = zx + zy$,

(iii) $(\lambda x)y = \lambda(xy)$, $x(\lambda y) = \lambda(xy)$,

(iv) $\|xy\| \leq \|x\|\|y\|$,

for all $x, y, z \in E$ and $\lambda \in \mathbb{F}$. Show that, if X is given the structure of a normed vector space as in Lemma 14.9, then we can find a map $M : X^2 \to X$ such that $M(x, y) = M_E(x, y)$ for all $x, y \in E$ and M has properties (i) to (iv) (with E replaced by X). Show that, if $M_E(x, y) = M_E(y, x)$ for all $x, y \in E$, then $M(x, y) = M(y, x)$ for all $x, y \in X$. Show that, if there exists an $e \in E$ such that $M_E(x, e) = M_E(e, x) = x$ for all $x \in E$, then $M(x, e) = M(e, x) = x$ for all $x \in X$.

K.330. [14.1, T] This neat proof of the existence part of Theorem 14.17, which states that every metric space (E, d) can be completed, is due to Kuratowski[17].

(i) Choose $e_0 \in E$. For each $e \in E$ define $f_e(t) = d(e, t) - d(e_0, t)$ $[t \in E]$. Show that $f_e \in \mathcal{C}(E)$, where $\mathcal{C}(E)$ is the space of bounded continuous functions $g : E \to \mathbb{R}$.

(ii) Give $\mathcal{C}(E)$ the usual uniform norm. Show that

$$\|f_u - f_v\| = d(u, v)$$

for all $u, v \in E$.

(iii) Let Y be the closure of $\{f_e : e \in E\}$. By using Theorem 11.39, or otherwise, show that Y with the metric $\tilde{d}$ inherited from $\mathcal{C}(E)$ is complete. Show that (E, d) has a completion by considering the map $\theta : E \to Y$ given by $\theta(e) = f_e$.

K.331. [14.1, P] Results like Lemma 14.9 rely on a strong link between the algebraic operation and the metric. From one point of view this exercise consists of simple results dressed up in jargon, but I think they shed some light on the matter.

(i) (This just sets up a bit of notation.) Suppose that (X, d) is a metric space. Show that $d_2 : E^2 \to \mathbb{R}$ given by

$$d_2((x, y), (x', y')) = d(x, x') + d(y, y')$$

[17] I take it from from a book [**16**] crammed with neat proofs.

defines a metric on X^2. Show that, if (X, d) is complete, so is (X^2, d_2).

(ii) Consider $X = [0, \infty)$ with the usual Euclidean metric d and $E = (0, \infty)$. Show that (X, d) is complete, E is a dense subset of X, and that, if we write $M_E(x, y) = xy$ (that is, if M is ordinary multiplication), then (E, M) is a group and $M_E : (E^2, d_2) \to (E, d)$ is continuous. Show, however, that there does not exist a continuous map $M : (X^2, d_2) \to (X, d)$ with $M(x, y) = M_E(x, y)$ for all x, $y \in E$ such that (X, M) is a group.

(iii) Consider $X = [0, \infty)$ with the usual Euclidean metric d and $E = (0, \infty) \cap \mathbb{Q}$. Show that (X, d) is complete, E is a dense subset of X, and, if we write $M_E(x, y) = xy$, then (E, M) is a group and $M_E : (E^2, d_2) \to (E, d)$ is continuous. Show, however, that there does not exist a continuous map $M : (X^2, d_2) \to (X, d)$ with $M(x, y) = M_E(x, y)$ for all x, $y \in E$ such that (X, M) is a group.

(iv) Consider $X = (0, \infty)$. Show that, if we write

$$d(x, y) = |\log x - \log y|,$$

then (X, d) is a metric space. Let $E = (0, \infty) \cap \mathbb{Q}$. Show that (X, d) is complete, E is a dense subset of X, and, if we write $M_E(x, y) = xy$, then (E, M) is a group and $M_E : (E^2, d_2) \to (E, d)$ is continuous. Show that there exists a continuous map $M : (X^2, d_2) \to (X, d)$ with $M(x, y) = M_E(x, y)$ for all x, $y \in E$ such that (X, M) is a group.

K.332. [14.1, P] (i) Observe that $\mathbb{R}$ is a group under addition. If E is a subgroup of $\mathbb{R}$ which is also a closed set with respect to the Euclidean metric, show that either $E = \mathbb{R}$ or

$$E = \{n\alpha : n \in \mathbb{Z}\}$$

for some $\alpha \in \mathbb{R}$.

(ii) Observe that

$$S^1 = \{\lambda \in \mathbb{C} : |\lambda| = 1\}$$

is a group under multiplication. What can you say about subgroups E of S^1 which are closed with respect to the usual metric?

(iii) Observe that $\mathbb{R}^2$ is a group under vector addition. What can you say about subgroups E of $\mathbb{R}^2$ which are closed with respect to the Euclidean metric? Generalise your result to $\mathbb{R}^m$.

K.333. [14.1, T, ↑↑] Exercise K.57 is not important in itself, but the method of its proof is. Extend the result and proof to $f : E \to \mathbb{R}$, where E is a dense subspace of a metric space (X, d).

Can the result be extended to $f : E \to Y$, where E is a dense subspace of a metric space (X, d) and (Y, ρ) is a metric space? (Give a proof or counterexample.) If not, what natural extra condition can we place on (Y, ρ) so that the result can be extended?

K.334. [14.1, P, S, ↑] Suppose that $f : \mathbb{R} \to \mathbb{R}$ satisfies

$$|f(x) - f(y)| \leq (x - y)^2 \text{ for all } x, y \in \mathbb{R}.$$

Show that f is constant.

Show that the result remains true if we replace $\mathbb{R}$ by $\mathbb{Q}$. Explain why this is consistent with examples of the type given in Example 1.3

K.335. [14.1, P, S] Let (X, d) be a metric space which has the the Bolzano–Weierstrass property. Show that, given any $\epsilon > 0$, we can find a finite set of points $x_1, x_2, \ldots x_n$ such that the open balls $B(x_j, \epsilon)$ with centre x_j and radius ϵ cover X (that is to say, $\bigcup_{j=1}^{n} B(x_j, \epsilon) = X$). (This result occurs elsewhere both in the main text and in the exercises, but it will do no harm to reprove it.) Deduce that (X, d) has a countable dense subset.

Give an example of a complete metric space which does not have the Bolzano–Weierstrass property but does have a countable dense subset. Give an example of a metric space which is not complete but does have a countable dense subset. Give an example of a complete metric space which does not have a countable dense subset.

K.336. [14.1, P] (i) By observing that every open interval contains a rational number, or otherwise, show that every open subset of $\mathbb{R}$ (with the standard Euclidean metric) can be written as the countable union of open intervals.

(ii) Let (X, d) be a metric space with a countable dense subspace. Show that every open subset of X can be written as the countable union of open balls.

(iii) Give an example of a complete metric space which has an open set that cannot be written as the countable union of open balls.

(iv) Let (X, d) be a metric space with a countable dense subspace. Show that every open ball can be written as the countable union of closed balls. Show that every open set can be written as the countable union of closed balls. Show that, if U is an open set, we can find bounded closed sets K_j with $K_{j+1} \subseteq K_j$ $[1 \leq j]$ and $\bigcup_{j=1}^{\infty} K_j = U$. [This result is useful for spaces like $\mathbb{R}^n$ with the usual metric, where we know, in addition, that bounded closed sets have the Bolzano–Weierstrass property.]

K.337. [14.1, P] The previous Exercise K.336 dealt with general metric spaces. Apart from the last part, this exercise deals with the particular space $\mathbb{R}$ with the usual metric. We need the notion of an equivalence relation.

(i) Let U be an open subset of $\mathbb{R}$. If $x, y \in U$, write $x \sim y$ if there is an open interval $(a, b) \subset U$ with $x, y \in (a, b)$. Show that $\sim$ is an equivalence relation on U.

(ii) Write $[x] = \{y \in U : y \sim x\}$ for the equivalence class of some $x \in U$. If $[x]$ is bounded, show, by considering the infimum and supremum of $[x]$, or otherwise, that $[x]$ is an open interval. What can we say about $[x]$ if it is bounded below but not above? Prove your answer carefully. What can we say about $[x]$ if it is bounded above but not below? What can we say if $[x]$ is neither bounded above nor below?

(iii) Show that U is the disjoint union of a collection $\mathcal{C}$ of sets of the form (a, b), (c, ∞), $(-\infty, c)$ and $\mathbb{R}$.

(iv) Suppose that U is the disjoint union of a collection $\mathcal{C}'$ of sets of the form (a, b), (c, ∞), $(-\infty, c)$ and $\mathbb{R}$. If $J \in \mathcal{C}'$, explain why there exists an $I \in \mathcal{C}$ with $I \supseteq J$. Explain why, if a is an end point of J which is not an end point of I, there must exist a $J' \in \mathcal{C}'$ with $a \in J'$. Hence, or otherwise, show that $J = I$. Conclude that $\mathcal{C}' = \mathcal{C}$.

(v) Show that $\mathcal{C}$ is countable.

(vi) We saw in part (iv) that $\mathcal{C}$ is uniquely defined, and this raises the possibility of defining the 'length' of U to be the sum of the lengths of the intervals making up $\mathcal{C}$. However, this approach fails in higher dimensions. The rest of this exercise concerns $\mathbb{R}^2$ with the usual metric.

Show that the open square $(-a, a) \times (-a, a)$ is not the union of disjoint open discs. [It may be helpful to look at points on the boundary of a disc forming part of such a putative union.]

Show that the open disc $\{(x, y) : x^2 + y^2 < 1\}$ is not the union of disjoint open squares.

K.338. [14.1, P, ↑↑] Recall that we say that a function $f : [0, 1] \to \mathbb{R}$ is of bounded variation if there exists a constant K such that, whenever $0 = x_0 \le x_1 \le \cdots \le x_n = 1$, we have

$$\bigstar \qquad \sum_{j=1}^{n} |f(x_j) - f(x_{j-1})| \le K.$$

We write $BV([0, 1])$ for the space of all such functions and, if $f \in BV([0, 1])$, we write $V(f)$ for the infimum of the set of K for which ★ holds.

(i) Show that, if we write

$$\|f\|_{BV} = |f(0)| + V(f),$$

then $(BV([0, 1]), \| \ \|_{BV})$ is a complete normed space.

(ii) Show that $C([0, 1])$ is not a subset of $BV([0, 1])$ and $BV([0, 1])$ is not a subset of $C([0, 1])$. Show that $C([0, 1]) \cap BV([0, 1])$ is a closed subspace of $(BV([0,1]), \| \|_{BV})$. Show that $C^1([0,1])$, the set of all continuously differentiable functions on $[0,1]$, is a subspace of $(BV([0,1]), \| \|_{BV})$ but is not closed.

(iii) If f is the 'devil's staircase' function constructed in Exercise K.253 and $g \in C^1([0,1])$, show that

$$V(f+g) = V(f) + V(g).$$

Deduce that $C^1([0,1])$ is not dense in $C([0,1]) \cap BV([0,1])$ equipped with the norm $\| \|_{BV}$.

K.339. [14.1, T] We say that metric spaces (X,d) and (Y,ρ) are homeomorphic if there exists a bijective map $f : X \to Y$ such that f and f^{-1} are continuous. We say that f is a homeomorphism between X and Y.

(i) Show that homeomorphism is an equivalence relation on metric spaces.

(ii) If $f : X \to Y$ is a homeomorphism, show that U is open in (X,d) if and only if $f(U)$ is open in (Y,ρ).

(iii) By constructing an explicit homeomorphism, show that $\mathbb{R}$ with the usual metric is homeomorphic to the open interval $(-1,1)$ with the usual metric. Deduce that the property of completeness is not preserved under homeomorphism.

(iv) By constructing an explicit homeomorphism, show that $I = (-1,1)$ with the usual metric is homeomorphic to

$$J = \{z \in \mathbb{C} : |z| = 1, \ z \neq 1\}$$

with the usual metric. Show that $[-1,1]$ with the usual metric is a completion of I. Find a completion of J.

Explain briefly why the completion of I must contain two extra points.

K.340. [14.1, T] (i) Suppose (X,d) is a metric space with the Bolzano–Weierstrass property and (Y,ρ) is any metric space. If $f : X \to Y$ is a bijective continuous function, show that $f^{-1} : Y \to X$ is uniformly continuous. (Thus we have shown that, in the language of Exercise K.339, (X,d) and (Y,ρ) are homeomorphic.) Show also that (Y,ρ) has the Bolzano–Weierstrass property.

(ii) Look briefly at Exercise 5.66. Which results (if any) of that exercise can can be obtained using (i)?

(iii) Consider $\mathbb{R}$ with the usual metric. Give an example of a uniformly continuous bijective map $f : \mathbb{R} \to \mathbb{R}$ with f^{-1} not uniformly continuous.

(iv) Let d be the usual metric on $\mathbb{R}$ and ρ the discrete metric on $\mathbb{R}$. Let $f : (\mathbb{R},\rho) \to (\mathbb{R},d)$ be given by $f(x) = x$. Show that f is is a bijective continuous function but f^{-1} is not continuous.

K.341. [14.1, T, ↑] Suppose that (X, d) is a metric space with the Bolzano–Weierstrass property. Explain (by referring to Exercise K.335, if necessary) why we can find a countable dense subset $\{x_1, x_2, x_3, \ldots\}$, say, for X. Consider l^2 with its usual norm (see Exercise K.208). Show that the function $\mathbf{f} : X \to l^2$ given by

$$\mathbf{f}(x) = (d(x, x_1), 2^{-1}d(x, x_2), 2^{-2}d(x, x_3), \ldots)$$

is well defined, continuous and injective. Deduce that $\mathbf{f}(X)$ is homeomorphic to X. [Thus l^2 contains subsets homeomorphic to any given metric space with the Bolzano–Weierstrass property.]

K.342. [14.1, P] Suppose that (X, d) and (Y, ρ) are metric spaces and $f : X \to Y$ is a continuous surjective map.

(i) Suppose that $\rho(f(x), f(x')) \leq Kd(x, x')$ for all x, $x' \in X$ and some $K > 0$. If (X, d) is complete, does it follow that (Y, ρ) is complete? If (Y, ρ) is complete, does it follow that (X, d) is complete? Give proofs or counterexamples as appropriate.

(ii) Suppose that $\rho(f(x), f(x')) \geq Kd(x, x')$ for all x, $x' \in X$ and some $K > 0$. If (X, d) is complete, does it follow that (Y, ρ) is complete? If (Y, ρ) is complete, does it follow that (X, d) is complete? Give proofs or counterexamples as appropriate.

K.343. [14.1, P, S] Let X be the space of open intervals $\alpha = (a, b)$ with $a < b$ in $\mathbb{R}$. If α, $\beta \in X$, then the symmetric difference $\alpha \triangle \beta$ consists of the empty set, one open interval or two disjoint open intervals. We define $d(\alpha, \beta)$ to be the total length of the intervals making up $\alpha \triangle \beta$. Show that d is a metric on X.

Show that the completion of (X, d) contains precisely one further point.

K.344. [14.1, P] Consider the set $\mathbb{N}$ of non-negative integers. If $n \in \mathbb{N}$ and $n \neq 0$, then there exist unique r, $s \in \mathbb{N}$ with s an odd integer and $n = s2^r$. Write $\lambda(n) = r$. If n, $m \in \mathbb{N}$ and $n \neq m$, we write

$$d(n, m) = 2^{-\lambda(|n-m|)}.$$

We take $d(n, n) = 0$.

(i) Show that d is a metric on $\mathbb{N}$.

(ii) Show that the open ball with centre 1 and radius 1 is closed.

(iii) Show that no one-point set $\{n\}$ is open.

(iv) Show that the function $f : \mathbb{N} \to \mathbb{N}$ given by $f(x) = x^2$ is continuous.

(v) Show that the function $g : \mathbb{N} \to \mathbb{N}$ given by $f(x) = 2^x$ is not continuous at any point of $\mathbb{N}$.

(vi) Show that d is not complete. (Be careful. You must show that your Cauchy sequence does not converge to any point of $\mathbb{N}$.)

Reflect on what the completion might look like. (You are not asked to come to any conclusion.)

K.345. [Appendix C, P] Show that there exists a continuous function $G : \mathbb{R}^2 \to \mathbb{R}$ which is infinitely differentiable at every point except $\mathbf{0}$ and which has directional derivative zero in all directions at $\mathbf{0}$ but which is not differentiable at $\mathbf{0}$.

Bibliography

1. F. S. Acton, *Numerical methods that work*, Harper and Row, 1970.
2. A. F. Beardon, *Limits. a new approach to real analysis*, Springer, 1997.
3. R. Beigel, *Irrationality without number theory*, American Mathematical Monthly **98** (1991), 332–335.
4. B. Belhoste, *Augustin-Louis Cauchy. a biography*, Springer, 1991, Translated from the French by F. Ragland, but the earlier French publication of the same author is a different book.
5. D. Berlinski, *A tour of the calculus*, Pantheon Books, New York, 1995.
6. P. Billingsley, *Probability and measure*, Wiley, 1979.
7. E. Bishop and D. Bridges, *Constructive analysis*, Springer, 1985.
8. R. P. Boas, *Lion hunting and other mathematical pursuits*, Dolciani Mathematical Expositions, vol. 15, MAA, 1995.
9. J. R. Brown, *Philosophy of mathematics*, Routledge and Kegan Paul, 1999.
10. J. C. Burkill, *A first course in mathematical analysis*, CUP, 1962.
11. R. P. Burn, *Numbers and functions*, CUP, 1992.
12. W. S. Churchill, *My early life*, Thornton Butterworth, London, 1930.
13. D. R. Dickinson, *Calculus*, George G. Harrap and Co., London, 1958.
14. J. Dieudonné, *Foundations of modern analysis*, Academic Press, 1960.
15. ———, *Infinitesimal calculus*, Kershaw Publishing Company, London, 1973, Translated from the French *Calcul Infinitésimal* published by Hermann, Paris in 1968.
16. R. M. Dudley, *Real analysis and probability*, second ed., Wadsworth and Brooks, CUP, 2002.
17. J. Fauvel and J. Gray (eds.), *History of mathematics: A reader*, Macmillan, 1987.
18. W. L. Ferrar, *A textbook of convergence*, OUP, 1938.
19. J. E. Gordon, *Structures, or why things don't fall down*, Penguin, 1978.
20. A. Grünbaum, *Modern science and Zeno's paradoxes*, George Allen and Unwin, 1968.
21. P. R. Halmos, *Naive set theory*, Van Nostrand, 1960.

22. ______, *I want to be a mathematician*, Springer, 1985.

23. G. H. Hardy, *Collected papers*, vol. V, OUP, 1908.

24. ______, *A course of pure mathematics*, CUP, 1908, Still available in its 10th edition.

25. ______, *Divergent series*, OUP, 1949.

26. G. H. Hardy and E. M. Wright, *An introduction to the theory of numbers*, CUP, 1938.

27. P. T. Johnstone, *Notes on logic and set theory*, CUP, 1987.

28. Y. Katznelson and K. Stromberg, *Everywhere differentiable, nowhere monotone, functions*, American Mathematical Monthly **81** (1974), 349–54.

29. F. Klein, *Elementary mathematics from an advanced standpoint (part 1)*, Dover, 1924, Third edition. Translated by E. R. Hedrick and C. A. Hedrick.

30. M. Kline, *Mathematical thought from ancient to modern times*, OUP, 1972.

31. A. N. Kolmogorov and S. V. Fomin, *Introductory real analysis*, Prentice Hall, 1970, Translated from the Russian and edited by R. A. Silverman.

32. T.W. Körner, *The behavior of power series on their circle of convergence*, Banach Spaces, Harmonic Analysis, and Probability Theory, Lecture Notes in Mathematics, vol. 995, Springer, 1983, pp. 56–94.

33. ______, *Differentiable functions on the rationals*, Bulletin of the London Mathematical Society **23** (1991), 557–62.

34. ______, *Butterfly hunting and the symmetry of mixed partial derivatives*, American Mathematical Monthly **105** (1998), 756–8.

35. E. Laithwaite, *Invitation to engineering*, Blackwell, 1984.

36. J. E. Littlewood, *A mathematician's miscellany*, 2nd ed., CUP, 1985, (Editor B. Bollobás).

37. P. Mancosu and E. Vilati, *Torricelli's infinitely long horn and its philosophical reception in the seventeeenth century*, Isis **82** (1991), 50–70.

38. J. C. Maxwell, *Treatise on electricity and magnetism*, OUP, 1873.

39. ______, *Scientific letters and papers*, vol. II, CUP, 1995.

40. D. J. Newman, *A problem seminar*, Springer, 1982.

41. G. M. Phillips, *Two millennia of mathematics. from Archimedes to Gauss*, CMS books in mathematics **6**, Springer, 2000.

42. Plato, *Parmenides*, Routledge and Kegan Paul, 1939, Translated by F. M. Cornford.

43. F. Poundstone, *Labyrinths of reason*, Penguin, 1991.

44. M. J. D. Powell, *Approximation theory and methods*, CUP, 1988.

45. T. A. Ptaclusp, *Episodes from the lives of the great accountants*, Pyramid Press, Tsort, 1970.

46. M. C. Reed, *Fundamental ideas of analysis*, Wiley, 1998.

47. P. L. Roe, *The best shape for a tin can*, The Mathematical Gazette **75** (1991), no. 472, 147–50.

48. K. R. Stromberg, *Introduction to classical real analysis*, Wadsworth, 1981.

49. S. Wagon, *The Banach-Tarski paradox*, CUP, 1985.

50. E. T. Whittaker and G. N. Watson, *A course of modern analysis*, CUP, 1902, Still available in its 4th edition.

51. P. Whittle, *Optimization under constraints*, Wiley, 1971.

Index

You may find Appendix J useful for tracking down particular ideas.

Titles in This Series

62 **T. W. Körner,** A companion to analysis: A second first and first second course in analysis, 2004

61 **Thomas A. Ivey and J. M. Landsberg,** Cartan for beginners: Differential geometry via moving frames and exterior differential systems, 2003

60 **Alberto Candel and Lawrence Conlon,** Foliations II, 2003

59 **Steven H. Weintraub,** Representation theory of finite groups: algebra and arithmetic, 2003

58 **Cédric Villani,** Topics in optimal transportation, 2003

57 **Robert Plato,** Concise numerical mathematics, 2003

56 **E. B. Vinberg,** A course in algebra, 2003

55 **C. Herbert Clemens,** A scrapbook of complex curve theory, second edition, 2003

54 **Alexander Barvinok,** A course in convexity, 2002

53 **Henryk Iwaniec,** Spectral methods of automorphic forms, 2002

52 **Ilka Agricola and Thomas Friedrich,** Global analysis: Differential forms in analysis, geometry and physics, 2002

51 **Y. A. Abramovich and C. D. Aliprantis,** Problems in operator theory, 2002

50 **Y. A. Abramovich and C. D. Aliprantis,** An invitation to operator theory, 2002

49 **John R. Harper,** Secondary cohomology operations, 2002

48 **Y. Eliashberg and N. Mishachev,** Introduction to the h-principle, 2002

47 **A. Yu. Kitaev, A. H. Shen, and M. N. Vyalyi,** Classical and quantum computation, 2002

46 **Joseph L. Taylor,** Several complex variables with connections to algebraic geometry and Lie groups, 2002

45 **Inder K. Rana,** An introduction to measure and integration, second edition, 2002

44 **Jim Agler and John E. McCarthy,** Pick interpolation and Hilbert function spaces, 2002

43 **N. V. Krylov,** Introduction to the theory of random processes, 2002

42 **Jin Hong and Seok-Jin Kang,** Introduction to quantum groups and crystal bases, 2002

41 **Georgi V. Smirnov,** Introduction to the theory of differential inclusions, 2002

40 **Robert E. Greene and Steven G. Krantz,** Function theory of one complex variable, 2002

39 **Larry C. Grove,** Classical groups and geometric algebra, 2002

38 **Elton P. Hsu,** Stochastic analysis on manifolds, 2002

37 **Hershel M. Farkas and Irwin Kra,** Theta constants, Riemann surfaces and the modular group, 2001

36 **Martin Schechter,** Principles of functional analysis, second edition, 2002

35 **James F. Davis and Paul Kirk,** Lecture notes in algebraic topology, 2001

34 **Sigurdur Helgason,** Differential geometry, Lie groups, and symmetric spaces, 2001

33 **Dmitri Burago, Yuri Burago, and Sergei Ivanov,** A course in metric geometry, 2001

32 **Robert G. Bartle,** A modern theory of integration, 2001

31 **Ralf Korn and Elke Korn,** Option pricing and portfolio optimization: Modern methods of financial mathematics, 2001

30 **J. C. McConnell and J. C. Robson,** Noncommutative Noetherian rings, 2001

29 **Javier Duoandikoetxea,** Fourier analysis, 2001

28 **Liviu I. Nicolaescu,** Notes on Seiberg-Witten theory, 2000

27 **Thierry Aubin,** A course in differential geometry, 2001

26 **Rolf Berndt,** An introduction to symplectic geometry, 2001

25 **Thomas Friedrich,** Dirac operators in Riemannian geometry, 2000

24 **Helmut Koch,** Number theory: Algebraic numbers and functions, 2000
23 **Alberto Candel and Lawrence Conlon,** Foliations I, 2000
22 **Günter R. Krause and Thomas H. Lenagan,** Growth of algebras and Gelfand-Kirillov dimension, 2000
21 **John B. Conway,** A course in operator theory, 2000
20 **Robert E. Gompf and András I. Stipsicz,** 4-manifolds and Kirby calculus, 1999
19 **Lawrence C. Evans,** Partial differential equations, 1998
18 **Winfried Just and Martin Weese,** Discovering modern set theory. II: Set-theoretic tools for every mathematician, 1997
17 **Henryk Iwaniec,** Topics in classical automorphic forms, 1997
16 **Richard V. Kadison and John R. Ringrose,** Fundamentals of the theory of operator algebras. Volume II: Advanced theory, 1997
15 **Richard V. Kadison and John R. Ringrose,** Fundamentals of the theory of operator algebras. Volume I: Elementary theory, 1997
14 **Elliott H. Lieb and Michael Loss,** Analysis, 1997
13 **Paul C. Shields,** The ergodic theory of discrete sample paths, 1996
12 **N. V. Krylov,** Lectures on elliptic and parabolic equations in Hölder spaces, 1996
11 **Jacques Dixmier,** Enveloping algebras, 1996 Printing
10 **Barry Simon,** Representations of finite and compact groups, 1996
9 **Dino Lorenzini,** An invitation to arithmetic geometry, 1996
8 **Winfried Just and Martin Weese,** Discovering modern set theory. I: The basics, 1996
7 **Gerald J. Janusz,** Algebraic number fields, second edition, 1996
6 **Jens Carsten Jantzen,** Lectures on quantum groups, 1996
5 **Rick Miranda,** Algebraic curves and Riemann surfaces, 1995
4 **Russell A. Gordon,** The integrals of Lebesgue, Denjoy, Perron, and Henstock, 1994
3 **William W. Adams and Philippe Loustaunau,** An introduction to Gröbner bases, 1994
2 **Jack Graver, Brigitte Servatius, and Herman Servatius,** Combinatorial rigidity, 1993
1 **Ethan Akin,** The general topology of dynamical systems, 1993